Pierre Pichat (Ed.)

Photocatalysis:
Fundamentals, Materials and Potential

MDPI

This book is a reprint of the Special Issue that appeared in the online, open access journal, *Molecules* (ISSN 1420-3049) from September 2014–August 2015 (available at: http://www.mdpi.com/journal/molecules/special_issues/photocatalysis).

Guest Editor
Pierre Pichat
CNRS/ Ecole Centrale de Lyon (STMS)
France

Editorial Office
MDPI AG
Klybeckstrasse 64
Basel, Switzerland

Publisher
Shu-Kun Lin

Managing Editor
Ran Dang

1. Edition 2016

MDPI • Basel • Beijing • Wuhan • Barcelona

ISBN 978-3-03842-183-2 (Hbk)
ISBN 978-3-03842-184-9 (PDF)

Table of Contents

Chapter 1: Fundamentals: Photon Absorption, Active Species, Mechanisms, Reaction Pathways, Efficiency Evaluation

Chapter 2: UV and Visible-Light Sensitive Photocatalysts: Efficiency Effects of Nature, Composition, Preparation, Structure and Texture

Chapter 3: Air, Water and Surface Decontamination

Chapter 4: Photocatalysis and Photoelectrochemistry for Production of Energy and Chemicals

List of Contributors

Gopal Achari: Department of Civil Engineering, University of Calgary, 2500 University Dr. NW, Calgary, AB T2N 1N4, Canada.

Angelo Albini: PhotoGreen Lab, Department of Chemistry, University of Pavia, via Taramelli 12, Pavia 27100, Italy.

Rose Amal: Particles and Catalysis Research Group, School of Chemical Engineering, The University of New South Wales, Sydney, NSW 2052, Australia.

Beata Bajorowicz: Department of Chemical Technology, Faculty of Chemistry, Gdansk University of Technology, ul. G. Narutowicza 11/12, Gdansk 80-233, Poland.

Zuzana Barbieriková: Institute of Physical Chemistry and Chemical Physics, Faculty of Chemical and Food Technology, Slovak University of Technology in Bratislava, Radlinského 9, Bratislava SK-812 37, Slovakia.

Bernardí Bayarri: Departament d'Enginyeria Química, Universitat de Barcelona, C/Martí i Franquès 1, Barcelona 08028, Spain.

Kaustava Bhattacharyya: Department of Chemistry, Institute for Catalysis in Energy Processes, Northwestern University, Evanston, IL 60208, USA.

Vlasta Brezová: Institute of Physical Chemistry and Chemical Physics, Faculty of Chemical and Food Technology, Slovak University of Technology in Bratislava, Radlinského 9, Bratislava SK-812 37, Slovakia.

Michaela Brunclikova: Department of Inorganic Technology, University of Chemistry and Technology, Prague, Technická 5, Prague 16628, Czech Republic.

John Anthony Byrne: Nanotechnology and Integrated BioEngineering Centre, Ulster University, Newtownabbey, Northern Ireland BT37 0QB, UK.

Yong Cai: Department of Chemistry and Biochemistry, Florida International University, Miami, FL 33199, USA.

Yanke Che: Beijing National Laboratory for Molecular Sciences, Key Laboratory of Photochemistry, Institute of Chemistry, The Chinese Academy of Sciences, Beijing 100190, China.

Chuncheng Chen: Beijing National Laboratory for Molecular Sciences, Key Laboratory of Photochemistry, Institute of Chemistry, The Chinese Academy of Sciences, Beijing 100190, China.

Zhong Chen: Energy Research Institute @ NTU, 1 CleanTech Loop, Clean Tech One, Singapore 637141, Singapore; School of Materials Science and Engineering, Nanyang Technological University, 50 Nanyang Avenue, Singapore 639798, Singapore.

José Colina-Márquez: Program of Chemical Engineering, Universidad de Cartagena Piedra de Bolívar Campus, Av. El Consulado 48-152, A.A. 130015, Cartagena de Indias, Colombia.

Anna Cybula: Department of Chemical Technology, Faculty of Chemistry, Gdansk University of Technology, ul. G. Narutowicza 11/12, Gdansk 80-233, Poland.

Dionysios D. Dionysiou: Environmental Engineering and Science Program, University of Cincinnati, Cincinnati, OH 45221-0012, USA.

Patrick Stuart Morris Dunlop: Nanotechnology and Integrated BioEngineering Centre, Ulster University, Newtownabbey, Northern Ireland BT37 0QB, UK.

Dana Dvoranová: Institute of Physical Chemistry and Chemical Physics, Faculty of Chemical and Food Technology, Slovak University of Technology in Bratislava, Radlinského 9, Bratislava SK-812 37, Slovakia.

Terry A. Egerton: School of Chemical Engineering and Advanced Materials, University of Newcastle, Newcastle NE1 7RU, UK.

Santiago Esplugas: Departament d'Enginyeria Química, Universitat de Barcelona, C/Martí i Franquès 1, Barcelona 08028, Spain.

Pilar Fernández-Ibáñez: Plataforma Solar de Almería—CIEMAT, PO Box 22, 04200 Tabernas, Almería, Spain.

Leanne Fisher: School of Healthcare Science, Manchester Metropolitan University, Manchester M1 5GD, UK.

Akira Fujishima: Kanagawa Academy of Science and Technology, KSP East 407, 3-2-1 Sakado, Takatsu-ku, Kawasaki, Kanagawa 213-0012, Japan; Photocatalysis International Research Center, Tokyo University of Science, 2641 Yamazaki, Noda, Chiba 278-8510, Japan.

Jaime Giménez: Departament d'Enginyeria Química, Universitat de Barcelona, C/Martí i Franquès 1, Barcelona 08028, Spain.

Óscar González: Departament d'Enginyeria Química, Universitat de Barcelona, C/Martí i Franquès 1, Barcelona 08028, Spain.

Kimberly A. Gray: Department of Civil and Environmental Engineering, Institute for Catalysis in Energy Processes, Northwestern University, Evanston, IL 60208, USA.

Jeremy William John Hamilton: Nanotechnology and Integrated BioEngineering Centre, Ulster University, Newtownabbey, Northern Ireland BT37 0QB, UK.

Stephen O. Hay: United Technologies Research Center (ret.), 35 Weigel Valley Drive, Tolland, CT 06082, USA.

Yuji Hayashi: PACT World Co. Ltd., Domicile 301, 1-3-1 Sugeinadazutsumi, Tama-ku, Kawasaki, Kanagawa 214-0003, Japan.

Junkai He: Institute of Materials Science, University of Connecticut, U-3060, 91 North Eagleville Road, Storrs, CT 06269-3060, USA.

James Highfield: Heterogeneous Catalysis, Institute of Chemical & Engineering Sciences (ICES, A * Star), 1 Pesek Road, Jurong Island, 627833, Singapore.

Satoshi Horikoshi: Department of Materials and Life Sciences, Faculty of Science and Technology, Sophia University, 7-1 Kioicho, Chiyodaku, Tokyo 102-8554, Japan.

Zdenek Hubicka: Institute of Physics, Academy of Sciences of the Czech Republic, Na Slovance 2, Prague 14800, Czech Republic.

Erina Ichihashi: PACT World Co. Ltd., Domicile 301, 1-3-1 Sugeinadazutsumi, Tama-ku, Kawasaki, Kanagawa 214-0003, Japan.

Yasushi Ishiguro: Division of Materials and Manufacturing Science, Graduate School of Engineering, Osaka University, 2-1 Yamadaoka, Suita, Osaka 565-0871, Japan.

Maryam Izadifard: Department of Chemistry University of Calgary, 2500 University Dr. NW, Calgary, AB T2N 1N4, Canada.

Hongwei Ji: Beijing National Laboratory for Molecular Sciences, Key Laboratory of Photochemistry, Institute of Chemistry, The Chinese Academy of Sciences, Beijing 100190, China.

Ting Jiang: Department of Chemical and Bimolecular Engineering, University of Connecticut, U-3222, 191 Auditorium Road, Storrs, CT 06269-3060, USA.

Wenjun Jiang: Department of Chemistry and Biochemistry, Florida International University, Miami, FL 33199, USA.

Ryota Jono: Department of Chemical System Engineering, School of Engineering, The University of Tokyo, 7-3-1 Hongo, Bunkyo-ku, Tokyo 113-8656, Japan.

Takashi Kamegawa: Division of Materials and Manufacturing Science, Graduate School of Engineering, Osaka University, 2-1 Yamadaoka, Suita, Osaka 565-0871, Japan; Nanoscience and Nanotechnology Research Center, Osaka Prefecture University, 1-2 Gakuencho, Nakaku, Sakai, Osaka 599-8570, Japan.

Pushkar Kanhere: Energy Research Institute @ NTU, 1 CleanTech Loop, Clean Tech One, Singapore 637141, Singapore; School of Materials Science and Engineering, Nanyang Technological University, 50 Nanyang Avenue, Singapore 639798, Singapore.

Peter J. Kelly: Surface Engineering Group, Dalton Research Institute, Manchester Metropolitan University, Manchester M1 5GD, UK.

Ryota Kido: Division of Materials and Manufacturing Science, Graduate School of Engineering, Osaka University, 2-1 Yamadaoka, Suita, Osaka 565-0871, Japan.

Tomasz Klimczuk: Department of Solid State Physics, Faculty of Applied Physics and Mathematics, Gdansk University of Technology, ul. G. Narutowicza 11/12, Gdansk 80-233, Poland.

Stepan Kment: Joint Laboratory of Optics, Palacky University, RCPTM, 17. listopadu 12, Olomouc 77146, Czech Republic.

Ewa Kowalska: Catalysis Research Center, Hokkaido University, Sapporo 001-0021, Japan.

Denis Kozlov: Boreskov Institute of Catalysis, pr. Ak. Lavrentieva 5, Novosibirsk 630090, Russia; Novosibirsk State University, st. Pirogova 2, Novosibirsk 630090, Russia; Research and Educational Centre for Energoefficient Catalysis (NSU), st. Pirogova 2, Novosibirsk 630090, Russia.

Josef Krysa: Department of Inorganic Technology, University of Chemistry and Technology, Prague, Technická 5, Prague 16628, Czech Republic.

Aleksandra Kułagowska: Institute of Chemical and Environment Engineering, West Pomeranian University of Technology, ul. Pułaskiego 10, 70–322 Szczecin, Poland.

Cooper Langford: Department of Chemistry, University of Calgary, 2500 University Dr. NW, Calgary, AB T2N 1N4, Canada.

Gianluca Li Puma: Environmental Nanocatalysis & Photoreaction Engineering, Department of Chemical Engineering, Loughborough University, Loughborough LE11 3TU, UK.

Panagiotis Lianos: Department of Chemical Engineering, University of Patras, University Campus, Patras 26500, Greece; FORTH/ICE-HT, Stadiou Str., Platani, P.O. Box 1414, Patras 26504, Greece.

Sanly Liu: Particles and Catalysis Research Group, School of Chemical Engineering, The University of New South Wales, Sydney, NSW 2052, Australia.

Zhu Luo: Institute of Materials Science, University of Connecticut, U-3060, 91 North Eagleville Road, Storrs, CT 06269-3060, USA.

Wanhong Ma: Beijing National Laboratory for Molecular Sciences, Key Laboratory of Photochemistry, Institute of Chemistry, The Chinese Academy of Sciences, Beijing 100190, China.

Tadashi Machida: Tanashin Denki Co. Ltd., 8-19-20 Fukasawa, Setagaya-ku, Tokyo 158-0081, Japan.

Fiderman Machuca-Martínez: School of Chemical Engineering, Universidad del Valle, Campus Meléndez, Calle 13 No. 100-00, A.A. 25360 Cali, Colombia.

Sixto Malato: Plataforma Solar de Almería (CIEMAT), Carretera de Senes, km 4, Tabernas, Almería 04200, Spain.

Yongtao Meng: Department of Chemistry, University of Connecticut, U-3060, 55 North Eagleville Road, Storrs, CT 06269-3060, USA.

Robert Michal: Department of Chemical Engineering, University of Patras, University Campus, Patras 26500, Greece; Permanent address: Department of Inorganic Chemistry, Faculty of Natural Sciences, Comenius University, Mlynská Dolina, 84215 Bratislava, Slovakia.

Antoni W. Morawski: Institute of Chemical and Environment Engineering, West Pomeranian University of Technology, ul. Pułaskiego 10, 70–322 Szczecin, Poland.

Yuko Morito: Photocatalysis International Research Center, Tokyo University of Science, 2641 Yamazaki, Noda, Chiba 278-8510, Japan; VIX Corporation, 2-14-8 Midorigaoka, Meguro-ku, Tokyo 152-0034, Japan.

Sylwia Mozia: Institute of Chemical and Environment Engineering, West Pomeranian University of Technology, ul. Pułaskiego 10, 70–322 Szczecin, Poland.

Steven C. Murphy: Department of Chemistry, University of Connecticut, U-3060, 55 North Eagleville Road, Storrs, CT 06269-3060, USA.

Nobuaki Negishi: Research Institute for Environmental Management Technology, National Institute of Advanced Industrial Science and Technology, 16-1 Onogawa, Tsukuba 305-8569, Japan.

Naoki Nishida: Tanashin Denki Co. Ltd., 8-19-20 Fukasawa, Setagaya-ku, Tokyo 158-0081, Japan.

Masami Nishikawa: Nagaoka University of Technology, 1603-1 Kamitomioka, Nagaoka 940-2188, Japan.

Yoshio Nosaka: Nagaoka University of Technology, 1603-1 Kamitomioka, Nagaoka 940-2188, Japan.

Atsuko Y. Nosaka: Nagaoka University of Technology, 1603-1 Kamitomioka, Nagaoka 940-2188, Japan.

Timothy Obee: United Technologies Research Center (ret.), 351 Foster Street, South Windsor, CT 06074, USA.

Tsuyoshi Ochiai: Kanagawa Academy of Science and Technology, KSP East 407, 3-2-1 Sakado, Takatsu-ku, Kawasaki, Kanagawa 213-0012, Japan; Photocatalysis International Research Center, Tokyo University of Science, 2641 Yamazaki, Noda, Chiba 278-8510, Japan.

Bunsho Ohtani: Catalysis Research Center, Hokkaido University, Sapporo 001-0021, Japan.

Kevin E. O'Shea: Department of Chemistry and Biochemistry, Florida International University, Miami, FL 33199, USA.

Soheyla Ostovarpour: Surface Engineering Group, Dalton Research Institute, Manchester Metropolitan University, Manchester M1 5GD, UK; School of Healthcare Science, Manchester Metropolitan University, Manchester M1 5GD, UK.

Xibin Pang: Beijing National Laboratory for Molecular Sciences, Key Laboratory of Photochemistry, Institute of Chemistry, The Chinese Academy of Sciences, Beijing 100190, China.

Sagi Pasternak: Department of Chemical Engineering, Technion—Israel Institute of Technology, Haifa 3200003, Israel.

Yaron Paz: Department of Chemical Engineering, Technion—Israel Institute of Technology, Haifa 3200003, Israel.

José Peral: Departament de Química, Edifici Cn, Universitat Autònoma de Barcelona, Bellaterra, Cerdanyola del Vallès 08193, Spain.

Pierre Pichat: "Photocatalyse et Environnement", CNRS/Ecole Centrale de Lyon (STMS), 69134 Ecully CEDEX, France.

Inmaculada Polo-López: Plataforma Solar de Almería—CIEMAT, PO Box 22, 04200 Tabernas, Almería, Spain.

Stefano Protti: PhotoGreen Lab, Department of Chemistry, University of Pavia, via Taramelli 12, Pavia 27100, Italy.

Emad Radwan: Department of Water Pollution Research, National Research Centre, Cairo 12311, Egypt.

Mamun Rashid: Department of Chemistry and Biochemistry, Florida International University, Miami, FL 33199, USA.

Marina Ratova: School of Chemistry and Chemical Engineering, Queen's University Belfast, Belfast BT9 5AG, UK.

Davide Ravelli: PhotoGreen Lab, Department of Chemistry, University of Pavia, via Taramelli 12, Pavia 27100, Italy.

Malka Rochkind: Department of Chemical Engineering, Technion—Israel Institute of Technology, Haifa 3200003, Israel.

Taizo Sano: Research Institute for Environmental Management Technology, National Institute of Advanced Industrial Science and Technology, 16-1 Onogawa, Tsukuba 305-8569, Japan.

Jun Sawada: Department of Materials Science and Technology, Tokyo University of Science, 6-3-1 Niijuku, Katsushika-ku, Tokyo 125-8585, Japan.

Jason Scott: Particles and Catalysis Research Group, School of Chemical Engineering, The University of New South Wales, Sydney, NSW 2052, Australia.

Hiroshi Segawa: Research Center for Advance Science and Technology, The University of Tokyo, 4-6-1, Komaba, Meguro-ku, Tokyo 153-8904, Japan.

Dmitry Selishchev: Boreskov Institute of Catalysis, pr. Ak. Lavrentieva 5, Novosibirsk 630090, Russia; Novosibirsk State University, st. Pirogova 2, Novosibirsk 630090, Russia; Research and Educational Centre for Energoefficient Catalysis (NSU), st. Pirogova 2, Novosibirsk 630090, Russia.

Nick Serpone: PhotoGreen Laboratory, Dipartimento di Chimica, Università di Pavia, via Taramelli 12, Pavia 27100, Italy.

Stavroula Sfaelou: Department of Chemical Engineering, University of Patras, University Campus, Patras 26500, Greece.

Preetam Kumar Sharma: Nanotechnology and Integrated BioEngineering Centre, Ulster University, Newtownabbey, Northern Ireland BT37 0QB, UK.

Steven Suib: Institute of Materials Science, University of Connecticut, U-3060, 91 North Eagleville Road, Storrs, CT 06269-3060, USA; Department of Chemistry, University of Connecticut, U-3060, 55 North Eagleville Road, Storrs, CT 06269-3060, USA; Department of Chemical Engineering, University of Connecticut, U-3060, 91 North Eagleville Road, Storrs, CT 06269-3060, USA.

Yoshitsugu Uchida: Tanashin Denki Co. Ltd., 8-19-20 Fukasawa, Setagaya-ku, Tokyo 158-0081, Japan.

Ashlene Sarah Margaret Vennard: Nanotechnology and Integrated BioEngineering Centre, Ulster University, Newtownabbey, Northern Ireland BT37 0QB, UK.

Joanna Verran: School of Healthcare Science, Manchester Metropolitan University, Manchester M1 5GD, UK.

Baiju K. Vijayan: Department of Civil and Environmental Engineering, Institute for Catalysis in Energy Processes, Northwestern University, Evanston, IL 60208, USA.

Zhishun Wei: Catalysis Research Center, Hokkaido University, Sapporo 001-0021, Japan.

Eric Weitz: Department of Chemistry, Institute for Catalysis in Energy Processes, Northwestern University, Evanston, IL 60208, USA.

Glen T. West: Surface Engineering Group, Dalton Research Institute, Manchester Metropolitan University, Manchester M1 5GD, UK.

Yossy Wicaksana: Particles and Catalysis Research Group, School of Chemical Engineering, The University of New South Wales, Sydney, NSW 2052, Australia.

Michal J. Winiarski: Department of Solid State Physics, Faculty of Applied Physics and Mathematics, Gdansk University of Technology, ul. G. Narutowicza 11/12, Gdansk 80-233, Poland.

Weiqiang Wu: Department of Chemistry, Institute for Catalysis in Energy Processes, Northwestern University, Evanston, IL 60208, USA.

Hiromi Yamashita: Division of Materials and Manufacturing Science, Graduate School of Engineering, Osaka University, 2-1 Yamadaoka, Suita, Osaka 565-0871, Japan; Elements Strategy Initiative for Catalysts and Batteries (ESICB), Kyoto University, Katsura, Kyoto 615-8520, Japan.

Koichi Yamashita: Department of Chemical System Engineering, School of Engineering, The University of Tokyo, 7-3-1 Hongo, Bunkyo-ku, Tokyo 113-8656, Japan.

Sayaka Yanagida: Department of Materials Science and Technology, Tokyo University of Science, 6-3-1 Niijuku, Katsushika-ku, Tokyo 125-8585, Japan; Photocatalysis International Research Center, Research Institute for Science and Technology, Tokyo University of Science, 2641 Yamazaki, Noda-shi, Chiba 278-8510, Japan.

Shozo Yanagida: Frontier Research Institute, Osaka University, 2-1, Yamada-oka, Suita, Osaka 565-0871, Japan.

Susumu Yanagisawa: Department of Physics and Earth Sciences, Faculty of Science, University of the Ryukyus, 1, Senbaru, Nishihara, Okinawa 903-0213, Japan.

Atsuo Yasumori: Department of Materials Science and Technology, Tokyo University of Science, 6-3-1 Niijuku, Katsushika-ku, Tokyo 125-8585, Japan; Photocatalysis International Research Center, Research Institute for Science and Technology, Tokyo University of Science, 2641 Yamazaki, Noda-shi, Chiba 278-8510, Japan.

Adriana Zaleska: Department of Chemical Technology, Faculty of Chemistry, Gdansk University of Technology, ul. G. Narutowicza 11/12, Gdansk 80-233, Poland.

Sifani Zavahir: School of Chemistry, Physics and Mechanical Engineering, Science and Engineering Faculty, Queensland University of Technology, Brisbane QLD 4001, Australia.

Jincai Zhao: Beijing National Laboratory for Molecular Sciences, Key Laboratory of Photochemistry, Institute of Chemistry, The Chinese Academy of Sciences, Beijing 100190, China.

Shan Zheng: Department of Chemistry and Biochemistry, Florida International University, Miami, FL 33199, USA.

Huaiyong Zhu: School of Chemistry, Physics and Mechanical Engineering, Science and Engineering Faculty, Queensland University of Technology, Brisbane QLD 4001, Australia.

Martin Zlamal: Department of Inorganic Technology, University of Chemistry and Technology, Prague, Technická 5, Prague 16628, Czech Republic.

About the Guest Editor

Pierre Pichat, as "Directeur de Recherche de 1ère classe" (first-class) with the CNRS (National Center for Scientific Research, France), has been active in heterogeneous photocatalysis for many years. He has founded a laboratory dealing with both basic investigations on this field and applications regarding self-cleaning materials and purification of air or water. He has published numerous research papers and several reviews of the domain. He has edited a book and special issues. He is a frequent invited lecturer at Conferences. He is a member of the International Scientific Committees of most of the International Conferences on photocatalysis. Over the years, he has served on CNRS-related Committees on diverse aspects of chemistry; he has been the coordinator or advisor of European Community projects on photocatalysis; he has evaluated projects on environmental chemistry for various countries. He has received an International Appreciation Award acknowledging his pioneering contributions to heterogeneous photocatalysis.

Preface

Following an initiative by the Multidisciplinary Digital Publishing Institute (MDPI), a series of articles on heterogeneous photocatalysis were published in Molecules from September 2014 to August 2015 after rigorous peer-review. These articles are freely accessible online. Nevertheless, it was thought that a printed book gathering them in an organized manner will be very useful. The book format allows one to browse through the articles in a much easier way. Anybody in a laboratory can have the printed book at hand for consulting at any time. Attention of potential readers to the existence of a book can be drawn readily in libraries and online. A book is also more appropriate for storage than a pile of copies!

In the present case, another argument in favor of the book format was that the articles as a whole provide a timely picture of heterogeneous photocatalysis. This picture includes both retrospective analyses and proper examples of diverse aspects of the current research, so that this book presents an outstanding overview of the field. I would also like to emphasize that such an ensemble of as well-known authors has rarely been brought together in previous books on this topic. Undoubtedly, this will allow the community of senior scientists and students to possess a book of great significance for a low price.

This book contains a total of thirty-one articles among which six are feature articles and seven are reviews. It is structured in four sections. The contents of each section are summarized hereafter.

As detailed in its title, **Section 1** covers many basic aspects of photocatalysis. The feature article by T. Egerton addresses a fundamental aspect that is too rarely considered in sufficient detail, viz. the photon absorption which is obviously critical for photocatalytic reactions and depends on several parameters that are not always taken into account properly. Combined with the review from the Y. Paz's group on the too frequent use of dyes for evaluating photocatalytic efficiency, this feature article also shows the importance of "good practices" in photocatalysis for valid comparisons of experimental results. A review from J. Zhao's group draws the attention on the great interest of employing O-labeled dioxygen and water to unravel the photocatalytic basic mechanisms of the hydroxylation or ring-opening of benzene derivatives and the decarboxylation of aliphatic acids. Three other papers concern the appropriate use of spectroscopic techniques to obtain information on the fundamental steps in photocatalysis; both the review from Y. Nosaka's group and the article from V. Brezova's group draw attention on the issues and illustrate the interest of using electron paramagnetic resonance to detect some active species and to determine their role in photocatalytic mechanisms and pathways, while the article from K. Gray's group emphasizes the pertinence of the use of infrared spectroscopy to investigate the interactions of molecules with photocatalysts, as shown by the examples of H_2O and CO_2.

Section 2 devoted to photocatalysts contains the highest number of articles in this book, which reflects the importance of the material aspect. Unsurprisingly, with the exception of three, these articles deal with TiO_2 as the unique or principal photo-active component of the photocatalysts. The feature article from P. Kelly's and J. Verran's groups presents the advantages of magnetron sputtering as a versatile, flexible coating method (including of thermally sensitive materials) that allows easy doping and can be industrialized. The effects of the texture and the structure of pure TiO_2 upon the rates of photocatalytic reactions are discussed in a short review by P. Pichat relative to nanotubes and an article from B. Ohtani's group relative to anatase particles differing by shape and ratios of exposed facets; these reports show the complexity of the issue. Several articles refer to the combination of TiO_2 with another non-semiconducting adsorbent (zeolites, a clay, silica and carbon) or with WO_3 in attempts to increase the photocatalytic efficiency. The effects have been assessed in a variety of photocatalytic reactions principally in the gas phase (H. Yamashita's, D. Kozlov's and A. Yasumori's groups), and also in the aqueous phase (C. Langford's group) or in a liquid organic phase (H. Zhu's group), and interpretations for these effects are suggested. Section 2 also reports on the use of semiconductors other than TiO_2, the main goal being to explore the potentialities of materials more sensitive to visible-light irradiation for solar applications. A comprehensive review (from Z. Chen's group with 132 references) deals with perovskite-based photocatalysts and two articles concern WO_3 (R. Amal's group) or $KTaO_3$, CdS, and MoS_2 either pristine or as composites (A. Zaleska's group). These reports provide the state-of-the-art and insights into the potential development of these semiconductors in photocatalysis.

The articles gathered in **Section 3** focus especially on the relevance of photocatalysis for decontamination, unlike those in Section 2 that concentrate on the materials used in photocatalysis, even though these materials were most often also tested for the removal of pollutants. One feature article from S. Hay's and S. Suib's groups discusses the viability of photocatalysis on the basis of field experiments and the authors conclude that the photocatalyst lifetime needs to be extended for cost-effective purification of indoor air. The other feature article from group led by J. Gimenez addresses the question of the environmental impact of water photocatalytic treatment with an original comparison with other advanced oxidation processes, particularly the photo-Fenton process. A detailed review (169 references) from groups led by J. Byrne is devoted to the state-of-the-art inactivation of microorganisms in water and on surfaces and considers the perspectives. Another review by S. Horikoshi and N. Serpone demonstrates that concomitant use of microwave and photocatalysis can be beneficial and the origins of the improvement are discussed. Two articles present results of field experiments for the photocatalytic treatment of either outdoor air under solar light (by N. Negishi and T. Sano) or indoor air in combination with a plasma reactor (from groups led by T. Ochiai and A. Fujishima) by use of newly conceived demonstration devices. Two other articles report on the potential use of photocatalysis to tackle two serious environmental problems, viz. the removal from water of toxic cations in mixture (from K. O'Shea's and D. Dionysiou's groups) or endocrine disruptors (J. Colina's and G. Li Puma's groups).

Whereas the potentialities of photocatalysis for decontamination are well-established, the viability of photocatalysis regarding the production of energy and chemical compounds is still debated because these processes are based on uphill (or endergonic) reactions. These potentialities and the hurdles to their development are reviewed and illustrated in **Section 4**. One very detailed (370 references) feature article by J. Highfield critically reviews advances in photocatalysis and photoelectrocatalysis for producing H_2 from water, reforming bio-oxygenates, and synthesizing organic fuels through CO_2 reduction. Another feature article from A. Albini's group elaborates comprehensively on the potential development of photocatalysis and photochemistry in organic synthesis, which would offer the advantage of operating under mild conditions. An article by A. Morawski's group indicates the formation, in both aerobic and anaerobic conditions, of H_2 and C_1-C_3 alkanes from C_1-C_2 alcohols, acetic acid or glucose, though with low efficiency, over TiO_2 modified with 20 wt% Fe. A general article from P. Lianos' group describes in detail under which conditions a photoelectrochemical cell comprising a TiO_2 photoanode (which can be made sensitive to visible light) and a carbon black cathode with Pt particles can produce electricity or H_2 from fuels. Photoelectrochemical water splitting is considered in two other articles from either the material view or the theoretical view. One from J. Krysa's group underlines the importance for the efficiency not only of the type of semiconductor (TiO_2 or Fe_2O_3) but also of the way it is elaborated. In the other article from groups led by S. Yanagida, simulation of adsorbed H_2O clusters via density functional theory is shown to allow one to interpret the photoelectrochemical mechanism of water oxidation using nanocrystalline TiO_2 electrodes.

In conclusion, this book will be helpful to the beginners who would like to learn more about heterogeneous photocatalysis and its issues, as well as to the senior scientists who will find reviews and articles allowing them to refresh or update their knowledge of some aspects of this multidisciplinary field.

I sincerely thank the contributors for their response to my solicitation. Initially, none of them knew who would respond positively and they were just confident in me for being able to accomplish this venture. I think they do not regret their decision, given the exceptional group of eminent experts who authored these articles published in Molecules and now gathered in this book. Naturally, I thank them, above all those who wrote long feature articles and reviews, for their time. I am also very grateful to the many reviewers who accepted to evaluate the manuscripts and to write constructive comments. Obviously, hearty thanks are also due to Ms. Ran Dang, Managing Editor at MDPI, with whom I always had excellent and efficient email relationships, and to the Assistant Editors who helped me with high competence to speed up the reviewing process.

Pierre Pichat
Guest Editor
CNRS/Ecole Centrale de Lyon, France

Chapter 1:
Fundamentals: Photon Absorption, Active
Species, Mechanisms, Reaction Pathways,
Efficiency Evaluation

UV-Absorption—The Primary Process in Photocatalysis and Some Practical Consequences

Terry A. Egerton

Abstract: TiO$_2$ photochemistry studies generally address reactions of photogenerated charge-carriers at the oxide surface or the recombination reactions which control the proportion of charge carriers that reach the surface. By contrast, this review focuses on UV absorption, the first photochemical step in semiconductor photocatalysis. The influence of particle size on absorption and scattering of light by small TiO$_2$ particles is summarized and the importance of considering, the particle size in the application, not the BET or X-ray line broadening size, is emphasized. Three different consequences of UV absorption are then considered. First, two commercially important systems, pigmented polymer films and paints, are used to show that TiO$_2$ can protect from direct photochemical degradation. Then the effect of UV absorption on the measured photocatalytic degradation of aqueous solutions of organics is considered for two separate cases. Firstly, the consequences of UV absorption by TiO$_2$ on the generation of hydroxyl radicals from H$_2$O$_2$ are considered in the context of the claimed synergy between H$_2$O$_2$ and TiO$_2$. Secondly, the effect of altered UV absorption, caused by changed effective particle size of the catalyst, is demonstrated for photocatalysis of propan-2-ol oxidation and salicylic acid degradation.

Reprinted from *Molecules*. Cite as: Egerton, T.A. UV-Absorption—The Primary Process in Photocatalysis and Some Practical Consequences. *Molecules* **2014**, *19*, 18192-18214.

1. Introduction

Semiconductor photocatalysis by TiO$_2$ has been widely studied for 50 years. Potential uses include destruction of bacteria [1], the oxidation of pollutants [2], e.g., dye residues [3], and removal of organic films from glass and polymer substrates [4]. However, early work emphasized undesirable aspects, e.g., photocatalytic degradation of TiO$_2$ pigmented paint films [5] or textile fibres [6]. Commercial research to minimize TiO$_2$ photocatalysis continues, but little is published in the open literature. Both objectives have driven research into the photocatalytic mechanism [7,8]. It is generally agreed that UV absorption excites an electron from the valence band to the conduction band of the semiconductor. The resulting excited electrons, in the otherwise empty conduction band, and the "positive holes" in the valence band allow charge transfer to the TiO$_2$ surface which facilitates oxidation of surrounding molecules. Sometimes direct charge-transfer causes the oxidation [9]. Alternatively, hydroxyl radicals, formed by the "positive holes" in the valence band accepting electrons from hydroxyl ions, are the catalytically active intermediates [8,10].

The *Grotthus-Draper Law*, formulated in 1817 and rediscovered in 1841, has been stated as *"Only radiations which are absorbed by the reacting system are effective in producing chemical change."* [11]. However, discussion of light absorption in semiconductor photocatalysis is generally restricted to extending absorption to longer wavelengths (to harvest a greater proportion of solar radiation) [12] and the possibility of photo-sensitizing reactions by exploiting electron transfer to the semiconductor from excited states of adsorbed dyes [13]. UV-photocatalytic papers mainly address topics such as the activity of new TiO_2 preparations [14,15], the intermediates formed in specific reactions [16], the extent to which transition metals modify charge-carrier recombination [17,18], and whether specialized procedures such as the application of an electric field increase photocatalytic activity [19]. Often, potential changes in UV-absorption of the particle dispersion resulting from, e.g., the preparation of novel TiO_2's, or from supporting catalysts on inert supports or from inadvertently changing particle aggregation during surface modification or doping, are overlooked. Therefore this survey focuses on ways in which unconsidered aspects of UV-absorption may modify either undesirable or desirable photocatalysis.

2. A Summary of Small Particle Optics (Where Diameter, d, Is Comparable with the Radiation Wavelength, λ)

2.1. Background Theory

The optical properties of particulate dispersions, unlike those of molecular solutions, depend on absorption and scattering. For particulate suspensions the well known extinction equation may be written as:

$$\log_{10}(I_o/I_t) = q.c.l = (q_{abs} + q_{sca}) \tag{1}$$

where c is the concentration of particles in suspension and l is the path length and q, the extinction coefficient, is the sum of the coefficients of scattering, q_{sca}, and absorption, q_{abs}. Scattering of radiation, of wavelength λ, by small particles, of diameter d ($d/\lambda \ll 0.1$), was analyzed by Rayleigh [20]. Later Mie's treatment of scattering by larger spherical particles took into account interference between light scattered from different points on the particle surface [21]. The difference between Rayleigh scattering and Mie scattering of 555 nm radiation by isolated, spherical, rutile particles is shown in Figure 1 (both theories describe optically dilute systems, *i.e.*, neither treats interference between light scattered by *different* particles.)

Mie showed that light scattering depends on both the ratio d/λ, and on the refractive index of the medium, m_m, and the particle, m_p. Refractive index is a complex quantity with real and imaginary components and, as shown for rutile spheres, in Figure 2 both of these vary with wavelength. Light *scattering* is controlled by the real component; light *absorption* by particles is controlled by both real and imaginary components and (unlike absorption by molecules) also by particle size. It is often convenient to express both scattering and absorption on a unit-volume (or unit-mass) basis and typical results are shown in Figure 3a,b [23].

Figure 1. Rayleigh (▲, dashed line) and Mie scattering (◊ and ■, full lines) of 555 nm radiation by isolated rutile spheres as a function of particle diameter. The curves plotted through ▲ and ◊, points assume that all particles are of identical. The curve plotted through the ■ points assumes a log-normal distribution of particle size.

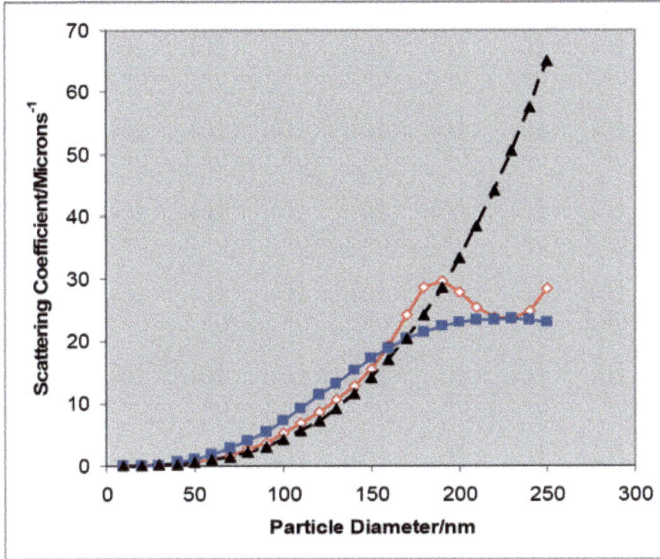

Figure 2. Variation of the real, ◊, and imaginary, ●, components of the refractive index of rutile between 275 and 400 nm based on results of Vos and Krusmeyer [22]. Above 400 nm the imaginary component, which controls light absorption, is negligible even though the real component is not. Pure rutile crystals are transparent in the visible region of the spectrum but can scatter visible radiation.

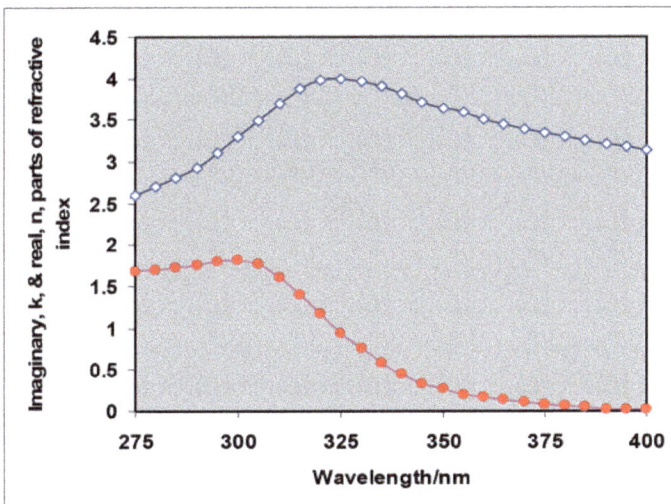

Figure 3. Mie theory calculations of (**a**) q_{sca}, scattering and (**b**) q_{abs}, absorption per unit volume for 20 nm, ×; 50 nm, Δ; 100 nm ■; and 220 nm ♦ rutile particles dispersed in an organic medium, plotted as a function of mean size for a log normal particle distribution with $\sigma = 1.33$. Reproduced with permission from Egerton & Tooley, International Journal of Cosmetic Science **2012**, *34*, 117–122 [23] published by Society of Cosmetic Sciences; Société Française de Cosmétologie and Blackwell Publishing.

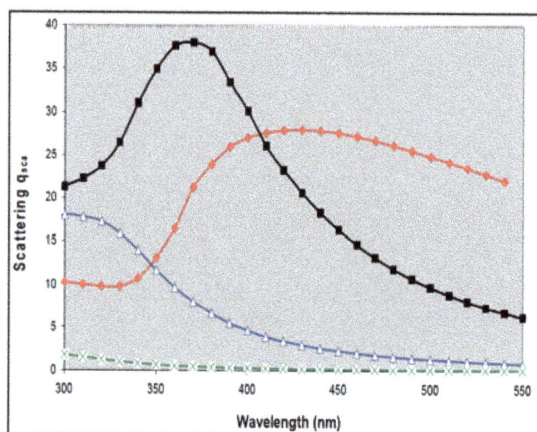

(**a**)

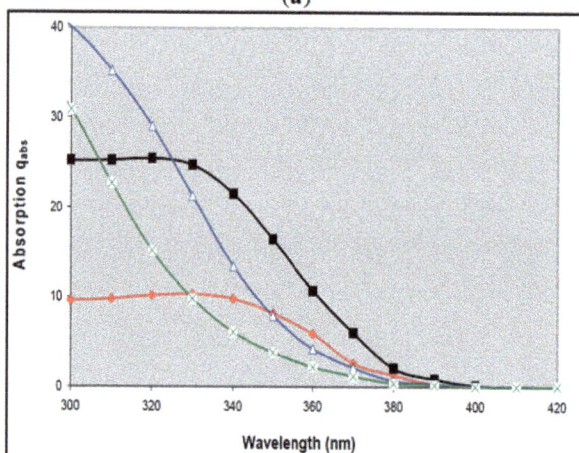

(**b**)

Calculated scattering and absorption coefficients of a particulate suspension are shown in Figure 3a,b. The extinction coefficient may be calculated by simply summing the scattering and absorption coefficients. Inspection of these graphs shows that in an organic medium the percentage contributions of absorption to the extinction coefficient of for mean sizes of 20, 50, 100 and 220 nm particles are 94% (93%), 66% (63%), 53% (52%) and 49% (46%) at 310 nm and are 81% (50%), 30% (17%), 22% (14%) and 26% (30%) at 360 nm. The results in brackets are the results of earlier calculations published by Robb, Simpson and Tunstall [24] (the differences between the two sets are probably a consequence of small differences in the input refractive indices and the size-distributions in the two sets of calculations). Except for 20 nm particles at a wavelength of 360 nm the results of the two sets of calculations are in reasonable agreement. Both sets show that the contribution of absorption is greater at 310 than at 360 nm for all sizes in the range 20 to 220 nm, but even at 310 nm, the absorption contribution becomes progressively less important as the mean particle size is increased. For 220 nm particles, the size of a typical TiO_2 pigment, scattering and absorption make approximately equal contributions to attenuation of 310 nm radiation but at 360 nm the most important contribution to attenuation is scattering. Because both the absorption and scattering are particle-size dependent absorption and scattering coefficients from, e.g., measurements on single crystals [25], on CVD films [26], or from study of different TiO_2 particles [27] cannot be used to calculate the radiation flux in a photocatalytic study.

2.2. Comparison of Theory and Experiment

The variation in absorption and scattering coefficients with the ratio of d/λ is paralleled by changes in the angular distribution of the scattered light [28]. Consequently the experimental estimation of the amount of radiation absorbed by a particulate suspension is not straightforward. In a typical "optical-density" type transmission measurement, as used in most photocatalytic studies, the measured attenuation is the sum of the absorption and the variable fraction of light not incident on the detector because it is scattered out of the beam. A proper measure of the radiation absorbed by the TiO_2 requires the amount total amount of scattered light to be measured—by the use of an integrating sphere [23,29]. Figure 4 compares calculated (from the results in Figure 3a,b) and experimental coefficients for 50 nm rutile and shows how the experimental result changes as size (measured by x-ray sedimentation) varies from 35 to 145 nm [23]. A comparison with the absorption properties of sunscreen formulations has also been made [23]. Satuf *et al.* have used diffuse reflectance and transmittance spectrophotometric measurements of suspensions to derive curves for Aldrich TiO_2, Degussa (now Evonik) P25 and Hombikat UV 100. At 310 and 360 nm the derived values of absorption coefficient are ca 3.1 and 4.6 $\times$ 10^4 $cm^2 \cdot g^{-1}$ respectively for P25 and 0.75 and 0.5 for Hombikat UV 100 [29]. When the scattering and absorption coefficients are known, they may be used to identify optimum catalyst loading for reactors of specified geometry and radiation input [30]. Li Puma *et al.* have developed a 6-flux model, to estimate the flux throughout a photocatalytic reactor and have used their treatment to model the effect of catalyst loading on the overall volumetric rate of photon absorption, and hence on the effective quantum yield [31,32].

Although conventional reflectance spectra on pressed discs of powder may usefully demonstrate the position of the absorption edge (or band-gap) [12], and confirm, for example, that anatase absorbs at shorter wavelengths (has a larger band-gap) than rutile, they do not effectively show differences in absorption at wavelengths shorter than the band gap by suspensions of particles. This is illustrated in Figure 5a which shows the very similar diffuse reflectance spectra of compacts of three rutile samples with mean sizes of 35, 50 and 145 nm, and the size distributions shown in Figure 5b, whose extinction coefficients in suspension are shown in Figure 4. Even though there are large differences between the transmission spectra of suspensions of the three samples, the differences between the reflectance spectra are negligible. Little information about the optics of particle suspensions is contained in the reflectance spectra of discs made from powders from which the suspensions were prepared.

Figure 4. Experimental extinction coefficients for 35 ■, 50 ▲ and 145 nm ● rutile particles and the calculated coefficients for 50 nm △, derived from Figure 3a,b.

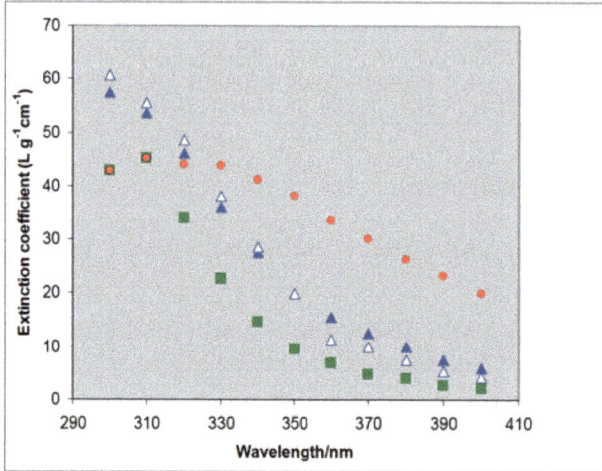

Although conventional reflectance spectra on pressed discs of powder may usefully demonstrate the position of the absorption edge (or band-gap) [12], and confirm, for example, that anatase absorbs at shorter wavelengths (has a larger band-gap) than rutile, they do not effectively show differences in absorption at wavelengths shorter than the band gap by suspensions of particles. This is illustrated in Figure 5a which shows the very similar diffuse reflectance spectra of compacts of three rutile samples with mean sizes of 35, 50 and 145 nm, and the size distributions shown in Figure 5b, whose extinction coefficients in suspension are shown in Figure 4. Even though there are large differences between the transmission spectra of suspensions of the three samples, the differences between the reflectance spectra are negligible. Little information about the optics of particle suspensions is contained in the reflectance spectra of discs made from powders from which the suspensions were prepared.

At least two factors contribute to the difference between the reflectance and transmission spectra. Firstly, the reflectance spectra depend on the ratio, K/S, of the absorption coefficients (K) and the scattering coefficient (S) of the powder. Since both of these depend on the ratio d/λ the effect of changes in particles-size, d, are minimized (by contrast, the transmission spectrum of Figure 4 depends on the sum of K and S). Secondly, because of interference effects, the scattering of particles is strongly influence by particle-particle distance, which is clearly quite different in dilute suspensions and pressed powder compacts [33].

Figure 5. (a) The reflectance spectra (measured on a Jasco 670 spectrometer fitted with an integrating sphere) of pressed discs made from the three different rutile samples of mean size 35 —; 50 — and 145 — whose suspension spectra are shown in Figure 4 and whose size distributions are shown in Figure 5b; (b) The particle-size distributions measured by X-ray size sedimentation (Brookhaven X-ray disc sedimentometer) of three rutile samples (reprinted with permission from Egerton & Tooley, International Journal of Cosmetic Science **2012**, *34*, 117–122 [23] published by Society of Cosmetic Sciences; Société Française de Cosmétologie and Blackwell Publishing).

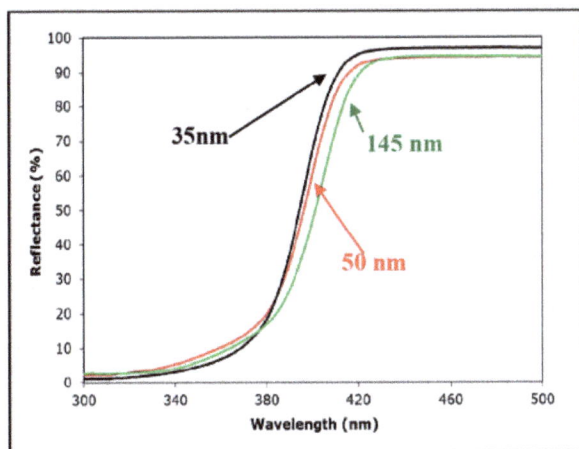

(a)

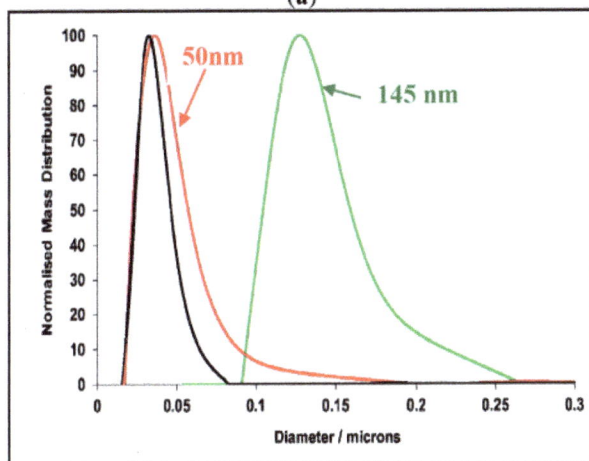

(b)

2.3. The Major Problem Associated with the Calculation of Attenuation by Semi-Conductor Dispersions

The previous sections have shown that scattering and absorption by TiO_2 particles depends on their size in suspension. The size of the fundamental, or primary, particles may be measured by X-ray line broadening (if the particles are crystalline) or inferred from BET surface area measurements, using area gm^{-1}, A = 6 d/ρ), and are usually in reasonable agreement (Figure 6) but it is rare for the TiO$_2$ to exist as isolated primary particles in suspension.

Instead, as represented schematically by Figure 7 and, for the case of P25, by the transmission micrographs of Figure 6 of reference [34], the particles in typical photocatalytic suspensions are, if great care has been taken, aggregates, but more usually, weakly-bound agglomerates. (By agglomerates is meant a secondary particle composed of flocculated or coagulated particles which can be broken down by changes in such factors as suspension pH and mechanical forces. By Aggregate is meant a more strongly bound secondary particle—perhaps formed by sintering during the particle preparation process—and much less susceptible to mechanical disruption.)

Figure 6. A comparison of published XRD sizes with the BET-derived sizes of (mainly) anatase samples identified from the publications listed in the caption to figure 15 of Egerton, T.A.; Tooley, I.R., *Intl. J. Cosmetic Sci.* **2014**, *36*, 195–206 [35]. The dashed line corresponds to S = 6 D/ρ. where S is the surface area, D the particle diameter and ρ the particle density.

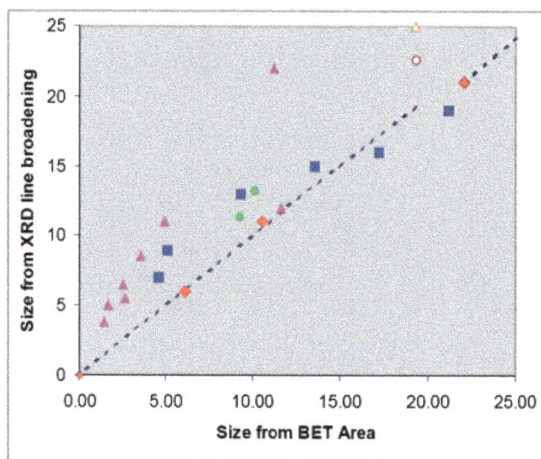

Since the gap between the primary particles is normally small in comparison with the radiation wavelength, the optical behaviour is controlled by the size of the aggregates or agglomerates in suspension as explicitly recognized by both Martin, Baltanas and Cassano [34] and by Egerton and Tooley [36]. If the pH is near the isoelectric point there is negligible particle-particle electrostatic repulsion but significant particle attraction because of TiO$_2$'s high Hamaker constant. On the basis of electron microscopy Martin *et al.* concluded that "*Any degree of stirring (be it mechanical agitation, pump recycling or gentle gas bubbling) produces an important change in the optical properties*" [34]. However, because the primary particle size *is* relevant to calculations of adsorption and surface-coverage, and perhaps because the primary particle size is easier to measure, particle characterization continues to be mainly by surface area and XRD derived sizes. Therefore, it is appropriate to summarize some of the results which show that there is normally significant particle agglomeration in suspension, even at pH's at which there is significant surface charge.

Figure 7. Schematic representation of the aggregation and agglomeration of titanium dioxide nanoparticles. Primary particles may flocculate to form weakly bound agglomerates or sinter to form much stronger aggregates. Agglomerates may sinter to form more strongly bound aggregates. Agglomerates break down more easily than aggregates.

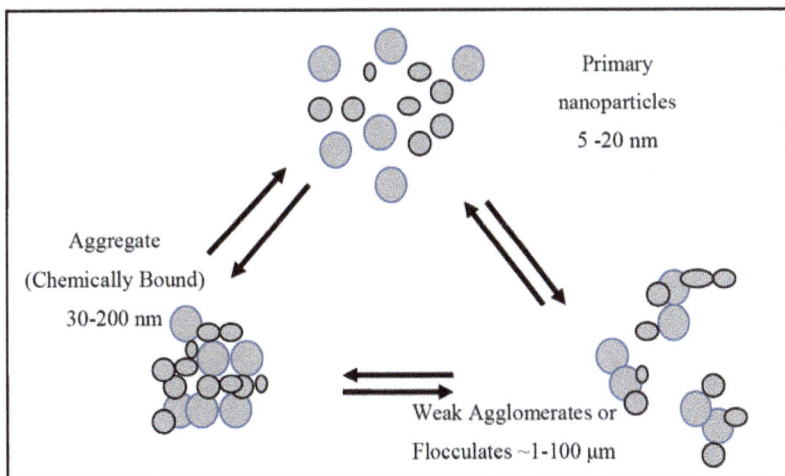

Ridley and coworkers studied 0.15 g·dm^{-3} aqueous dispersions of a crystalline anatase (Ishihara ST-01) at a pH of 2.7, well below the p.z.c. of 6.85 [37]. The TEM size was 4.6 nm, the BET equivalent sphere diameter was 5.0 nm, and the XRD crystallite size was 7 nm. However, laser diffraction methods showed the dispersed particles to have a relatively small mode centred near 100 nm and to consist predominantly of microscale aggregates with a median diameter of 2329 nm (99.6% of the distribution volume was characterized by diameters greater than 400 nm). At this pH, ultrasonication did not significantly alter the measured size distribution. French and co-workers used dynamic light scattering (DLS) at a sample concentration of ~40 mg·dm^{-3} to show that 4–5 nm sol-gel derived anatase particles form stable agglomerates with an average diameter of 50–60 nm at pH ~ 4.5 in 0.0045 M NaCl [38]. When the ionic strength was increased to 0.0165 M micron-sized agglomerates formed within 15 min. At all other pH values tested (5.8–8.2), micron-sized agglomerates formed in less than 5 min even at low ionic strength (0.0084–0.0099 M NaCl). DLS measurements were used by Lee *et al.* to show that in suspension particles with a TEM size below 100 nm, increased in size from ~100 to 500 nm as their concentration increased from 5 to 80 mg·dm^{-3} [39]. Egerton and Tooley compared XRD line-broadening and, BET equivalent sphere diameters with sizes measured by X-ray sedimentation and DLS of three surface-treated rutile samples [35]. In each case the DLS size was and 5–10 times larger than the XRD or BET sizes. The sedimentation size and DLS sizes were comparable. A comparison of XRD, BET, sedimentation and Laser measurements is shown in Table 1.

The size-measurement results are, with hindsight, supported by two widely used experimental procedures—filtration of catalysts from reaction mixtures and the practice of stirring suspensions of photocatalysts to prevent sedimentation. If the particles were dispersed, it would be impossible to remove 20 nm particles of P25 catalyst particles from reaction mixtures by 0.1 μm Millipore filters, or by guard columns, as is commonly done prior to chromatographic analysis [3,36]. Also, since Stokes' equation shows that primary particles would not sediment significantly over the time of a typical catalytic experiment, there would be no need of the common experimental practice of stirring suspensions of photocatalysts to prevent sedimentation [2,40]. Despite this problem, it is still both useful and important, to consider the effect of absorption when interpreting the results of photocatalytic experiments by dispersions of TiO_2. The rest of this paper will give some examples of how this may be the case.

Table 1. A comparison of particle-sizes of nano-particulate TiO_2 measured by X-ray line broadening or by nitrogen adsorption (BET) on the dry powders and by X-ray sedimentation and laser diffraction or laser scattering on aqueous dispersions of TiO_2.

Base Crystal & Surface-Treatment	XRD Line Broadening	Equivalent Sphere Diameter from BET Area	Sedimentation Brookhaven	Dynamic Light Scattering (DLS) or Laser Diffraction
1. Ishihara ST-01 Anatase Untreated [37]	7	5	-	2329 Beckman Coulter LS-230 Laser Diffraction
2. Rutile-A silica-alumina [35]	15	12	91 ± 10	133 (DLS)
3: Rutile-A Stearate [35]	15	23	53 ± 10	124 (DLS)
4: Rutile-B Stearate [35]	9	48	160 ± 19	201 (DLS)

3. The Need to Consider UV Absorption when Interpreting Photocatalytic Results

3.1. Protection against Photodegradation of Organic Materials by Dispersed TiO₂

3.1.1. Photodegradation of Polyethylene Films

Academic studies of photocatalysis usually select systems in which oxidative degradation is negligible in the absence of TiO_2. However, many commercial systems—e.g., most polymers and paints are degraded by UV even when TiO_2 is not present and Figure 8 shows the increasing infrared signature of the carbonyl oxidation products of an unpigmented polyethylene (PE) film exposed to UV for increasing times [41]. Ultimately the film is totally oxidized to carbon dioxide and water.

If TiO_2 is present, it has normally been added to opacify the product—e.g., to pigment a polythene film. Commercial pigments, often rutile because its higher refractive index, enhance opacity, and are frequently surface-treated, e.g., with an amorphous layer of hydrous silica and/or alumina and or/zirconia (see Figure 9) to minimize photocatalysis, and lengthen the lifetime of the product [5,42–46].

Figure 8. The development of the infrared absorption characteristic of carbonyl oxidation products in an unpigmented polyethylene film as the UV exposure in QUV accelerated weathering equipment, fitted with UVA-340 tubes and operated at 40 °C, increases from 125 to 1348 h. (reprinted from Polymer Degradation and Stability. **2007**, *92*, 2163–2172 [41] with permission).

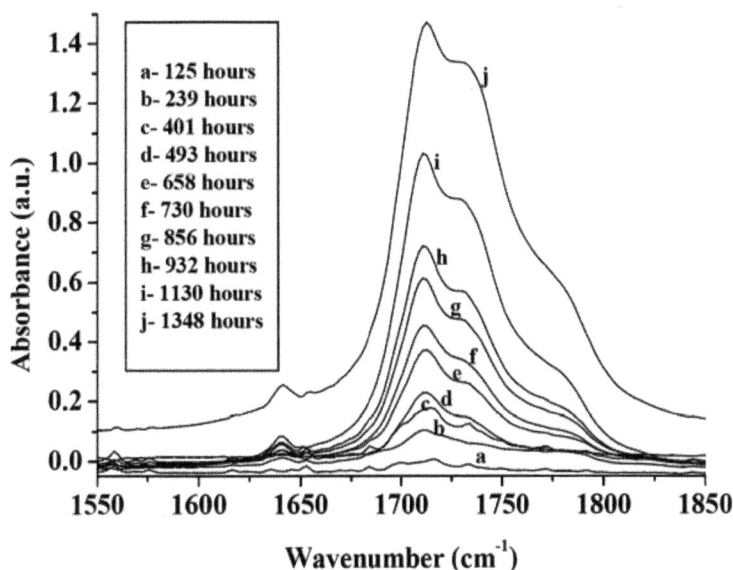

In polymers pigmented with a surface treated rutile, the absorption of UV by added TiO_2 may reduce the photochemical oxidation and extend the life of the polymer and this is shown in Figure 10 where the development of the carbonyl absorption in 100 μm thick films of low density polyethylene in unpigmented (PE-U1) and films prepared with different TiO_2 pigments at a loading of 5 p.h.r. (parts per hundred resin by weight). For the anatase (PE-A1) and lightly coated rutile (PE-R1-1) films incorporation of pigment has increased the photodegradation—*i.e.*, photocatalysis by the TiO_2 is more important than any UV absorption by the pigment. For the remaining pigments (R3-1 and R4-1), in which photocatalysis has been suppressed by a heavier surface coating UV absorption, the absorption by UV dominates so that the total degradation is less than that of the unpigmented film. This reduction occurs even though the total path length is only 100 μm and at wavelengths greater than 250 nm the absorbance is less than 0.2 for the unpigmented film. Figure 10b shows that the same pattern is observed if total oxidation of the polymer to carbon dioxide is measured instead of the formation of the intermediate carbonyl groups [41]. A similar pattern is also observed in PVC for which Worsley *et al.* have noted that the activity of P25 is two orders of magnitude higher than that of a Al/Si/Zr coated rutile pigment [45].

Figure 9. Transmission electron micrographs showing (**a**) the uncoated TiO2 crystals and (**b**) surface treated (coated) TiO2 rutile pigment. The ZrO2/Al2O3 coating, with a thickness of 3–10 nm, shows as a less dense outline to the images (reprinted from J. Mater. Sci. **2002**, *37*, 4901–4909 [47] with permission).

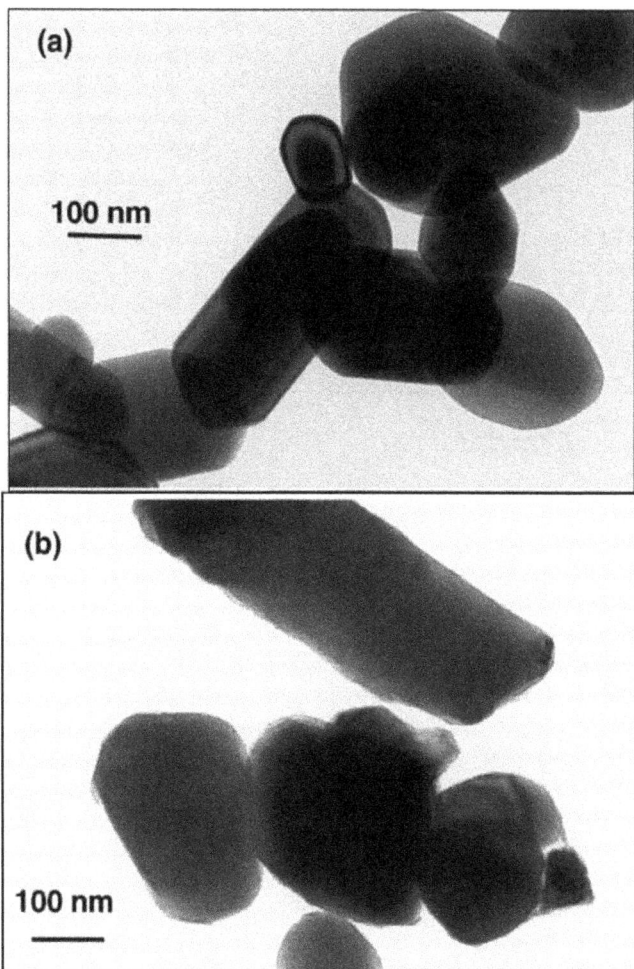

3.1.2. Photodegradation of Alkyd Paint Films

Alkyd films, of the type used in oil-based paints, absorb more strongly between 300 and 400 nm (Figure 11a) than PE films, and the pigment volume fraction (p.v.c.) of TiO_2 in an alkyd paint may be ten times that in a polymer film. The oxidative degradation of these paints to CO_2 may be measured by monitoring their loss in weight when exposed to UV and Figure 11b shows results for a series of such paints exposed to UV in carbon-arc Marr weathering equipment [47]. For the uncoated rutile the weight-loss increases as the p.v.c. increases from 5% to 40%. The much lower weight loss from paints made with coated pigment confirms that the surface coating reduces the pigment photoactivity and therefore has increased the relative importance of protective UV absorption. The decrease in weight loss as the p.v.c. of the coated pigments increases from 5% to 40% demonstrates that the protective role of TiO_2 can also occur when the TiO_2 loading is much higher than that in polymer films.

Figure 10. (a) The development of carbonyl absorption in unpigmented and pigmented polythene films (PE A-1, ■; PE R1-1, ▲, PE U-1, ●; PE R2-1, ♦; and PE R3-1, ▼ and PE R4-1, ╋. as a function of exposure in QUV accelerated weathering equipment. After recording each IR spectrum the disc was returned to the exposure unit for further UV exposure); **(b)** CO_2 evolution from the photo-oxidation of the same films as used for the carbonyl development measurements in Figure 10a but exposed to irradiation from a xenon lamp [46].

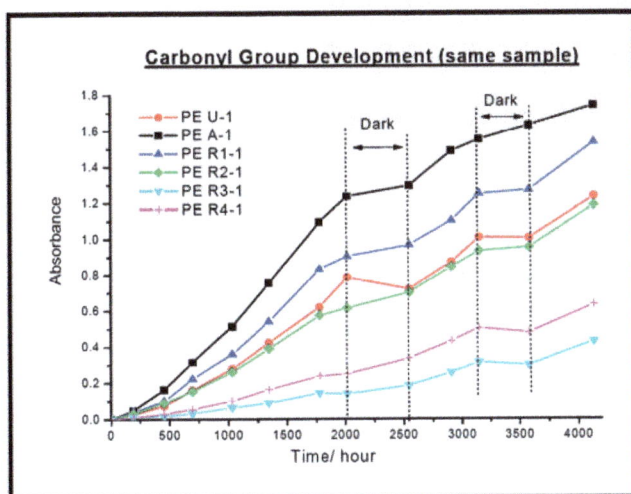

(a)

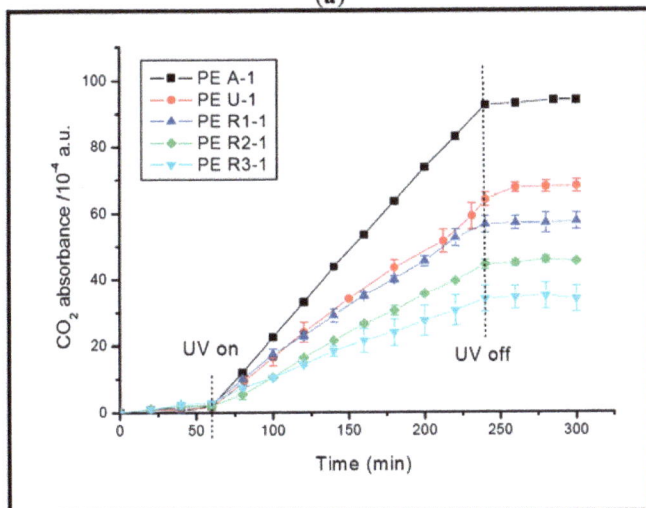

(b)

Figure 11. (a) UV absorption of a 90 μm unpigmented alkyd film compared with the spectral distribution of the carbon arc lamps used in an "accelerated weathering test". LH scale, arc lamp distribution; RH scale, Film Absorbance (reprinted from *J. Mater. Sci.* **2002**, *37*, 4901–4909 [47] with permission); (b) The degradation (weight loss as a function of time exposed to radiation from carbon arcs) of paint films of long-oil soya alkyd opacified with either uncoated rutile (solid points) or coated rutile (empty points) by 40 (■,□), 25 (●○) and 5 (◆◇) volume % pigment.

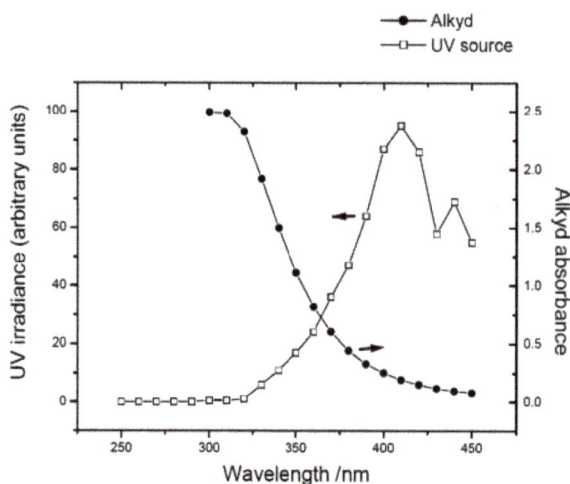

(a)

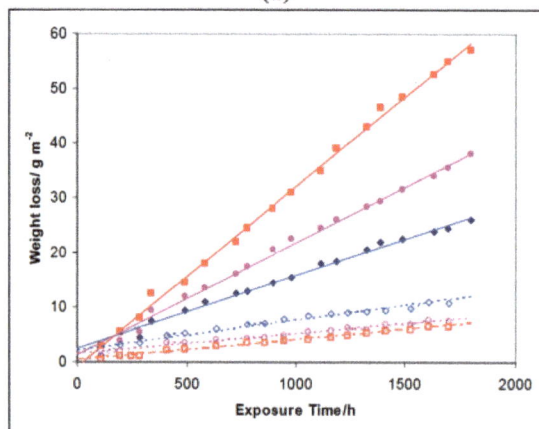

(b)

3.1.3. Disinfection by UV-C

Dillert, Siemon and Bahnemann compared the disinfection of: (a) pre-treated municipal wastewater and (b) a model pollutant, *E. coli.* [48] by two different lamps, A and B. In the wavelength ranges < 280 nm, 280–315 nm, and 315–400 nm the output of lamp A (Solidex glass envelope) was 5, 150 and 220 mEinstein·h^{-1} whilst that of B (Jena glass envelope) was 0.2, 18 and 390 mEinstein·h^{-1}, respectively. *i.e.*, the short wavelength output of A was much greater than that of B. A summary of their results (Table 2) shows that although, without TiO$_2$, both lamps disinfected, A, because of its higher UV-B and UV-C output, was more effective than B in both systems. However, addition of 5 g·dm^{-3} P25 TiO$_2$ reduced the effectiveness of lamp A (disinfection times increased to more than 360 min. for waste-water and to more than 270 min. for E. coli) but greatly increased the effectiveness of lamp B for disinfection of E. coli and this was attributed to photocatalysis by the TiO$_2$. By contrast with this increase, the effectiveness of lamp B for wastewater treatment was decreased. Dillert et al. attributed the significant reduction of the effectiveness of lamp A to the "shadowing effect" of the TiO$_2$ particles. Their conclusion that photon absorption by TiO$_2$ can reduce the number of photons available for direct photochemical not only demonstrates the importance of UV absorption in the photocatalytic treatment of practical systems, but also implies that this absorption may be more important with respect to some wavelengths (in this case short wavelengths) than for others.

Table 2. A comparison, derived from the results of Dillert et al. of the disinfection achieved by two different UV lamps, A and B [48]. The results show the potential importance of UV absorption by TiO$_2$ in reducing the disinfection rates in systems for which direct UV treatment is effective.

Conditions	Pre-Treated Wastewater		E. coli	
	Initial Number of Colony Forming Units (c.f.u.)	C.f.u after 60 min Treatment	Initial Number of Colony Forming Units (c.f.u.)	C.f.u after 60 min Treatment
Lamp A	~6 × 10^4	Not detectable	3 × 10^6	Not detectable
Lamp A+TiO$_2$	3 × 10^5	1.5 × 10^4	6 × 10^6	1 × 10^5
Lamp B	1 × 10^4	2 × 10^2	~8 × 10^6	~6.5 × 10^5
Lamp B+TiO$_2$	3 × 10^4	6 × 10^3	~8 × 10^6	1.5 × 10^1

3.2. Reduced Photocatalytic Activity of Other Solution Species because of UV Absorption by Nano-Particulate TiO$_2$

UV-C photocatalysed dye-decolouration by hydrogen peroxide/anatase mixtures has been widely reported to be faster than photocatalytic decolouration by anatase alone. For Tropaeoline and Reactive Red the reaction rate is doubled [49,50]; for Safira HEXL anionic azo dye the rate is enhanced ×4.5, at pH 7, and ×9, at pH 5 [51]. For Acid Red 14 a 20 fold enhancement has been reported [52]. However results such as those shown in Figure 12, for the UV-C decolouration by anatase PC500 of the azo-dye Reactive Orange 16, suggest that the true picture is less simple [53]. Experiments with both 2 mM and 20 mM H$_2$O$_2$ showed that although decolouration by UV-C+TiO$_2$+H$_2$O$_2$ is faster than decolouration by UV-C+TiO$_2$, it is slower than decolouration by UV-C+H$_2$O$_2$ in the absence of anatase. If the reaction is carried out in water TiO$_2$ the rate is decreased to 45% of that H$_2$O$_2$ alone. In 0.1 M NaCl the reduction is by 43% and In 0.1 Na$_2$SO$_4$ the reduction is 64%. UV-C photolyses H$_2$O$_2$ to hydroxyl radicals with a quantum efficiency that can approach unity [54] but the quantum efficiencies for the TiO$_2$-photocatalysed processes are << 5%. Therefore, the reduction in decolouration rate when TiO$_2$ is added to H$_2$O$_2$ shows that UV-C photon absorption by the TiO$_2$ reduces the number of photons which would otherwise be available to efficiently photolyse H$_2$O$_2$.

Figure 12. UV-C decolouration of 0.05 mM RO16 in the presence of TiO$_2$ (● UV-C only: ▲ 2 g·dm^{-3} TiO$_2$: ♦ 2 g·dm^{-3} TiO$_2$, 20 mM H$_2$O$_2$: ◊ No TiO$_2$, 20 mM H$_2$O$_2$). Reprinted from Egerton, T.A.; Purnama, H. *Dyes Pigments* **2014**, *101*, 280–285 [53] with permission.

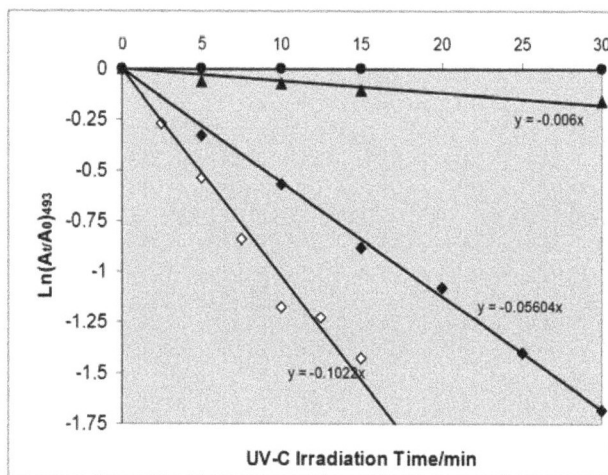

3.3. Changes in UV Absorption Caused by Agglomeration or Flocculation Alter the Measured Photocatalytic Activity of Nano-Particles

3.3.1. Predicted Change in Photocatalytic Activity Resulting from an Increase in UV Absorption

As stressed in Section 2.3, because of flocculation/agglomeration the effective size of TiO$_2$ particles in suspension is larger than that of their constituent primary particles. As illustrated by the large flocculate sizes measured by DLS and other optical techniques (Table 1), it is the size of the secondary particles that control the optics (because UV photons have a much larger wavelength, ~300 nm, than the spaces between the primary particles (10–20 nm). However, because charge transfer between primary particles is limited, the charge-carrier recombination, unlike the optical behaviour, is limited mainly by the size of primary particles. Consequently, changes in the degree of particle agglomeration, such as brought about by milling, should change the measured photocatalytic activity.

The absorption curves of Figure 3b demonstrate that UV-A absorption increases significantly as rutile particle size is decreased from 200 to 100 nm. Assuming that the reactor geometry is such that all the UV is absorbed, the effect of increased absorption is that the same amount of UV is absorbed in a shorter path-length—by fewer primary particles, as illustrated by Figure 13.

The effect of this difference in absorption may now be considered. In a suspension each primary particles acts as micro-reactor and the flux of incident photons is usually sufficiently large for recombination kinetics to dominate, *i.e.*, the rate is proportional to the square root of incident intensity [8,29,55–57].

Figure 13. A schematic depiction of the effects of increased UV absorption associated with improved dispersion of TiO$_2$ particles. The second and third rows represent changes in the transmission spectrum and the attenuation of the incident UV beam as the particle dispersion is altered in the way depicted in the top row.

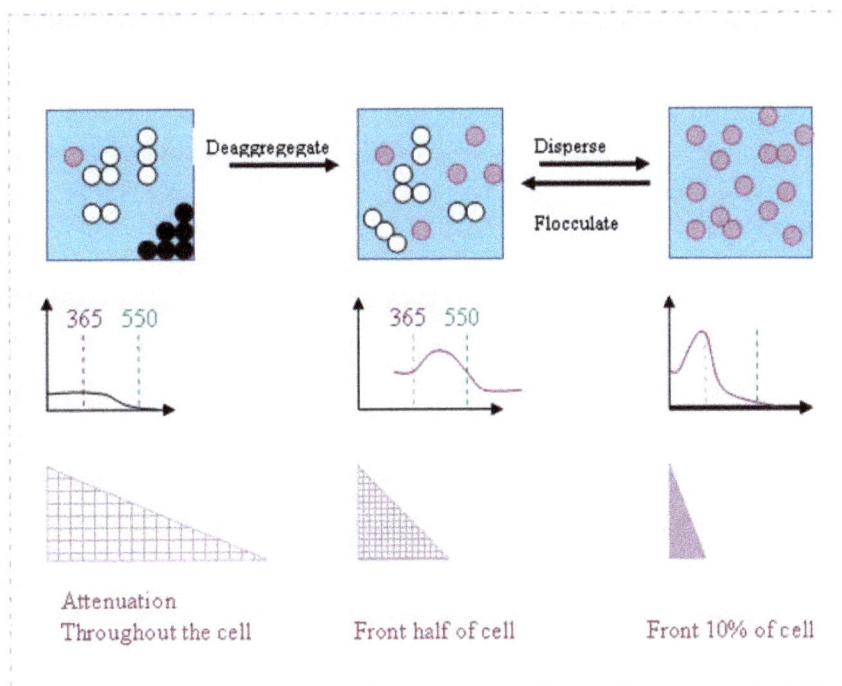

Then, if all of the P incident photons are absorbed by n of the more weakly absorbing 200 nm particles, generating P/n electron-hole pairs in each, the observed rate, R$_{200}$, is proportional to the product of the number of micro-reactors and the rate in each of them:

$$R_{200} = k \times n \times (P/n)^{0.5} = k \times P^{0.5} n^{0.5} \tag{2}$$

If the 100 nm particles absorb twice as strongly as the 200 nm particles all of the photons will be absorbed in only n/2 particles and the number of electron-hole pairs in each of these will be P/(n/2) = 2P/n. This leads to an observed rate, R$_{100}$, given by:

$$R_{100} = k \times n/2 \times (2P/n)^{0.5} = k \times P^{0.5} n^{0.5} 2^{-0.5} \tag{3}$$

The effect of the *increased* absorption has been to *decrease* the measured rate because recombination has meant that absorption in fewer particles is not compensated by the increased number of electron-hole pairs that are generated in each. Thus, on the basis of the optics alone more strongly absorbing particles are predicted to be less active than less strongly absorbing, larger particles.

3.3.2. Effect of Milling on Photocatalytic Activity for Propan-2-ol Oxidation

The conclusion that improving catalyst dispersion, and decreasing the mean size of the agglomerated particles, reduces measured photocatalytic activity has been demonstrated for propan-2-ol oxidation using a high area rutile, 140 $m^2 \cdot g^{-1}$, with a line broadening size of 7–10 nm [36]. X-ray disc sedimentation measurements demonstrated that despite the XRD size of 7–10 nm the particle size of unmilled particles in suspension was greater than 40 nm and that vigorous milling was unable to reduce the size below 20 nm. The optical changes measured from 300 to 700 nm on diluted suspensions prepared by small-scale sand milling (probably less vigorous than used for the size-measurements) for increasing times are shown in Figure 14a. The significant turbidity between 400 and 700 nm is due to scattering and confirms that many particles must be of the order of 100 nm or more. The decrease in the turbidity at, e.g., 550 nm indicates that some of these large particles are broken down as milling times increase from 15 to 30 minutes. Correspondingly, the absorption below 400 nm increases as the milling time is extended. The effect of these changes in particle size are shown in Figure 14b. Similar results, a halving of the measured activity, have been measured for a rutile with a much larger primary particle size; ~200 nm by TEM and a surface area of 8 $m^2 \cdot g^{-1}$. A decrease in activity with milling has also been measured for P25 (Millenium), 10 $m^2 \cdot g^{-1}$ anatase (Huntsman), a low area rutile (Huntsman) surface treated with rutile, a 76 $m^2 \cdot g^{-1}$ anatase (Tayca) and high area rutile doped with iron [40].

Figure 14. (a) Optical transmission curves measured on (A) unmilled 140 $m^2 \cdot g^{-1}$ rutile, and the same rutile milled for (B) 7.5 minute, (C) 15 minutes and (E) 30 minutes. Curve D was measured on a 15-minute milled and diluted sample that had been left to stand for the duration of a typical oxidation experiment prior to making the measurement; **(b)** The time dependence of propanone formation during the photocatalytic oxidation of propan-2-ol by the same 140 $m^2 \cdot g^{-1}$ rutile before sand-milling ●, and after milling for 7.5 ■, 15 ▲ and 30 ×, minutes. Reprinted from Egerton & Tooley, *J. Phys. Chem. B* **2004**, *108*, 5066–5072 [36] with permission.

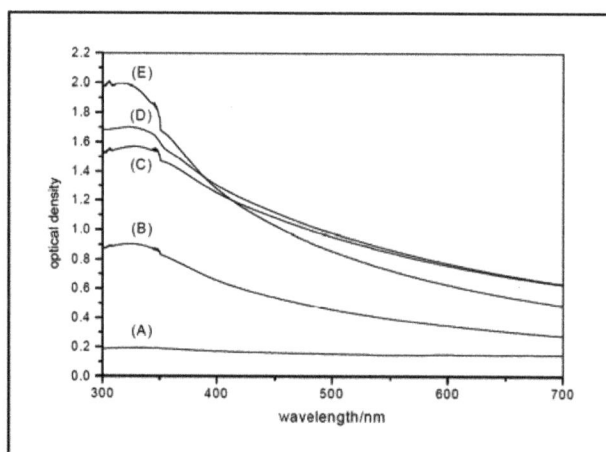

(a)

Figure 14. *Cont.*

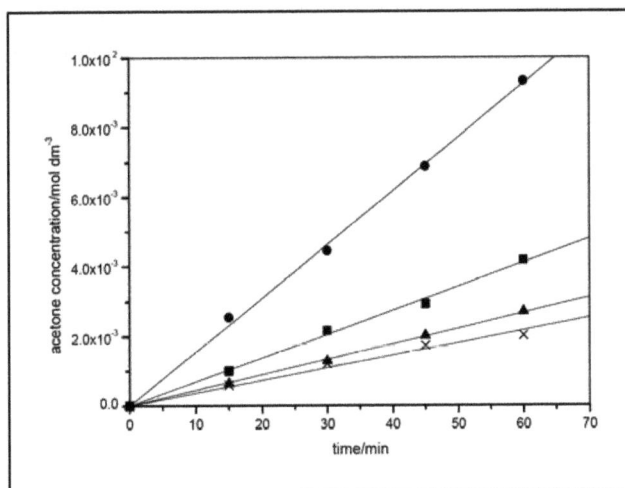

(b)

3.3.3. Effect of Milling on Photocatalytic Degradation of Salicylic Acid

Propan-2-ol oxidation is believed to proceed by a hydroxyl radical mechanism and so the effect of milling on the photocatalytic degradation a reaction of salicylic acid has been investigated. (Preliminary experiments using neutral density filters confirmed the rate of this reaction varied as the square root of the UV intensity.) Once again, milling decreased the rate of reaction, as shown in Figure 15 [40].

Figure 15. The effect increasing milling times on the first order rate constant for the degradation by high area rutile of 0.36 mM salicylic acid at pH 4. Reprinted from *J. Photochem. Photobiol. A* **2010**, *216*, 268–274 [40] with permission.

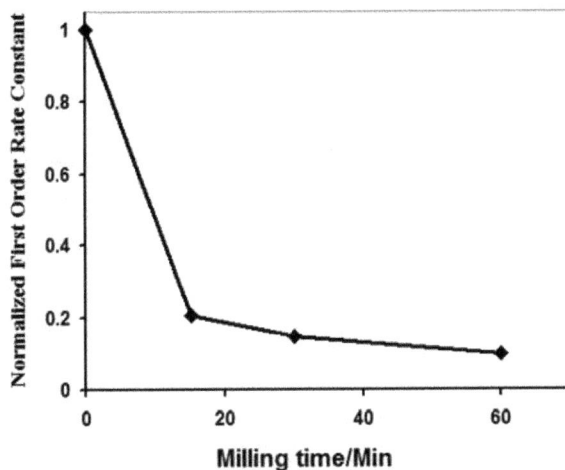

3.3.4. Effect of Milling on Photocatalytic Degradation of Dichloroacetic Acid

The UV attenuation of suspensions of TiO$_2$ in dichloroacetate increased with milling, as with propanol and salicylic acid. However, the rate of dichloroacetate degradation increased linearly with UV intensity, unlike the I$^{0.5}$ dependency of rate found for propan-2-ol oxidation and salicylic acid degradation. Therefore the conclusion of Section 3.3.1, that more strongly UV-absorbing particles should be less active, is not necessarily valid for this system. The results shown in Figure 16 show that the photocatalytic activity for dichloroacetic acid degradation at pH 3 did *not* decrease as the catalyst suspension was milled. Therefore increased UV absorption by individual crystal should compensate for the reduction in the number of micro-reactors. For P25 rate was independent of milling time and for high area rutile of the rate actually increased [40].

Figure 16. Effect of milling time on the DCA oxidation rate by a high area rutile ♦ and by P25 ■. Reprinted from *J. Photochem. Photobiol. A* **2010**, *216*, 268–274 [40] with permission.

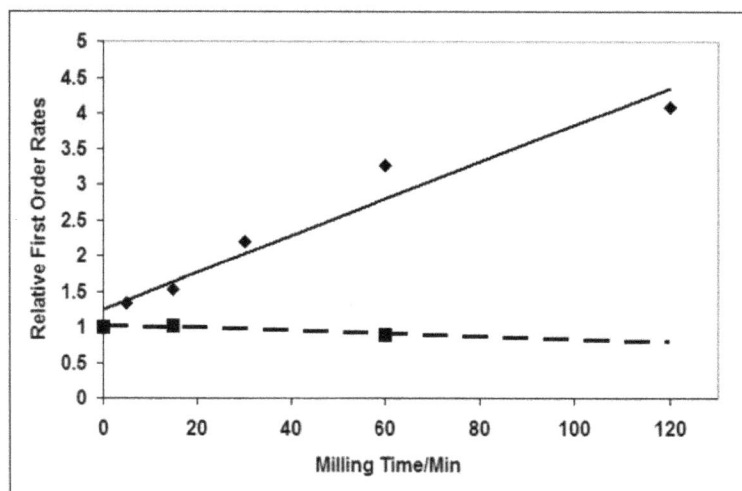

4. Conclusions and Implications

UV absorption should be considered explicitly when considering TiO$_2$ photocatalysis, because, in broad-band semiconductors, UV absorption initiates charge-carrier generation Despite many problems in their application, theories of light scattering and absorption have been of practical use in the design of inorganic UV–blocks for cosmetics and of opacifying pigments for paints. Both scattering and absorption depend on TiO$_2$ particle size. However, the size that is relevant to photocatalysis in aqueous dispersions of TiO$_2$ is *not* the size inferred from X-ray line-broadening or from surface area measurements. This is because these methods measure the size of primary particles, but particles in suspension are almost always aggregates or agglomerates of primary particles. It is the aggregated or agglomerated particles that control the optics.

The *protection* of commercial pigmented products such as polymers or paints, from UV is one demonstration of the practical significance of UV absorption by TiO$_2$. Degradation is reduced because the reduction in the number of active photons is more important than the residual photocatalysis of surface-treated TiO_2. Even for photocatalytic grades of TiO_2, UV absorption reduces the number of active photons and can negate the benefits that have been claimed to result from the addition of photochemically active species to TiO_2 suspensions. E.g., TiO_2 can reduce the decolouration of dye residues by H_2O_2. Such effects should be explicitly considered in studies of hoped-for synergy.

Increased absorption by TiO_2 particles may *decrease* measured photocatalytic rates. This is because charge-carrier recombination within particles may lead to the absorption of n photons by one photocatalyst particle being less effective than the absorption of 1 photon by each of n particles. Therefore, milling of photocatalysts to improve photocatalytic dispersion may reduce the measured rate of photocatalysis—as shown for the degradation of propanol, and salicylic acid. By contrast, milling does not decrease the measured rate of photocatalytic degradation of dichloroacetate; the difference is attributed to the fact that this reaction does not have an $I^{0.5}$ dependence on the intensity, I, of the incident UV.

For reactions that do have an $I^{0.5}$ intensity dependence, the effects of changes in dispersion may need to be considered. Sometimes, as when catalysts are immobilized on papers or fibres, these dispersion changes are deliberately induced. In many more cases they are overlooked, e.g., (i) if particles produced by a new synthetic route have a different degree of aggregation; or (ii) if deliberate changes in suspension pH alter surface potential and therefore flocculation; or (iii) if heat-treatment during doping increases aggregation. However, all such changes in particle dispersion may change the measured photocatalytic activity for reasons that are purely a consequence of the changed optics and suitable control experiments should be used.

Acknowledgments

It is a pleasure to thank the students and colleagues whose work has been reviewed in this paper, and especially to thank Dave Tunstall (formerly of Tioxide) and Ian Tooley (Croda) for many contributions and discussions.

Conflicts of Interest

The authors declare no conflict of interest.

References

1. Matsunaga, T.; Okochi, M. TiO_2-Mediated Photochemical Disinfection of *Escherichia coli* Using Optical Fibers. *Environ. Sci. Technol.* **1995**, *29*, 501–505.
2. Maillard, C.; Guillard, C.; Pichat, P. Comparative effects of the TiO_2-UV, H_2O_2-UV H_2O_2-Fe^{2+} systems on The disappearance Rate of benzamide and 4-hydroxybenzamide in water. *Chemosphere* **1992**, *24*, 1085–1094.
3. Mills, A.; Belghazi, A.; Davies, R.H.; Morris, S.A. Kinetic study of the bleaching of Rhodamine 6G photosensitized by titanium dioxide. *J. Photochem. Photobiol. A* **1994**, *79*, 131–139.

4. Fateh, R.; Dillert, R.; Bahnemann, D. Self-cleaning properties, mechanical stability, and adhesion strength of transparent photocatalytic TiO_2-ZnO coatings on polycarbonate. *ACS Appl. Mater. Interfaces* **2014**, *6*, 2269–2277.

5. Hughes, W. Phodegradation of paint films containing TiO_2 pigments. In *Xth FATIPEC Congress 1970*; Verlag Chemie GmbH: Weinheim/Bergst, Germany, 1970; pp. 67–82.

6. Allen, N.S.; McKellar, J.F.; Phillips, G.O.; Chapman, C.B. The TiO_2 photosensitized degradation of nylon 6, 6: Stabilizing action of manganese ions. *J. Polym. Sci. C Polym. Lett.* **1974**, *12*, 723–727.

7. Turchi, C.S.; Ollis, D.F. Photocatalytic degradation of organic-water contaminants—Mechanisms involving hydroxyl radical attack. *J. Catal.* **1990**, *122*, 178–192.

8. Egerton, T.A.; King, C.J. The Influence of Light Intensity on Photoactivity in TiO_2 pigmented systems. *J. Oil Colour Chem. Assoc.* **1979**, *62*, 386–391.

9. Bard, A.J. Heterogeneous photocatalytic decomposition of saturated carboxylic acids on TiO_2 powder decarboxylative route to alkanes. *J. Am. Chem. Soc.* **1978**, *100*, 5985–5992.

10. Mills, A.; LeHunte, S. An overview of semiconductor photocatalysis. *J. Photochem. Photobiol. A* **1997**, *108*, 1–35.

11. Glasstone, S. *Textbook of Physical Chemistry*, 2nd ed.; Macmillan: London, UK, 1962; p. 1155.

12. Asahi, R.; Morikawa, T.; Ohwaki, T.; Aoki, K.; Taga, Y. Visible-light photocatalysis in nitrogen-doped titanium oxides. *Science* **2001**, *293*, 269–271.

13. Vinodgopal, K.; Wynkoop, D.E.; Kamat, P.V. Environmental photochemistry on semiconductor surfaces: Photosensitized degradation of a textile azo dye, acid orange 7, on TiO_2 particles using visible light. *Environ. Sci. Technol.* **1996**, *30*, 1660–1666.

14. Chen, Y.S.; Crittenden, J.C.; Hackney, S.; Sutter, L.; Hand, D.W. Preparation of a novel TiO_2-based p-n junction nanotube photocatalyst. *Environ. Sci. Technol.* **2005**, *39*, 1201–1208.

15. Morawski, A.W.; Grzechulska, J.; Kalucki, K. A new method for the preparation of potassium-pillared layered titanate applied in photocatalysis. *J. Phys. Chem. Solids* **1996**, *57*, 1011–1017.

16. Pelizzetti, E.; Carlin, V.; Minero, C.; Pramauro, E.; Vincenti, M. Degradation pathways of atrazine under solar light and in the presence of TiO_2 colloidal particles. *Sci. Total Environ.* **1992**, *123*, 161–169.

17. Wang, L.; Egerton, T.A. The influence of chromium on photocatalysis of propan-2-ol and octadecanoic acid oxidation by rutile TiO_2. *J. Photochem. Photobiol. A* **2013**, *252*, 211–215.

18. Litter, M.I. Heterogeneous photocatalysis-transition metal ions in photocatalytic systems. *Appl. Catal. B Environ.* **1999**, *23*, 89–114.

19. Egerton, T. Does photoelectrocatalysis by TiO_2 work? *J. Chem. Technol. Biotechnol.* **2011**, *86*, 1024–1031.

20. Rayleigh, L. On the light from the sky, its polarization and colour. *Philos. Mag.* **1871**, *41*, 107–120, 274–279.

21. Mie, G. Articles on the optical characteristics of turbid tubes, especially colloidal metal solutions. *Ann. Phys.* **1908** *25*, 377–445.

22. Vos, K.; Krusemeyer, H.J. Reflectance and electroreflectance of TiO_2 single crystals. *J. Phys. C Solid State Phys.* **1977**, *10*, 3893.

23. Egerton, T.A.; Tooley, I.R. UV absorption and scattering properties of inorganic-based sunscreens. *Int. J. Cosmet. Sci.* **2012**, *34*, 117–122.

24. Robb, J.L.; Simpson, L.A.; Tunstall, D.F. Titanium dioxide and UV radiation. *Drug Cosmet. Ind.* **1994**, *154*, 32–39.

25. Ghosh, A.K.; Maruska, H.P. Photoelectrolysis of water in sunlight on sensitized semiconductor electrodes. *J. Electrochem. Soc.* **1977**, *124*, 1516–1522.

26. Mollers, F.; Tolle, H.J.; Memming, R. On the origin of photocatalytic deposition of noble metals on TiO_2. *J. Electrochem. Soc.* **1974**, *121*, 1160–1167.

27. Martin, C.A.; Baltanas, M.A.; Cassano, A.E. Photocatalytic reactors II. Quantum efficiencies allowing for scattering effects. An experimental approximation. *J. Photochem. Photobiol. A* **1996**, *94*, 173–189.

28. Romero, R.L.; Alfano, O.M.; Cassano, A.E. Cylindrical Photocatalytic Reactors. Radiation absorption and scattering effects produced by suspended fine particles in an annular space. *Ind. Eng. Chem. Res.* **1997**, *36*, 3094–3109.

29. Satuf, M.A.; Brandi, R.J.; Cassano, A.E.; Orlando, M.A. Experimental method to evaluate the optical properties of aqueous titanium dioxide suspensions. *Ind. Eng. Chem. Res.* **2005**, *44*, 6643–6649.

30. Marugan, J.; van Grieken, R.; Pablos, C.; Satuf, M.L.; Cassano, A.E. Modelling of bench-scale photocatalytic reactor for water disinfection from laboratory-scale data. *Chem. Eng. J.* **2013**, *2245*, 39–45.

31. Toepfer, B.; Gora, A.; Li Puma, G. Photocatalytic oxidation of multicomponent solutions of herbicides Reaction kinetics with explicit photon absorption effects *Appl. Catal. B Environ.* **2006**, *68*, 171–180.

32. Angel Mueses, M.; Machuca-Martinez, F.; Li Puma, G. Effective quantum yield and reaction rate model for evaluation of photocatalytic degradation of water contaminants in heterogeneous pilot-scale solar photoreactors. *Chem. Eng. J.* **2013**, *215–216*, 937–947.

33. Tunstall, D.F.; Hird, M.J. Effect of particle crowding on scattering power of TiO_2 pigments. *J. Paint Technol.* **1974**, *46*, 33–40.

34. Martin, C.A.; Baltanas, M.A.; Cassano, A.E. Photocatalytic reactors I. Optical behaviour of titanium oxide particulate suspensions. *J. Photochem. Photobiol. A* **1993**, *76*, 199–208.

35. Egerton, T.A.; Tooley, I.R. Physical characterization of titanium dioxide nanoparticles. *Int. J. Cosmet. Sci.* **2014**, *36*, 195–206.

36. Egerton, T.A.; Tooley, I.R. Effect of changes in TiO_2 dispersion on its measured photocatalytic activity. *J. Phys. Chem. B* **2004**, *108*, 5066–5072.

37. Ridley, M.K.; Hackley, V.A.; Machesky, M.L. Characterization and Surface Reactivity of Nanocrystalline Anatase in Aqueous Solutions. *Langmuir* **2006**, *22*, 10972–10982.

38. French, R.A.; Jacobson, A.R.; Kim, B.; Isley, S.L.; Penn, R.L.; Baveye, P.C. Influence of Ionic Strength, pH, and Cation Valence on Aggregation Kinetics of Titanium Dioxide Nanoparticles. *Environ. Sci. Technol.* **2009**, *43*, 1354–1359.

39. Lee, B.C.; Kim, K.T.; Cho, J.G.; Lee, J.W.; Ryu, T.K.; Yoon, J.H.; Lee, S.H.; Duong, C.N.; Eom, I.C.; Ki, P.J.; *et al.* Oxidative stress in juvenile common carp (Cyprinus carpio) exposed to TiO2 nanoparticles. *Mol. Cell. Toxicol.* **2012**, *8*, 357–366.

40. Egerton, T.A.; Harrison, R.W.; Hill, S.E.; John, A.; Mattinson, J.A.; Purnama, H. Effects of particle dispersion on the measurement of semi-conductor photocatalytic activity. *J. Photochem. Photobiol. A* **2010**, *216*, 268–274.

41. Fernando, S.S.; Christensen, P.A.; Egerton, T.A.; White, J.R. Carbon dioxide evolution and carbonyl group development during photodegradation of polyethylene and polypropylene. *Polym. Degrad. Stab.* **2007**, *92*, 2163–2172.

42. Egerton, T.A. The modification of fine powders by inorganic coatings. *Kona* **1998**, *16*, 46–59.

43. Egerton, T.A. Titanium Compounds, Inorganic. In *Kirk-Othmer Encyclopedia of Chemical Technology*, 4th ed.; John Wiley & Sons Inc.: New York, NY, USA, 1997; Volume 24, pp. 225–274.

44. Egerton, T.A.; Tooley, I.R. The surface characterization of coated titanium dioxide by FTIR spectroscopy of adsorbed nitrogen. *J. Mater. Chem.* **2002**, *12*, 1111–1117.

45. Worsley, D.A.; Searle, J.R. Photoactivity test for TiO2 pigment photocatalysed polymer degradation. *Mater. Sci. Technol. Lond.* **2002**, *18*, 681–684.

46. Jin, C. FTIR Studies of TiO2-pigmented Polymer Photodegradation. Ph.D. Thesis, University of Newcastle, Newcastle, UK, 1974.

47. Christensen, P.A.; Dilks, A.; Egerton, T.A.; Lawson, E.J.; Temperley, J. Photocatalytic oxidation of alkyd paint films measured by FTIR analysis of UV generated carbon dioxide. *J. Mater. Sci.* **2002**, *37*, 4901–4909.

48. Dillert, R.; Siemon, U.; Bahnemann, D. Photocatalytic disinfection of municipal wastewater. *Chem. Eng. Technol.* **1998**, *21*, 356–360.

49. Gupta, V.K.; Jain, R.; Agarwal, S.; Shrivastava, M. Kinetics of photocatalytic degradation of tropaeoline 000 using UV/TiO2 in a UV reactor. *Colloid Surf. A* **2011**, *378*, 22–26.

50. Wu, C.H. Effects of operational parameters on the decolourization of C.I. reactive red 198 in UV/TiO2 systems. *Dyes Pigment.* **2008**, *77*, 31–38.

51. Sauer, T.; Neto, C.; Jose, H.J.; Moreira, R.F.P.M. Kinetics of photocatalytic degradation of reactive dyes in a TiO2 slurry reactor. *J. Photochem. Photobiol. A* **2002**, *149*, 147–154.

52. Daneshvar, N.; Salari, D.; Khataee, A.R. Photocatalytic Degradation of azo acid red 14 in water: Investigation of the effect of operational parameters. *J. Photochem. Photobiol. A* **2003**, *157*, 111–116.

53. Egerton, T.A.; Purnama, H. Does hydrogen peroxide really accelerate TiO2 UV-C photocatalyzed decolouration of azo dyes such as Reactive Orange 16? *Dyes Pigment.* **2014**, *101*, 280–285.

54. Baxendale, J.H.; Wilson, J.A. Photolysis of hydrogen peroxide at high light intensities. *Trans. Faraday Soc.* **1957**, *53*, 344–356.

55. Kormann, C.; Bahnemann, D.W.; Hoffmann, M.R. Photolysis of Chloroform and other organic molecules in aqueous TiO2 suspensions. *Environ. Sci. Technol.* **1991**, *25*, 494–500.

56. Blake, D.M.; Webb, J.; Turchi, C.; Magrine, K. Kinetic and Mechanistic Overview of TiO$_2$. *Sol. Energy Mater.* **1991**, *24*, 584–593.
57. Okamoto, K.; Yamamoto, Y.; Tanaka, H.; Itaya, A. Kinetics of heterogeneous photocatalytic decomposition of phenol over anatase powder. *Bull. Chem. Soc. Jpn.* **1985**, *58*, 2023–2028.

Using Dyes for Evaluating Photocatalytic Properties: A Critical Review

Malka Rochkind, Sagi Pasternak and Yaron Paz

Abstract: This brief review aims at analyzing the use of dyestuffs for evaluating the photocatalytic properties of novel photocatalysts. It is shown that the use of dyes as predictors for photocatalytic activity has its roots in the pre visible-light activity era, when the aim was to treat effluents streams containing hazardous dyes. The main conclusion of this review is that, in general, dyes are inappropriate as model compounds for the evaluation of photocatalytic activity of novel photocatalysts claimed to operate under visible light. Their main advantage, the ability to use UV-Vis spectroscopy, is severely limited by a variety of factors, most of which are related to the presence of other species. The presence of a second mechanism, sensitization, diminishes the generality required from a model contaminant used for testing a novel photocatalyst. While it is recommended not to use dyes for general testing of novel photocatalysts, it is still understandable that a model system consisting of a dye and a semiconductor can be of large importance if the degradation of a specific dye is the main aim of the research, or, alternatively, if the abilities of a specific dye to induce the degradation of a different type of contaminant are under study.

Reprinted from *Molecules*. Cite as: Rochkind, M.; Pasternak, S.; Paz, Y. Using Dyes for Evaluating Photocatalytic Properties: A Critical Review. *Molecules* **2015**, *20*, 88-110.

1. Introduction

Over the years enormous numbers of manuscripts have been published on the application of photocatalysts for water and air decontamination, as well as for maintaining clean and superhydrophilic surfaces. As part of this scientific endeavor thousands of compounds have been tested [1], demonstrating the versatility of photocatalysis and its inherent non-preferential nature, which is closely connected to the radical mechanism involved in the photocatalytic degradation process.

Among the many compounds that served, and still serve, to evaluate photocatalytic activity, are organic dyes. Dyes are usually categorized according to their chromophores. Table 1 presents the main categories of dyes and an estimation regarding the number of publications on their photocatalytic degradation under UV and under visible light. The estimation is based on the SciFinder™ data source, taking the words "photocatalysis" "visible light" (or "UV light") and "a specific name of dye (in both its formal and its commercial name)" as keywords. The table is based on data obtained for 250 different dyes. The most studied dyes are the thiazine dyes (with a dominance of methylene blue: 37% of all papers on thiazines), second to them are the xanthenes (with a dominance of rhodamine B: 30% of all manuscripts on xanthenes). Azo dyes, despite their dominance in global production (50%–70% of the market), hence their dominance in contributing to the environmental challenge, come only fourth. From the table it is evident that for all dye categories the number of manuscripts on visible light photocatalysis of dyes is larger than that of the number of manuscripts on UV light photocatalysis. The higher ratio between the number of manuscripts on visible light photocatalysis and UV light photocatalysis is found for xanthenes (2.18) and thiazines (1.80). The ratio is in particular low for azo dyes (1.56). In what follows a rationalization for this difference in the ratios is proposed.

Table 1. An estimation regarding the number of publications on the photocatalytic degradation of dyes under UV light and under visible light, organized according to dye categories. The estimation is based on SciFinder[TM] data source, taking the words "photocatalysis" "visible light" (or "UV light") and "a specific name of dye" as keywords.

Class	UV	Visible
Anthraquinones	238	390
Azo dyes	1285	2006
Natural dyes	187	303
Thiazines	7496	13471
Triarylmethanes	1439	2758
Xanthenes	5625	12244
Others	303	557

Monitoring the photocatalytic degradation of dyes as a tool for demonstrating the technological benefits of photocatalysis is quite common today. Indeed, recent years have shown a tremendous increase in the number of manuscripts describing the photocatalytic degradation of dyes. And yet, a sense of dissatisfaction from this situation appears in private communications, in scientific conferences and during conversations among peers. It is for this reason that we decided to dedicate this manuscript to analyze the sources of the use of dyes in photocatalysis and to summarize the pros and cons in taking the photocatalytic degradation of dyes as a probe for the general properties of photocatalysts.

2. The Mechanisms

The first two decades of research on the photocatalytic degradation of dyes were characterized by the dominance of un-doped titanium dioxide, a UV active photocatalyst. The general degradation scheme of dyes by the UV-active titanium dioxide (Figure 1A) consisted of photon absorption by the photocatalyst, charge separation and the generation of active species on the surface of the photocatalyst. Generally speaking, the main active species under this mechanism are OH radicals formed by oxidation of water molecules by the photogenerated holes, hence the primary attack of the dye molecules is oxidative [2,3]. Evidence for direct oxidative attack by holes was also recorded. Likewise, irreversible reductive decolarization initiated by electrons or by superoxides formed on the photocatalyst's surface can be quite efficient, as was found for azo dyes [4,5].

Decolorization can take place also by a self-sensitization mechanism (Figure 1B). Here, the light is absorbed by the dye molecule. Charge transfer from the excited dye molecule to the conduction band of the semiconductor results in the formation of an unstable dye cation radical and in parallel an active specie on the semiconductor surface that attacks the destabilized dye molecule. One of the first demonstrations of this mechanism, published as early as 1977, described highly efficient N-deethylation of rhodamine B adsorbed on CdS [6]. Likewise, the fact that de-coloring kinetics of methylene blue under solar light in the presence of (undoped) TiO$_2$ was faster than de-coloring kinetics under UV light was explained by this self-sesnsitization mechanism [7].

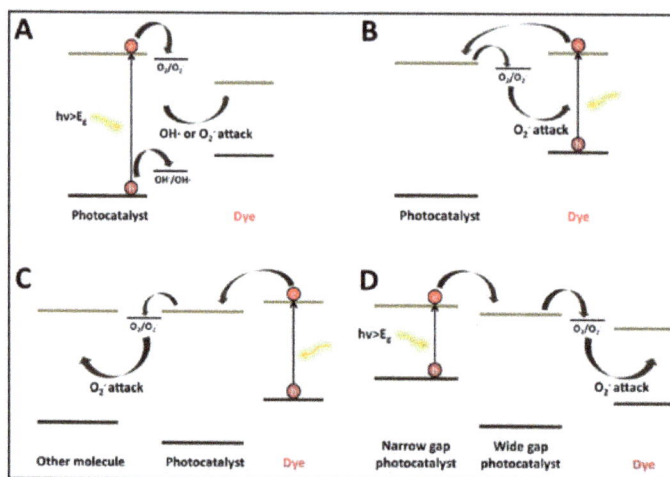

Figure 1. The mechanisms of light-induced degradation of dyes (**A**) photocatalysis (**B**) dye sensitization followed by dye degradation (**C**) dye sensitization followed by reduction of a second molecule; (**D**) degradation by coupled semiconductors under visible light.

An in-depth insight into the mechanism involved in self-sensitization was presented by Liu *et al.*, who studied the photooxidation of alizarin red in TiO$_2$ under visible light, combining ESR spin-trapping technique with molecular orbitals calculations [8]. It was found that the main active species was O$_2^-$ or OOH. Of large interest is their claim that the electron transferred from the dye to the semiconductor is likely to arrive from the atom having the largest electron density in the ground state. Later, this atom becomes the site where the attack by the superoxide anions radicals, formed at the surface of the semiconductor, takes place. Therefore, one may identify two major differences between the photocatalytic mechanism (Figure 1A) and the photosensitization mechanism (Figure 1B): the type of active species and the existence of a preferential location of attack in the sensitization scenario.

Of particular interest is a third photocatalytic mechanism (Figure 1C). Here, absorption takes place on the dye, as in scheme B, however the electrons transferred to the semiconductor are utilized to reduce another molecule [9]. As an example, thionine and eosineY adsorbed on TiO$_2$ were able to photoinduce the degradation of phenol, chlorophenol and 1,2-dichloroethane [10,11]. A study on the degradation of aromatics, trichloroethylene and surfactants by rhodamine B and methylene blue–adsorbed TiO$_2$ suggested that the active species formed upon charge injection from the dye to the semiconductor are *O$_2^-$/*HO$_2$ [12]. It should be noted that some degradation of the sensitizers may occur in parallel to the degradation of the non-absorbing contaminants. This undesired process becomes more important at low concentrations of the non-absorbing contaminants.

One of the main benefits of using a dye sensitizer attached to a semiconductor is the ability to induce chemical changes in a controlled, selective, manner. Such a control can be achieved by utilizing the fact that the location of the HOMO–LUMO levels relative to that of the photocatalyst depends on the specific dye that is being used. For example, TiO$_2$ sensitized by phthalocyanines was able to degrade, upon absorbing visible light, phenol, thiophenol, 4-chlorophenol and hydroquinone, but not oxalic acid, benzoquinone, or EDTA [13].

A narrow bandgap semiconductor coupled to a wide band gap semiconductor may serve as a sensitizer, provided that the conduction band of the wide bandgap semiconductor is more anodic than that of the sensitizer (Figure 1D). An example is the degradation of acid orange II in a coupled CdS/TiO$_2$ photocatalysts under visible light [14]. A question mark might be raised justly regarding the possible role of light absorption by the dye. Comparing the kinetics to that in the presence of only one semiconductor may assist in understanding the role of the dye. For the specific system of acid orange II/CdS/TiO$_2$ the slower kinetics measured in the presence of only one type of semiconductor, convinced the authors that light absorption by the CdS played the dominant role.

Our discussion of the photoinduced degradation mechanisms of dyes will not be completed without mentioning direct photolysis, $i.e.$, the degradation of the dye due to absorption of photons, without a need for a photocatalyst [15,16]. Testing of direct photolysis of dyes is performed routinely as part of the developing of a dye or a dye formulation, and assessing of product stability under weathering conditions. Likewise, photolysis tests are routinely performed and reported, as control-experiments, in almost all manuscripts dealing with photocatalysis, whether under UV light or under visible light.

Total mineralization can be achieved with most dyes [17,18]. This is not the case for dyes containing the triazine group, where the highly stable cyanuric acid is formed [19]. In azo dyes color disappearance usually reflects an attack on the azo bond (C-N=N-) [20]. This usually precedes the opening of the aromatic rings [21,22]. Therefore, aromatic amines or phenolic compounds are often observed as intermediate products. The opening of the aromatic (in other cases naphthalene rings) yields a variety of carboxylic acids, which eventually decarboxylate by the "photo-Kolbe" reaction to yield CO$_2$. It should be noted that azo dyes containing a phenyl azo substitution (naphtol blue, chromotrope 2R, $etc.$) are likely to generate benzene as an end product when degraded by hydroxyl radicals [23].

It was established that dyes containing sulfur atoms are mineralized into sulfate ions [19,24]. The kinetics of sulfate formation was found to be only slightly slower than that of decolorization. Chlorinated dye molecules release chloride ions, already at the beginning of the photocatalytic process. Since chlorinated compounds cause problems in biological treatment, it was claimed that the early release of chlorides advocates for the use of a combined AOP- biological treatment, where a short photocatalytic step precedes an activated sludge treatment step [25]. Dyes containing nitrogen may release NH$_4^+$, NO$_3^-$ and even N$_2$, depending on the initial oxidation state of the nitrogen atoms. Generally speaking, amino groups, consisting of nitrogen in its -3 oxidation state, produce NH$_4^+$. Once formed, the ammonium ions may slowly be photocatalytically oxidized into nitrate ions [22]. In contrast, N$_2$ is the favorite end-product in the degradation of azo bonds, where each nitrogen atom is in its +1 oxidation state.

3. Studying Photoinduced Dye Degradation: The Motivation

3.1. The Early Days

Originally, the motivation to study the photocatalytic degradation of dyestuffs was explained by an actual need to treat contaminated wastewater released by the textile industry [25–27] claimed to be responsible for the release of as much as 15% of the total world production of dyes [28]. Other industries that release considerable amounts of dyestuff are the leather tanning industry [29], the paper industry [30], the hair-coloring industry [31] and the food industry [32]. While many of the dyes are considered toxic by themselves, the toxicity of the raw materials used for their synthesis (in particular aromatic amines) might be of larger concern [33]. Apart from toxicity, one needs to add the obvious (non)aesthetic properties of the streams; properties that are assets to photographers, journalists and politicians.

This rationalization, i.e., the need to degrade large number of different contaminants paved the way for the study the photocatalytic degradation of a large number of dyes. The dominance of un-doped TiO_2, active only under UV light in the photocatalytic arena, was clearly mapped onto the research performed on photocalytically induced dye degradation. Accordingly, almost all published data refers to scheme A, and was dominated by those dyes that pose an environmental problem. The research was mainly concentrated in monitoring the kinetics of disappearance of the primary substrate (Langmuir- Hinshelwood kinetics was mostly reported, manifesting the role of adsorption [34,35]) as well as studying the factors affecting the photocatalytic degradation rates. These included initial dye concentration (an optimal concentration was found in many cases [36,37]), solution's pH [2,38] (which affects adsorption of charged dye molecules), light intensity [39,40], addition of oxidants such as H_2O_2 and $S_2O_8^{2-}$ [24,41], and the presence of co-existing ions [42].

The notion in the early days that the main aim in studying the photocatalytic degradation of dyes was to treat water that had been contaminated by dye-stuff and not as a means to evaluate performance of photocatalysts was echoed also by the large number of manuscripts examining the toxicity of the intermediates and that of the end-products. Methods included the monitoring of the bioluminescence of the bacteria *Vibrio fisheri* [43] or, alternatively, by following the inhibition of bacterial respiration [44,45]. Results varied according to the specific dye. For example, Reactive Black 5 emitted toxic by-products [46], whereas the intermediate products of the triazine dyes MX-5B and K-2G were found to be non-toxic [19,39].

The introduction of self-cleaning surfaces, i.e., self-cleaning glass [47,48], self-cleaning cementitious materials [49,50], and self- cleaning fabrics [51] marked a change in the role that dyes played in pohotocatalysis. For these applications, the dyes no longer served as actual contaminants that had to be degraded in order to protect from their own toxicity, but rather as a tool for demonstrating the potential of photocatalysis as a means to protect aesthetic assets. Indeed, the Japanese industrial standard JIS-R-1703-2:2007 utilizes methylene blue for evaluation of self-cleaning surfaces [52]. Choosing dyes for such demonstrations was a natural choice. After all, the main purpose was to preserve the original color and to prevent color change due to weathering. Intermediate products, mineralization, adsorption were no longer of importance. The only important characteristic was now the kinetics of de-coloring. This focus change was manifested by a change in the scientific tools monitoring the photocatalytic activity. Infrared spectroscopy, HPLC, GCMS and NMR became almost obsolete for these applications. Instead, visible light spectrophotometery, a non-expensive, easy-to-use technique became the technique of choice. A direct step further was the use of human eyes as indicators for decoloration of dyes, thus presenting photocatalysis to laymen, and, no less important, to policy makers. Often, the surfaces of large objects like walls or pavements were colored with dyes and exposed selectively, thus stressing the photocatalytic effect to the human eye, which, by nature, is very sensitive to slight differences in color and tint at domain boundaries.

3.2. The Second Phase- Developing New Materials Operating under Visible Light

The introduction of photocatalysts that respond to visible light marked a significant change in the study of photocatalytic degradation of dyes. Originally, visible light response was achieved by manipulating the activity of titanium dioxide either by doping with non-metallic elements (in particular nitrogen [53,54], carbon [55], sulfur [56,57], fluorine [58], and their mixtures [59]), by doping with transition metals [60], and by coupling the photocatalyst with other semiconductors [61]. In parallel to the manipulation of titanium dioxide a zeal for developing new photocatalysts having narrow bandgaps has been noticed. Apparently, this trend occurred (actually still occurs) in parallel with the developing of the Dye Sensitized Solar Cells (DSSC) and with the renewed interest in hydrogen production by photoinduced water splitting. Many of the proposed materials are ternary and quaternary oxides of bismuth [62–64].

A very common means to evaluate the properties of the new photocatalysts is reflection spectroscopy, from which apparent bandgap values can be calculated based on Tauc plots [65]. Without getting into whether doping or any other photocatalyst's manipulation is manifested by band narrowing or by introduction of energy levels within the bandgap, it is quite obvious that absorbing visible light does not necessarily imply photocatalytic response to that light. This is partly because of high recombination rates and partly because of energy mismatch that thermodynamically prevents the formation of critical species such as hydroxyl radicals or superoxides. For this reason there is a constant need for generic method to evaluate the visible light activity of the new photocatalysts.

In light of this need, the photoinduced degradation of dyes (and in particular rhodamine B, methylene blue) became in recent years the preferred way to evaluate the photoactivity of the novel, apparently visible-light-active materials. In other words, the main reason for studying the photocatalytic degradation of dyes was no longer the need to treat contaminated water but rather their potential use as a probe for the activity of the novel photocatalysts.

That the popularity of using dyes has increased since the introduction of photocatalysts operating in the visible part of the spectrum is demonstrated in Figure 2. Figure 2A presents, on an annual basis, the ratio between the number of manuscripts retrieved by SciFinder™ database upon using the keywords "Visible light" and "Photocatalysis" and the number of manuscripts retrieved by using only "Photocatalysis" as a keyword. An abrupt increase in that ratio is observed around years 2002–2004, marking the change in focus from photocatalysts operating under UV light (mostly undoped titania) to photocatalysts claimed to operate under visible light. Likewise, Figure 2B depicts the ratio between the number of manuscripts retrieved by the SciFinder™ database upon using the keywords "dye" and "Photocatalysis" and the number of manuscripts retrieved by using only "Photocatalysis" as a keyword. A striking similarity between the two graphs is noticed, supporting the linkage made by us between the developing of photocatalysts operating under visible light and the use of dyes as model contaminants.

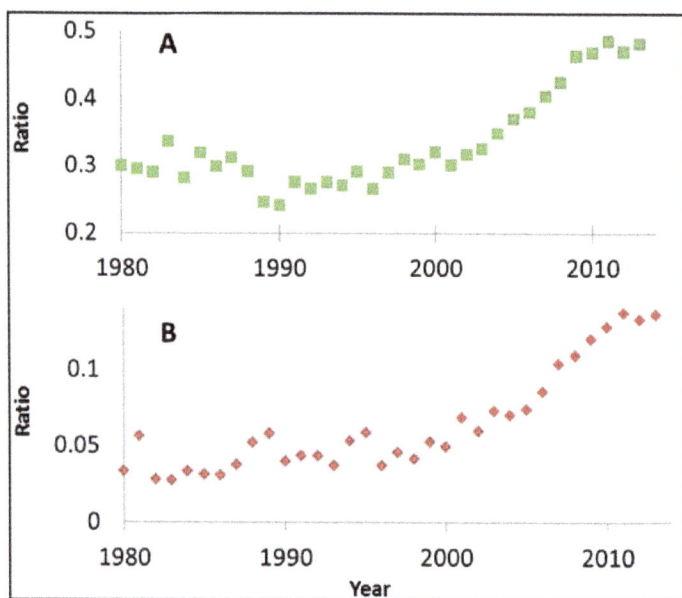

Figure 2. (A) The ratio between the number of manuscripts retrieved by SciFinder™ database upon using the keywords "Visible light" and "Photocatalysis" and the number of manuscripts retrieved by using only "Photocatalysis" as a keyword; **(B)** The ratio between the number of manuscripts retrieved by the SciFinder™ database upon using the keywords "dye" and "Photocatalysis" and the number of manuscripts retrieved by using only "Photocatalysis" as a keyword.

Using any model contaminant as a probe for effectiveness of a new photocatalyst requires that the degradation kinetics, whatever they are, will be, as much as possible, generic, in a sense that comparative studies performed on a set of photocatalysts will have a relevance (at least partially) for the degradation of other contaminants. In practice, most manuscripts do not report that the degradation they measured under visible light could be due to photosensitization rather than photocatalysis. In many cases, even when sensitization is mentioned as the governing mechanism, the (ir)relevance for other contaminants is only rarely mentioned.

One of the first studies pointing out the inadequacy of dyes as a probe molecule for semiconductor photocatalysis (methylene blue in this case) was presented by Yan *et al.* [66]. Here, the action spectra of S-doped TiO_2 measured during the degradation of methylene blue revealed activity at 580–650 nm (in correlation with the absorption spectrum of MB), whereas no activity at this range of the spectrum was observed towards acetic acid. The same group repeated its concern about the use of dyes for evaluating activity also at a later manuscript [67]. Recently, a clear recommendation not to use dye tests for activity assessment of visible light photocatalysts was presented, based on experiments with six visible- light photocatalysts degrading five organic dyes [68].

Basic ethics of integrity require that comparative studies, which are most likely relevant to one contaminant only, will be reported as such, and will not be publicized in a generalized manner that misleads the readers. It is quite unfortunate that, generally speaking, these basic ethics are not strictly obeyed. Was this behavior a matter of misinterpretation of data? Of innocent misunderstanding of the role that dye sensitization played in the obtained kinetics? Was it due to a sloppy scientific community that did not emphasize enough the difference between sensitization and "true" photocatalysis? This is not for us to judge. What we can do (and we do it hereby) is to analyze the factors that render dyes inappropriate for serving as generalized indicators for photocatalytic activity, and to discuss possible remedies for this situation.

This manuscript is basically a mini-review manuscript. As such, it relies on published articles and on work made by large number of research groups. While relying on the work of others, we thought it would be beneficial to perform a set of experiments that would demonstrate our claims and would augment our conclusions. In doing so we aimed at providing common grounds to the presented claims, knowing that while there are many reviews on the photocatalytic degradation of dyes, the published reviews basically summarize the kinetic and mechanistic results and are quite silent in giving a critical view on the fundamentals of studying the photocatalytic degradation of dyes.

4. Rhodamine B and BiOCl

The photocatalytic degradation of rhodamine B (N,N,N',N'-tetraethylrodamine) under both UV light and visible light in the presence of bismuth oxychloride (BiOCl) was used as a model system. Rhodamine was chosen based on its popularity among research groups studying visible light activity. Figure 3 presents, on an annual basis, the ratio between the number of SciFinder™ hits for rhodamine B and visible light photocatalysis and the number of hits for visible light and photocatalysis. Although this ratio should not be regarded as more than a rough estimation, the monotonic increase as a function of time and the high values obtained (0.13 in 2013, for example), are indicative of the important role of this dye in evaluating the photoactivity of the so-called visible light photocatalysts.

Rhodamine B belongs to the oxygen-containing heterocyclic xanthene dyes family. It is neither degraded in the dark in the presence of a photocatalyst nor under illumination in the absence of a photocatalyst. Under visible light, and in the presence of an appropriate semiconductor, rhodamine B degrades via an efficient N-deethylation sensitization mechanism. The process is initiated by sensitization of the dye, charge transport to the semiconductor and formation of active superoxide ions by reduction of oxygen pre-adsorbed on the semiconductor. Consequently, in this scenario the degradation of rhodamine B goes through the formation of triethylrhodamine, diethylrhodamine, ethylrhodamine and rhodamine having different λ_{max} at 555, 539, 522, 510, 498 nm, respectively and different relative extinction coefficients of 1:0.48:6.3:5.3:0.73, respectively [6]. At a later stage the nascent rhodamine is mineralized. Some photocatalysts, which absorb visible light, such as Bi_2O_3 (Eg = 2.85 eV) and $BiVO_4$ (Eg = 2.44 eV) tend to degrade RhB under visible light, not by a sensitization mechanism but by a photocatalytic mechanism which is manifested by a direct attack on the chromophore. This behavior was explained by the weak interaction between these photocatalysts and the dye, which prevents charge transport from the dye to the photocatalyst [69].

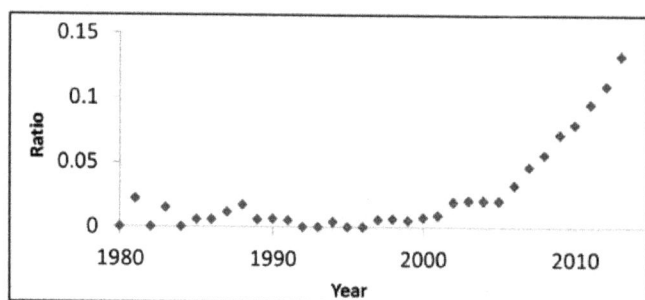

Figure 3. The ratio between the number of SciFinder™ hits for "rhodamine B" + "visible light" + "photocatalysis" and the number of hits for "visible light" + "photocatalysis".

The photocatalyst used by us in the present work was BiOCl, consisting of tetragonal Bi_2O_2 slabs, "sandwiched" between two chlorine ions slabs to form a $Bi_2O_2Cl_2$ layer along the c-axis. This structure forms internal electric fields between the Bi_2O_2 positive slabs and the halide anionic slabs that promote the separation of photogenerated electron-hole pairs thus improving the photocatalytic activity of the catalyst [70–72]. Unlike other bismuth oxyhalides the bandgap of BiOCl is large (3.2–3.5 eV) [73], thus it does not absorb in the visible part of the spectrum. For this reason any degradation induced by visible light can be regarded as originating from sesnsitization, whereas, by virtue of the small absorption cross section of the dye at 365 nm, degradation by UV light is likely to occur predominantly due to photocatalysis. Unlike TiO_2, which hardly physisorbs the rhodamine dye, BiOCl is known to chemically adsorb the dye [74], facilitating easy charge transport from the dye to the semiconductor.

4.1. Preparation of BiOCl

BiOCl was prepared by co-precipitation. 1.3 gram of $BiCl_3$ were dissolved in 20.5 mL of 1.23 M HCl to obtain bismuth solution concentration of 0.23 M. 18 ml of NaOH 2 M solution were added to the bismuth solution to obtain a pH of 11.5. The solution was then stirred at room temperature for 24 h. The particles were then collected by vacuum filtration, washed with water and dried in air. The material was characterized by XRD and SEM.

4.2. Photocatalytic Tests

BiOCl was used to study the photocatalytic and photosensitized degradation kinetics of Rhodamine B (RhB) in aqueous solutions. In a typical experiment, 30 mg of the photocatalyst were dispersed in 50 mL solution containing 1.25×10^{-5} M (in other cases 3.1×10^{-5} M) of RhB, in a reaction vessel under continuous stirring. Prior to exposure, the system was left in the dark for 90 min to obtain an adsorption/desorption equilibrium. Upon reaching equilibrium the solution was exposed to light at a specific wavelength in the range of 365–515 nm. A series of Light Emitting Diodes (LED-R Ltd., Ben-Shemen, Israel) having a typical 15 nm FWHM were used. Equal photon flux (3.17×10^{15} photons/(s·cm²)), rather than equal energy flux, was kept in all experiments. At given time intervals, 0.67 mL aliquots were sampled, centrifuged to remove the particles and measured by UV-vis spectroscopy (Lambda 40, Perkin-Elmer, Waltham, MA, USA).

5. Using Dyes for Evaluating Photocatalytic Properties

5.1. Monitoring the Degradation Kinetics

There is a general acceptance of the notion that the most important characteristic of a photocatalyst is its ability to degrade molecules of interest as fast as possible, at a given number of impinging photons and with a given amount of photocatalyst. This, almost automatically, raises a question regarding the definition of "degrading". Is it the kinetics of decoloring or, alternatively, the kinetics of mineralization?

As mentioned above, in the early days of photocatalytic dye degradation the main aim was the remediation of wastewater containing the specific dyes that were tested. Consequently, both de-coloring kinetics, types and formation/degradation kinetics of intermediate products, and mineralization rates were studied. Apparently, altering the focus into using the degradation of dyes as an indicator for the photoactivity of new photocatalysts operating under visible light reduced the importance of studying the fate of intermediates. Indeed, if one surveys the literature reporting on the developing of novel photocatalysts one finds mainly decoloring experiments and hardly any data on intermediates and their toxicity as well as on the rates of mineralization. Such an approach might be problematic, as explained below.

Dye molecules are usually comprised of relatively large number of functional groups. This may lead to numerous types of mechanisms, each releasing different intermediate products. Therefore, in comparing two different photocatalysts the rates of decolorization do not necessarily correlate with mineralization rates. Usually, the absorption spectrum of these intermediates partially overlaps with that of the dye; the extent of interference depends on the type of the intermediates, or, in other words, on the degradation mechanism. Since the mechanism is photocatalyst-dependent, estimating the rates of decoloration by monitoring a specific absorption peak inherently gives erroneous results that are influenced by the type of photocatalyst in use. This statement was clearly demonstrated in the study of visible-light photocatalytic degradation of methyl orange on carbon-doped TiO_2 and on Pt/WO$_3$ [68]. Here, degradation on carbon-doped TiO_2 that proceeded via demethylation yielded intermediates that absorbed light at higher energies (hence caused the blue-shifting of the peak [75]), yet partially interfered with the main peak in the absorption spectrum of the dye (505 nm). In contrast, degradation on the Pt/WO$_3$ photocatalyst did not yield any intermediate products that absorbed light in the visible part of the spectrum.

Figure 4 demonstrates the above statements by presenting the spectrum of rhodamine B following sensitized degradation under visible light by BiOCl (trace B) as well as by La_2BiNbO_7 (trace C). While both traces have the same absorbance at 554 nm (52% of the original) the two spectra are quite different. In fact, at this stage the maximum absorbance with La_2BiNbO_7 remained at 554 nm, whereas with BiOCl the maximum was located at 544 nm.

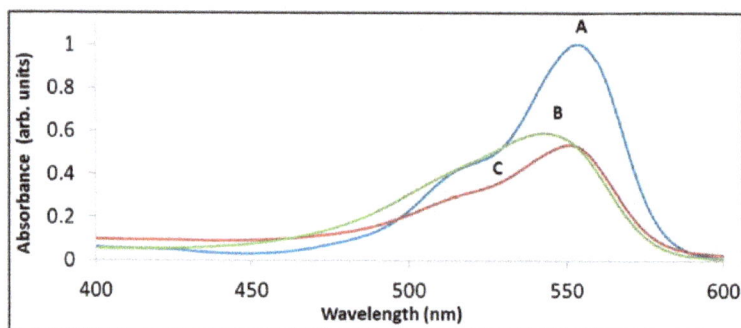

Figure 4. The degradation of rhodamine B with BiOCl (B) and with La₂BiNbO₇ (C) under visible light. Trace (A) presents the spectrum of RhB prior to degradation, peaked at 554 nm. Traces (B) and (C) present the spectrum at a point where 48% of the initial absorbance at at 554 nm has disappeared.

Hence, comparing between two photocatalysts by monitoring changes in the height of their main absorption peak might lead to flawed conclusions. The fact that, in general, the extinction coefficients of the partially overlapping intermediate products differ from that of the dye adds to the complexity of obtaining reliable kinetic data based on single wavelength Beer's law.

Plotting the absorbance at the center of the absorption peak of a dye is no doubt the easiest way to describe the decoloration kinetics. Such a representation reveals an exponential decay curve, in almost all cases, which enables to describe the kinetics by a single number, *i.e.*, by a first order rate constant. While we have our reservations as for the validity of these values in describing the so-called "activity" of the semiconductor (see below), we are not able to suggest a better way of representation. Hence, in what follows, we will not avoid using "k", the apparent first order rate constant of decoloration as a representative parameter of the process.

As discussed above, decolorization can be achieved by a reductive mechanism. The reduced form is then further degraded (usually by an oxidative pathway) to form the end-products. In certain cases, for example in the degradation of methylene blue, and under appropriate conditions (absence of light, high pH, high concentration of dissolved oxygen) the photoreduced specie might be re-oxidized back to colored methylene blue [76,77]. This observed (at least partial) reversibility, and in particular its dependence on the experimental conditions, marks methylene blue (and any other dye which behaves similarly) as inadequate for testing photocatalytic activity.

5.2. Sensitization Versus Photocatalysis

In light of the possible co-existing of a photocatalytic mechanism and a sensitization mechanism in the degradation of dyes it is required to analyze the relative contribution of dye sensitization *versus* that of "true" photocatalysis. Comparing the absorption spectrum of the photocatalyst, the absorption spectrum of the dye and the action spectra during the degradation of the dye is often suggested as a tool for obtaining such analysis. The expectation is that sensitization occurs at energies that are not absorbed by the photocatalyst and that the sensitization activity correlates with the absorption spectrum of the dye.

The fact that sensitization involves charge transport from the dye molecule to the semiconductor implies that adsorption should play a critical role in the process. The type of interaction (physisorption *versus* chemisorption) and (in the case of chemisorption) even the specific interaction between the adsorbate and adsorbent may significantly affect the efficiency of the process [78].

Charge transport from the photosensitized dye to the photocatalyst (and in the other direction as well) depends on the strength of interaction between the dye and the surface of the photocatalyst [79]. For this reason, the kinetic barrier for one electron transfer from an eosin Y dye to the conduction band of titanium dioxide, is lower than that for triethanolamine (TEOA), although the latter is a stronger reductant (under light and in the absence of a semiconductor triethanolamine can reduce free eosin Y) [12,80]. As a consequence, visible light illumination of TiO_2 in the presence of both eosin Y and TEOA degrades the triethanolamine by a mechanism that involves electron transport from the TEOA to the cation form of the adsorbed dye.

Therefore, surface treatments that promote this interaction (for example pre-treatment of TiO_2 with water [81], or altering the bridging group by fluorinating the surface of titanium dioxide [82]) are likely to improve sensitization-induced degradation. The strength of the interaction is likely to vary from one photocatalyst to the other. Hence, the relative contribution of sensitization by a specific dye might depend on the specific photocatalyst. Indeed, the higher rates of degradation of rhodamine B with BiOCl co-doped with niobium and iron compared with the rates with pristine BiOCl were explained by specific interaction between the dye molecules and the co-doped photocatalyst [64].

It could have been expected that sensitization would be manifested (at illumination energies above some threshold) by a correlation between the absorption spectrum of the dye and its action spectra (activity *versus* irradiated wavelength under a constant photon flux). This expectation is based on the pre-assumption that the quantum yield of the absorbed photons is independent of wavelength, as a result of fast intramolecular de-excitation to the bottom of the excited state band, *i.e.*, absence of hot-electrons effects. In practice, the correlation between the action spectrum and the absorption spectrum of the dye can be quite weak. For example, methylene blue has a very wide absorption peak. The extinction coefficient rises monotonically until it peaks at 660 nm. Yet, the degradation rate obeyed the following order: 570 = 640 = 670 < 540 < 605 < 555 < 585 < 525 < 620 < 655 nm [66].

One obvious explanation for the weak correlation between the absorption spectrum of the dye and its action spectrum stems from the fact that the parameter usually (but not always [66]) taken in the literature for this correlation is the absorption spectrum of the dye in solution instead of the absorption spectrum of the adsorbed dye. Figure 5 presents the absorption spectrum of an aqueous solution (15 mg/L) containing rhodamine B, together with the absorption spectrum of rhodamine B pre-adsorbed from the same concentration solution onto BiOCl. The latter spectrum was inferred from reflection measurements. For clarity both spectra were normalized according to their maximal absorption. It is evident that the two spectra differ in the width of their absorption peak (the adsorbed dye had a much broader peak), and, less significantly, in the position of their absorption peak (556 nm *versus* 554 nm for the adsorbed and the non-adsorbed rhodamine, respectively).

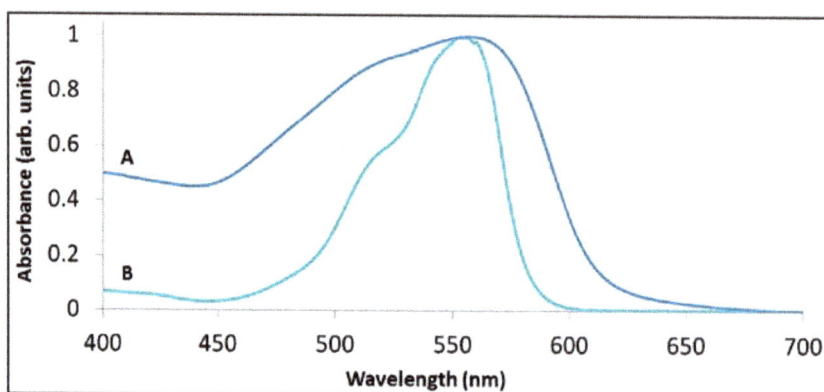

Figure 5. The absorption spectrum of rhodamine B (A) adsorbed on BiOCl (B) in aqueous solution.

It is noteworthy however that using the spectrum of the adsorbed dye as a source for correlation with the action spectrum still does not guarantee a high fidelity correlation. Figure 6 presents the decolorization rate constants obtained from a series of photoinduced degradation measurements of RhB (concentration: 6 mg/L) in the presence of BiOCl. Each data point represents exposure to light having a specific wavelength. The two data points at k = 0.1 sec^{-1} and k = 0.11 sec^{-1} represent exposure to photons in the UV range (365 nm and 375 nm, respectively) where the degradation mechanism is not photosensitization but photocatalysis, hence is the higher rate constant. A very weak correlation (if at all) between the absorption spectrum at the surface and the rate constant can be observed. This weak correlation disappeared upon degrading solutions with higher (15 mg/L) concentration (not shown here).

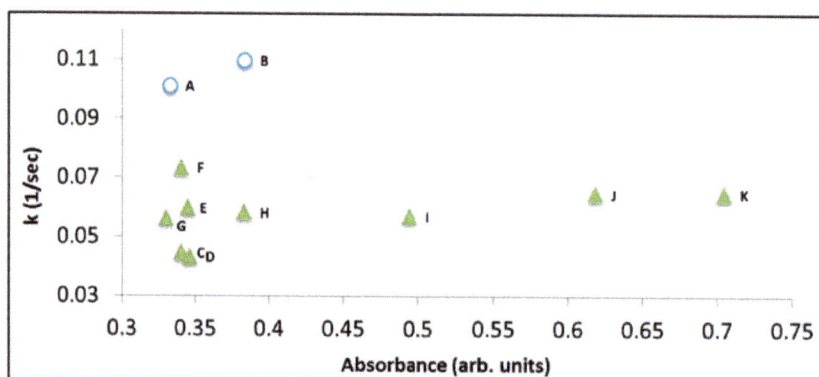

Figure 6. The decolorization rate constant *versus* relative absorption of RhB on the surface of BiOCl. Each point represents exposure to light of different wavelength (but same photon flux). The nominal concentration of the dye was 6 mg/L. (A) 365 nm; (B) 375 nm; (C) 395 nm; (D) 405 nm; (E) 410 nm; (F) 420 nm; (G) 435 nm; (H) 470 nm; (I) 490 nm; (J) 505 nm; (K) 515 nm.

5.3. Effect of Dye Concentration

Another deleterious effect of utilizing dye molecules for testing visible-light active photocatalysts is the dependence of the measured kinetics on the concentration of the dye. As mentioned above, working with UV-active photocatalysts to degrade molecules that do not absorb UV light, gives, for most cases, a Langmuir- Hinshelwood type of kinetics, where the rate increases upon increasing the concentration of the contaminant until rate-saturation is achieved. In that case, the apparent kinetics reflect the number of molecules that are adsorbed on the surface of the photocatalyst. The situation is totally different in the case of a system comprising of a visible—light absorbing photocatalyst and a visible light absorbing contaminant (dye). Here, increasing the concentration of the dye might reduce the rate of degradation, due to increased absorption of light by free dye molecules in the solution [25,83]. Along this line, the higher the concentration of the dye is, the weaker is the correlation between the action spectrum and the absorption spectrum of the dye.

High concentration might also cause aggregation and even surface dimerization. In certain cases, dimerization might have a deleterious effect on the degradation rates, as was observed in the case of phthalocyanines, where the dimers are considered to be photochemically inactive [13], or in the case of indigo carmine on N-doped TiO_2. In other cases (acid orange 7 on N-doped TiO_2) dimerization increases the decoloration rate [68].

Aggregation may affect the apparent decoloration rates also indirectly. Figure 7 presents the absorption spectra of rhodamine B adsorbed on BiOCl from solutions having different concentrations. A broadening of the absorption peak of the adsorbed dye to the blue upon increasing the concentration of the dye in the solution is clearly observed. Blue shifting in the absorption spectrum upon surface aggregation was reported also for Methylene Blue [52]). In that case, the higher the concentration of the dye during adsorption was, the more blue-shifting was observed. This means that the apparent degradation kinetics, often deduced based on single wavelength absorbance, are erroneous. Moreover, since the adsorpticity of a specific dye depends on the characteristics of the photocatalyst, it is obvious that the extent of concentration-dependence effects become photocatalyst-specific.

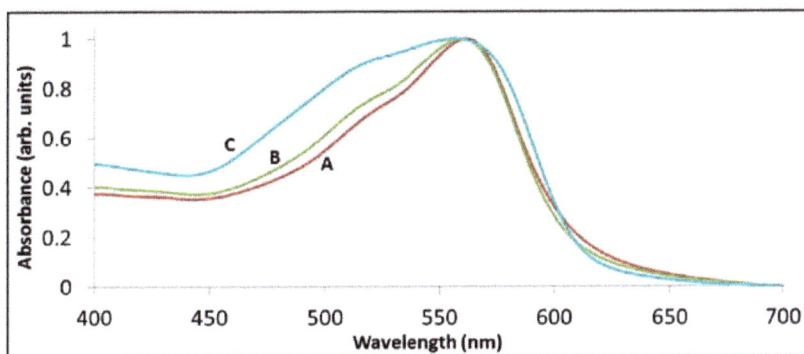

Figure 7. Changes in the absorption spectrum of rhodamine B pre-adsorbed on BiOCl from aqueous solutions containing various dye concentrations. For clarity, all graphs were normalized. (A) 0.5 mg/L; (B) 1.5 mg/L; (C) 15 mg/L.

As mentioned above, more and more researchers tend to criticize reports on degradation of dyes under visible light in the presence of a visible light absorbing semiconductors claiming that they do not reflect "true" photocatalysis but sensitization. A common counter-argument is based on Total Organic Carbon (TOC) measurements showing that the degradation proceeds all the way to mineralization. It is often claimed that observing mineralization can be regarded as an evidence that the semiconductor handles not only dyes, but also the intermediate products, including intermediate products that do not absorb visible light, hence cannot induce visible-light sensitization. We believe that this argumentation does not stand on solid grounds, for two reasons. One reason is that most intermediate products are destabilized chemical species hence continue to degrade quite easily. The second reason has to do with the ability of dye molecules that are attached to the semiconductor to photoinduce degradation of other molecules (Figure 1C). Thus, observing mineralization of intermediates does not necessarily imply that these intermediates may be degraded without the presence of sensitizing molecules.

5.4. Monitoring Intermediate Products

As mentioned before, organic dye molecules are characterized by the presence of a chromophore having several delocalized functional groups and side-functional groups that are instrumental in defining the energetic difference between the LUMO and the HOMO orbitals. As a consequence, light-induced degradation of the dyes is often characterized by large number of intermediate steps, on way to complete mineralization. The outcome is a gradual change in the spectrum of the solution that is characterized by a decrease in the absorption at the summit of the absorption peak of the dye and, in parallel, the appearance of a broad absorption envelope, representing the intermediates. Therefore, the kinetics of degradation can be defined by several ways: (1) following changes in the absorption peak of the dye (2) by following changes in the location of the superimposed peak (3) by following changes in the amount of intermediates. Calculating the amount and type of intermediates without separating the species is very cumbersome due to a strong partial overlap between the spectra of the intermediates and that of the dye. This problem is aggravated if the extinction coefficient of the species is different.

Figure 8 presents the location of the superimposed peak of rhodamine B and its intermediates during its photodegradation by BiOCl *versus* the absorbption at 554 nm, which is the wavelength of maximum absorption for pristine RhB. Each symbol in the figure represents exposure at a different wavelength. That way, the reaction progresses from right to left. This kind of representation of the degradation progression enables to compare between the two measurable parameters in the spectrum that are changed during degradation. The figure clearly shows high level of correlation between the location of the superimposed peak and the decrease in the absorption at 554 nm, suggesting, without the need of elaborate separation equipment, that the distribution of intermediates, at a specific percentage of decolorization of the dye is the same, regardless of the wavelength of exposure.

5.5. Effect of Light Intensity

Another way to describe the progression of degradation is by plotting the calculated (actually estimated) amount of intermediates *versus* the decrease in the concentration of the dye. Figure 9 demonstrates this attitude by comparing the progression in the degradation of rhodamine B on BiOCl under 515 nm light, at two light intensities: 0.3 mW/cm² and 1.19 mW/cm². As in Figure 8, the intact RhB concentration was calculated based on the absorption at 554 nm. The amount of intermediates was estimated by taking the absorption at 530 nm and subtracting the contribution to the 530 nm signal from non-degraded RhB, based on the 530 nm/554 nm absorption ratio in the RhB solution prior to exposure. Care was made to take into account, while calculating the amounts of RhB and the intermediates, the difference in the extinction coefficients. According to the figure, increasing the light intensity reduces the maximal concentration of intermediates. Since the rate constant for *N*-deethylation (producing the intermediates) also increases with intensity, the lower maximum should be explained by a faster degradation of the *N*-deethylated species upon increasing light intensity that more than compensates for the increase in the rate of *N*-deethylation. It is possible that this increase in the degradation rate of the intermediates is a consequence of a sensitization process by the intermediates. Therefore, it can be concluded that secondary degradation processes might mask the primary process, thus limit the ability to deduce conclusion on the activity by following the kinetics of disappearance of a single dye. One way to detour the problem is to irradiate at a wavelength that is absorbed only by the dye, however, this solution is cumbersome.

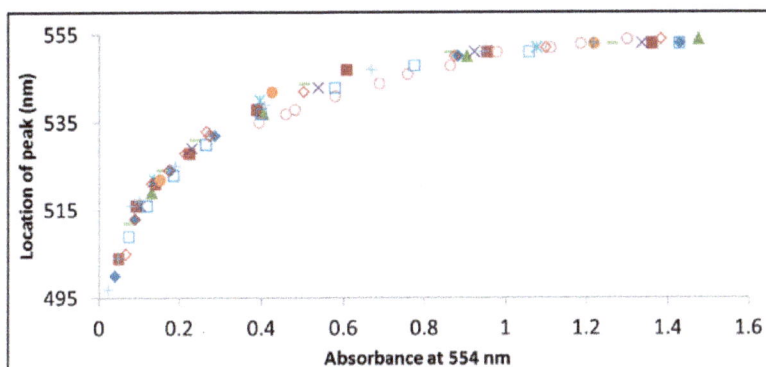

Figure 8. The location of the superimposed peak of rhodamine B and its intermediates during photodegradation by BiOCl *versus* the absorbption at 554 nm, which is the location of the RhB peak. Each symbol in the figure represents exposure at a different wavelength: 365 nm (filled diamonds), 375 nm (filled squares), 395 nm (filled triangles), 405 nm (× symbols), 410 nm (asterisks), 420 nm (filled circles), 435 nm (+ symbols), 470 nm (empty circles), 490 nm (− symbols), 505 nm (empty squares), 515 nm (empty diamonds).

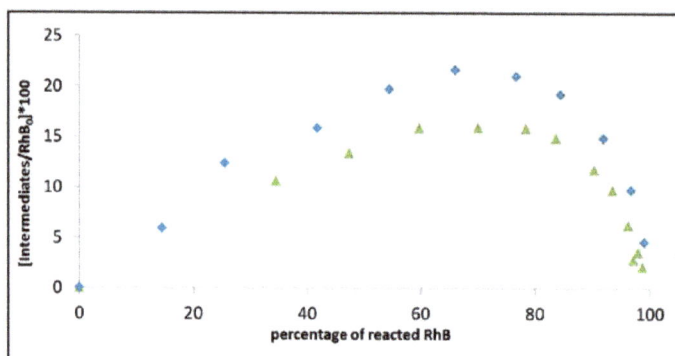

Figure 9. The progression of the photoinduced degradation of rhodamine B on BiOCl under 515 nm light, at two light intensities: 0.3 mW/cm^2 (diamonds) and 1.19 mW/cm^2 (triangles). The figure presents the calculated amount of intermediates *versus* the decrease in absorption at 554 nm. The initial concentration of RhB in both cases was 6 mg/L.

The sensitivity of the degradation mechanism in dyes to the experimental conditions does not stop at concentration, irradiance, or wavelength. In fact, the same photocatalyst can yield different mechanisms, if prepared in a different manner. The dye resazurin (Rz) is known to photocatalytically reduced to resorufin (Rf), which further degrades to non-absorbing species. Figure 10 presents the progression in the degradation of Rz upon exposure to 420 nm light (3 mW/cm^2) in the presence of BiYWO$_6$ prepared by a hydrothermal method and by a sol-gel method.

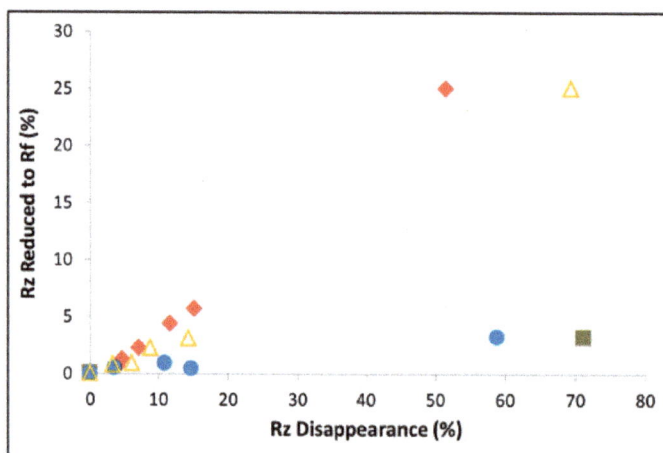

Figure 10. The disappearance of resazurin and the production of resorufin during photoinduced degradation of the dye with BiYWO6 prepared by various techniques: Hydrothermally prepared sample (filled squares), hydrothermally prepared sample calcined at 800 °C (filled circles), sol-gel sample calcined at 600 °C (filled rhombuses) and at 800 °C (empty triangles).

In the figure, the percentage of Rz disappearance is plotted against the percentage of formed Rf. A distinct difference in the amount of formed Rf at a given Rz disappearance is portrayed between the sol-gel-prepared BiYWO₆ and a hydrothermally-prepared BiYWO₆, manifested by a higher concentration of the intermediate Rf with the sol-gel prepared photocatalyst. Since the kinetics of Rz disappearance were similar for the two photocatalysts, the above difference has to be due to slower degradation kinetics of Rf with the sol-gel photocatalyst.

6. Conclusions

This brief review aims at analyzing the use of dye stuff for evaluating the photocatalytic properties of novel photocatalysts. As shown above, this use of dyes as predictors for photocatalytic activity has its roots in the pre-visible light activity era, when the aim was to treat effluents streams containing hazardous dyes. While dyes may be appropriate (with some limitations) for evaluating photocatalytic processes that take place under UV light and with UV-active photocatalysts, they are definitively problematic under visible light and with visible light-active photocatalysts, as discussed above.

The main conclusion of this review is that dyes in general are inappropriate as model systems for the evaluation of photocatalytic activity of novel photocatalysts claimed to operate under visible light. Their main advantage, the ability to use UV-vis spectroscopy, a non-expensive, easy to use technique, to measure the kinetics of their photocatalytic degradation, is severely limited by a variety of factors, most of which related to the presence of other species (intermediate products in the solution, dye aggregates on the surface, *etc.*). Exposure to visible light carries with it the most problematic aspect of using dyes, the presence of a second mechanism, sensitization, that diminishes the generality required from a model contaminant testing a novel photocatalyst. This conclusion is further supported by the low purity of commercially available dyes (70%–90%), that inhibits the obtaining of reliable data, and, more important, prevents making comparisons between results measured by different groups [68].

The observation that dye tests are highly specific to a dye/photocatalyst combination led to the recommendation to rely on a set of different dyes rather than on a single dye for evaluating the activity of a specific photocatalyst [68]. We believe this is definitively a step in the right direction from the current situation where novel photocatalysts are tested by their ability to degrade a single dye. The recommendation to use multitude of dyes is connected with the pH dependence of the adsorption of the dyes, however such dependence is not specific for dyes.

While it is recommended not to use dyes for general testing of novel photocatalysts, it is still understandable that a model system consisting of a dye and a semiconductor can be of large importance if the degradation of a specific dye is the main aim of the research, or, alternatively, if the abilities of a specific dye to induce the degradation of a different type of contaminant are under study. In both cases, scientific integrity demands that the relevant objectives of using dyes will be clarified by the reporting researchers. Furthermore, that sensitization will be reported as such, and that there will be no broad claims giving the readers false impression of a "true" visible light activity, when such activity does not exist. Above all, the paper calls for looking at a system containing both a specific dye and a semiconductor as a one integral system rather than a photocatalyst operating on a dye.

Acknowledgments

The authors are grateful to The Adelis Foundation, The Grand Technion Energy Program (GTEP) and The Russell Berrie Nanotechnology Institute (RBNI) for their support. The assistance of Idan Cohen in the data mining is gratefully acknowledged.

Conflicts of Interest

The authors declare no conflict of interest.

References

1. Blake, D.M. *Bibliography of Work on the Heterogeneous Photocatalytic Removal of Hazardous Compounds from Water and Air*; National Renewable Energy Laboratory: Golden, CO, USA, 1994.
2. Hustert, K.; Zepp, R.G. Photocatalytic Degradation of Selected Azo Dyes. *Chemosphere* **1992**, *24*, 335–342.
3. Ojani, R.; Raoof, J.; Zarei, E. Electrochemical Monitoring of Photoelectrocatalytic Degradation of Rhodamine B using TiO_2 Thin Film Modified Graphite Electrode. *J. Solid State Electrochem.* **2012**, *16*, 2143–2149.
4. Vinodgopal, K.; Bedja, I.; Hotchandani, S.; Kamat, P.V. A Photocatalytic Approach for the Reductive Decolorization of Textile Azo Dyes in Colloidal Semiconductor Suspensions. *Langmuir* **1994**, *10*, 1767–1771.
5. Vinodgopal, K.; Kamat, P.V. Photochemistry of Textile Azo Dyes. Spectral Characterization of Excited State, Reduced and Oxidized Forms of Acid Orange 7. *J. Photochem. Photobiol. A* **1994**, *83*, 141–146.
6. Watanabe, T.; Takizawa, T.; Honda, K. Photocatalysis through Excitation of Adsorbates. 1. Highly Efficient N-Deethylation of Rhodamine B Adsorbed to Cadmium Sulfide. *J. Phys. Chem.* **1977**, *81*, 1845–1851.
7. Kuo, W.; Ho, P. Solar Photocatalytic Decolorization of Methylene Blue in Water. *Chemosphere* **2001**, *45*, 77–83.
8. Liu, G.; Li, X.; Zhao, J.; Horikoshi, S.; Hidaka, H. Photooxidation Mechanism of Dye Alizarin Red in TiO_2 Dispersions under Visible Illumination: An Experimental and Theoretical Examination. *J. Mol. Catal. A* **2000**, *153*, 221–229.
9. Taqui Khan, M.; Chatterjee, D.; Bala, M. Photocatalytic Reduction of N_2 to NH_3 Sensitized by the [Ru III-Ethylenediaminetetraacetate-2,2'-Bipyridyl]−Complex in a Pt-TiO_2 Semiconductor Particulate System. *J. Photochem. Photobiol. A.* **1992**, *67*, 349–352.
10. Chatterjee, D.; Mahata, A. Photoassisted Detoxification of Organic Pollutants on the Surface Modified TiO_2 Semiconductor Particulate System. *Catal. Commun.* **2001**, *2*, 1–3.
11. Chatterjee, D.; Mahata, A. Demineralization of Organic Pollutants on the Dye Modified TiO_2 Semiconductor Particulate System using Visible Light. *Appl. Catal. B* **2001**, *33*, 119–125.
12. Chatterjee, D.; Mahata, A. Visible Light Induced Photodegradation of Organic Pollutants on Dye Adsorbed TiO_2 Surface. *J. Photochem. Photobiol. A* **2002**, *153*, 199–204.

13. San Romaen, E.; Navío, J.; Litter, M. Photocatalysis with Fe/TiO$_2$ Semiconductors and TiO$_2$ Sensitized by Phthalocyanines. *J. Adv. Oxid. Technol.* **1998**, *3*, 261–269.

14. Bessekhouad, Y.; Chaoui, N.; Trzpit, M.; Ghazzal, N.; Robert, D.; Weber, J. UV–vis Versus Visible Degradation of Acid Orange II in a Coupled CdS/TiO$_2$ Semiconductors Suspension. *J. Photochem. Photobiol. A.* **2006**, *183*, 218–224.

15. Bandara, J.; Kiwi, J. Fast Kinetic Spectroscopy, Decoloration and Production of H2O2 Induced by Visible Light in Oxygenated Solutions of the Azo Dye Orange II. *New J. Chem.* **1999**, *23*, 717–724.

16. Madhusudan, P.; Ran, J.; Zhang, J.; Yu, J.; Liu, G. Novel Urea Assisted Hydrothermal Synthesis of Hierarchical BiVO$_4$/Bi$_2$O$_2$CO$_3$ Nanocomposites with Enhanced Visible-Light Photocatalytic Activity. *Appl. Catal. B* **2011**, *110*, 286–295.

17. Lachheb, H.; Puzenat, E.; Houas, A.; Ksibi, M.; Elaloui, E.; Guillard, C.; Herrmann, J. Photocatalytic Degradation of various Types of Dyes (Alizarin S, Crocein Orange G, Methyl Red, Congo Red, Methylene Blue) in Water by UV-Irradiated Titania. *Appl. Catal. B* **2002**, *39*, 75–90.

18. Stylidi, M.; Kondarides, D.I.; Verykios, X.E. Pathways of Solar Light-Induced Photocatalytic Degradation of Azo Dyes in Aqueous TiO$_2$ Suspensions. *Appl. Catal. B* **2003**, *40*, 271–286.

19. Hu, C.; Yu, J.C.; Hao, Z.; Wong, P.K. Photocatalytic Degradation of Triazine-Containing Azo Dyes in Aqueous TiO$_2$ Suspensions. *Appl. Catal. B* **2003**, *42*, 47–55.

20. Zhang, F.; Zhao, J.; Shen, T.; Hidaka, H.; Pelizzetti, E.; Serpone, N. TiO$_2$-Assisted Photodegradation of Dye Pollutants II. Adsorption and Degradation Kinetics of Eosin in TiO$_2$ Dispersions under Visible Light Irradiation. *Appl. Catal. B* **1998**, *15*, 147–156.

21. Tang, W.; Zhang, Z.; An, H.; Quintana, M.; Torres, D. TiO$_2$/UV Photodegradation of Azo Dyes in Aqueous Solutions. *Environ. Technol.* **1997**, *18*, 1–12.

22. Tanaka, K.; Padermpole, K.; Hisanaga, T. Photocatalytic Degradation of Commercial Azo Dyes. *Water Res.* **2000**, *34*, 327–333.

23. Spadaro, J.T.; Isabelle, L.; Renganathan, V. Hydroxyl Radical Mediated Degradation of Azo Dyes: Evidence for Benzene Generation. *Environ. Sci. Technol.* **1994**, *28*, 1389–1393.

24. Augugliaro, V.; Baiocchi, C.; Bianco Prevot, A.; García-López, E.; Loddo, V.; Malato, S.; Marcí, G.; Palmisano, L.; Pazzi, M.; Pramauro, E. Azo-Dyes Photocatalytic Degradation in Aqueous Suspension of TiO$_2$ Under Solar Irradiation. *Chemosphere* **2002**, *49*, 1223–1230.

25. Konstantinou, I.K.; Albanis, T.A. TiO$_2$-Assisted Photocatalytic Degradation of Azo Dyes in Aqueous Solution: Kinetic and Mechanistic Investigations: A Review. *Appl. Catal. B* **2004**, *49*, 1–14.

26. Teh, C.M.; Mohamed, A.R. Roles of Titanium Dioxide and Ion-Doped Titanium Dioxide on Photocatalytic Degradation of Organic Pollutants (Phenolic Compounds and Dyes) in Aqueous Solutions: A Review. *J. Alloys Compd.* **2011**, *509*, 1648–1660.

27. Davis, R.J.; Gainer, J.L.; O'Neal, G.; Wu, I. Photocatalytic Decolorization of Wastewater Dyes. *Water Environ. Res.* **1994**, *66*, 50–53.

28. Houas, A.; Lachheb, H.; Ksibi, M.; Elaloui, E.; Guillard, C.; Herrmann, J. Photocatalytic Degradation Pathway of Methylene Blue in Water. *Appl. Catal. B* **2001**, *31*, 145–157.

29. Paschoal, F.M.M.; Anderson, M.A.; Zanoni, M.V.B. Simultaneous Removal of Chromium and Leather Dye from Simulated Tannery Effluent by Photoelectrochemistry. *J. Hazard. Mater.* **2009**, *166*, 531–537.

30. Crini, G. Non-Conventional Low-Cost Adsorbents for Dye Removal: A Review. *Bioresour. Technol.* **2006**, *97*, 1061–1085.

31. Nohynek, G.J.; Fautz, R.; Benech-Kieffer, F.; Toutain, H. Toxicity and Human Health Risk of Hair Dyes. *Food Chem. Toxicol.* **2004**, *42*, 517–543.

32. Alves, S.P.; Brum, D.M.; Branco de Andrade, É.C.; Pereira Netto, A.D. Determination of Synthetic Dyes in Selected Foodstuffs by High Performance Liquid Chromatography with UV-DAD Detection. *Food Chem.* **2008**, *107*, 489–496.

33. Brown, M.A.; de Vito, S.C. Predicting Azo Dye Toxicity. *Crit. Rev. Environ. Sci. Technol.* **1993**, *23*, 249–324.

34. Tang, W.Z.; An, H. UV/TiO$_2$ Photocatalytic Oxidation of Commercial Dyes in Aqueous Solutions. *Chemosphere* **1995**, *31*, 4157–4170.

35. Galindo, C.; Jacques, P.; Kalt, A. Photodegradation of the Aminoazobenzene Acid Orange 52 by Three Advanced Oxidation Processes: UV/H$_2$O$_2$, UV/TiO$_2$ and VIS/TiO$_2$: Comparative Mechanistic and Kinetic Investigations. *J. Photochem. Photobiol. A* **2000**, *130*, 35–47.

36. Saquib, M.; Muneer, M. TiO$_2$-Mediated Photocatalytic Degradation of a Triphenylmethane Dye (Gentian Violet), in Aqueous Suspensions. *Dyes Pigments* **2003**, *56*, 37–49.

37. Sakthivel, S.; Neppolian, B.; Shankar, M.; Arabindoo, B.; Palanichamy, M.; Murugesan, V. Solar Photocatalytic Degradation of Azo Dye: Comparison of Photocatalytic Efficiency of ZnO and TiO$_2$. *Sol. Energy Mater. Sol. Cells* **2003**, *77*, 65–82.

38. Guillard, C.; Lachheb, H.; Houas, A.; Ksibi, M.; Elaloui, E.; Herrmann, J. Influence of Chemical Structure of Dyes, of pH and of Inorganic Salts on their Photocatalytic Degradation by TiO$_2$ Comparison of the Efficiency of Powder and Supported TiO$_2$. *J. Photochem. Photobiol. A* **2003**, *158*, 27–36.

39. So, C.; Cheng, M.Y.; Yu, J.; Wong, P. Degradation of Azo Dye Procion Red MX-5B by Photocatalytic Oxidation. *Chemosphere* **2002**, *46*, 905–912.

40. Sauer, T.; Cesconeto Neto, G.; Jose, H.; Moreira, R. Kinetics of Photocatalytic Degradation of Reactive Dyes in a TiO$_2$ Slurry Reactor. *J. Photochem. Photobiol. A* **2002**, *149*, 147–154.

41. Daneshvar, N.; Salari, D.; Khataee, A. Photocatalytic Degradation of Azo Dye Acid Red 14 in Water on ZnO as an Alternative Catalyst to TiO$_2$. *J. Photochem. Photobiol. A* **2004**, *162*, 317–322.

42. Epling, G.A.; Lin, C. Investigation of Retardation Effects on the Titanium Dioxide Photodegradation System. *Chemosphere* **2002**, *46*, 937–944.

43. Reutergådh, L.B.; Iangphasuk, M. Photocatalytic Decolourization of Reactive Azo Dye: A Comparison between TiO$_2$ and Us Photocatalysis. *Chemosphere* **1997**, *35*, 585–596.

44. Poulios, I.; Micropoulou, E.; Panou, R.; Kostopoulou, E. Photooxidation of Eosin Y in the Presence of Semiconducting Oxides. *Appl. Catal. B* **2003**, *41*, 345–355.

45. Gomes de Moraes, S.; Sanches Freire, R.; Duran, N. Degradation and Toxicity Reduction of Textile Effluent by Combined Photocatalytic and Ozonation Processes. *Chemosphere* **2000**, *40*, 369–373.

46. Gouvea, C.A.; Wypych, F.; Moraes, S.G.; Duran, N.; Nagata, N.; Peralta-Zamora, P. Semiconductor-Assisted Photocatalytic Degradation of Reactive Dyes in Aqueous Solution. *Chemosphere* **2000**, *40*, 433–440.

47. Paz, Y.; Luo, Z.; Rabenberg, L.; Heller, A. Photooxidative Self-Cleaning Transparent Titanium Dioxide Films on Glass. *J. Mater. Res.* **1995**, *10*, 2842–2848.

48. Paz, Y.; Heller, A. Photo-Oxidatively Self-Cleaning Transparent Titanium Dioxide Films on Soda Lime Glass: The Deleterious Effect of Sodium Contamination and its Prevention. *J. Mater. Res.* **1997**, *12*, 2759–2766.

49. Lackhoff, M.; Prieto, X.; Nestle, N.; Dehn, F.; Niessner, R. Photocatalytic Activity of Semiconductor-Modified Cement—Influence of Semiconductor Type and Cement Ageing. *Appl. Catal. B* **2003**, *43*, 205–216.

50. Benedix, R.; Dehn, F.; Quaas, J.; Orgass, M. Application of Titanium Dioxide Photocatalysis to Create Self-Cleaning Building Materials. *Lacer* **2000**, *5*, 157–168.

51. Hashimoto, K.; Irie, H.; Fujishima, A. TiO2 Photocatalysis: A Historical Overview and Future Prospects. *Jpn. J. Appl. Phys.* **2005**, *44*, doi:10.1143/JJAP.44.8269.

52. Murugan, K.; Rao, T.N.; Gandhi, A.S.; Murty, B. Effect of Aggregation of Methylene Blue Dye on TiO2 Surface in Self-Cleaning Studies. *Catal. Commun.* **2010**, *11*, 518–521.

53. Sakthivel, S.; Janczarek, M.; Kisch, H. Visible Light Activity and Photoelectrochemical Properties of Nitrogen-Doped TiO2. *J. Phys. Chem. B* **2004**, *108*, 19384–19387.

54. Asahi, R.; Morikawa, T.; Ohwaki, T.; Aoki, K.; Taga, Y. Visible-Light Photocatalysis in Nitrogen-Doped Titanium Oxides. *Science* **2001**, *293*, 269–271.

55. Sakthivel, S.; Kisch, H. Daylight Photocatalysis by Carbon-modified Titanium Dioxide. *Angew. Chem. Int. Ed.* **2003**, *42*, 4908–4911.

56. Ohno, T.; Mitsui, T.; Matsumura, M. Photocatalytic Activity of S-Doped TiO2 Photocatalyst under Visible Light. *Chem. Lett.* **2003**, *32*, 364–365.

57. Ho, W. Low-Temperature Hydrothermal Synthesis of S-Doped TiO 2 with Visible Light Photocatalytic Activity. *J. Solid State Chem.* **2006**, *179*, 1171–1176.

58. Li, D.; Haneda, H.; Labhsetwar, N.K.; Hishita, S.; Ohashi, N. Visible-Light-Driven Photocatalysis on Fluorine-Doped TiO2 Powders by the Creation of Surface Oxygen Vacancies. *Chem. Phys. Lett.* **2005**, *401*, 579–584.

59. Liu, H.; Gao, L. (Sulfur, Nitrogen)-Codoped Rutile-Titanium Dioxide as a Visible-Light-Activated Photocatalyst. *J. Am. Ceram. Soc.* **2004**, *87*, 1582–1584.

60. Choi, W.; Termin, A.; Hoffmann, M.R. The Role of Metal Ion Dopants in Quantum-Sized TiO2: Correlation between Photoreactivity and Charge Carrier Recombination Dynamics. *J. Phys. Chem.* **1994**, *98*, 13669–13679.

61. Sant, P.A.; Kamat, P.V. Interparticle Electron Transfer between Size-Quantized CdS and TiO2 Semiconductor Nanoclusters. *Phys. Chem. Chem. Phys.* **2002**, *4*, 198–203.

62. Zou, Z.; Ye, J.; Arakawa, H. Optical and Structural Properties of Solid Oxide Photocatalyst Bi$_2$FeNbO$_7$. *J. Mater. Res.* **2001**, *16*, 35–37.

63. Luan, J.; Pan, B.; Paz, Y.; Li, Y.; Wu, X.; Zou, Z. Structural, Photophysical and Photocatalytic Properties of New Bi$_2$SbVO$_7$ under Visible Light Irradiation. *Phys. Chem. Chem. Phys.* **2009**, *11*, 6289–6298.

64. Nussbaum, M.; Shaham-Waldmann, N.; Paz, Y. Synergistic Photocatalytic Effect in Fe, Nb-Doped BiOCl. *J. Photochem. Photobiol. A* **2014**, *290*, 11–21.

65. Tauc, J.; Grigorovici, R.; Vancu, A. Optical Properties and Electronic Structure of Amorphous Germanium. *Phys. Status Solidi (b)* **1966**, *15*, 627–637.

66. Yan, X.; Ohno, T.; Nishijima, K.; Abe, R.; Ohtani, B. Is Methylene Blue an Appropriate Substrate for a Photocatalytic Activity Test? A Study with Visible-Light Responsive Titania. *Chem. Phys. Lett.* **2006**, *429*, 606–610.

67. Ohtani, B. Photocatalysis A to Z—What we Know and what we do Not Know in a Scientific Sense. *J. Photochem. Photobiol. C* **2010**, *11*, 157–178.

68. Bae, S.; Kim, S.; Lee, S.; Choi, W. Dye Decolorization Test for the Activity Assessment of Visible Light Photocatalysts: Realities and Limitations. *Catal. Today* **2014**, *224*, 21–28.

69. Saison, T.; Chemin, N.; Chanéac, C.; Durupthy, O.; Ruaux, V.; Mariey, L.; Maugé, F.; Beaunier, P.; Jolivet, J. Bi$_2$O$_3$, BiVO$_4$, and Bi$_2$WO$_6$: Impact of Surface Properties on Photocatalytic Activity under Visible Light. *J. Phys. Chem. C* **2011**, *115*, 5657–5666.

70. Wang, W.; Huang, F.; Lin, X. x-BiOI-(1-x)BiOCl as Efficient Visible-Light-Driven Photocatalysts. *Scr. Mater.* **2007**, *56*, 669–672.

71. Zhang, K.; Liu, C.; Huang, F.; Zheng, C.; Wang, W. Study of the Electronic Structure and Photocatalytic Activity of the BiOCl Photocatalyst. *Appl. Catal. B* **2006**, *68*, 125–129.

72. Jiang, J.; Zhao, K.; Xiao, X.; Zhang, L. Synthesis and Facet-Dependent Photoreactivity of BiOCl Single-Crystalline Nanosheets. *J. Am. Chem. Soc.* **2012**, *134*, 4473–4476.

73. Chen, F.; Liu, H.; Bagwasi, S.; Shen, X.; Zhang, J. Photocatalytic Study of BiOCl for Degradation of Organic Pollutants under UV Irradiation. *J. Photochem. Photobiol. A* **2010**, *215*, 76–80.

74. Wang, D.; Gao, G.; Zhang, Y.; Zhou, L.; Xu, A.; Chen, W. Nanosheet-Constructed Porous BiOCl with Dominant {001} Facets for Superior Photosensitized Degradation. *Nanoscale* **2012**, *4*, 7780–7785.

75. Dai, K.; Chen, H.; Peng, T.; Ke, D.; Yi, H. Photocatalytic Degradation of Methyl Orange in Aqueous Suspension of Mesoporous Titania Nanoparticles. *Chemosphere* **2007**, *69*, 1361–1367.

76. Mills, A.; Wang, J. Photobleaching of Methylene Blue Sensitised by TiO$_2$: An Ambiguous System? *J. Photochem. Photobiol. A* **1999**, *127*, 123–134.

77. De Tacconi, N.R.; Carmona, J.; Rajeshwar, K. Reversibility of Photoelectrochromism at the TiO$_2$/Methylene Blue Interface. *J. Electrochem. Soc.* **1997**, *144*, 2486–2490.

78. Zhao, Y.; Swierk, J.R.; Megiatto, J.D., Jr.; Sherman, B.; Youngblood, W.J.; Qin, D.; Lentz, D.M.; Moore, A.L.; Moore, T.A.; Gust, D.; *et al.* Improving the Efficiency of Water Splitting in Dye-Sensitized Solar Cells by using a Biomimetic Electron Transfer Mediator. *Proc. Natl. Acad. Sci. USA* **2012**, *109*, 15612–15616.

79. Zhao, J.; Wu, T.; Wu, K.; Oikawa, K.; Hidaka, H.; Serpone, N. Photoassisted Degradation of Dye Pollutants. 3. Degradation of the Cationic Dye Rhodamine B in Aqueous Anionic Surfactant/TiO$_2$ Dispersions under Visible Light Irradiation: Evidence for the Need of Substrate Adsorption on TiO$_2$ Particles. *Environ. Sci. Technol.* **1998**, *32*, 2394–2400.

80. Abe, R.; Hara, K.; Sayama, K.; Domen, K.; Arakawa, H. Steady Hydrogen Evolution from Water on Eosin Y-Fixed TiO$_2$ Photocatalyst using a Silane-Coupling Reagent under Visible Light Irradiation. *J. Photochem. Photobiol. A* **2000**, *137*, 63–69.

81. Pan, L.; Zou, J.; Zhang, X.; Wang, L. Water-Mediated Promotion of Dye Sensitization of TiO$_2$ under Visible Light. *J. Am. Chem. Soc.* **2011**, *133*, 10000–10002.

82. Wang, Q.; Chen, C.; Zhao, D.; Ma, W.; Zhao, J. Change of Adsorption Modes of Dyes on Fluorinated TiO$_2$ and its Effect on Photocatalytic Degradation of Dyes under Visible Irradiation. *Langmuir* **2008**, *24*, 7338–7345.

83. Neppolian, B.; Choi, H.; Sakthivel, S.; Arabindoo, B.; Murugesan, V. Solar Light Induced and TiO$_2$ Assisted Degradation of Textile Dye Reactive Blue 4. *Chemosphere* **2002**, *46*, 1173–1181.

Unraveling the Photocatalytic Mechanisms on TiO$_2$ Surfaces Using the Oxygen-18 Isotopic Label Technique

Xibin Pang, Chuncheng Chen, Hongwei Ji, Yanke Che, Wanhong Ma and Jincai Zhao

Abstract: During the last several decades TiO$_2$ photocatalytic oxidation using the molecular oxygen in air has emerged as a promising method for the degradation of recalcitrant organic pollutants and selective transformations of valuable organic chemicals. Despite extensive studies, the mechanisms of these photocatalytic reactions are still poorly understood due to their complexity. In this review, we will highlight how the oxygen-18 isotope labeling technique can be a powerful tool to elucidate complicated photocatalytic mechanisms taking place on the TiO$_2$ surface. To this end, the application of the oxygen-18 isotopic-labeling method to three representative photocatalytic reactions is discussed: (1) the photocatalytic hydroxylation of aromatics; (2) oxidative cleavage of aryl rings on the TiO$_2$ surface; and (3) photocatalytic decarboxylation of saturated carboxylic acids. The results show that the oxygen atoms of molecular oxygen can incorporate into the corresponding products in aqueous solution in all three of these reactions, but the detailed incorporation pathways are completely different in each case. For the hydroxylation process, the O atom in O$_2$ is shown to be incorporated through activation of O$_2$ by conduction band electrons. In the cleavage of aryl rings, O atoms are inserted into the aryl ring through the site-dependent coordination of reactants on the TiO$_2$ surface. A new pathway for the decarboxylation of saturated carboxylic acids with pyruvic acid as an intermediate is identified, and the O$_2$ is incorporated into the products through the further oxidation of pyruvic acid by active species from the activation of O$_2$ by conduction band electrons.

Reprinted from *Molecules*. Cite as: Pang, X.; Chen, C.; Ji, H.; Che, Y.; Ma, W.; Zhao, J. Unraveling the Photocatalytic Mechanisms on TiO$_2$ Surfaces Using the Oxygen-18 Isotopic Label Technique. *Molecules* **2014**, *19*, 16291-16311.

1. Introduction

Heterogeneous photocatalytic oxidation (HPO) based on TiO$_2$ as photocatalyst and solar light has emerged as a promising route for the degradation of persistent organic pollutants [1–3]. During photocatalytic oxidation, illumination of TiO$_2$ with light energy larger than the band gap generates conduction band electrons (e$^-_{cb}$) and valence band holes (h$^+_{vb}$), which are the initial "reactive reagents" of TiO$_2$ photocatalysis. The h$^+_{vb}$ and e$^-_{cb}$ can react with the H$_2$O solvent and the dissolved molecular oxygen, respectively, to produce various reactive oxygen species, such as hydroxyl radicals (·OH), hydrogen peroxide (H$_2$O$_2$), superoxide radicals (O$_2$·$^-$) or hydroperoxyl radicals (·OOH) [4,5]. However, the roles of H$_2$O and O$_2$, and the details of their photocatalytic reaction pathways are still elusive. Generally, the degradation reaction of organic pollutants is believed to be initiated by the ·OH radical, which is formed through the oxidation of H$_2$O by h$^+_{vb}$ and can oxidize almost all organic compounds [6]. O$_2$, the final oxidant of the whole photocatalytic oxidation process was exclusively considered as a scavenger of e$^-_{cb}$ to depress the recombination of photogenerated h^+_{vb}/e^-_{cb} pairs and regenerate the photocatalyst. According to this mechanism, the trapping of the e^-_{cb} by O_2 will only determine the rate of the photocatalytic reaction, but not change the reaction pathway and mechanism. However, recent studies have indicated that the participation of O_2 in the photocatalytic reaction would greatly influence the degradation product distribution. For example, in the photoelectrochemical degradation of 4-chlorophenol, when the e^-_{cb} is removed by using an appropriate bias, instead of by dioxygen [7], the mineralization of 4-chlorophenol cannot occur any more. This observation suggests that molecular oxygen may play an important role in the degradation mechanism such as the opening of aromatic rings and the subsequent mineralization, and not just act as an electron acceptor. Nevertheless, detailed mechanisms for the roles of molecular oxygen in photocatalysis are not fully understood so far.

One important reason that hinders the understanding of the mechanisms is the complexity of the photocatalytic process. In the photocatalytic reaction, the h^+_{vb}-induced oxidation half reaction and the e^-_{cb}-induced reduction half reaction proceed on the surface of one photocatalyst particle (usually of a nano size) at the same time, which makes it difficult to distinguish them in space and time. Moreover, the photocatalytic reaction involves a series of active free radical species and processes. It is challenging to investigate these species and processes with steady-state techniques. The isotopic labeling method is one of the most powerful techniques to unravel complicated reaction mechanisms [8]. Stable isotope marking, especially by $^{13}C/^{12}C$, H/D (D = ^2H) and $^{18}O/^{16}O$, is a versatile analytical tool across many realms of science [9]. In the TiO_2 photocatalytic system, the main reaction components O_2, H_2O and TiO_2 all contain oxygen atoms. Accordingly, oxygen atom isotopic labeling can be the most direct and reliable method to trace the O-atom origin of products and distinguish the role and pathways of these components in the different photocatalytic reactions. Another advantage of oxygen isotopic labeling technique is its flexibility, *i.e.*, each component such as ^{18}O-labeled $^{18}O_2$ [10–14], $H_2^{18}O$ [15–17], $Ti^{18}O_2$ [18–23] and ^{18}O-labeled substrate [24–26] can be labeled.

On the TiO_2 surface, the ^{18}O-labeled method has been frequently used in oxygen isotopic exchange measurements to study the stability of surface oxygen in thermally activated catalytic reactions [27,28]. More often, this method was employed in gas phase TiO_2 photocatalytic systems to investigate the photoinduced oxygen isotopic exchange with the aim of understanding the evolution of the intermediate species on the TiO_2 surface [29–36]. However, the photoinduced oxygen isotopic exchange and oxygen isotopic exchange on TiO_2 surface hinder the application of the isotopic labeling method in aerated aqueous TiO_2 photocatalysis systems, because the isotope scrambling among reaction components can make the assignment of the origin of the intermediates and products uncertain. On the other hand, many researchers have also reported that, in the gas phase, the adsorption of water and organic species would inhibit the progress of the photoinduced oxygen isotopic exchange [37–40]. All this suggests that, by deliberate selection of the appropriate photocatalytic systems and conditions, the photoinduced oxygen isotopic exchange can be largely avoided, even if the reaction is carried out in aerated aqueous solutions. In fact, the oxygen isotopic exchange and photoinduced oxygen isotopic exchange between the O_2 and TiO_2 or between O_2 and H_2O were shown to be rather slow in the aqueous TiO_2 photocatalysis systems, compared to the photocatalytic oxidation reaction in the gas phase [41–45]. Accordingly, in many situations, ^{18}O-

labeled methods can be employed to trace the pathway of oxygen-involved reaction in aqueous photocatalytic systems.

Aromatic rings are basic constituents of many kinds of organic pollutants, such as dyes, explosives, pesticides and pharmaceuticals. The release of these compounds could greatly affect the environment and human health [46,47]. Accordingly, aromatic compounds are the most frequently used model substrates to investigate photocatalytic mechanisms and to test the activity of photocatalysts. Before the complete mineralization of the aromatic compounds into CO_2 and H_2O, the photocatalytic degradation would proceed through many main intermediates with different functional groups. For example, hydroxylation, in which the hydrogen on the aromatics is replaced by the electron-donating hydroxyl (Scheme 1 process I), is regarded as an important process in the degradation of aromatic contaminants, especially at the beginning stage of the reaction [6,48–51]. On the way of mineralization, the cleavage of the aromatic ring, sometimes the hydroxylated one, to aliphatic compounds represents another critical process (Scheme 1 process II). The most stable intermediates after cleavage of aromatic ring should be the aliphatic carboxylates. The oxidative decarboxylation of these intermediates would lead to the formation of CO_2 and H_2O (Scheme 1 process III). Accordingly, this review is organized along these three processes: (1) hydroxylation of aromatics; (2) oxidative cleavage of aromatic rings; and (3) decarboxylation of saturated carboxylic acids, and tries to shed light on the application of ^{18}O-labeling methods to the study of photocatalytic mechanisms. We will first introduce how to distinguish the O-source and the investigation of the O_2-incorporation mechanism in the photocatalytic hydroxylation of aromatics by ^{18}O-labeling methods [41,52]. The application of ^{18}O-labeling methods in the study of the TiO_2-photocatalytic aryl ring-opening mechanism is illustrated by photocatalytic degradation of 3,5-di-*tert*-butylcatechol (DTBC) in aqueous solution [53]. Finally, we will focus on ^{18}O-labeling studies of the decarboxylation pathways of saturated mono- [44] and dicarboxylic acids [43]. These studies revealed that molecular oxygen can incorporate into the products to a different extent during all three of these processes, which means that O_2 is not just a conduction band electron scavenger, but also plays a crucial role in the degradation and mineralization of organic pollutants. Moreover, we also give a detailed picture of how molecular oxygen incorporates into the products. These studies also demonstrate that ^{18}O-isotope labeling is a very reliable and powerful method to study aspects of the mechanism of photocataytic oxidations, such as the role and reaction pathway of molecular oxygen and the solvent (water) in the photocatalytic reaction.

2. TiO$_2$ Photocatalytic Hydroxylation of Aromatics

During the TiO_2 photocatalytic degradation of aromatic compounds, hydroxylation products are always the main detected intermediates. It is also accepted that the hydroxylation process is the primary one, and sometimes even the rate-determining step of the whole photocatalytic degradation reaction of aromatic compounds. Usually, the hydroxylation of aromatics is believed to be initiated by direct oxidation by h_{vb}^+ followed by hydrolysis or the attack of ·OH (formed from the oxidation of water by h_{vb}^+) in the photocatalytic systems. According to both pathways, the O-atom of the hydroxyl group in the hydroxylated product should come from H_2O. However, while using the ^{18}O-isotope labeling method ($H_2^{18}O$ and $^{18}O_2$) to investigate the process of photocatalytic oxidation of benzene to CO_2 in aqueous solution, Matsumura and coworkers [51] found that the oxygen atoms of molecular oxygen were introduced into the phenol hydroxylation products (Scheme 2).

Scheme 1. The three main processes of the TiO_2 photocatalytic oxidation of aromatic compounds.

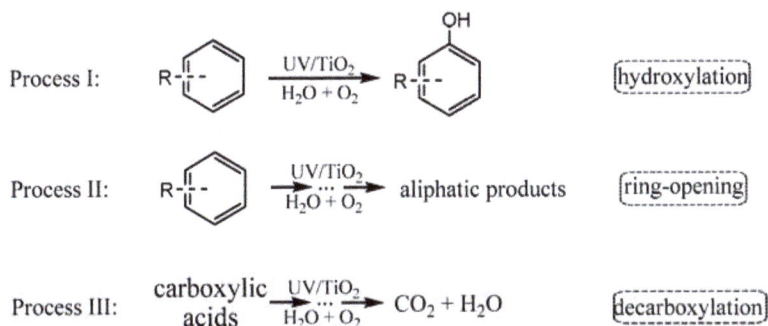

Process I: $R- \xrightarrow[H_2O + O_2]{UV/TiO_2} R-$ (with OH product) [hydroxylation]

Process II: $R- \xrightarrow[H_2O + O_2]{UV/TiO_2}$ aliphatic products [ring-opening]

Process III: carboxylic acids $\xrightarrow[H_2O + O_2]{UV/TiO_2} CO_2 + H_2O$ [decarboxylation]

Scheme 2. TiO_2 photocatalytic oxidation of benzene to phenol in aerated aqueous solution [51].

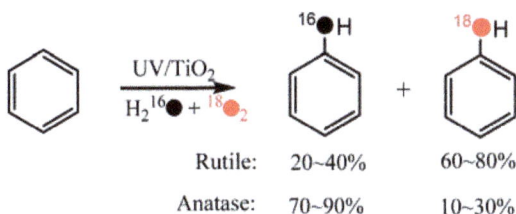

$$\text{benzene} \xrightarrow[H_2{}^{16}O + {}^{18}O_2]{UV/TiO_2} \text{phenol-}{}^{16}OH + \text{phenol-}{}^{18}OH$$

	${}^{16}OH$	${}^{18}OH$
Rutile:	20~40%	60~80%
Anatase:	70~90%	10~30%

Moreover, the incorporation of O atoms from O_2 exhibited great differences between the anatase and rutile systems, i.e., 10%–30% and 60%–80% of the O atoms in the phenol are from O_2 in the anatase and rutile systems, respectively. This observation has at least two implications: (1) dioxygen can participate directly in the oxidation process and incorporate it's O-atom into the hydroxylation products; (2) the extent of the O_2-incorporation is dependent on the photocatalytic conditions, such the crystal phase of the photocatalyst.

In order to unravel the detailed pathway of oxygen incorporation from O_2 in photocatalytic hydroxylation of aromatic compounds, the photocatalytic hydroxylation of several model substrates, such as benzoic acid, benzene, nitrobenzene, and benzonitrile were investigated by the isotope labeling method to trace the origin of the O atoms (from oxidant O_2 or solvent H_2O) in the hydroxyl groups of their corresponding hydroxylation products (Scheme 3) [41]. The results showed that, as reported by Matsumura et al., the O atoms in the hydroxyl groups of the hydroxylation products can originate from both H_2O and O_2, and their contributions are comparable to each other in the hydroxylation process. More importantly, the percentage of hydroxylation products from O_2 was found to depend markedly on the reaction conditions [41], such as the irradiation time and substrate concentration. The percentage of O_2-incorporation in hydroxybenzoic acid (BA-OH), for example, increased from 33.0% to 40.1% as the irradiation time increased from 1 h to 2 h at an initial benzoic acid concentration of 3 mM. In addition, when the initial concentration of benzoic acid increased from 3 mM to 25 mM, the percentage of O_2-incorporation in BA-OH decreased from 33.0% to 13.4% for the same irradiation time. Such a dependence of the percentage of O_2-incorporation in hydroxylation products on the irradiation time and initial substrate concentration was also observed in the photocatalytic hydroxylation of nitrobenzene and benzonitrile.

Scheme 3. The photocatalytic hydroxylation of benzene derivatives on TiO_2 (P25) in aerated aqueous solution [41].

R: -H, -COOH, -NO$_2$, -CN

It was also observed that when benzoic acid and benzene, which have different adsorption abilities on TiO_2, coexisted in the same reaction system, the percentage of O_2-incorporation in their hydroxylation products was different [41]. The addition of benzene to benzoic acid lowered slightly the percentage of O_2-incorporation in BA-OH, while the addition of benzoic acid in benzene notably increased the percentage of O_2-incorporation into its phenol hydroxylation product. The condition-dependence of the percentage of O_2-incorporation provides an excellent opportunity for us to probe further into the role and the mechanism of O_2 in the photocatalytic reaction.

Further, by selective removal of the reactive species generated from the water oxidation by h^+_{vb} or from O_2 reduction by e^-_{cb}, the effect of every reactive species on the isotope distribution of the hydroxylated product was investigated systematically. When formic acid was used to selectively remove the h^+_{vb}, the hydroxylation reaction still occurred. Moreover, nearly all O atoms in the hydroxyl groups of the hydroxylated products of benzoic acid came from O_2 in this situation (Scheme 4a). On the other hand, by employing the benzoquinone to scavenge e^-_{cb} or oxidize O_2^- back to O_2, the hydroxyl O atoms almost all originated from the H_2O solvent (Scheme 4b). Such a sharp contrast in the isotope distribution indicates that h^+_{vb} is indispensable to H_2O incorporation and H_2O cannot participate in the hydroxylation of aromatic compounds if the oxidation of H_2O by h^+_{vb} is blocked, while e^-_{cb} is indispensable to O_2 incorporation. These observations do not support the earlier proposed mechanism that the direct reaction between O_2 and the substrate radical species formed by hole oxidation or HO·-adduct is the main O_2-incorporation pathway in the hydroxylation process, but imply that the O_2 has to be activated by e^-_{cb} before its incorporation. This argument is further confirmed by the isotope experiment of directly oxidation benzoic acid to its cation radical by the one-electron oxidant SO_4^{--}, in which the hydroxyl O atoms were observed to nearly all from H_2O (Scheme 4c).

In addition, the accumulation of H_2O_2 was observed to be dependent on the concentration of substrates. The higher concentration of benzoic acid led to the slower consumption of H_2O_2, which is attributed to the competitive adsorption on TiO_2 between H_2O_2 and benzoic acid. The different percentages of O_2-incorporation in their hydroxylation products between benzoic acid and benzene, which have different adsorption abilities on TiO_2, also demonstrate that the adsorption is a key factor that determines the O_2-incorporation. All the experimental results indicate that the reaction on the surface tends to incorporate the more O atoms from H_2O into the product, while the hydroxylation of the unabsorbed substrate leads to the formation of the product containing the more O_2-derived O atoms. Therefore, both the pathways of the direct oxidation by h^+_{vb} with further hydrolysis (only from H_2O) on the surface and the addition of $\cdot OH$ (from both H_2O and O_2) in bulk solution are important in the TiO_2 photocatalytic hydroxylation of aromatics.

Scheme 4. (a) The hydroxlyation of benzoic acid initiated by e^-_{cb}; (b) the hydroxlylation of benzoic acid initiated by h^+_{vb}; (c) the hydroxlylation of benzoic acid via the path of one-electron oxidation by SO_4^{-} [41].

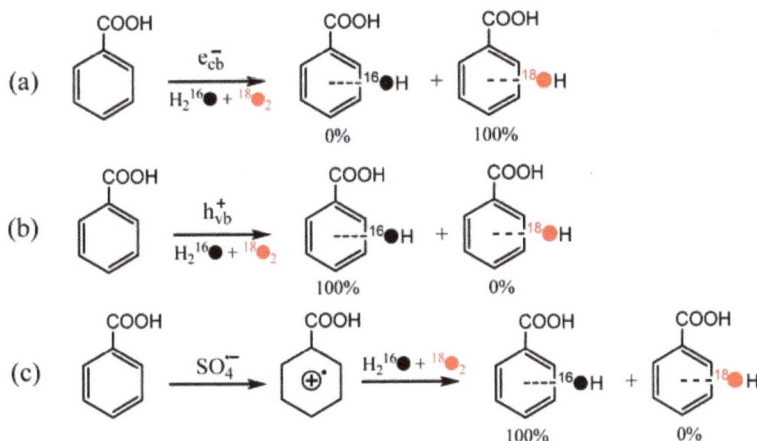

For substituted aromatic rings, different regioisomeric hydroxylated products, which usually exhibit different biological toxicity and secondary reactivity, can be formed during the photocatalytic hydroxylation. By detailed analysis of the monohydroxylated products of benzoic acid, it was found that three isomers of BA-OH, i.e., *meta*-(*m*-), *para*- (*p*-), and *ortho*- (*o*-) BA-OH, were all formed during the phtotocatalytic hydroxylation reaction (Scheme 5) [52]. The isotopic labeling experiments to trace the change of the oxygen source of the formed isomeric hydroxylated intermediates showed that the proportions of oxygen atom of the three hydroxylated isomers were remarkably different. The proportion of O atom of *m*-BA-OH from H_2O was higher than those of *p*- and *o*-BA-OHs. These observations are somewhat unexpected, because the hydroxylation reaction should have the same oxygen source if the same active species and hydroxylated mechanism are responsible for the hydroxylation in the same photocatalytic system. In addition, the difference in the isotopic abundance of product isomers increased stably with the decrease of partial pressure of O_2 (P_{O2}) and with the increase of substrate concentrations. The analysis of the monohydroxylated products of benzoic acid indicated that three isomers of BA-OH have different yields, and that the yield distributions of these three monohydroxylated products changed with P_{O2} and substrate concentration. The formation of *m*-BA-OH was depressed relative to the *p*- and *o*-BA-OH with the increase of P_{O2}, while the high substrate concentration favored the formation of *m*-BA-OH and disfavored the formation of *p*- and *o*-BA-OH, which is consistent with the changes in the isotope abundance.

Scheme 5. The formation of the three regioisomeric monohydroxylated products of benzoic acid: *meta*-hydroxyl benzoic acid (*m*-BA-OH), *para*-hydroxyl benzoic acid (*p*-BA-OH), and *ortho*- hydroxyl benzoic acid (*o*-BA-OH) [52].

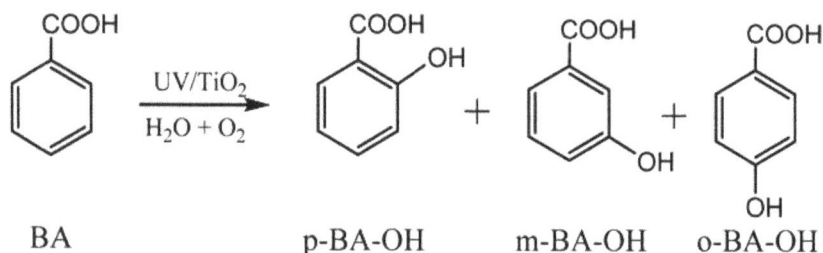

BA   p-BA-OH   m-BA-OH   o-BA-OH

The theoretical calculation indicated that the standard reduction potentials ($E°$ *vs.* NHE) of the HO-adduct benzoic acid radicals at *p*- and *o*-positions are −0.22 and −0.27 V, respectively, while *m*-HO-BA radical has an $E°$ of −0.66 V. The redox potentials of formed *p*- and *o*-BA-OH adduct radicals are below the bottom of conduction band of TiO_2 (−0.29 V), and these adduct radicals can be easily reduced by e^-_{cb}. In contrast, the reduction of *m*-BA-OH radical by e^-_{cb} is impossible because of the more negative reduction potential of *m*-BA-OH adduct radical. Confirming this observation, the redistribution of electron density in the presence of extra e^-_{cb} indicates that TiO_2 with adsorbed HO-adduct radical in the the *p*- and *o*- positions of benzoic acid, the added electron distributes mainly on the HO-BA radical, whereas this electron spreads predominantly over the d-orbits of Ti-atoms, which make up the conduction band of TiO_2, when *m*-HO-BA radical is adsorbed on the TiO_2 cluster [52]. Evidently, the reduction of *m*-BA-OH radical by e^-_{cb} occurs on the surface of TiO_2, where the O atoms of hydroxylated products predominantly come from H_2O. However, in the bulk solution where no reductive e^-_{cb} is available, all three HO-adduct radicals would transform into the corresponding hydroxylated products. Accordingly, any factor that can influence the adsorption of substrates and the accumulation of e^-_{cb} would change the yield distribution and O-origin of three isomeric hydroxylated products. For example, the lower P_{O2} and higher substrate concentration all would exaggerate the accumulation of e^-_{cb}, which favors the reduction of *p*- and *o*-BA-OH radicals. As a result, the relative ratio of *m*-BA-OH will increase. The concept that the e^-_{cb} can selectively recombine the formed surface HO-adduct radicals back to the original substrate (Scheme 6) implicates that we can modulate the hydroxylated intermediates distribution in the photocatalytic degradation of aromatic pollutants by tuning the Fermi level of the photocatalyst or controlling the accumulation amount of e^-_{cb} or surface modification.

Another example for the isotope-labeling study on photocatalytic mechanism comes from the modulation of O_2 reduction pathway by pendant proton relay [42]. In the photocatalytic oxidation of benzoic acid and benzene, it was found that the O_2-incorporation in the hydroxylation products was markedly depressed by the addition of phosphate, which can adsorb strongly on the surface of TiO_2. Further, the isotope-labeling analysis ($H_2^{16}O$ and $^{18}O_2$) indicated that nearly 50% of the O-atoms of hydroxyl groups in BA-OH were derived from O_2 for pristine TiO_2, while almost no O-atoms from O_2 were detected for the phosphate system with 2 mM phosphates. Similarly, for the photocatalytic oxidation of the weak-adsorbed benzene the significantly decreased O-atom incorporation from O_2 into phenol was also observed upon addition of phosphate (from about 50% to 10%) [42]. Since the O_2-incorporation is attributed to the sequential reduction of O_2 by e^-_{cb}, as mentioned above, the depression in O_2-incorporation is an indication of new reduction pathway of O_2 by e^-_{cb} in the presence of phosphates. Indeed, a detailed examination on the formation and decomposition of H_2O_2 during photoctalytic oxidation showed that the little H_2O_2 was generated in the presence of phosphates, while plenty of H_2O_2 was accumulated in the pristine TiO_2 system. The following electrochemical experiments confirmed that O_2 in the phosphate systems is reduced to H_2O via direct four electrons reduction (Equation (1)), while the pathway of the sequential single electron reduction to H_2O_2 (Equations (2–5)) is bypassed. Such a change in the O_2 reduction pathway is proposed to result from the management of surface proton by formation of "pendant proton relay structure" in which the surface-adsorbed phosphate has a pendant acid/base site [42]:

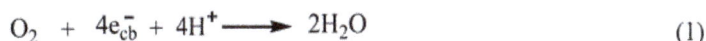

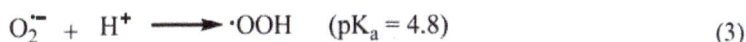

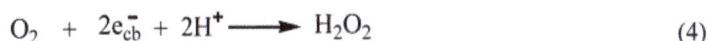

$$O_2 \ + \ 4e^-_{cb} \ + \ 4H^+ \longrightarrow \ 2H_2O \tag{1}$$

$$O_2 \ + \ e^-_{cb} \longrightarrow \ O_2^{-\cdot} \tag{2}$$

$$O_2^{-\cdot} \ + \ H^+ \longrightarrow \ \cdot OOH \quad (pK_a = 4.8) \tag{3}$$

$$O_2 \ + \ 2e^-_{cb} \ + \ 2H^+ \longrightarrow \ H_2O_2 \tag{4}$$

$$H_2O_2 + \ e^-_{cb} \longrightarrow \ \cdot OH \ + \ OH^- \tag{5}$$

Scheme 6. The selective reduction of HO-adduct radicals by conduction band electrons (e^-_{cb}) back to the original substrate [52].

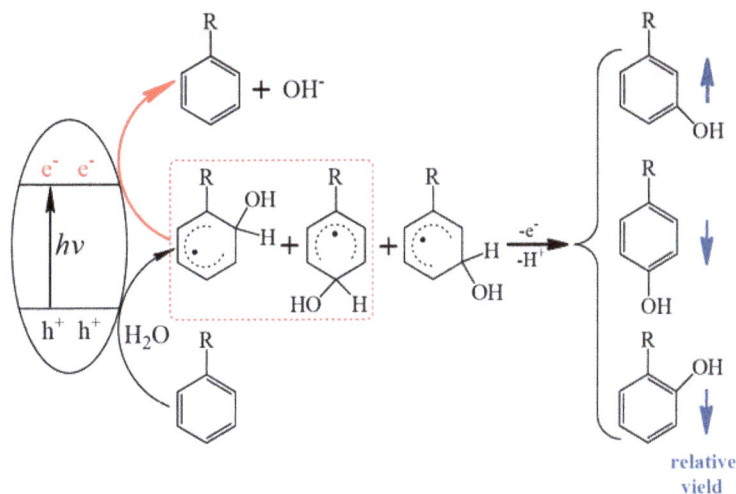

3. Photocatalytic Cleavage of Aryl-Ring on TiO_2 Surface

Another essential step on the way to the complete mineralization of aromatic pollutants into CO_2 is ring-opening, which involves the C-C bond cleavage of aryl rings and is more complex than the hydroxylation process. Accordingly, our understanding on the mechanism of cleavage of aryl ring is even poorer. Many researchers still believe that the $\cdot OH$ radicals are the main active oxygen species in the ring-opening step due to their high oxidative ability [6]. However, recent electrochemical studies suggested that the aromatic rings cannot be efficiently cleaved by the ·OH radicals in the absence of O_2 [7], indicating the ·OH radicals are not a good active species for the cleavage of aryl rings. Other reports [54–59] also argued that the reaction of the O_2-derived species, such as superoxide or singlet oxygen, with aromatics is the main pathway in photocatalytic ring-opening step. It is also considered that both the ·OH radicals and O_2 are both important in the aromatic ring cleavage, i.e., ·OH radicals attack the aromatic ring, and the cleavage results from the reaction of the radicals formed by OH-attack and O_2 [55,60]. As a matter of fact, whether ·OH radical or O_2 ultimately breaks the C-C bond of aromatic ring is still experimentally under debate and the mechanism of the ring-opening reaction in most reports remains somewhat speculative. Another popular mechanism in the ring-opening step for TiO_2 photocatalysis is that the superoxide radical anion, generated from the reduction of O_2 by e_{cb}^-, reacts with substrate radical cation which is formed from oxidation of the substrate by h_{vb}^+, to form a dioxetane intermediate, then homolytic cleavage of the C-C bond leads to the formation the muconaldehyde (Scheme 7) [33,55,58,61–64].

Scheme 7. The previously proposed mechanism for TiO_2 photocatalytic cleavage of aryl-ring via a dioxetane intermediate.

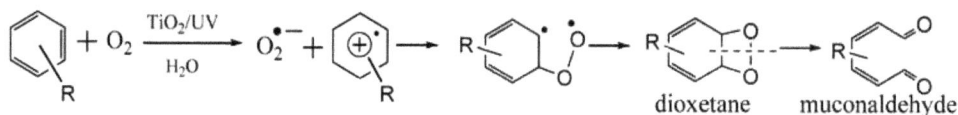

However, no direct evidence was provided so far to support this proposal. In aqueous solution, the detected initial ring-opening products are always carboxylic acids or carboxylic acid derivatives, instead of the expected dialdehyde, probably because the aldehyde is an unstable intermediate and it can be rapidly converted into a stable carboxylic acid under photocatalytic conditions. Another possibility for the low yield of the aldehyde was attributed to the hydroxylation of the aromatic ring before the cleavage, while the ring cleavage still occurs via a dioxetane intermediate process. Recently, Matsumura et al. [65] indeed detected the muconaldehyde intermediate during the cleavage of benzene rings by TiO_2 photocatalyst although its yield is also very low in aqueous solution. Unfortunately, they could not track the origin of O atoms introduced into the muconaldehyde due to the fast oxygen exchange between carbonyl groups of the muconaldehyde and water. In order to track the ring-open pathway by avoiding the oxygen exchange between the products and the solvent, the ortho-dihydroxybenzenes were used as model substrates, because the oxidative cleavage of ortho-dihydroxy-benzenes would form directly the muconic acid which cannot exchange its oxygen atom with H_2O. Therefore, the behaviours of oxygen in the ring-opening products can be definitely tracked by the ^{18}O-labeling methods in the aqueous TiO_2 photocatalytic system. Further, among the numerous ortho-dihydroxyl substituted benzenes, 3,5-di-tert-butylcatechol (DTBC) is an ideal molecular probe for the ring-cleavage, because its substituents can distinguish the ring-opening position and the steric effect of bulky t-butyl groups also can induce the primary ring-opening reaction with unexpectedly high yields [53]. Moreover, the pathway for oxidative cleavage of aromatic ring of DTBC has been extensively studied in the systems of catechol oxygenases and its artificially synthesised iron-containing analogues.

As illustrated in Scheme 8, the main oxidation products are generally divided into three groups: (1) products 2 and 3 are the primary ring-opening products without the loss of a carbon atom, and they are generally intradiol products, which means that the O-atom was inserted into the C-C bond between the two *ortho* hydroxyls; (2) products 4 and 5 have a loss of one carbon relative to their matrix and are generally extradiol products because the O-atom is inserted into the C-C bond out of the two hydroxyls; and (3) products 6–8 are the products by simple oxidation of the DTBC in which no aromatic C-C bond is completely broken. The yields of these products are known to change over a wide range, dependent on the reaction conditions, oxidation systems and reaction mechanisms.

Scheme 8. The structures of the main intermediate products of photocatalytic cleavage of 3,5-di-tert-butylcatechol (DTBC) by TiO₂ (P25) in aerated, water/acetonitrile mixed solution. The main products are divided into three groups and the values in brackets indicate the highest yields of the corresponding products [53].

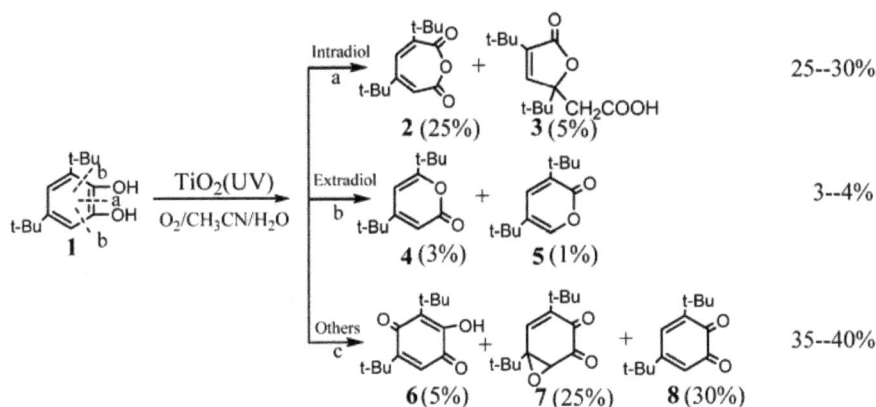

Under the optimized reaction conditions [53], the total yield of the identified products during photocatalytic oxidation of DTBC could account for nearly 75% of the initial DTBC, and the initial ring-opening products (2+3+4+5) had a yield of 30% (Scheme 8), which means that most of the products were recovered after the photocatalytic reaction. In addition, the isotope exchange experiment of the initial ring-opening products 2–5 with H₂¹⁸O or ¹⁸O₂ showed that the oxygen isotopic exchange or the photoinduced oxygen isotopic exchange process was slow and can be ignored under the present experimental conditions. All these characteristics of the oxidation products of DTBC make it possible to conveniently determine the oxygen sources and accurately quantify the ring-opening products by oxygen-18 isotope labeling experiments. Among these products, products 2 and 3 are the intermediates that most directly reflect the aromatic ring-opening mechanism. The product 2 had the highest yield among all of the cleavage products (~25%). In addition, it has been widely accepted in the literature and confirmed by our experiments that product 2 is the precursor of product 3, through the hydrolysis and subsequently addition reactions.

Therefore, product **2** bears the direct information of the primary ring-opening process and is the most desirable intermediates to trace the oxygen source by isotope method.

The isotope-labeling result showed that the inserted O atom (the bridging oxygen in anhydride functional group) in the ring of product **2** was from the O_2 in both $^{18}O_2/H_2^{16}O$ and $^{16}O_2/H_2^{18}O$ systems (Figure 1), which provides the most direct information on how the O_2 cleaves the aromatic ring in the photocatalytic reaction [53].

Figure 1. The oxygen-isotope distribution of product **2** under the various isotope conditions. In each panel, the horizontal axis represents the three isotope conditions: (1) Natural $^{16}O_2$ and $H_2^{16}O$; (2) $^{18}O_2$ and $H_2^{16}O$; (3) $^{16}O_2$ and $H_2^{18}O$; the vertical axis represents the oxygen-isotope distribution ratio (%); M, M+2, M+4 denote products including 0, 1, 2 atoms of ^{18}O in place of ^{16}O [53].

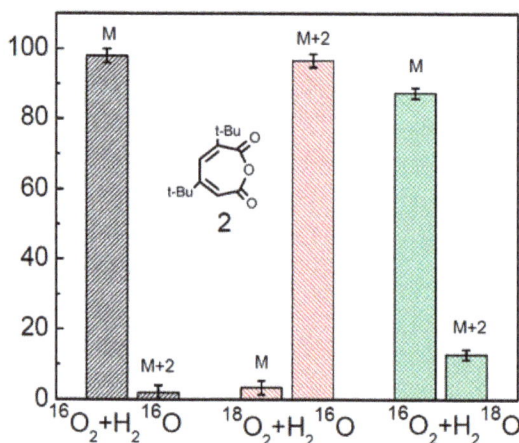

This observation seems not to support the earlier proposal of the usual ring opening pathway via a dioxetane intermediate (Scheme 7). If the TiO_2 photocleavage of aromatics is through the hemolysis of dioxetane intermediate, the major ring-opening products should be a diacid (muconic acid) or acid-aldehydes. However, in the primary ring-opening products, only product **2** was detected in the photocatalytic reaction. No evidence for muconic acid formation could be obtained, neither by HPLC-ESI nor GC-MS detection of the compound itself or of its silylated derivatives. Two possibilities may explain this observation: muconic acid is formed but rapidly converts to product **2**, or muconic acid is not formed at all and product **2** is directly derived from the insertion of an O atom in the C-C bond of DTBC. The former can be excluded because only one O atom was labeled in product **2** (Scheme 9a). If the product **2** were derived from lactonization of muconic acid, it is impossible for product **2** to be completely preserved with one labeled O atom, because the two oxygen atoms in the carboxylate anions are equal (Scheme 9b). Therefore, the isotopic labeling experiments indicate that the inserting O-atom of O_2 into the C-C bond can lead to the cleavage of the aromatic rings in photocatalytic systems.

This single oxygen insertion from O_2 was also found to be dominant in the formation of products **4** and **5** by using $^{16}O_2/H_2^{18}O$ or $^{18}O_2/H_2^{16}O$. These results imply that the O_2-incorporation into the aryl ring is through a single-oxygen insertion rather than the insertion of both oxygen atoms of the O_2 as proposed earlier [33,55,58,61–64]. This incorporation of a single oxygen atom from O_2 is analogous to the reaction of oxygenases in biological systems (Scheme 9c), which leads to distinct intradiol or extradiol ring-opening products, respectively, depending on the initial sites of O_2 coordination [66–69]. It is expected that the activation and insertion of O_2 into aromatic ring in the present case is dependent on its coordination to the Ti–sites on the TiO_2 surface, just as in the active centre of oxygenases.

Scheme 9. The oxygen-isotope distribution of product **2** via the **(a)** single oxygen insertion process or **(b)** molecular oxygen 1,2-addition process. **(c)** the single oxygen incorporation processes in biological systems [53].

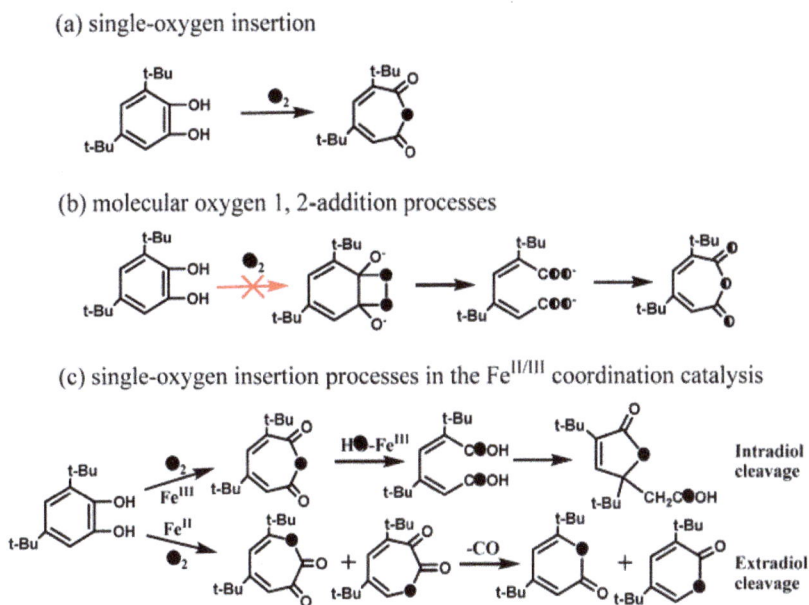

(a) single-oxygen insertion

(b) molecular oxygen 1, 2-addition processes

(c) single-oxygen insertion processes in the $Fe^{II/III}$ coordination catalysis

Also interestingly, it was found that the different TiO_2 particles with identical crystal structures and similar defect concentrations in bulk, but different exposed Ti-site coordination on the surface exhibited different the distribution in the intradiol products (**2 and 3**) and extradiol products (**4 and 5**) [53]. The ratio of intradiol to extradiol products had a good correlation with the ratio of Ti-**4c** to Ti-**3c** sites. Since the relative amounts of the exposed Ti-**4c** to Ti-**3c** sites would be largely dependent on the TiO_2 particle size, the ratio can be determined by the inherent geometric size of the particles. The most noteworthy fact was that even for the smallest size of TiO_2 particles (~9.7 nm), the formation of intradiol products still exceeds that of extradiol products. This was consistent with the geometrical proportions of Ti-**4c** and Ti-**3c** sites on the surface of any size of TiO_2 nanocrystals. Thus, it was proposed that particle size determines the surface Ti-site coordination state, and then determines the chemoselectivity via the different activation pathways of O_2 [53]. The high proportion of the intradiol products must be yielded on the steps (Ti-**4c**) or kinks, where the O_2 should be activated and incorporated by the anchored DTBC radicals since there is no available

Ti-site left for O_2 coordination and reduction (Scheme 10a). Similarly, a small proportion of the extradiol products should be delivered from the corners (Ti-3c) or partial oxygen vacancies (also Ti-3c) with the smallest distribution proportion on the surface, and the activation of O_2 was performed by Ti^{3+} (or e_{cb}^-) because there is an available Ti-site left for O_2 coordination (Scheme 10b). Finally, the Ti center channels the decomposition of these proxy adducts and leads to the ring-opening products by inducing the cleavage of O-O bond (Scheme 10). The mechanism suggests that it is the molecular oxygen that breaks the C-C bond of the aromatics. The surface-mediated aromatic ring cleavage mechanism appears in TiO_2 photocatalytic system shows the site coordination, steric hindrance and stereoelectronic effects play more important roles than the single oxidation potential. It also provides further understanding of the essence of heterogeneous photocatalytic oxidation, namely the final conduction band electron or Ti-sites activating dioxygen.

Scheme 10. Proposed mechanism for singly O-atom incorporation in the photocatalytic cleavage of catechol by TiO_2. **(a)** intradiol cleavage via the anchored DTBC radicals active dioxygen at the step or edge, **(b)** extradiol cleavage via Ti-sites active dioxygen on the corner [53].

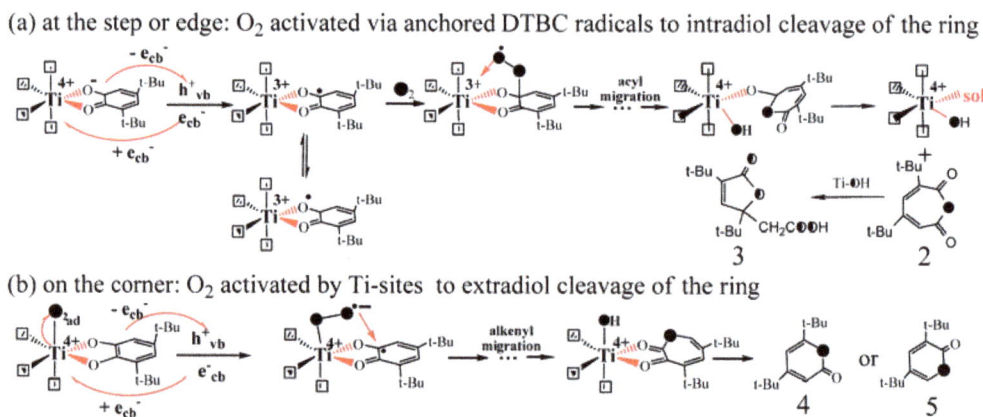

(a) at the step or edge: O_2 activated via anchored DTBC radicals to intradiol cleavage of the ring

(b) on the corner: O_2 activated by Ti-sites to extradiol cleavage of the ring

4. TiO$_2$ Photocatalytic Decarboxylation of Carboxylic Acids

As discussed above, the observed initial ring-opening products of aromatic compounds in aerated aqueous solutions are always carboxylic acids or carboxylic acid derivatives. Thus decarboxylation of these carboxylic acids is one of the most important steps for complete mineralization of organic pollutants. The most accepted mechanism for decarboxylation is through the Photo-Koble reactions [70], which is initiated by the hole oxidation of the carboxylic acids to release the CO_2 and alkyl radicals. In the presence of O_2, the alkyl radicals react with O_2 to form the peroxyl radicals, and then decompose to hydroxylated and carbonylated intermediates via quaternion peroxide intermediates (Russell mechanism) [71]. However, the experimental results for the degradation of saturated monocarboxylic acids (from acetic acid (C_2) to valeric (C_5)) in aerated aqueous solution by TiO_2 photocatalysis indicated that the photocatalytic oxidation of the acids led to the release of CO_2 and the formation of carboxylate acid with one less carbon, that is, a C_5 acid sequentially formed C_4 products, then C_3 and so forth (Scheme 11) [44,72], but little hydroxylated and carbonylated intermediates were detected. This means that there are other possible mechanisms for the TiO_2 photocatalytic decarboxylation of carboxylic acids other than the Photo-Koble process and Russell mechanism.

Scheme 11. Stepwise cleavages of C^1-C^2 bonds in carboxylic acids.

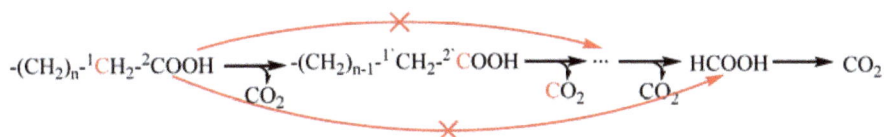

Since this decarboxylation breaks the C^1-C^2 bond of the original acid and establishes a new carboxyl group in the immediately smaller acid, it is feasible to directly trace these processes by using the 18-oxygen isotopic labeling method. The photocatalytic decarboxylation of propionic acid was carried out in $H_2^{18}O$ solution to observe the ^{18}O profile of the produced acetic acid [44]. The isotope-labeling results showed that both the oxygen atoms of O_2 and H_2O can incorporate into the acetic acid. The percentage of the O_2-incorporation can reach as much as 42% at a conversion of 25.1%. To obtain information about the intermediates generated during decarboxylation, diffuse reflectance FTIR measurements (DRIFTS) were employed to monitor *in-situ* the oxidative decarboxylation process of propionic acid. During irradiation, an absorption peak of C=O stretch of α-keto group of pyruvic acid was observed. This peak shifted from 1772 cm^{-1} to 1726 cm^{-1} when the 18-oxygen isotope labeling of $H_2^{18}O$ was introduced, indicating that pyruvic acid is the intermediate of decarboxylation of propionic acid, and the oxygen atom of α-keto group comes from H_2O (Equation (6)) [44]:

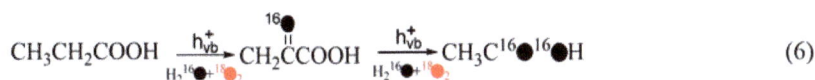

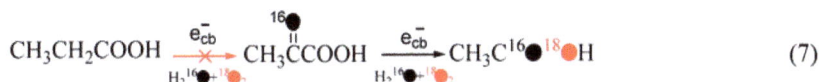

$$CH_3CH_2COOH \xrightarrow[H_2^{16}O + {}^{18}O_2]{h_{vb}^+} CH_3\overset{{}^{16}O}{\overset{\|}{C}}COOH \xrightarrow[H_2^{16}O + {}^{18}O_2]{h_{vb}^+} CH_3C^{16}O^{16}OH \qquad (6)$$

$$CH_3CH_2COOH \xrightarrow[H_2^{16}O + {}^{18}O_2]{e_{cb}^-} CH_3\overset{{}^{16}O}{\overset{\|}{C}}COOH \xrightarrow[H_2^{16}O + {}^{18}O_2]{e_{cb}^-} CH_3C^{16}O^{18}OH \qquad (7)$$

The isotope labeling results on the decarboxylation of propionic acid showed the O atom in O_2 was incorporated into the product acetic acid, while O_2 did not participate into the formation of the intermediate pyruvic acid (Equation (7)) [44]. Thus O_2 should be introduced during the oxidation of pyruvic acid to acetic acid. By using pyruvic acid as the model substrate, the 18-oxygen labeling experiment in $^{18}O_2/ H_2^{16}O$ was carried out to examine the transformation of pyruvic acid to acetic acid by TiO_2 photocatalysis. The oxygen atom of O_2 was largely incorporated into the acetic acid, and at least one oxygen atom of the substrate pyruvic acid was preserved in the carboxyl group of the formed acetic acid. The proportion of O_2-incorporation greatly depended on the reaction conditions, because the O_2-involved decarboxylation competes with the hole/OH radical-promoted decarboxylation process. The electrochemical experiment further confirmed that, in the absence of holes/OH radicals, O_2 could independently cleave the C^1-C^2 bond of pyruvic acid to generate acetic acid with 100% selectivity at a negative bias, while no reaction was observed in the case of propionic acid [44]. A reasonable explanation for such a difference is the different modes of coordination of propionic acid and pyruvic acid on the TiO_2 surface. An α-keto acid adsorbs on the TiO_2 surface via bidentate coordination with the Ti sites, which is favorable for the incorporation of O_2 into the C^1-C^2 bond, likely through a Crigee rearrangement (Scheme 12). However, for propionic acids, it can only chemisorb via monodentate coordination which is not favorable for O_2-incorporation.

Scheme 12. A possible pre-coordination mechanism for dioxygen incorporation into the product during the decarboxylation of α-keto acids [44].

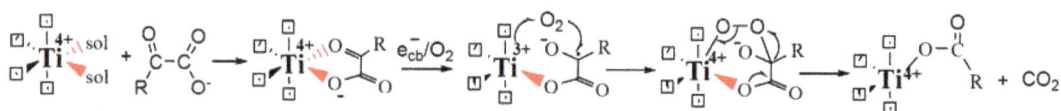

For the decarboxylation of dicarboxylic acids, the process was also found to proceed through stepwise loss fo carbon atoms [43]. More interestingly, the dicarboxylic acids with an even number of carbon atoms (e-DAs) always degraded more slowly than those acids with an odd number of carbon atoms (o-DAs) (Figure 2).

Figure 2. Average rates for both full conversion (■) and TOC removal (▲) by TiO2-based photocatalysis for the five dicarboxylic acids as a function of their carbon number [43].

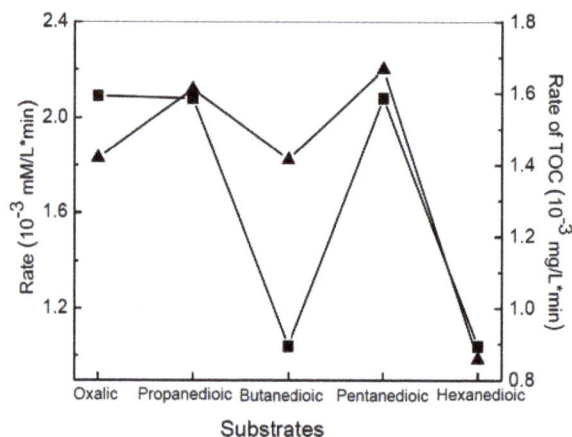

The attenuated total reflection FTIR (ATR-FTIR) combined with the ^{13}C labeling method showed that both carboxyl groups of the e-DAs acids coordinate to TiO_2 through bidentate chelating forms. However, for the o-DAs acids, only one carboxyl group coordinates to TiO_2 in a bidentate chelating manner and the other carboxyl group interacts with the surface in a monodentate mode. Further 18-oxygen labeling experiments showed that the photocatalytic oxidation of o-DAs and e-DAs had different oxygen sources in the carboxyl group of the formed decarboxylated products of dicarboxylic acid [43]. For o-DAs, which have two adsorption different modes: the asymmetrical bidentate and monodentate chelating mode, only the oxygen from H_2O was incorporated into the initial decarboxylated products (Scheme 13a). In contrast, for the e-DAs, which exhibit symmetrical bidentate chelating mode on TiO_2 surface both O_2 and H_2O contributed to their decarboxylated products (Scheme 13b). All these results indicated that the coordination patterns of the substrates on the TiO_2 surface are very important in the TiO_2 photocatalytic decarboxylation of saturated moncarboxylic or dicarboxylic acids and are sometimes the main factor that determines the active species, such as $h^+{}_{vb}/\cdot OH$ and $e^-{}_{cb}/O_2$, to cleave the C-C or C-H bond of the substrates.

Scheme 13. Schematic diagrams of photoctalytic decarboxylation of dicarboxylic acids. **(a)** Pentanedioic acid represents dicarboxylic acids with an odd number of carbon atoms (o-DAs); **(b)** butanedioic acid represents dicarboxylic acids with an even number of carbon atoms (e-DAs) [43].

(a) Asymmetrical coordination: only $h^+{}_{vb}/H_2O$ break the C-C bond

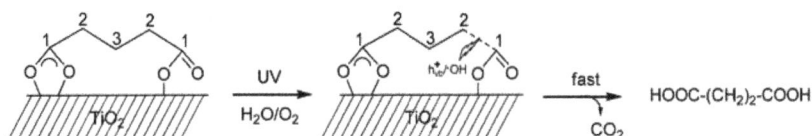

(b) Symmetrical coordination: both $h^+{}_{vb}/H_2O$ and $e^-{}_{cb}/O_2$ can equally break the C-C bond

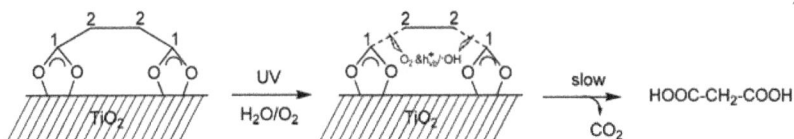

5. Conclusions

In this review, we have summarized our recent work on the oxygen-18 labeling method to unravel the TiO_2-photocatalytic mechanisms, focusing mainly on the mechanisms of hydroxylation of aromatics, cleavage of aryl rings and decarboxylation of saturated carboxylic acids. The incorporation of oxygen atoms from O_2 or H_2O into the initial products is tracked using oxygen-18 labeled $^{18}O_2$ or $H_2{}^{18}O$ in these oxidation processes and could help us understand the essential mechanisms of TiO_2 photocatalytic oxidation. The isotope results indicated that the mineralization of organic compounds needs the cooperation of $H_2O/h^+{}_{vb}$ and $O_2/e^-{}_{cb}$ in the TiO_2 photocatalytic oxidation. The activation and incorporation of molecular oxygen which is the final oxidant of TiO_2 photocatalytic oxidation are mainly affected by the coordination environment of Ti sites on the TiO_2 surface. This site coordination plays a more important role than the single oxidation potential in the photocleavage of aromatic rings. It also showed that ^{18}O-isotope labeling is a very efficient and powerful method for unraveling the TiO_2-photocatalytic mechanisms which are the basis of incorporation of oxygen into products. However, one must be mindful of the processes of oxygen isotopic exchange and photoinduced oxygen isotopic exchange when using the oxygen-18 labeling method in TiO_2-photocatalysis systems

Acknowledgments

This work was supported by 973 project (Nos.2013CB632405, 2010CB933503), NSFC (Nos. 21137004, 21322701, 21221002, 21277147) and the "Strategic Priority Research Program" of the Chinese Academy of Sciences (No. XDA09030200).

Author Contributions

Xibin Pang and Jincai Zhao conceived and designed the review; Xibin Pang and Chuncheng Chen drafted the manuscript; Hongwei Ji, Yanke Che, and Wanhong Ma discussed the results and commented on the manuscript. All authors read and approved the final manuscript.

Conflicts of Interest

The authors declare no conflict of interest.

References

1. Hoffmann, M.R.; Martin, S.T.; Choi, W.; Bahnemann, D.W. Environmental applications of semiconductor photocatalysis. *Chem. Rev.* **1995**, *95*, 69–96.
2. Gaya, U.I.; Abdullah, A.H. Heterogeneous photocatalytic degradation of organic contaminants over titanium dioxide: A review of fundamentals, progress and problems. *J. Photochem. Photobiol. C Photochem. Rev.* **2008**, *9*, 1–12.
3. Pichat, P. Photocatalysis and water purification: From fundamentals to recent applications. In *Photocatalysis and Water Purification*; Pichat, P., Ed.; Wiley-VCH: Weinheim, Germany, 2013.
4. Fox, M.A.; Dulay, M.T. Heterogeneous photocatalysis. *Chem. Rev.* **1993**, *93*, 341–357.
5. Linsebigler, A.L.; Lu, G.; Yates, J.T. Photocatalysis on TiO_2 surfaces: Principles, mechanisms, and selected results. *Chem. Rev.* **1995**, *95*, 735–758.
6. Wang, J.L.; Xu, L.J. Advanced oxidation processes for wastewater treatment: Formation of hydroxyl radical and application. *Crit. Rev. Environ. Sci. Technol.* **2012**, *42*, 251–325.
7. Yang, J.; Dai, J.; Chen, C.C.; Zhao, J.C. Effects of hydroxyl radicals and oxygen species on the 4-chlorophenol degradation by photoelectrocatalytic reactions with TiO_2-film electrodes. *J. Photochem. Photobiol. A* **2009**, *208*, 66–77.
8. Holmes, J.L.; Jobst, K.J.; Terlouw, J.K. Isotopic labelling in mass spectrometry as a tool for studying reaction mechanisms of ion dissociations. *J. Label. Compd. Radiopharm.* **2007**, *50*, 1115–1123.
9. Lloyd-Jones, G.C.; Munoz, M.P. Isotopic labelling in the study of organic and organometallic mechanism and structure: An account. *J. Label. Compd. Radiopharm.* **2007**, *50*, 1072–1087.
10. Almeida, A.R.; Moulijn, J.A.; Mul, G. Photocatalytic oxidation of cyclohexane over TiO_2: Evidence for a Mars-van Krevelen mechanism. *J. Phys. Chem. C* **2011**, *115*, 1330–1338.
11. Epling, W.S.; Peden, C.H.F.; Henderson, M.A.; Diebold, U. Evidence for oxygen adatoms on TiO_2 (110) resulting from O_2 dissociation at vacancy sites. *Surf. Sci.* **1998**, *412–413*, 333–343.

12. Kim, H.H.; Ogata, A.; Schiorlin, M.; Marotta, E.; Paradisi, C. Oxygen isotope (O-18(2)) evidence on the role of oxygen in the plasma-driven catalysis of VOC oxidation. *Catal. Lett.* **2011**, *141*, 277–282.

13. Liao, L.F.; Lien, C.F.; Shieh, D.L.; Chen, M.T.; Lin, J.L. Ftir study of adsorption and photoassisted oxygen isotopic exchange of carbon monoxide, carbon dioxide, carbonate, and formate on TiO_2. *J. Phys. Chem. B* **2002**, *106*, 11240–11245.

14. Thompson, T.L.; Diwald, O.; Yates, J.T. Molecular oxygen-mediated vacancy diffusion on TiO_2 (110)—new studies of the proposed mechanism. *Chem. Phys. Lett.* **2004**, *393*, 28–30.

15. Wu, T.P.; Kaden, W.E.; Anderson, S.L. Water on rutile TiO_2 (110) and Au/ TiO_2 (110): Effects on a mobility and the isotope exchange reaction. *J. Phys. Chem. C* **2008**, *112*, 9006–9015.

16. Suprun, W.; Sadovskaya, E.M.; Rudinger, C.; Eberle, H.J.; Lutecki, M.; Papp, H. Effect of water on oxidative scission of 1-butene to acetic acid over V_2O_5-TiO_2 catalyst. Transient isotopic and kinetic study. *Appl. Catal. A Gen.* **2011**, *391*, 125–136.

17. Bui, T.D.; Yagi, E.; Harada, T.; Ikeda, S.; Matsurnura, M. Isotope tracing study on oxidation of water on photoirradiated TiO_2 particles. *Appl. Catal. B Environ.* **2012**, *126*, 86–89.

18. Civis, S.; Ferus, M.; Kubat, P.; Zukalova, M.; Kavan, L. Oxygen-isotope exchange between CO_2 and solid (TiO_2)-O-18. *J. Phys. Chem. C* **2011**, *115*, 11156–11162.

19. Montoya, J.F.; Ivanova, I.; Dillert, R.; Bahnemann, D.W.; Salvador, P.; Peral, J. Catalytic role of surface oxygens in TiO_2 photooxidation reactions: Aqueous benzene photooxidation with (TiO_2)-O-18 under anaerobic conditions. *J. Phys. Chem. Lett.* **2013**, *4*, 1415–1422.

20. Frank, O.; Zukalova, M.; Laskova, B.; Kurti, J.; Koltai, J.; Kavan, L. Raman spectra of titanium dioxide (anatase, rutile) with identified oxygen isotopes (16,17,18). *PCCP* **2012**, *14*, 14567–14572.

21. Civis, S.; Ferus, M.; Zukalova, M.; Kavan, L.; Zelinger, Z. The application of high-resolution ir spectroscopy and isotope labeling for detailed investigation of TiO_2/gas interface reactions. *Opt. Mater.* **2013**, *36*, 159–162.

22. Kavan, L.; Zukalova, M.; Ferus, M.; Kurti, J.; Koltai, J.; Civis, S. Oxygen-isotope labeled titania: $Ti^{18}O_2$. *PCCP* **2011**, *13*, 11583–11586.

23. Choi, J.; Kang, D.; Lee, K.H.; Lee, B.; Kim, K.J.; Hur, N.H. Evidence for light-induced oxygen exchange in the oxidation of liquid hydrocarbons on oxygen 18-labelled titanium dioxide. *RSC Adv.* **2013**, *3*, 9402–9407.

24. Li, X.J.; Jenks, W.S. Isotope studies of photocatalysis: Dual mechanisms in the conversion of anisole to phenol. *J. Am. Chem. Soc.* **2000**, *122*, 11864–11870.

25. Zhang, X.G.; Ke, X.B.; Zheng, Z.F.; Liu, H.W.; Zhu, H.Y. TiO_2 nanofibers of different crystal phases for transesterification of alcohols with dimethyl carbonate. *Appl. Catal. B Environ.* **2014**, *150*, 330–337.

26. Zhang, M.; Wang, Q.; Chen, C.C.; Zang, L.; Ma, W.H.; Zhao, J.C. Oxygen atom transfer in the photocatalytic oxidation of alcohols by TiO_2: Oxygen isotope studies. *Angew. Chem. Int. Ed.* **2009**, *48*, 6081–6084.

27. Sato, S. Hydrogen and oxygen isotope exchange-reactions over illuminated and nonilluminated TiO_2. *J.Phys. Chem.* **1987**, *91*, 2895–2897.

28. Yanagisawa, Y.; Sumimoto, T. Oxygen-exchange between CO_2 adsorbate and TiO_2 surfaces. *Appl. Phys. Lett.* **1994**, *64*, 3343–3344.

29. Mikhaylov, R.V.; Lisachenko, A.A.; Titov, V.V. Investigation of photostimulated oxygen isotope exchange on TiO_2 Degussa P25 surface upon UV-Vis irradiation. *J. Phys. Chem. C* **2012**, *116*, 23332–23341.

30. Avdeev, V.I.; Bedilo, A.F. Molecular mechanism of oxygen isotopic exchange over supported vanadium oxide catalyst VOx/TiO_2. *J. Phys. Chem. C* **2013**, *117*, 2879–2887.

31. Montoya, J.F.; Peral, J.; Salvador, P. Surface chemistry and interfacial charge-transfer mechanisms in photoinduced oxygen exchange at O_2- TiO_2 interfaces. *ChemPhysChem* **2011**, *12*, 901–907.

32. Courbon, H.; Herrmann, J.M.; Pichat, P. Effect of platinum deposits on oxygen-adsorption and oxygen isotope exchange over variously pretreated, ultraviolet-illuminated powder TiO_2. *J. Phys. Chem.* **1984**, *88*, 5210–5214.

33. Pichat, P.; Courbon, H.; Enriquez, R.; Tan, T.T.Y.; Amal, R. Light-induced isotopic exchange between O_2 and semiconductor oxides, a characterization method that deserves not to be overlooked. *Res. Chem. Intermed.* **2007**, *33*, 239–250.

34. Avdeev, V.I.; Bedilo, A.F. Electronic structure of oxygen radicals on the surface of VOx/TiO_2 catalysts and their role in oxygen isotopic exchange. *J. Phys. Chem. C* **2013**, *117*, 14701–14709.

35. Sato, S.; Kadowaki, T.; Yamaguti, K. Photocatalytic oxygen isotopic exchange between oxygen molecule and the lattice oxygen of TiO_2 prepared from titanium hydroxide. *J. Phys. Chem.* **1984**, *88*, 2930–2931.

36. Kumthekar, M.W.; Ozkan, U.S. Nitric oxide reduction with methane over Pd/TiO_2 catalysts. 2. Isotopic labeling studies using N-15, O-18, and C-13. *J. Catal.* **1997**, *171*, 54–66.

37. Muggli, D.S.; Falconer, J.L. Uv-enhanced exchange of O_2 with H_2O adsorbed on TiO_2. *J. Catal.* **1999**, *181*, 155–159.

38. Civis, S.; Ferus, M.; Zukalova, M.; Kubat, P.; Kavan, L. Photochemistry and gas-phase FTIR spectroscopy of formic acid interaction with anatase $Ti^{18}O_2$ nanoparticles. *J. Phys. Chem. C* **2012**, *116*, 11200–11205.

39. Henderson, M.A. Formic acid decomposition on the {110}-microfaceted surface of TiO_2 (100): Insights derived from ^{18}O-labeling studies. *J. Phys. Chem.* **1995**, *99*, 15253–15261.

40. Courbon, H.; Formenti, M.; Pichat, P. Study of oxygen isotopic exchange over ultraviolet irradiated anatase samples and comparison with the photooxidation of isobutane into acetone. *J. Phys. Chem.* **1977**, *81*, 550–554.

41. Li, Y.; Wen, B.; Yu, C.L.; Chen, C.C.; Ji, H.W.; Ma, W.H.; Zhao, J.C. Pathway of oxygen incorporation from O_2 in TiO_2 photocatalytic hydroxylation of aromatics: Oxygen isotope labeling studies. *Chem. Eur. J.* **2012**, *18*, 2030–2039.

42. Sheng, H.; Ji, H.W.; Ma, W.H.; Chen, C.C.; Zhao, J.C. Direct four-electron reduction of O_2 to H_2O on TiO_2 surfaces by pendant proton relay. *Angew. Chem. Int. Ed.* **2013**, *52*, 9686–9690.

43. Sun, Y.; Chang, W.; Ji, H.; Chen, C.; Ma, W.; Zhao, J. An unexpected fluctuating reactivity for odd and even carbon numbers in the TiO$_2$-based photocatalytic decarboxylation of C2–C6 dicarboxylic acids. *Chem. Eur. J.* **2014**, *20*, 1772–1772.

44. Wen, B.; Li, Y.; Chen, C.C.; Ma, W.H.; Zhao, J.C. An unexplored O$_2$-involved pathway for the decarboxylation of saturated carboxylic acids by TiO$_2$ photocatalysis: An isotopic probe study. *Chem. Eur. J.* **2010**, *16*, 11859–11866.

45. Zhao, Y.B.; Ma, W.H.; Li, Y.; Ji, H.W.; Chen, C.C.; Zhu, H.Y.; Zhao, J.C. The surface-structure sensitivity of dioxygen activation in the anatase-photocatalyzed oxidation reaction. *Angew. Chem. Int. Ed.* **2012**, *51*, 3188–3192.

46. Jones, K.C.; de Voogt, P. Persistent organic pollutants (POPS): State of the science. *Environ. Pollut.* **1999**, *100*, 209–221.

47. Schwarzenbach, R.P.; Gschwend, P.M.; Imboden, D.M. Chemical transformations i: Hydrolysis and reactions involving other nucleophilic species. In *Environmental Organic Chemistry*; John Wiley & Sons, Inc.: Hoboken, NJ, USA, 2005; pp. 489–554.

48. Andino, J.M.; Smith, J.N.; Flagan, R.C.; Goddard, W.A.; Seinfeld, J.H. Mechanism of atmospheric photooxidation of aromatics: A theoretical study. *J. Phys. Chem.* **1996**, *100*, 10967–10980.

49. Nicolaescu, A.R.; Wiest, O.; Kamat, P.V. Mechanistic pathways of the hydroxyl radical reactions of quinoline. 1. Identification, distribution, and yields of hydroxylated products. *J. Phys. Chem. A* **2005**, *109*, 2822–2828.

50. Li, X.; Cubbage, J.W.; Tetzlaff, T.A.; Jenks, W.S. Photocatalytic degradation of 4-chlorophenol. 1. The hydroquinone pathway. *J. Org. Chem.* **1999**, *64*, 8509–8524.

51. Thuan, D.B.; Kimura, A.; Ikeda, S.; Matsumura, M. Determination of oxygen sources for oxidation of benzene on TiO$_2$ photocatalysts in aqueous solutions containing molecular oxygen. *J. Am. Chem. Soc.* **2010**, *132*, 8453–8458.

52. Li, Y.; Wen, B.; Ma, W.H.; Chen, C.C.; Zhao, J.C. Photocatalytic degradation of aromatic pollutants: A pivotal role of conduction band electron in distribution of hydroxylated intermediates. *Environ. Sci. Technol.* **2012**, *46*, 5093–5099.

53. Pang, X.; Chang, W.; Chen, C.; Ji, H.; Ma, W.; Zhao, J. Determining the TiO$_2$-photocatalytic aryl-ring-opening mechanism in aqueous solution using oxygen-18 labeled O$_2$ and H$_2$O. *J. Am. Chem. Soc.* **2014**, *136*, 8714–8721.

54. Szabo-Bardos, E.; Markovics, O.; Horvath, O.; Toro, N.; Kiss, G. Photocatalytic degradation of benzenesulfonate on colloidal titanium dioxide. *Water Res.* **2011**, *45*, 1617–1628.

55. Muneer, M.; Qamar, M.; Bahnemann, D. Photoinduced electron transfer reaction of few selected organic systems in presence of titanium dioxide. *J. Mol. Catal. A:Chem.* **2005**, *234*, 151–157.

56. Lang, K.; Mosinger, J.; Wagnerova, D.M. Progress in photochemistry of singlet oxygen. *Chem. Listy* **2005**, *99*, 211–221.

57. Chen, C.; Lei, P.; Ji, H.; Ma, W.; Zhao, J.; Hidaka, H.; Serpone, N. Photocatalysis by titanium dioxide and polyoxometalate/TiO$_2$ cocatalysts. Intermediates and mechanistic study. *Environ. Sci. Technol.* **2003**, *38*, 329–337.

58. Fox, M.A.; Chen, C.C.; Younathan, J.N.N. Oxidative cleavage of substituted naphthalenes induced by irradiated semiconductor powders. *J. Org. Chem.* **1984**, *49*, 1969–1974.

59. Wahab, H.S.; Bredow, T.; Aliwi, S.M. Computational investigation of the adsorption and photocleavage of chlorobenzene on anatase TiO_2 surfaces. *Chem. Phys.* **2008**, *353*, 93–103.

60. Loeff, I.; Stein, G. 489. The radiation and photochemistry of aqueous solutions of benzene. *J. Chem. Soc. (Resumed)* **1963**, 2623–2633.

61. Maillard-Dupuy, C.; Guillard, C.; Courbon, H.; Pichat, P. Kinetics and products of the TiO_2 photocatalytic degradation of pyridine in water. *Environ. Sci. Technol.* **1994**, *28*, 2176–2183.

62. Li, X.; Cubbage, J.W.; Jenks, W.S. Photocatalytic degradation of 4-chlorophenol. 2. The 4-chloro-catechol pathway. *J. Org. Chem.* **1999**, *64*, 8525–8536.

63. Cermenati, L.; Albini, A.; Pichat, P.; Guillard, C. TiO_2 photocatalytic degradation of haloquinolines in water: Aromatic products gm-ms identification. Role of electron transfer and superoxide. *Res. Chem. Intermed.* **2000**, *26*, 221–234.

64. Cermenati, L.; Pichat, P.; Guillard, C.; Albini, A. Probing the TiO_2 photocatalytic mechanisms in water purification by use of quinoline, photo-Fenton generated OH radicals and superoxide dismutase. *J. Phys. Chem.* **1997**, 101, 2650–2658.

65. Bui, T.D.; Kimura, A.; Higashida, S.; Ikeda, S.; Mafsurnura, M. Two routes for mineralizing benzene by TiO_2-photocatalyzed reaction. *Appl. Catal. B Environ.* **2011**, *107*, 119–127.

66. Bugg, T.D.H.; Lin, G. Solving the riddle of the intradiol and extradiol catechol dioxygenases: How do enzymes control hydroperoxide rearrangements? *Chem. Commun.* **2001**, 941–952.

67. Costas, M.; Mehn, M.P.; Jensen, M.P.; Que, L. Dioxygen activation at mononuclear nonheme iron active sites: Enzymes, models, and intermediates. *Chem. Rev.* **2004**, *104*, 939–986.

68. Xin, M.T.; Bugg, T.D.H. Evidence from mechanistic probes for distinct hydroperoxide rearrangement mechanisms in the intradiol and extradiol catechol dioxygenases. *J. Am. Chem. Soc.* **2008**, *130*, 10422–10430.

69. Jo, D.H.; Que, J.L. Tuning the regiospecificity of cleavage in feiii catecholate complexes: Tridentate facial *versus* meridional ligands. *Angew. Chem. Int. Ed.* **2000**, *39*, 4284–4287.

70. Kraeutler, B.; Bard, A.J. Heterogeneous photocatalytic decomposition of saturated carboxylic acids on titanium dioxide powder. Decarboxylative route to alkanes. *J. Am. Chem. Soc.* **1978**, *100*, 5985–5992.

71. Russell, G.A. Deuterium-isotope effects in the autoxidation of aralkyl hydrocarbons—Mechanism of the interaction of peroxy radicals. *J. Am. Chem. Soc.* **1957**, *79*, 3871–3877.

72. Serpone, N.; Martin, J.; Horikoshi, S.; Hidaka, H. Photocatalyzed oxidation and mineralization of C1–C5 linear aliphatic acids in UV-irradiated aqueous titania dispersions—kinetics, identification of intermediates and quantum yields. *J. Photochem. Photobiol. A Chem.* **2005**, *169*, 235–251.

Spectroscopic Investigation of the Mechanism of Photocatalysis

Yoshio Nosaka, Masami Nishikawa and Atsuko Y. Nosaka

Abstract: Reaction mechanisms of various kinds of photocatalysts have been reviewed based on the recent reports, in which various spectroscopic techniques including luminol chemiluminescence photometry, fluorescence probe method, electron spin resonance (ESR), and nuclear magnetic resonance (NMR) spectroscopy were applied. The reaction mechanisms elucidated for bare and modified TiO_2 were described individually. The modified visible light responsive TiO_2 photocatalysts, *i.e.*, Fe(III)-deposited metal-doped TiO_2 and platinum complex-deposited TiO_2, were studied by detecting paramagnetic species with ESR, $\cdot O_2^-$ (or H_2O_2) with chemiluminescence photometry, and OH radicals with a fluorescence probe method. For bare TiO_2, the difference in the oxidation mechanism for the different crystalline form was investigated by the fluorescence probe method, while the adsorption and decomposition behaviors of several amino acids and peptides were investigated by ^{1}H-NMR spectroscopy.

Reprinted from *Molecules*. Cite as: Nosaka, Y.; Nishikawa, M.; Nosaka, A.Y. Spectroscopic Investigation of the Mechanism of Photocatalysis. *Molecules* **2014**, *19*, 18248-18267.

1. Introduction

TiO_2 photocatalysts have been widely utilized for the oxidation of organic pollutants [1–4]. For further practical applications, the improvement in the photocatalytic efficiency and the extension of the effective wavelength to visible region are desired. To develop photocatalysts, understanding of the detailed photocatalytic mechanisms is prerequisite. Recently, the reaction mechanisms of TiO_2 photocatalysis have been extensively reviewed [5] and the authors also reviewed the reports published up to 2011 from the view of the detection of active oxygen species [6]. In this manuscript, recent development in the reaction mechanism mainly reported by our group was reviewed. Main techniques used were ESR spectroscopy for the state of photoinduced electron and holes, fluorescence probe method for the formation of OH radical and NMR spectroscopy for the adsorption and decomposition of biological molecules in solution.

2. Spectroscopic Methods for Investigating Photocatalysis

2.1. ESR (Electron Spin Resonance) Spectroscopy

ESR spectroscopy is conventionally used to detect unpaired electrons. Photocatalytic reactions proceed by the two following reactions: reduction of reactants with photoexcited electrons and oxidation of reactants with holes. Therefore, it is important to examine the generation behavior of these active species. In TiO_2 photocatalytic systems, two kinds of active species (photoexcited electron and hole) are generated on absorbing photons. Some of the electrons and holes are trapped at Ti and O atoms, to become Ti^{3+} and O^-, respectively. Therefore, by detecting these trapped electrons and holes using ESR spectroscopy under light irradiation of different wavelengths, the generation behavior of excited species can be examined. Moreover, the electron transfer between photocatalysts and co-catalysts can be also examined because if the electron transfer occurs, the amount of the unpaired electron in the co-catalyst should change before and after light irradiation. Therefore, ESR spectroscopy is very useful to elucidate photocatalytic reaction mechanism.

2.2. Chemiluminescence Photometry

Reduced oxygen molecules such as superoxide radical ($\cdot O_2^-$) and H_2O_2 can be detected by chemiluminescence with luminol (LH$_2$, aminodiazabenzoquinone). The one electron oxidized state of luminol ($\cdot L^-$) reacts with $\cdot O_2^-$ to form the excited state of 3-aminophthalic acid to emit fluorescence in alkaline solution [7], where $\cdot L^-$ is formed from LH$_2$ by the oxidation with $\cdot O_2^-$ [8]. Since $\cdot O_2^-$ is rather stable in alkaline solution, after the irradiation on photocatalyst was stopped, luminol is injected to measure the amount of $\cdot O_2^-$ by the chemiluminescence intensity. The same chemiluminescence was obtained from H_2O_2 by the reaction with L that is two-electron oxidized state of LH$_2$ [9]. To measure the amount of H_2O_2 in solution, after mixing luminol, hemoglobin was added to oxidize luminol, because L is rather unstable [8]. Luminol chemiluminescence method has some problems. It is available only in alkaline solution, and luminol emits light with SiO$_2$ in the absence of $\cdot O_2^-$ and H_2O_2. Therefore, in the case of SiO$_2$ deposited TiO$_2$, instead of luminol, MCLA and lucigenin were employed for the detection of $\cdot O_2^-$ and H_2O_2, respectively, by means of chemiluminescence photometry [10].

2.3. Florescence Probe Method

Hydroxyl radical ($\cdot OH$) has been recognized as a key active species in the oxidation mechanism in photocatalysis [9,11]. For the detection of $\cdot OH$ we employed coumarin. It reacts with $\cdot OH$ to produce 7-OH coumarin (umbelliferone) which emits strong fluorescence [12]. After the irradiation of a coumarin aqueous solution containing photocatalysts powder for a given time, the fluorescence intensity of the fluorescent products (umbelliferone) in the solution was measured. The $\cdot OH$ concentration could be calculated from the concentration of umbelliferone with the aid of data of radiation chemistry [12]. Since carboxyl group is known to adsorb on TiO$_2$, the similar experiments were performed for 3-carboxylic acid derivative of coumarin (CCA, Figure 1), and ensured the reaction with $\cdot OH$ to form OH-CCA as illustrated in Figure 1 [13].

2.4. NMR (Nuclear Magnetic Resonance) Spectroscopy

^{1}H-NMR spectroscopy has been recognized as an effective technique to investigate the behaviors of the reactant molecules in the photocatalytic systems. The adsorption and the decomposition of biomolecules such as amino acids and peptides in the aqueous suspension of photocatalysts can be investigated with ^{1}H-NMR spectroscopy with relatively feasible experimental procedures [14] as follows. Firstly, ^{1}H-NMR of organic molecules dissolved in the solvent are measured. Then certain amount of the photocatalysts is added to the solution. From the initial decrease in the intensity of ^{1}H-NMR peaks of the corresponding reactant molecules the amount of the adsorption can be estimated. Then, by measuring the decrease in the intensities of reactant molecules for various photoirradiation times, one could estimate the photodecomposition rates of the reactants [15].

Figure 1. Probing reaction of OH radical with CCA (coumarin 3-carboxy acid) to form fluorescent molecule OH-CCA (7-hydroxy coumarin 3-carboxy acid).

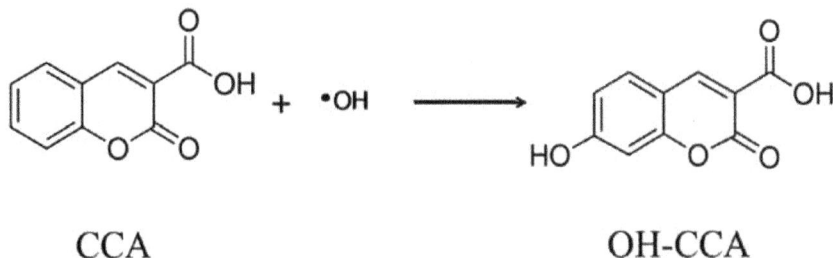

CCA   OH-CCA

3. Mechanism of Photocatalysis

3.1. Bare-TiO$_2$ and Visible-Light Responsive TiO$_2$ Photocatalysts

General scheme of photocatalysis applied for the oxidation of pollutant is shown in Figure 2. Light absorption in semiconductor corresponds to the formation of an electron (e$^-$) in the conduction band (CB) and a hole (h$^+$) in the valence band (VB). Usually e$^-$ reduces O$_2$ in air to form •O$_2^-$ and H$_2$O$_2$.

The photocatalytic oxidation of organic compounds is accelerated with oxygen [16]. The consumption of O$_2$ at the oxidation site of the photocatalyst has been suggested from the experiment of electrochemical probe reactions at the surface of illuminated TiO$_2$ photoelectrode [17]. Therefore, the generalized oxidation mechanism of organic molecules (RH) can be illustrated as shown in Figure 2. Organic reactants RH will degrade by losing one carbon atom by releasing CO$_2$ through the intermediates like aldehyde R'CHO or carboxylate R'COO$^-$. Although •OH has been often regarded to play an important role in the actual oxidation mechanism of photocatalytic reactions, •OH is not involved in the main oxidation process for organic compounds. In place of it, the surface trapped holes play the role of oxidation, which may be acknowledged as the surface adsorbed •OH in the de-protonated form as stated below.

For the extension of the practical applications of photocatalysts, the utilization of visible light has been intensively promoted. Figure 3 shows the energy levels of several representative visible-light responsive photocatalysts. Since the one-electron reduction potential of O$_2$ is very close to that of the CB bottom of TiO$_2$ and the energy level of VB has sufficient oxidation ability, the shift of the VB by doping N (or, C and S) anions has been attempted to absorb visible light (b). In this case, photogenerated holes at the donor level should have the oxidation ability similarly to that of bare TiO$_2$ [9].

Figure 2. General reaction processes for the photocatalytic oxidation of organic molecules.

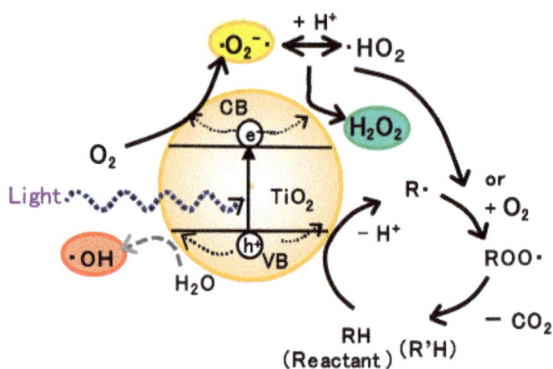

Figure 3. Classification of visible-responsive photocatalysts by the primary reaction mechanism. **(a)** Unmodified TiO_2; **(b)** Nitrogen doped TiO_2; **(c)** Fe(III) grafted TiO_2; **(d)** Fe(III) grafted metal doped TiO_2; **(e)** Platinum complex deposited TiO_2. Adopted with permission from [18]. © 2013 American Chemical Society.

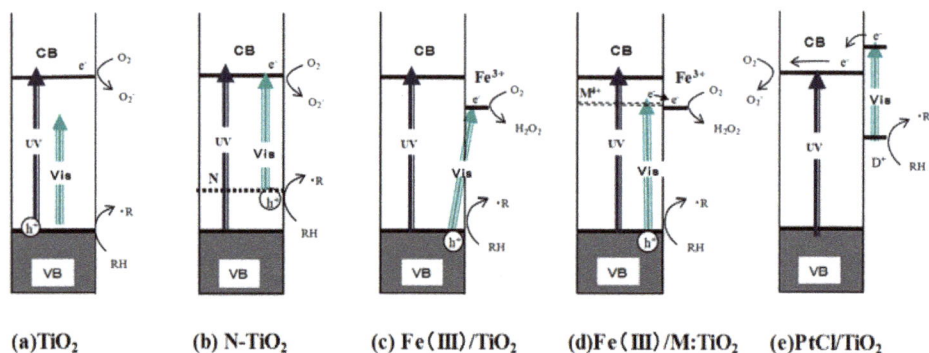

(a)TiO_2 (b) N-TiO_2 (c) Fe(III)/TiO_2 (d)Fe(III)/M:TiO_2 (e)PtCl/TiO_2

Recently, interfacial charge transfer (IFCT) type absorption originating from the excitation of VB electrons to deposited (or grafted) metal ions has been proposed (c). In this case, if the deposited compound has a catalytic ability of O_2 reduction, the efficiency is expected to be increased [19,20]. Since the absorbance of IFCT is very small, to increase the absorption, the transferring of the excited electron to the graft level by doping of metal ions was proposed (d). Photocatalysts of photosensitization type were also proposed, in which the stable metal complex such as $PtCl_6^{2-}$ is deposited as a sensitizer (e). The deposited compound absorbs the visible light and transfers the excited electron to produce a cation radical D^+, which can oxidize organic pollutant molecules. In this case, the enough oxidation power with good stability is required for the oxidized sensitizer D^+ [21,22]. To confirm the suggested reaction mechanism, several spectroscopic methods have been applied to the detection of the paramagnetic species produced on the catalysts along with the primary products ($\cdot O_2^-$, H_2O_2, $\cdot OH$).

3.1.1. Fe(III) Grafted TiO₂ Based Photocatalysts

Fe(III) grafted TiO₂ (Fe/TiO₂) showed the photocatalytic activity under visible light irradiation. The quantum efficiency of Fe/TiO₂ prepared under optimized condition was reported to be 22% [20]. We examined the photocatalytic reaction mechanism of the Fe/TiO₂ using ESR spectroscopy [23]. As shown in Figure 4A, under visible light irradiation, the ESR signal assigned to Fe^{3+} (g = 4.3) was decreased and the ESR signal assigned to trapped holes (g = 2.01) at the TiO₂ host was observed. In the case of TiO₂ without the grafting of Fe^{3+}, the trapped hole signal was scarcely observed as compared to the Fe/TiO₂ under visible light irradiation. This means that electrons at VB are directly transferred to the grafted Fe^{3+} rather than CB (Figure 4B). Using ESR spectroscopy, we could reveal for the first time that the direct electron transfer from the VB of TiO₂ to the Fe^{3+} is the origin of the visible light response.

Figure 4. (A) ESR spectra of **(a)** Fe^{3+} and **(b)** holes for Fe/TiO₂ and **(c)** holes for TiO₂ before and after light irradiation with different wavelengths; **(B)** Photocatalytic reaction mechanism for Fe/TiO₂ under visible light irradiation. Reprinted with permission from [23]. © 2012 American Chemical Society.

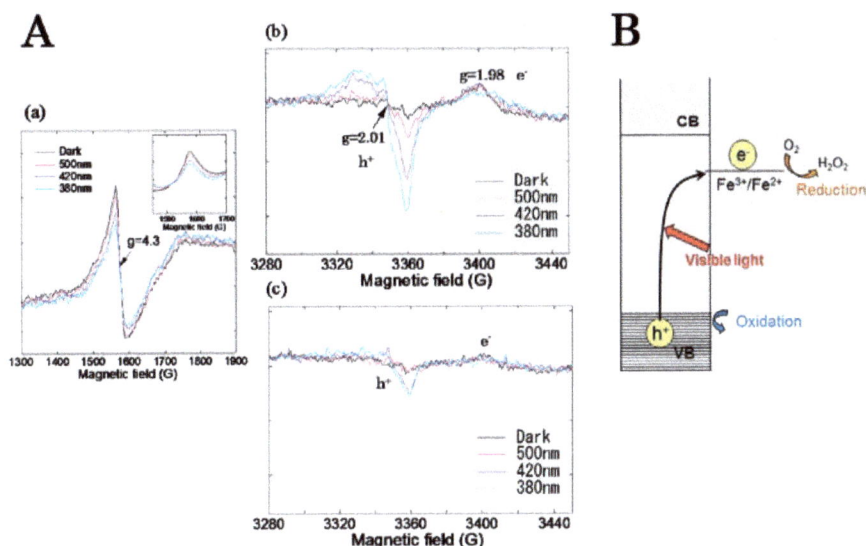

Photocatalytic reactions cannot proceed when the photogenerated electrons do not react with molecular oxygen, which is the only molecule to be reduced in ambient atmosphere even though the redox potential of the photogenerated hole is positive enough to decompose organic compounds. Therefore, it is important to confirm the reduction of O₂ into $\cdot O_2^-$ by one-electron or H₂O₂ by two-electron reductions. Under visible light irradiation, for Fe/TiO₂, the production of H₂O₂ was dominant as compared to $\cdot O_2^-$. This means that the excited Fe^{2+} can reduce O₂ to H₂O₂ through two-electron process. Since the electrons having a potential energy of +0.695 V (*vs.* SHE at pH = 0) can reduce O₂ to H₂O₂ by two-electron process [24], the redox potential of the grafted Fe^{3+} to Fe^{2+} was equal to or less than +0.695 V (*vs.* SHE at pH = 0). Since photogenerated electrons were consumed by the reduction of O_2 to H_2O_2, holes remained at valence band could decompose organic substances efficiently, resulting in the high performance.

Moreover, after the grafting of Fe^{3+} on the TiO_2 doped with metal (M) ions such as Ru, Ir, Rh or Cr ions ($Fe/M:TiO_2$), the photocatalytic activities were enhanced compared to the Fe/TiO_2 as shown in Figure 5A. The visible light response was increased in the order Ir > Cr > Ru > Rh. In the case of the $Fe/Ru:TiO_2$, based on the measurements by ESR spectroscopy, Ru ions were doped as tetravalent and played a role as an acceptor level. Then, the photoexcited Ru ions, by receiving electrons from the VB, immediately transfer electrons to the grafted Fe^{3+} under visible light irradiation. This indirect electron transfer from the VB to the Fe^{3+} via the doped Ru ions occurred in addition to the direct electron transfer from the VB to the Fe^{3+}, leading to the enhancement of photocatalytic activity.

Figure 5. (A) Photocatalytic activities of the $Fe/M:TiO_2$ against gaseous acetaldehyde under visible light irradiation of (a) $\lambda = 470$ nm and (b) $\lambda = 625$ nm; (B) The amount of (a) $\cdot O_2^-$ and (b) H_2O_2 generated on the $M:TiO_2$ and the $Fe/M:TiO_2$, respectively under visible light irradiation of $\lambda = 625$ nm; (C) Schematic energy level diagram for the $Fe/M:TiO_2$. Reprinted with permission from [25]. © 2014 Elsevier B. V.

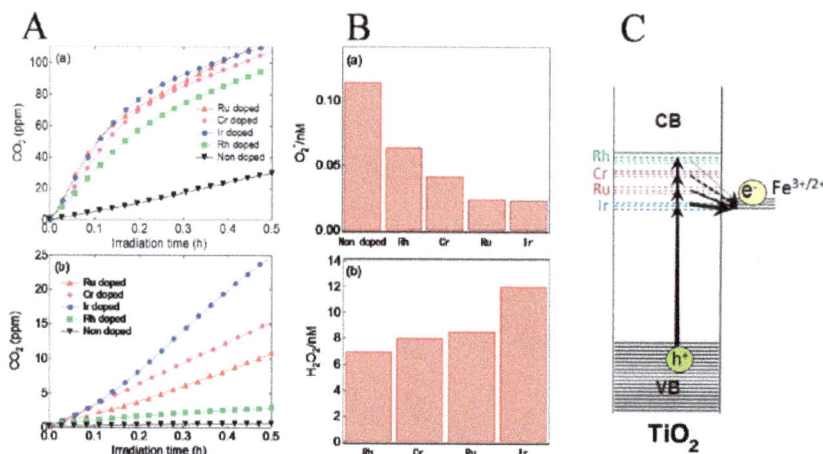

Furthermore, we examined the desirable character of doped metal ions for photocatalytic performance in detail. Figure 5B shows the amounts of $\cdot O_2^-$ and H_2O_2 generated on the $M:TiO_2$ and the $Fe/M:TiO_2$, respectively, under visible light irradiation. The electrons having a potential energy of $+0.38$ V (vs. SHE at pH = 0) can reduce O_2 into $\cdot O_2^-$ under the experimental conditions (pH = 11.5) [24]. Therefore, by measuring the amount of $\cdot O_2^-$, the redox potential of dopants can be relatively estimated. For the non-doped TiO_2, the generation amount of $\cdot O_2^-$ was larger than that of the $M:TiO_2$. For the non-doped TiO_2, O_2 was reduced to $\cdot O_2^-$ by an electron excited at conduction band from defect level under visible light irradiation. This result indicated that all kinds of dopant used played a role as acceptor, because if they play a role as donor, the generation amount of $\cdot O_2^-$ should increase due to electron excitations from dopants to CB. Among the $M:TiO_2$, the generation amount of $\cdot O_2^-$ was decreased in the order of Rh > Cr > Ru > Ir. Since the order of the redox level of dopants should be consistent with that of the $\cdot O_2^-$ amount, their redox levels would be more negative in the order of Rh > Cr > Ru > Ir. Secondly, when electrons transfer to the grafted Fe^{3+}, the excited Fe^{2+} can reduce O_2 to H_2O_2 by a two-electron process as mentioned below. Therefore, we can determine the degree of electron transfer to the Fe^{3+} from the dopants by the measurement of amount of H_2O_2. The generation amount of H_2O_2 was decreased in the order of Ir > Ru > Cr > Rh (Figure 5B). This H_2O_2 generation tendency was opposite to that of $\cdot O_2^-$. This means that through the dopant with more positive redox potential, electrons can transfer more easily to the Fe^{3+} as illustrated in Figure 5C. This is due to the small energy loss of electrons when the redox potential of dopant was close to that of the Fe^{3+}. Therefore, we concluded that the high photocatalytic activity of the Fe/Ir:TiO_2 under visible light irradiation was attributable to the acceptor level due to Ir^{4+} formed close to the redox potential of the grafted Fe^{3+} (Figure 5C).

In the case of TiO_2 codoped with Rh and Sb ions, the efficiency of the indirect electron transfer to the Fe^{3+} was lowered compared to the TiO_2 doped with Rh ions alone (Figure 6A). By codoping with Sb ions, Rh^{4+} was reduced to Rh^{3+} and the formed Rh^{3+} played a role as donor [26,27]. This indicated that the efficiency of the indirect electron transfer of $Rh^{3+} \rightarrow CB \rightarrow Fe^{3+}$ was lower than that of $VB \rightarrow Rh^{4+} \rightarrow Fe^{3+}$ (Figure 6B).

Figure 6. (A) The co-doped effect on photocatalytic activities for CO_2 formation from gaseous acetaldehyde under visible light irradiation of $\lambda = 470$ nm. **(a)** Before and **(b)** after grafting of Fe^{3+}; **(B)** Indirect electron transfer paths to the Fe^{3+} via **(a)** donor level and **(b)** acceptor level. Reprinted with permission from [25]. © 2014 Elsevier B. V.

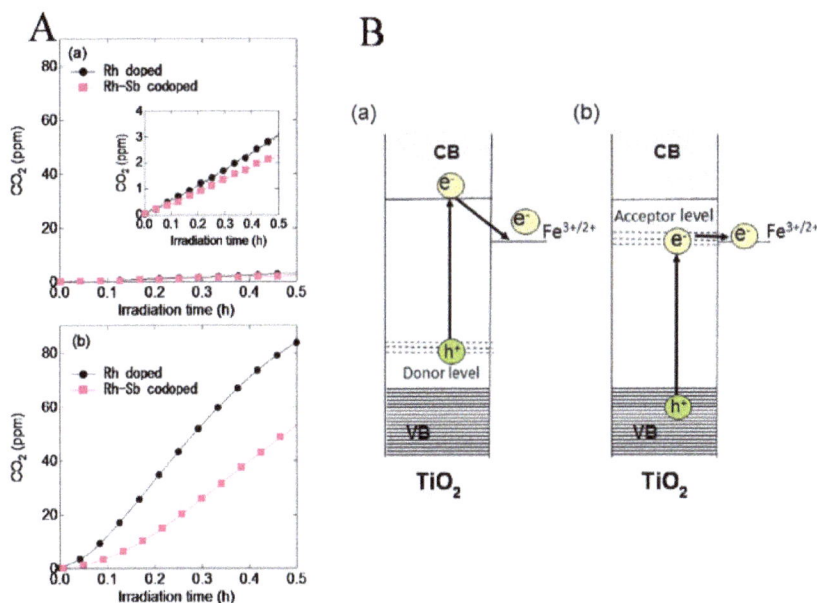

The energy gap of the redox potential between conduction band and the grafted Fe^{3+} was larger than that between the doped Rh^{4+} and the grafted Fe^{3+} and therefore electrons photoexcited at CB could not effectively transfer to the Fe^{3+} because of the large energy loss. From these results, forming acceptor level closed to the redox potential of the grafted Fe^{3+} was important for high performance of Fe/TiO_2 based photocatalysts under visible light irradiation [25]. Recently, a better energy level matching in $Fe/M:TiO_2$ was achieved by employing Fe^{3+} as a doping metal ion [28].

3.1.2. Pt Chloride Deposited TiO₂ Photocatalysts

Pt^{4+} chloride deposited TiO_2 ($PtCl/TiO_2$) also showed a photocatalytic activity under visible light irradiation and its quantum efficiency was 9.8% [20]. In the past, Kisch *et al.*, reported a mechanistic hypothesis to explain $PtCl/TiO_2$ activity [21,29]. The proposed hypothesis was that the PtCl undergoes homolytic Pt-Cl cleavage by absorbing of light, generating a Pt^{3+} intermediate and a chlorine atom, the Pt^{3+} injects an electron to the conduction band of TiO_2, and then the Cl radical oxidizes organic compounds. However, it is not clear whether the Pt-Cl cleavage in the $PtCl/TiO_2$ system would occur. In addition, there is no sufficient evidence to support the injection of electron from Pt^{3+} to the conduction band of TiO_2. Therefore, we clarified the charge transfer between the PtCl and TiO_2 under visible-light irradiation using ESR spectroscopy [30].

For a bare TiO_2 without deposition of PtCl, under visible light irradiation, both ESR signals assigned to trapped electrons and holes were not observed (Figure 7A(a)). For the $PtCl/TiO_2$, a signal assigned to Pt^{3+} was observed. This means that Pt^{4+} chloride complexes were charge-separated into Pt^{3+} and Cl radicals. Then in the TiO_2 host, trapped electrons (g ≈ 1.98) were observed (Figure 7A(b)). These results proved that TiO_2 could receive electrons from excited Pt^{3+} as well as the hypothetical mechanism. However, unlike the hypothetical mechanism, trapped hole signal (g = 2.01) was also observed. Based on the results, some electrons in the VB of TiO_2 would be excited to the orbital of the Cl radicals similarly to the case of direct electron transfer from the VB of TiO_2 to the grafted Fe^{3+} for the Fe/TiO_2 photocatalysts. Since the redox potential (+3.0 V *vs.* SHE at pH = 0) of the VB of rutile TiO_2 is more positive than that (2.47 V *vs.* SHE at pH = 0) of Cl/Cl^- [31], the high photocatalytic activity of $PtCl/TiO_2$ would be owing to the generation of holes in the TiO_2 host.

Generation behaviors of $•O_2^-$ and H_2O_2 under visible light irradiation were also examined for the $PtCl/TiO_2$ as shown in Figure 6B. $•O_2^-$ was predominantly generated compared to H_2O_2. This was opposite behavior to the Fe/TiO_2 for which H_2O_2 was dominantly generated rather than $•O_2^-$. This means that photoexcited electrons have a higher potential energy than +0.38 V (*vs.* SHE at pH = 0) and therefore the reduction of O_2 to $•O_2^-$ was produced by the electrons photoexcited at CB of TiO_2, which supported the ESR results [30].

A plausible reaction mechanism for the $PtCl/TiO_2$ photocatalyst is illustrated in Figure 7C. Photoexcited Pt^{3+} generated by ligand- metal charge transfer in deposited PtCl complex by adsorption of visible light gives an electron to the TiO_2 CB and then the electron is consumed by reduction of O_2 into $•O_2^-$. Some of the photogenerated Cl radicals can decompose organic substances and the other receive electrons by photo-excitation from the valence band of TiO_2, resulting the generation of holes in TiO_2. The organic substances can be efficiently decomposed by the generated holes in TiO_2 with strong oxidation ability.

Figure 7. (**A**) ESR spectra of electrons and holes for (**a**) TiO$_2$ and (**b**) PtCl/TiO$_2$; (**B**) (**a**) •O$_2^-$ and (**b**) H$_2$O$_2$ generated on the PtCl/TiO$_2$ under visible light irradiation. (**C**) Photocatalytic reaction mechanism of PtCl/TiO$_2$. Reprinted with permission from [30]. © 2012 American Chemical Society.

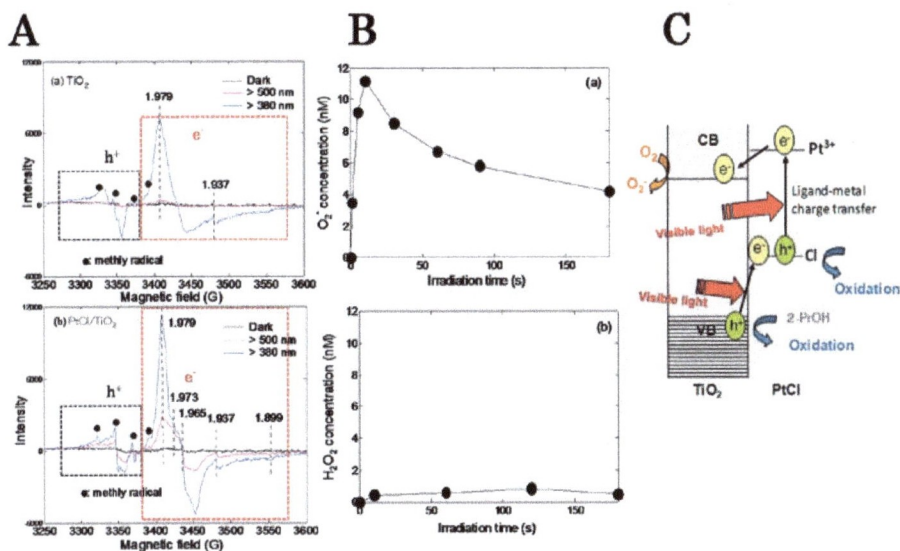

3.1.3. Comparison of the Visible-Light Responsive TiO$_2$ Photocatalysts

The reaction mechanisms of various modified TiO$_2$ were investigated by detecting •OH quantitatively by means of a coumarin fluorescence probe method [18]. The photocatalysts investigated were nitrogen-doped, Fe(III)-grafted, Fe(III)-grafted Ru-dopedTiO$_2$, and Pt-complex-deposited, whose diffuse reflectance spectra are shown in Figure 8A. On the irradiation with 470 nm light in the presence of coumarin, the concentration of umbelliferone was increased (Figure 8B). From the slope, the formation rate of •OH was calculated. Then, the •OH quantum yield was calculated with the absorbed light intensity which was evaluated from the absorption and irradiance spectra in Figure 8A. The quantum yield ranged from 10^{-5} for N-TiO$_2$ to 4×10^{-4} for Fe/TiO$_2$ [18]. In the presence of 0.14 mM H$_2$O$_2$, the •OH yield decreased for N-TiO$_2$ while it increased for Fe/TiO$_2$. The increase for Fe/TiO$_2$ suggests that H$_2$O$_2$ is a reaction intermediate for producing •OH.

The photocatalytic activity was evaluated by the rate of CO$_2$ generation associated with acetaldehyde decomposition and then it was plotted in Figure 9A as a function of the •OH formation rate for each photocatalyst. The CO$_2$ generation rates of the photocatalysts were positively correlated with those of the •OH formation. However, the formation rates of CO$_2$ were extremely larger (10^3 times) than those of •OH. This finding indicates that the oxidation reaction predominantly takes place at the photocatalyst surface with the trapped holes. The good correlation in the figure suggests that •OH in the bulk solution is equilibrated with trapped holes (Equation (1)), but the equilibrium is significantly shifted to the surface trapped holes.

$$\text{•OH} + \text{Ti}^{4+}[\text{TiO}_2] \longleftrightarrow \text{•O}^-\text{Ti}^{4+}[\text{TiO}_2] + \text{H}^+ \tag{1}$$

Figure 8. (A) Absorption spectra as the complement of the reflectance (1-R) and the irradiance spectra of LED used in the study; **(B)** Concentration of umbelliferone generated under 470-nm irradiation in aqueous coumarin solution was plotted as a function of the irradiation time. Reprinted with permission from [18]. © 2013 American Chemical Society.

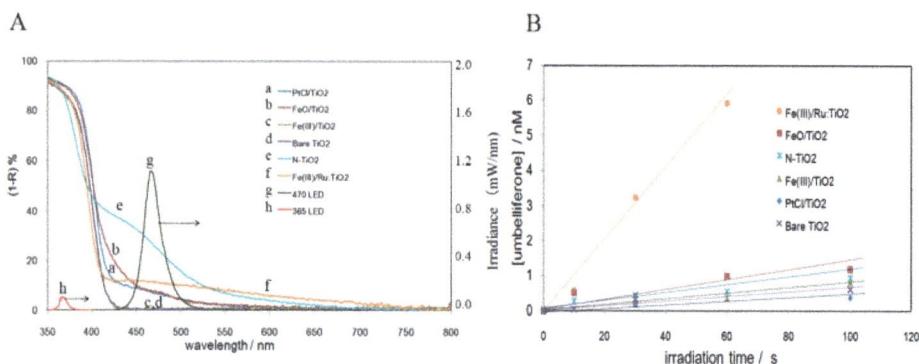

The highest photocatalytic activity in the suspension system was obtained for Fe(III)-deposited Ru-doped TiO_2 (Fe(III)/Ru:TiO_2) whose reaction mechanism is shown in Figure 9B. On the basis of the ESR and chemiluminescence experiments mentioned above [23], the CB electrons are formed by two step excitation with visible light irradiation and O_2 is reduced to H_2O_2. The grafted Fe^{3+} is reduced by Ru^{3+} or by IFCT and then the formed Fe^{2+} produces •OH from H_2O_2. The •OH produced in solution is adsorbed on the TiO_2 surface to form trapped holes which could oxidize organic compounds, such as acetaldehyde, leading to CO_2.

Figure 9. (A) Relationship between the formation rates of CO_2 and •OH under the irradiation of 470 nm LED. The CO_2 formation rate is a measure of the photocatalytic reaction rate in the acetaldehyde decomposition in aqueous suspension system; **(B)** Schematic illustration of reaction mechanism of Fe(III)-grafted Ru-doped TiO_2 photocatalyst based on the detection of •OH and CO_2. Key: IFCT, interfacial charge transfer; h^+_{tr}, surface trapped hole. Reprinted with permission from [18]. © 2013 American Chemical Society.

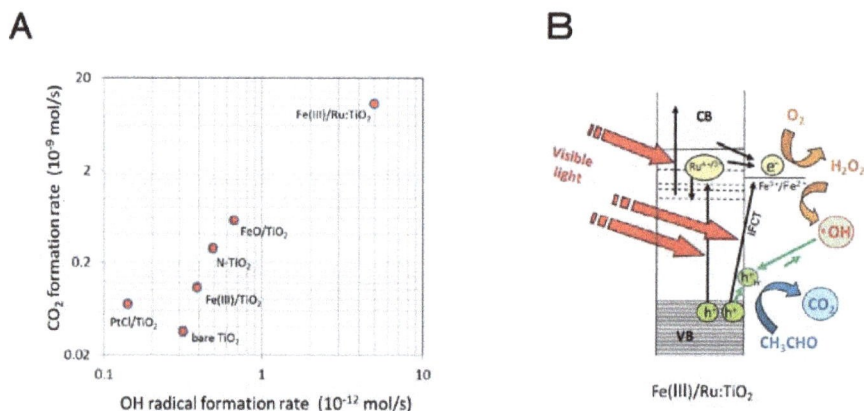

3.2. Photocatalysis with Bare TiO_2

3.2.1. Reactivities of Rutile and Anatase Surfaces

The photogeneration of molecular oxygen at rutile TiO_2 electrode is a famous historical reaction [32]. To investigate the oxidation mechanism, •OH formation was measured by employing three electrodes of rutile TiO_2 (100), (110), and (001) [33]. Figure 10A shows the amount of the produced umbelliferone, which is normalized to the number of charges used in the reaction. For all electrodes, the photocurrent efficiency of •OH was less than 1%, while that of O_2 was about 100%. This observation implies that the conventionally proposed mechanism to produce O_2 via •OH formation is not a major mechanism in water oxidation at TiO_2 surface. Figure 10B shows the plausible reaction steps in the formation of O_2 and •OH through surface peroxo (Ti-O-O-Ti). By cleaving Ti-O bond in the peroxo, O_2 is formed as shown in Figure 10B(a) [34]. When O-O bond is cleaved instead of Ti-O bond, •OH is formed as a byproduct. The •OH formation in Figure 10A increases in the order of (001) < (110) < (100), which can be explained by the strength of Ti-O bond deduced from the surface structure [33].

Figure 10. (A) The amount of produced umbelliferone normalized for the current charge is plotted against the coumarin concentration in solution for the rutile (100), (110), and (001) TiO_2 electrodes; **(B)** Plausible reaction steps starting from peroxo to form **(a)** O_2 and **(b)** •OH at the TiO_2 surface. Reprinted with permission from [33]. © 2013 American Chemical Society.

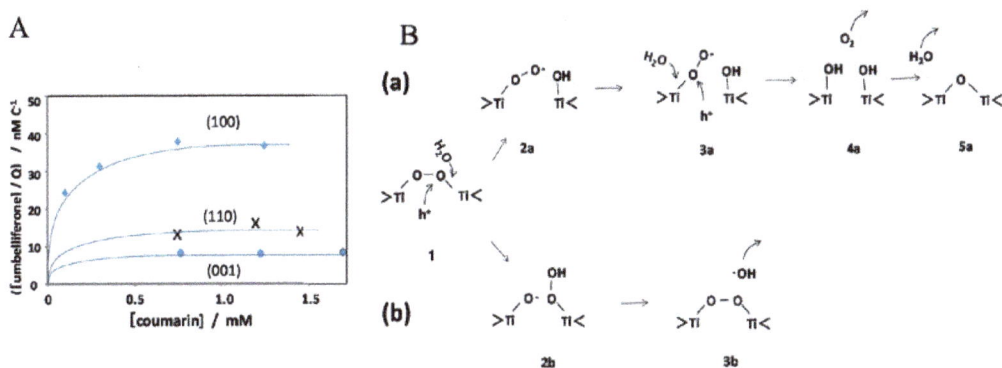

Though rutile TiO_2 shows high activity for O_2 evolution, anatase TiO_2 is known to have a higher activity in the photocatalytic oxidation of organic molecules [35]. The difference in the generation process of •OH between rutile and anatase was investigated by using CCA and coumarin [36]. Figure 11A shows the quantum yields of •OH generation measured with coumarin and CCA together with the adsorbed fraction of CCA. As shown in Figure 11A, anatase and anatase-contained TiO_2 (ST-01, P25, and F1) generated •OH in the substantial yields. The quantum yield for OH-CCA was much larger than that for umbelliferone, indicating that •OH is formed at the TiO_2 surface and diffused into bulk solution. Furthermore, this observation indicates that the active site is different from the adsorption site of $-COO^-$ group. Since H_2O_2 is produced in photocatalysis, the effect of H_2O_2 on the •OH generation was investigated. Figure 11B shows the effect of the addition of H_2O_2 on the formation rate for (a) OH-CCA and (b) umbelliferone. On the addition of H_2O_2, the •OH generation for pure anatase TiO_2 decreased but increased for rutile and rutile-contained TiO_2. This phenomenon has been reported previously for other several TiO_2 powders [37]. The amount of •O_2^- was significantly increased with the addition of H_2O_2 [37]. Although the formation of •OH from H_2O_2 by CB electrons is commonly suggested in the •OH generation mechanism, the fact that the •O_2^- was significantly increased with H_2O_2 denied the one-electron reduction of H_2O_2.

Figure 11. (A) Quantum yields of •OH using the different probe molecules, coumarin (blue) and CCA (brown), and the fraction of adsorbed CCA (green) for four kinds of TiO_2 powders; **(B)** Effect of the addition of 0.14 mM H_2O_2 on the formation rates of **(a)** OH-CCA and **(b)** umbelliferone. Reprinted with permission from [36]. © 2014 American Chemical Society.

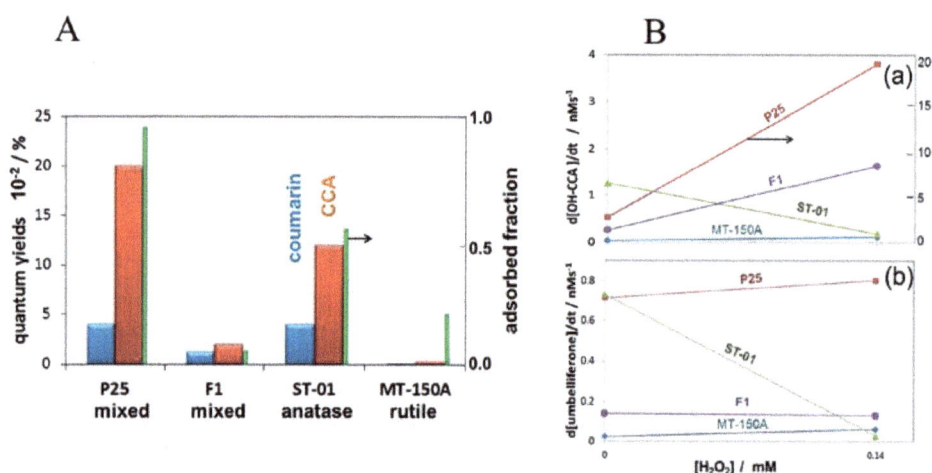

Since the increase is remarkable for anatase than rutile TiO_2, with the addition of H_2O_2, the generation of •OH at anatase surface was replaced by the oxidation of H_2O_2 to form •O_2^-, as illustrated in Figure 12A.

For rutile TiO_2 adsorbed H_2O_2 is equivalent to the surface peroxo, Ti-O-O-Ti and promotes the formation of •OH as stated above. The detailed generation mechanism of •OH on anatase and rutile TiO_2 surfaces can be proposed as shown in Figure 12B. On the anatase surface, photogenerated valence band holes, h^+, are trapped at the surface oxygen to form trapped holes (Ti−O•) that can be regarded as the adsorbed •OH in the deprotonated form (•O^-) [18] then an •OH is released to the solution as represented by Equation (1). On the other hand, for rutile TiO_2, since the crystalline structure is packed more tightly than that for anatase, the stability of the surface trapped holes may be different. By trapping of h^+ predominantly near the trapped hole, Ti-peroxo is formed. As described above in Figure 10B(b), •OH radical is produced by h^+ from H_2O with Ti-peroxo, which plays the role of a catalyst. Thus, the increase of the •OH generation with H_2O_2 for rutile TiO_2 can be explained.

Figure 12. (A) Photocatalytic processes at the conduction band (C.B.) and the valence band (V.B.) of TiO₂ with anatase and rutile crystalline types in the absence and the presence of H₂O₂. The thickness of arrows expresses the degree of the reaction rate; (B) Plausible mechanisms of •OH generation at anatase TiO₂ (upper part) and rutile TiO₂ (lower part). Reprinted with permission from [36]. © 2014 American Chemical Society.

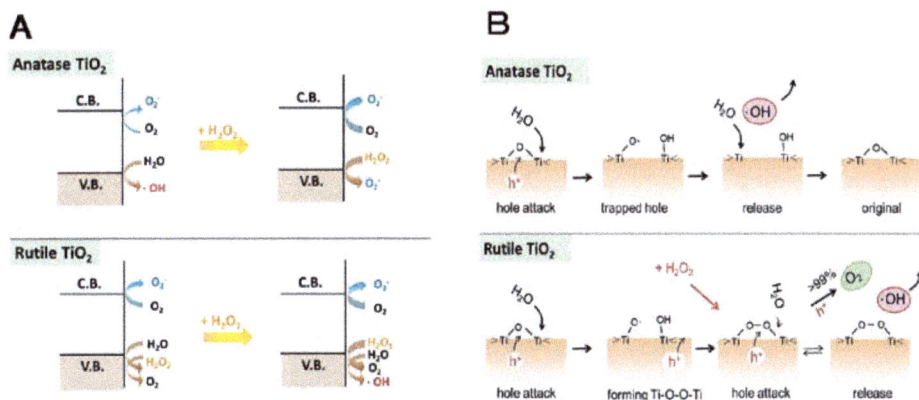

3.2.2. Adsorption and Decomposition of Glycine Related Peptides

The application of photocatalysts to biological fields for their antibacterial effect and in medical treatments for diseases, including cancer, has been proceeding extensively [38,39]. It is believed that the active oxygen species generated on the photocatalysts such as H₂O₂, •OH, and singlet oxygen are involved in the attack to kill various kinds of virus and bacteria [40]. However, the mechanism underlying the photobiological activity is not yet well understood. Since the photocatalytic process is expected to occur at the interface between the photocatalysts and the liquid medium, the interface between protein molecules and inorganic materials has recently received much attention.

Proteins and peptides are composed of various kinds of amino acids. For a proper understanding of the adsorptive and photocatalytic interactions between the surface of the photocatalysts and proteins/peptides, fundamental knowledge on the adsorption and photocatalytic reactivity of individual constituent amino acids would be necessary.

TiO₂ is widely used for practical applications as a photocatalyst. The surface of TiO₂ is amphiphilic, which consists of hydrophobic and hydrophilic parts [41]. The hydrophilic parts involve two kinds of hydroxyl group, that is, the acidic bridged hydroxyl group and the basic terminal hydroxyl group. Both groups can be adsorptive and/or photocatalytic active sites, depending on the kinds of titanium dioxides which are characterized by different particle size, surface area, and crystal forms such as anatase, rutile and brookite. The photocatalyst with different characteristic surface shows different adsorbability and photocatalytic activity [40].

It was demonstrated that both hydrophilic and hydrophobic sites are adsorptive sites but that only hydrophobic sites are photocatalytically active for ST-01 TiO₂ (100% anatase crystal form with a BET surface area of 320 m²·g⁻¹ and a particle size of 9 nm; Ishihara Sangyo Ltd., Osaka,

Japan) [15]. After the calcinations at 973 K hydrophilic parts of the surface of ST-01 can be eliminated and a highly hydrophobic surface (designated as HT-TiO$_2$) is created without changing the crystal form [15]. By employing these characteristics, the adsorption and decomposition sites of the simplest amino acid glycine, whose adsorbability on the TiO$_2$ surface is still controversial [42], and its homopeptides (Gly-Gly and Gly-Gly-Gly) were investigated by ^{1}H-NMR spectroscopy [43]. For Gly-Gly and Gly-Gly-Gly the carboxylic group and the peptide bond were assigned as the adsorptive sites of the peptides on the surface of ST-01. The adsorption feature of Gly-Gly-Gly on TiO$_2$ (ST-01) are illustrated in Figure 13; the peptide would adsorb by the C-terminal carboxyl group most probably with the terminal hydroxyl group at 5-coordinated Ti of TiO$_2$ as is generally believed [42]. The photo decomposition took place by the weak adsorption of the peptide bonds on the surface of TiO$_2$ (ST-01).

Figure 13. Schematic presentation of the adsorption of the peptide (Gly-Gly-Gly) on the hydrophilic and hydrophobic surface of TiO$_2$ (ST-01). Reprinted with permission from [43] © 2014 American Chemical Society.

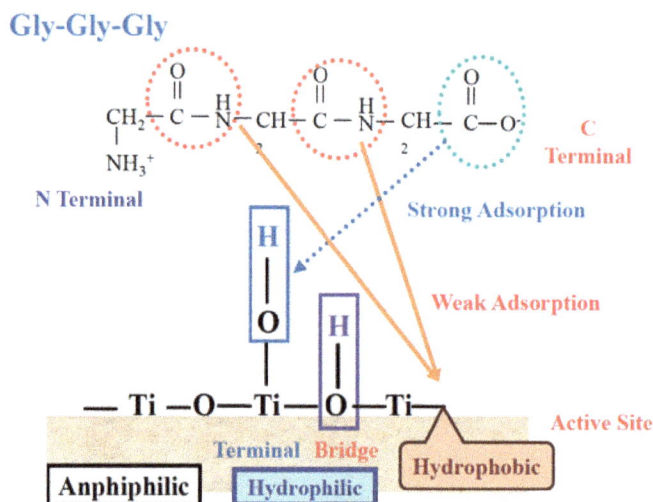

On the other hand, when a hydrophobic side chain Leu is incorporated, in addition to the peptide bonds and the carboxylic group, the adsorption of hydrophobic leucyl residue on the hydrophobic parts of TiO$_2$ surface would take place. As shown in Figure 14, for HT-TiO$_2$, the adsorption of the Leu- containing peptides increased with the increase of the number of the peptide bond that is, Leu < Leu-Gly, Gly-Leu < Leu-Gly-Gly (Figure 14C).

However, the decomposition rates are almost the same (Figure 14D). These facts suggest that both the peptide bond and leucyl side chain could adsorb on the hydrophobic surface of TiO$_2$ but photocatalytic decomposition should take place through the adsorption of the leucyl side chain which would adsorb preferably on the photocatalytic active part of the hydrophobic TiO$_2$ surface. Thus leucyl residue would adsorb preferably on the active site of the hydrophobic part of TiO$_2$ instead of the peptide bonds and photocatalysis proceeds. The adsorption feature of Leu-Gly-Gly, on TiO$_2$ (ST-01) are illustrated in Figure 15.

Figure 14. Equilibrium adsorption and rate of decomposition under the UV irradiation measured at 297 K for Leu, Leu-Gly, Gly-Leu, and Leu-Gly-Gly in the aqueous suspensions of TiO₂; (**A,B**) for untreated TiO₂ and (**C,D**) for HT-TiO₂ (TiO₂ calcined at 973 K). Reprinted with permission from [43] © 2014 American Chemical Society.

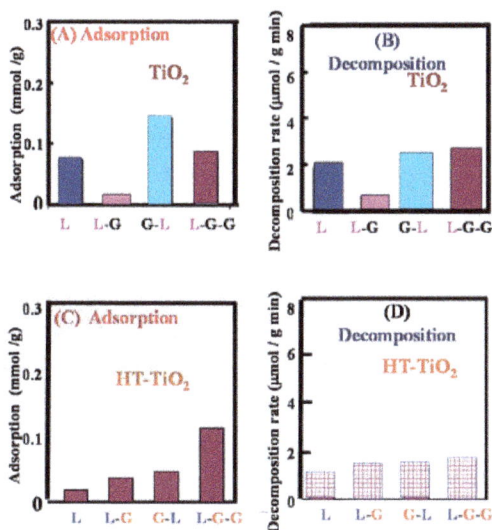

Figure 15. Schematic presentation of the plausible adsorption of the peptide (Leu-Gly-Gly) on the hydrophilic and hydrophobic surface of TiO₂ (ST-01). Reprinted with permission from [43] © 2014 American Chemical Society.

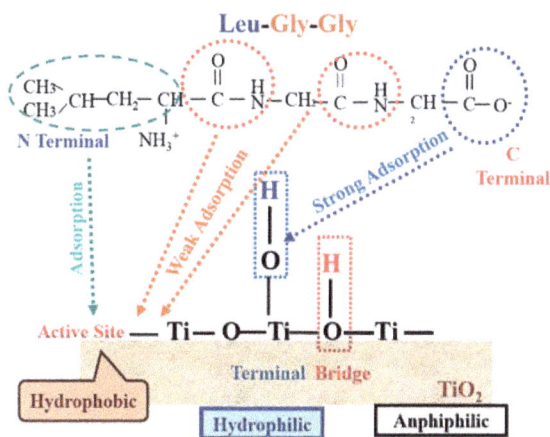

However, as shown in Figure 14A,B it was found that Leu-Gly showed remarkably low adsorbability and decomposition rate as compared to Gly-Leu due to the specific conformation, in which the positively charged amino group and negatively charged carboxyl group interact strongly by electrostatic force [43]. Thus, when a peptide or proteins take a specific conformation, photocatalysis does not work effectively. For the effective use of TiO₂ it would be necessary to acquire information on the surface conformation of the corresponding proteins/peptides to access the surface of the photocatalysts. By combining the information about the surface conformation of proteins/peptides and the active sites of TiO₂ (hydrophobic or hydrophilic), we could design the TiO₂ effective to diminish the specific virus, bacteria or environmental hazardous materials.

3.2.3. Glutathione and Related Amino Acids

With increased applications of TiO₂ nanoparticles, the concerns about their potential human toxicity and their environmental impact have also increased. Although details of human biological responses to TiO₂ exposure are still unavailable, numerous *in vitro* examinations concerning cellular responses induced by TiO₂ have been reported [44–46].

Glutathione is a tri-peptide capable of diminishing active oxygen species in living cells. In spite of the importance of glutathione in defense against oxidative stress, its actual affects and the mechanism for the TiO₂-induced cytotoxicity and genotoxicity have not been completely elucidated yet.

The photocatalytic decomposition of glutathione and related amino acids in TiO₂ suspension was investigated with ^{1}H NMR spectroscopy [47]. The results suggest, that as shown in Figure 16A, both glutathione in reduced (GSH) and oxidative forms (GSSG) are adsorbed on the TiO₂ surface by carboxyl or amino groups but not by the thiol group (SH) of the side chain which plays a crucial role in the glutathione cycle (Scheme 1), to be degraded. This suggests that the function of glutathione cycle should be deteriorated in living cells by the adsorption. However, the decomposition rates are considerably slow as compared with those of the constituent amino acids (Glu, Cys and Gly) as shown in Figure 16B, possibly reflecting the self-defensive property against active oxygen species.

Figure 16. (A) Schematic presentation of the plausible adsorption of glutathione on the surface of TiO₂ (ST-01); (B) (a) Adsorption and (b) decomposition rates of glutathione (GSH and GSSG) and the constituent amino acids (Glu, Cys, and Gly) in aqueous suspension of TiO₂ (5 mg/0.4 mL D₂O) under UV irradiation at 297 K. Reprinted with permission from [47] © 2012 American Chemical Society.

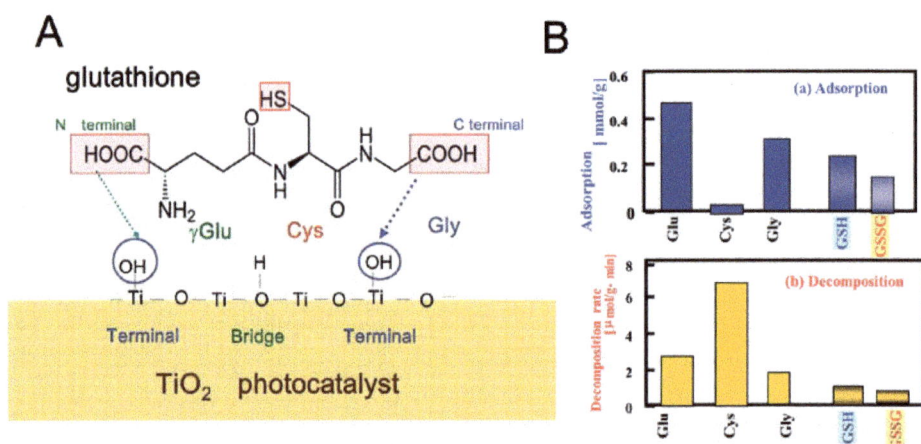

Scheme 1. Glutathione cycle.

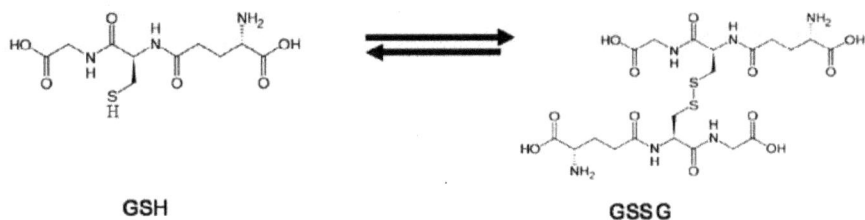

GSH   GSSG

4. Conclusions

TiO$_2$ photocatalysts have been utilized for the oxidation of organic pollutants. For the development of further practical applications, the improvement of the activity with the aid of an understanding of the detailed mechanism(s) of action is a prerequisite. The primary process of photocatalysis reported in the literatures still have some confusion. To clarify the reaction mechanism, the proper and reliable detection of primary active species, such as trapped electrons, trapped holes, •O$_2^-$ and •OH, in photocatalytic systems is required. By employing various spectroscopic techniques we have succeeded in elucidating some of the mechanisms of important photocatalytic reactions. Further investigations are proceeding in our laboratory.

Author Contributions

Main contributions of each author follow; Yoshio Nosaka; Sections 2.2, 2.3, 3.1, 3.1.3, 3.2.1, Masami Nishikawa; Sections 2.1, 3.1.1, 3.1.2, and Atsuko Y. Nosaka; Sections 2.4, 3.2.2, 3.2.3. All of the authors participated in the revision and corrections of the manuscript.

Conflicts of Interest

The authors declare no conflict of interest.

References

1. Kaneko, M., Ohkura, I., Eds. *Photocatalysis Science and Technology*; Kodansha Ltd.: Tokyo, Japan; Springer: New York, NY, USA, 2002.
2. Anpo, M.; Kamat, P.V. (Eds.) *Environmentally Benign Photocatalysts: Applications of Titanium Oxide-Based Materials*; Springer: New York, NY, USA, 2010.
3. Pichat, P. (Ed.) *Photocatalysis and Water Purification: From Fundamentals to Recent Applications*; Wiley-VCH: Weinheim, Germany, 2013.
4. Fujishima, A.; Zhang, X.; Tryk, D.A. TiO$_2$ Photocatalysis and related surface phenomena. *Surf. Sci. Rep.* **2008**, *63*, 515–582.
5. Henderson, M.A. A surface science perspective on TiO$_2$ photocatalysis. *Surf. Sci. Rep.* **2011**, *66*, 185–297.

6. Nosaka, Y.; Nosaka, A.Y. Identification and roles of the active species generated on various photocatalysts. In *Photocatalysis and Water Purification*; Pichat, P., Ed.; Wiley-VCH: Weinheim, Germany, 2013; pp. 3–24.

7. Nosaka, Y.; Yamashita, Y.; Fukuyama, H. Application of chemiluminescent probe to monitoring superoxide radicals and hydrogen peroxide in TiO_2 photocatalysis. *J. Phys. Chem. B* **1997**, *101*, 5822–5827.

8. Koizumi, Y.; Nosaka, Y. Kinetics simulation of luminol chemiluminescence based on quantitative analysis of photons generated in electrochemical oxidation. *J. Phys. Chem. A* **2013**, *117*, 7705–7711.

9. Hirakawa, T.; Nosaka, Y. Selective production of superoxide ions and hydrogen peroxide over nitrogen- and sulfur-doped TiO_2 photocatalysts with visible light in aqueous suspension systems. *J. Phys. Chem. C* **2008**, *112*, 15818–15823.

10. Oguma, J.; Kakuma, Y.; Murayama, S.; Nosaka, Y. Effects of silica coating on photocatalytic reactions of anatase titanium dioxide studied by quantitative detection of reactive oxygen species. *Appl. Catal. B* **2013**, *129*, 282–286.

11. Nosaka, Y.; Komori, S.; Yawata, K.; Hirakawa, T.; Nosaka, A.Y. Photocatalytic •OH radical formation in TiO_2 aqueous suspension studied by several detection methods. *Phys. Chem. Chem. Phys.* **2003**, *5*, 4731–4735.

12. Louit, G.; Foley, S.; Cabillac, J.; Coffigny, H.; Taran, F.; Valleix, A.; Renault, J.P.; Pin, S. The reaction of coumarin with the OH radical revisited: Hydroxylation product analysis determined by fluorescence and chromatography. *Radiat. Phys. Chem.* **2005**, *72*, 119–124.

13. Gerald, L.N.; Jamie, R.M. Fluorescence detection of hydroxyl radicals. *Radiat. Phys. Chem.* **2006**, *75*, 473–478.

14. Matsushita, M.; Tran, H.; Nosaka, A.Y.; Nosaka, Y. Photo-oxidation mechanism of L-alanine in TiO_2 photocatalytic systems. *Catal. Today* **2007**, *120*, 240–244.

15. Nosaka, A.Y.; Nishino, J.; Fujiwara, T.; Yagi, H.; Akutsu, H.; Nosaka, Y. Effects of thermal treatments on the recovery of adsorbed water and photocatalytic activities of TiO_2 photocatalytic systems. *J. Phys. Chem. B* **2006**, *110*, 8380–8385.

16. Maldotti, A.; Molinari, A.; Amadelli, R. Photocatalysis with organized systems for the oxofunctionalization of hydrocarbons by O_2. *Chem. Rev.* **2002**, *102*, 3811–3836.

17. Ikeda, K.; Sakai, H.; Ryo, R.; Hashimoto, K.; Fujishima, A. Photocatalytic reactions involving radical chain reactions using microelectrodes. *J. Phys. Chem. B* **1997**, *101*, 2617–2620.

18. Zhang, J.; Nosaka, Y. Quantitative detection of OH radicals for investigating the reaction mechanism of various visible-light TiO_2 photocatalysts in aqueous suspension. *J. Phys. Chem. C* **2013**, *117*, 1383–1391.

19. Irie, H.; Miura, S.; Kamiya, K.; Hashimoto, K. Efficient visible light-sensitive photocatalysts: Grafting Cu(II) ions onto TiO_2 and WO_3 photocatalysts. *Chem. Phys. Lett.* **2008**, *457*, 202–205.

20. Yu, H.; Irie, H.; Shimodaira, Y.; Hosogi, Y.; Kuroda, Y.; Miyauchi, M.; Hashimoto, K. An efficient visible-light-sensitive Fe(III)-grafted TiO_2 photocatalyst. *J. Phys. Chem. C* **2010**, *114*, 16481–16487.

21. Macyk, W.; Kisch, H. Photosensitization of crystalline and amorphous titanium dioxide by platinum(IV) chloride surface complexes. *Chem. Eur. J.* **2001**, *7*, 1862–1867.

22. Ishibai, Y.; Sato, J.; Nishikawa, T.; Miyagishi, S. Synthesis of visible-light active TiO_2 photocatalyst with Pt-modification: Role of TiO_2 substrate for high photocatalytic activity. *Appl. Catal. B* **2008**, *79*, 117–121.

23. Nishikawa, M.; Hiura, S.; Mitani, Y.; Nosaka, Y. Photocatalytic reaction mechanism of Fe(III)-grafted TiO_2 studied by means of ESR spectroscopy and chemiluminescence photometry. *J. Phys. Chem. C* **2012**, *116*, 14900–14907.

24. Nosaka, Y.; Takahashi, S.; Sakamoto, H.; Nosaka, A.Y. Reaction mechanism of Cu(II)-grafted visible-light responsive TiO_2 and WO_3 photocatalysts studied by means of ESR spectroscopy and chemiluminescence photometry. *J. Phys. Chem. C* **2011**, *115*, 21283–21290.

25. Nishikawa, M.; Takanami, R.; Nakagoshi, F.; Suizu, H.; Nagai, H.; Nosaka, Y. Dominated factors for high performance of Fe^{3+} grafted metal doped TiO_2 based photocatalyst. *Appl. Catal. B* **2014**, *160–161*, 722–729.

26. Niishiro, R.; Konta, R.; Kato, H.; Chun, W.J.; Asakura, K.; Kudo, A. Photocatalytic O_2 evolution of rhodium and antimony-codoped rutile-type TiO_2 under visible light irradiation. *J. Phys. Chem. C* **2007**, *111*, 17420–17426.

27. Oropeza, F.E.; Egdell, R.G. Control of valence states in Rh-doped TiO_2 by Sb co-doping: A study by high resolution X-ray photoemission spectroscopy. *Chem. Phys. Lett.* **2011**, *515*, 249–253.

28. Liu, M.; Qiu, X.; Miyauchi, M.; Hashimoto, K. Energy-level matching of Fe(III) ions grafted at surface and doped in bulk for efficient visible-light photocatalysts. *J. Am. Chem. Soc.* **2013**, *135*, 10064–10072.

29. Burgeth, G.; Kisch, H. Photocatalytic and photoelectrochemical properties of titania-chloroplatinate(IV). *Coord. Chem. Rev.* **2002**, *230*, 41–47.

30. Nishikawa, M.; Sakamoto, H.; Nosaka, Y. Reinvestigation of photocatalytic reaction mechanism for Pt-complex-modified TiO_2 under visible-light irradiation by means of ESR Spectroscopy and chemiluminescence photometry. *J. Phys. Chem. A* **2012**, *116*, 9674–9679.

31. Bard, A.J.; Parsons, R.; Jordan, J. (Eds.) *Standard Potentials in Aqueous Solution*; Marcel Dekker: New York, NY, USA, 1985.

32. Fujishima, A.; Honda, K. Electrochemical photolysis of water at a semiconductor electrode. *Nature* **1972**, *238*, 37–38.

33. Nakabayashi, Y.; Nosaka, Y. OH radical formation at distinct faces of rutile TiO_2 crystal in the procedure of photoelectrochemical water oxidation. *J. Phys. Chem. C* **2013**, *117*, 23832–23839.

34. Imanishi, A; Fukui, K. Atomic-scale surface local structure of TiO_2 and its influence on the water photooxidation process. *J. Phys. Chem. Lett.* **2014**, *5*, 2108–2117.

35. Ohno, T.; Sarukawa, K.; Matsumura, M. photocatalytic activities of pure rutile particles isolated from TiO_2 powder by dissolving the anatase component in HF Solution. *J. Phys. Chem. B* **2001**, *105*, 2417–2420.

36. Zhang, J.; Nosaka, Y. Mechanism of the OH radical generation in photocatalysis with TiO_2 of different crystalline types. *J. Phys. Chem. C* **2014**, *118*, 10824–10832.

37. Hirakawa, T.; Yawata, K.; Nosaka, Y. Photocatalytic reactivity for O_2^- and OH radical formation in anatase and rutile TiO_2 suspension as the effect of H_2O_2 addition. *Appl. Catal. A* **2007**, *325*, 105–111.

38. Dadjour, M.F.; Ogino, C.; Matsumura, S.; Nakamura, S.; Shimizu, N. Disinfection of legionella pneumophila by ultrasonic treatment with TiO_2. *Water Res.* **2006**, *40*, 1137–1142.

39. Ishiguro, H.; Nakano, R.; Yao, Y.; Kajioka, A.; Fujishima, A.; Sunada, K.; Minoshima, M.; Hashimoto, K.; Kubota, Y. Inactivation of Qβ bacteriophage by photocatalysis using TiO_2 thin film under weak with long wavelength UV irradiation. *Photochem. Photobiol. Sci.* **2011**, *10*, 1825–1829.

40. Diebold, U. The surface science of titanium dioxide. *Surf. Sci. Rep.* **2003**, *48*, 53–229.

41. Mastikhin, V.M.; Mudrakovsky, I.L.; Nosov, A.V. ^{1}H-NMR magic angle spinning (MAS) studies of heterogeneous catalysis. *Prog. NMR Spectrosc.* **1991**, *23*, 259–299.

42. Köppen, S.; Bronkalla, O.; Langel, W. Molecular simulation of protein-surface interactions. *J. Phys. Chem. C* **2008**, *112*, 13600–13606.

43. Nosaka, A.Y.; Tanaka, G.; Nosaka, Y. Study by use of ^{1}H-NMR spectroscopy of the adsorption and decompoaition of glycine, leucine, and derivatives in TiO_2 photocatalysis. *J. Phys. Chem. B* **2014**, *118*, 7561–7567.

44. Fenoglio, I.; Greco, G.; Livraghi, S.; Fubini, B. Non-UV-induced radical reactions at the surface of TiO_2 nanoparticles that may trigger toxic responses. *Chem. Eur. J.* **2009**, *15*, 4614–4621.

45. Petković, J.; Žegura, B.; Filipič, M. Influence of TiO_2 nanoparticles on cellular antioxidant defense and its involvement in genotoxicity in HepG2 cells. *J. Phys. Conf. Ser.* **2011**, *304*, 1–8.

46. Horie, M.; Kato, H.; Fujita, K.; Endoh, S.; Iwahashi, H. *In vitro* evaluation of cellular response induced by manufactured nanoparticles. *Chem. Res. Toxicol.* **2012**, *25*, 605–619.

47. Nosaka, A.Y.; Tanaka, G.; Nosaka, Y. The behaviors of glutathione and related Amino Acids in TiO_2 photocatalytic system. *J. Phys. Chem. B* **2012**, *116*, 11098–11102.

Radical Intermediates in Photoinduced Reactions on TiO₂ (An EPR Spin Trapping Study)

Dana Dvoranová, Zuzana Barbieriková and Vlasta Brezová

Abstract: The radical intermediates formed upon UVA irradiation of titanium dioxide suspensions in aqueous and non-aqueous environments were investigated applying the EPR spin trapping technique. The results showed that the generation of reactive species and their consecutive reactions are influenced by the solvent properties (e.g., polarity, solubility of molecular oxygen, rate constant for the reaction of hydroxyl radicals with the solvent). The formation of hydroxyl radicals, evidenced as the corresponding spin-adducts, dominated in the irradiated TiO₂ aqueous suspensions. The addition of ^{17}O-enriched water caused changes in the EPR spectra reflecting the interaction of an unpaired electron with the ^{17}O nucleus. The photoexcitation of TiO₂ in non-aqueous solvents (dimethylsulfoxide, acetonitrile, methanol and ethanol) in the presence of 5,5-dimethyl-1-pyrroline *N*-oxide spin trap displayed a stabilization of the superoxide radical anions generated via electron transfer reaction to molecular oxygen, and various oxygen- and carbon-centered radicals from the solvents were generated. The character and origin of the carbon-centered spin-adducts was confirmed using nitroso spin trapping agents.

Reprinted from *Molecules*. Cite as: Dvoranová, D.; Barbieriková, Z.; Brezová, V. Radical Intermediates in Photoinduced Reactions on TiO₂ (An EPR Spin Trapping Study). *Molecules* **2014**, *19*, 17279-17304.

1. Introduction

Among the materials previously studied as potential photocatalysts, titanium dioxide meets the criteria for industrial-scale utilization. Stability, low cost, relatively low toxicity and appropriate photocatalytic activity predispose TiO₂ to a wide range of applications in various areas (gas sensors, photocatalysts, solar cells, thin film capacitors, self-cleaning surfaces, *etc.*) [1–6]. Especially attractive are nowadays the prospects of titania photocatalysts applications in the remediation of polluted water, soil and air, or in unconventional organic syntheses [7–13]. Consequently, all titanium dioxide polymorphs (anatase, brookite, rutile) have been intensively studied regarding their ability to produce, upon UVA photoexcitation, electron (e⁻) and hole (h⁺) pairs further involved in the consecutive chemical reactions [1,2,5,14–16]. In general, the photoactivity of TiO₂ is determined by the processes of electron/hole pair generation, recombination, interfacial transfer and by the surface reactions of these charge carriers with the species adsorbed on the surface of the photocatalyst [1,2,5,10,15–17]. The photoinduced processes on TiO₂ nanoparticles upon ultra-band gap irradiation are also well influenced by the bulk structure, surface properties and the electronic structure of the photocatalyst [5]. The reactions of photogenerated holes with the adsorbed hydroxide anions and water molecules lead to the formation of highly reactive hydroxyl radicals, which, together with the hole itself, can initiate the oxidative degradation of organic pollutants down to water and carbon dioxide [1,2,5,10,16].

The efficient production of hydroxyl radicals and their non-selective reactions with organic and inorganic pollutants represent a crucial point considering the application of photocatalytic processes in water and air purification [1]. Recent investigations have revealed different mechanisms on anatase and rutile surfaces [18], as well as the role of surface-bridging oxygens of TiO_2 on the $^{\bullet}OH$ formation associated with the oxidation of surface hydroxide anions and water molecules by the photogenerated holes [19,20]. The presence of molecular oxygen also plays a substantial role in the photoinduced processes on irradiated TiO_2 surfaces, as it enables an effective charge carriers separation. The electrons trapped transiently on the surface or on the next-to-surface defects can react with the adsorbed oxygen molecules [21–23]. The consecutive reactions of the so generated $O_2^{\bullet-}$ are influenced by the solvent properties [24]. Although the superoxide radical anion is quite stable in the aprotic solvents [25], in aqueous solutions the reaction with protons is favorable, and hydrogen peroxide is formed and involved in further photocatalytic processes, Equations (1)–(6) [26]:

$$O_2^{\bullet-} + H^+ \rightarrow {}^{\bullet}O_2H \tag{1}$$

$$2\,{}^{\bullet}O_2H \rightarrow H_2O_2 + O_2 \tag{2}$$

$$H_2O_2 + O_2^{\bullet-} \rightarrow {}^{\bullet}OH + O_2 + OH^- \tag{3}$$

$$H_2O_2 \xrightarrow{h\nu} 2\,{}^{\bullet}OH \tag{4}$$

$$H_2O_2 + e^- \rightarrow {}^{\bullet}OH + OH^- \tag{5}$$

$$H_2O_2 + h^+ + OH^- \rightarrow H_2O + {}^{\bullet}O_2H \tag{6}$$

Consequently, superoxide radical anion and hydrogen peroxide as the most important products of the molecular oxygen reduction play an important role in the complex mechanism of Reactive Oxygen Species (ROS, e.g., $^{\bullet}OH$, $^{\bullet}O_2H$ or singlet oxygen) generation on the irradiated TiO_2 surfaces [1,18,26–28].

EPR spectroscopy occupies an exclusive position in the investigation of titania photocatalysts, providing a characterization of paramagnetic centers produced via the trapped photogenerated electrons and holes [29–36], and of titanium dioxide materials with transition-metal ions doping [17,34]. A majority of the research exploiting EPR spectroscopy deals with the investigation of reactive radical intermediates produced in the irradiated TiO_2 particulate systems where the application of an indirect spin trapping technique is inevitable [37–45]. This method is based on the chemical reaction of a diamagnetic spin trap (ST) with a short-lived radical, producing a more stable nitroxide radical, i.e., spin-adduct, using nitrones, N-oxides and nitroso compounds as the spin trapping agents (Figure 1). Spin traps possessing N-oxide and nitrone groups are mainly applied in the identification of hydroxyl radicals generation, as well as other oxygen-, nitrogen- and sulfur-centered reactive radicals, however the information on the structure of carbon-centered radicals trapped with these agents is limited, and the application of nitroso spin traps is necessary to bring the knowledge on other nuclei in the vicinity of the trapped carbon [46,47]. Successful assignment of measured EPR spectra of spin-adducts requires a thorough interpretation of the acquired data and careful choice of the spin trapping agent for the specific experimental conditions [48].

Figure 1. Overview of the spin trapping agents applied in *in situ* EPR investigations.

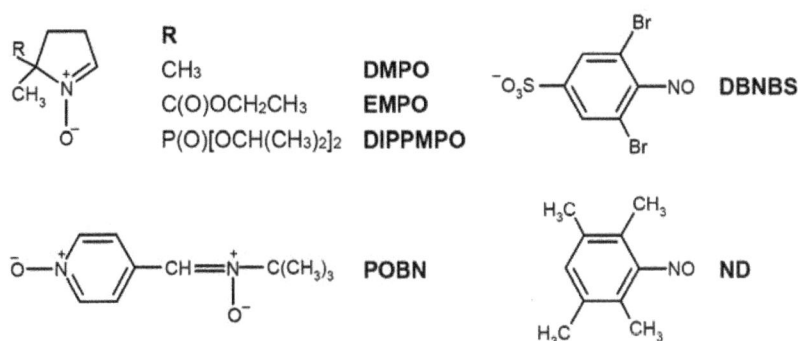

	R	
	CH3	DMPO
	C(O)OCH2CH3	EMPO
	P(O)[OCH(CH3)2]2	DIPPMPO

In the literature hydroxyl radicals are frequently declared as the most important reactive species generated upon TiO2 irradiation, in spite of the addition of organic co-solvent to the reaction systems, which effects the character and amount of radicals formed. The main aim of our study was to point on the formation of radical intermediates in aqueous TiO2 suspensions, as well as in suspensions prepared in an organic solvent (DMSO, acetonitrile, methanol, ethanol) and thus provide a straightforward comparison of the radical species detected under UVA exposure of TiO2 suspended in water and in aprotic and protic polar solvents exploiting the *in situ* EPR spin trapping technique. The oxidation of sterically hindered amines to the corresponding nitroxide radicals via ROS photogenerated in the aqueous and acetonitrile TiO2 suspensions was also monitored by *in situ* EPR spectroscopy.

2. Results and Discussion

2.1. Spin Trapping in the Aqueous TiO2 Suspensions

Despite the fact that the detection of hydroxyl radicals upon UVA irradiation of the aerated aqueous TiO2 suspensions in the presence of 5,5-dimethyl-1-pyrroline N-oxide (DMPO) spin trap represents a frequently applied EPR technique [38–45,49], in order to bring the complete information on the radical species generated in the irradiated TiO2 in different media we also report the results of EPR spin trapping experiments in aqueous TiO2 suspensions using different spin trapping agents. As expected, immediately after the irradiation started, a typical four-line EPR signal attributed to ˙DMPO-OH spin-adduct with spin Hamiltonian parameters (a_N = 1.497 mT, a_H = 1.477 mT; g = 2.0057 [50]) was generated in the system TiO2/DMPO/H2O/air, as is shown in Figure 2a. The concentration profile of ˙DMPO-OH during the *in situ* EPR spin trapping experiments is strongly influenced by TiO2 loading, UVA radiation dose, as well as by the initial oxygen and spin trap concentrations [38,44].

The primary source of the ˙OH radicals in the irradiated aqueous TiO2 suspensions is the oxidation of OH⁻ and H2O by the photogenerated holes, however further reactions of the reactive oxygen species generated in the system leading to ˙OH cannot be excluded (Equations (1)–(6)). The EPR spin trapping technique is assumed to detect the photogenerated hydroxyl radicals on the photocatalysts' surfaces, based on the previous comparison with the quantification of ˙OH via fluorescence detection using the hydroxylation of terephthalic acid, by which the bulk 'OH are detected [43].

Figure 2. The sets of individual EPR spectra (magnetic field sweep width, $SW = 8$ mT) monitored upon continuous UVA irradiation ($\lambda_{max} = 365$ nm; irradiance 15 mW·cm^{-2}) of aerated TiO_2 P25 suspensions in the presence of spin trapping agent DMPO: (**a**) water; (**b**) mixed solvent water/dimethylsulfoxide (5:1 v:v); (**c**) acetonitrile. TiO_2 concentration 0.167 mg·mL^{-1}, $c_{0,DMPO} = 0.035$ M.

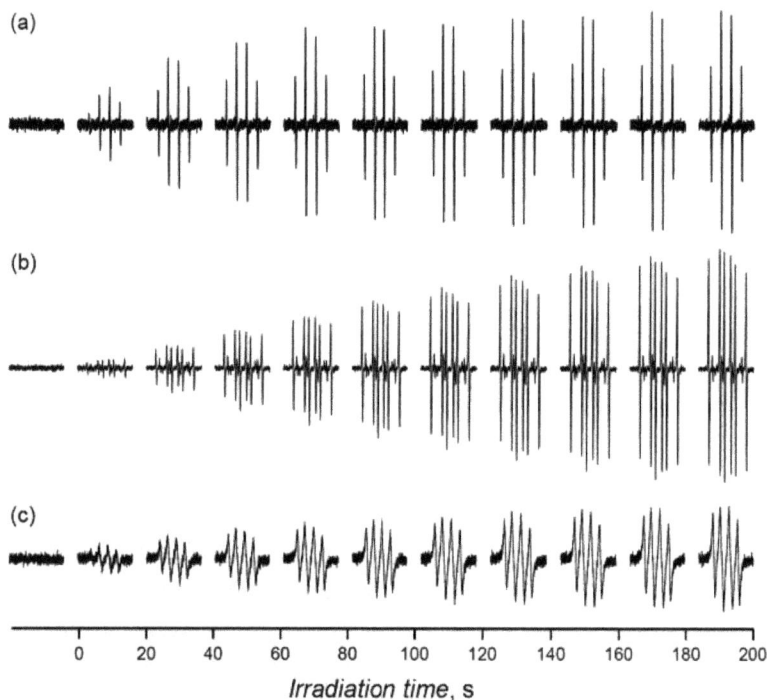

Figure 3 illustrates experimental and simulated EPR spectra of 'DMPO-OH measured in the TiO_2 suspensions prepared either using ordinary water, or water enriched with the magnetically active ^{17}O (13%–17% atom.) nucleus. The EPR spectrum depicted in Figure 3b is fully compatible with the presence of both spin-adducts, *i.e.*, 'DMPO-OH and 'DMPO-^{17}OH [51], unambiguously identifying the adsorbed and close-to-surface water molecules as the source of hydroxyl radicals. Recently, we conducted EPR spin trapping experiments using aerated aqueous suspensions of $Ti^{17}O_2$ (containing up to 90% atom. ^{17}O) with DMPO, and the EPR spectra of 'DMPO-OH corresponded to the interaction of one nitrogen nucleus ($a_N = 1.492$ mT) and one hydrogen nucleus ($a_H = 1.476$ mT) with an unpaired electron [52]. No evidence of a hyperfine coupling from ^{17}O was found, consequently the lattice oxygens from TiO_2 were excluded as the source of hydroxyl radicals trapped by DMPO under the given experimental conditions [52].

Figure 3. Experimental **(black)** and simulated **(red)** EPR spectra ($SW = 8$ mT) obtained upon irradiation ($\lambda_{max} = 365$ nm; irradiance 15 mW·cm^{-2}; exposure 400 s) of the aerated aqueous TiO$_2$ P25 suspensions in the presence of spin trapping agent DMPO (TiO$_2$ concentration 0.167 mg·mL^{-1}, $c_{0,DMPO} = 0.035$ M): **(a)** ordinary water; **(b)** water enriched with H$_2$^{17}O (13%–17% atom.). Simulation parameters (hfcc in mT): (a) ˙DMPO–OH ($a_N = 1.497$, $a_H = 1.477$; $g = 2.0057$); (b) linear combination of spin-adducts, *i.e.*, ˙DMPO–OH (relative concentration in %, 82) and ˙DMPO-^{17}OH ($a_N = 1.494$, $a_H = 1.480$, $a_{17O} = 0.467$; $g = 2.0057$; 18).

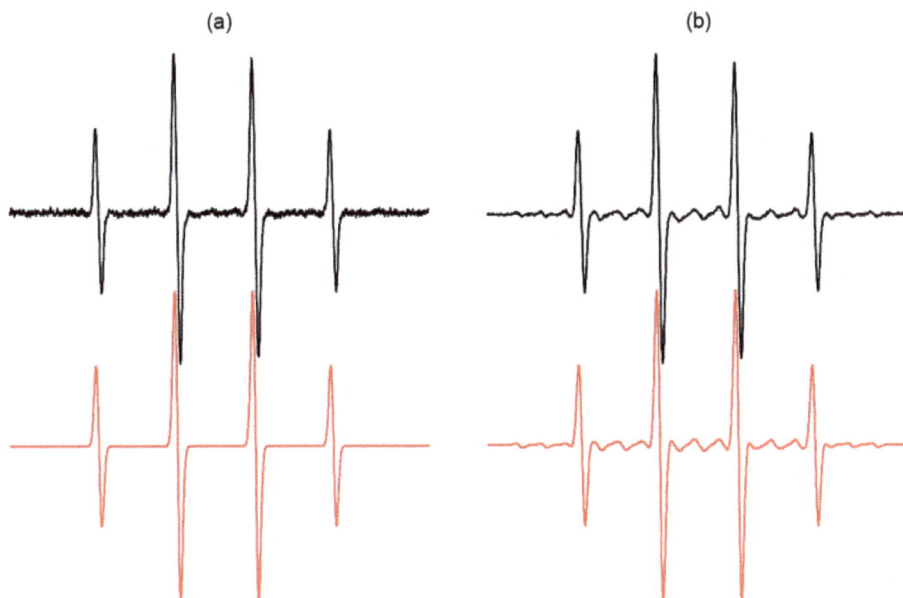

Previously, the possibility of a direct oxidation of the spin trapping agent DMPO via photogenerated holes to a radical cation DMPO˙$^+$, which subsequently reacts with water molecules forming a so-called imposter spin-adduct ˙DMPO-OH, was supposed [41,43,44]. In addition, the degradation of the low-stability ˙DMPO-O$_2$H spin-adduct, theoretically also generated in the studied system, results in the ˙DMPO-OH formation [53]. However, the generation of the surface hydroxyl radicals can be evidenced by the addition of dimethylsulfoxide (DMSO) to the aqueous TiO$_2$ suspensions, since the rapid reaction of hydroxyl radicals with DMSO (Table 1) produces methyl radicals [54], detectable in the reaction with spin trap (ST) as the corresponding carbon-centered spin-adduct, Equations (7) and (8):

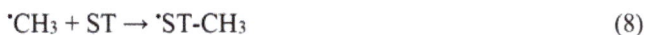

$$(CH_3)_2SO + {}^{\bullet}OH \rightarrow CH_3(OH)SO + {}^{\bullet}CH_3 \tag{7}$$

$$^{\bullet}CH_3 + ST \rightarrow {}^{\bullet}ST\text{-}CH_3 \tag{8}$$

Table 1. Bimolecular rate constants for the reaction of hydroxyl radical with the selected solvents [54], decay constants and lifetime of singlet oxygen ($^1\Delta_g$) in selected solvents [55].

Solvent	$k_{\cdot OH}$, $M^{-1} \cdot s^{-1}$	$k^1_{O_2}$, s^{-1}	τ ($1/k^1_{O_2}$), μs
Water	–	2.4×10^5	4.2
Dimethylsulfoxide [#]	7.0×10^9	5.2×10^4	19
Acetonitrile [#]	2.2×10^7	1.4×10^4	71
Methanol [#]	8.3×10^8	1.1×10^5	9
Ethanol [#]	2.2×10^9	7.9×10^4	13

[#]: determined in aqueous solutions.

Figure 2b shows the set of EPR spectra monitored during the exposure of TiO$_2$/DMPO/air in mixed solvent water/DMSO (5:1 v:v). The EPR spectra obtained are more complex compared to those found in water suspensions (Figure 2a), and represent a superposition of the dominant six-line signal attributed to ˙DMPO-CH$_3$ and the low-intensity signal of ˙DMPO-OH with slightly modified hyperfine coupling constants (hfcc) caused by the DMSO presence in the system [56]. The experimental and simulated EPR spectra found upon 400 s exposure are depicted in Figure 4a and the corresponding spin Hamiltonian parameters elucidated from the simulated spectra are summarized in Table 2.

Further experiments using 3,5-dibromo-4-nitrosobenzene sulfonate (DBNBS) spin trapping agent in aerated aqueous TiO$_2$ suspensions containing DMSO or DMSO-d_6 (water/DMSO, 5:1 v:v) unambiguously confirmed the generation of methyl radicals via the reaction of hydroxyl radicals with the solvent, as the EPR spectra monitored upon UVA irradiation are well-matched to ˙DBNBS-CH$_3$ or ˙DBNBS-CD$_3$ spin-adducts (Figure 4b,c), respectively, with the spin Hamiltonian parameters well correlated with literature data (Table 2).

Figure 4. Experimental (**black**) and simulated (**red**) EPR spectra obtained upon irradiation (λ_{max} = 365 nm; irradiance 15 mW·cm^{-2}; exposure 400 s) of the aerated TiO$_2$ P25 suspensions in mixed solvent water/DMSO (5:1 v:v) in the presence of various spin trapping agents (TiO$_2$ concentration 0.167 mg·mL^{-1}): (**a**) DMPO (SW = 8 mT; $c_{0,DMPO}$ = 0.035 M); (**b**) DBNBS (SW = 10 mT; $c_{0,DBNBS}$ = 0.008 M); (**c**) DBNBS using DMSO-d_6. The spin Hamiltonian parameters of corresponding spin-adducts are listed in Table 2. (a) ˙DMPO-CH$_3$ (relative concentration in %, 94), ˙DMPO-OH (6); (b) ˙DBNBS–CH$_3$ (100); (c) ˙DBNBS-CD$_3$ (100).

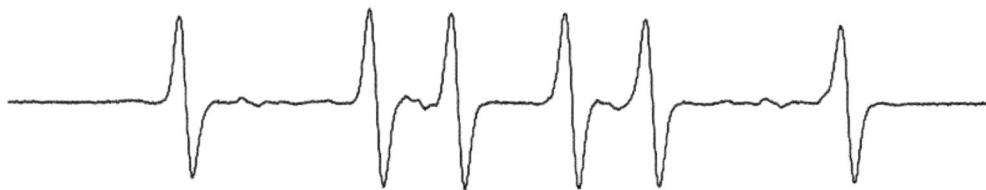

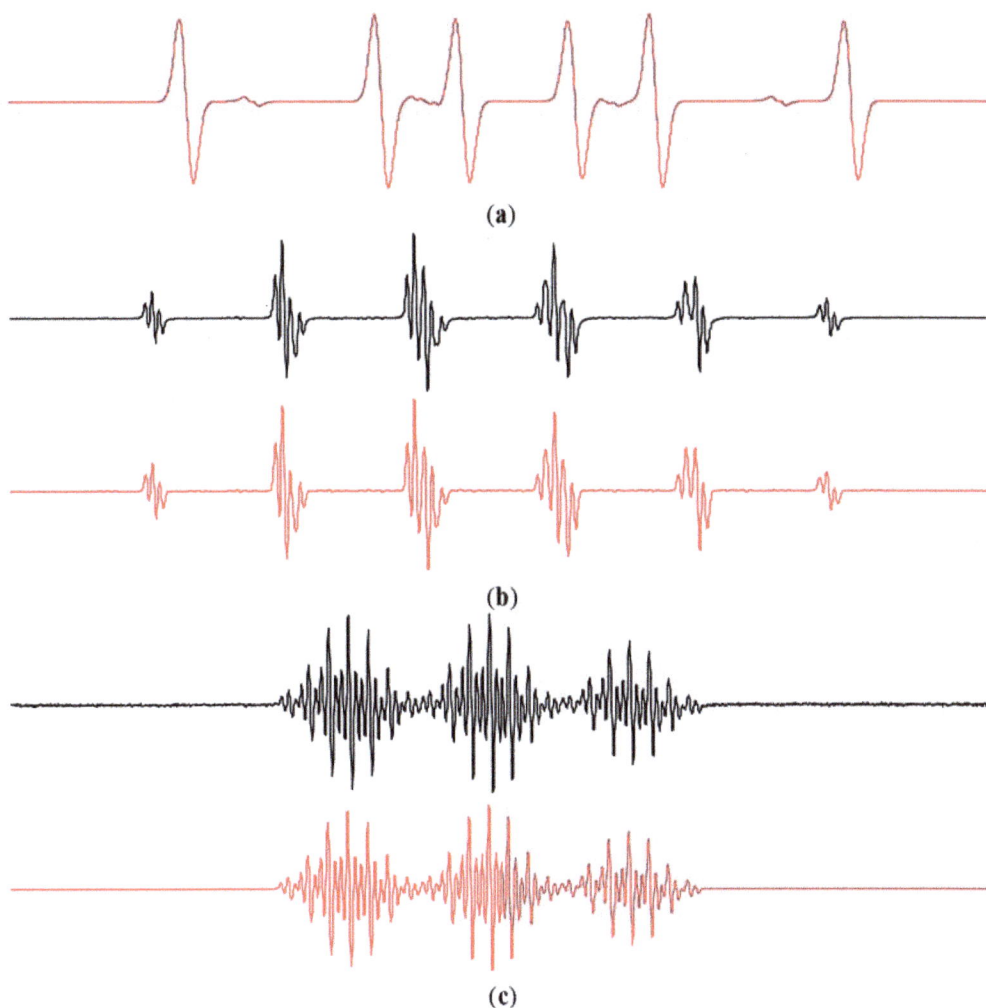

(a)

(b)

(c)

The application of 5-(ethoxycarbonyl)-5-methyl-1-pyrroline *N*-oxide (EMPO), 5-(diisopropoxyphosphoryl)-5-methyl-1-pyrroline *N*-oxide (DIPPMPO) and α-(4-pyridyl-1-oxide)-*N*-*tert*-butylnitrone (POBN) spin trapping agents suitable for the detection of oxygen-centered radicals in the aerated aqueous TiO_2 suspensions upon exposure confirmed the dominant generation of hydroxyl radical spin-adducts (Figure 5). The chiral centre in EMPO and DIPPMPO molecules may result in the production of *trans* and *cis* spin-adduct diastereoisomers. Indeed, the simulation analysis of the corresponding experimental EPR spectra summarized in Table 2, revealed the superposition of two individual EPR signals belonging to hydroxyl radical spin-adduct diastereoisomers. Additionally, low-intensity EPR signals of radical intermediates originating from the spin traps decomposition were detected as the carbon-centered spinadducts (˙EMPO$_{degr}$, ˙DIPPMPO$_{degr}$) or the four-line signal of hydroxy *tert*-butylnitroxide (˙POBN$_{degr}$) [61].

Table 2. Spin Hamiltonian parameters (hyperfine coupling constants and g-values) of the spin-adducts elucidated from simulations of the experimental EPR spectra obtained upon UVA irradiation ($\lambda_{max} = 365$ nm) of aerated TiO$_2$ P25 suspensions in water and water/dimethylsulfoxide mixed solvent (5:1 v:v) in the presence of the corresponding spin trapping agents.

Spin-Adduct	Hyperfine Coupling Constants (mT)		g-Value	Reference
	a_{NO}	a_i		
		Water		
˙DMPO–OH	1.497	$a_H^\beta = 1.477$	2.0057	[50]
˙DMPO–^{17}OH	1.494	$a_H^\beta = 1.480$; $a_{17O} = 0.469$	2.0057	[51,57]
˙DMPO–N$_3$	1.481	$a_H^\beta = 1.426$; $a_N = 0.314$	2.0057	[50]
trans-˙EMPO–OH	1.410	$a_H^\beta = 1.278$; $a_H^\gamma = 0.066$; $a_H^\gamma = 0.043$	2.0056	[58,59]
cis-˙EMPO–OH	1.410	$a_H^\beta = 1.542$	2.0056	[58,59]
˙EMPO$_{degr}$	1.514	$a_H^\beta = 2.187$	2.0056	–
trans-˙DIPPMPO–OH	1.410	$a_H^\beta = 1.319$; $a_P = 4.692$	2.0055	[59,60]
cis-˙DIPPMPO–OH	1.646	$a_H^\beta = 1.236$; $a_P = 3.572$	2.0055	[59,60]
˙DIPPMPO$_{degr}$	1.469	$a_H^\beta = 2.147$; $a_P = 4.830$	2.0055	–
˙POBN–OH	1.508	$a_H^\beta = 0.169$	2.0057	[44]
˙POBN$_{degr}$	1.461	$a_H^\beta = 1.413$	2.0055	[61]
		Water/DMSO (5:1 v:v)		
˙DMPO–OH	1.469	$a_H^\beta = 1.358$; $a_H^\gamma = 0.067$	2.0057	[56]
˙DMPO–CH$_3$	1.588	$a_H^\beta = 2.250$	2.0055	[50]
˙DBNBS–CH$_3$	1.434	$a_H(3H) = 1.331$; $a_H(2H''') = 0.069$; $a_{13C} = 0.929$	2.0063	[50,62]
˙DBNBS–CD$_3$	1.434	$a_D(3D) = 0.201$; $a_H(2H''') = 0.070$	2.0063	[50]

Symbols β and γ denote the position of the interacting hydrogen nuclei.

Previously, the photoinduced generation of O$_2^{\cdot-}$ on the irradiated TiO$_2$ nanoparticles was confirmed by low temperature EPR measurements below 160 K [25,31,32,34,52]. Even though the spin trapping agents EMPO and DIPPMPO were specially designed for the detection of superoxide radical anion in aqueous media and biological systems [60,63–66], the EPR signals reflecting the presence of ˙EMPO-O$_2^-$/O$_2$H and ˙DIPPMPO-O$_2^-$/O$_2$H spin-adducts were not found in the irradiated aqueous TiO$_2$ suspensions (Figure 5). Most probably, under the given experimental conditions, the superoxide radical anions are preferably transformed to hydrogen peroxide by a disproportionation with protons [24,26,67]. Moreover, a significantly lower rate constants for the addition of O$_2^{\cdot-}$/˙O$_2$H to the nitrone spin traps may cause the limited production of spin-adducts [46,48]. The rapid transformation of superoxide radical anions, as well as their very slow reaction with nitrone spin traps caused that at room temperature in aerated aqueous TiO$_2$ suspensions superoxide detection using conventional spectroscopic techniques failed, and only chemiluminescence with luminol or luciferin analog was applied as a suitable experimental method [68–70].

Figure 5. Experimental (**black**) and simulated (**red**) EPR spectra obtained upon irradiation (λ_{max} = 365 nm; irradiance 15 mW·cm^{-2}; exposure 400 s) of the aerated aqueous TiO$_2$ P25 suspensions in the presence of various spin trapping agents (TiO$_2$ concentration 0.167 mg·mL^{-1}): (**a**) EMPO (*SW* = 8 mT; $c_{0,EMPO}$ = 0.05 M); (**b**) DIPPMPO (*SW* = 16 mT; $c_{0,DIPPMPO}$ = 0.035 M); (**c**) POBN (*SW* = 8 mT; $c_{0,POBN}$ = 0.05 M). Simulations represent linear combinations of corresponding spin-adducts (hfcc parameters listed in Table 2): (a) *trans*-˙EMPO–OH (relative concentration in %, 68), *cis*-˙EMPO-OH (28), ˙EMPO$_{degr}$ (4); (b) *trans*-˙DIPPMPO-OH (77), *cis*-˙DIPPMPO-OH (10), ˙DIPPMPO$_{degr}$ (13); (c) ˙POBN-OH (92), ˙POBN$_{degr}$ (8).

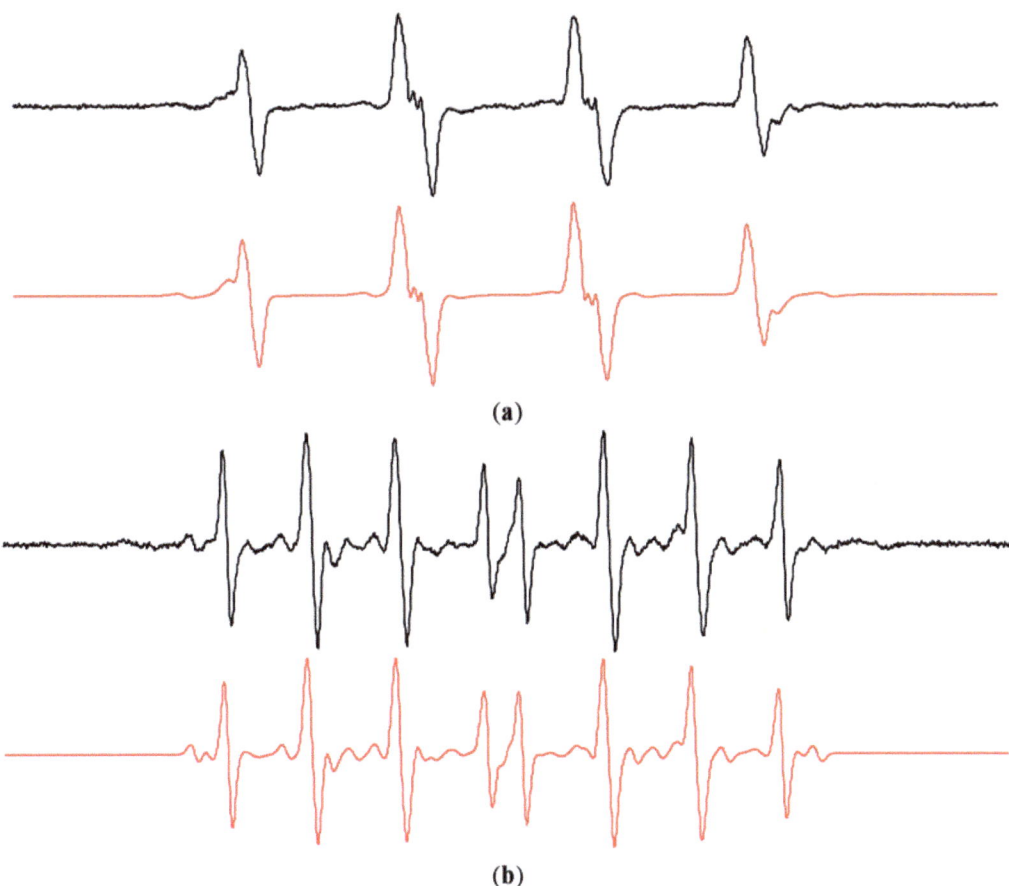

(**a**)

(**b**)

Figure 5. *Cont.*

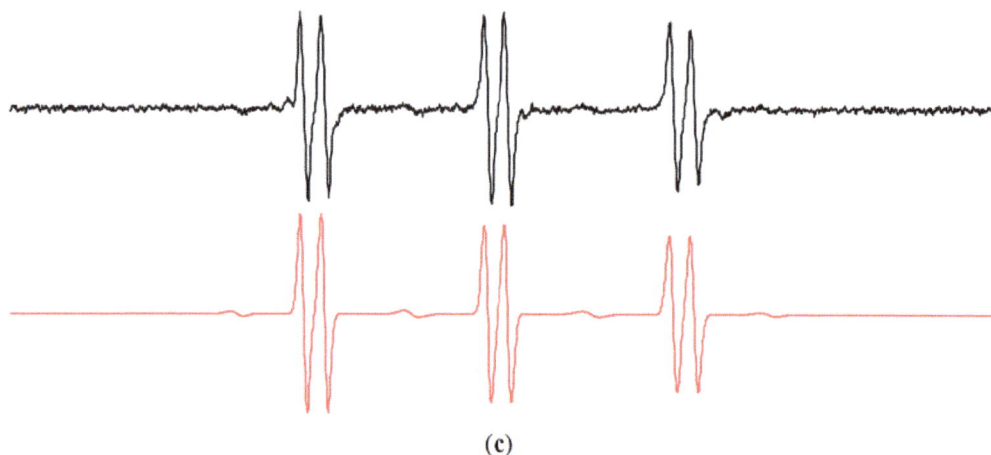

(c)

2.2. Spin Trapping in Non-Aqueous TiO₂ Suspensions

EPR spin trapping investigations of reactive radicals produced in the irradiated TiO$_2$ dispersions so far were focused mainly on aqueous systems and on the detection of hydroxyl radicals [20,31,35–37,49]. Analogous experiments performed in organic solvents may provide interesting information concerning the radical intermediates generated [13,71–74], consequently we carried out spin trapping experiments with TiO$_2$ nanoparticles dispersed in DMSO, acetonitrile (ACN), methanol and ethanol. The generation of electron-hole pairs upon TiO$_2$ irradiation and their consecutive reactions resulting in the free radicals formation are substantially influenced by the solvent properties [54,75,76]. The increased solubility of molecular oxygen plays an important role in these processes (Table 3), together with the stabilization effect of the aprotic solvents on the superoxide radical anions and the reactivity of holes and hydroxyl radicals with the solvents (Table 1).

Table 3. Solubility of molecular oxygen in various solvents at 25 °C.

Solvent	c_{O2}, mM	Reference
Water	1.0	[77]
Dimethylsulfoxide	2.1	[78]
Acetonitrile	8.1	[78]
Methanol	9.4–10.3	[79]
Ethanol	7.5–11.6	[79]

2.2.1. Dimethylsulfoxide

The EPR spectra monitored upon UVA photoexcitation of TiO$_2$ suspensions in aerated DMSO (Figure 6a) differ from those found when water was used as a solvent (Figure 3a), and the dominating signals represent spin-adducts ˙DMPO-O$_2$¯ and ˙DMPO-OCH$_3$ with the spin Hamiltonian parameters summarized in Table 4. The superoxide radical anion stabilization in the aprotic solvent explains its favourable generation and consequently also trapping [24,25]. The production of 'DMPO-OCH₃ adduct is initiated by the oxidation of hydroxide anions or water molecules adsorbed on the titanium dioxide surface producing reactive hydroxyl radicals, which immediately attack the DMSO solvent forming methyl radicals (Equation (7)), as shown above in the mixed water/DMSO solvent (Figure 4). The rapid reaction of methyl radicals with molecular oxygen results in the generation of peroxomethyl radicals serving as a source of 'DMPO–OCH₃ spin-adducts (Equations (9)–(13)) [73]. Further low-intensity oxygen-centered spin-adduct assigned to 'DMPO–OR originates from the solvent and most probably represents 'DMPO–OCH₂S(O)CH₃:

$$\cdot CH_3 + O_2 \rightarrow CH_3OO\cdot \tag{9}$$

$$DMPO + CH_3OO\cdot \rightarrow {}^\cdot DMPO\text{–}OOCH_3 \tag{10}$$

$$2\, {}^\cdot DMPO\text{–}OOCH_3 \rightarrow O_2 + 2\, {}^\cdot DMPO\text{–}OCH_3 \tag{11}$$

$$2\, CH_3OO\cdot \rightarrow 2\, CH_3O\cdot + O_2 \tag{12}$$

$$DMPO + CH_3O\cdot \rightarrow {}^\cdot DMPO\text{–}OCH_3 \tag{13}$$

The photoinduced generation of methyl radicals upon the irradiation of titania-DMSO dispersions was confirmed by the nitroso spin trapping agent DBNBS, suitable for the identification of carbon-centered radicals, as typical signal of 'DBNBS-CH₃ and 'DBNBS-CD₃ (using DMSO-d_6) were detected (Figure 6b,c). Depending on the experimental conditions the spin trap decomposition may occur during the photocatalytic processes demonstrated by the generation of 'DBNBS-SO₃⁻ (Table 4).

Figure 6. Experimental (**black**) and simulated (**red**) EPR spectra (SW = 8 mT) obtained upon irradiation (λ_{max} = 365 nm; irradiance 15 mW·cm⁻²; exposure 400 s) of the aerated TiO₂ P25 DMSO suspensions in the presence of various spin trapping agents (TiO₂ concentration 0.167 mg·mL⁻¹): (**a**) DMPO ($c_{0,DMPO}$ = 0.035 M); (**b**) DBNBS ($c_{0,DBNBS}$ = 0.008 M); (**c**) DBNBS using DMSO-d_6 (83% vol.). The spin Hamiltonian parameters of corresponding spin-adducts are listed in Table 4. (**a**) 'DMPO–O₂⁻ (relative concentration in %, 46), 'DMPO–OCH₃ (46), 'DMPO–OR (6), 'DMPO–CH₃ (2); (**b**) 'DBNBS–CH₃ (90), 'DBNBS–SO₃⁻ (10); (**c**) 'DBNBS–CD₃ (86), 'DBNBS–CH₃ (14).

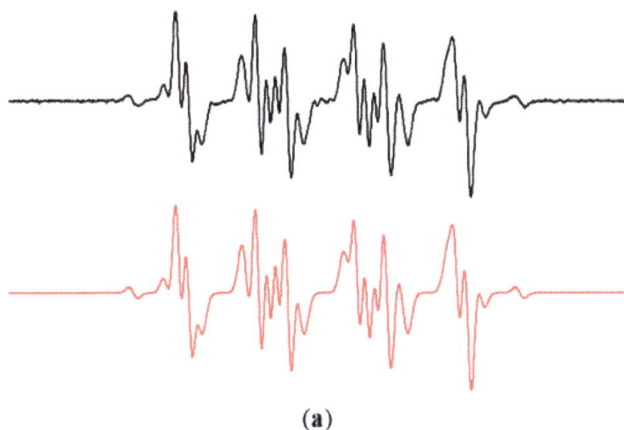

(**a**)

Figure 6. *Cont.*

(b)

(c)

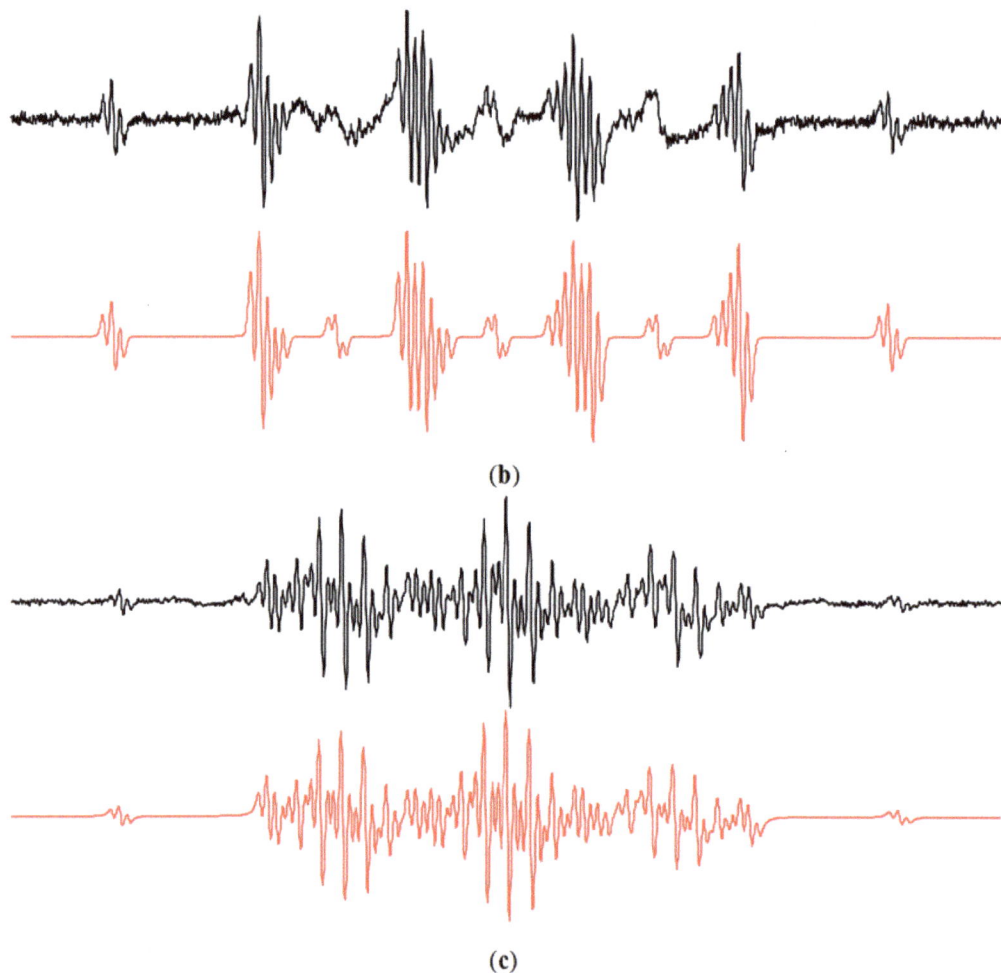

2.2.2. Acetonitrile

Due to the increased solubility of molecular oxygen in ACN (Table 3) the EPR spectra monitored in the irradiated systems TiO$_2$/DMPO/ACN/air represent a four-line EPR signal with significantly broadened spectral lines (Figure 2c, Figure 7 blue line). Consequently, to obtain spectra suitable for the identification of spin-adduct parameters, the saturation of the exposed sample with argon is necessary. The experimental EPR spectrum recorded immediately after a post-radiation Ar-saturation and EPR spectrometer re-tuning, along with its simulation is shown in Figure 7 (black and red lines). Acetonitrile as an aprotic solvent stabilizes superoxide radical anions, consequently the spin-adduct ˙DMPO-O$_2$⁻ dominates the EPR spectrum. The use of dried ACN solvent indicates that the hydroxyl radicals are generated by the oxidation of OH⁻/H$_2$O adsorbed on the TiO$_2$ surface via the photogenerated holes [80]. A lower reactivity of the photogenerated hydroxyl radicals towards acetonitrile (Table 1) allows the hydroxyl radicals to be trapped by DMPO and the ˙DMPO-OH was found in the spectra (Figure 7). The formation of

˙DMPO-OCH₃ most probably relates to the interaction of hydroxyl radicals with the solvent [81] producing CH₃OO˙ radicals trapped as the ˙DMPO-OCH₃ spin-adducts [73]. The spin Hamiltonian parameters of the individual spin-adducts elucidated by the simulation of experimental EPR spectra obtained in TiO₂/DMPO/ACN/air are summarized in Table 4.

Figure 7. Experimental (**blue**) EPR spectra (SW = 8 mT) obtained upon irradiation (λ_{max} = 365 nm; irradiance 15 mW·cm^{-2}; exposure 400 s) of the aerated TiO₂ P25 (0.167 mg·mL^{-1}) suspensions in ACN containing DMPO spin trap ($c_{0,DMPO}$ = 0.035 M), along with EPR spectra measured after post-radiation saturation with argon (**black**) and their simulations (**red**). The spin Hamiltonian parameters of corresponding spin-adducts are listed in Table 4. ˙DMPO-O₂$^-$ (rel. conc. in %, 60), ˙DMPO–OH (9), ˙DMPO-OCH₃ (23), ˙DMPO$_{degr}$ (8).

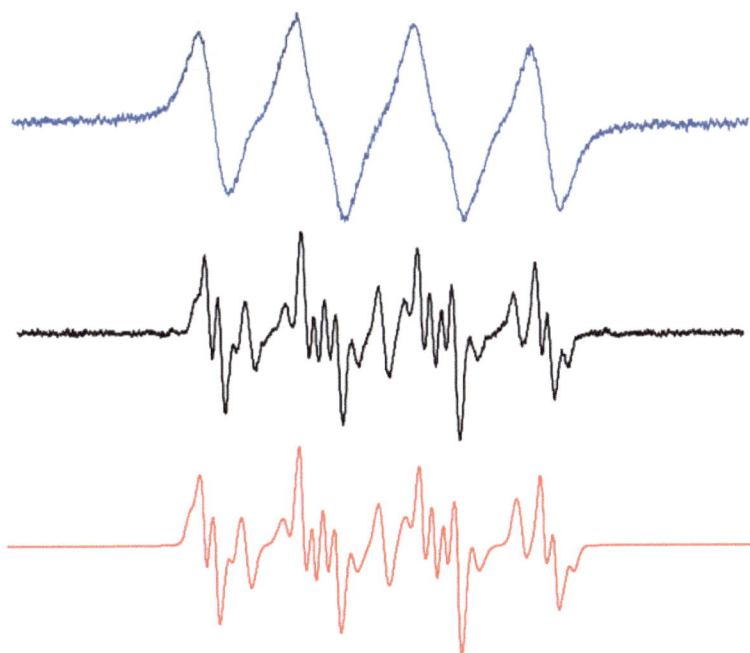

Additional experiments with nitrosodurene (ND) spin trap were performed in order to identify the structure of the carbon-centered radicals produced during the exposure of TiO₂/ACN/air. The EPR spectra measured upon continuous irradiation (spectra not shown) revealed the presence of a nine-line signal of ˙ND–CH₂CN, produced via the interaction of ˙OH with ACN [54], and a broad, three-line signal of ND˙$^+$ generated by the spin trap oxidation (Table 4).

2.2.3. Methanol and Ethanol

The protic solvents methanol and ethanol are characterized with increased concentration of dissolved molecular oxygen (Table 3), and these solvents are well-known as efficient scavengers of photogenerated holes [82]. By their application radical intermediates created via interaction with both photogenerated charge carriers may be observed, *i.e.*, electrons are scavenged by molecular oxygen forming $O_2^{•-}$ and holes react with alcohols producing the primary alkoxy radical species, $•OCH_3$ and $•OCH_2CH_3$ (Equations (14) and (15)) [12,13,62].

$$CH_3OH + h^+ \rightarrow CH_3O^• + H^+ \qquad (14)$$

$$CH_3CH_2OH + h^+ \rightarrow CH_3CH_2O^• + H^+ \qquad (15)$$

Table 4. Spin Hamiltonian parameters (hyperfine coupling constants and *g*-values) of spin-adducts elucidated from the simulations of experimental EPR spectra obtained upon UVA irradiation (λ_{max} = 365 nm) of aerated TiO$_2$ P25 suspensions in organic solvents in the presence of spin traps.

Spin-Adduct	Hyperfine Coupling Constants (mT)		*g*-Value	Reference
	a_{NO}	a_i		
DMSO				
$•$DMPO–O$_2^-$	1.287	a_H^β = 1.041; a_H^γ = 0.139	2.0057	[25,50,83]
$•$DMPO–OCH$_3$	1.329	a_H^β = 0.808; a_H^γ = 0.164	2.0057	[50,84]
$•$DMPO–OR	1.301	a_H^β = 1.464	2.0057	[84]
$•$DMPO–CH$_3$	1.462	a_H^β = 2.093	2.0056	[50]
$•$DBNBS–CH$_3$	1.337	a_H(3H) = 1.211; a_H(2H''') = 0.067	2.0064	[50]
$•$DBNBS–CD$_3$	1.334	a_D(3D) = 0.183; a_H(2H''') = 0.067	2.0064	[50]
$•$DBNBS–SO$_3^-$	1.295	a_H(2H''') = 0.054	2.0064	[85]
ACN				
$•$DMPO–O$_2^-$ #	1.296	a_H^β = 1.044; a_H^γ = 0.133	2.0057	[50,74]
$•$DMPO–OH #	1.382	a_H^β = 1.200; a_H^γ = 0.080	2.0057	[74]
$•$DMPO–OCH$_3$ #	1.312	a_H^β = 0.796; a_H^γ = 0.179	2.0057	[74]
$•$DMPO$_{degr}$ #	1.479		2.0056	–
$•$ND–CH$_2$CN	1.342	a_H(2H) = 0.977	2.0057	[86,87]
ND$^{•+}$	2.608		2.0057	[50]
Methanol				
$•$DMPO–O$_2^-$ #	1.376	a_H^β = 0.963; a_H^γ = 0.132	2.0057	[50]
$•$DMPO–OCH$_3$ #	1.363	a_H^β = 0.775; a_H^γ = 0.167	2.0057	[50]
$•$DMPO–OCH$_2$OH #	1.414	a_H^β = 1.266; a_H^γ = 0.075	2.0057	[50]
$•$DMPO–CH$_2$OH #	1.506	a_H^β = 2.116	2.0056	[50]
$•$DMPO$_{degr}$ #	1.523		2.0056	–
$•$ND–CH$_2$OH	1.387	a_H(2H) = 0.771	2.0057	[62,86,87]
Ethanol				
$•$DMPO–O$_2^-$ #	1.322	a_H^β = 1.050; a_H^γ = 0.133	2.0057	[50]
$•$DMPO–OCH$_2$CH$_3$ #	1.356	a_H^β = 0.761; a_H^γ = 0.174	2.0057	[50]
$•$DMPO–OR #	1.470	a_H^β = 1.094; a_H^γ = 0.090	2.0057	[50]
$•$DMPO–CR$_1$ #	1.481	a_H^β = 2.195	2.0056	[50]
$•$DMPO–CR$_2$ #	1.534	a_H^β = 2.215	2.0056	[50]
$•$ND–CH(CH$_3$)OH	1.398	a_H = 0.702	2.0057	[62,86,87]
$•$ND–CH$_3$	1.452	a_H(3H) = 1.345	2.0057	[62,86,87]

#: post-radiation saturation with argon. Symbols β and γ denote the position of the interacting hydrogen nuclei.

These oxygen-centered radical species can be easily identified using the DMPO spin trap. Due to the higher concentration of dissolved oxygen in these solvents, the EPR spectra of spin-adducts measured in aerated methanol and ethanol TiO$_2$ suspensions are characterized by a significant line broadening, which hinders a detailed simulation analysis (Figure 8a,b blue lines). However, the post-radiation saturation of the TiO$_2$ suspensions with argon, and subsequent measurement of EPR spectra provide the EPR signals of sufficient quality (Figure 8a,b black lines).

Figure 8. Experimental (**blue**) EPR spectra ($SW = 8$ mT) obtained upon irradiation ($\lambda_{max} = 365$ nm; irradiance 15 mW·cm^{-2}; exposure 400 s) of the aerated TiO$_2$ P25 (0.167 mg·mL^{-1}) suspensions in organic solvents containing DMPO spin trap ($c_{0,DMPO} = 0.035$ M), along with EPR spectra measured after post-radiation saturation with argon (**black**) and their simulations (**red**): (**a**) methanol; (**b**) ethanol. The spin Hamiltonian parameters of corresponding spin-adducts are listed in Table 4. (a) $^\bullet$DMPO–O$_2^-$ (rel. conc. in %, 33), $^\bullet$DMPO–OCH$_3$ (35), $^\bullet$DMPO–OCH$_2$OH (21), $^\bullet$DMPO–CH$_2$OH (7), $^\bullet$DMPO$_{degr}$ (4); (b) $^\bullet$DMPO–O$_2^-$ (17), $^\bullet$DMPO–OCH$_2$CH$_3$ (59), $^\bullet$DMPO–OR (10), $^\bullet$DMPO–CR$_1$ (9), $^\bullet$DMPO–CR$_2$ (5).

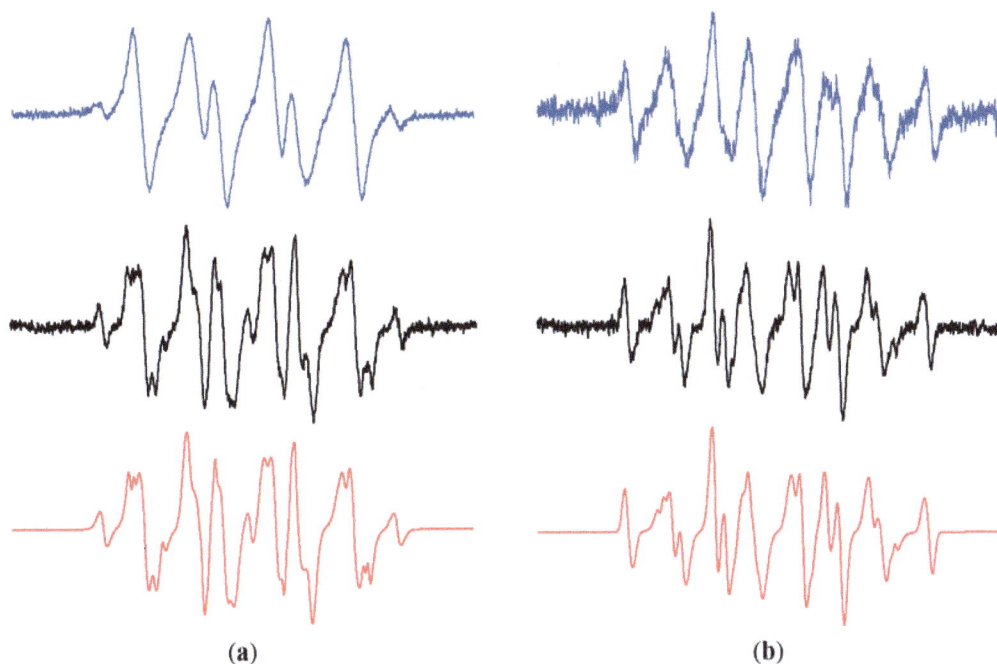

(**a**)   (**b**)

Simulations of the EPR spectra measured using this experimental procedure in the system TiO$_2$/DMPO/methanol revealed the presence of individual DMPO spin-adducts corresponding to $^\bullet$DMPO-O$_2^-$, $^\bullet$DMPO-OCH$_3$, $^\bullet$DMPO-CH$_2$OH, $^\bullet$DMPO-OCH$_2$OH and a triplet signal assigned to the DMPO degradation product (Figure 8a red line). The spin Hamiltonian parameters of the identified spin-adducts are gathered in Table 4. We assume that the radical species $^\bullet$CH$_2$OH and $^\bullet$OCH$_2$OH are produced via complex reactions of $^\bullet$OCH$_3$ radicals with methanol molecules and molecular oxygen [62], an alternative mechanism of ˙CH$_2$OH generation represents the hydrogen abstraction from methanol via hydroxyl radicals (Table 1, Equations (16) and (17)).

$$CH_3O^\bullet + CH_3OH \rightarrow CH_3OH + {}^\bullet CH_2OH \tag{16}$$

$$CH_3OH + {}^\bullet OH \rightarrow {}^\bullet CH_2OH + H_2O \tag{17}$$

Simulation of the EPR spectra obtained in the irradiated suspensions TiO$_2$/DMPO/ethanol after the saturation with argon evidenced the presence of spin-adducts characteristic for ˙DMPO-O$_2^-$, ˙DMPO-OCH$_2$CH$_3$, ˙DMPO-OR, and two carbon-centered spin-adducts with slightly differing hyperfine coupling constants (Figure 8b, Table 4). Nitrosodurene spin trapping agent was used in the analogous experiments to identify the carbon-centered radicals in the irradiated methanol or ethanol TiO$_2$ suspensions (spectra not shown). In methanol only the generation of ˙ND-CH$_2$OH was evidenced. The EPR spectra monitored in TiO$_2$/ethanol suspensions are compatible with ˙ND-CH(OH)CH$_3$ and ˙ND-CH$_3$ spin-adducts in good accordance with the reactions of ethoxy, 1-hydroxyethyl and 2-hydroxyethyl radicals in ethanol (Equations (18)–(20)) published previously [62], and also with two DMPO carbon-centered spin-adducts detected:

$$CH_3CH_2O^\bullet + CH_3CH_2OH \rightarrow CH_3CH_2OH + CH_3{}^\bullet CHOH \tag{18}$$

$$CH_3{}^\bullet CHOH + CH_3CH_2OH \rightarrow {}^\bullet CH_2CH_2OH + CH_3CH_2OH \tag{19}$$

$$CH_2{}^\bullet CH_2OH \rightarrow {}^\bullet CH_3 + CH_2O \tag{20}$$

2.3. Oxidation of Sterically Hindered Amine in TiO$_2$ Suspensions

The irradiation of titanium dioxide nanoparticles in the presence of molecular oxygen results in the generation of singlet oxygen, but the specific mechanism of ${}^1O_2$ formation is not straightforward and alternative reaction pathways have been suggested [88–90]. The direct detection of ${}^1O_2$ is based on the phosphorescence measurement at 1270 nm corresponding to the radiative transition $O_2({}^1\Delta_g) \rightarrow O_2({}^3\Sigma_g)$ [89,91]. The principle of the indirect techniques of ${}^1O_2$ monitoring is the specific reaction with an organic compound generating a product detectable by a suitable method [92], supported by the application of ${}^1O_2$ scavengers and traps, or using the effect of deuterated solvents. The photoinduced formation of singlet oxygen in the homogeneous systems is frequently monitored also by EPR spectroscopy, detecting the generation of nitroxide radicals derived from 4(R)-2,2,6,6-tetramethylpiperide N-oxyl (R = hydroxy, oxo) produced by the oxidation of corresponding sterically hindered amines (SHA) [84,93,94]. Although this method is widely used for the singlet oxygen detection, many questions arise concerning its selectivity [95]. Particular problems may appear when a numerous ROS or other reactive species are formed in the studied system and their interaction with SHA cannot be excluded, e.g., in the irradiated TiO$_2$ suspensions. The detailed analysis of paramagnetic species generated in homogeneous ACN solutions and TiO$_2$ suspensions in the presence SHA and ROS was performed previously in our laboratory [74].

The concentration of molecular oxygen in aqueous TiO$_2$ suspensions play an important role during the oxidation of 4-oxo-2,2,6,6-tetramethylpiperidine (TMPO) to the radical product 2,2,6,6-tetramethylpiperidine N-oxyl (Tempone; $a_N = 1.617$ mT, $a_{13C}(4^{13}C) = 0.610$ mT;

$g = 2.0054$). In the photoexcited system TiO$_2$/TMPO/water/air the concentrations of Tempone was very low, and the prolonged irradiation led to a total disappearance of the EPR signal (data not shown). However, the saturation of the aqueous TiO$_2$ suspension by oxygen led to the continuous growth of the EPR signal of Tempone ($a_N = 1.479$ mT; $g = 2.0057$) as shown in Figure 9a. The addition of sodium azide, a widely used water-soluble singlet oxygen quencher, to the TiO$_2$/TMPO/water/O$_2$ systems completely suppressed the Tempone generation. However this result should be very carefully analyzed, since besides the singlet oxygen, azide anions also react very fast with the hydroxyl radicals ($k = 1.4 \times 10^{10}$ M^{-1} s^{-1} [55]) producing the azide radical ˙N$_3$ [96] detected here as the corresponding spin-adduct ˙DMPO-N$_3$ in the photoexcited system TiO$_2$/DMPO/water/NaN$_3$/air (Table 2). Despite the limited water solubility of β-carotene, an analogous inhibition of TMPO photooxidation was observed also when β-carotene as an effective singlet oxygen quencher [97] was added to the TiO$_2$ suspensions (Figure 9b). However, due to the lack of specificity the alternative reaction pathways of β-carotene with the radical species generated in the irradiated titania suspensions must be considered [98]. The role of hydroxyl radicals in the SHA oxidation was further demonstrated in the mixed solvent containing DMSO, where the total inhibition of Tempone formation was found, due to the effective scavenging of hydroxyl radicals by DMSO (Figure 9c). The increased concentration of dissolved molecular oxygen in acetonitrile (Table 3), as well as longer lifetime of ^{1}O$_2$ in this solvent (Table 1) resulted in the effective oxidation of TMPO to Tempone. The higher ^{3}O$_2$ concentration in the systems TiO$_2$/TMPO/ACN/air is reflected also in the spectral linewidth growth with not-resolved ^{13}C-satellites (Figure 9d).

Figure 9. The sets of individual EPR spectra ($SW = 8$ mT) monitored upon continuous UVA irradiation ($\lambda_{max} = 365$ nm; irradiance 15 mW·cm^{-2}) of aerated TiO$_2$ P25 suspensions in the presence of sterically hindered amine TMPO: (**a**) oxygenated water; (**b**) oxygenated water saturated with β-carotene; (**c**) oxygenated mixed solvent water/DMSO (5:1 v:v); (**d**) aerated ACN. TiO$_2$ concentration 0.167 mg·mL^{-1}, $c_{0,TMPO} = 0.008$ M.

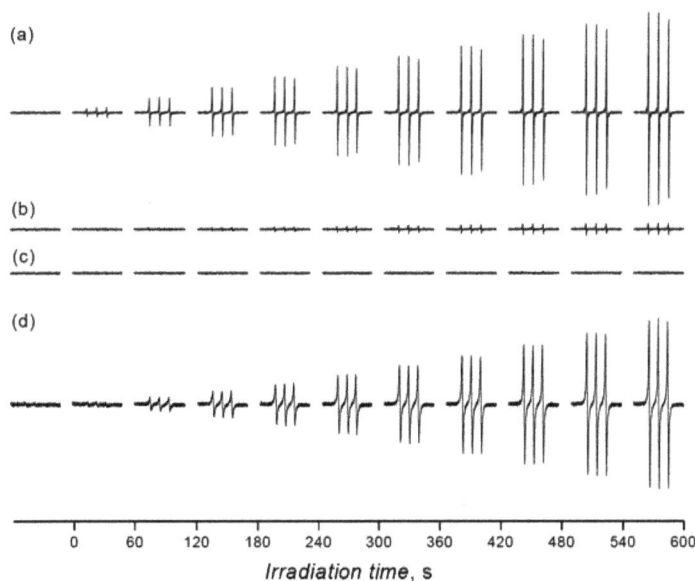

3. Experimental Section

The commercial titanium dioxide Aeroxide® P25 (Evonic Degussa, Essen, Germany) was used and stock suspensions containing 1 mg TiO_2 mL^{-1} were prepared in redistilled water, dimethylsulfoxide (Merck, Darmstadt, Germany, SeccoSolv®, max. 0.025% H_2O), acetonitrile (Merck, SeccoSolv®, max. 0.005% H_2O), methanol (spectroscopic grade, Lachema, Brno, Czech Republic), and ethanol (for UV spectroscopy, MikroChem, Pezinok, Slovak Republic). The isotopically enriched water-^{17}O (20%–24.9% atom. ^{17}O) and deuterated DMSO-d_6, both from Sigma-Aldrich (Buchs, Switzerland), were used as co-solvents. The stock TiO_2 suspensions were homogenized for 1 min using ultrasound (Ultrasonic Compact Cleaner TESON 1; Tesla, Piešťany, Slovak Republic). The spin trapping agent 5,5-dimethyl-1-pyrroline N-oxide (DMPO, Sigma-Aldrich) was distilled prior to use. 5-(Diisopropoxyphosphoryl)-5-methyl-1-pyrroline N-oxide (DIPPMPO, Enzo Life Sciences, Farmingdale, NY, USA), 5-(ethoxycarbonyl)-5-methyl-1-pyrroline N-oxide (EMPO; Enzo Life Sciences), α-(4-pyridyl-1-oxide)-N-tert-butylnitrone (POBN; Janssen Chimica, Geel, Belgium), 2,3,5,6,-tetramethylnitrosobenzene (nitrosodurene, ND, Sigma-Aldrich) and 3,5-dibromo-4-nitrosobenzene sulfonate (DBNBS, Sigma-Aldrich) were used without extra purification. All spin traps were stored at −18 °C. The stock solutions of the spin trapping agents were prepared in studied solvents, apart from the ND, characteristic with a limited solubility in polar solvents, which was applied in a saturated suspension directly before the specific experiments. The concentrations of spin traps applied were chosen in order to minimize the undesired photochemical reactions of the spin traps and to gain the effective trapping of photogenerated radical species. The sterically hindered amine 4-oxo-2,2,6,6-tetramethylpiperidine (TMPO, Merck-Schuchardt, Hohenbrunn, Germany) was used as supplied. Sodium azide (analytical grade, Sigma-Aldrich) and β-carotene (UV grade, Sigma-Aldrich) were applied as the singlet oxygen quenchers. Concentrations of the photogenerated paramagnetic species were determined using solutions of 4-oxo-2,2,6,6-tetramethylpiperidine N-oxyl (Tempone, Sigma-Aldrich) as the calibration standards.

The TiO_2 P25 suspensions containing the spin trapping agent or the TMPO was mixed and carefully saturated with air or oxygen using a slight gas stream immediately before the EPR measurement. So prepared samples were transferred to a small quartz flat cell (WG 808-Q, optical cell length 0.04 cm; Wilmad-LabGlass, Vineland, NJ, USA) optimized for the TE_{102} cavity (Bruker, Rheinstetten, Germany) of the spectrometer X-band EPR spectrometer (EMXplus, Bruker). During the EPR photochemical experiments the samples were irradiated at 295 K directly in the EPR resonator, and the EPR spectra were recorded in situ during a continuous photoexcitation or after a defined exposure. As an irradiation source a UV LED monochromatic radiator (λ_{max} = 365 nm; Bluepoint LED, Hönle UV Technology, Gräfelfing/München, Germany) was used. The irradiance value (λ_{max} = 365 nm; 15 mW·cm^{-2}) within the EPR cavity was determined using a UVX radiometer (UVP, Upland, CA, USA). In some cases, argon saturation needed to be applied after the irradiation of the aerated suspensions prior to the subsequent EPR experiment to get better resolved spectra by suppressing the line-broadening effect of molecular oxygen.

Typical EPR spectrometer settings in a standard photochemical experiment were: microwave frequency, ~9.424 GHz; microwave power, 10.53 mW; center field, 335.6 mT; sweep width, 8–16 mT;

gain, 1×10^5 to 1×10^6; modulation amplitude, 0.05–0.1 mT; scan, 20 s; time constant, 10.24 ms. The g-values (± 0.0001) were determined using a built-in magnetometer. The EPR spectra so obtained were analyzed and simulated using the Bruker software WinEPR and SimFonia and the Winsim2002 [99].

4. Conclusions

The EPR spin trapping experiments using a variety of spin trapping agents (DMPO, EMPO, DIPPMPO, POBN, DBNBS and ND) were performed to identify reactive intermediates formed upon irradiation of TiO_2 suspended in water and organic solvents. The role of water in the photoinduced generation of the hydroxyl radical spin-adduct ˙DMPO-OH in aerated aqueous TiO_2 systems was evidenced using ^{17}O-enriched water. Application of a water-soluble nitroso spin trapping agent DBNBS confirmed the production of methyl radicals when DMSO was added to the aqueous TiO_2 suspensions and the addition of DMSO-d_6 revealed also the origin of these radicals. The photoexcitation of TiO_2 in non-aqueous solvents (DMSO, ACN, methanol and ethanol) in the presence of spin trapping agents showed the stabilization of superoxide radical anions generated via electron transfer reaction to molecular oxygen, as well as the production of various oxygen- and carbon-centered radicals from the solvents. The oxidation of sterically hindered amine TMPO to radical Tempone via ROS was monitored in aqueous and acetonitrile TiO_2 suspensions.

The results obtained demonstrate that indirect EPR spectroscopy techniques represent valuable tools for the characterization of radical intermediates generated in irradiated TiO_2 suspensions. However, a careful selection of the experimental conditions and a precise analysis of the experimental EPR spectra considering alternative reaction pathways is an important aspect of any successful application of these indirect techniques in the characterization of TiO_2 photoactivity.

Supplementary Materials

Supplementary materials can be accessed at: http://www.mdpi.com/1420-3049/19/11/17279/s1.

Acknowledgments

This work was financially supported by Scientific Grant Agency of the Slovak Republic (Project VEGA 1/0289/12) and Slovak University of Technology in Bratislava Young Researcher Grant (Z. Barbieriková).

Author Contributions

Dana Dvoranová and Vlasta Brezová designed experimental research; Dana Dvoranová, Zuzana Barbieriková and Vlasta Brezová performed analysis of experimental data and wrote the paper. All authors read and approved the final manuscript.

Conflicts of Interest

The authors declare no conflict of interest.

References

1. Minero, C.; Maurino, V.; Vione, D. Photocatalytic mechanisms and reaction pathways drawn from kinetic and probe molecules. In *Photocatalysis and Water Purification: From Fundamentals to Recent Applications*, 1st ed.; Pichat, P., Ed.; Wiley-VCH: Weinheim, Germany, 2013; pp. 53–72.

2. Fujishima, A.; Zhang, X.; Tryk, D. TiO_2 photocatalysis and related surface phenomena. *Surf. Sci. Rep.* **2008**, *63*, 515–582.

3. McCullagh, C.; Robertson, J.; Bahnemann, D.; Robertson, P. The application of TiO_2 photocatalysis for disinfection of water contaminated with pathogenic micro-organisms: A review. *Res. Chem. Intermed.* **2007**, *33*, 359–375.

4. Agrios, A.; Pichat, P. State of the art and perspectives on materials and applications of photocatalysis over TiO_2. *J. Appl. Electrochem.* **2005**, *35*, 655–663.

5. Carp, O.; Huisman, C.; Reller, A. Photoinduced reactivity of titanium dioxide. *Prog. Solid State Chem.* **2004**, *32*, 33–177.

6. Shi, H.; Magaye, R.; Castranova, V.; Zhao, J. Titanium dioxide nanoparticles: A review of current toxicological data. *Part. Fibre Toxicol.* **2013**, *10*, 15.

7. Kanakaraju, D.; Glass, B.; Oelgemoller, M. Titanium dioxide photocatalysis for pharmaceutical wastewater treatment. *Environ. Chem. Lett.* **2014**, *12*, 27–47.

8. Lang, X.; Ma, W.; Chen, C.; Ji, H.; Zhao, J. Selective aerobic oxidation mediated by TiO_2 photocatalysis. *Acc. Chem. Res.* **2014**, *47*, 355–363.

9. Ahmed, S.; Rasul, M.; Martens, W.; Brown, R.; Hashib, M. Advances in heterogeneous photocatalytic degradation of phenols and dyes in wastewater: A review. *Water Air Soil Pollut.* **2011**, *215*, 3–29.

10. Gaya, U.; Abdullah, A. Heterogeneous photocatalytic degradation of organic contaminants over titanium dioxide: A review of fundamentals, progress and problems. *J. Photochem. Photobiol. C Rev.* **2008**, *9*, 1–12.

11. McCullagh, C.; Skillen, N.; Adams, M.; Robertson, P. Photocatalytic reactors for environmental remediation: A review. *J. Chem. Technol. Biotechnol.* **2011**, *86*, 1002–1017.

12. Amadelli, R.; Samiolo, L.; Maldotti, A.; Molinari, A.; Gazzoli, D. Selective photooxidation and photoreduction processes at TiO_2 surface-modified by grafted vanadyl. *Int. J. Photoenergy* **2011**, *2011*, doi:10.1155/2011/259453.

13. Molinari, A.; Montoncello, M.; Rezala, H.; Maldotti, A. Partial oxidation of allylic and primary alcohols with O_2 by photoexcited TiO_2. *Photochem. Photobiol. Sci.* **2009**, *8*, 613–619.

14. Henderson, M. A surface science perspective on TiO_2 photocatalysis. *Surf. Sci. Rep.* **2011**, *66*, 185–297.

15. Diebold, U. Structure and properties of TiO_2 surfaces: A brief review. *Appl. Phys. A: Mater. Sci. Process.* **2003**, *76*, 681–687.

16. Pelaez, M.; Nolan, N.; Pillai, S.; Seery, M.; Falaras, P.; Kontos, A.; Dunlop, P.; Hamilton, J.; Byrne, J.; O'Shea, K.; *et al.* A review on the visible light active titanium dioxide photocatalysts for environmental applications. *Appl. Catal. B* **2012**, *125*, 331–349.

17. Diebold, U. The surface science of titanium dioxide. *Surf. Sci. Rep.* **2003**, *48*, 53–229.

18. Zhang, J.; Nosaka, Y. Mechanism of the OH radical generation in photocatalysis with TiO_2 of different crystalline types. *J. Phys. Chem. C* **2014**, *118*, 10824–10832.

19. Montoya, J.; Ivanova, I.; Dillert, R.; Bahnemann, D.; Salvador, P.; Peral, J. Catalytic role of surface oxygens in TiO_2 photooxidation reactions: Aqueous benzene photooxidation with $Ti^{18}O_2$ under anaerobic conditions. *J. Phys. Chem. Lett.* **2013**, *4*, 1415–1422.

20. Salvador, P. On the nature of photogenerated radical species active in the oxidative degradation of dissolved pollutants with TiO_2 aqueous suspensions: A revision in the light of the electronic structure of adsorbed water. *J. Phys. Chem. C* **2007**, *111*, 17038–17043.

21. Green, J.; Carter, E.; Murphy, D. An EPR investigation of acetonitrile reactivity with superoxide radicals on polycrystalline TiO_2. *Res. Chem. Intermed.* **2009**, *35*, 145–154.

22. Carter, E.; Carley, A.; Murphy, D. Evidence for O_2^- radical stabilization at surface oxygen vacancies on polycrystalline TiO_2. *J. Phys. Chem. C* **2007**, *111*, 10630–10638.

23. Berger, T.; Sterrer, M.; Diwald, O.; Knozinger, E.; Panayotov, D.; Thompson, T.; Yates, J. Light-induced charge separation in anatase TiO_2 particles. *J. Phys. Chem. B* **2005**, *109*, 6061–6068.

24. Sawyer, D.T.; Valentine, J.S. How super is superoxide? *Acc. Chem. Res.* **1981**, *14*, 393–400.

25. Harbour, J.R.; Hair, M.L. Detection of superoxide ions in nonaqueous media. Generation by photolysis of pigment dispersions. *J. Phys. Chem.* **1978**, *82*, 1397–1399.

26. Nosaka, Y.; Nosaka, A.Y. Identification and roles of the active species generated on various photocatalysts. In *Photocatalysis and Water Purification: From Fundamentals to Recent Applications*, 1st ed.; Pichat, P., Ed.; Wiley-VCH: Weinheim, Germany, 2013; pp. 3–24.

27. Hirakawa, T.; Yawata, K.; Nosaka, Y. Photocatalytic reactivity for $O_2^{•-}$ and $^•OH$ radical formation in anatase and rutile TiO_2 suspension as the effect of H_2O_2 addition. *Appl. Catal. A* **2007**, *325*, 105–111.

28. Hirakawa, T.; Daimon, T.; Kitazawa, M.; Ohguri, N.; Koga, C.; Negishi, N.; Matsuzawa, S.; Nosaka, Y. An approach to estimating photocatalytic activity of TiO_2 suspension by monitoring dissolved oxygen and superoxide ion on decomposing organic compounds. *J. Photochem. Photobiol. A Chem.* **2007**, *190*, 58–68.

29. Wang, Z.; Ma, W.; Chen, C.; Ji, H.; Zhao, J. Probing paramagnetic species in titania-based heterogeneous photocatalysis by electron spin resonance (ESR) spectroscopy—A mini review. *Chem. Eng. J.* **2011**, *170*, 353–362.

30. Micic, O.; Zhang, Y.; Cromack, K.; Trifunac, A.; Thurnauer, M. Trapped holes on TiO_2 colloids studied by electron-paramagnetic-resonance. *J. Phys. Chem.* **1993**, *97*, 7277–7283.

31. Nakaoka, Y.; Nosaka, Y. ESR investigation into the effects of heat treatment and crystal structure on radicals produced over irradiated TiO_2 powder. *J. Photochem. Photobiol. A Chem.* **1997**, *110*, 299–305.

32. Coronado, J.; Maira, A.; Conesa, J.; Yeung, K.; Augugliaro, V.; Soria, J. EPR study of the surface characteristics of nanostructured TiO_2 under UV irradiation. *Langmuir* **2001**, *17*, 5368–5374.

33. Dimitrijevic, N.; Saponjic, Z.; Rabatic, B.; Poluektov, O.; Rajh, T. Effect of size and shape of nanocrystalline TiO2 on photogenerated charges. An EPR study. *J. Phys. Chem. C* **2007**, *111*, 14597–14601.

34. Kokorin, A.I. Electron Spin Resonance of nanostructured oxide semiconductors. In *Chemical Physics of Nanostructured Semiconductors*; Kokorin, A.I., Bahnemann, D.W., Eds.; VSP BV: Utrecht, The Netherlands, 2003; pp. 203–263.

35. Ghiazza, M.; Alloa, E.; Oliaro-Bosso, S.; Viola, F.; Livraghi, S.; Rembges, D.; Capomaccio, R.; Rossi, F.; Ponti, J.; Fenoglio, I. Inhibition of the ROS-mediated cytotoxicity and genotoxicity of nano-TiO2 toward human keratinocyte cells by iron doping. *J. Nanopart. Res.* **2014**, *16*, 2263.

36. Chiesa, M.; Paganini, M.C.; Livraghi, S.; Giamello, E. Charge trapping in TiO2 polymorphs as seen by electron paramagnetic resonance spectroscopy. *Phys. Chem. Chem. Phys.* **2013**, *15*, 9435–9447.

37. Li, M.; Yin, J.J.; Wamer, W.G.; Lo, Y.M. Mechanistic characterization of titanium dioxide nanoparticle-induced toxicity using electron spin resonance. *J. Food Drug Anal.* **2014**, *22*, 76–86.

38. Grela, M.A.; Coronel, M.E.J.; Colussi, A.J. Quantitative spin-trapping studies of weakly illuminated titanium dioxide sols. Implications for the mechanism of photocatalysis. *J. Phys. Chem.* **1996**, *100*, 16940–16946.

39. Jaeger, C.D.; Bard, A.J. Spin trapping and electron spin resonance detection of radical intermediates in the photodecomposition of water at TiO2 particulate systems. *J. Phys. Chem.* **1979**, *83*, 3146–3152.

40. Dvoranová, D.; Brezová, V.; Mazúr, M.; Malati, M.A. Investigations of metal-doped titanium dioxide photocatalysts. *Appl. Catal. B* **2002**, *37*, 91–105.

41. Taborda, A.V.; Brusa, M.A.; Grela, M.A. Photocatalytic degradation of phthalic acid on TiO2 nanoparticles. *Appl. Catal. A* **2001**, *208*, 419–426.

42. Brezová, V.; Staško, A.; Biskupič, S.; Blažková, A.; Havlínová, B. Kinetics of hydroxyl radical spin trapping in photoactivated homogeneous (H_2O_2) and heterogeneous (TiO_2, O_2) aqueous systems. *J. Phys. Chem.* **1994**, *98*, 8977–8984.

43. Nosaka, Y.; Komori, S.; Yawata, K.; Hirakawa, T.; Nosaka, A. Photocatalytic ˙OH radical formation in TiO2 aqueous suspension studied by several detection methods. *Phys. Chem. Chem. Phys.* **2003**, *5*, 4731–4735.

44. Brezová, V.; Dvoranová, D.; Staško, A. Characterization of titanium dioxide photoactivity following the formation of radicals by EPR spectroscopy. *Res. Chem. Intermed.* **2007**, *33*, 251–268.

45. Brezová, V.; Billik, P.; Vrecková, Z.; Plesch, G. Photoinduced formation of reactive oxygen species in suspensions of titania mechanochemically synthesized from TiCl4. *J. Mol. Catal. A Chem.* **2010**, *327*, 101–109.

46. Hawkins, C.L.; Davies, M.J. Detection and characterisation of radicals in biological materials using EPR methodology. *Biochim. Biophys. Acta Gen. Subj.* **2014**, *1840*, 708–721.

47. Spasojevic, I. Free radicals and antioxidants at a glance using EPR spectroscopy. *Crit. Rev. Clin. Lab. Sci.* **2011**, *48*, 114–142.

48. Alberti, A.; Macciantelli, D. Spin Trapping. In *Electron Paramagnetic Resonance: A Practitioner's Toolkit*; Brustolon, M., Giamelo, E., Eds.; John Wiley & Sons: Hoboken, NJ, USA, 2009; pp. 287–323.

49. Dodd, N.J.F.; Jha, A.N. Photoexcitation of aqueous suspensions of titanium dioxide nanoparticles: An electron spin resonance spin trapping study of potentially oxidative reactions. *Photochem. Photobiol.* **2011**, *87*, 632–640.

50. Buettner, G.R. Spin trapping: ESR parameters of spin adducts. *Free Radic. Biol. Med.* **1987**, *3*, 259–303.

51. Lloyd, R.V.; Hanna, P.M.; Mason, R.P. The origin of the hydroxyl radical oxygen in the Fenton reaction. *Free Radic. Biol. Med.* **1997**, *22*, 885–888.

52. Brezová, V.; Barbieriková, Z.; Zukalová, M.; Dvoranová, D.; Kavan, L. EPR study of ^{17}O-enriched titania nanopowders under UV irradiation. *Catal. Today* **2014**, *230*, 112–118.

53. Finkelstein, E.; Rosen, G.M.; Rauckman, E.J. Production of hydroxyl radical by decomposition of superoxide spin-trapped adducts. *Mol. Pharmacol.* **1982**, *21*, 262–265.

54. Buxton, G.; Greenstock, C.; Helman, W.; Ross, A. Critical-review of rate constants for reactions of hydrated electrons, hydrogen-atoms and hydroxyl radicals ($^{\bullet}$OH/$^{\bullet}$O^{-}) in aqueous-solution. *J. Phys. Chem. Ref. Data* **1988**, *17*, 513–886.

55. Wilkinson, F.; Helman, W.; Ross, A. Rate constants for the decay and reactions of the lowest electronically excited singlet-state of molecular-oxygen in solution—An expanded and revised compilation. *J. Phys. Chem. Ref. Data* **1995**, *24*, 663–1021.

56. Zalibera, M.; Rapta, P.; Staško, A.; Brindzová, L.; Brezová, V. Thermal generation of stable SO$_4^{\bullet-}$ spin trap adducts with super-hyperfine structure in their EPR spectra: An alternative EPR spin trapping assay for radical scavenging capacity determination in dimethylsulphoxide. *Free Radic. Res.* **2009**, *43*, 457–469.

57. Mottley, C.; Connor, H.D.; Mason, R.P. [^{17}O]oxygen hyperfine structure for the hydroxyl and superoxide radical adducts of the spin traps DMPO, PBN and 4-POBN. *Biochem. Biophys. Res. Commun.* **1986**, *141*, 622–628.

58. Stolze, K.; Rohr-Udilova, N.; Rosenau, T.; Hofinger, A.; Kolarich, D.; Nohl, H. Spin trapping of C- and O-centered radicals with methyl-, ethyl-, pentyl-, and phenyl-substituted EMPO derivatives. *Bioorg. Med. Chem.* **2006**, *14*, 3368–3376.

59. Culcasi, M.; Rockenbauer, A.; Mercier, A.; Clément, J.L.; Pietri, S. The line asymmetry of electron spin resonance spectra as a tool to determine the *cis:trans* ratio for spin-trapping adducts of chiral pyrrolines *N*-oxides: The mechanism of formation of hydroxyl radical adducts of EMPO, DEPMPO, and DIPPMPO in the ischemic-reperfused rat liver. *Free Radic. Biol. Med.* **2006**, *40*, 1524–1538.

60. Chalier, F.; Tordo, P. 5-Diisopropoxyphosphoryl-5-methyl-1-pyrroline *N*-oxide, DIPPMPO, a crystalline analog of the nitrone DEPMPO: Synthesis and spin trapping properties. *J. Chem. Soc. Perkin Trans. 2* **2002**, 2110–2117, doi:10.1039/B206909C.

61. Huling, S.G.; Arnold, R.G.; Sierka, R.A.; Miller, M.R. Measurement of hydroxyl radical activity in a soil slurry using the spin trap α-(4-pyridyl-1-oxide)-*N-tert*-butylnitrone. *Environ. Sci. Technol.* **1998**, *32*, 3436–3441.

62. Brezová, V.; Tarábek, P.; Dvoranová, D.; Staško, A.; Biskupič, S. EPR study of photoinduced reduction of nitroso compounds in titanium dioxide suspensions. *J. Photochem. Photobiol. A Chem.* **2003**, *155*, 179–198.

63. Clement, J.; Gilbert, B.; Ho, W.; Jackson, N.; Newton, M.; Silvester, S.; Timmins, G.; Tordo, P.; Whitwood, A. Use of a phosphorylated spin trap to discriminate between the hydroxyl radical and other oxidising species. *J. Chem. Soc. Perkin Trans. 2* **1998**, 1715–1718, doi:10.1039/A804098B.

64. Patel, A.; Rohr-Udilova, N.; Rosenau, T.; Stolze, K. Synthesis and characterization of 5-alkoxycarbonyl-4-hydroxymethyl-5-alkyl-pyrroline *N*-oxide derivatives. *Bioorg. Med. Chem.* **2011**, *19*, 7643–7652.

65. Stolze, K.; Rohr-Udilova, N.; Hofinger, A.; Rosenau, T. Spin trapping properties of aminocarbonyl- and methylamino-carbonyl-substituted EMPO derivatives. *Free Radic. Res.* **2009**, *43*, 81–81.

66. Abbas, K.; Hardy, M.; Poulhès, F.; Karoui, H.; Tordo, P.; Ouari, O.; Peyrot, F. Detection of superoxide production in stimulated and unstimulated living cells using new cyclic nitrone spin traps. *Free Radic. Biol. Med.* **2014**, *71*, 281–290.

67. Halliwel, B.; Gutteridge, J. *Free Radicals in Biology and Medicine*, 3rd ed.; Oxford University Press: Oxford, UK, 1999; p. 60.

68. Hirakawa, T.; Nakaoka, Y.; Nishino, J.; Nosaka, Y. Primary passages for various TiO_2 photocatalysts studied by means of luminol chemiluminescent probe. *J. Phys. Chem. B* **1999**, *103*, 4399–4403.

69. Nosaka, Y.; Yamashita, Y.; Fukuyama, H. Application of chemiluminescent probe to monitoring superoxide radicals and hydrogen peroxide in TiO_2 photocatalysis. *J. Phys. Chem. B* **1997**, *101*, 5822–5827.

70. Nosaka, Y.; Fukuyama, H. Application of chemiluminescent probe to the characterization of TiO_2 photocatalysts in aqueous suspension. *Chem. Lett.* **1997**, *26*, 383–384.

71. Marino, T.; Molinari, R.; García, H. Selectivity of gold nanoparticles on the photocatalytic activity of TiO_2 for the hydroxylation of benzene by water. *Catal. Today* **2013**, *206*, 40–45.

72. Molinari, A.; Maldotti, A.; Amadelli, R. Probing the role of surface energetics of electrons and their accumulation in photoreduction processes on TiO_2. *Chem. Eur. J.* **2014**, *20*, 7759–7765.

73. Brezová, V.; Gabčová, S.; Dvoranová, D.; Staško, A. Reactive oxygen species produced upon photoexcitation of sunscreens containing titanium dioxide (An EPR study). *J. Photochem. Photobiol. B Biol.* **2005**, *79*, 121–134.

74. Barbieriková, Z.; Mihalíková, M.; Brezová, V. Photoinduced oxidation of sterically hindered amines in acetonitrile solutions and titania suspensions (An EPR study). *Photochem. Photobiol.* **2012**, *88*, 1442–1454.

75. Lide, D.R., Ed. *CRC Handbook of Chemistry and Physics*, 86th ed.; CRC Press: Boca Raton, FL, USA, 2005.

76. Mitroka, S.; Zimmeck, S.; Troya, D.; Tanko, J. How solvent modulates hydroxyl radical reactivity in hydrogen atom abstractions. *J. Am. Chem. Soc.* **2010**, *132*, 2907–2913.

77. Turro, N.J.; Ramamurthy, V.; Scaiano, J.C. *Modern Molecular Photochemistry of Organic Molecules*; University Science Books: Sausalito, CA, USA, 2010; p. 1008.

78. Wadhawan, J.; Welford, P.; McPeak, H.; Hahn, C.; Compton, R. The simultaneous voltammetric determination and detection of oxygen and carbon dioxide—A study of the kinetics of the reaction between superoxide and carbon dioxide in non-aqueous media using membrane-free gold disc microelectrodes. *Sens. Actuat. B* **2003**, *88*, 40–52.

79. Golovanov, I.; Zhenodarova, S. Quantitative structure-property relationship: XXIII. Solubility of oxygen in organic solvents. *Russ. J. Gen. Chem.* **2005**, *75*, 1795–1797.

80. Di Paola, A.; Bellardita, M.; Palmisano, L.; Barbieriková, Z.; Brezová, V. Influence of crystallinity and OH surface density on the photocatalytic activity of TiO_2 powders. *J. Photochem. Photobiol. A Chem.* **2014**, *273*, 59–67.

81. Addamo, M.; Augugliaro, V.; Coluccia, S.; di Paola, A.; García-López, E.; Loddo, V.; Marcì, G.; Martra, G.; Palmisano, L. The role of water in the photocatalytic degradation of acetonitrile and toluene in gas-solid and liquid-solid regimes. *Int. J. Photoenergy* **2006**, *2006*, doi:10.1155/IJP/2006/39182.

82. Micic, O.; Zhang, Y.; Cromack, K.; Trifunac, A.; Thurnauer, M. Photoinduced hole transfer from TiO_2 to methanol molecules in aqueous-solution studied by electron-paramagnetic-resonance. *J. Phys. Chem.* **1993**, *97*, 13284–13288.

83. Pieta, P.; Petr, A.; Kutner, W.; Dunsch, L. *In situ* ESR spectroscopic evidence of the spin-trapped superoxide radical, $O_2^{\cdot-}$, electrochemically generated in DMSO at room temperature. *Electrochim. Acta* **2008**, *53*, 3412–3415.

84. Barbieriková, Z.; Bella, M.; Kučerák, J.; Milata, V.; Jantová, S.; Dvoranová, D.; Veselá, M.; Staško, A.; Brezová, V. Photoinduced superoxide radical anion and singlet oxygen generation in the presence of novel selenadiazoloquinolones (An EPR study). *Photochem. Photobiol.* **2011**, *87*, 32–44.

85. Guo, R.; Davies, C.; Nielsen, B.; Hamilton, L.; Symons, M.; Winyard, P. Reaction of the spin trap 3,5-dibromo-4-nitrosobenzene sulfonate with human biofluids. *Biochim. Biophys. Acta Gen. Subj.* **2002**, *1572*, 133–142.

86. Konaka, R.; Terabe, S.; Mizuta, T.; Sakata, S. Spin trapping by use of nitrosodurene and its derivatives. *Can. J. Chem.* **1982**, *60*, 1532–1542.

87. Terabe, S.; Kuruma, K.; Konaka, R. Spin trapping by use of nitroso-compounds. Part VI. Nitrosodurene and other nitrosobenzene derivatives. *J. Chem. Soc.* **1973**, *9*, 1252–1258.

88. Daimon, T.; Hirakawa, T.; Nosaka, Y. Monitoring the formation and decay of singlet molecular oxygen in TiO_2 photocatalytic systems and the reaction with organic molecules. *Electrochemistry* **2008**, *76*, 136–139.

89. Daimon, T.; Nosaka, Y. Formation and behavior of singlet molecular oxygen in TiO_2 photocatalysis studied by detection of near-infrared phosphorescence. *J. Phys. Chem. C* **2007**, *111*, 4420–4424.

90. Daimon, T.; Hirakawa, T.; Kitazawa, M.; Suetake, J.; Nosaka, Y. Formation of singlet molecular oxygen associated with the formation of superoxide radicals in aqueous suspensions of TiO_2 photocatalysts. *Appl. Catal. A* **2008**, *340*, 169–175.

91. Nakamura, K.; Ishiyama, K.; Ikai, H.; Kanno, T.; Sasaki, K.; Niwano, Y.; Kohno, M. Reevaluation of analytical methods for photogenerated singlet oxygen. *J. Clin. Biochem. Nutr.* **2011**, *49*, 87–95.

92. Wu, H.; Song, Q.; Ran, G.; Lu, X.; Xu, B. Recent developments in the detection of singlet oxygen with molecular spectroscopic methods. *TrAC Trends Anal. Chem.* **2011**, *30*, 133–141.

93. Barbieriková, Z.; Bella, M.; Sekeráková, L.; Lietava, J.; Bobeničová, M.; Dvoranová, D.; Milata, V.; Sádecká, J.; Topoľská, D.; Heizer, T.; *et al.* Spectroscopic characterization, photoinduced processes and cytotoxic properties of substituted *N*-ethyl selenadiazoloquinolones. *J. Phys. Org. Chem.* **2013**, *26*, 565–574.

94. Lion, Y.; Gandin, E.; van de Vorst, A. On the production of nitroxide radicals by singlet oxygen reaction: An EPR study. *Photochem. Photobiol.* **1980**, *31*, 305–309.

95. Nosaka, Y.; Natsui, H.; Sasagawa, M.; Nosaka, A. Electron spin resonance studies on the oxidation mechanism of sterically hindered cyclic amines in TiO_2 photocatalytic systems. *J. Phys. Chem. B* **2006**, *110*, 12993–12999.

96. Maldotti, A.; Amadelli, R.; Carassiti, V. An electron spin resonance spin trapping investigation of azide oxidation on TiO_2 powder suspensions. *Can. J. Chem.* **1988**, *66*, 76–80.

97. Konovalova, T.A.; Lawrence, J.; Kispert, L.D. Generation of superoxide anion and most likely singlet oxygen in irradiated TiO_2 nanoparticles modified by carotenoids. *J. Photochem. Photobiol. A Chem.* **2004**, *162*, 1–8.

98. Jomová, K.; Valko, M. Health protective effects of carotenoids and their interactions with other biological antioxidants. *Eur. J. Med. Chem.* **2013**, *70*, 102–110.

99. Duling, D.R. Simulation of multiple isotropic spin-trap EPR spectra. *J. Magn. Reson. B* **1994**, *104*, 105–110. Available online: http://www.niehs.nih.gov/research/resources/software/tox-pharm/tools/ (accessed on sss22 October 2014).

Sample Availability: Not available.

Probing Water and CO_2 Interactions at the Surface of Collapsed Titania Nanotubes Using IR Spectroscopy

Kaustava Bhattacharyya, Weiqiang Wu, Eric Weitz, Baiju K. Vijayan and Kimberly A. Gray

Abstract: Collapsed titania nanotubes (cTiNT) were synthesized by the calcination of titania nanotubes (TiNT) at 650 °C, which leads to a collapse of their tubular morphology, a substantial reduction in surface area, and a partial transformation of anatase to the rutile phase. There are no significant changes in the position of the XPS responses for Ti and O on oxidation or reduction of the cTiNTs, but the responses are more symmetric than those observed for TiNTs, indicating fewer surface defects and no change in the oxidation state of titanium on oxidative and/or reductive pretreatment. The interaction of H_2O and CO_2 with the cTiNT surface was studied. The region corresponding to OH stretching absorptions extends below 3000 cm^{-1}, and thus is broader than is typically observed for absorptions of the OH stretches of water. The exchange of protons for deuterons on exposure to D_2O leads to a depletion of this extended absorption and the appearance of new absorptions, which are compatible with deuterium exchange. We discuss the source of this extended low frequency OH stretching region and conclude that it is likely due to the hydrogen-bonded OH stretches. Interaction of the reduced cTiNTs with CO_2 leads to a similar but smaller set of adsorbed carbonates and bicarbonates as reported for reduced TiNTs before collapse. Implications of these observations and the presence of proton sources leading to hydrogen bonding are discussed relative to potential chemical and photochemical activity of the TiNTs. These results point to the critical influence of defect structure on CO_2 photoconversion.

Reprinted from *Molecules*. Cite as: Bhattacharyya, K.; Wu, W.; Weitz, E.; Vijayan, B.K.; Gray, K.A. Probing Water and CO_2 Interactions at the Surface of Collapsed Titania Nanotubes Using IR Spectroscopy. *Molecules* **2015**, *20*, 15469-15487.

1. Introduction

Nano-scale TiO_2 materials have attracted scientific interest due to their unusual physico-chemical properties such as high specific surface area, ion-exchangeability, and photocatalytic activity. Interest in nano-scale titania materials has extended to one-dimensional structures including nanotubes, nanorods, and nanowires, which can now be synthesized by relatively standard methods [1–6].

In the late 1990s, Kasuga *et al.* reported the first hydrothermal synthesis of a nano-tubular structure of titania [1,2]. Researchers subsequently showed that these structures consist of hydrated dititanate ($H_2Ti_2O_5$), trititanate ($H_2Ti_3O_7$) [3–5], and $H_2Ti_4O_9 \cdot xH_2O$ stoichiometries [6], as well as their Na salts. These structures can be thermally treated to improve phase crystallinity, although there is still no clear consensus on the specific conditions under which thermal treatment leads to the transformation of titania/titanates in titania nanotubes (TiNTs) to TiO_2 (B), rutile, or brookite. Nevertheless, it is clear that the morphology of the TiNTs starts to undergo a change around 550 °C. Poudel *et al.* [7] observed a change from nanotubes to nanowire morphology at an annealing temperature of 650 °C. Tsai and Teng [8], Yoshida *et al.* [9], and Vijayan *et al.* [10] all reported a drop in surface area and pore volume with increasing temperature. The surface area of the TiNTs steadily decreases as a function of temperature for as long as the nanotube morphology is retained and then decreases sharply upon further elevation of the temperature, which results in the loss of the tubular morphology [11]. Vijayan *et al.* characterized these morphological changes to TiNTs over a range of calcination temperatures from 200 to 800 °C and, using EPR, they detailed how these changes altered charge trapping behavior and photocatalytic reactivity [10].

There is an enormous body of literature on the photocatalytic reactivity, particularly oxidation reactions, of titania and doped titania materials of various morphologies. Comparatively less is known about the details of the photocatalytic reduction of CO_2 by titania materials, especially the molecular level mechanism(s) for this complex process. Our prior study of the adsorption of CO_2 on the surface of TiNTs and platinized TiNTs focused on changes in adsorptive chemistry that took place when the TiNTs were subjected to reductive and/or oxidative pretreatments. We found that differences in the surface species formed, which included carbonates and bicarbonates, could be correlated with the Lewis acidity and basicity of the Ti and O on the TiNT surface [12].

Calcination temperature affects the chemistry taking place on TiNTs. Vijayan *et al.* [10] reported maximum CO_2 conversion for TiNT calcined at 400 °C and dramatically decreasing CO_2 photo-reduction with increasing calcination temperature, particularly above 600 °C. The collapse of the tubular nanostructure is accompanied by a diminished number of under-coordinated Ti sites and marked changes in charge trapping as characterized by EPR. Differences in the respective reactivity of TiNT and collapsed TiNT (cTiNT) with CO_2 may be explained by differences in CO_2 binding at the catalyst surface [10,12]. Water has been found to be a critical participant in the reduction of CO_2. The presence of water has been shown to facilitate CO_2 photoreductive chemistry. This is not surprising since water can provide the protons needed to go from CO_2 to hydrocarbons and the coordination of water can potentially lower the barrier for reactions to desired hydrocarbon products such as formic acid and methanol.

The objective of the research reported herein is to spectroscopically probe the interactions of water and CO_2 at the cTiNT surface with the ultimate goal of identifying the surface characteristics of photocatalytic materials capable of driving CO_2 conversion. Specifically, we report the effect of higher temperature annealing on the phase, structure, and morphological distortion, which leads to changes in the chemical interactions of CO_2 with cTiNT. We also address an issue in the literature with regard to the assignment of low frequency absorptions in the OH stretching region of the infrared. These absorptions are assigned to OH stretches that form on exposure of the nanotubes to water. We speculate that hydrogen bonding between H-containing moieties within the structure of the nanotubes may be the sources of these low frequency OH absorptions.

2. Results

2.1. X-ray Diffraction Data

The interpretation of the X-ray diffraction data (Figure 1) for the titania nanotubes that have been heated to 350 °C has been discussed previously in detail [12]. These data confirm that the TiNTs are essentially pure anatase. Thermal annealing at 650 °C leads to formation of a rutile phase along with the characteristic anatase phase. As seen in Figure 1, after annealing at 650 °C, a series of diffraction peaks appear which are consistent with anatase TiO_2 (JCPDS 75-1537): (101), 25.7°; (200), 47.9°; (105), 53.7°; (211), 55.0°; (213), 61.9°; (204), 62.8°; (116), 68.3°; (220), 69.7°; and (301), 75.2°. An additional diffraction peak characteristic of the rutile phase is seen at 27.3° (110) (JCPDS-77-0443, 77-0444). This new peak indicates that there is a partial transformation between anatase and rutile as a result of heating to 650 °C. Though anatase is metastable in the bulk, it has been reported that the anatase phase is more stable as nano-scale TiO_2 than as bulk TiO_2 [13–17]. A prior report for TiNT prepared by anodization of Ti foil indicated that above 550–600 °C, rutile is the dominant phase (>50%) [18,19]. In contrast, Albu *et al.* found that double-walled anodized TiNTs were extraordinarily stable, retained their structural integrity even when annealed to temperatures higher than 600 °C, and showed only traces of rutile [20]. Similarly, we determine a rutile composition of only ~5.5%. The smaller amount of rutile formed is probably due to the enhanced stability of anatase in nano-scale titania materials, but the transformation could also be kinetically limited, possibly as a result of the hydrothermal synthesis process. However, the peak at a 2θ value of 13.3° is completely absent in the XRD pattern, indicating the loss of the titania nanotubular morphology for samples calcined at 650 °C.

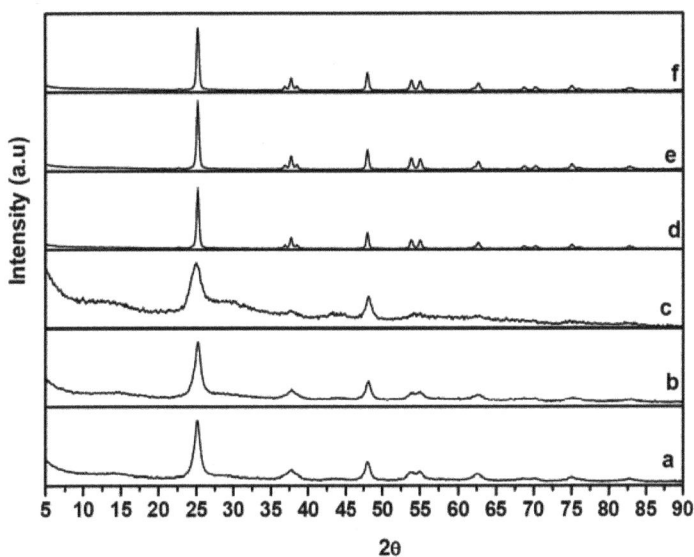

Figure 1. XRD profiles of (**a**) TiNT; (**b**) TiNT-O$_2$; (**c**) TiNT-O$_2$-H$_2$; (**d**) cTiNT-; (**e**) cTiNT-O$_2$; (**f**) cTiNT-O$_2$-H$_2$. Note that traces **a–c** have been included here for completeness and are reproduced from Figure 2 in reference [12]. Reprinted with permission from: Kaustava Bhattacharyya, Alon Danon, Baiju K. Vijayan, Kimberley A. Gray, Peter C. Stair, and Eric Weitz, *The Journal of Physical Chemistry C* **117** (2013), 12661–12678, Copyright 2013, American Chemical Society.

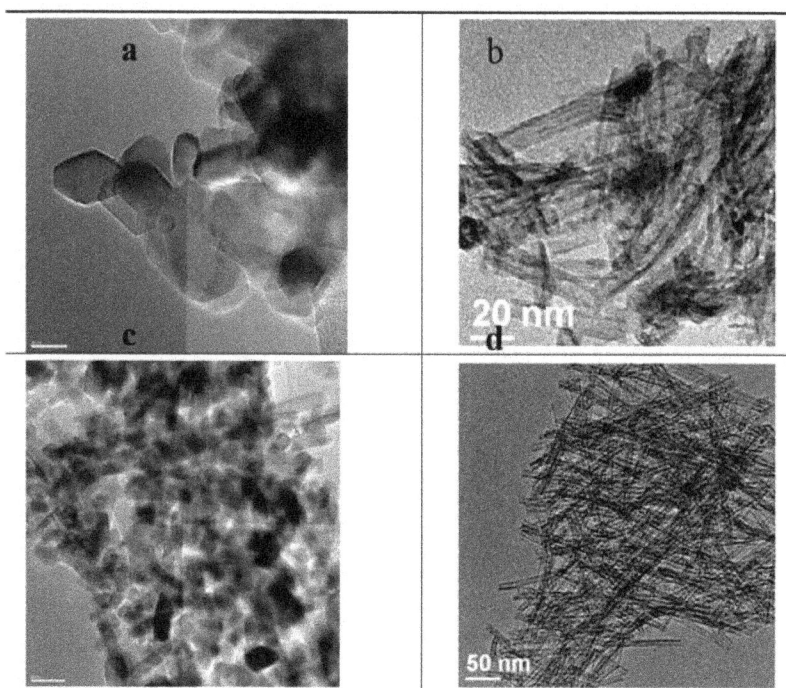

Figure 2. TEM images of the TiNT and the cTiNT: (**a**) cTiNT with a magnification scale of 20 nm; (**b**) TiNT with a magnification scale of 20 nm; (**c**) cTiNT with a magnification scale of 50 nm; (**d**) TiNT with a magnification scale of 50 nm. Note that panels (**b,d**) are included here for completeness and are from Figure 1 in Reference [12]. Reprinted with permission from: Kaustava Bhattacharyya, Alon Danon, Baiju K. Vijayan, Kimberley A. Gray, Peter C. Stair, and Eric Weitz, *The Journal of Physical Chemistry* C **117** (2013), 12661–12678, Copyright 2013, American Chemical Society.

2.2. Transmission Electron Microscopy

The interpretation of the electron microscopy data for the titania nanotubes heated to 350 °C has been discussed in detail previously [12]. Briefly, the data in Figures 2–4 compare HRTEM images of the TiNTs calcined at 350 °C and 650 °C. The conclusion from the HRTEM data is that the nanotubular morphology of the TiNTs is intact for the TiNTs calcined at 350 °C and after pretreatment under an oxygen atmosphere, and subsequently under H_2. However, when these TiNT are calcined at 650 °C, the tubular morphology is lost and the tubes transform into a variety of other nanoscale shapes, although nanorods are the predominant shape.

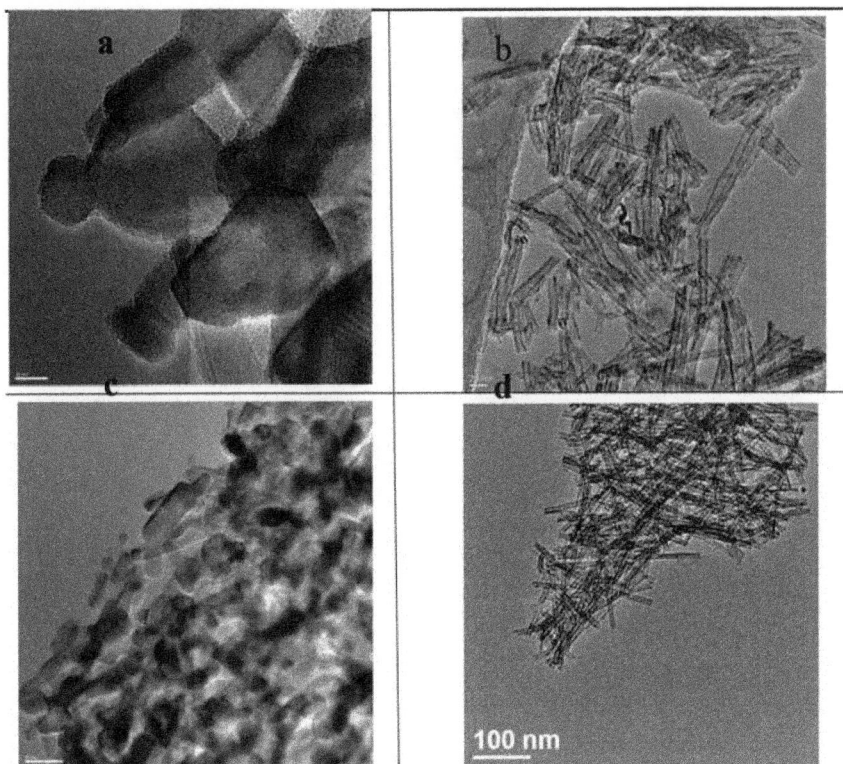

Figure 3. TEM images of the TiNT-O_2 and the cTi-NT-O_2: (**a**) cTiNT-O_2 with a magnification scale of 20 nm; (**b**) TiNT-O_2 with a magnification scale of 20 nm; (**c**) cTiNT-O_2 with a magnification scale of 50 nm; (**d**) TiNT-O_2 with a magnification scale of 100 nm.

2.3. BET Surface Area

Measurement involving a determination of the BET surface area and pore size distribution of TiNTs has been previously reported [10,12]. Briefly, the as-prepared TiNTs have a surface area of ~235 m^2/g, which is almost 20 times that of the TiO$_2$ anatase powder from which the tubes were prepared. Upon calcination at 350 °C the surface area decreases to ~200 m^2/gm. When these TiNTs are subject to either oxidative and/or reductive pretreatments at 350 °C the surface area decreases further, into the ~180–195 m^2/gm range, which is still comparable to that before pretreatment. Upon calcinations of the TiNTs at 650 °C the surface area decreases substantially to ~90 m^2/g. The surface areas do not change significantly for the cTiNT samples as a result of oxidization and subsequent reduction. Thus, as the temperature increases toward 650 °C, the collapse of the nanotubes leads to formation of coarser grain particles and as the crystallite size grows, so does the inter-particle pore size.

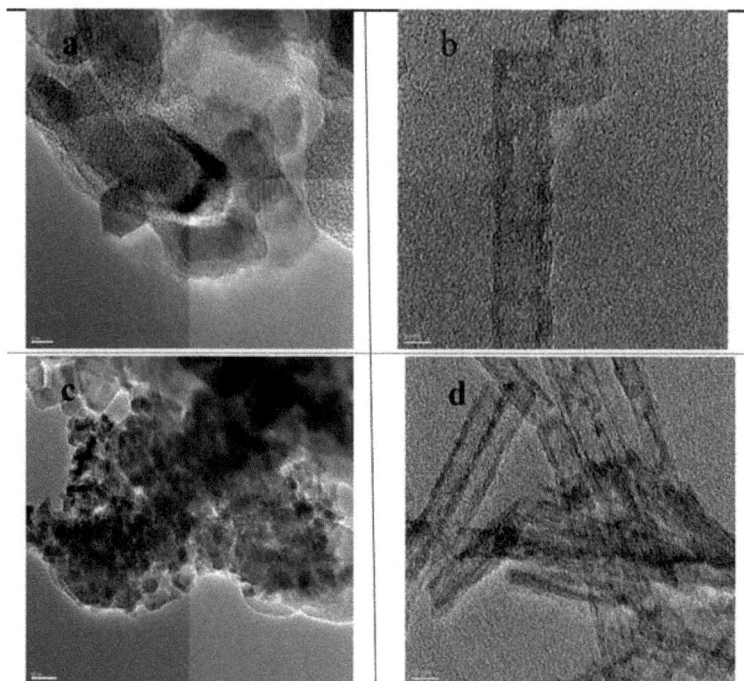

Figure 4. TEM images of the TiNT–O_2-H_2 and the cTiNT-O_2-H_2: (**a**) cTiNT-O_2-H_2 with a magnification scale of 20 nm; (**b**) TiNT with a magnification scale of 20 nm; (**c**) cTiNT-O_2-H_2 with a magnification scale of 50 nm; (**d**) TiNT with a magnification scale of 50 nm. Note that panel 2.3d is from Figure 1 in reference [12]. It has been included here for completeness and is reprinted with permission from: Kaustava Bhattacharyya, Alon Danon, Baiju K. Vijayan, Kimberley A. Gray, Peter C. Stair, and Eric Weitz, *The Journal of Physical Chemistry* C **117** (2013), 12661–12678, Copyright 2013, American Chemical Society.

2.4. XPS

The top panel in Figure 5 shows the XPS Ti 2p spectra of the cTiNT subjected to oxidative and reductive pretreatments. All of the cTiNT samples have Ti $2p_{3/2}$ and Ti $2p_{1/2}$ peaks at 458.6 and 464.3 eV, respectively, in their Ti XPS spectra. These peak positions are typical of anatase TiO_2 nano-powder samples [21,22]. There is no observed change in these peaks as a function of oxidation or reduction. The lack of change in these peaks on oxidation and/or reduction is in marked contrast to what is observed for TiNT samples, where upon reduction there is formation of the Ti^{3+} and splitting of the Ti^{4+} peak [12].

The O 1s XPS spectra shown in the bottom panel in Figure 5 for the samples calcined at 650 °C (cTiNT) exhibits analogous behavior. Each sample shows an O 1s XPS peak at 531.66 eV, which is typical of bulk anatase TiO_2 samples. These O 1s peaks are symmetric and possess a smaller FWHM than the TiNT samples calcined at 350 °C, which is indicative of fewer O defects and fewer OH groups. Thus, we can infer based on the XPS data [21,23] that the electronic environment of surface

Ti and O atoms in the nanotubular morphology is more sensitive to oxidation and/or reduction than their counterparts in the nanotubes that have collapsed as a result of annealing. Overall, there is no observable shift in either the Ti or the O XPS peaks upon oxidation of the cTiNT, and upon reduction there is neither formation of Ti^{3+} for this sample nor is there any splitting of the Ti^{4+} peak, as was observed for the corresponding TiNTs that were annealed at 350 °C [12].

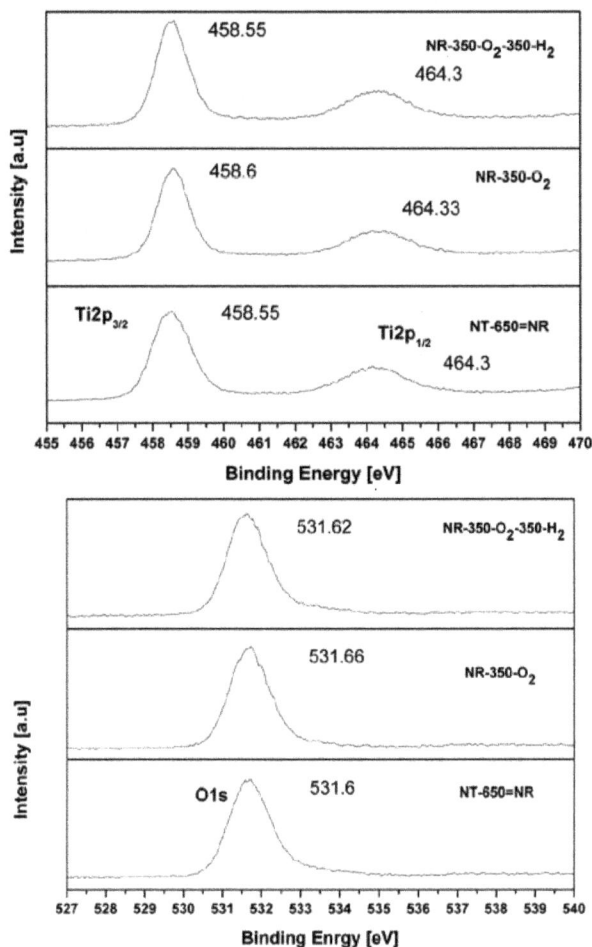

Figure 5. The XPS spectra for (**top panel**) Ti 2p and (**bottom panel**) O 1s for the cTiNTs which have been subjected to different pre-treatments: (**bottom trace**) cTiNT; (**middle trace**) cTiNT-O_2; (**top trace**) cTiNT-O_2-H_2.

2.5. FT-IR Spectroscopy

Figure 6 shows a spectrum of the cTiNTs after they have been collapsed by calcining to 650 °C and then cooled to room temperature. There is an absorption that stretches from ~3700 cm^{-1} down to ~2700 cm^{-1}. The absorption appears to consist of at least two overlapped peaks as a result of the increase in intensity near 3000 cm^{-1}. There are no observable absorptions due to Ti-OH stretches, which are expected near 3700 cm^{-1} [24]. The insert shows the cTiNT-O$_2$ after exposure to water vapor at room temperature and subsequent to evacuation of the water vapor.

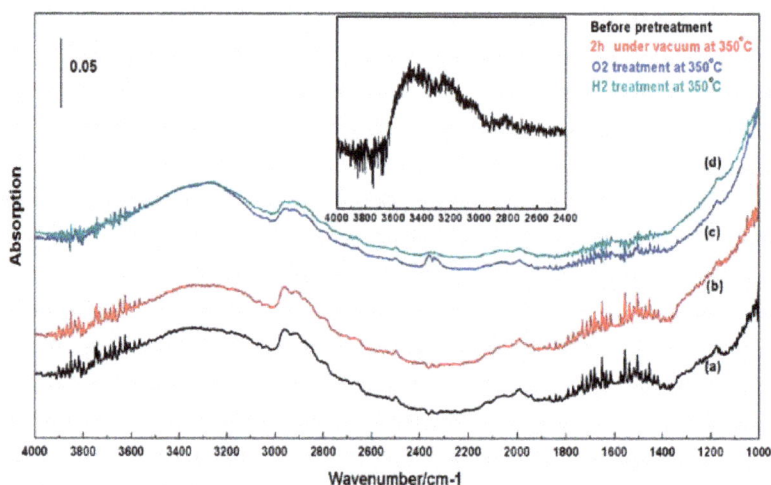

Figure 6. *In situ* FT-IR spectra of the c-TiNT. The collapsed NTs were prepared by calcination at 650 °C in a furnace and were allowed to cool to room temperature and introduced into the IR cell and an IR spectrum was recorded (**a**). Spectrum **b** was recorded after these cTiNTs were exposed to vacuum for 2 h at 350 °C. Spectrum **c** was recorded after the same cTiNTs were heated in O$_2$ for 3 h at 350 °C to produce cTiNT-O$_2$. Spectrum **d** was recorded after the oxygen was evacuated from the cell subsequent to obtaining spectrum **c** and the oxidized cTiNTs were heated in H$_2$ for 3 h at 350 °C to produce cTiNT-O$_2$-H$_2$. The inset shows a cTiNT-O$_2$ sample that has been exposed to 10 Torr of water vapor for 15 min at room temperature. The background for all spectra was the cell with the tungsten wire grid in place but no sample on the grid.

To probe the source of these absorptions we exposed the sample to D$_2$O. Figure 7 shows that exposure to 10 Torr of gas-phase D$_2$O produces a broad absorption between ~2700 and ~1950 cm^{-1}. A "negative absorption" is also seen between ~3500 and ~2700 cm^{-1}. We note that the spectra shown are referenced to background spectra of the cTiNTs. Thus, depletion of a species that is present and absorbs in the background spectrum will appear as a "negative absorption" in the sample spectrum. The upper trace also has absorptions due to OD stretching modes of D$_2$O in the ~2900 to ~2500 cm^{-1} range, which show the rotational structure characteristic of gas-phase absorptions, as do the absorptions below 1600 cm^{-1}, which are due to the bending mode of D$_2$O. The lower trace demonstrates that evacuation removes the gas-phase D$_2$O. Evacuation also leads to a decrease in intensity of the 1950 to 2700 cm^{-1} absorption, with a larger decrease in intensity at higher frequencies, where weakly bound D$_2$O would be expected to absorb. There is also a sharp peak at 2350 cm^{-1} that is superimposed on the broad absorption. This is where gas-phase CO$_2$ absorbs [25], and this absorption is likely due to a small change in the CO$_2$ content of the purged housing that takes place over the timescale of the experiment.

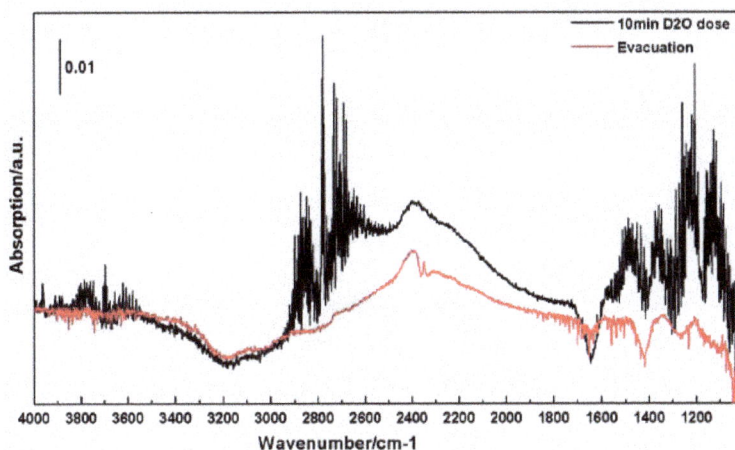

Figure 7. The upper spectrum is the *in situ* FT-IR spectra of the cTiNT-O_2-H_2 after exposure to D_2O vapor for 10 min at room temperature. The lower spectrum is of the same sample after exposure to vacuum at room temperature. The background for these spectra is the c-TiNT-O_2-H_2s.

Figure 8 shows the spectrum obtained on exposure of the reduced cTiNTs to 20 Torr of CO_2. The very intense absorption centered at 2350 cm^{-1} is due to the asymmetric stretch of CO_2 and the absorptions near ~3610 and ~3715 cm^{-1} are due to combination bands of CO_2 [25]. Carbonates and bicarbonates have characteristic absorptions below ~1700 cm^{-1} [26]. As shown in the insert, exposure to CO_2 also leads to a small "negative" absorption in the OH stretching region. The nature of these absorptions along with the absorption at ~1718 cm^{-1} is discussed in more detail below.

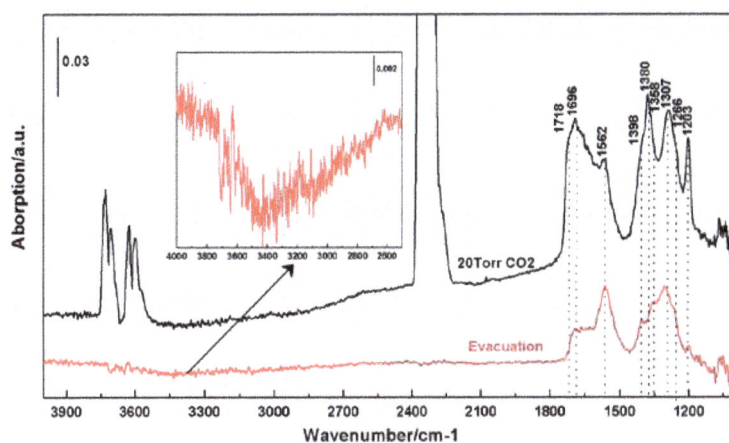

Figure 8. The upper trace is an *in situ* FT-IR spectrum of the cTiNT-O_2-H_2 exposed to 20 Torr of CO_2 for 30 min at room temperature. The lower trace is the same sample after exposure to vacuum at room temperature. The background for these spectra is the c-TiNT-O_2-H_2. The insert is a blow-up of the 4000 to ~2400 cm^{-1} region.

3. Discussion

3.1. Structure of the NTs

It is clear from the XRD results that as the TiNTs are annealed to 650 °C, the anatase TiNTs are partially transformed to the rutile phase. The TEM results show that the tubular morphology is completely lost upon calcining at 650 °C. In contrast to XPS data showing significant changes in the electronic environment around the Ti and O sites on oxidation and/or reduction of the TiNTs heated to 350 °C [12], the XPS spectra shown in Figure 5 reveal that there are no significant changes in the electronic environment around the Ti and O surface sites on oxidation and/or reduction of the cTiNT. All of the XPS spectra for the cTiNTs exhibit Ti $2p_{3/2}$ and Ti $2p_{1/2}$ peaks at 458.6 and 464.3 eV, respectively. They also have very similar O 1s XPS spectra, which are not affected by further oxidation or reduction. Each spectrum shows an O 1s XPS peak at 531.6 eV that is typical of bulk anatase TiO_2 samples. These O1s peaks [27] are symmetric and possess a smaller FWHM than seen for the TiNT samples that were calcined at 350 °C [12]. The narrowing and the symmetry of this peak are likely the result of either fewer surface –OH groups, O defect sites, non-stoichiometric O sites, or some combination of the three [23,27]. One of the more interesting observations with the TiNT samples is that reduction subsequent to oxidation leads to the appearance of a XPS peak that is characteristic of Ti^{3+} [12]. No Ti^{3+} signals are seen for any treatment of the cTiNT samples that we have studied.

3.2. Interactions of H_2O and D_2O with c-TiNT

Titania nanotubes are formed from the dehydration of titanates, such as $H_2Ti_2O_5$, that is driven by calcination at elevated temperatures. Figure 6 shows the IR spectrum of cTiNT (calcined at 650 °C in an oven, allowed to cool to room temperature, and then transferred to our IR cell). Traces are shown for the cTiNTs after being transferred into the cell, after 2 h under vacuum at room temperature, after oxidation by heating in oxygen at 350 °C for 3 h, and after reduction in hydrogen at 350 °C for 3 h (subsequent to oxygen pretreatment and evacuation of oxygen). The high energy portion of the four spectra are dominated by a broad absorption starting near ~2700 cm^{-1} with a feature with maximum amplitude near 2950 cm^{-1} and another broad feature that peaks in the 3300 to 3400 cm^{-1} region.

There have been many studies of water adsorption on metal oxides and on titania in particular. Suda and Morimoto report three features above 3000 cm^{-1} on rutile titania powder [28]. One is a sharp absorption near 3660 cm^{-1}, which is split into a doublet at elevated temperature. This feature is assigned to surface OH groups, and the two components of the peak are attributed to monodentate and bidentate OH binding. The remainder of the absorption above 3000 cm^{-1} is assigned to surface-bound water. They assign the feature at 3520 cm^{-1} to hydrogen-bonded OH groups and the broad feature at ~3400 cm^{-1} to molecular water. The interaction of water with a thin titania film was studied by Nakamura et al., and a broad absorption due to physisorbed water is reported centered at 3500 cm^{-1} [29]. This water can be removed by evacuation at room temperature. More strongly absorbed water is reported to be centered at 3270 cm^{-1}. This assignment parallels those for other metal oxides where physisorbed water is reported at higher frequencies than more strongly adsorbed water and both absorb above 3000 cm^{-1} [24]. Ti-OH absorptions, if present, are typically reported around 3700 cm^{-1} [24]. Interestingly, a spectrum reported by Tsuchiya *et al.* [30] for anodic titania NTs shows a broad absorption centered at 3450 cm^{-1} that extends below 3000 cm^{-1}. Chen *et al.* [5] also present a spectrum of titania NTs that shows a broad absorption from just above 2500 cm^{-1} to ~ 3400 cm^{-1}

The spectra reported by Tsuchiya *et al.* [30] and Chen *et al.* [5] are qualitatively similar in extent and shape to the water absorption shown in Figure 6. There is no absorption characteristic of OH groups in Figure 6 (or in the two references cited above). This is consistent with the removal of OH groups from the surface when the TiNTs are heated to 650 °C. Presumably this can occur as a result of a reaction to form water. This water then either desorbs into the gas phase or adsorbs on the cTiNTs. Water can also adsorb on the cTiNTs as they cool in air. Gas-phase water has its asymmetric stretching vibration centered at 3756 cm^{-1}, its symmetric vibration centered at 3651 cm^{-1}, and its bending mode centered at 1595 cm^{-1} [25]. The manifold of small peaks between 3900 and 3500 cm^{-1} and between 1400 and 1900 cm^{-1} in Figure 6 are due to absorptions of gas-phase water. These absorptions are likely due to small changes in water content in the purged spectrometer housing taking place over the timescale of hours. Inspection of Figure 6 clearly demonstrates that adsorbed water absorbs at a lower frequency than gas-phase water. The observation that there is little change in the shape of the higher energy portion of the water absorption upon evacuation at an elevated temperature (350 °C) is consistent with the dominant water species that is present being strongly adsorbed water, which absorbs at a lower frequency than weakly bound water [24]. We now turn to the lower frequency portion of the absorption in this region.

Absorptions at lower frequencies than seen for adsorbed water on anatase powder or thin films were noted in the literature [5,30] and in a prior study of TiNTs in our group [31]. In that study, TiNTs decorated with Pt nanoparticles were exposed to supra-bandgap light [31]. Pt nanoparticles can act as a "sink" for photo-generated electrons, extending the lifetime of photo-generated holes. The longer hole lifetime leads to greater oxidation of water. On illumination of the platinized TiNTs, the intensity in the region near 3000 cm^{-1} decreases below the baseline. Again, we note that a "negative absorption" is characteristic of a depletion of a species that is present in the background spectrum. This depletion on illumination of the platinized TiNTs is preferentially in the low frequency region of the water absorption and suggests that this lower frequency region is dominated by a different type of water than water absorbing at higher frequencies. Since the holes are primarily expected to be generated within the Pt-TiNTs, the water that they react with leading to preferential depletion of the low frequency region of the absorption near 3000 cm^{-1} was referred to as "structural water". This term is meant to indicate water within the structure of the TiNT, where it may be more proximate to the location of hole generation than surface water. The depletion of structural water is accompanied by an increase in the absorption at higher frequencies as more surface-bound water is produced as a result of H and OH thus generated with other OH groups [31]. Thus, our findings are consistent with previous work and suggest that there are two different OH stretching regions.

We note that the shape of the spectrum in the water region in Figure 6 is also consistent with two absorption regions: one that dominates the high frequency region and the other that dominates the low frequency region. The lower frequency band has a maximum near 2900 cm^{-1} and a relatively sharp drop-off near 3000 cm^{-1}. The partitioning of amplitude between these two bands is affected by the treatment and history of the cTiNTs. Though we have not studied this effect in detail, as discussed below, there are reports in the literature that protons are incorporated in the TiNT structure. This is expected based on the exchange of Na$^+$ with protons during the TiNT synthesis. This could result in a change in the effective pH of the interior of the TiNTs and lead to a change in the hydrophilicity of the interior, making it a more favorable environment for "structural water". On collapse, some of this water and/or protons could be trapped in the interior of the cTiNTs.

The data in Figure 7 show the spectrum of the collapsed oxidized cTiNTs after exposure to gas-phase D$_2$O. New absorptions as a result of interaction with D$_2$O are seen in the 1950–2700 cm^{-1} region and loss of absorption is seen in the 2700–3500 cm^{-1} region. It is difficult to provide an exact delineation of the extent of either region because of uncertainties in the baseline. However, a "negative absorption" is seen between ~3550 and ~2700 cm^{-1}. These spectra are referenced against a background spectrum of the cTiNT-O$_2$-H$_2$s. A negative absorption indicates that a species that was present in the background is no longer present in the displayed spectrum. New absorptions are seen between ~2700 and ~1950 cm^{-1}. A new absorption with rotational structure is seen between ~2500 and ~2900 cm^{-1}. The asymmetric stretching mode of gas-phase D$_2$O is centered at 2789 cm^{-1} and the symmetric stretch is centered at 2666 cm^{-1}. A Q branch is apparent at 2789 cm^{-1}. Thus, the new absorption with rotational structure centered at ~2789 cm^{-1} is due to stretching modes of gas-phase D$_2$O. The ratio of the frequencies of the vibrational modes of gas-phase D$_2$O and their protonated counterparts is calculated from their spectra to be between ~0.73 and 0.74 [25]. A value of ~0.73 would be predicted from the ideal harmonic oscillator model for an OH stretch. There is a wider spread in these ratios for water absorptions in solid hydrates [32]. The ratio of the frequencies of the high and low energy edges of the loss in absorbance (negative absorption) and the new absorption is ~0.76 and ~0.72. Given the uncertainties as to where the two regions begin and end, and potential perturbations due to the difference in the environment in these experiments and the gas phase, we conclude that these frequency ratios are consistent with deuterated water exchanging with protonated water. The fact that the entire absorption band above 2700 cm^{-1} is depleted on interaction with D$_2$O is strong evidence that this absorption is due to moieties whose protons are readily exchanged by exposure to D$_2$O and, thus, this observation is consistent with absorptions being due to OH stretching modes.

The question remains, though, why do we see absorptions from OH stretching modes at a significantly lower frequency than the OH stretches in water adsorbed on the surface of titania powders? We do not have a definitive answer, but there is considerable evidence in the literature that interactions in the solid state can lead to shifts in OH stretching frequencies. The OH and OD stretching frequencies for solid hydrates can be significantly red-shifted [32]. Hydrogen bonding can lead to the appearance of an absorption that is at a lower frequency than a typical OH absorption. The length of the hydrogen bond has been correlated with its absorption frequency, and has been reported to be as low as 2000 cm^{-1} [33]. The exact structure of TiNTs is still unresolved, and may be influenced by the synthesis procedure and subsequent treatments. The cTiNTs have been subjected to much less detailed study. However, evidence has been presented in the literature to show that TiNTs can contain water and protons [34,35]. This opens up the possibility for hydrogen bonding within the structure of TiNTs. The structure of the cTiNTs is different than those of the multi-walled TiNTs; however, the similarity between what we observe and what is observed in prior studies [5,30] suggests similar factors leading to a lower frequency band in the region of the spectrum in which OH absorptions appear. This hydrogen bonding would involve water and/or protons. It is possible that the cTiNTs trap water and/or protons within their structure on collapse of the TiNTs. It is also likely that the cTiNTs adsorb atmospheric water as they cool after being collapsed in an oven.

3.3. Interactions of CO_2 with cTiNT

Figure 8 shows the spectrum taken after exposure of the cTiNT-O_2-H_2 to CO_2. The new absorptions in the 1700 to 1200 cm^{-1} region are characteristic of carbonates and bicarbonates [12,26,31]. Though the exact nature of the observed carbonate or bicarbonate species is not of major significance in the current study, for completeness we provide the assignments we have deduced for the observed absorptions with the caveat that some of these assignments are still a subject of discussion [12,26,31].

- Bidentate carbonate (in cm^{-1}): 1690, 1562, 1380, 1358, 1307
- Monodentate carbonate (in cm^{-1}): 1266
- Bicarbonate (in cm^{-1}): 1404, 1398, 1203
- The 1290 cm^{-1} absorption is likely a convolution of peaks including a bidentate carbonate reported at 1278 cm^{-1}.

We note that many of the absorption features are lost on evacuation, which is consistent with weakly bound species. The peak frequencies of the remaining absorptions may then shift slightly due to overlap with other absorptions that disappeared on evacuation. Evacuation results in bidentate carbonate absorptions at 1563, 1358, and 1307 cm^{-1}, a bicarbonate at 1404 cm^{-1}, and a monodentate carbonate at 1266 cm^{-1}. The absorption at 1718 cm^{-1} is too high in energy to be due to a carbonate or bicarbonate. It is in a region in which carbonyl stretching absorptions can appear [36]. The carbonyl stretch of gas-phase formic acid absorbs at 1740 cm^{-1} [25] and an absorption due to formic acid has been assigned previously for an uncollapsed platinized TiNT at ~1725 cm^{-1} [31]. Thus, we assign the 1718 cm^{-1} absorption to the carbonyl stretch of formic acid. The observation of formic acid is interesting since there are reports of dissociative adsorption of formic acid to formate for anatase titania, particularly for the (101) plane [37]. A search of the literature does not provide a definitive explanation for our observation of formic acid. However, we note that formate and formic acid are expected to be in equilibrium. The observation of formate on exposure to formic acid suggests that in those systems, the equilibrium is heavily shifted toward formate. Since titania provides an acidic surface [12] and our infrared studies provide evidence for the presence of protons and/or water, we hypothesize that in our system, the presence of proton sources is sufficient to shift the equilibrium for the nanotubes towards formic acid. Other factors could also contribute to our observation of formic acid. One suggested pathway for formic acid formation involving isolated hydroxyls [38] would not be expected to be significant in our system due to the absence of isolated OH moieties. It is also possible that the efficient sites for dissociation of formic acid in the presence of water, such as the (101) plane [37,39], have been annealed as a result of high temperature calcination of the samples under study.

Clearly there is less adsorption due to carbonates and bicarbonates than seen with the uncollapsed TiNTs. A decrease in absorbance would be expected due to a decrease in surface area of the cTiNTs *versus* the uncollapsed TiNTs. There is also a difference in the pattern of the absorptions seen relative to that for the reduced TiNTs. However, what is perhaps most surprising is the degree of similarity in the absorptions between the similarly treated collapsed and uncollapsed reduced TiNTs, though there are more species absorbing for the reduced uncollapsed TiNTs. The observation of formic acid suggests that these cTiNT-O_2-H_2s are capable of inducing C-H bond formation. We note that similarly prepared cTiNTs have been shown to be photochemically active for carbon dioxide reduction, albeit at much lower yields [10]. How, then, do we rationalzie this qualitative similarity in adsorption behavior but disparity in photoreduction of CO_2? As shown by XPS results, defects on cTiNTs are substantially less than on the TiNTs. Defects such as oxygen vacancies and associated Ti^{3+} sites promote significant surface carboxylate formation due to CO_2 adsorption, which is, in turn, postulated to be a critical intermediate in the photoconversion of CO_2 [12].

Figure 8 shows a blow-up of the 1800–4000 cm^{-1} region which exhibits a weak "negative absorption" between ~2500 and ~3500 cm^{-1}. Bicarbonate formation must involve the interaction of CO_2 with an H-containing moiety. The negative absorption due to the depletion of moieties present in the background spectrum is consistent with the formation of bicarbonates and indicates a depletion of surface adsorbed water. The lower frequency OH stretch region is due to OH stretches of H-containing moieties within the cTiNTs. It is also possible that CO_2 and water compete for the same surface sites and this competition leads to the depletion of OH bond-containing moieties that absorb between ~2500 and 3500 cm^{-1}.

4. Experimental Section

4.1. Synthesis of TiNT, cTiNT, and in Situ Pretreatments

Titania nanotubes were prepared by a modified hydrothermal method that has been previously reported [1,2,40]. In a typical experiment, 2 g of anatase titania powder (purity 99%, Sigma Aldrich Chemicals, St. Louis, MO, USA) were stirred with 50 mL of 10 M NaOH solution (purity 97%, BDH Chemicals, Radnor, PA, USA) in a closed 125 mL Teflon cup. The Teflon cup is sealed inside a stainless steel outer vessel and placed in an oven for 48 h at 120 °C. The resulting precipitate was washed with 1 M HCl (purity 38%, EMD Chemicals, Billerica, MA, USA) followed by several washings with deionized water to attain a pH between 6 and 7. The TiNT powder thus formed was dried overnight in an oven held at 110 °C. Some of the TiNT samples were then calcined at a temperature of 650 °C to probe the effect of temperature on the morphology of these materials. The TiNTs were initially heated *in situ* under an O_2 atmosphere at 350 °C [O_2 ~2 Torr] to clean the surface, and were then allowed to cool to room temperature under vacuum. The TiNTs that are only heated to 350 °C are labeled by the pretreatment (*i.e.*, -TiNT-O_2 for oxidized TiNTs) while those heated to 650 °C are labeled as cTiNT and the pretreatment (*i.e.*, -cTiNT-O_2). After oxidative pretreatment, some of the TiNTs were treated in a H_2 atmosphere at 350 °C [H_2 ~2 Torr] for 3 h and then allowed to cool to room temperature under vacuum. The TiNTs heated to 350 °C that were reduced are referred to as TiNT-O_2-H_2 while those heated to 650 °C and then reduced after oxidation are referred to as cTiNT-O_2-H_2.

4.2. Characterization of the TiNTs

The morphology of titania nanomaterials was probed by transmission electron microscopy (TEM, STEM JEOL-2100F, Peabody, MA, USA), with an accelerating voltage of 200 kV. To establish crystallinity and phase purity, the powder X-ray diffraction (XRD) patterns of the TiNT and the cTiNT were recorded on a Rigaku Domex diffractometer (The Woodlands, TX, USA) using Cu Kα radiation, a continuous scan, and a scintillation-type detector for 2θ from 5° to 90°. X-ray photoelectron spectra (XPS) were obtained with an Omicron (Houston, TX, USA) ESCA-2000-125-based spectrometer using an Al Kα radiation source (1486.6 eV, 30 mA × 8 kV). These spectra provided data on the oxidation states of the ions in the TiNT samples. The C 1s response at 284.6 eV was used as an internal reference for the absolute binding energy.

4.3. In Situ FT-IR Spectroscopy

In situ FT-IR spectra were recorded with a Nicolet (Madison, WI, USA) 6700 FTIR spectrometer equipped with both a mercury cadmium telluride (MCT) and a DTGS (Deuterated Triglycine Sulfate) detector. Each spectrum was obtained by averaging 64 scans at a resolution of 4 cm^{-1}. The custom fabricated infrared cell, which was designed to study highly scattering powder samples in a transmission mode, has been described previously [31]. Briefly, it consists of a stainless steel cube with two CaF$_2$ windows positioned on opposite sides of the cube. For this study, samples were pressed onto a highly transmissive tungsten wire grid held between two nickel jaws. The grid, which is resistively heated to a temperature measured by a Chromel-Alumel thermocouple attached to its center, provides a support for highly scattering samples so that very thin samples can be studied in transmission mode. The vacuum system was pumped using a Turbo pump backed by a mechanical forepump to achieve a base pressure of 1×10^{-5} Torr. The infrared beam was directed out of the spectrometer, allowed to pass through the cell windows and the sample on the wire grid, and was detected with the MCT detector. The cell and detector were contained in an enclosure that was purged with boiled off nitrogen before acquisition of spectra. Unless otherwise stated, background spectra are of the samples cooled to ambient temperature, under vacuum, after pretreatment. A Baratron capacitance manometer was used to monitor the pressure of CO$_2$. D$_2$O, used in some of the IR experiments, was specified as 99% atomic purity.

5. Conclusions

This research probes the interaction of water on the cTiNTs and we propose an explanation for the unusual low frequency OH stretching region. On exposure to water the OH stretching region of c-TiNT-O_2, illustrated in Figure 6, is broader than that reported for titania powder and what is typical for other metal oxides [41,42], extending below 3000 cm^{-1}. Interaction with D$_2$O demonstrates that the OH region contains exchangeable protons, and absorptions due to the deuterated analogs of the species are observed with absorptions between ~2700 and ~1950 cm^{-1}. Taking into account that variations in the baseline of the spectra produce uncertainty in the extent of these absorptions, the shifts in the absorptions on deuteration are consistent with expectations based on typical frequency ratios for OH *vs.* OD stretches. Based on these observations and data in the literature, we suggest that the unusually low frequency OH stretching modes are best explained by hydrogen bonding involving H-containing moieties within the structure of the TiNTs. It is recognized that the addition of water can increase the efficiency of photoinduced reduction reactions taking place on titania [43], and proton sources are needed for the formation of hydrocarbon products. We note that the present results suggest that proton sources can be sequestered on and within the structure of the TiNTs and c-TiNTs

Exposure of the c-TiNT-O_2-H_2 to CO_2 leads to the formation of new species on the c-TiNTs, which are identified as carbonates and bicarbonates, as well as formic acid. Though there are differences in the number and nature of the carbon-containing species observed on the surface of the c-TiNT-O_2-H_2 *versus* the TiNT-O_2-H_2, many moieties are observed on both types of TiNTs. The large decrease in surface area that occurs on the collapse of the TiNTs to c-TiNTs would be expected to contribute to a smaller amount of surface carbonates and bicarbonates. However, the fact that many of the surface moieties are the same for the c-TiNT-O_2-H_2 and the TiNT-O_2-H_2 suggests that many of the sites of interaction on TiNT-O_2-H_2 remain on c-TiNT-O_2-H_2, but despite adsorption, CO_2 photoreduction is low on the cTiNTs. We suggest that adsorption mediated by specific defects (O-vacancy and Ti^{3+}), which are largely absent on cTiNT, is required for the significant photoreduction of CO_2. An implication of these findings, then, is that a material's defect structure is crucial for CO_2 photoconversion.

Acknowledgments

This work was supported by the Chemical Sciences, Geosciences, and Biosciences Division, Office of Basic Energy Sciences, Office of Science, U.S. Department of Energy (Award No. DE-FG02-03-ER15457). The XPS work was performed in the Keck-II facility of NUANCE Center at Northwestern University. NUANCE Center is supported by NSF-NSEC, NSF-MRSEC, Keck Foundation, the State of Illinois, and Northwestern University.

Author Contributions

K. Bhattacharyya, W. Wu, and B. K. Vijayan synthesized the materials and conducted the experimental work. The research and manuscript preparation were supervised by E. Weitz and K.A. Gray.

Conflicts of Interest

The authors declare no conflict of interest.

References

1. Kasuga, T.; Hiramatsu, M.; Hoson, A.; Sekino, T.; Niihara, K. Titania Nanotubes Prepared by Chemical Processing. *Adv. Mater.* **1999**, *11*, 1307–1311.

2. Kasuga, T.; Hiramatsu, M.; Hoson, A.; Sekino, T.; Niihara, K. Formation of titanium oxide nanotube. *Langmuir* **1998**, *14*, 3160–3163.

3. Du, G.H.; Chen, Q.; Che, R.C.; Yuan, Z.Y.; Peng, L.M. Preparation and structure analysis of titanium oxide nanotubes. *Appl. Phys. Lett.* **2001**, *79*, 3702–3704.

4. Chen, Q.; Zhou, W.Z.; Du, G.H.; Peng, L.M. Trititanate nanotubes made via a single alkali treatment. *Adv. Mater.* **2002**, *14*, 1208–1211.

5. Chen, Q.; Du, G.H.; Zhang, S.; Peng, L.M. The structure of trititanate nanotubes. *Acta Crystallogr. Sect. B Struct. Sci.* **2002**, *58*, 587–593.

6. Nakahira, A.; Kato, W.; Tamai, M.; Isshiki, T.; Nishio, K.; Aritani, H. Synthesis of nanotube from a layered $H_2Ti_4O_9 \cdot H_2O$ in a hydrothermal treatment using various titania sources. *J. Mater. Sci.* **2004**, *39*, 4239–4245.

7. Poudel, B.; Wang, W.Z.; Dames, C.; Huang, J.Y.; Kunwar, S.; Wang, D.Z.; Banerjee, D.; Chen, G.; Ren, Z.F. Formation of crystallized titania nanotubes and their transformation into nanowires. *Nanotechnology* **2005**, *16*, 1935–1940.

8. Tsai, C.C.; Teng, H.S. Regulation of the physical characteristics of Titania nanotube aggregates synthesized from hydrothermal treatment. *Chem. Mater.* **2004**, *16*, 4352–4358.

9. Yoshida, R.; Suzuki, Y.; Yoshikawa, S. Effects of synthetic conditions and heat-treatment on the structure of partially ion-exchanged titanate nanotubes. *Mater. Chem. Phys.* **2005**, *91*, 409–416.

10. Vijayan, B.; Dimitrijevic, N.M.; Rajh, T.; Gray, K. Effect of Calcination Temperature on the Photocatalytic Reduction and Oxidation Processes of Hydrothermally Synthesized Titania Nanotubes. *J. Phys. Chem. C* **2010**, *114*, 12994–13002.

11. Suzuki, Y.; Yoshikawa, S. Synthesis and thermal analyses of TiO_2-derived nanotubes prepared by the hydrothermal method. *J. Mater. Res.* **2004**, *19*, 982–985.

12. Bhattacharyya, K.; Danon, A.; Vijayan, B.K.; Gray, K.A.; Stair, P.C.; Weitz, E. Role of the Surface Lewis Acid and Base Sites in the Adsorption of CO_2 on Titania Nanotubes and Platinized Titania Nanotubes: An in Situ FT-IR Study. *J. Phys. Chem. C* **2013**, *117*, 12661–12678.

13. Zhang, J.; Li, M.J.; Feng, Z.C.; Chen, J.; Li, C. UV Raman spectroscopic study on TiO_2. I. Phase transformation at the surface and in the bulk. *J. Phys. Chem. B* **2006**, *110*, 927–935.

14. Zhang, H.Z.; Banfield, J.F. Thermodynamic analysis of phase stability of nanocrystalline titania. *J. Mater. Chem.* **1998**, *8*, 2073–2076.

15. Muscat, J.; Swamy, V.; Harrison, N.M. First-principles calculations of the phase stability of TiO_2. *Phys. Rev. B* **2002**, *65*, 22412–22415.

16. Kumar, K.N.P. Growth of Rutile Crystallites during the Initial-Stage of Anatase-to-Rutile Transformation in Pure Titania and in Titania-Alumina Nanocomposites. *Scr. Metall. Mater.* **1995**, *32*, 873–877.

17. Ovenstone, J.; Yanagisawa, K. Effect of hydrothermal treatment of amorphous titania on the phase change from anatase to rutile during calcination. *Chem. Mater.* **1999**, *11*, 2770–2774.

18. Regonini, D.; Jaroenworaluck, A.; Stevens, R.; Bowen, C.R. Effect of heat treatment on the properties and structure of TiO$_2$ nanotubes: Phase composition and chemical composition. *Surf. Interface Anal.* **2010**, *42*, 139–144.

19. Schulte, K.L.; DeSario, P.A.; Gray, K.A. Effect of crystal phase composition on the reductive and oxidative abilities of TiO$_2$ nanotubes under UV and visible light. *Appl. Catal. B Environ.* **2010**, *97*, 354–360.

20. Albu, S.P.; Ghicov, A.; Aldabergenova, S.; Drechsel, P.; LeClere, D.; Thompson, G.E.; Macak, J.M.; Schmuki, P. Formation of Double-Walled TiO$_2$ Nanotubes and Robust Anatase Membranes. *Adv. Mater.* **2008**, *20*, 4135–4139.

21. Bhattacharyya, K.; Varma, S.; Tripathi, A.K.; Bharadwaj, S.R.; Tyagi, A.K. Effect of Vanadia Doping and Its Oxidation State on the Photocatalytic Activity of TiO$_2$ for Gas-Phase Oxidation of Ethene. *J. Phys. Chem. C* **2008**, *112*, 19102–19112.

22. Sodergren, S.; Siegbahn, H.; Rensmo, H.; Lindstrom, H.; Hagfeldt, A.; Lindquist, S.E. Lithium intercalation in nanoporous anatase TiO$_2$ studied with XPS. *J. Phys. Chem. B* **1997**, *101*, 3087–3090.

23. Li, J.; Zeng, H.C. Preparation of monodisperse Au/TiO$_2$ nanocatalysts via self-assembly. *Chem. Mater.* **2006**, *18*, 4270–4277.

24. Finnie, K.S.; Cassidy, D.J.; Bartlett, J.R.; Woolfrey, J.L. IR Spectroscopy of Surface Water and Hydroxyl Species on Nanocrystalline TiO$_2$ Films. *Langmuir* **2001**, *17*, 816–820.

25. Herzberg, G. *Infrared and Raman Spectra*; Van Nostrand Rheinhold Company Inc: New York, NY, USA, 1945.

26. Busca, G.; Lorenzelli, V. Infrared Spectroscopic Identification os Species arising from reactive adsorption of carbon oxides on metal-oxide surfaces. *Mater. Chem.* **1982**, *7*, 89–126.

27. Li, J.; Tang, S.; Lu, L.; Zeng, H.C. Preparation of nanocomposites of metals, metal oxides, and carbon nanotubes via self-assembly. *J. Am. Chem. Soc.* **2007**, *129*, 9401–9409.

28. Suda, Y.; Morimoto, T. Molecularly Adsorbed H$_2$O on the Bare Surface of TiO$_2$ (Rutile). *Langmuir* **1987**, *3*, 786–788.

29. Nakamura, R.; Ueda, K.; Sato, S. *In Situ* Observation of the Photoenhanced Adsorption of Water on TiO$_2$ Films by Surface-Enhanced IR Absorption Spectroscopy. *Langmuir* **2001**, *17*, 2298–2300.

30. Tsuchiya, H.; Macak, J.M.; Müller, L.; Kunze, J.; Müller, F.; Greil, P.; Virtanen, S.; Schmuki, P. Hydroxyapatite growth on anodic TiO$_2$ nanotubes. *J. Biomed. Mater. Res. Part A* **2006**, *77A*, 534–541.

31. Wu, W.; Bhattacharyya, K.; Gray, K.; Weitz, E. Photoinduced Reactions of Surface-Bound Species on Titania Nanotubes and Platinized Titania Nanotubes: An *in Situ* FTIR Study. *J. Phys. Chem. C* **2013**, *117*, 20643–20655.

32. Berglund, B.; Lindgren, J.; Tegenfeldt, J. O-H and O-D Stretching Vibrations in Isotopically Dilute HDO Molecules in Some Solid Hydrates. *J. Mol. Struct.* **1978**, *43*, 169–177.

33. Nakamoto, K.; Margoshes, M.; Rundle, R.E. Stretching Frequencies as a Function of Distances in Hydrogen Bonds. *J. Am. Chem. Soc.* **1955**, *77*, 6480–6486.

34. Ferrari, A.M.; Lessio, M.; Szieberth, D.; Maschil, L. On the Stability of Dititanate Nanotubes: A Density Functional Theory Study. *J. Phys. Chem. C* **2010**, *114*, 21219–21225.

35. Izawa, H.; Kikkawa, B.; Koizumi, M. Ion Exchange and Dehydration of Layered Titanates: $Na_2Ti_3O_7$ and $K_2Ti_4O_9$. *J. Phys. Chem.* **1982**, *86*, 5023–5026.

36. Banwell, C.N. *Fundamentals of Molecular Spectroscopy*; McGraw Hill Book Company Ltd: London, UK, 1972.

37. Miller, K.L.; Faconer, J.L.; Medin, J.W. Effect of water on the adsorbed structure of formic acid on TiO_2 anatase (101). *J. Catal.* **2011**, *278*, 321–328.

38. Nanayakkara, C.E.; Dillon, J.K.; Grassian, V.H. Surface Adsorption and Photochemistry of Gas-Phase Formic Acid on TiO_2 Nanoparticles: The Role of Adsorbed Water in Surface Coordination, Adsorption Kinetics, and Rate of Photoproduct Formation. *J. Phys. Chem. C* **2014**, *118*, 25487–25495.

39. Kim, S.Y.; van Duin, A.C.T.; Kubicki, J.D. Molecular dynamics simulations of the interactions between TiO_2 nanoparticles and water with Na^+ and Cl^-, methanol, and formic acid using a reactive force field. *J. Mater. Res.* **2012**, *28*, 513–520.

40. Baiju, K.V.; Shukla, S.; Biju, S.; Reddy, M.L.P.; Warrier, K.G.K. Hydrothermal processing of dye-adsorbing one-dimensional hydrogen titanate. *Mater. Lett.* **2009**, *63*, 923–926.

41. Chung, J.S.; Miranda, R.; Bennett, C. Study of Methanol and Water Chemisorbed on Molybdenum Oxide. *J. Chem. Soc. Faraday Trans.* **1985**, *81*, 19–36.

42. Hind, A.A.; Grassian, V.H. FT-IR Study of Water Adsorption on Aluminum Oxide Surfaces. *Langmuir* **2003**, *19*, 341–347.

43. Dimitrijevic, N.D.; Vijayan, B.K.; Poluektov, O.G.; Rajh, R.; Gray, K.A.; He, H.; Zapol, P. Role of Water and Carbonates in Photocatalytic Transformation of CO_2 to CH_4 on Titania. *J. Am. Chem. Soc.* **2011**, *133*, 3964–3971.

Sample Availability: The methods by which material samples were synthesized are available from the authors.

Chapter 2:
UV and Visible-Light Sensitive Photocatalysts:
Efficiency Effects of Nature, Composition,
Preparation, Structure and Texture

Structural Formation and Photocatalytic Activity of Magnetron Sputtered Titania and Doped-Titania Coatings

Peter J. Kelly, Glen T. West, Marina Ratova, Leanne Fisher, Soheyla Ostovarpour and Joanna Verran

Abstract: Titania and doped-titania coatings can be deposited by a wide range of techniques; this paper will concentrate on magnetron sputtering techniques, including "conventional" reactive co-sputtering from multiple metal targets and the recently introduced high power impulse magnetron sputtering (HiPIMS). The latter has been shown to deliver a relatively low thermal flux to the substrate, whilst still allowing the direct deposition of crystalline titania coatings and, therefore, offers the potential to deposit photocatalytically active titania coatings directly onto thermally sensitive substrates. The deposition of coatings via these techniques will be discussed, as will the characterisation of the coatings by XRD, SEM, EDX, optical spectroscopy, *etc.* The assessment of photocatalytic activity and photoactivity through the decomposition of an organic dye (methylene blue), the inactivation of *E. coli* microorganisms and the measurement of water contact angles will be described. The impact of different deposition technologies, doping and co-doping strategies on coating structure and activity will be also considered.

Reprinted from *Molecules*. Cite as: Kelly, P.J.; West, G.T.; Ratova, M.; Fisher, L.; Ostovarpour, S.; Verran, J. Structural Formation and Photocatalytic Activity of Magnetron Sputtered Titania and Doped-Titania Coatings. *Molecules* **2014**, *19*, 16327-16348.

1. Introduction

Photocatalytic titania-based surfaces and coatings have many potential applications, including "self-cleaning" windows, anti-fogging screens or lenses, air cleaning and water purification devices and "self-sterilizing" antibacterial tiles [1–6]. Although it is relatively straightforward to demonstrate the effectiveness of these coatings in a laboratory environment, producing highly photoactive coatings in a commercially viable process is more challenging, and this has limited the exploitation of this technology to date. Titania can be produced in nanoparticle form for incorporation into paints and other building products [7], or as slurries and suspensions for water treatment [8,9]. Whilst the latter arrangement provides high surface areas of active material, there is usually a requirement for downstream filtration of the particles, limiting its practicality. In other applications, such as windows, lenses or tiles, a titania thin film or coating is the preferred option, where the reduced surface area is compensated for by high transparency and durability.

There are a number of physical and chemical deposition techniques that can be used to produce titania and doped-titania coatings. These include pulsed laser deposition [10], magnetron sputtering [11–13], reactive evaporation [13], ion beam assisted deposition [14], chemical vapour deposition [15], sol-gel [16], dip-coating [17], hydrothermal synthesis [18] and atomic layer deposition [19,20]. The characteristics of each process have a major bearing on deposition parameters, such as substrate temperature (and thereby, choice of substrate material) and throughput and coating properties, such as adhesion, crystallinity, grain size, lattice defects, transparency and surface roughness, and in general, the performance of the coating is inextricably linked to the choice of deposition process. The production of photoactive titania coatings is further complicated by the requirement for the coating to be predominantly in the anatase crystal form (mixed phase anatase/rutile structures have also been reported as being effective [10,20]). Titania coatings deposited at ambient temperature tend to be amorphous [12], though and the formation of anatase structures usually requires elevated temperatures (~400 °C) during deposition or post-deposition annealing, which imposes additional processing costs and restricts the use of thermally sensitive substrate materials.

Of the deposition techniques available, magnetron sputtering is widely used for the production of high quality coatings for applications ranging from Low-E and solar control glazing products, tool coatings, micro- and opto-electronic components, data storage media and thin film photovoltaics. Indeed, the scalability and versatility of the magnetron sputtering process and the uniformity and repeatability of the resulting coatings has made this the process of choice for many commercial applications [21].

The magnetron sputtering process has been described in detail elsewhere [21] and the finer nuances of magnetron design and process control are beyond the scope of this paper. In simple terms, though, it is a physical vapour deposition process in which positively charged ions from a glow discharge plasma are accelerated towards a negatively biased target plate of the material to be deposited, which is mounted on the magnetron body. The incident ions remove or "sputter" atoms from the surface of the target through a momentum exchange mechanism. The process takes place in a reduced pressure (typically 0.1 to 0.5 Pa) atmosphere, usually of argon, in which the plasma can be readily maintained. The sputtered atoms diffuse across the chamber and condense on the substrate as a thin film. Reactive gases, such as oxygen or nitrogen can be introduced with the argon in order to form compound films of oxides or nitrides. However, during the deposition of dielectric materials, such as oxides, the build-up of positive charges on the target can result in arc events, which are detrimental to the stability of the process and the quality of the coating. This problem can be negated by powering the magnetron in the mid-frequency (20–350 kHz) pulsed DC mode, where the polarity of the target alternates rapidly between positive and negative voltages. Again, this process has been described elsewhere [22].

Another variant of pulsed sputtering is the recently introduced HiPIMS (high power impulse magnetron sputtering) technique, which utilises lower pulse frequencies (50–1000 Hz), higher peak voltages (−500 to −1000 V) and very high peak currents (up to 1000 A). This results in similar time-averaged powers, but at much lower duty cycles, compared to pulsed DC magnetron sputtering, giving very high current densities at the target and leading to significant ionisation of the deposition flux. HiPIMS has been reported to enhance the film structure and make possible the deposition of crystalline thin films, including titania, without additional heat treatment [23,24]. Furthermore, the present authors have demonstrated that the thermal energy flux delivered to the substrate during HiPIMS deposition is several times lower than for DC or pulsed DC magnetron sputtering at the same time-averaged power [25]. This work was extended to demonstrate for the first time that photocatalytically active titania coatings can be deposited directly onto polymeric substrates by HiPIMS in a single stage process [26].

Sputtering systems can be configured with multiple magnetrons fitted with different target materials in order to deposit doped coatings, in which the dopant level is controlled by the relative power delivered to each magnetron. Alternatively, in a single magnetron system an alloy target can be used to produce doped coatings directly, although the dopant level in this case is fixed to that of the target material.

The ability to produce doped coatings is of great importance in this context, because the relatively high band gap of anatase (3.2 eV) means that it requires UV light (<390 nm) for activation. Photocatalytic activity can be both increased and extended into the visible range, though, by doping with different metallic elements (e.g., W, Mo, Nb, Ta) or non-metallic elements (e.g., N, C, S). Doping titanium dioxide with non-metal atoms narrows the band gap due to a mixing of the dopant p-states with the p-states of oxygen forming the valence band of titanium dioxide [27]. Of the range of possible non-metal dopants, nitrogen is one of the most described in literature for improving the photocatalytic activity of titanium dioxide [28–30] and extending its activity into the visible range. The nitrogen atom has a size comparable with the size of an oxygen atom, thus it can be easily introduced into the titania structure in either substitutional or interstitial positions [31].

Doping with transition metal ions is reported to create impurity levels near the conduction band that may perform as trapping centres, which extend the lifetime of photogenerated electrons and holes [32]. It is reported that the best results for transition metal doping can be achieved when the ionic radius of the doping metal is close to that of titanium [33] to enable incorporation into the titania lattice. Of the variety of candidate metals described in the literature, transition metals such as tungsten [34], chromium [35], vanadium [36] and molybdenum [37] are mentioned as efficient dopants for shifting the activity to the visible range.

Both of these doping strategies, and the idea of simultaneously co-doping titania with metallic and non-metallic elements, have been extensively investigated by many researches in the past few years [32,38]. Despite this, at present, there is no uniform theory explaining the optimum choice of dopant element(s) and doping level to maximise the photocatalytic properties in the visible range. This paper gives an overview of studies of doping and co-doping strategies conducted by the authors on magnetron sputtered titania coatings [25,39–43]. The influence of different elements on structural formation is considered and the production of as-deposited anatase coatings using the HiPIMS process is also described. Attempts to optimise the photoactivity of the coatings under UV, fluorescent and visible light irradiation are discussed.

2. Experimental Section

2.1. Coating Deposition Process

All the coatings described here were deposited by reactive magnetron sputtering in a Teer Coatings Ltd. (Droitwich, UK) UDP 450 system (Figure 1). Up to three 300 mm × 100 mm unbalanced planar magnetrons were installed vertically opposed through the chamber walls. Depending on the experimental array the system was configured with either two magnetrons fitted with titanium targets (99.5% purity) and one with a metallic dopant target (W, Mo, Ta or Nb—all

99.9% purity) [39–41], or for the HiPIMS array, a single magnetron was used with either a titanium target or a 5 at% W-doped Ti target installed [25,42,43].

Figure 1. Schematic representation of the Teer Coatings Ltd. UDP450 sputtering rig with three planar magnetrons installed.

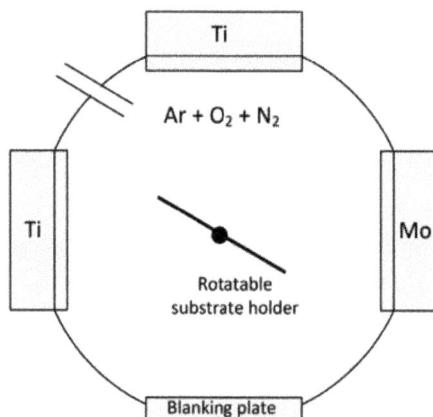

For the multiple magnetron configuration, the magnetrons with the titanium targets were driven in mid-frequency pulsed DC mode using a dual channel Advanced Energy Pinnacle Plus supply at a frequency of 100 kHz and a duty of 50% (in synchronous mode) at a constant time-averaged power of 1 kW per channel. In order to vary the doping level, the magnetron with the dopant target was driven at powers in the range 100–180 W in continuous DC mode using an Advanced Energy MDX power supply. The reactive sputtering process was carried out in an argon:oxygen atmosphere at 0.3 Pa, and was controlled by optical emissions monitoring using an operating set point (15% of the full metal signal) previously found to produce stoichiometric TiO_2 coatings [44]. The substrates (microscope slides initially, but later 20×10 mm^2 304 2B stainless steel coupons were also coated for antimicrobial testing) were ultrasonically pre-cleaned in propanol and placed onto the electrically floating substrate holder, which was rotated continuously during the deposition process at 4 rpm at a distance of 100 mm from the magnetrons. During the nitrogen and co-doping experiments, the nitrogen flow was controlled using a mass flow controller in the range from 0 to 10 sccm to vary dopant levels [41]. Coating thicknesses were in the range 500 nm to 1 μm. Initial experiments showed that the as-deposited pulsed DC coatings were amorphous. Therefore, these coatings were post-deposition annealed in air at either 400 or 600 °C for 30 min and then allowed to cool in air.

For the HiPIMS experiments, the magnetron was driven at time-averaged powers of 600 W and 880 W using a Huettinger HMP1/1_P2 HiPIMS power supply. The working pressure was varied in the range of 0.13 to 0.93 Pa. Pulse frequency (100–300 Hz) and pulse width (50–200 μs) were used as two other process variables. Sputtering was carried out in an argon:oxygen atmosphere of 2:3 for all deposition runs (10 sccm of Ar and 15 sccm of O_2), which corresponded to the poisoned mode for this system. The thresholds of these variable parameters were chosen to maintain stable plasma discharge conditions and, thereby, control over the deposition process. The coatings were initially deposited onto soda-lime glass substrates. Coating thickness measurements were obtained by means of surface profilometry. All coatings deposited in this mode were of the order of 100 nm. Optimised operating conditions were then used to deposit coatings onto 100 μm PET (polyethylene terephthalate) web and PC (polycarbonate) substrates. The HiPIMS coatings were analysed in the as-deposited condition and were not annealed.

2.2. Coating Characterization

The coatings were typically analyzed by Raman spectroscopy (Renishaw Invia, 514 nm laser) and X-ray diffraction (XRD) in θ–2θ mode (Philips PW1729 diffractometer with CuKα1 radiation at 0.154 nm) to ascertain their crystalline structure. Composition was investigated by energy dispersive X-ray spectroscopy (EDX—Edax Trident, installed on a Zeiss Supra 40 VP-FEG-SEM). The surface roughness and surface areas of the coatings were determined using a MicroXAM white light surface profilometer. Finally, values of the optical band gaps of the coatings were calculated using the Tauc plot method [45], by plotting $(\alpha h v)^{1/2}$ vs. hv and extrapolating the linear region to the abscissa (where α is absorbance coefficient, h is Plank's constant, v is the frequency of vibration).

2.3. Assessment of Photocatalytic Activity and Hydrophilicity

The determination of photocatalytic activity was carried out using the methylene blue (MB) degradation test. MB is an organic dye with molecular formula $C_{16}H_{18}ClN_3S$, and is often used as an indicating organic compound to measure the activity of photocatalysts. In fact, ISO10678 confirms the use of methylene blue as a model dye for surface photocatalytic activity determination in aqueous medium [46].

An aqueous solution of MB shows strong optical absorption at approximately 665 nm wavelength. Changes in the absorption peak height are used for monitoring the concentration of MB, and hence its degradation in contact with a photocatalytic surface.

Prior to the photocatalytic measurements, coating samples of equal size (15 × 25 mm^2) were immersed in a conditioning solution of methylene blue for pre-absorption of MB on the test surfaces to exclude the effect of absorption during the photocatalytic experiment. The photocatalytic measurements were carried out for 1 h in continuous mode. The absorption peak height of the methylene blue solution was measured with an Ocean Optics USB 2000+ spectrometer with continuous magnetic stirring.

Each coating was tested both under UV and fluorescent light sources; 2 × 15 W 352 nm Sankyo Denki BLB lamps were used as the UV light source (integrated power flux to the sample = 4 mW/cm^2) and 2 × 15 W Ushio fluorescent lamps as the fluorescent light source (integrated power flux to the sample = 6.4 mW/cm^2). Selected coatings were additionally tested under a visible light source. The visible light source was simulated by combining a fluorescent light source with a Knight Optical 395 nm long pass UV filter. The natural decay rate of methylene blue (without the photocatalyst present) under each type of light source was measured for reference purposes, as well as the degradation rate of methylene blue in contact with photocatalytic surface but without light irradiation (i.e., in the dark). In both cases the decay rate of methylene blue was of zero order and, thus was neglected in the following calculations, meaning any changes in the absorption peak height could be attributed to the photocatalytic activity [37]. The experimental setup for the MB tests is shown schematically in Figure 2.

Figure 2. Schematic of methylene blue photocatalytic testing equipment.

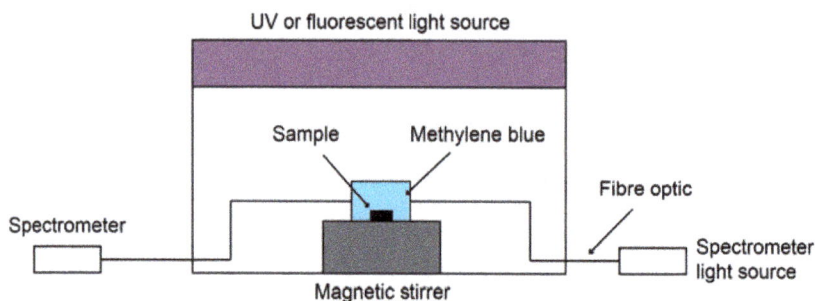

According to the Lambert-Beer law, the concentration of dye, c, is proportional to the absorbance value:

$$A = \varepsilon\, cl \tag{1}$$

where A is absorbance, ε is the molar absorbance coefficient; l is the optical length of the cell where the photocatalyst is immersed into MB.

The photocatalytic decomposition of MB was approximated to first order kinetics, as shown in the equation:

$$\ln\left[\frac{C_0}{C}\right] = k_a t \tag{2}$$

where C_0 and C are the concentrations of MB solution at time 0 and time t of the experiment, respectively. If the ratio of absorption decay is proportional to the concentration decay, the first order reaction constant, k_a can be found from the slope of the plot $\ln(A_0/A)$ against time.

The hydrophilic properties of the coatings were estimated via measurements of contact angles of deionised water droplets on the surface of the coating made with a Kruss goniometer.

2.4. Assessment of Antimicrobial Properties

Escherichia coli (ATCC 8739) was used as a model organism in these experiments. Measurements of the antimicrobial activity of selected coatings deposited onto 304 2B stainless steel substrates were performed using ISO 27447:2009 as guidance (with minor modifications) [47]. Stainless steel was selected because it is the material of choice in the food and beverage production industries and the coatings were developed for field trials in industrial facilities [48,49]. In brief, 50 µL of suspension containing approximately 10^5 colony forming units (cfu) per mL of bacterial cells were placed on the surfaces and a polyethylene film was placed over the bacterial suspension to ensure even distribution. Surfaces were illuminated (wavelength range of 300–700 nm) in a 20 °C incubator (Gallenkamp, Loughborough, UK) fitted with six fluorescent lamps (Sylvania, ON, Canada) with an energy output of 6.4 mW/cm². At selected time points (0, 12, 24 and 48 h), surfaces were removed and vortexed for 1 min in neutralizing broth (20 g·L^{-1} Soya Lectin (Holland and Barrett, Nuneaton, UK) and 30 g·L^{-1} Tween 80 (Sigma Aldrich, Gillingham, UK) to remove any surviving bacteria. Bacteria were enumerated by plate counts. All tests were carried out in triplicate. Stainless steel was used as a light control and a set of coated surfaces were also kept in dark conditions to serve as further controls.

3. Results

3.1. Transition Metal-Doped Pulsed DC Coatings

3.1.1. Structures and Compositions

The coatings produced by pulsed DC sputtering had dense, defect-free structures, with relatively smooth surfaces. A typical example is shown in Figure 3, which is a SEM micrograph showing the fracture section and surface topography of a Mo-doped (2.44 at%) coating after annealing at 400 °C. The sputtering rates of the dopant metals investigated increased in order Nb < Mo < Ta < W. Thus, the dopant content increased in this order when the same given power was applied to the dopant target (see Table 1), meaning some calibration of the process is required if coatings with the same dopant content are required. However, that was not the overriding concern with these experiments, which were more focused on structural formation and photocatalytic activity.

Figure 3. SEM micrograph of the fracture section of 2.44 at% Mo-doped titania coating deposited onto a glass substrate.

As mentioned above, the as-deposited coatings were assumed to be amorphous on the basis of analysis by XRD and Raman spectroscopy. This concurs with previous work, which showed that for pure titania coatings, strongly crystalline anatase structures formed for coatings annealed at 400 °C and that this structure persisted up to 600 °C before evidence of rutile was observed [50]. For doped titania coatings, the dopant element has an important influence on structural formation during the annealing of these coatings. This is illustrated in Figures 4 and 5, which show XRD spectra of selected doped-titania coatings annealed at 400 and 600 °C, respectively. The dopant compositions are indicated in Table 1. For Mo-, Ta- and Nb- doped coatings, a strong anatase structure has clearly evolved at 400 °C, whereas, doping with W appears to suppress the formation of this structure. Annealing at 600 °C results in the formation of an anatase structure for all the dopants investigated, but in the case of tungsten, broad rutile peaks were also detected in the Raman spectra for these samples (Figure 6), indicating a mixed-phase structure. This finding also highlights the different sensitivities of Raman spectroscopy and XRD for thin film analysis.

Table 1. Dopant levels and band gap values for transition metal-doped titania coatings deposited on glass substrates and annealed.

Dopant	Power on Dopant Target W	Dopant Content at%	Coating Thickness nm	Annealing Temperature °C	Band Gap eV	$k_a \times 10^{-5}$, s^{-1} UV light	$k_a \times 10^{-5}$, s^{-1} fluor. light	Crystal Structure
none (pure TiO$_2$)	-	-	586	400	3.12	1.0	0.5	anatase
				600	3.15	1.7	0.6	anatase
	100	0.74	607	400	3.16	2.0	0.5	anatase
				600	3.16	0.6	0.3	anatase
Nb	150	1.94	697	400	3.15	1.8	0.5	anatase
				600	3.15	0.6	0.4	anatase
	180	2.67	712	400	3.13	1.5	0.9	anatase
				600	3.13	0.4	0.0	anatase
	100	2.44	685	400	3.17	4.0	1.9	anatase
				600	3.00	2.8	1.8	anatase
	150	5.37	727	400	3.11	0.6	0.3	anatase
Mo				600 *	2.95	-	-	anatase
	180	6.96	754	400	3.09	0.5	0.2	amorphous
				600 *	2.97	-	-	anatase
	100	10.03	814	400	3.22	0.6	0.4	amorphous
				600	3.02	2.2	1.6	anatase
W	150	13.87	889	400	3.22	0.6	0.7	amorphous
				600	3.00	1.4	0.8	anatase/rutile
	180	15.84	896	400	3.22	0.4	0.3	amorphous
				600	2.98	0.9	0.6	rutile
	100	3.07	594	400	3.08	0.6	0.4	anatase
				600	3.09	1.3	0.6	anatase
Ta	150	9.10	786	400	3.20	0.7	0.0	amorphous
				600	3.16	1.0	0.0	anatase
	180	13.51	943	400	3.28	0.9	0.0	amorphous
				600	3.24	0.7	0.0	rutile

* coatings delaminated from substrate and were not tested.

Figure 4. XRD analysis of selected doped-titania coatings deposited on glass substrates and annealed at 400 °C.

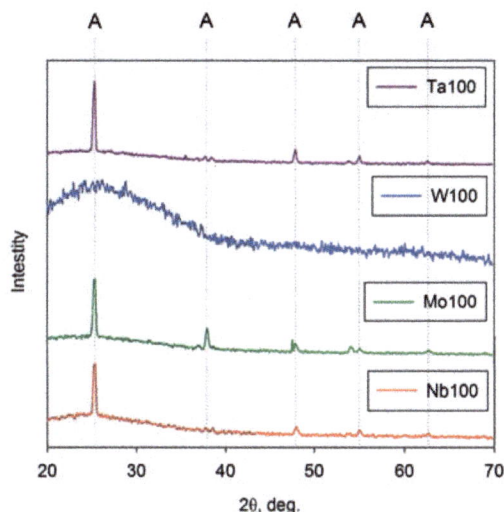

Figure 5. XRD analysis of selected doped titania coatings on glass substrates and annealed at 600 °C.

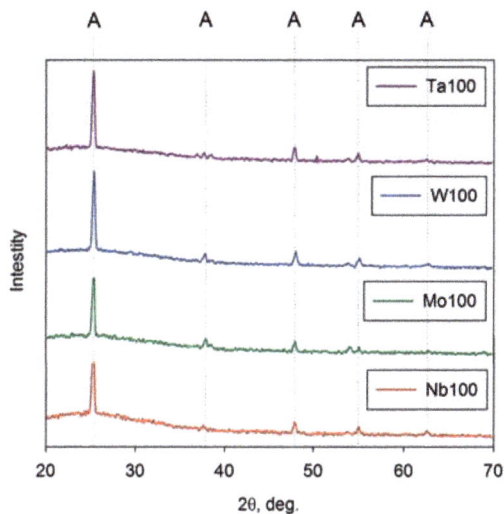

The choice of dopant material and annealing temperature also influenced the band gap of the resulting coating. Again, some typical examples are given in Table 1. Annealing at 400 °C produced very small red shifts for Mo-doped coatings, but small blue shifts for the other dopants. In contrast, annealing at 600 °C resulted in more significant red shifts for most of the combinations tested, and particularly for the Mo- and W- doped coatings (up to 0.2 eV).

Figure 6. Raman spectra showing structural variations as a function of W-content for W-doped titania coatings after annealing at 600 °C.

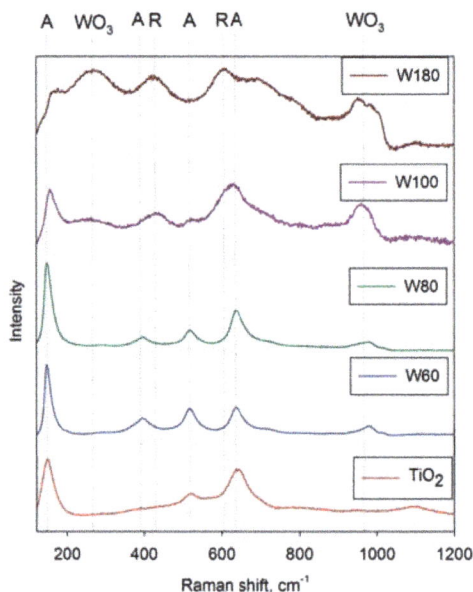

3.1.2. Photocatalytic Activity

As a benchmark for the doped titania coatings, the rate constants for the decomposition of methylene blue for pure titania coatings annealed at 400 °C and 600 °C sources were $1.0 \times 10^{-5} \cdot s^{-1}$ and $1.7 \times 10^{-5} \cdot s^{-1}$ under UV radiation and $0.5 \times 10^{-5} \cdot s^{-1}$, and $0.6 \times 10^{-5} \cdot s^{-1}$ under fluorescent light, respectively. As might be expected from the structural data shown in Figure 4, for the coatings annealed at 400 °C, Nb-doped coatings (best result: $k_a = 2.0 \times 10^{-5} \cdot s^{-1}$ with 0.7 at% Nb) and Mo-doped coatings (best result: $k_a = 4.0 \times 10^{-5} \cdot s^{-1}$ at 2.4 at% Mo) proved most effective at increasing photocatalytic activity under UV radiation. Only the 2.4 at% Mo-doped coating showed any notable improvement in fluorescent light activity ($k_a = 2.8 \times 10^{-5} \cdot s^{-1}$), which again would be expected from the observed band gap shifts. Ta- and W- doped coatings showed a reduction in activity under both light sources.

For the coatings annealed at 600 °C, now both Nb and Ta proved ineffective as dopant elements, with reduced activities compared to pure titania. In this case, Mo-doped coatings and W-doped coatings showed the greatest increases in activity. The best rate constants obtained with 2.4 at% Mo were $2.8 \times 10^{-5} \cdot s^{-1}$ and $1.8 \times 10^{-5} \cdot s^{-1}$ for UV and fluorescent light radiation, respectively. The equivalent values for coatings with 10.0 at% W were $2.2 \times 10^{-5} \cdot s^{-1}$ and $1.6 \times 10^{-5} \cdot s^{-1}$.

3.1.3. Optimisation of Tungsten Dopant Level

Although the W-doped coatings showed enhanced activity, it was recognised that the initial experimental conditions had produced relatively high levels of tungsten in the coatings (10–15 at%). Thus, a second series of W-doped coatings were produced where the power to the dopant target was varied over a lower range of values (60–90 W), to produce lower W dopant levels, with a view to optimising the activity level.

Other than the range of dopant target powers, the additional W-doped coatings were deposited under identical conditions to the initial batch of coatings, as described in Section 2.1. The coatings were then annealed at 600 °C. The dopant content and thickness of these and the previous W-doped coatings are given in Table 2. After annealing, these coatings showed a transition from anatase structures at low-W levels, through a mixed phase structure to rutile structures at higher W levels. Evidence for the formation of tungsten oxides was also identified for the higher W dopant levels. This structural transition with dopant content is illustrated in Figure 6, which shows selected Raman spectra for these coatings.

Table 2. Compositional data, band gap values and photocatalytic activity results for tungsten-doped titania coatings deposited on glass substrates and annealed at 600 °C.

Sample ID	Power on Dopant Target W	Dopant Content at%	Coating Thickness nm	Band Gap eV	$k_a \times 10^{-5}$, s^{-1} UV light	$k_a \times 10^{-5}$, s^{-1} fluor. light	Crystal Structure
TiO$_2$	-	-	586	3.15	1.7	0.6	anatase
W60	60	3.83	702	3.12	4.0	1.0	anatase
W70	70	4.64	746	3.09	5.6	1.2	anatase
W80	80	5.89	758	3.09	9.9	2.7	anatase
W90	90	7.09	793	3.05	6.4	2.1	anatase
W100	100	10.03	814	3.02	2.2	1.6	anatase
W150	150	13.87	889	3.00	1.4	0.8	anatase/rutile
W180	180	15.84	896	2.98	0.9	0.6	rutile

Once again, photocatalytic activity was assessed in terms of the degradation rate of methylene blue and band gap shifts were calculated from Tauc plots [40]. The results are also included in Table 2, together with surface area measurements calculated from white light profilometer scans. The rate constants obtained from the MB tests under UV and fluorescent light sources are also shown graphically as a function of W content in Figure 7, together with the surface area values for both sets of coatings. A clear, sharp peak in activity occurred at 5.9 at% W, with the k_a values showing an approximately five-fold increase compared to the values for pure titania coatings. Further increases in W content beyond 5.9 at% lead to a rapid fall off in activity levels. The peak in activity appears to almost coincide with the peak in surface area of these coatings (as determined by white light profilometry). Whilst increased surface area would be expected to contribute to an increase in activity, due to the greater area in contact with the MB, consideration of the data presented shows that the maximum to minimum variation in surface area is only around 2%, which cannot alone account for the 500% increase in activity. Furthermore, the band gap of the coatings decreased progressively with W-content, from 3.12 at 3.8 at% to 2.98 at 15.8 at%, implying that the increased activity is not linked in this instance to a reduction in band gap energy.

A mechanism has been forward by a number of authors to account for the increase in activity at specific tungsten dopant levels [51]. When the photocatalyst is irradiated, the photogenerated electrons will be transferred into the tungsten oxide conduction band, which is located lower than the corresponding band of titanium dioxide (2.5–2.8 eV). Conversely, the holes will accumulate on the valence band of titania, promoting efficient charge separation. In the case of coatings with higher W content, excessive levels of dopant act as recombination centres for photogenerated electrons and holes. Additionally, the formation of a separate phase of tungsten oxide reduces the surface area of titanium dioxide, as proved by the surface morphology results, and thus reduces the area of contact between the pollutant and the photocatalyst. These factors result in a significant loss of photocatalytic activity for higher W-content coatings.

Figure 7. Variation in MB degradation rates for UV and fluorescent light irradiation and surface area, as functions of W content for W-doped titania coatings after annealing at 600 °C.

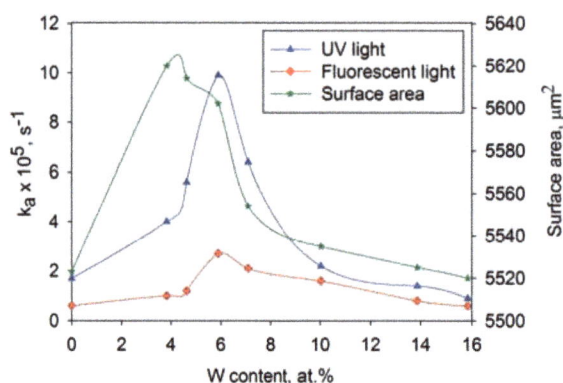

3.1.4. Synergistic Effects of Co-Doping with Molybdenum and Nitrogen

To investigate the potential of co-doping with two elements, a batch of Mo-doped coatings were produced, which also incorporated varying levels of nitrogen. Coatings with the previously determined optimum Mo dopant level of 2.4 at% were used for these experiments. Coatings doped only with nitrogen were also produced for comparison purposes. Details of the coating compositions are given in Table 3. Once again, these coatings were post-deposition annealed at 600 °C prior to testing. Despite using the same range of flow rates, it can be seen that the nitrogen content was significantly lower in the N-doped only coatings, compared to the co-doped coatings. Indeed, the nitrogen content in coatings N1–N5 was too low to be quantified with the techniques used here (XPS and EDX). Co-doping with N and Mo is known to increase the solubility limits of both N and Mo in TiO_2 [31]. This effect is described as being more pronounced in the case of nitrogen, as the solubility of N in titania is usually very low. The data presented here are in good agreement with this finding. Further detailed interpretation of the XPS analyses has been given elsewhere [41].

Raman analysis of the coatings confirmed an anatase structure for the annealed coatings, which XRD indicated had a strong (1 0 1) texture (not shown here) [41]. Band gap values and MB degradation rates are listed in Table 4. For the N-doped only coatings, samples N1 and N3 show some increase in UV activity, compared to the undoped titania coating, but apart from these two results, the effect of N-doping alone is negligible. However, the results for the co-doped coatings show a progressive increase in UV and fluorescent activity, with coating MoN7 demonstrating the highest activity under both light sources. This coating showed an increase in UV light activity of >4× and an increase in fluorescent light activity of >9× that of the pure titania coating. The equivalent values when compared to the Mo-only doped coating are both approximately a 3× increase in activity. Furthermore, visible light testing (using the 395 nm long pass filter) demonstrated that these coatings also exhibited some activity under this light source, while for undoped/N-doped titania coatings no activity was recorded.

Table 3. Compositional properties and thickness of titania coatings doped with nitrogen ("N" series) and co-doped with molybdenum/nitrogen ("MoN" series).

Dopant	Sample ID	Flow of Nitrogen, Sccm	Content of Nitrogen, at%	Coating Thickness, nm
Mo	TiO_2 + Mo	-	-	685
	N1	1	<1%	654
	N3	3	<1%	657
N	N5	5	<1%	654
	N7	7	1.09	658
	N10	10	3.67	661
	MoN1	1	1.22	760
	MoN3	3	3.08	764
Mo + N	MoN5	5	4.95	758
	MoN7	7	7.13	761
	MoN10	10	9.12	766

Table 4. Band gap values and MB degradation rate constants for N-doped and Mo-N co-doped titania coatings.

Sample ID	Band Gap eV	Band Gap Shift (Compared to TiO_2)	$k_a \times 10^{-5}, s^{-1}$ UV Light	$k_a \times 10^{-5}, s^{-1}$ Fluor. Light	$k_a \times 10^{-5}, s^{-1}$ Vis. Light
TiO_2	3.15	-	1.7	0.6	0
TiO_2 + Mo	3.00	−0.15	2.8	1.8	0.6
N1	3.22	+0.07	3.6	0.9	0
N3	3.22	+0.07	2.9	1.6	0
N5	3.20	+0.05	1.7	1.1	0
N7	3.14	−0.01	1.7	1.0	0
N10	3.08	−0.07	1.4	0.9	0
MoN1	3.09	−0.06	1.0	0.6	0
MoN3	3.09	−0.06	4.9	1.7	0.4
MoN5	3.05	−0.10	6.9	2.1	0.7
MoN7	3.04	−0.11	7.5	5.6	1.2
MoN10	3.07	−0.08	5.6	3.7	1.0

The results of photocatalytic tests showed that doping with nitrogen only had at best a moderately positive effect on photocatalytic activity, while co-doping with nitrogen and molybdenum resulted in significant improvements in photocatalytic activity. The efficiency of N-doped coatings under UV light, compared to that of undoped titania, was higher by a factor of 2 at most and generally lower than this. However, despite widely published information about N-doping as an efficient method of improving the photocatalytic properties under fluorescent/visible light [30,52], the N-doped titania coatings studied in this work had only a marginally higher efficiency of MB degradation under the fluorescent light source. As no noticeable band gap shift towards the visible range was observed, the increased photocatalytic activity under the fluorescent light source could only be attributed to improved electron-hole separation and the extended lifetime of charge carriers, as a result of nitrogen incorporation.

The observed increase in activity of the co-doped coatings can be assumed to be a result of more efficient electron-hole separation, compared to undoped or singly Mo- or N-doped titania coatings, due to the synergistic effect of Mo-N co-doping. A mechanism of explaining more efficient charge carrier separation was proposed by Cheng *et al.*, who observed similar results for Mo-N co-doped coatings prepared by a hydrolysis-precipitation method [38]. In the proposed mechanism nitrogen and molybdenum create local energy levels within the titania band gap, and therefore several ways of charge carrier excitation are available, and consequently more photo-induced charge carriers can be efficiently separated to participate in the photocatalytic process. Co-doped coatings with optimum content of nitrogen and molybdenum demonstrate significantly higher photocatalytic activity, due to more efficient charge carrier separation and their extended lifetimes. A shift of the band gap towards the visible range, compared to undoped titania, enables the presence of photocatalytic activity under fluorescent and visible light sources.

3.2. Characterisation of As-Deposited HiPIMS Titania Coatings

In a preliminary study using the HiPIMS process, coatings were deposited from a single titanium target and a Ti target containing 5 at% W [42]. Analysis by Raman spectroscopy of the as-deposited coatings indicated that the pure titania samples had formed a mixed anatase/rutile structure, whereas the W-doped coatings had only a weakly crystalline anatase structure. Example spectra are shown in Figure 8. These spectra imply that the presence of tungsten in the coating has suppressed the formation of a crystalline structure, as observed previously for the coatings deposited by pulsed DC sputtering. The band gap data and photocatalytic activity rate constants for these coatings are listed in Table 5. A number of interesting points emerge from these data. Firstly, the UV light rate constants (k_a values ~2.0 to 2.4 × $10^{-5} \cdot s^{-1}$) are noticeably higher than those obtained for pulsed DC titania coatings after annealing (typically $k_a = 1.7 × 10^{-5} \cdot s^{-1}$). It was also observed that the presence of W reduced the band gap of these coatings quite considerably (by 0.14–0.15 eV), which in turn lead to a 2 to 4 fold improvement in the level of fluorescent light activity for the doped coatings. Finally, the W-doped HiPIMS titania coatings displayed visible light activity levels very close to the values measured under fluorescent light sources.

Figure 8. Raman spectra of as-deposited W-doped (TiO$_2$-W) and undoped titania (TiO$_2$) coatings deposited by HiPIMS on glass substrates.

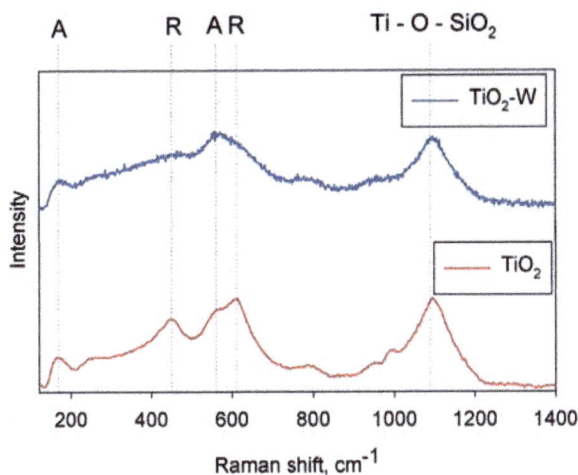

Table 5. Overview of deposition conditions and properties of titania and W-doped titania coatings deposited on glass substrates by HiPIMS.

Coating ID	Mean Sputtering Power kW	W at%	Band Gap eV	Surface Roughness μm	Surface Area μm^2	$k_a \times 10^{-5}$ UV Light, s^{-1}	$k_a \times 10^{-5}$ Fluorescent Light, s^{-1}	$k_a \times 10^{-5}$ Visible Light, s^{-1}
TiO$_2$	44	-	3.21	0.084	5521	2.0	0.3	-
TiO$_2$W	44	5.67	3.04	0.092	5524	2.1	1.2	0.9

Having demonstrated the potential of the HiPIMS process, further studies were carried out to attempt to optimise the performance of the coatings produced by this technique [43]. Process variables including deposition pressure, pulse width (*i.e.*, duration of the power pulse delivered to the target) and pulse frequency were varied and their impact on photocatalytic activity and water contact angle was investigated. Of these variables, deposition pressure emerged as the most influential. This is clearly illustrated in Figure 9, which compares the variation with coating pressure of contact angle and the photocatalytic activity rate constants for MB degradation under UV light irradiation. Following these experiments, coatings were deposited directly onto PET and polycarbonate (PC) substrates under optimised conditions. Within experimental accuracy, the same values of first order rate constants were obtained for these coatings, independent of the substrate materials tested (glass, PET and PC).

3.3. Antimicrobial Activity

The antimicrobial activity against *E. coli* of selected Mo- and W- doped titania coatings was assessed and compared to pure titania coatings and stainless steel controls. As mentioned earlier, 304 2B stainless steel was chosen for its compatibility with industrial processing. The coatings were deposited by pulsed DC magnetron sputtering, using the conditions described earlier, and then annealed at 600 °C. Interestingly, for this choice of substrate material, a higher molybdenum content (6.9 at%) was found to provide the greatest photocatalytic activity from MB tests (compared with 2.44 at% for glass), so this dopant content was used, along with coatings doped with 3.8 at% W, which were also found to be optimal in this case. This finding concurs with other studies that have shown that the choice of substrate material (particularly whether it is electrically conductive or not) influences photocatalytic activity [53].

Figure 9. Variation in water contact angle and UV light photocatalytic rate constant as a function of deposition pressure for titania coatings deposited on glass substrates by HiPIMS. Images of the water droplets are included for samples deposited at 0.16 and 0.93 Pa.

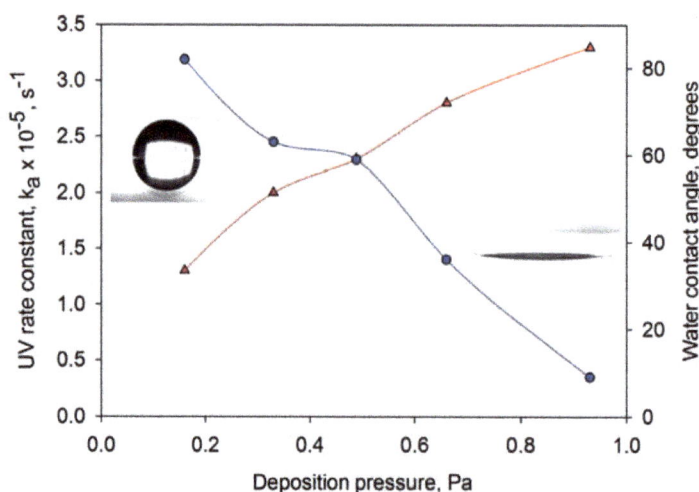

The stainless steel controls did not reduce the number *E. coli* colony forming units in light or dark conditions in a 48 h period (Figure 10). The pure titania coatings showed only a weak photocatalytic effect in reducing the number of colony forming units by 2-logs in this time period. However, coatings doped with Mo eradicated *E. coli* within 24 h in both light and dark conditions. The activity displayed in the dark suggests that the surface is dual functioning, being both photocatalytic (as determined by the degradation of MB) and innately antimicrobial. W-doped coatings also reduced microbial counts by 5-logs within 48 h in the light but not the dark, *i.e.*, only photocatalytic behaviour was observed. All doped surfaces displayed an ability to inactivate *E. coli* when tested under visible light and, in the Mo case, when in the dark, highlighting the potential use of such surfaces for indoor applications, allowing a choice between a coating with an active antimicrobial function (Mo), or one which is inert unless irradiated (W), depending on requirements and regulations.

Figure 10. Antimicrobial effect of TiO$_2$, TiO$_2$-Mo and TiO$_2$-W surfaces on *Escherichia coli* Stainless steel surfaces were used as controls and light and dark conditions were investigated.

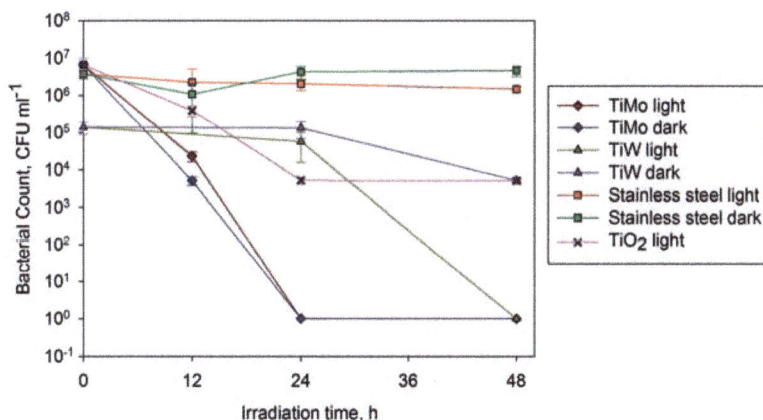

4. Discussion

This paper has considered a number of deposition and doping strategies for the production of titania-based photocatalytically active thin films. Reactive magnetron sputtering is a versatile, flexible technique for the production of high quality, fully dense coatings. When operating in pulsed DC mode, it provides a stable, arc-free process for the deposition of dielectric materials, such as titania. Furthermore, the coatings can be readily doped via transition metals, or via non-metal gaseous species, or a combination of both. In each case, control of the dopant level is straightforward. However, when operating in this mode, the as-deposited coatings were found to be amorphous and, therefore, showed no activity. Effective annealing temperatures for structural formation varied with dopant element. Mo-doped coatings annealed at 400 °C were found to demonstrate significantly higher activities than pure titania coatings annealed at the same temperature, whereas a temperature of 600 °C was required to achieve the same result for the W-doped coatings. Figure 11 additionally shows that the synergistic effect obtained by co-doping with Mo and N also produced coatings with a UV activity, close to that of the W-doped coatings, and a noticeably higher activity in fluorescent light. The UV and fluorescent light activities of the W-doped and MoN co-doped coatings also exceed the values shown in Figure 11 obtained for a sample of Pilkington's Activ, which is a commercially available product. Direct comparisons with this sample should be avoided, because it is produced via a chemical vapour deposition pyrolysis route and is significantly thinner than the sputtered coatings. However, as there is a dearth of "standard samples" in this field, it still serves as a useful guide to relative activity levels. A mixed anatase/rutile phase was detected for the samples with the highest levels of tungsten doping in the pulsed DC study, although the best photocatalytic results were still found for anatase coatings. In contrast, for the pure titania deposited via HiPIMS, a mixed phase structure gave superior photocatalytic activity [43]. This, of course, is not a new finding and several researchers have proposed that the mixed phase structure is optimal for photocatalytic activity [10,20].

Figure 11. Bar chart of maximum rate constants obtained for the decomposition of methylene blue under UV and fluorescent light sources for pure titania coatings, transition metal doped titania coatings, Mo and N co-doped coatings deposited by pulsed DC magnetron sputtering (annealed at 600 °C) and titania and W-doped titania coatings deposited by HiPIMS (as-deposited).

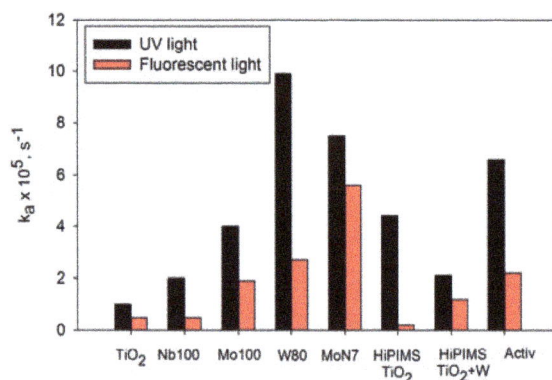

The HiPIMS process is still in the development stage and there remain issues with power supply stability and process control. Nevertheless, the potential of this process to produce, at least, semi-crystalline coatings in the as-deposited state is a clear advantage over other deposition processes. Furthermore, the low net deposition temperature makes it a suitable technique for deposition onto thermally sensitive materials, as demonstrated here with PET and PC substrates. The data presented in Figure 11 indicates that, when optimised, the HiPIMS pure titania coatings could achieve approximately twice the UV activity rate of the annealed pulsed DC coatings. The W-doped HiPIMS showed a reduction in UV activity, attributed to the weaker crystalline structure, but higher fluorescent light activity, attributed to a substantial band gap shift.

The capacity to break down organic compounds, as modelled here with methylene blue, is just one of the phenomena associated with photocatalytic coatings. The inactivation of microorganisms is another important ability. Numerous researchers have claimed antimicrobial activity for their coatings, but care must be taken in assessing these claims. The test method for antibacterial activity of photocatalytic materials is complex, requiring specific experimental conditions to be met and multiple repeat experiments if results are to be tested for reproducibility and compared to other published data. The results presented here are a case in point. A limited number of replicates were tested and only one microorganism was used; the Gram-negative *E. coli*. Ideally, more replicates would be tested and a Gram-positive microorganism, such as *Staphylococcus aureus*, would also be investigated. Despite this, the doped titania coatings showed the ability to eradicate *E coli* within 24 to 48 h. There was also an interesting distinction between the dopant elements, with the Mo-doped coatings being effective in light and dark and the W-doped coatings only being effective in the light. These results certainly merit more detailed investigation in the future. The recent introduction of antibacterial testing under indoor lighting (ISO 17094:2014) [54] has now allowed for visible light active photocatalytic surfaces to be tested more precisely, however, a more rapid antimicrobial testing method which could be performed by non-microbiologist would still be valuable.

5. Conclusions

Reactive magnetron sputtering techniques have been used to produce a range of titania and doped titania coatings. Choice of deposition technique (pulsed DC sputtering or HiPIMS) and choice of dopant element had a significant influence on structural formation and, subsequently, photocatalytic activity for these coatings. Pulsed DC coatings were amorphous in the as-deposited state, with no measurable activity against methylene blue, whereas the HiPIMS coatings were weakly crystalline as-deposited with moderate levels of activity. The benefits of this technique were further demonstrated by depositing active coatings onto polymeric substrates in a single stage process. Of the transition metals investigated as dopant elements, molybdenum and tungsten were the most effective. The highest UV activity recorded in these experiments was achieved by coatings doped with 5.9 at% W after annealing at 600 °C. This was slightly higher than the UV activity of MoN-doped coatings after annealing, but the co-doped coatings showed a higher level of activity under fluorescent light irradiation. Although only limited tests were performed, the Mo- and W- doped coatings also demonstrated the ability to inactivate *E. coli*. In the former case, the coatings were both antimicrobial (active in the dark) and photocatalytic (active in the light), whereas the W-doped coatings only displayed photocatalytic activity.

Acknowledgments

Funding for some of the antimicrobial aspects of this work from the Technology Strategy Board (UK) is gratefully acknowledged. The authors would also like to thank Xiaohong Xia and Yun Gao from the Faculty of Materials Science and Engineering, Hubei University, China for providing XPS analysis of selected coatings.

Author Contributions

Kelly was the main author of this paper and part of the supervisory team for the experimental programmes described here. West and Verran were also part of the project supervisory teams. Ratova conducted the majority of the experimental work on the production and testing of the photocatalytic coatings. Fisher and Ostovarpour were responsible for the production and testing of the antimicrobial coatings.

Conflicts of Interest

The authors declare no conflict of interest.

References

1. Paz, Y. Application of TiO₂ photocatalysis for air treatment: Patents' overview. *Appl. Catal. B Environ.* **2010**, *99*, 448–460.
2. Vidal, A. Developments in solar photocatalysis for water purification. *Chemosphere* **1998**, *36*, 2593–2606.

3. Fujishima, A.; Zhang, X. Titanium dioxide photocatalysis: Present situation and future approaches. *CR Chim.* **2006**, *9*, 750–760.

4. Markowska-Szczupak, A.; Ulfig, K.; Morawski, A.W. The application of titanium dioxide for deactivation of bioparticulates: An overview. *Catal. Today* **2011**, *169*, 249–257.

5. Fujishima, A.; Rao, T.N.; Tryk, D.A. Titanium dioxide photocatalysis. *J. Photochem. Photobiol. C Photochem. Rev.* **2000**, *1*, 1–21.

6. Yang, Y.; Li, X.J.; Chen, J.T.; Wang, L.Y. Effect of doping mode on the photocatalytic activities of Mo/TiO$_2$. *J. Photochem. Photobiol. A* **2004**, *163*, 517–522.

7. Allen, N.S.; Edge, M.; Verran, J.; Stratton, J.; Maltby, J.; Bygott, C. Photocatalytic titania based surfaces: Environmental benefits. *Polym. Degrad. Stab.* **2008**, *93*, 1632–1646.

8. Braham, R.J.; Harris, A.T. Review of major design and scale-up considerations for solar photocatalytic reactors. *Ind. Eng. Chem. Res.* **2009**, *48*, 8890–8905.

9. McCullagh, C.; Robertson, P.K.J.; Adams, M.; Pollard, P.M.; Mohammed, A. Development of a slurry continuous flow reactor for photocatalytic treatment of industrial waste water. *J. Photochem. Photobiol. A* **2010**, *211*, 42–46.

10. Zhao, L.; Han, M.; Lian, J. Photocatalytic activity of TiO$_2$ films with mixed anatase and rutile structures prepared by pulsed laser deposition. *Thin Solid Films* **2008**, *516*, 3394–3398.

11. Musil, J.; Herman, D.; Sicha, J. Low-temperature sputtering of crystalline TiO$_2$ films. *J. Vac. Sci. Technol.* **2008**, *24*, 3793–3800.

12. Šícha, J.; Musil, J.; Meissner, M.; Čerstvý, R. Nanostructure of photocatalytic TiO$_2$ films sputtered at temperatures below 200 °C. *Appl. Surf. Sci.* **2008**, *254*, 3793–3800.

13. Frach, P.; Glöß, D.; Metzner, C.; Modes, T.; Scheffel, B.; Zywitzki, O. Deposition of photocatalytic TiO$_2$ layers by pulse magnetron sputtering and by plasma-activated evaporation. *Vacuum* **2006**, *80*, 679–683.

14. Zhang, F.; Wolf, G.K.; Wang, X.; Liu, X. Surface properties of silver doped titanium oxide films. *Surf. Coat. Technol.* **2001**, *148*, 65–70.

15. Yates, H.M.; Brook, L.A.; Ditta, I.B.; Evans, P.; Foster, H.A.; Sheel, D.W.; Steele, A. Photo-induced self-cleaning and biocidal behaviour of titania and copper oxide multilayers. *J. Photochem. Photobiol. A* **2008**, *197*, 197–205.

16. Yaghoubi, H.; Taghavinia, N.; Alamdari, E.K. Self-cleaning TiO$_2$ coating on polycarbonate: Surface treatment, photocatalytic and nanomechanical properties. *Surf. Coat. Technol.* **2010**, *204*, 1562–1568.

17. Yang, J.H.; Han, Y.S.; Choy, J.H. TiO$_2$ thin-films on polymer substrates and their photocatalytic activity. *Thin Solid Films* **2006**, *495*, 266–271.

18. Wei, F.; Ni, L.; Cui, P. Preparation and characterization of N-S-codoped TiO$_2$ photocatalyst and its photocatalytic activity. *J. Hazard. Mater.* **2008**, *156*, 135–140.

19. Lee, C.S.; Kim, J.; Son, J.Y.; Choi, W.; Kim, H. Photocatalytic functional coatings of TiO$_2$ thin films on polymer substrate by plasma enhanced atomic layer deposition. *Appl. Catal. B Environ.* **2009**, *91*, 628–633.

20. Kääriäinen, M.L.; Cameron, D.C. The importance of the majority carrier polarity and p-n junction in titanium dioxide films to their photoactivity and photocatalytic properties. *Surf. Sci.* **2012**, *606*, L22–L25.

21. Kelly, P.J.; Arnell, R.D. Magnetron sputtering: A review of recent developments and applications. *Vacuum* **2000**, *56*, 159–172.

22. Kelly, P.J.; Bradley, J.W. Pulsed magnetron sputtering—Process overview and applications. *J. Optoelectron. Adv. Mater.* **2009**, *11*, 1101–1107.

23. Lundin, D.; Sarakinos, K. Smoothing of discharge inhomogeneities at high currents in gasless high power impulse magnetron sputtering. *J. Mater. Res.* **2012**, *27*, 780–792.

24. West, G.; Kelly, P.; Barker, P.; Mishra, A.; Bradley, J. Measurements of deposition rate and substrate heating in a HiPIMS discharge. *Plasma Process. Polym.* **2009**, *6*, S543–S547.

25. Sarakinos, K.; Alami, J.; Konstantinidis, S. High power pulsed magnetron sputtering: A review on scientific and engineering state of the art. *Surf. Coat. Technol.* **2010**; *204*, 1661–1684.

26. Kelly, P.J.; Barker, P.M.; Ostovarpour, S.; Ratova, M.; West, G.T.; Iordanova, I.; Bradley, J.W. Deposition of photocatalytic titania coatings on polymeric substrates by HiPIMS. *Vacuum* **2012**, *86*, 1880–1882.

27. Asahi, R.; Morikawa, T.; Ohwaki, T.; Aoki, K.; Taga, Y. Visible-light photocatalysis in nitrogen-doped titanium oxides. *Science* **2001**, *293*, 269–271.

28. Tavares, C.J.; Marques, S.M.; Viseu, T.; Teixeira, V.; Carneiro, J.O.; Alves, E.; Barradas, N.P.; Munnik, F.; Girardeau, T.; Rivière, J.P. Enhancement of the photocatalytic nature of nitrogen-doped PVD-grown titanium dioxide thin films. *J. Appl. Phys.* **2009**, *106*, 113535.

29. Chiu, S.M.; Chen, Z.S.; Yang, K.Y.; Hsu, Y.L.; Gan, D. Photocatalytic activity of doped TiO$_2$ coatings prepared by sputtering deposition. *J. Mater. Process. Technol.* **2007**, *192–193*, 60–67.

30. Wong, M.S.; Pang, C.H.; Yang, T.S. Reactively sputtered N-doped titanium oxide films as visible-light photocatalyst. *Thin Solid Films* **2006**, *494*, 244–249.

31. Pelaez, M.; Nolan, N.T.; Pillai, S.C.; Seery, M.K.; Falaras, P.; Kontos, A.G.; Dunlop, P.S.M.; Hamilton, J.W.J.; Byrne, J.A.; O'Shea, K.; *et al.* A review on the visible light active titanium dioxide photocatalysts for environmental applications. *Appl. Catal. B Environ.* **2012**, *125*, 331–349.

32. Liu, H.; Lu, Z.; Yue, L.; Liu, J.; Gan, Z.; Shu, C.; Zhang, T.; Shi, J.; Xiong, R. Mo-N co-doped TiO$_2$ for enhanced visible-light photoactivity. *Appl. Surf. Sci.* **2011**, *257*, 9355–9361.

33. Choi, W.; Termin, A.; Hoffmann, M.R. The role of metal ion dopants in quantum-sized TiO$_2$: Correlation between photoreactivity and charge carrier recombination dynamics. *J. Phys. Chem.* **1994**, *98*, 13669–13679.

34. Li, X.Z.; Li, F.B.; Yang, C.L.; Ge, W.K. Photocatalytic activity of WO$_x$-TiO$_2$ under visible light irradiation. *J. Photochem. Photobiol. A* **2001**, *141*, 209–217.

35. Peng, Y.H.; Huang, G.F.; Huang, W.Q. Visible-light absorption and photocatalytic activity of Cr-doped TiO$_2$ nanocrystal films. *Adv. Powder Technol.* **2012**, *23*, 8–12.

36. Akbarzadeh, R.; Umbarkar, S.B.; Sonawane, R.S.; Takle, S.; Dongare, M.K. Vanadia–titania thin films for photocatalytic degradation of formaldehyde in sunlight. *Appl. Catal. A Gen.* **2010**, *374*, 103–109.

37. Štengl, V.; Bakardjieva, S.; Murafa, N. Preparation and photocatalytic activity of rare earth doped TiO₂ nanoparticles. *Mater. Chem. Phys.* **2009**, *114*, 217–226.

38. Heng, X.; Yu, X.; Xing, Z. Characterization and mechanism analysis of Mo–N-co-doped TiO₂ nano-photocatalyst and its enhanced visible activity. *J. Colloid Interface Sci.* **2012**, *372*, 1–5.

39. Ratova, M.; Kelly, P.J.; West, G.T.; Iordanova, I. Enhanced properties of magnetron sputtered photocatalytic coatings via transition metal doping. *Surf. Coat. Technol.* **2013**, *228*, S544–S549.

40. Ratova, M.; West, G.T.; Kelly, P.J. Optimization study of photocatalytic tungsten-doped coatings deposited by reactive magnetron co-sputtering. *Coatings* **2013**, *3*, 194–207.

41. Ratova, M.; West, G.T.; Kelly, P.J.; Xia, X.; Gao, Y. Synergistic effect of doping with nitrogen and molybdenum on the photocatalytic properties of thin titania films. *Vacuum* **2014**, in press.

42. Ratova, M.; West, G.T.; Kelly, P.J. HiPIMS deposition of tungsten-doped titania coatings for photocatalytic applications. *Vacuum* **2014**, *102*, 48–50.

43. Ratova, M.; West, G.T.; Kelly, P.J. Optimisation of hipims photocatalytic titania coatings for low temperature deposition. *Surf. Coat. Technol.* **2014**, *250*, 7–13.

44. Onifade, A.A.; Kelly, P.J. The influence of deposition parameters on the structure and properties of magnetron-sputtered titania coatings. *Thin Solid Films* **2006**, *494*, 8–12.

45. Tauc, J.; Grigorovici, R.; Vancu, A. Optical properties and electronic structure of amorphous germanium. *Phys. Status Solidi* **1966**, *15*, 627–637.

46. *ISO10678. Fine Ceramics, Advanced Technical Ceramics—Determination of Photocatalytic Activity of Surfaces in an Aqueous Medium by Degradation of Methylene Blue*; International Organisation for Standards: Geneva, Switzerland, 2010.

47. *ISO27447. Fine Ceramics (Advanced Ceramics, Advanced Technical Ceramics). Test Methods for Antibacterial Activity of Semiconducting Photocatalytic Materials*; International Organisation for Standards, Geneva, Switzerland, 2009.

48. Fisher, L.; Ostovarpour, S.; Kelly, P.; Whitehead, K.A.; Cooke, K.; Stogråds, E.; Verran, J. Molybdenum doped titanium dioxide photocatalytic coatings for use as hygienic surfaces—The effect of soiling on antimicrobial activity. *Biofouling* **2014**, *30*, 911–919.

49. Navabpour, P.; Ostovarpour, S.; Tattershall, C.; Cooke, K.; Kelly, P.; Verran, J.; Whitehead, K.; Hill, C.; Raulio, M.; Priha, O. Photocatalytic TiO₂ and Doped TiO₂ Coatings to Improve the Hygiene of Surfaces Used in Food and Beverage Processing—A Study of the Physical and Chemical Resistance of the Coatings. *Coatings* **2014**, *4*, 433–449.

50. Kulczyk-Malecka, J.; Kelly, P.J.; West, G.; Clarke, G.C.; Ridealgh, J.A. Characterisation Studies of the Structure and Properties of As-Deposited and Annealed Pulsed Magnetron Sputtered Titania Coatings. *Coatings* **2013**, *3*, 166–176.

51. Lorret, O.; Francová, D.; Waldner, G.; Stelzer, N. W-doped titania nanoparticles for UV and visible-light photocatalytic reactions. *Appl. Catal. B Environ.* **2009**, *91*, 39–46.

52. Qin, H.L.; Gu, G.B.; Liu, S. Preparation of nitrogen-doped titania with visible-light activity and its application. *CR Chim.* **2008**, *11*, 95–100.

53. Yao, M.; Chen, J.; Zhao, C.; Chen, Y. Photocatalytic activities of ion doped TiO$_2$ thin films when prepared on different substrates. *Thin Solid Films* **2009**, *517*, 5994–5999.

54. *ISO17094. Fine Ceramics (Advanced Ceramics, Advanced Technical Ceramics). Test Methods for Antibacterial Activity of Semiconducting Photocatalytic Materials under Indoor Lighting Environments*; International Organisation for Standards: Geneva, Switzerland, 2014.

Are TiO₂ Nanotubes Worth Using in Photocatalytic Purification of Air and Water?

Pierre Pichat

Abstract: Titanium dioxide nanotubes (TNT) have mainly been used in dye sensitized solar cells, essentially because of a higher transport rate of electrons from the adsorbed photo-excited dye to the Ti electrode onto which TNT instead of TiO₂ nanoparticles (TNP) are attached. The dimension ranges and the two main synthesis methods of TNT are briefly indicated here. Not surprisingly, the particular and regular texture of TNT was also expected to improve the photocatalytic efficacy for pollutant removal in air and water with respect to TNP. In this short review, the validity of this expectation is checked using the regrettably small number of literature comparisons between TNT and commercialized TNP referring to films of similar thickness and layers or slurries containing an equal TiO₂ mass. Although the irradiated geometrical area differed for each study, it was identical for each comparison considered here. For the removal of toluene (methylbenzene) or acetaldehyde (ethanal) in air, the average ratio of the efficacy of TNT over that of TiO₂ P25 was about 1.5, and for the removal of dyes in water, it was around 1. This lack of major improvement with TNT compared to TNP could partially be due to TNT texture disorders as seems to be suggested by the better average performance of anodic oxidation-prepared TNT. It could also come from the fact that the properties influencing the efficacy are more numerous, their interrelations more complex and their effects more important for pollutant removal than for dye sensitized solar cells and photoelectrocatalysis where the electron transport rate is the crucial parameter.

Reprinted from *Molecules*. Cite as: Pichat, P. Are TiO₂ Nanotubes Worth Using in Photocatalytic Purification of Air and Water? *Molecules* **2014**, *19*, 15075-15087.

1. Introduction

Research on titanium dioxide nanotubes (TNT) began about 15 years ago. It was incited by earlier studies on carbon nanotubes which were mainly aimed at augmenting their adsorption properties. TNT have been employed principally and successfully in dye sensitized solar cells [1–3] in attempts to increase the efficacy of conversion of light to electricity.

Regarding the photocatalytic removal of pollutants, several assertions can be found in the literature claiming the likely superiority of TNT relative to layers of titanium dioxide nanoparticles (TNP), e.g., "The optimum TNT structures outperform over standard TNP P25 films rendering them very promising for outdoor photocatalytic applications." [4]; "TNT can have a higher photocatalytic reactivity than a comparable nanoparticulate layer." [5]; "As TNT are very resilient during heat treatment, their application as catalysts in photocatalysis is very attractive." [6]. Although these sentences contain some restrictive terms, they clearly present an optimistic view of the applicability of TNT for the photocatalytic removal of pollutants.

The aim of the present article was to thoroughly determine whether these assertions were essentially based on theoretical expectations or whether they were actually supported by appropriate comparisons. Unfortunately, in many reports regarding the photocatalytic removal of pollutants, TNT were compared to one another to determine the effect of the preparation procedures and were not compared to TNP. In some other reports the TNP used to rank the TNT photocatalytic efficacy were home-made so that the comparisons cannot be referred to. The only studies taken into account here are those in which the photocatalytic efficacy was measured for TNT and commercialized TNP (P25 from Evonik/Degussa except in one case) under the same conditions. For the removal of gaseous pollutants, this means that TNT and TNP films of similar thickness or deposited layers containing an equal mass of TNT or TNP were used. For the removal of pollutants in aqueous phase, it was the same condition for films and an equal mass of TiO_2 in the case of slurries. The irradiation characteristics varied between studies as usual in photocatalysis. However, for each TNT-TNP comparison taken into account here, they were identical and the same geometrical area of film or the same volume of slurry was irradiated. The main part of this article is devoted to these comparisons. However, to orient and inform the reader, the theoretical expectations in favor of the use of TNT instead of TNP are first mentioned. Then, the dimension ranges of TNT are indicated and the two main synthesis methods of TNT are very briefly described. Some electron microscopy images of TNT are also shown. Because TiO_2 nanorods, TiO_2 nanowires, and titanate nanotubes correspond to either other textures or types of oxide, they are not included in this short review.

2. What Are the Main Expectations for Using TNT for Photocatalytic Pollutant Removal?

First, in the case of TNP, the texture is most often irregular so that a part of the TiO_2 surface is not easily accessed by the reactants. On the contrary, reactants are expected to diffuse easily in the straight tubes of TNT. Second, the recombination rate of photoproduced charges should be increased by interparticular connections which are more numerous in TNP layers than in TNT. In addition, higher light-harvesting by TNT is expected because of scattering within the tubes [7].

Indeed, these expectations have been put forward to explain results obtained in the field of dye- sensitized solar cells [1–3]. For instance, the half reaction time of conduction band electrons with a photo-excited dye (a bipyridyl ruthenium complex) was reduced from 180 µs to 18 µs for TiO_2 nanofibers compared with mesoporous TiO_2 layers of similar thickness [8].

3. Dimension Ranges of TNT

Figure 1 indicates the dimension ranges of TNT schematized as perfect cylinders. These ranges include the dimensions most frequently reported. Dimensions outside these ranges have sometimes been mentioned. On the average, the thickness, width and length can vary by factors of about 10, 100 and 1000, respectively. The tubes can be adjacent or separated by a distance of up to a few µm. Adjusting the numerous parameters of the various syntheses and/or changing the synthesis method allow one to modify all these dimensions [1–3,5,6], which should affect the TNT photocatalytic efficacy. Undoubtedly, this research field will be further explored to yield TNT that are more effective for air and water photocatalytic purification, which is the particular application considered in this paper. In other words, the comparisons presented here will have to be completed and the conclusions perhaps amended depending on future reports.

Figure 1. Scheme indicating the ranges of the most often reported dimensions of TNT.

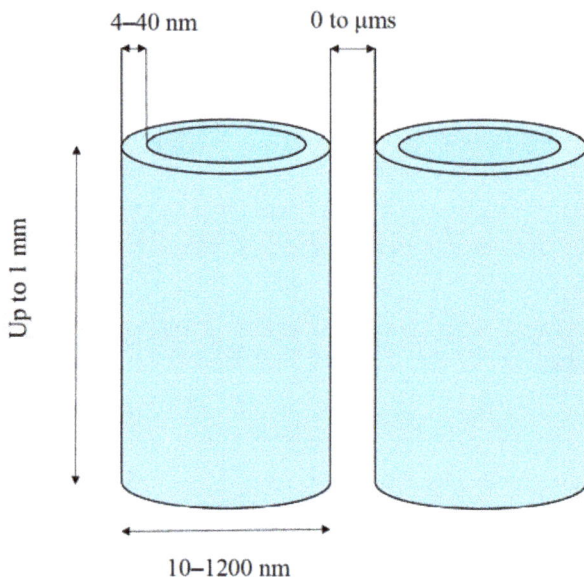

4. Preparation Methods of TNT

TNT can be synthesized in many ways: anodic oxidation of titanium, hydrothermal synthesis, sol-gel, seeded growth, alumina templating, surfactant-directed synthesis, chemical vapor deposition, *etc*. Presenting them in detail is beyond the scope of this article. Some information about anodic oxidation of titanium and hydrothermal synthesis—the two methods that seem to have been employed most often—is nonetheless provided here to show that numerous parameters can be varied to enable one to tailor the dimensions of the TNT, including the tube spacing, and also to change other characteristics such as the crystallinity and the allotropic form.

Anodic oxidation of titanium permits one to obtain TNT that can be directly used in dye-sensitized solar cells since they are attached to the titanium foil onto which they are formed. This synthesis is performed in a conventional electrochemical cell (Figure 2) [1]. The main parameters are the voltage—several voltages can be successively employed or alternately the voltage can be increased gradually—and the electrolyte: aqueous HF or NH_4F are often used; organic solvents, principally alcohols, are also utilized. Other parameters are the pH, the temperature and the duration. The TNT are then routinely calcined in the 673 K–773 K temperature range [1–3,5]. The use of supercritical carbon dioxide for drying has been shown to prevent deformations of the TNT [9].

Preparation of TNT via hydrothermal synthesis [6] is usually carried out in an autoclave containing TNP in an aqueous solution of NaOH (2 to 20 mol·L^{-1}) at a temperature above the water boiling point for several hours. That process breaks some of the Ti-O-Ti bonds and form Ti-O-Na bonds. Subsequent washing, generally with an aqueous solution of HCl, removes the Na$^+$ ions and produce TNT. The mechanisms of the effects of these successive treatments are still debated. In efforts to adjust the characteristics of the TNT, many factors can be changed: the starting TiO$_2$ or TiO$_2$ precursor, the concentration in NaOH, the temperature and duration of the hydrothermal process, the washing/cation exchange procedure, the final calcination; additional processes can also be used (e.g., ultrasound pretreatment of TNP, microwave heating). Because the synthesis mechanisms are presently less well understood for the hydrothermal method than for anodic oxidation, it may be more difficult to achieve a desirable TNT morphology.

Figure 2. Scheme of an electrochemical cell enabling the formation of TNT on a Ti anode. Reproduced with permission from reference [1].

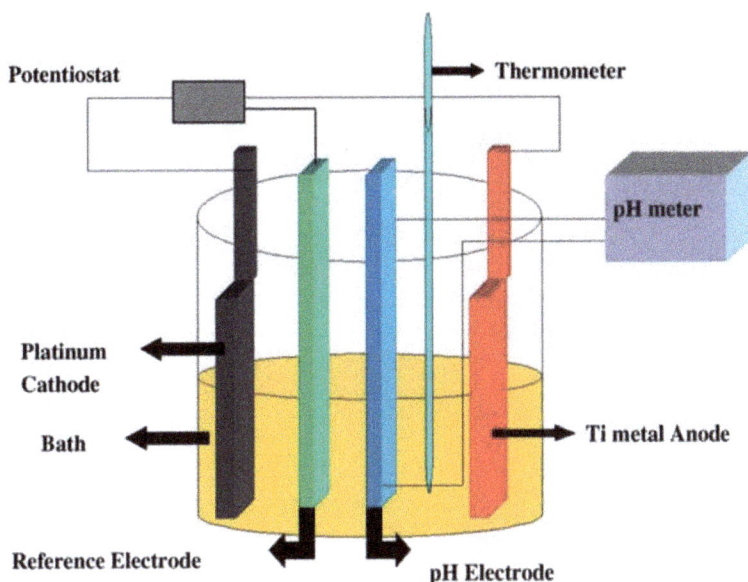

5. Electron Microscopy Images of TNT

Figure 3a–c show that TNT of regular dimensions, shape and arrangements can indeed be obtained. For example, the cylindrical TNT in Figure 3a were prepared by anodic oxidation of Ti at a fixed voltage. The conical TNT in Figure 3b resulted from a programmed variation of the voltage. However, disorders can also occur. A relatively minor disorder is the formation of bundles due to wall collapsing especially at the TNT top because the walls cannot carry their weight or the voltage was too high (Figure 3d). Much more pronounced disorders can result in grass-like aspects caused by too long anodization in a very acidic electrolyte (Figure 3e). These pictures emphasize the critical role of the experimenter in synthesizing well-ordered TNT.

Figure 3. Scanning electron microscopy images: (**a–c**) highly ordered NTT viewed from various angles; (**d**) bundled TNT within the white circles; (**e**) TNT with grass-like aspect ("nanograss") due to collapsing of tube walls. Reproduced with permission from references [1] (a,b) and [5] (c–e).

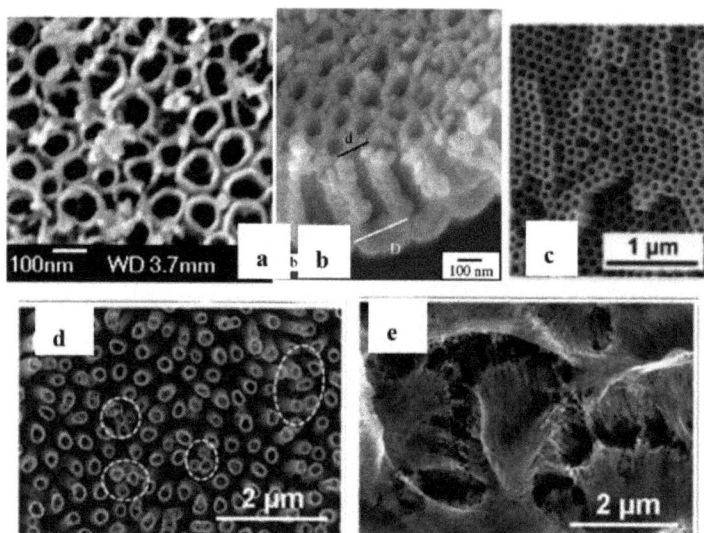

6. Removal of Gaseous Pollutants

As mentioned in the Introduction section, the only comparisons taken into account were those referring to TNT and layers of commercialized TNP of similar thickness and geometrical area. In the corresponding papers, only two pollutants were in fact used, *viz.* toluene (ethylbenzene) [4,10,11] and acetaldehyde (ethanal) [12,13]. They represent well the monocyclic aromatic hydrocarbons and aldehydes found in tropospheric air. Nevertheless, the use of other test pollutants is desirable to assess the effect on the efficacy ratios of various properties such as the molecular size and the presence of functional groups affecting the adsorption strength.

The ratio of the efficacy of various TNT over that of TiO_2 P25 (Figure 4) was comprised between 0.7 and 1.7. It referred to apparent first-order removal rates; this kinetic order resulted from the low vapor pressure of the tested pollutant. It is worth noting that the ratio of 0.7 corresponded to TNT synthesis by use of an alumina template [10] or a hydrothermal method [13], whereas the ratios of about 1 and 1.7 corresponded to TNT synthesis by anodic oxidation [4,11,12]. This might be an indication of the higher photocatalytic efficacy of TNT obtained via anodic oxidation, even though the number of reports is too small to draw a definitive conclusion. Anyhow, the main point is that this range of ratios does not appear to be in agreement with the efficacy enhancement anticipated from the use of straight regular tubes in place of TNP (*cf.* Section 2).

Figure 4. TNT/P25 TNP efficacy ratio in the photocatalytic removal of toluene [4,10,11] or acetaldehyde [12,13]. The corresponding references are indicated in brackets.

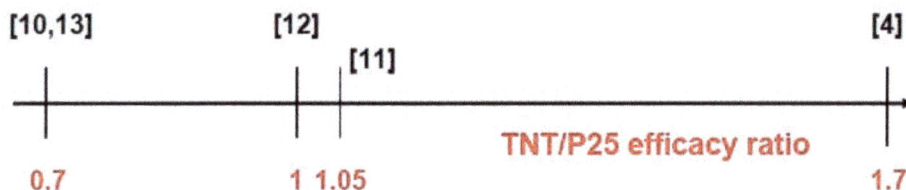

For each study, the efficacy ratios indicated here refer to the TNT presenting the highest efficacy. For example, in the case of acetaldehyde removal [13], Figure 5 shows the prominent role of the calcination temperature of TNT. The TNT calcined below 773 K were less efficient than P25, those calcined at 773 K were about as efficient, and those calcined at 873 K were markedly more efficient. This much higher efficacy was a priori unexpected because calcination at 873 K produced a drastic change in the texture: the tubes collapsed, forming nanorods. The supposed easier access of the reactants because of the tubular shape was accordingly suppressed, so that the efficacy should have decreased instead of increased. According to the authors [13], the decrease in surface area was more than compensated by the decrease in the recombination rate of photoproduced charges resulting from a higher crystallinity; the appearance of faceted anatase particles favoring the photocatalytic removal of acetaldehyde was also suggested (*cf.* Section 9). Indeed, a tradeoff between surface area and charge recombination rate, and possibly other factors, has also been proposed regarding the pollutant-dependent efficacy of various TNP in the removal of pollutants in liquid water [14–17].

Figure 5. Pseudo-first-order rate constant of CH_3CHO removal against the calcination temperature T_c of TNT with reference to P25 TiO_2. At $T_c > 400$ °C, the tubular morphology collapsed. Reproduced with permission from reference [13].

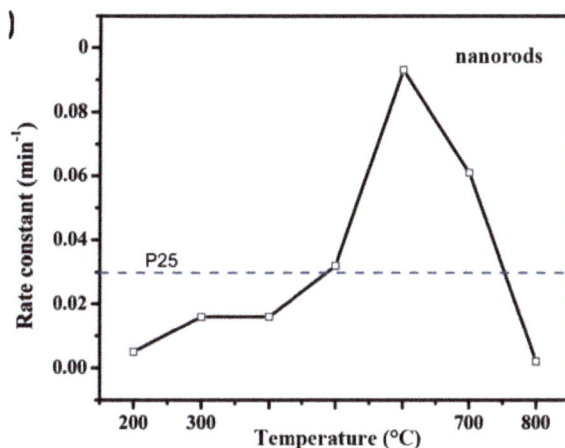

Also, the efficacy ratios indicated here refer to the initial rate of pollutant removal. In the case of acetaldehyde, Figure 6 shows that TNT and P25 had the same efficacy during the initial period of degradation (see the rectangle in red color) [12]. Then the TNT became more efficient. According to the authors, the higher surface area offered by the TNT restricted the competition between ethanal and its intermediate products of degradation and thereby enhanced the removal rate of ethanal. This advantage of TNT over TNP may even be expected to be more pronounced with organic compounds containing more C atoms than ethanal and hence producing more numerous intermediate products of degradation. The message here is that the comparisons of efficacy ratios also depend on the irradiation time. Obviously, it should additionally be affected by the nature and partial pressure of the pollutant.

These remarks emphasize the need for more investigations with several probe molecules. However, given the results reported so far [4,10–13], it would be surprising that a general substantial advantage of TNP for the photocatalytic removal of gaseous pollutants would be observed.

Figure 6. Kinetic variations in the concentrations of CH_3CHO and CO_2 during CH_3CHO removal when either TNP or a P25 TiO_2 film was the photocatalyst. Reproduced with permission from reference [12].

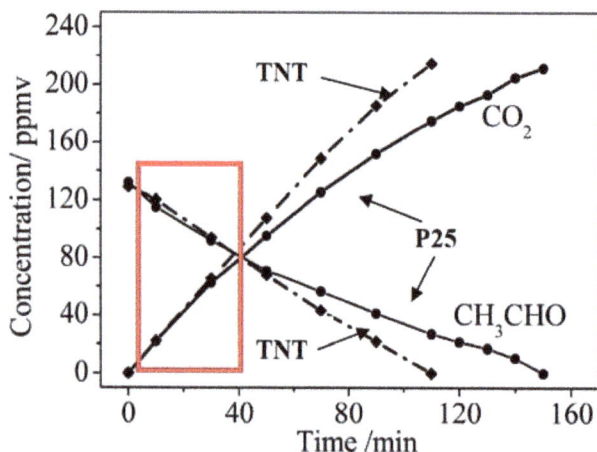

7. Removal of Pollutants in Aqueous Phase

Unfortunately, for the studies enabling comparisons of TNT and layers of commercialized TNP of similar thickness and geometrical area, the only probe compounds employed were dyes [18–22] (except in one case where it was phenol [23]). This is understandable because dye decolorization can be followed easily and rapidly using a commonly available spectrometer. However, photolysis, reductive bleaching and dye sensitization are phenomena that can cause decolorization, in addition to photocatalysis that is supposed to be exclusively monitored by the spectra [24–28]. Consequently, the use of a dye as a probe of photocatalytic efficacy implies that the presence of TNT or TNP does not affect or equally affects the phenomena other than photocatalysis. That is questionable because dye adsorption is likely different on the TNT and TNP investigated. Moreover, dyes are usually not pure; therefore, an additional bias could be the different role, with respect to TNT or TNP, of the compounds present as impurities. Accordingly, the use of a series of probe pollutants [14–17] both pure and non-photosensitive to the wavelengths employed to activate TiO_2 would be highly desirable in future works for a better comparison of TNT and TNP.

As in the case of gaseous pollutants, the efficacy ratios considered here are those referring to the TNT presenting the highest efficacy, since, in particular, the calcination temperature had a strong influence (e.g., reference [18]). They also corresponded to apparent first-order removal rates arising from low dye initial concentration. With P25 being the reference, the efficacy ratios were either 0.8 (four cases) or 1.3 (one case) depending on the TNT and the test dye (Figure 7). The ratio of 0.8 corresponded to TNT synthesized using a hydrothermal method [18,20,21] and to C/TiO_2 composite nanotubes obtained by electrospinning so that they were in fact hollow nanofibers [22]. The ratio of 1.3 corresponded to TNT prepared by anodic oxidation [19]. As for pollutant removal in air, this latter synthesis type seems to enable one to get TNT having a higher photocatalytic efficacy. However, given the small number of cases, no definitive conclusion can be drawn. Anyhow, whatever the TNT synthesis, no great difference in efficacy was observed between various TNT and P25. An efficacy ratio of 1.4 (Figure 7) was reported when Ishihara ST-01 was the reference for TNP instead of 0.8 when it was P25, the dye being rhodamine B [22]. This difference was not unexpected because highly porous TiO_2 like ST-01 (300 $m^2 \cdot g^{-1}$) is usually less efficient than non-porous P25 for photocatalytic dye removal.

Figure 7. TNT/TNP efficacy ratio in the photocatalytic removal of dyes: amaranth (A), rhodamine B (Rh), methyl orange (MO), and acid orange 7 (AO). The TNP were P25 except in one case as is shown. The corresponding references are indicated in brackets.

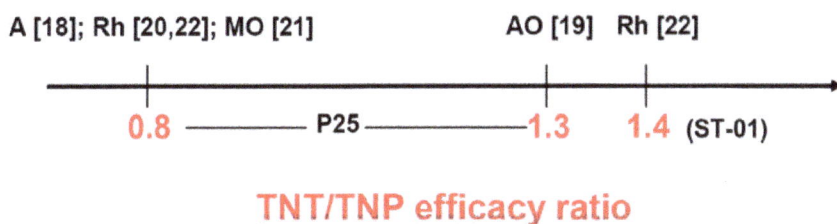

The situation is different when TNT are used in a photoelectrocatalytic device. The efficacy ratio of TNT and P25 films with similar thickness and geometrical area was found to increase from 0.85 (photocatalysis) to 1.85 (photoelectrocatalysis with a 0.6 V bias) for the initial rate constant of phenol removal [23]. This suggests better electron transport by TNT to the Ti electrode in the photoelectrochemical cell. Indeed, the initial photocurrent was higher by a factor of ~30 for TNT than for the P25 film. Clearly, this result is in favor of the use of TNT in photoelectrocatalysis compared to photocatalysis.

8. Modifications of TNT to Improve the Efficacy in the UV or to Extend It to the Visible

The same types of modifications have been applied to TNT as those commonly used for TNP to improve the photocatalytic efficacy under UV irradiation and to extend it to the visible spectral range. For example, they included: metal deposits (Pt [10], Ag [21,29,30], Au [30]); cation doping (Bi^{3+} [31], Nd^{3+} [32], Gd^{3+} [33], Zr^{4+} [34]); N "doping" [33,34]; formation of composites with chalcogenides (CdS [20], CuS [35], $CuInS_2$ [36], ZnTe [37]) or activated carbon [22] or graphene [30,38]. To measure the improvements, the following probe compounds were employed: toluene in the gas phase [10] and, in aqueous phase, most often dyes but also phenol [35], 4,4'-dibromobiphenyl [33], 2,4-dichlorophenoxyacetic acid (2,4-D) [30,36] and anthracene-9-carboxylic acid [37].

As was expected, significant or even substantial photocatalytic efficacy increases were observed in the UV spectral range. The optimum percentage of cation doping was the same for TNT as for TNP. Visible-light induced activity was found in the case of N "doping" [33,34]. However, to the best of our knowledge, no comparison with commercialized TiO_2 particles modified similarly was reported. Therefore these studies do not provide answers as to whether modified TNP are more efficient than similarly modified TNP in photocatalytic purification of water and air.

9. Why do TNT Appear not to be Decisively Attractive for Air and Water Photocatalytic Purification?

The expectations for high TNT photocatalytic efficacy (see Section 2) based mainly on: (i) easier access of the reactants; (ii) decreased recombination rate of photogenerated charges; and (iii) possibly a better use of the photons although this use depends strongly on the TNT morphology [7,39], seem to be *a priori* reasonable. Unfortunately, the efficacy ratios of TNT and layers of commercialized TNP of similar thickness and geometrical area collected in the present article (Figures 4 and 7) appear not to be decisively attractive for removal of both gaseous and aqueous pollutants and hence not in line with these expectations.

A possible explanation for this discrepancy between expected and observed efficacy ratios could arise from the occurrence of various disorders (cracks, distortions, formation of bundles, *etc.*) in the TNT texture and arrangements, whereas the expectations are based on TNT in the shape of perfect and parallel cylinders. Indeed, higher efficacy ratios (*cf.* Sections 6 and 7) corresponded to TNT prepared by anodic oxidation which usually generates TNT that are better ordered The unfavorable effect of TNT disorders has been shown for electron transfer in dye sensitized solar cells. For example, TNT films cleaned with water and then dried in air contained both clusters of bundles and microcracks, whereas those cleaned with ethanol and then dried using supercritical carbon dioxide did not (Figure 8) [9]. The reaction time of the conduction band electrons of TiO_2 with the photo-excited dye (a bipyridyl ruthenium complex) was increased by ~20% for the disordered TNT. For pollutant removal both hole-induced and electron-induced events are thought to be involved for both the formation of radicals, such as hydroxyl radicals, and direct charge transfer from TiO_2 to the pollutant. Consequently, higher decreases in efficacy might result from TNT disorders than in the case of the simpler reduction of a photo-excited dye.

Additionally, it must not be forgotten that easier access of the reactants and decreased recombination rate of photogenerated charges are not the only factors affecting the photocatalytic efficacy. Among these factors, the crystallinity, the fraction of under-coordinated surface atoms and the exposed planes of TiO_2 [40–42] have been shown to also play a role. For the TNT compared here with P25 layers, these interrelated factors could counteract the advantages expected from the TNT texture regarding the photocatalytic efficacy.

Figure 8. Scanning electron microscopy images of TNT films that were cleaned with either (**a**) ethanol or (**b**) water, and then either (a) supercritical CO_2-dried or (b) air-dried. Reproduced with permission from reference [9].

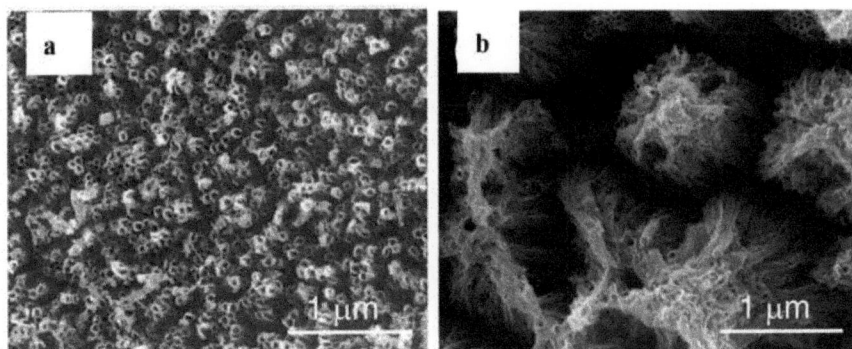

10. Conclusions

Reports allowing comparison of the photocatalytic efficacy of TNT and commercialized TNP under the strict conditions indicated in the Introduction are in fact scarce. Most papers include efficacy comparisons between differently prepared and/or treated TNT. In other papers, comparison with TNP does not fulfill the conditions selected here.

In the gas phase, the TNT/P25 TNP efficacy ratio derived from five reports ranges between 0.7 and 1.7 for toluene and acetaldehyde, the only pollutants tested. In the case of liquid water, this ratio ranges between 0.8 and 1.3 for the four dyes investigated in five reports and it is 0.85 for phenol, the only other probe compound conforming to the criteria. All these values illustrate that the TNT used in these comparisons do not show a decisive efficacy improvement, in contrast with the expectations based on the TNT texture (*cf.* Section 2).

Possible explanations for the discrepancy between the expectations and the actual results are discussed in Section 9. Disorders, such as cracks, distortions, formation of bundles, *etc.*, in the TNT morphology would obviously impair the anticipated advantages. However, it might also well be that the expectations are based on a too simple view of photocatalytic events. Multiple, interrelated factors that intervene in photocatalysis and are ignored in this simple view could counteract the positive roles expected from the TNT texture. To wit, the TNT/P25 ratio of the initial rate constant of phenol removal was very substantially decreased in photocatalysis relative to photoelectrocatalysis where the easier electron transport rate attributed to the tubular morphology is a dominant factor [23].

Clearly, more correctly-designed comparisons involving well-ordered TNT are necessary to reach better substantiated conclusions regarding the question used as the title of this article. As aforementioned, they must include a series of appropriately chosen probe molecules. The question of the stability of the photocatalytic efficacy of the TNT during the photocatalytic removal of pollutants should also be addressed; until now, only laboratory trials based on a few repetitions have been reported.

Conflicts of Interest

The author declares no conflict of interest.

References

1. Mor, G.K.; Varghese, O.K.; Paulose, M.; Shankar, K.; Grimes, C.A. A review on highly ordered vertically oriented TiO$_2$ nanotube arrays: Fabrication, material properties, and solar energy applications. *Sol. Energy Mater. Sol. Cells* **2006**, *90*, 2011–2075.

2. Yan, J.; Zhou, F. TiO$_2$ nanotubes: Structure optimization for solar cells. *J. Mater. Chem.* **2011**, *21*, 9406–9418.

3. Zhang, Q.; Cao, G. Nanostructured photoelectrodes for dye-sensitized solar cells. *Nano. Today* **2011**, *6*, 91–109.

4. Kontos, A.G.; Katsanaki, A.; Maggos, T.; Likodimos, V.; Ghicov, A.; Kim, D.; Kunze, J.; Vasilakos, C.; Schmuki, P.; Falaras, P. Photocatalytic degradation of gas pollutants on self-assembled titania nanotubes. *Chem. Phys. Lett.* **2010**, *490*, 58–62.

5. Roy, P.; Berger, S.; Schmuki, P. TiO$_2$ Nanotubes: Synthesis and Applications. *Angew. Chem. Int. Ed.* **2011**, *50*, 2904–2939.

6. Wong, C.L.; Tan, Y.N.; Mohamed, A.R. A review on the formation of titania nanotube photocatalysts by hydrothermal treatment. *J. Environ. Manag.* **2011**, *92*, 1669–1680.

7. Ma, Y.; Lin, Y.; Xiao, X.; Li, X.; Zhou, X. Synthesis of TiO$_2$ nanotubes and light scattering property. *Chin. Sci. Bull.* **2005**, *50*, 1985–1990.

8. Ghadiri, E.; Taghavinia, N.; Zakeeruddin, S.M.; Grätzel, M.; Moser, J.-E. Electron Collection Efficiency in Dye-Sensitized Solar Cells Based on Nanostructured TiO$_2$ Hollow Fibers. *Nano Lett.* **2010**, *10*, 1632–1638.

9. Zhu, K.; Vinzant, T.B.; Neale, N.R.; Frank, A.J. Removing Structural Disorder from Oriented TiO$_2$ Nanotube Arrays: Reducing the Dimensionality of Transport and Recombination in Dye-Sensitized Solar Cells. *Nano Lett.* **2007**, *7*, 3739–3746.

10. Chen, Y.; Crittenden, J.C.; Hackney, S.; Sutter, L.; Hand, D.W. Preparation of a Novel TiO$_2$-Based p-n Junction Nanotube Photocatalyst. *Environ. Sci. Technol.* **2005**, *39*, 1201–1207.

11. Wu, Z.; Guo, S.; Wang, H.; Liu, Y. Synthesis of immobilized TiO$_2$ nanowires by anodic oxidation and their gas phase photocatalytic properties. *Electrochem. Commun.* **2009**, *11*, 1692–1695.

12. Liu, Z.; Zhang, X.; Nishimoto, S.; Murakami, T.; Fujishima, A. Efficient Photocatalytic Degradation of Gaseous Acetaldehyde by Highly Ordered TiO$_2$ Nanotube Arrays. *Environ. Sci. Technol.* **2008**, *42*, 8547–8551.

13. Vijayan, B.; Dimitrijevic, N.M.; Rajh, T.; Gray, K. Effect of Calcination Temperature on the Photocatalytic Reduction and Oxidation Processes of Hydrothermally Synthesized Titania Nanotubes. *J. Phys. Chem. C* **2010**, *114*, 12994–13002.

14. Agrios, A.G.; Pichat, P. Recombination rate of photogenerated charges *versus* surface area: Opposing effects of TiO₂ sintering temperature on photocatalytic removal of phenol, anisole, and pyridine in water. *J. Photochem. Photobiol. A* **2006**, *180*, 130–135.

15. Enriquez, R.; Pichat, P. Different Net Effect of TiO₂ Sintering Temperature on the Photocatalytic Removal Rates of 4-Chlorophenol, 4-Chlorobenzoic acid and Dichloroacetic acid in Water. *J. Environ. Sci. Health A* **2006**, *41*, 955–966.

16. Enriquez, R.; Agrios, A.G.; Pichat, P. Probing multiple effects of TiO₂ sintering temperature on photocatalytic activity in water by use of a series of organic pollutant molecules. *Catal. Today* **2007**, *120*, 196–202.

17. Ryu, J.; Choi, W. Substrate-Specific Photocatalytic Activities of TiO₂ and Multiactivity Test for Water Treatment Application. *Environ. Sci. Technol.* **2008**, *42*, 294–300.

18. Qamar, M.; Yoon, C.R.; Oh, H.J.; Lee, N.H.; Park, K.; Kim, D.H.; Lee, K.S.; Lee, W.J.; Kim, S.J. Preparation and photocatalytic activity of nanotubes obtained from titanium dioxide. *Catal. Today* **2008**, *131*, 3–14.

19. Hou, Y.; Li, X.; Liu, P.; Zou, X.; Chen, G.; Yue, P.-L. Fabrication and photo-electrocatalytic properties of highly oriented titania nanotube arrays with {101} crystal face. *Sep. Purif. Technol.* **2009**, *67*, 135–140.

20. Zhou, Q.; Fu, M.-L.; Yuan, B.-L.; Cui, H.-J.; Shi, J.-W. Assembly, characterization, and photocatalytic activities of TiO₂ nanotubes/CdS quantum dots nanocomposites. *J. Nanopart. Res.* **2011**, *13*, 6661–6672.

21. Liu, Z.; Chena, J.; Zhang, Y.; Wu, L.; Li, X. The effect of sandwiched Ag in the wall of TiO₂ nanotube on the photo-catalytic performance. *Mater. Chem. Phys.* **2011**, *128*, 1–5.

22. Im, J.H.; Yang, S.J.; Yun, C.H.; Park, C.R. Simple fabrication of carbon/TiO₂ composite nanotubes showing dual functions with adsorption and photocatalytic decomposition of Rhodamine B. *Nanotechnology* **2012**, *23*, doi:10.1088/0957-4484/23/3/035604.

23. Liu, Z.; Zhang, X.; Nishimoto, S.; Jin, M.; Tryk, D.A.; Murakami, T.; Fujishima, A. Highly Ordered TiO₂ Nanotube Arrays with Controllable Length for Photoelectrocatalytic Degradation of Phenol. *J. Phys. Chem. C* **2008**, *112*, 25–259.

24. Julson, A.J.; Ollis, D.F. Kinetics of dye decolorization in an air–solid system. *Appl. Catal. B* **2006**, *65*, 315–325.

25. Yan, X.; Ohno, T.; Nishijima, K.; Abe, R.; Ohtani, B. Is methylene blue an appropriate substrate for a photocatalytic activity test? A study with visible-light responsive titania. *Chem. Phys. Lett.* **2006**, *429*, 606–610.

26. Chin, P.; Ollis, D.F. Decolorization of organic dyes on Pilkington Activ™ photocatalytic glass. *Catal. Today* **2007**, *123*, 177–188.

27. Mills, A. An overview of the methylene blue ISO test for assessing the activities of photocatalytic films. *Appl. Catal. B* **2012**, *128*, 144–149.

28. Ollis, D.; Silva, C.G.; Faria, J. Simultaneous photochemical and photocatalyzed liquid phase reactions: Dye decolorization kinetics. *Catal. Today* **2014**, in press.

29. Paramasivam, I.; Macak, J.M.; Schmuki, P. Photocatalytic activity of TiO₂ nanotube layers loaded with Ag and Au nanoparticles. *Electrochem. Commun.* **2008**, *10*, 71–75.

30. Tang, Y.; Luo, S.; Teng, Y.; Liu, C.; Xu, X.; Zhang, X.; Chen, L. Efficient removal of herbicide 2,4-dichlorophenoxyacetic acid from water using Ag/reduced graphene oxide co-decorated TiO_2 nanotube arrays. *J. Hazard. Mater.* **2012**, *241–242*, 323–330.

31. Xu, J.; Wang, W.; Shang, M.; Gao, E.; Zhang, Z.; Ren, J. Electrospun nanofibers of Bi-doped TiO_2 with high photocatalytic activity under visible light irradiation. *J. Hazard. Mater.* **2011**, *196*, 426–430.

32. Xu, Y.-H.; Chen, C.; Yang, X.-L.; Li, X.; Wang, B.-F. Preparation, characterization and photocatalytic activity of the neodymium-doped TiO_2 nanotubes. *Appl. Surf. Sci.* **2009**, *255*, 8624–8628.

33. Liu, H.; Liu, G.; Xie, G.; Zhang, M.; Hou, Z.; He, Z. Gd^{3+}, N-codoped trititanate nanotubes: Preparation, characterization and photocatalytic activity. *Appl. Surf. Sci.* **2011**, *257*, 3728–3732.

34. Liu, H.; Liu, G.; Fan, J.; Zhou, Q.; Zhou, H.; Zhang, N.; Hou, Z.; Zhang, M.; He, Z. Photoelectrocatalytic degradation of 4,4'-dibromobiphenyl in aqueous solution on TiO_2 and doped TiO_2 nanotube arrays. *Chemosphere* **2011**, *82*, 43–47.

35. Ratanatawanate, C.; Bui, A.; Vu, K.; Balkus, K.J., Jr. Low-Temperature Synthesis of Copper(II) Sulfide Quantum Dot Decorated TiO_2 Nanotubes and Their Photocatalytic Properties. *J. Phys. Chem. C* **2011**, *115*, 6175–6180.

36. Liu R.; Liu, Y.; Liu, C.; Luo, S.; Teng, Y.; Yang, L.; Yang, R.; Cai, Q. Enhanced photoelectrocatalytic degradation of 2,4-dichlorophenoxyacetic acid by $CuInS_2$ nanoparticles deposition onto TiO_2 nanotube arrays. *J. Alloys Compd.* **2011**, *509*, 2434–2440.

37. Liu, Y.; Zhang, X.; Liu, R.; Yang, R.; Liu, C.; Cai, Q. Fabrication and photocatalytic activity of high-efficiency visible-light-responsive photocatalyst ZnTe/TiO_2 nanotube arrays. *J. Solid State Chem.* **2011**, *184*, 684–689.

38. Perera, S.D.; Mariano, R.G.; Vu, K.; Nour, N.; Seitz, O.; Chabal, Y.; Balkus, K.J., Jr. Hydrothermal Synthesis of Graphene-TiO_2 Nanotube Composites with Enhanced Photocatalytic Activity. *ACS Catal.* **2012**, *2*, 949–956.

39. Atyaoui, A.; Bousselmi, L.; Cachet, H.; Pu, P.; Sutter, E.M.M. Influence of geometric and electronic characteristics of TiO_2 electrodes with nanotubular array on their photocatalytic efficiencies. *J. Photochem. Photobiol. A* **2011**, *224*, 71–79.

40. Ohtani, B. Design and Development of Active Titania and Related Photocatalysts. In *Photocatalysis and Water Purification: From Fundamentals to Recent Applications*; Pichat, P., Ed.; Wiley-VCH: Weinheim, Germany, 2013; pp. 75–102.

41. Liu, G.; Yu, J.C.; Lu, G.Q.; Cheng, H.-M. Crystal facet engineering of semiconductor photocatalysts: Motivations, advances and unique properties. *Chem. Commun.* **2011**, *47*, 6763–6783.

42. Jiang, Z.; Tang, Y.; Tay, Q.; Zhang, Y.; Malyi, O.I.; Wang, D.; Deng, J.; Lai, Y.; Zhou, H.; Chen, X. Understanding the Role of Nanostructures for Efficient Hydrogen Generation on Immobilized Photocatalysts. *Adv. Energy Mater.* **2013**, *3*, 1368–1380.

Influence of Post-Treatment Operations on Structural Properties and Photocatalytic Activity of Octahedral Anatase Titania Particles Prepared by an Ultrasonication-Hydrothermal Reaction

Zhishun Wei, Ewa Kowalska and Bunsho Ohtani

Abstract: The influence of changes in structural and physical properties on the photocatalytic activity of octahedral anatase particles (OAPs), exposing eight equivalent {101} facets, caused by calcination (2 h) in air or grinding (1 h) in an agate mortar was studied with samples prepared by ultrasonication (US; 1 h)–hydrothermal reaction (HT; 24 h, 433 K). Calcination in air at temperatures up to 1173 K induced particle shape changes, evaluated by aspect ratio (AR; d_{001}/d_{101} = depth vertical to anatase {001} and {101} facets estimated by the Scherrer equation with data obtained from X-ray diffraction (XRD) patterns) and content of OAP and semi-OAP particles, without transformation into rutile. AR and OAP content, as well as specific surface area (SSA), were almost unchanged by calcination at temperatures up to 673 K and were then decreased by elevating the calcination temperature, suggesting that calcination at a higher temperature caused dull-edging and particle sintering, the latter also being supported by the analysis of particle size using XRD patterns and scanning electron microscopic (SEM) images. Time-resolved microwave conductivity (TRMC) showed that the maximum signal intensity (I_{max}), corresponding to a product of charge-carrier density and mobility, and signal-decay rate, presumably corresponding to reactivity of charge carriers, were increased with increase in AR, suggesting higher photocatalytic activity of OAPs than that of dull-edged particles. Grinding also decreased the AR, indicating the formation of dull-edged particles. The original non-treated samples showed activities in the oxidative decomposition of acetic acid (CO_2 system) and dehydrogenation of methanol (H_2 system) comparable to and lower than those of a commercial anatase titania (Showa Denko Ceramics FP-6), respectively. The activities of calcined and ground samples for the CO_2 system and H_2 system showed almost linear relations with AR and I_{max}, respectively, suggesting that those activities may depend on different properties.

Reprinted from *Molecules*. Cite as: Wei, Z.; Kowalska, E.; Ohtani, B. Influence of Post-Treatment Operations on Structural Properties and Photocatalytic Activity of Octahedral Anatase Titania Particles Prepared by an Ultrasonication-Hydrothermal Reaction. *Molecules* **2014**, *19*, 19573-19587.

1. Introduction

Titanium(IV) oxide (titania) has been the most frequently used photocatalyst in various areas [1,2]. In almost all practical applications of photocatalysis, titania has been used as a photocatalyst because of its many advantages including low price, high photostability, nontoxicity and superior redox ability [3]. Although studies on titania photocatalysis [3] have suggested the desired structural and/or physical properties for high-level photocatalytic activity, e.g., anatase crystallites rather than rutile ones and smaller particles, *i.e.*, higher specific surface area, intrinsic interpretation for possible structure-activity correlations has not yet been obtained, at least to the authors' knowledge [4–6].

Particle morphology has recently attracted much attention from scientists as a possible key parameter for controlling the activity of photocatalyst particles [7–14]. Various methodologies have been developed to control particle morphology by using diverse treatments such as ultrasonication, grinding, washing, microwave irradiation, and thermal and pressure treatments [15–17], and by selective preparation of single-crystalline facetted photocatalyst particles [18–20]. Along this line, octahedral anatase particles (OAPs), exposing eight equivalent {101} facets, with sizes of several tens of nanometers, have been prepared by ultrasonication (US) of partially proton-exchanged potassium titanate nanowires followed by hydrothermal reaction (HT) [21,22]. As is suggested by the fact that natural anatase titania minerals have been found in an octahedral shape, the {101} facets are thermodynamically most stable and tend to be exposed when titania is prepared under equilibrium conditions such as HT. In the previous study of our group, it was shown that OAPs prepared by HT after US showed relatively high activity for oxidative decomposition of acetic acid under an aerobic atmosphere (CO_2 system) but low activity for methanol dehydrogenation under an argon atmosphere (H_2 system) [21–23].

As has been often and commonly observed for particulate photocatalysts, photocatalytic activities of OAP-containing samples prepared under different US and HT conditions have different physical/structural properties, morphologies and photocatalytic activities. However, it was accidentally observed that particles with almost the same structural activities, such as specific surface area, crystalline composition and particle size, except only for OAP content, could be prepared by changing only US time, and the photocatalytic activity, especially for a CO_2 system, was almost proportional to the OAP content; *i.e.*, the morphology of particles directly determines the photocatalytic activity [23].

In the present study, an OAP-containing sample was calcined or ground in an agate mortar in air to change mainly the particle morphology, and the influence of change in morphology on photocatalytic activity was examined in order to clarify the intrinsic reason for the morphology-dependent photocatalytic activity of anatase titania photocatalyst particles.

2. Results and Discussion

2.1. Preparation of Original OAP-Containing Sample

The previously reported US-HT process [21] led to conversion of partially protonated potassium titanate nanowires (TNWs) into anatase titania particles of predominantly octahedral shape. SEM images of the original sample (without post-treatment) are shown in Figure 1. The obtained particles were smaller than 50 nm and exhibited various morphologies, *i.e.*, OAP, semi-OAP and others (see the Experimental Section). In the XRD pattern, only peaks assigned to anatase crystallites appeared (Figure 2a). Images of high-resolution transmission electron microscopy (HRTEM) supported the presence of single crystals of anatase as shown in Figure 2b; *i.e.*, lattice fringes with a spacing of 0.35 nm and an angle between two kinds of fringes of 68.3° agreed well with the previously reported (101) lattice spacing and angle between (101) and (001) [21].

Figure 1. SEM images with (**a**) low magnification and (**b**) high magnification of the original OAP-containing sample. Scale bar: 100 nm.

Figure 2. (**a**) XRD pattern and (**b**) HRTEM image of the original OAP-containing sample.

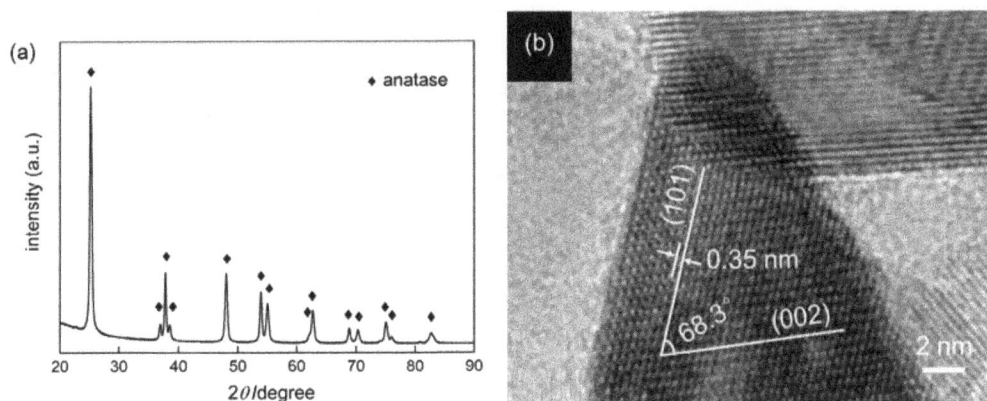

2.2. Influence of Calcination on the Structure of Particles

The influence of calcination on the structural properties of samples was studied by changing calcination temperatures (573–1173 K; 2 h) and keeping the other process conditions, *i.e.*, HT time (24 h), HT temperature (433 K), US time (1 h), TNW amount of titanate nanowires (267 mg) and Milli-Q water volume (80 mL) the same. Figure 3a shows the influence of calcination temperature on crystallinity (anatase; content of crystalline phases) and specific surface area (SSA). It should be noted that no peaks assigned to rutile crystalline were observed even with calcination at 1173 K. This feature will be discussed later. Figure 3b shows changes in crystallite size estimated from a 101 XRD peak ("XRD size"; same as in Figure 3a), particle size estimated by SEM observation ("SEM size"), and particle size expected from SSA ("SSA size") calculated with the assumption of spherical uniform-sized anatase particles [4]. It is expected that SEM size, the longest part of each particle, may be larger than XRD size, depth of the particle measured in the direction vertical to the {101} lattice plane. Furthermore, SEM size may be the size of aggregated particles if they are seen as a single crystal, while XRD size corresponds to the average size of a single crystalline part in aggregated particles. Considering those problems, use of XRD size seems better to discuss the change in particle size. As seen in Figure 3a, with elevation of calcination temperature, XRD size slightly increased at ≤973 K but was much larger in the temperature range higher than 973 K, while SSA decreased even at a lower temperature, *i.e.*, 773 K. Therefore, at a temperature higher than 773 K, a difference between XRD size and SSA size can be seen in Figure 3b, indicating partial sintering, with lattice mismatch, *i.e.*, formation of grain boundaries, of anatase crystallites by calcination. The decrease in crystallinity in the temperature range of 673–873 K is attributable to the formation of grain boundaries. (The increase in crystallinity at the lower temperature might be due to dehydration and/or crystallization of amorphous phase.) Thus, calcination at temperatures of 773–1073 K in air induced sintering of some of the particles without lattice matching as the difference between XRD size and SSA size shows; XRD size became comparable to or larger than that of SSA size by calcination at 1173 K, suggesting fusion of crystallites to a larger single crystal.

These findings indicate that the original OAP-containing particles (content of OAP and semi-OAP: 62%) are stable toward heat-induced crystal transformation at the temperature at lowest below 1073 K, presumably due to their exposure of ordered {101} facets with less defects, which may trigger crystal transformation into larger single crystals (fusion) or rutile. As was stated above, all of the samples, even those heated at 1173 K, reported here included no rutile phase, though it has been reported that conversion of anatase to rutile usually occurs at much lower temperatures, e.g., 823–973 K [24–27]. The high level of heat tolerance of the present OAP-containing particles is attributable to the above-mentioned particle morphology exposing ordered facets. It has been reported that rice-like titania nanorods were only stable enough to convert 2% of anatase into rutile at 1173 K [28]. Another possible reason for the heat tolerance is the presence of stabilizers, *i.e.*, potassium cations possibly contaminated from TNWs. It was reported that the presence of impurities enhanced the heat tolerance of anatase; e.g., lanthanum(III) oxide-modified titania underwent phase transformation by calcination at >1073 K [29]. Study on this heat tolerance is now under way and the details will be published elsewhere [30].

Calcination also changed the morphology of samples. As Figure 4 shows, sharply angulate particles became round-shaped particles by calcination accompanied by the above-mentioned sintering, and the higher the calcination temperature was, the higher was the extent of round edging. These observations were quantified using OAP (and semi-OAP) content and aspect ratio (AR) as shown in Figure 3c. The former was evaluated by counting numbers of OAPs, semi-OAPs and others in SEM images, and the latter was evaluated as the ratio of crystallite sizes calculated from 004 and 101 XRD-peak widths and might be closely related to the OAP content. With calcination at < 675 K, there was little change in either OAP content or AR, *i.e.*, morphology was not changed in this low temperature region. This is consistent with results of the above-mentioned particle-size analysis. In a higher temperature region, ≥773 K, both OAP content and AR were decreased to 3% and 0.72 at 1173 K, respectively, suggesting that sintering (or fusion) of crystallites induced a change in morphology to a round-edged shape. As Figure 3d shows, OAP and semi-OAP contents decreased linearly with decrease in AR > 1, indicating that both are measures of particle shape; the content of semi-OAPs, which include particles with round-edged summits, increased with decrease in AR. The x-intercept of the extrapolated linear part of plots was *ca.* 0.9.

Based on the assumption that an octahedral particle is changed to a decahedral particle that exposes eight {101} and two {001} facets, the AR value giving the lowest surface area/volume (weight) ratio is estimated to be ca. 0.95 [31], similar to the observed intercept. It seems that the calcination-induced AR decrease is accounted for by modification of particle shape to give smallest surface area/volume ratio. Considering the arbitrary property of OAP content, AR will be used when discussing shape dependence in the following sections.

Figure 3. (a) Correlations between crystalline size (XRD size), crystallinity, specific surface area and calcination temperature; **(b)** Comparison of XRD size, SEM size and SSA size of samples as functions of calcination temperature; **(c)** Changes in OAP/semi-OAP content and aspect ratio (AR) as functions of calcination temperature; **(d)** OAP/semi-OAP content as a function of AR.

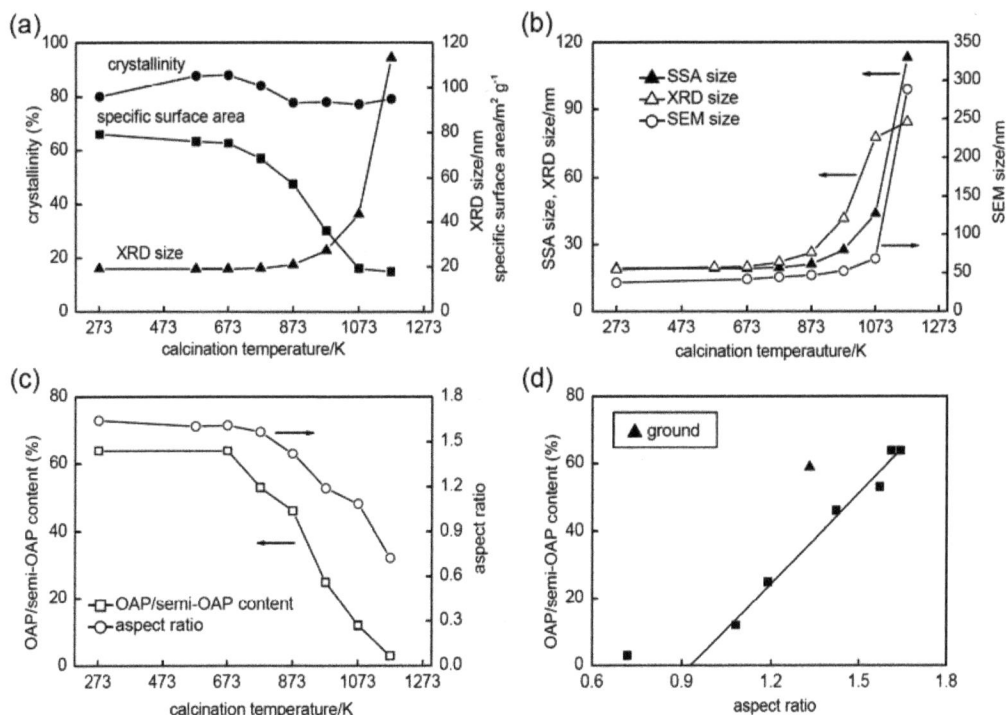

Figure 4. SEM images of OAPs with calcination temperatures of (**a**) 673 K; (**b**) 873 K; (**c**) 1073 K and (**d**) 1173 K. Scale bar: 100 nm.

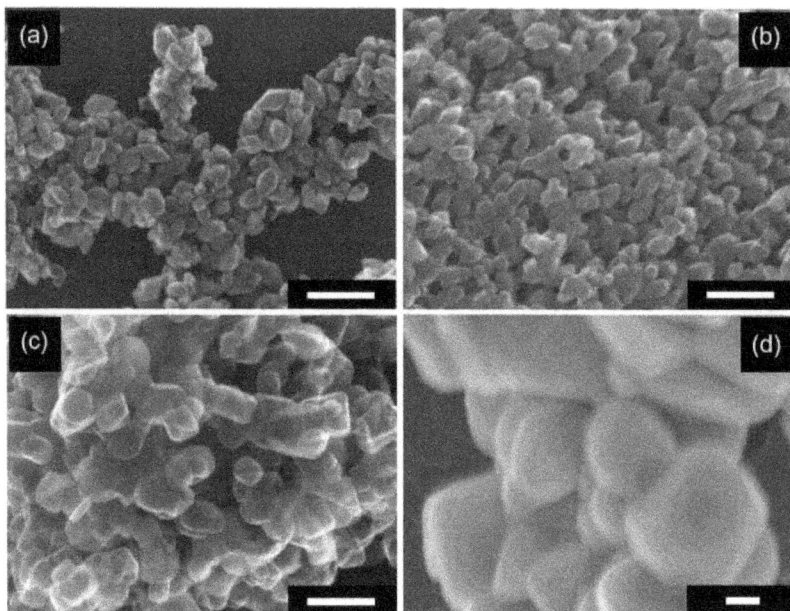

2.3. Influence of Calcination on Photocatalytic Activities

Photocatalytic activity for two reaction systems, *i.e.*, decomposition of acetic acid (CO_2 system) and methanol dehydrogenation (H_2 system), was examined. The dependence of photocatalytic activities on calcination temperature is shown in Figure 5a. The highest photocatalytic activity in the CO_2 system was obtained for the sample calcined at 673 K. The influence of calcination was less evident in the H_2 system than in the CO_2 system, and the sample calcined at 873 K showed 1.7 times higher photocatalytic activity than that of the original uncalcined sample. Although this activity trend can be explained, in a conventional manner, by a good balance of higher specific surface area and higher crystallinity [32], such an explanation does not give any intrinsic insights into the correlation between the physical/structural property and photocatalytic activity. In a previous study on photocatalytic activity of OAP-containing particles, it was found that photocatalytic activities for CO_2 and H_2 systems are governed only by the OAP content of samples; the activities were linearly increased with increase in OAP content for samples with almost the same other structural properties such as SSA, crystallinity, XRD size and total density of electron traps [23]. This tendency was also observed for the present samples, especially for activity for the CO_2 system as shown in Figure 5b as a function of AR, since AR seems to be a slightly better measure for content of OAPs as shown in Figure 3d. The plots (not shown) of photocatalytic activities in both CO_2 and H_2 systems as a function of SSA showed trends, increase with SSA, but had poor linearity. Thus, the activity of OAP-containing samples seems to be regulated by the shape of particles. The meaning of this shape-dependent activity and another dependence of activity in the H_2 system will be discussed later.

Figure 5. (a) Correlation between photocatalytic activity and calcination temperature (data of the original uncalcined sample shown at "273 K"); (b) Photocatalytic activities as functions of aspect ratio (AR). Plots for the original uncalcined sample are on a dotted line. Open (CO_2) and closed (H_2) triangles are of the ground sample.

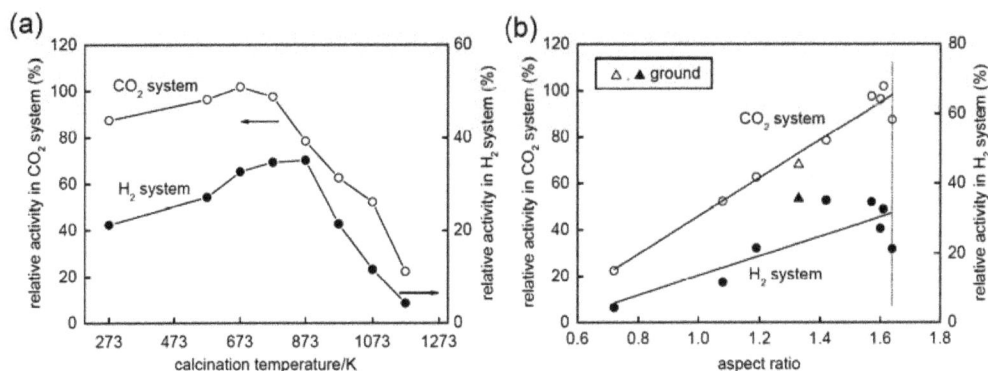

2.4. Time-Resolved Microwave Conductivity

As one of the physical properties closely related to photocatalytic activity, time-resolved microwave conductivity (TRMC) was measured for the present samples. Such carrier dynamics may be also studied by photocurrent measurements as was reported previously [33], but due to the possible complexity in interpretation of the measurements owing to the change in sample structural properties during the electrode preparation process, we used TRMC measurement, which can be performed in the form of powder. Intensity of the TRMC signal generally shows microwave absorption by migration of charge carriers and is considered to be a product of charge-carrier density and their mobility [34]. It is also thought that positive holes, one of the charge carriers, generated in titania particles, are quickly trapped in certain sites within the time of a nanosecond laser pulse to result in negligible migration. Therefore, the TRMC signal may reflect only migration of photoexcited electrons [34]. Figure 6 shows parts of time-course curves of TRMC signals; all samples exhibited a rise of the signal within a 355-nm laser-pulse duration (*ca.* 10 ns), and decay was observed in the μs time region. First, we assume that electrons photoexcited in a conduction band (CB) are trapped in traps, the energy level of which is lower than the bottom of the CB, within a ps time scale [35], *i.e.*, faster than a process detected in TRMC measurement. Then, charge migration giving maximum signal intensity (I_{max}) can be assigned to electron migration through the trap-detrap mechanism, *i.e.*, trapping in shallow traps and detrapping to the CB thermally, since electron hopping between deep traps is very slow [36] and thereby electrons in deep traps may have little involvement in the TRMC response. In such a circumstance, the higher the density of shallow traps is, the higher the mobility is. On the basis of these considerations, density of electrons trapped in shallow traps was slightly increased by calcination at <873 K and then decreased by calcination at >873 K. On the other hand, decay of the signal 10–20 ns after a laser pulse is attributable to trapping of electrons (once trapped in shallow traps) in deep traps, prohibiting further migration of electrons. The rate of this decay was evaluated, for convenience, by calculating the ratio of intensities at 4000 ns and the maximum (I_{4000}/I_{max}). The decay became slightly faster with calcination at ≤ 873 K, suggesting an increase in density of deep traps presumably caused by particle sintering, but, overall, the trends of I_{max} and I_{4000}/I_{max} were of mirror images, i.e., the higher I_{max} was, the lower was I_{4000}/I_{max}, indicating that the rate of trapping by deep traps depends also on the density of shallow traps.

Figure 6. (a) Influence of calcination temperature on TRMC signal; (b) I_{max} and I_{4000}/I_{max} as functions of AR; (c) Correlation between photocatalytic activities in CO_2 and H_2 systems and maximum intensity of the TRMC signal (I_{max}). Data for the original samples are plotted on dotted lines. Open (CO_2) and closed (H_2) triangles are of the ground sample.

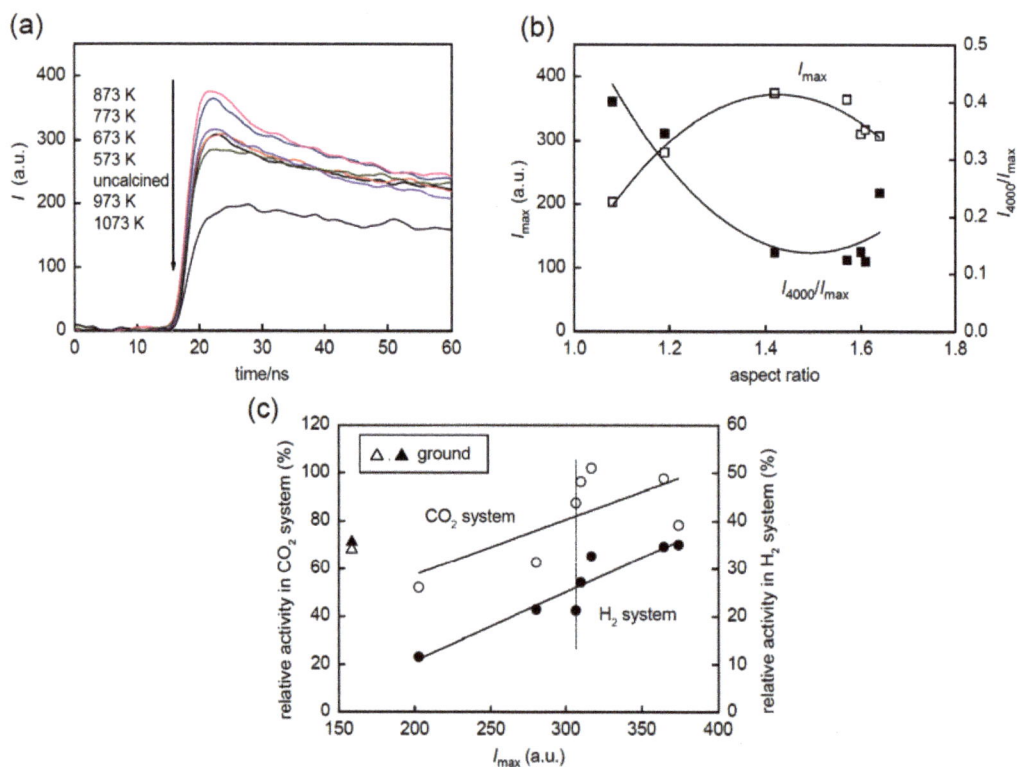

Figure 6c shows the dependence of photocatalytic activities for CO_2 and H_2 systems on I_{max}. A fairly linear relation was observed for activity in the H_2 system. In the H_2 system, platinum was in-situ deposited as a catalyst for hydrogen liberation by photoexcited electrons, and it has been observed that deposition of platinum reduces, in a short time region, the density of trapped electrons by capturing them [37]. Therefore, it seems reasonable that there is such a linear relation between activity in the H_2 system and I_{max} with the assumption that electrons that are able to be trapped in shallow traps migrate efficiently to platinum deposits to liberate hydrogen without being trapped by deep traps. On the other hand, oxygen as a possible electron acceptor in the CO_2 system is known to have a negligible influence on the nanosecond transient behavior of photoexcited electrons [38] and this is one of the reasons for the absence of a clear correlation with the TRMC data. Although the intrinsic reason is still ambiguous, the activity for the CO_2 system was governed by the particle morphology as is represented by AR (Figure 5b).

2.5. Influence of Grinding on Structure and Photocatalytic Activity of OAP-Containing Particles

Grinding is one of the most commonly used post-treatment operations to make samples homogeneous by separating loosely bound agglomerates. However, grinding may also change the structure of samples and their photocatalytic activity. In this regard, the influence of grinding was studied for OAP-containing samples. After 1 h grinding in an agate mortar, the morphology of OAP-containing samples was obviously changed to round-edged as shown in Figure 7; OAPs were changed to semi-OAPs and others, as schematically shown in Figure 7.

Figure 7. SEM images with (**a**) low magnification and (**b**) high magnification of the ground OAP-containing sample (scale bar: 100 nm) and schematic representation of change in morphology.

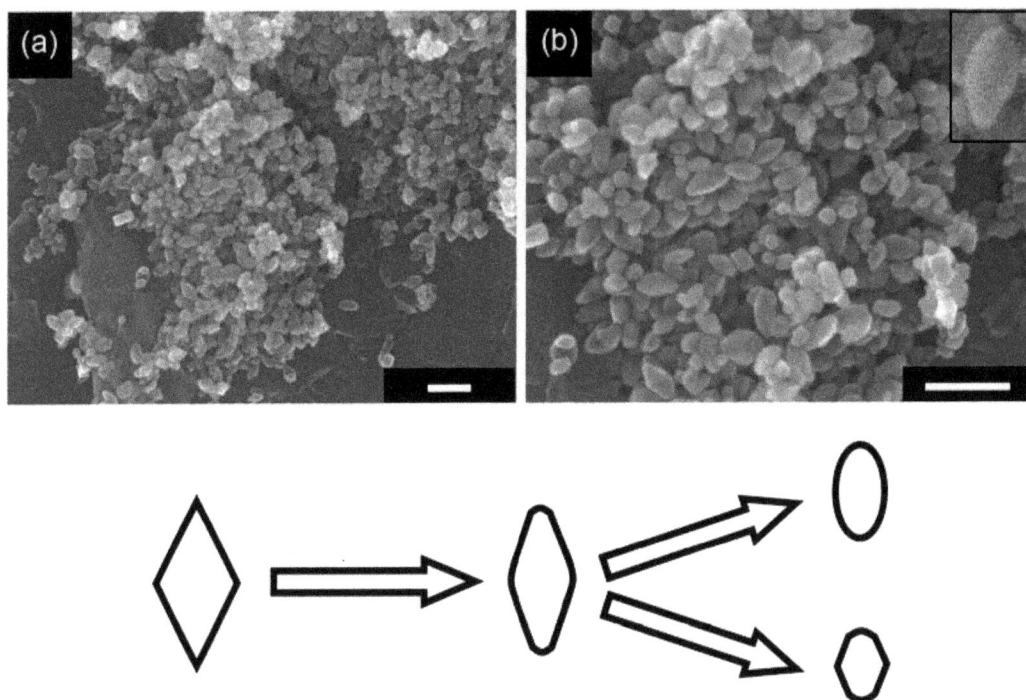

This SEM observation is consistent with the decrease in AR (1.33) from that of the original sample (1.64) as shown in Figure 3d. (A deviation of a plot for the ground sample from a linear relation in Figure 3d might be caused by the arbitrary property of counting the numbers of semi-OAPs and others in SEM images; 30% semi-OPA content is expected from the plots.)

Photocatalytic activities in CO_2 and H_2 systems are plotted in Figure 5b,c, respectively. Again, the activity in the CO_2 system of the ground sample seems to be a linear relation of activity with AR (Figure 5b) but not being explained by I_{max} dependence (Figure 6c). On the other hand, the activity in the H_2 system was not plotted in a linear relation with I_{max} (Figure 6c). It has been observed that grinding titania particles in an agate mortar for a long time (>24 h) caused the formation of deep traps, leading to negligible photocatalytic activity presumably due to enhanced recombination at the deep traps [39]. Assuming that similar deep traps were also produced in the present case with grinding for a relatively short time, decreases in I_{max} and I_{4000}/I_{max} evaluated in TRMC measurements were accounted for by fast trapping in such deep traps in the ground sample. However, loaded platinum in the H_2 system may capture photoexcited electrons before being trapped in the deep traps, resulting in unexpectedly higher photocatalytic activity of the ground sample in the H_2 system. Of course, the possibility that preferable platinum loading on the ground sample led to higher photocatalytic activity in the H_2 system cannot be excluded. Study to find the true activity-controlling structural property is now in progress.

3. Experimental Section

3.1. Preparation of OAPs Samples

Partially proton-exchanged potassium titanate nanowires (TNWs, 267 mg) were ultrasonically dispersed in Milli-Q water (40 mL) for 1 h at 298 K. The suspension was placed in a sealed Teflon bottle (100 mL), into which was poured an additional 40 mL of Milli-Q water. The bottle was heated for 24 h at 433 K in an oven without agitation. After cooling, titania as white precipitate was centrifuged, washed with RO (reverse osmosis) water, and dried under vacuum (353 K, 12 h). The thus-obtained OAP-containing samples were used as the original starting samples for further studies.

Two post-treatment operations, calcination and grinding in air, were performed. For calcination, a 1.2-g sample in an air-open ceramic crucible was placed in an oven, and the temperature was raised to a given temperature (573–1173 K) at the rate of 10 K·min^{-1}, kept at that temperature for 2 h, and cooled down to ambient temperature. For grinding, a 1.2 g sample was ground in an agate mortar for 1 h in air.

3.2. Characterization

Specific surface area of samples was evaluated by nitrogen adsorption at 77 K using the Brunauer–Emmett–Teller (BET) equation. The morphology was studied by scanning electron microscopy (SEM, JEOL JSM-7400F, Akishima, Japan), scanning transmission electron microscopy (STEM, HITACHI HD-2000, Tokyo, Japan) and transmission electron microscopy (TEM, JEOL JEM-2100F). Particles in a sample were classified into three groups based on the results of SEM analysis: (a) OAP, an octahedral particle without observable defects; (b) semi-OAP, an octahedral particle with a defect (defects) and (c) others, an irregular shaped non-octahedral particle. Composition of these particles in each sample was measured by counting at least 200 particles in several SEM images of a sample [23].

XRD analysis was performed using the SmartLab intelligent X-ray diffraction system (Rigaku, Akishima, Japan) equipped with a sealed tube X-ray generator (a copper target; operated at 40 kV and 30 mA), a D/teX high-speed position-sensitive detector system and an ASC-10 automatic sample changer. Data acquisition conditions were as follows: 2θ range, 10–90°; scan speed, $1.00° \cdot min^{-1}$; and scan step, 0.008°. The obtained XRD patterns were analyzed by Rigaku PDXL, a crystal structure analysis package including Rietveld analysis [40], installed in a computer controlling the diffractometer. Crystallite size (XRD size) was evaluated from corrected width of an anatase 101 diffraction peak using the Scherrer equation. Crystallinity of a sample was evaluated using an internal standard, highly crystalline nickel oxide (NiO). The standard (20.0 wt %) was mixed thoroughly with a sample (80.0 wt %) by braying in an agate mortar. Since Rietveld analyses give composition of each crystal among total crystal content, the composition of the standard (formally 20.0 wt %) is measured to be larger if the sample contains a non-crystalline component. Therefore, crystalline and non-crystalline compositions are estimated by re-calculation to make the standard composition to be 20.0 wt %. At present, the authors regard the non-crystalline part to be water, which can be estimated by thermogravimetry, and amorphous titania and/or titanates.

3.3. Photocatalytic Activity Test

Photocatalytic activities of samples were examined by measuring the amount of evolved carbon dioxide (CO_2) and hydrogen (H_2) from continuously stirred (1000 rpm) suspensions of a sample (50 mg) in an aerated aqueous acetic acid solution (5.0 mL, 5.0 vol %) (CO_2 system) and a deaerated aqueous methanol solution (5.0 mL, 50 vol %) containing chloroplatinic acid (corresponding to 2.0 wt % (as platinum) of a photocatalyst) for *in-situ* platinum photodeposition (H_2 system), respectively. Photoirradiation (>290 nm) was performed with a 400-W high-pressure mercury lamp (Eiko-sha, Osaka, Japan) at 298 K. Amounts of liberated CO_2 and H_2 in gas phase were measured by gas chromatography (TCD-GC). The photocatalytic activities are presented with reference to those of a commercial titania photocatalyst, Showa Denko Ceramics FP-6 (anatase, SSA: *ca.* 100 $m^2 \cdot g^{-1}$, XRD size: 15 nm). FP-6 is known to exhibit a high level of photocatalytic activity among commercial titania powders, similar to well-known Evonik (Degussa, Essen, Germany) P25 [41,42]. The average rates of FP-6 and P25 were, respectively, *ca.* 0.54 and 0.91 $\mu mol \cdot h^{-1}$ in the H_2 system and 0.039 and 0.046 $\mu mol \cdot h^{-1}$ in the CO_2 system [23]. Due to negligible solubility of H_2 in water and CO_2 in an aqueous solution of acetic acid, no correction was made for gas dissolution in reaction suspensions.

3.4. Time-Resolved Microwave Conductivity Measurements

Charge-carrier dynamics was studied by measuring time-resolved microwave conductivity (TRMC) induced by a ns time-scale laser pulse in Laboratory of Physical Chemistry, University of Paris-Sud. Incident 36.8-GHz microwaves and UV laser pulses were generated by a Gunn diode of the K_a band and third harmonic of a 1064-nm Nd:YAG laser (10 Hz) with full width at half maximum of *ca.* 10 ns, respectively [43].

4. Conclusions

Post-treatments, calcination and grinding, of OAP-containing particles changed structural properties, including OAP/semi-OAP content, aspect ratio (AR) and particle size, and TRMC responses, e.g., I_{max} and I_{4000}/I_{max} reflecting change in charge-carrier dynamics. The changes caused by calcination were interpreted by annealing of single-crystal OAPs and sintering to be aggregates of single crystals of OAPs, leading to, respectively, higher crystallinity/higher density of shallow traps and slightly lower crystallinity/lower aspect ratio, *i.e.*, lower OAP content/higher density of deep traps due to grain boundaries, while the changes caused by grinding were interpreted by round-edging, *i.e.*, decrease in OAP (and semi-OAP) content/higher density of deep traps. As a result, photocatalytic activities in CO_2 and H_2 systems were linearly increased or decreased depending on AR reflecting OAP content and maximum TRMC signal intensity (I_{max}), respectively. Although few supporting experimental results have been obtained, a working hypothesis is that photocatalytic activity is governed by electron traps, the density of which is influenced by the method of preparation and post-treatment [44], and that most electron traps are located on the surface of particles and their energy depends on the structure of exposed surfaces such as {101} facets on OAPs. Analyses of energy-resolved density of electron traps [45] in the present OAP-containing particles are now in progress and results will be published in the near future.

Acknowledgments

The authors are grateful to Christophe Colbeau-Justin, Jonathan Verrett and Hynd Remita for technical assistance and discussion on the results of TRMC analyses. This study was partly supported by CONCERT-Japan Program (Japan Science and Technology Agency) and a Grant-in-Aid (KAKENHI) from the Ministry of Education, Culture, Sports, Science and Technology (MEXT) of Japan (Grant No. 2510750303). One of the authors (Z.W.) appreciates China Scholarship Council (CSC) for support.

Author Contributions

ZW, EK and BO designed research, analyzed the data and wrote the paper; ZW performed experiments. All authors read and approved the final manuscript.

Conflicts of Interest

The authors declare no conflict of interest.

References and Notes

1. Diebold, U. The surface science of titanium dioxide. *Surf. Sci. Rep.* **2003**, *48*, 53–229.
2. Chen, X.; Mao, S.S. Titanium dioxide nanomaterials: Synthesis, properties, modifications, and applications. *Chem. Rev.* **2007**, *107*, 2891–2959.
3. Fujishima, A.; Zhang, X.; Tryk, D.A. TiO$_2$ photocatalysis and related surface phenomena. *Surf. Sci. Rep.* **2008**, *63*, 515–582.

4. Ohtani, B. Preparing articles on photocatalysis—Beyond the illusions, misconceptions and speculation. *Chem. Lett.* **2008**, *37*, 217–229.

5. Ohtani, B. Photocatalysis A to Z—What we know and what we do not know in a scientific sense. *J. Photochem. Photobiol. C* **2010**, *11*, 157–178.

6. Ohtani, B. Revisiting the fundamental physical chemistry in heterogeneous photocatalysis: Its thermodynamics and kinetics. *Phys. Chem. Chem. Phys.* **2014**, *16*, 1788–1797.

7. Kudo, A.; Miseki, Y. Heterogeneous photocatalyst materials for water splitting. *Chem. Soc. Rev.* **2009**, *38*, 253–278.

8. Zhang, H.M.; Han, Y.H.; Liu, X.L.; Liu, P.R.; Yu, H.; Zhang, S.Q.; Yao, X.D.; Zhao, H.J. Anatase TiO_2 microspheres with exposed mirror-like plane {001} facets for high performance dye-sensitized solar cells (DSSCs). *Chem. Commun.* **2010**, *46*, 8395–8397.

9. Roy, P.; Berger, S.; Schmuki, P. TiO_2 nanotubes: Synthesis and applications. *Angew. Chem. Int. Ed.* **2010**, *50*, 2904–2939.

10. Saggioro, E.M.; Oliveira, A.S.; Pavesi, T.; Maia, C.G.; Ferreira, L.F.V.; Moreira, J.C. Use of titanium dioxide photocatalysis on the remediation of model textile wastewaters containing azo dyes. *Molecules* **2011**, *16*, 10370–10386.

11. Jun, J.W.; Casula, M.F.; Sim, J.H.; Kim, S.Y.; Cheon, J.; Alivisatos, A.P. Surfactant-assisted elimination of a high energy facet as a means of controlling the shapes of TiO_2 nanocrystals. *J. Am. Chem. Soc.* **2003**, *125*, 15981–15985.

12. Barnard, A.S.; Curtiss, L.A. Prediction of TiO_2 nanoparticle phase and shape transitions controlled by surface chemistry. *Nano Lett.* **2005**, *5*, 1261–1266.

13. Murakami, N.; Kurihara, Y.; Tsubota, T.; Ohno, T. Shape-controlled anatase titanium(IV) oxide particles prepared by hydrothermal treatment of peroxo titanic acid in the presence of polyvinyl alcohol. *J. Phys. Chem. C* **2007**, *113*, 3062–3069.

14. Liu, G.; Sun, C.; Yang, H.G.; Smith, S.C.; Wang, L.; Lu, G.Q.; Cheng, H.M. Nanosized anatase TiO_2 single crystals for enhanced photocatalytic activity. *Chem. Commun.* **2010**, *46*, 755–757.

15. Liu, S.; Yu, J.; Wang, W. Effects of annealing on the microstructures and photoactivity of fluorinated N-doped TiO_2. *Phys. Chem. Chem. Phys.* **2010**, *12*, 12308–12315.

16. Wei, Z.; Liu, Y.; Wang, H.; Mei, Z.; Ye, J.; Wen, X.; Gu, L.; Xie, Y. A gas-solid reaction growth of dense TiO_2 nanowire arrays on Ti foils at ambient atmosphere. *J. Nanosci. Nanotechnol.* **2012**, *12*, 316–323.

17. Zaban, A.; Aruna, S.T.; Tirosh, S.; Gregg, B.A.; Mastai, Y. The effect of the preparation condition of TiO_2 colloids on their surface structures. *J. Phys. Chem. B* **2000**, *104*, 4130–4133.

18. Ohno, T.; Sarukawa, K.; Matsumura, M. Crystal faces of rutile and anatase TiO_2 particles and their roles in photocatalytic reactions. *New J. Chem.* **2002**, *26*, 1167–1170.

19. Yang, H.G.; Sun, C.H.; Qiao, S.Z.; Zou, J.; Liu, G.; Smith, S.C.; Cheng, H.M.; Lu, G.Q. Anatase TiO_2 single crystals with a large percentage of reactive facets. *Nature* **2008**, *453*, 638–642.

20. Amano, F.; Prieto-Mahaney, O.O.; Terada, Y.; Yasumoto, T.; Shibayama, T.; Ohtani, B. Decahedral single-crystalline particles of anatase titanium(IV) oxide with high photocatalytic activity. *Chem. Mater.* **2009**, *21*, 2601–260.

21. Amano, F.; Yasumoto, T.; Prieto-Mahaney, O.O.; Uchida, S.; Shibayama, T.; Ohtani, B. Photocatalytic activity of octahedral single-crystalline mesoparticles of anatase titanium(IV) oxide. *Chem. Commun.* **2009**, *17*, 2311–2313.

22. Ohtani, B.; Amano, F.; Yasumoto, T.; Prieto-Mahaney, O.O.; Uchida, S.; Shibayama, T.; Terada, Y. Highly active titania photocatalyst particles of controlled crystal phase, size, and polyhedral shape. *Top. Catal.* **2010**, *53*, 455–461.

23. Wei, Z.; Kowalska, E.; Ohtani, B. Enhanced photocatalytic activity by particle morphology—Preparation, characterization and photocatalytic activities of octahedral anatase titania particles. *Chem. Lett.* **2014**, *43*, 346–348.

24. Li, W.; Ni, C.; Lin, H.; Huang, C.P.; Ismat-Shah, S. Size dependence of thermal stability of TiO₂ nanoparticles. *J. Appl. Phys.* **2004**, *96*, 6663–6668.

25. Zhang, J.; Xu, Q.; Feng, Z.; Li, M.; Li, C. Importance of the relationship between surface phases and photocatalytic activity of TiO₂. *Angew. Chem. Int. Ed.* **2008**, *47*, 1766–1769.

26. Kominami, H.; Murakami, S.; Kera, Y.; Ohtani, B. Titanium(IV) oxide photocatalyst of ultra-high activity: A new preparation process allowing compatibility of high adsorptivity and low electron-hole recombination probability. *Chem. Lett.* **1998**, 125–125.

27. Kominami, H.; Kumamoto, H.; Kera, Y.; Ohtani, B. Photocatalytic decolorization and mineralization of malachite green in an aqueous suspension of titanium(IV) oxide nano-particles under aerated conditions: correlation between some physical properties and their photocatalytic activity. *J. Photochem. Photobiol. A: Chem.* **2003**, *160*, 99–104.

28. Grover, I.S.; Singh, S.; Pal, B. Stable anatase TiO₂ formed by calcination of rice-like titania nanorod at 800 °C exhibits high photocatalytic activity. *RSC Adv.* **2014**, *4*, 24704–24709.

29. Baiju, K.V.; Sibu, C.P.; Rajesh, K.; Krishna Pillaia, P.; Mukundan, P.; Warrier, K.G.K.; Wunderlich, W. An aqueous sol–gel route to synthesize nanosized lanthana-doped titania having an increased anatase phase stability for photocatalytic application. *Mater. Chem. Phys.* **2005**, *90*, 123–127.

30. Wei, Z.; Kowalska, E.; Nitta, A.; Ohtani, B. Highly heat tolerant octahedral anatase particles—Effect of surface structure on the phase anatase-rutile transformation and photocatalytic activity. *J. Mater. Chem. A* **2014**, in preparation.

31. Janczarek, M.; Kowalska, E.; Ohtani, B. Synthesis and characterization of decahedral-shape anatase titania photocatalyst particles in a coaxial flow gas-phase reaction. *Chem. Mater.* **2014**, in preparation.

32. Kominami, H.; Matsuura, T.; Iwai, K.; Ohtani, B.; Nishimoto, S.I.; Kera, Y. Ultra-highly active titanium(IV) oxide photocatalyst prepared by hydrothermal crystallization from titanium(IV) alkoxide in organic solvents. *Chem. Lett.* **1995**, 693–694.

33. Park, H.; Choi, W. Effects of TiO₂ surface fluorination on photocatalytic reactions and photoelectrochemical behaviors. *J. Phys. Chem. B* **2004**, *108*, 4086–4093.

34. Katoh, R.; Huijser, A.; Hara, K.; Savenije, T.J.; Siebbeles, L.D. A. Effect of the particle size on the electron injection efficiency in dye-sensitized nanocrystalline TiO$_2$ films studied by time-resolved microwave conductivity (TRMC) measurements. *J. Phys. Chem. C* **2007**, *111*, 10741–10746.

35. Colombo, D.P., Jr.; Bowman, R.M. Does interfacial charge transfer compete with charge carrier Recombination? A femtosecond diffuse reflectance investigation of TiO$_2$ nanoparticles. *J. Phys. Chem.* **1996**, *100*, 18445–18449.

36. Boettger, H.; Bryksin, V.V. Hopping conductivity in ordered and disordered solids (I). *Phys. Status Solidi B* **1976**, *78*, 9–56.

37. Katoh, R.; Furube, A.; Yamanaka, K.; Morikawa, T. Charge separation and trapping in N-doped TiO$_2$ photocatalysts: A time-resolved microwave conductivity study. *J. Phys. Chem. C* **2010**, *1*, 3261–3265.

38. Ohtani, B.; Bowman, R.M.; Colombo, D.P.; Kominami, H.; Noguchi, H.; Uosaki, K. Femtosecond diffuse reflectance spectroscopy of aqueous titanium(IV) oxide suspension: Correlation of electron-hole recombination kinetics with photocatalytic activity. *Chem. Lett.* **1998**, 579–580.

39. Ohtani, B.; Majima, T. Control of photocatalytic activity of titania particles by braying and calcination—Effect of electron traps on the photocatalytic activity. Catalysis Research Center, Hokkaido University, Sapporo 001-0021, Japan. Unpublished results, 2014.

40. Izumi, F.; Momma, K. Three-dimensional visualization in powder diffraction. *Solid State Phenom.* **2007**, *130*, 15–20.

41. Ohtani, B.; Prieto-Mahaney, O.O.; Li, D.; Abe, R. What is Degussa (Evonik) P25? Crystalline composition analysis, reconstruction from isolated pure particles and photocatalytic activity test. *J. Photochem. Photobiol. A* **2010**, *216*, 179–182.

42. Ohtani, B; Prieto-Mahaney, O.O.; Amano, F.; Murakami, N.; Abe, R. What are titania photocatalysts?—An exploratory correlation of photocatalytic activity with structural and physical properties. *J. Adv. Oxid. Technol.* **2010**, *15*, 247–261.

43. Kowalska, E.; Remita, H.; Colbeau-Justin, C.; Hupka, J.; Belloni, J. Modification of titanium dioxide with platinum ions and clusters: Application in photocatalysis. *J. Phys. Chem. C* **2008**, *112*, 1124–1131.

44. Ohtani, B. Titania photocatalysis beyond recombination: A critical review. *Catalysts* **2013**, *3*, 942–953.

45. Nitta, A.; Takase, M.; Ohtani, B. Energy-resolved measurement of electron traps in metal oxide particulate photocatalysts: Reversed double-beam photoacoustic spectroscopy. **2014**, in preparation.

Sample Availability: Samples of the compounds are not available from the authors.

Design of Composite Photocatalyst of TiO₂ and Y-Zeolite for Degradation of 2-Propanol in the Gas Phase under UV and Visible Light Irradiation

Takashi Kamegawa, Yasushi Ishiguro, Ryota Kido and Hiromi Yamashita

Abstract: Hydrophobic Y-zeolite (SiO_2/Al_2O_3 = 810) and TiO_2 composite photocatalysts were designed by using two different types of TiO_2 precursors, *i.e.*, titanium ammonium oxalate and ammonium hexafluorotitanate. The porous structure, surface property and state of TiO_2 were investigated by various characterization techniques. By using an ammonium hexafluorotitanate as a precursor, hydrophobic modification of the Y-zeolite surface and realizing visible light sensitivity was successfully achieved at the same time after calcination at 773 K in the air. The prepared sample still maintained the porous structure of Y-zeolite and a large surface area. Highly crystalline anatase TiO_2 was also formed on the Y-zeolite surface by the role of fluorine in the precursor. The usages of ammonium hexafluorotitanate were effective for the improvement of the photocatalytic performance of the composite in the degradation of 2-propanol in the gas phase under UV and visible light ($\lambda > 420$ nm) irradiation.

Reprinted from *Molecules*. Cite as: Kamegawa, T.; Ishiguro, Y.; Kido, R.; Yamashita, H. Design of Composite Photocatalyst of TiO₂ and Y-Zeolite for Degradation of 2-Propanol in the Gas Phase under UV and Visible Light Irradiation. *Molecules* **2014**, *19*, 16477-16488.

1. Introduction

Titanium dioxide (TiO_2)-based photocatalytic materials have been used for the decomposition of undesired and harmful organic compounds in the air and water [1–7]. Electron-hole pairs formed under light irradiation by using a suitable light source play significant roles in the degradation of organic compounds into CO_2 and H_2O. TiO_2-based photocatalytic materials are also continuously researched for their importance in relation to the utilization of light energy for the synthesis of chemicals, the production of clean energy, *etc.*, under carefully-controlled conditions [8–13]. Coating technologies for TiO_2-based photocatalytic materials open the way for the utilization of unique functions, such as photocatalytic activity, the self-cleaning effect and photoinduced superhydrophilicity [3,14–17]. By increasing the awareness of environmental issues, purification of diluted organic contaminants in water and air is becoming an increasingly important agenda. TiO_2-based photocatalytic materials have potential to solve the problem of air pollution by volatile organic compounds in our living spaces, *i.e.*, the cause of sick house syndrome emitted from interiors. However, photocatalytic performance of TiO_2 is insufficiently utilized in our living spaces, due to the limitation of the amount of light in the UV region. Bare TiO_2 can only absorb UV light, which also corresponds to *ca.* 3% of the energy of natural solar light. Therefore, many efforts have been devoted to the design of highly efficient photocatalysts, which can work under not only UV, but also visible, light irradiation. Doping of various heteroatoms, e.g., carbon, nitrogen, sulfur and transition metals (Cr, V, Fe, *etc.*), into TiO_2 was previously reported and enabled the use of light in the visible region [18–25]. The modification of the TiO₂ surface by metals, metal ions and chlorides is also another method for realizing the sensitivity to visible light [26–28]. Utilization of visible light is also achieved by the anchoring of phenolic compounds on the TiO₂ surface by the formation of surface complexes [12,13].

On the other hand, adsorbents, such as activated carbon, are often used as a disposable material for the removal of diluted organic contaminants in water and air. The combination of adsorbent and TiO₂ photocatalyst is intensively studied for the design of efficient photocatalytic systems with specific functions [6,29–36]. Adsorption and enrichment of organic contaminants around combined TiO₂ from air and water play crucial roles in the photocatalytic decomposition process and lead to the efficient removal of diluted organic contaminants in the air and water. The combination of TiO₂ and silicate materials, such as zeolite, mesoporous silica and clay minerals, is achieved by applying different methods, such as wet impregnation and sol-gel processes [29–36]. The physical and chemical properties of the composites, especially the surface properties, such as hydrophilicity/hydrophobicity or surface charge, strongly affect the photocatalytic performance. In our former works, surface modification from a hydrophilic to more hydrophobic state was successfully achieved by the grafting of the fluorine group containing silylation reagents on zeolite and the mesoporous silica surface [36]. The effect of the direct and selective modification of the TiO₂ surface on mesoporous silica by graphene on the photocatalytic performance in water purification was also shown in a previously reported paper [37]. In the case of composite systems, photocatalytic performance in water and air purification relies on the surface properties. In the present work, we designed a composite system of hydrophobic Y-zeolite ($SiO_2/Al_2O_3 = 810$) and TiO₂ by using two different types of TiO₂ precursors (titanium ammonium oxalate ($(NH_4)_2[TiO(C_2O_4)_2]$) and ammonium hexafluorotitanate ($(NH_4)_2[TiF_6]$)). We mainly focus on the application of the composite in the decomposition of 2-propanol in the gas phase as a model contaminant of air under UV and visible light ($\lambda > 420$ nm) irradiation.

2. Results and Discussion

2.1. Characterization of TiO₂/Y-Zeolite Composite Photocatalysts

Figure 1 shows the UV-Vis absorption spectra of AO-TiO₂/Y and AF-TiO₂/Y, which were prepared by a combination of Y-zeolite and $(NH_4)_2[TiO(C_2O_4)_2]$ or $(NH_4)_2[TiF_6]$ as a TiO₂ source, respectively [35]. Both samples exhibited the typical absorption in the UV light region corresponding to the band gap energy of TiO₂ particles loaded on Y-zeolite. The absorption band edge of TiO₂ was obviously changed by the quantum-size effect [38]. The blue shifts of absorption spectra suggest that TiO₂ nanoparticles are successfully loaded on the Y-zeolite surface with a dispersed state. The small differences in the absorption spectra of AO-TiO₂/Y and AF-TiO₂/Y indicate the differences of the TiO₂ particle size and crystallinity formed on the Y-zeolite surface [39,40]. The powder color of AO-TiO₂/Y was a simple white, while AF-TiO₂/Y was a clear yellow powder. As shown in Figure 1 (inset), AF-TiO₂/Y exhibited visible light absorption from 400 to 500 nm. This visible light absorption was estimated to be induced by the doping of nitrogen and fluorine into TiO₂ during the decomposition of $(NH_4)_2[TiF_6]$ in the calcination process. No visible light absorption was attained by use of $(NH_4)_2[TiO(C_2O_4)_2]$, although both precursors contain a nitrogen source (ammonium cation). The existence of fluorine might induce the encapsulation of nitrogen within TiO_2 formed on the Y-zeolite surface by keeping a charge balance of each component (Ti^{4+}, O^{2-}, N^{3-} and F^-) during the calcination in the air [41,42].

Figure 1. UV-Vis absorption spectra of (**a**) AO-TiO₂/Y and (**b**) AF-TiO₂/Y.

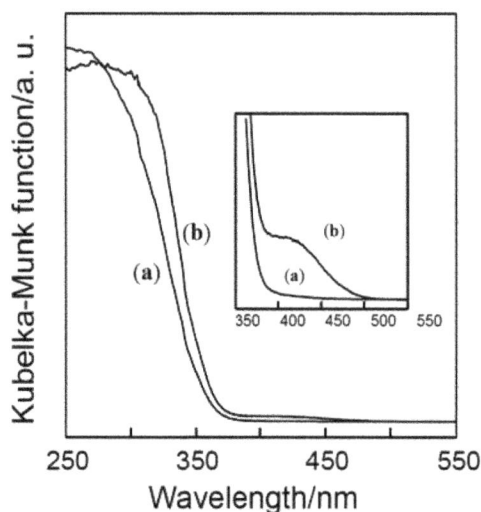

The state of TiO_2 in each sample was investigated by Ti K-edge X-ray absorption fine structure (XAFS) measurements. Figure 2A–C shows the X-ray absorption near edge structure (XANES) of AO-TiO₂/Y, AF-TiO₂/Y and the reference sample (anatase TiO_2). The XANES spectrum of anatase TiO_2 exhibited several pre-edge peaks (A_1, A_2, and A_3) from 4960 to 4975 eV. The A_1 peak is assigned to an exciton band or the 1s to $1t_{1g}$ transition. The A_2 and A_3 peak is attributed to the 1s to 3d transition, as well as being also assigned to the 1s to $2t_{2g}$ and 1s to 3d transitions, respectively [25,43,44]. AO-TiO₂/Y and AF-TiO₂/Y exhibited well-defined three pre-edge peaks, which were similar to those for anatase TiO_2 as a reference. These results indicated that anatase TiO_2 was formed on the Y-zeolite surface without relying on the differences of TiO_2 precursors. In the Fourier transforms of extended X-ray absorption fine structure (EXAFS) spectra (Figure 2a–c), peaks due to the existence of oxygen neighbors (Ti-O) and the Ti neighbors (Ti-O-Ti) were observed at *ca.* 1.8 and between 2.0 and 3.0 Å (without phase-shift correction), respectively [39,40]. The clear peak due to the existence of the Ti-O-Ti bond indicated the formation of aggregated large TiO_2 particles with octahedral coordination. AF-TiO₂/Y exhibited a relatively intense peak compared to that of AO-TiO₂/Y, showing the formation of TiO_2 with relatively high crystallinity by using $(NH_4)_2[TiF_6]$ as a precursor.

Figure 2. (A–C) XANES and (a–c) Fourier transforms of EXAFS spectra of (A,a) AO-TiO₂/Y, (B,b) AF-TiO₂/Y and (C,c) TiO₂ (anatase).

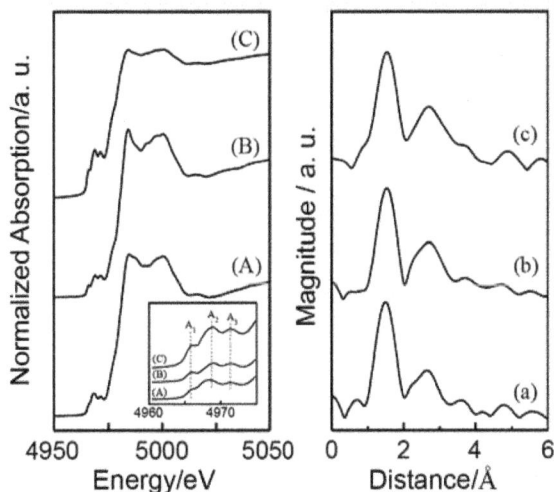

The XRD patterns of AO-TiO₂/Y and AF-TiO₂/Y are shown in Figure 3. The diffraction peaks attributed to the framework structure of Y-zeolite (5° < 2θ < 45°) were clearly observed in both samples. The typical peak assigned to the (101) reflection of the TiO₂ anatase phase is observed at around 2θ = 25° in the XRD measurement. AF-TiO₂/Y showed an intense peak in this region, while AO-TiO₂/Y only showed a small peak, except for the peaks due to the Y-zeolite. The crystallinity of TiO₂ formed on the Y-zeolite surface was affected to a large degree by the differences of the TiO₂ precursors. It has been reported that crystallization of TiO₂ was enhanced by the role of the fluorine ion. The addition of the fluorine ion in the hydrolysis process of titanium isopropoxide realized the good crystallinity of TiO₂ [40,41]. AF-TiO₂/Y have thus relatively high crystallinity by the effect of the contained fluorine in (NH₄)₂[TiF₆] used as a precursor.

Figure 3. XRD patterns of (a) AO-TiO₂/Y and (b) AF-TiO₂/Y.

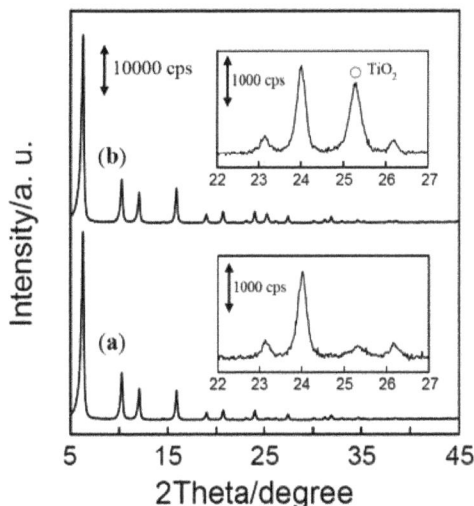

The textural properties of AO-TiO$_2$/Y and AF-TiO$_2$/Y were investigated by the measurement of nitrogen adsorption-desorption isotherms at 77 K. As shown in Figure 4A, AO-TiO$_2$/Y and AF-TiO$_2$/Y exhibited typical type I isotherms with a steep increase in the adsorbed amount of nitrogen at the low relative pressure region (P/P$_0$ < 0.01). The BET surface area of samples was determined to be 664 m^2/g (AO-TiO$_2$/Y) and 645 m^2/g (AF-TiO$_2$/Y), respectively. The pore size distribution curve also showed a peak at around 0.75 nm. These kinds of TiO$_2$ precursors hardly affect the structure of Y-zeolite.

Figure 4. (**A**) Nitrogen adsorption-desorption isotherms at 298 K and (**B**) water adsorption isotherms at 298 K of (a) AO-TiO$_2$/Y and (b) AF-TiO$_2$/Y. The insets of (A) and (B) show the pore size distribution curves and the adsorbed amount of water at around P/P$_0$ = 0.2, respectively.

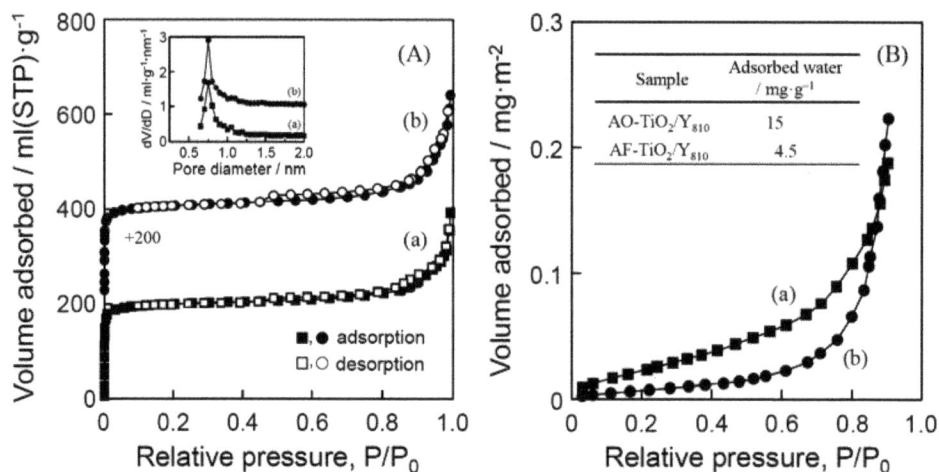

The investigations of the surface hydrophilic and hydrophobic nature of both samples were also carried out by measurement of water adsorption isotherms at 298 K. Figure 4B shows the water adsorption isotherms of AO-TiO$_2$/Y and AF-TiO$_2$/Y. The adsorbed amount of water on AF-TiO$_2$/Y was quite small up to relative pressure, P/P$_0$ = 0.8. The inset of Figure 4B shows the adsorbed amount of water at around P/P$_0$ = 0.2 on both samples. The adsorbed amount of water on AF-TiO$_2$/Y was less than a third of that on AO-TiO$_2$/Y, showing the good surface hydrophobic property of AF-TiO$_2$/Y.

In the case of zeolite, the surface hydrophilic and hydrophobic nature depends strongly on the SiO$_2$/Al$_2$O$_3$ ratio of samples. As the SiO$_2$/Al$_2$O$_3$ ratio of zeolite increases, the adsorbed amount of water becomes small. Y-zeolite with high SiO$_2$/Al$_2$O$_3$ ratio (SiO$_2$/Al$_2$O$_3$ = 810), which has a good hydrophobic nature in the series of commercially available Y-zeolite, was adopted for the preparation of AO-TiO$_2$/Y and AF-TiO$_2$/Y. The surface property of samples significantly changed depending on the kinds of TiO$_2$ precursors. By using (NH$_4$)$_2$[TiF$_6$] as a precursor, the improvement of the surface hydrophobic property of the TiO$_2$-zeolite composite was achieved in a brief preparation process. Accompanying the generation of hydrofluoric acid and ammonia gas, the decomposition of (NH$_4$)$_2$[TiF$_6$] to TiO$_2$ gradually occurs above 473 K [45]. Generated hydrofluoric acid gas might be reacting with the surface hydroxyl groups of Y-zeolite and the formation of fluorinated groups, which make the surface of AF-TiO$_2$/Y quite hydrophobic. The presence of fluorine moieties (Si-F) on the surface of AF-TiO$_2$/Y were confirmed by F$_{1s}$ XPS analysis. As shown in Figure 5, AF-TiO$_2$/Y exhibited a weak peak at around 690 eV, while no peak was observed in the case of AO-TiO$_2$/Y. This peak was assigned to the covalent F atoms, indicating the formation of Si-F through the reaction of surface hydroxyl groups and generated hydrofluoric acid gas [36]. The peak attributed to the presence of the F$^-$ ion doped into TiO$_2$ is also observed in the same region [42]. However, no peak was observed in the N$_{1s}$ XPS analysis of AF-TiO$_2$/Y. The concentration of nitrogen in AF-TiO$_2$/Y was below the detection limitation. Considering these obtained results, the amount of nitrogen and fluorine doped into TiO$_2$ on the Y-zeolite surface was estimated to be so small, while AF-TiO$_2$/Y exhibited clear absorption in the visible light region. These results suggested that a large part of fluorine exists on the Y-zeolite surface and leads to the good surface hydrophobicity. In fact, AF-TiO$_2$/Y showed small a peak in FT-IR spectrum at around 3740 cm^{-1} as compared to that of AO-TiO$_2$/Y. This peak is assigned to the surface hydroxyl group of Y-zeolite.

Figure 5. F$_{1s}$ XPS spectra of (a) AO-TiO$_2$/Y and (b) AF-TiO$_2$/Y.

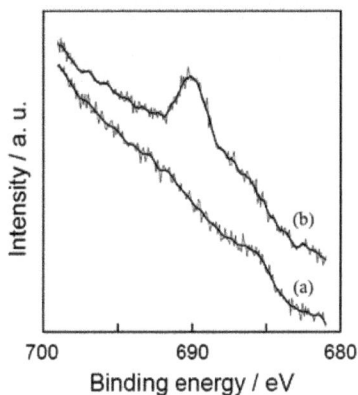

2.2. Photocatalytic Performance of TiO$_2$/Y-Zeolite Composite Photocatalysts

The photocatalytic performance of AO-TiO$_2$/Y and AF-TiO$_2$/Y was evaluated in the degradation of 2-propanol gas diluted in the air under UV and visible light ($\lambda > 420$ nm) irradiation. The degradation of 2-propanol in the gas phase was adopted as a model reaction. It is well known that 2-propanol was decomposed to CO$_2$ and H$_2$O via the formation of acetone. In the initial reaction stage, acetone was mainly formed in the gas phase. The formed acetone is fully decomposed into CO$_2$ and H$_2$O by the progress of the reaction time [28,36,39]. As shown in Figure 6A, AF-TiO$_2$/Y exhibited a two-times higher photocatalytic activity than that on AO-TiO$_2$/Y under UV light irradiation. As is obvious from the XRD measurement, AF-TiO$_2$/Y contained TiO$_2$ with good crystallinity, leading to an enhancement of photocatalytic performance. The crystallinity of TiO$_2$ is related to various properties, such as electron conductivity, hole mobility and the electron-hole recombination probability. The good crystallinity of TiO$_2$ reduces the recombination probability and enhances the photocatalytic reactions [46,47]. In the case of hydrophilic zeolite, H$_2$O molecules are easily adsorbed on the surface and filled inside of pores. Organic molecules are thus not preferentially adsorbed on the surface and diffuse into the pores, resulting in lower photocatalytic activity. The noticeable improvement on AF-TiO2/Y was attained by the combinational effect of the hydrophobicity of the Y-zeolite support and the good crystallinity of the formed TiO2. In response to the light absorption property, AF-TiO2/Y also showed photocatalytic activity under visible light (λ > 420 nm) irradiation (Figure 6B). Utilization of $(NH_4)_2[TiF_6]$ is effective for realizing the improvement of the hydrophobicity of the support, the good crystallinity of TiO2, as well as the visible light sensitivity in a simple preparation process.

Figure 6. Conversion in the photocatalytic degradation of 2-propanol gas diluted in the air on AO-TiO2/Y and AF-TiO2/Y under (**A**) UV and (**B**) visible light (λ > 420 nm) irradiation (reaction time: (A) 1 h; (B) 12 h).

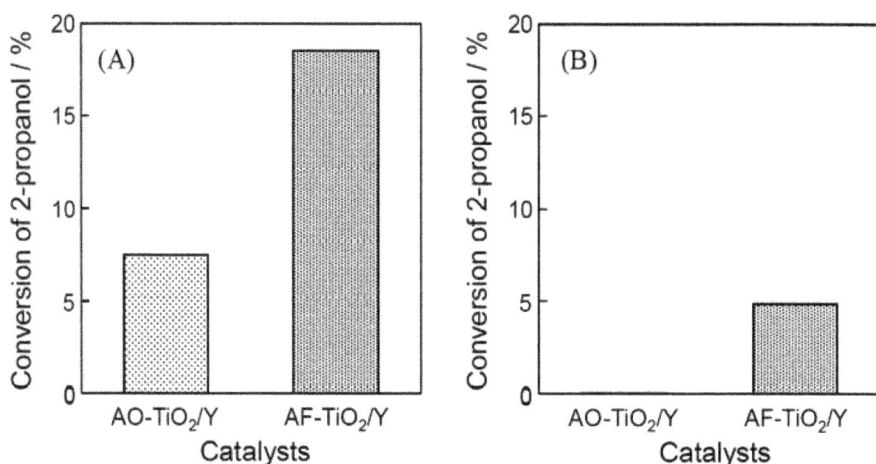

3. Experimental Section

3.1. Materials

Proton type Y-zeolite (SiO_2/Al_2O_3 = 810) used as a support of the TiO2 photocatalyst was supplied by Tosoh Co. (Tokyo, Japan). Titanium ammonium oxalate ($(NH_4)_2[TiO(C_2O_4)_2]$) and ammonium hexafluorotitanate ($(NH_4)_2[TiF_6]$) were purchased from Kishida Chemicals (Osaka, Japan). 2-propanol was purchased from Nacalai Tesque Inc (Kyoto, Japan). All chemicals were used without further purification.

3.2. Sample Preparation

The combination of TiO2 and Y-zeolite (SiO_2/Al_2O_3 = 810) was carried out by a conventional impregnation method [35,37]. Y-zeolite was suspended in an aqueous solution of $(NH_4)_2[TiO(C_2O_4)_2]$ or $(NH_4)_2[TiF_6]$ and was stirred at 323 K for 1 h. Water was then evaporated at 343 K under reduced pressure. The obtained powder was dried at 373 K for 12 h and then calcined at 773 K for 5 h in the air (heating rate: *ca.* 2.5 K/min). The content of TiO2 was adjusted to 10 wt %

in both samples. The thus obtained samples were denoted as AO-TiO$_2$/Y and AF-TiO$_2$/Y, which were prepared by using an aqueous solution of (NH$_4$)$_2$[TiO(C$_2$O$_4$)$_2$] and (NH$_4$)$_2$[TiF$_6$], respectively.

3.3. Characterization Techniques

Nitrogen adsorption-desorption isotherms at 77 K, as well as water adsorption isotherms at 298 K were measured by a BEL-SORP max (BEL Japan, Inc., Osaka, Japan). Prior to the measurements of isotherms, each sample was degassed under vacuum at 473 K for 2 h. Diffuse reflectance UV-Vis spectra were recorded at 298 K with a Shimadzu UV-2450A double-beam digital spectrophotometer. FT-IR measurements were carried out at 298 K in transmission mode with a resolution of 4 cm^{-1} using a JASCO FT/IR-6100. Prior to FT-IR measurements, self-supporting pellets of samples were degassed at 673 K for 1 h. XPS measurements were carried out by using a Shimadzu ESCA-3200 using Mg Kα radiation. The powder XRD measurements were performed using a Rigaku Ultima IV X-ray diffractometer with Cu Kα radiation. Ti K-edge X-ray absorption fine structure (XAFS) measurements were carried out at the BL-7C facility of the Photon Factory (high energy acceleration research organization, Tsukuba, Japan). XAFS spectra were measured at 298 K in the fluorescence mode. A Si(111) double crystal was used to monochromatize the synchrotron radiation from the 2.5 GeV electron storage ring. The obtained data were examined using the analysis program (Rigaku REX2000). Fourier transformations were performed on k^3-weighted EXAFS oscillations in the range 3–10 Å^{-1} to obtain the radial structure function. Thermogravimetry-differential thermal analyses for determining the decomposition temperature of (NH$_4$)$_2$[TiF$_6$] was performed using a TG-DTA2000S (MAC Science Co. Ltd., Tokyo, Japan) from RT to 1073 K at a heating rate of 10 K/min under an air flow (50 mL/min).

3 4. Photocatalytic Reaction

The photocatalytic activity of samples was evaluated by monitoring the decomposition of 2-propanol gas in the air under UV and visible light irradiation. A sample (10 mg) was fixed on a glass filter in accordance with previous papers [27,28]. After pretreatment for the removal of residual organics on the surface, the gas phase in a glass reactor equipped with a flat quartz window was replaced with artificial air. 2-propanl gas (0.13 mmol) was then injected into the reactor. The irradiation of UV light was carried out by using a 200-W mercury xenon lamp (UVF-204S, San-ei Electric Co., Ltd., Osaka, Japan) under controlled light intensity (5 mW/cm^2 at around 360 nm). Visible light (λ > 420 nm) irradiation was also performed by using the same light source through a colored filter (HOYA; L-42). The progress of the reaction was monitored by gas chromatography analysis (Shimadzu GC-14B with FID and TCD, Kyoto, Japan).

4. Conclusions

Comparative studies were carried out by using composite photocatalysts (AO-TiO$_2$/Y and AF-TiO$_2$/Y) prepared by a combination of Y-zeolite (SiO$_2$/Al$_2$O$_3$ = 810) and two different types of precursors ((NH$_4$)$_2$[TiO(C$_2$O$_4$)$_2$] and (NH$_4$)$_2$[TiF$_6$]). AF-TiO$_2$/Y, having visible light sensitivity, good crystallinity of TiO$_2$ and a highly hydrophobic surface property, was successfully achieved by using

$(NH_4)_2[TiF_6]$ as a precursor. Instead of the use of $(NH_4)_2[TiO(C_2O_4)_2]$ as a precursor, these additional and advanced functions were realized at the same time in a simple preparation process. Based on these functions, AF-TiO$_2$/Y exhibited a good photocatalytic performance in the decomposition of 2-propanol in the gas phase under not only UV, but also visible, light ($\lambda > 420$ nm) irradiation.

Acknowledgments

This study was supported by a Grant-in-Aid for Scientific Research (KAKENHI) from the Ministry of Education, Culture, Sports, Science and Technology (MEXT), Japan (No. 26420786, 26220911). The X-ray absorption measurements were performed at the BL-7C facility of the Photon Factory at the National Laboratory for High-Energy Physics, Tsukuba, Japan (2012G126).

Author Contributions

T.K. and H.Y. designed research; T.K., Y.I., and R.K. performed research and analyzed the data; T.K. and H.Y. wrote the paper. All authors read and approved the final manuscript.

Conflicts of Interest

The authors declare no conflict of interest.

References

1. Fox, M.A.; Dulay, M.T. Heterogeneous Photocatalysis. *Chem. Rev.* **1993**, *93*, 341–357.
2. Hoffmann, M.R.; Martin, S.T.; Choi, W.Y.; Bahnemann, D.W. Environmental Applications of Semiconductor Photocatalysis. *Chem. Rev.* **1995**, *95*, 69–96.
3. Fujishima, A.; Rao, T.N.; Tryk, D.A. Titanium Dioxide Photocatalysis. *J. Photochem. Photobiol. C: Photochem. Rev.* **2000**, *1*, 1–21.
4. Chen, X.; Mao, S.S. Titanium Dioxide Nanomaterials: Synthesis, Properties, Modifications, and Applications. *Chem. Rev.* **2007**, *107*, 2891–2959.
5. Kuwahara, Y.; Kamegawa, T.; Mori, K.; Yamashita, H. Design of New Functional Titanium Oxide-based Photocatalysts for Degradation of Organics Diluted in Water and Air. *Curr. Org. Chem.* **2010**, *14*, 616–629.
6. Qian, X.; Fuku, K.; Kuwahara, Y.; Kamegawa, T.; Mori, K.; Yamashita, H. Design and Functionalization of Photocatalytic Systems within Mesoporous Silica. *ChemSusChem* **2014**, *7*, 1528–1536.
7. Kubacka, A.; Fernández-García, M.; Colón, G. Advanced Nanoarchitectures for Solar Photocatalytic Applications. *Chem. Rev.* **2012**, *112*, 1555–1614.
8. Lang, X.; Ma, W.; Chen, C.; Ji, H.; Zhao J. Selective Aerobic Oxidation Mediated by TiO$_2$ Photocatalysis. *Acc. Chem. Res.* **2014**, *47*, 355–363.
9. Palmisano, G.; García-López, E.; Marcí, G.; Loddo, V.; Yurdakal, S.; Augugliaro, V.; Palmisano, L. Advances in Selective Conversions by Heterogeneous Photocatalysis. *Chem. Commun.* **2010**, *46*, 7074–7089.

10. Shiraishi, Y.; Sugano, Y.; Tanaka, S.; Hirai, T. One-Pot Synthesis of Benzimidazoles by Simultaneous Photocatalytic and Catalytic Reactions on Pt@TiO₂ Nanoparticles. *Angew. Chem. Int. Ed.* **2010**, *49*, 1656–1660.

11. Kominami, H.; Yamamoto, S.; Imamura, K.; Tanaka, A.; Hashimoto, K. Photocatalytic Chemoselective Reduction of Epoxides to Alkenes Along with Formation of Ketones in Alcoholic Suspensions of Silver-loaded Titanium(IV) Oxide at Room Temperature without the Use of Reducing Gasses. *Chem. Commun.* **2014**, *50*, 4558–4560.

12. Kamegawa, T.; Seto, H.; Matsuura, S.; Yamashita, H. Preparation of Hydroxynaphthalene-Modified TiO₂ via Formation of Surface Complexes and their Applications in the Photocatalytic Reduction of Nitrobenzene under Visible-Light Irradiation. *ACS Appl. Mater. Interfaces* **2012**, *4*, 6635–6639.

13. Kamegawa, T.; Matsuura, S.; Seto, H.; Yamashita, H. A Visible-Light-Harvesting Assembly with a Sulfocalixarene Linker between Dyes and a Pt-TiO₂ Photocatalyst. *Angew. Chem. Int. Ed.* **2013**, *52*, 916–919.

14. Wang, R.; Hashimoto, K.; Fujishima, A.; Chikuni, M.; Kojima, E.; Kitamura, A.; Shimohigoshi, M.; Watanabe, T. Light-induced Amphiphilic Surfaces. *Nature* **1997**, *388*, 431–432.

15. Wang, R.; Hashimoto, K.; Fujishima, A.; Chikuni, M.; Kojima, E.; Kitamura, A.; Shimohigoshi, M.; Watanabe, T. Photogeneration of Highly Amphiphilic TiO₂ Surfaces. *Adv. Mater.* **1998**, *10*, 135–138.

16. Kamegawa, T.; Suzuki, N.; Yamashita, H. Design of Macroporous TiO₂ Thin Film Photocatalysts with Enhanced Photofunctional Properties. *Energy Environ. Sci.* **2011**, *4*, 1411–1416.

17. Kamegawa, T.; Shimizu, Y.; Yamashita, H. Superhydrophobic Surfaces with Photocatalytic Self-Cleaning Properties by Nanocomposite Coating of TiO₂ and Polytetrafluoroethylene. *Adv. Mater.* **2012**, *24*, 3697–3700.

18. Choi, W.; Termin, A.; Hoffmann, M.R. The Role of Metal Ion Dopants in Quantum-Sized TiO₂: Correlation between Photoreactivity and Charge Carrier Recombination Dynamics. *J. Phys. Chem.* **1994**, *98*, 13669–13679.

19. Yamashita, H.; Harada, M.; Misaka, M.; Takeuchi, M.; Neppolian, B.; Anpo, M. Photocatalytic Degradation of Organic Compounds Diluted in Water Using Visible Light-responsive Metal Ion-Implanted TiO₂ Catalysts: Fe Ion-implanted TiO₂. *Catal. Today* **2003**, *84*, 191–196.

20. Yamashita, H.; Harada, M.; Misaka, J.; Takeuchi, M.; Ikeue, K.; Anpo, M. Degradation of Propanol Diluted in Water under Visible Light Irradiation using Metal Ion-implanted Titanium Dioxide Photocatalysts. *J. Photochem. Photobiol. A-Chem.* **2002**, *148*, 257–261.

21. Kamegawa, T.; Sonoda, J.; Sugimura, K.; Mori, K.; Yamshita, H. Degradation of Isobutanol Diluted in Water over Visible Light Sensitive Vanadium Doped TiO₂ Photocatalyst. *J. Alloys Compd.* **2009**, *486*, 685–688.

22. Asahi, R.; Morikawa, T.; Ohwaki, T.; Aoki, K.; Taga, Y. Visible-Light Photocatalysis in Nitrogen-Doped Titanium Oxides. *Science* **2001**, *293*, 269–271.

23. Sakthivel, S.; Kisch, H. Daylight Photocatalysis by Carbon-Modified Titanium Dioxide. *Angew. Chem. Int. Ed.* **2003**, *42*, 4908–4911.

24. Ohno, T.; Akiyoshi, M.; Umebayashi, T.; Asai, K.; Mitui, T.; Matsumura, M. Preparation of S-doped TiO_2 Photocatalysts and their Photocatalytic Activities under Visible Light. *Appl. Catal. A: Gen.* **2004**, *265*, 115–121.

25. Maeda, K.; Shimodaira, Y.; Lee, B.; Teramura, K.; Lu, D.; Kobayashi, H.; Domen, K. Studies on $TiN_xO_yF_z$ as a Visible-Light-Responsive Photocatalyst. *J. Phys. Chem. C* **2007**, *111*, 18264–19270.

26. Kisch, H.; Zang, L.; Lange, C.; Maier, W.F.; Antonius, C.; Meissner, D. Modified, Amorphous Titania-A Hybrid Semiconductor for Detoxification and Current Generation by Visible Light. *Angew. Chem. Int. Ed.* **1998**, *37*, 30345–3036.

27. Irie, H.; Kamiya, K.; Shibanuma, T.; Miura, S.; Tryk, D.A.; Yokouyama, T.; Hashimoto, K. Visible Light-Sensitive Cu(II)-Grafted TiO_2 Photocatalysts: Activities and X-ray Absorption Fine Structure Analyses. *J. Phys. Chem. C* **2009**, *113*, 10761–10766.

28. Kitano, S.; Hashimoto, K.; Kominami, H. Photocatalytic Degradation of 2-propanol over Metal-ion-loaded Titanium(IV) Oxide under Visible Light Irradiation: Effect of Physical Properties of Nano-crystalline Titanium(IV) Oxide. *Appl. Catal. B: Environ.* **2011**, *101*, 206–211.

29. Corma, A.; Garcia, H. Zeolite-based Photocatalysts. *Chem. Commun.* **2004**, 1443–1459.

30. Qian, X.F.; Kamegawa, T.; Mori, K.; Li, H.X.; Yamashita, H. Calcium Phosphate Coatings Incorporated in Mesoporous TiO_2/SBA-15 by a Facile Inner-pore Sol-gel Process toward Enhanced Adsorption-photocatalysis Performances. *J. Phys. Chem. C* **2013**, *117*, 19544–19551.

31. Yamashita, H.; Nose, H.; Kuwahara, Y.; Nishida, Y.; Yuan, S.; Mori, K. TiO_2 Photocatalyst Loaded on Hydrophobic Si_3N_4 Support for Efficient Degradation of Organics Diluted in Water. *Appl. Catal. A: Gen.* **2008**, *350*, 164–168.

32. Kuwahara, Y.; Aoyama, J.; Miyakubo, K.; Eguchi, T.; Kamegawa, T.; Mori, K.; Yamashita, H. TiO_2 Photocatalyst for Degradation of Organic Compounds in Water and Air Supported on Highly Hydrophobic FAU Zeolite: Structural, Sorptive, and Photocatalytic Studies. *J. Catal.* **2012**, *285*, 223–234.

33. Inumaru, K.; Kasahara, T.; Yasui, M.; Yamanaka, S. Direct Nanocomposite of Crystalline TiO_2 Particles and Mesoporous Silica as a Molecular Selective and Highly Active Photocatalyst. *Chem. Commun.* **2005**, 2131–2133.

34. Torimoto, T.; Okawa, Y.; Takeda, N.; Yoneyama, H. Effect of Activated Carbon Content in TiO_2-loaded Activated Carbon on Photodegradation Behaviors of Dichloromethane. *J. Photochem. Photobiol. A: Chem.* **1997**, *103*, 153–157.

35. Kamegawa, T.; Kido, R.; Yamahana, D.; Yamashita, H. Design of TiO_2-zeolite Composites with Enhanced Photocatalytic Performances under Irradiation of UV and Visible Light. *Microporous Mesoporous Mater.* **2013**, *165*, 142–147.

36. Kuwahara, Y.; Maki, K.; Matsumura, Y.; Kamegawa, T.; Mori, K.; Yamashita, H. Hydrophobic Modification of a Mesoporous Silica Surface Using a Fluorine-Containing Silylation Agent and Its Application as an Advantageous Host Material for the TiO_2 Photocatalyst. *J. Phys. Chem. C* **2009**, *113*, 1552–1559.

37. Kamegawa, T.; Yamahana, D.; Yamashita, H. Graphene Coating of TiO$_2$ Nanoparticles Loaded on Mesoporous Silica for Enhancement of Photocatalytic Activity. *J. Phys. Chem. C* **2011**, *114*, 15049–15053.

38. Lassaletta, G.; Fernandez, A.; Espinos, J.P.; Gonzalez-Elipe, A.R. Spectroscopic Characterization of Quantum-Sized TiO$_2$ Supported on Silica: Influence of Size and TiO$_z$-SiO$_z$ Interface Composition. *J. Phys. Chem.* **1995**, *99*, 1484–1490.

39. Yamashita, H.; Honda, M.; Harada, M.; Ichihashi, Y.; Anpo, M.; Hirao, T.; Itoh, N.; Iwamoto, N. Preparation of Titanium Oxide Photocatalysts Anchored on Porous Silica Glass by a Metal Ion-Implantation Method and Their Photocatalytic Reactivities for the Degradation of 2-Propanol Diluted in Water. *J. Phys. Chem. B* **1998**, *102*, 10707–10711.

40. Liu, Z.; Davis, R.J. Investigation of the Structure of Microporous Ti-Si Mixed Oxides by X-ray, UV Reflectance, FT-Raman, and FT-IR Spectroscopies. *J. Phys. Chem.* **1994**, *98*, 1253–1261.

41. Du, X.; He, J.; Zhao, Y. Facile Preparation of F and N Codoped Pinecone-Like Titania Hollow Microparticles with Visible Light Photocatalytic Activity. *J. Phys. Chem. C* **2009**, *113*, 14151–14158.

42. Yu, J.C.; Yu, J.; Ho, W.; Jiang, Z.; Zhang, L. Effects of F⁻ Doping on the Photocatalytic Activity and Microstructures of Nanocrystalline TiO$_2$ Powders. *Chem. Mater.* **2002**, *14*, 3808–3816.

43. Greegor, R.B.; Lytle, F.W.; Sandstrom, D.R.; Wong, J.; Schultz, P. Investigation of TiO$_2$SiO$_2$ Glasses by X-ray Absorption Spectroscopy. *J. Non-Cryst. Solids* **1983**, *55*, 27–43.

44. Chen, L.X.; Rajh, T.; Wang, Z.; Thurnauer, M.C. XAFS Studies of Surface Structures of TiO$_2$ Nanoparticles and Photocatalytic Reduction of Metal Ions. *J. Phys. Chem. B* **1997**, *101*, 10688–10697.

45. Chen, D.; Jiang, Z.; Geng, J.; Zhu, J.; Yang, D. A Facile Method to Synthesize Nitrogen and Fluorine Co-doped TiO$_2$ Nanoparticles by Pyrolysis of (NH$_4$)$_2$TiF$_6$. *J. Nanopart. Res.* **2009**, *11*, 303–313.

46. Furube, A.; Asahi, T.; Masuhara, H.; Yamashita, H.; Anpo, M. Charge Carrier Dynamics of Standard TiO$_2$ Catalysts Revealed by Femtosecond Diffuse Reflectance Spectroscopy. *J. Phys. Chem. B* **1999**, *103*, 3120–3127.

47. Kominami, H.; Murakami, S.; Kato, J.; Kera, Y.; Ohtani, B. Correlation between Some Physical Properties of Titanium Dioxide Particles and Their Photocatalytic Activity for Some Probe Reactions in Aqueous Systems. *J. Phys. Chem. B* **2002**, *106*, 10501–10507.

Sample Availability: Samples are available from the authors.

Photocatalytic Oxidation of Diethyl Sulfide Vapor over TiO₂-Based Composite Photocatalysts

Dmitry Selishchev and Denis Kozlov

Abstract: Composite TiO₂/activated carbon (TiO₂/AC) and TiO₂/SiO₂ photocatalysts with TiO₂ contents in the 10 to 80 wt. % range were synthesized by the TiOSO₄ thermal hydrolysis method and characterized by AES, BET, X-ray diffraction and FT-IR ATR methods. All TiO₂ samples were in the anatase form, with a primary crystallite size of about 11 nm. The photocatalytic activities of the TiO₂/AC and TiO₂/SiO₂ samples were tested in the gas-phase photocatalytic oxidation (PCO) reaction of diethyl sulfide (DES) vapor in a static reactor by the FT-IR *in situ* method. Acetaldehyde, formic acid, ethylene and SO₂ were registered as the intermediate products which finally were completely oxidized to the final oxidation products – H₂O, CO₂, CO and SO₄²⁻ ions. The influence of the support on the kinetics of DES PCO and on the TiO₂/AC and TiO₂/SiO₂ samples' stability during three long-term DES PCO cycles was investigated. The highest PCO rate was observed for TiO₂/SiO₂ photocatalysts. To evaluate the activity of photocatalysts the turnover frequency values (TOF) were calculated for three photocatalysts (TiO₂, TiO₂/AC and TiO₂/SiO₂) for the same amount of mineralized DES. It was demonstrated that the TOF value for composite TiO₂/SiO₂ photocatalysts was 3.5 times higher than for pure TiO₂.

Reprinted from *Molecules*. Cite as: Selishchev, D.; Kozlov, D. Photocatalytic Oxidation of Diethyl Sulfide Vapor over TiO₂-Based Composite Photocatalysts. *Molecules* **2014**, *19*, 21424-21441.

1. Introduction

Volatile organic compounds containing N, S, P or Cl heteroatoms are often highly toxic and very dangerous for human health [1–3], and some of them could be used as chemical warfare agents (CWA) [4]. One of the best known CWAs is bis(2-chloroethyl) sulfide or mustard gas (HD). This species is a highly toxic vesicant which causes destruction of cell membranes and nucleic acids. It binds with nucleophilic groups like sulphur atoms in the SH-groups of proteins and nitrogen atoms in the nitrogen bases of DNA [5]. The relative toxicity (LD_{50}) for HD inhalation is about 1.5 mg·min/L and this value is the highest among vesicants [6]. In this way the development of effective methods for HD neutralization is an important task to ensure human safety.

The main chemical methods of HD detoxification include nucleophilic substitution or oxidation, which result in the cleavage of C-S or C-Cl bonds and partial or even complete oxidation of the target molecule [7]. HD can be destroyed by photolysis or oxidized by photogenerated ozone under UV light irradiation [8]. At the same time the method of photocatalytic oxidation using TiO₂ as the photocatalyst is regarded as one of the promising methods of CWA disposal due to the high oxidative ability of TiO₂ under UV irradiation [9]. PCO makes it possible to destroy dangerous compounds completely with the formation of CO_2, H_2O, NO_3^-, SO_4^{2-}, PO_4^{3-} and Cl^- as final products [10].

In the view of the high toxicity of HD, researchers usually use for laboratory investigations simulants such as 2-chloroethyl ethyl sulfide [11], 2-chloroethyl methyl sulfide [12], 2-phenethyl 2-

chloroethyl sulfide [13], diethyl sulfide (DES) [14] and dimethyl sulfide [15]. These simulants are safer due to their lower toxicity, but at the same time they simulate well the chemical behavior of HD.

In the current work we focused on the investigation of DES vapor PCO. In our previous works we demonstrated that DES can be easily decontaminated under the UV irradiation using TiO_2 as the photocatalyst with the formation of CO_2, H_2O and surface sulfate and carbonate species as the final PCO products [14,16–18]. Acetaldehyde, ethanol, ethylene, SO_2 and other trace products were detected as the gas-phase intermediates, while polysulfides, diethyl sulfone, diethyl sulfoxide were detected as the surface intermediates. All intermediates were completely oxidized to the final products after long-term irradiation [14,16]. Analysis of intermediates and products allowed the authors to propose the main routes of the DES PCO, which include C-S bond cleavage, and oxidation of sulfur and carbon atoms.

To enhance the rate of air purification from DES vapor in a closed chamber a TiO_2 aerosol generated by a sonic method could be applied [19]. Aerosol spraying led to the fast adsorption of DES vapor and its further photocatalytic oxidation under UV irradiation.

The main problem during the long-term DES oxidation is the deactivation of the TiO_2 photocatalyst. An increase of the time required for the complete mineralization of DES was clearly seen from the kinetic curves of CO_2 accumulation during several oxidation cycles in a batch reactor [14]. FT-IR analysis demonstrated that the accumulation of non-volatile organic intermediates like polysulfides, diethyl sulfone, diethyl sulfoxide and sulfate species on the surface of photocatalyst are responsible for its deactivation during the long-term experiments. The positive influence of using the composite TiO_2/adsorbent photocatalysts was also discussed in our previously published paper devoted to the computer simulation of the kinetics of photocatalytic reactions [20] where we demonstrated an increase of the rate of substrate removal for TiO_2/adsorbent photocatalysts.

In recent years composite photocatalysts in which TiO_2 is deposited onto the surface of a porous support like activated carbon (AC), silica or zeolite were actively investigated in the PCO processes of various pollutants, both in the gas and liquid phases [21–26]. In addition to the increase of the adsorption capacity in some cases the increase of the PCO rate and the decrease of deactivation degree were observed for such composite photocatalysts [27–29].

Concerning mustard gas, several research groups have studied PCO of HD simulants using composite photocatalysts. Cr-modified TiO_2-loaded MCM-41 silica photocatalyst was studied in the oxidation of DES vapor in a batch reactor under UV irradiation [30]. It was demonstrated that the TiO_2 deposition on the Cr-MCM-41 support increases the DES removal rate, but decreases the CO_2 formation rate if compares with the commercial Hombifine N TiO_2 (Sachtleben Chemie GmbH, Duisburg, Germany).

Panayotov and co-workers investigated the PCO of 2-chloroethyl ethyl sulfide (CEES) and DES on a mixed oxide TiO_2-SiO_2 photocatalyst [31–33]. They revealed that the CEES adsorbs on the surface of the composite photocatalyst through both the chlorine and sulfur atoms by bonding to isolated OH groups. The authors also demonstrated that the presence of the Cl atom in the CEES molecule does not significantly influence the PCO rate if compared with the DES molecule. Partially or fully oxidized products were observed during the photooxidation of both tested molecules over the composite photocatalyst. Partially oxidized products have been demonstrated to block OH groups on the surface of photocatalysts and to prevent further adsorption of target molecules and to thus reduce the rate of photooxidation. Unfortunately, no comparison between pure TiO_2 and composite TiO_2-SiO_2 photocatalyst was done. The main drawback of the previous works is the absence of systematic investigations of the behavior of composite photocatalysts during the long-term PCO of HD simulants.

In spite of the fact that AC is the most frequently used TiO_2 support in composite photocatalysts, the SiO_2 material is also promising due to its higher hydrophilicity, transparency and quantity of OH-groups. In this connection, the main objective of the current study was to investigate the PCO of DES in the gas-phase over composite photocatalysts in which TiO_2 was deposited onto AC or SiO_2 surfaces. We investigated the effect of the porous support on the kinetics of DES PCO and on the composite photocatalyst activity in multiple long-term experiments. Finally, a comparison between pure TiO_2 and TiO_2/adsorbent photocatalysts was made.

2. Results and Discussion

2.1. Characterization of the Synthesized Photocatalysts

Synthesis of TiO_2, TiO_2/AC and TiO_2/SiO_2 samples was performed by the $TiOSO_4$ thermal hydrolysis method. This method has some advantages in comparison with the popular sol-gel method which utilizes titanium alkoxides because titanyl sulfate is a cheaper precursor. Synthesized TiO_2 samples were of anatase crystal structure with a high surface area and good crystallinity. As a result the TiO_2 photocatalyst synthesized by this procedure usually have high photocatalytic activity in the oxidation of volatile organic compounds [34].

In our previous work TiO_2/AC samples with TiO_2 contents higher than 60 wt. % demonstrated high photocatalytic activity [22]. Therefore in this work we prepared TiO_2/AC samples with estimated TiO_2 contents equal to 65, 70 and 80 wt. %. The TiO_2/SiO_2 photocatalysts have high photoactivity even at a relatively low TiO_2 content, so we prepared several TiO_2/SiO_2 samples with estimated TiO_2 contents in the range from 10 to 80 wt. %. Varying the TiO_2 content in the series of TiO_2/adsorbent samples allows us to choose a photocatalyst with high adsorption capacity and at the same time with high photocatalytic activity for further investigation of DES PCO. The results of AES and BET analysis are presented in the Table 1.

It follows from the Table 1 that the synthesized TiO_2 sample has a high surface area (208 m^2/g) and pore volume (0.15 cm^3/g). The corresponding values for composite TiO_2/AC and TiO_2/SiO_2 photocatalysts are higher because of the higher porosity of AC and SiO_2. Moreover, the lower the TiO_2 content is the higher specific surface area and pore volume of the composite photocatalyst are (Figure 1). Figure 1 demonstrates that the specific surface area and pore volume of the composite photocatalyst are slightly lower than the algebraic sum of the corresponding values of TiO_2 and adsorbent (AC or SiO_2). This means that a partial blocking of the support surface by TiO_2 nanoparticles occurs.

Table 1. TiO₂ content and textural properties of the samples.

Series	Sample *	TiO₂ Content, wt. %	Surface Area, m²/g	Pore Volume, cm³/g
Supports	AC	--	825	0.54
	SiO₂	--	442	0.78
TiO₂	TiO₂	100	208	0.15
TiO₂/AC	80-TC	77.5	299	0.22
	70-TC	68.5	367	0.27
	65-TC	62.5	401	0.26
TiO₂/SiO₂	80-TS	76.5	237	0.27
	60-TS	57.5	270	0.37
	40-TS	39.7	298	0.48
	20-TS	22.2	351	0.58
	10-TS	12.3	396	0.68

* Number in the sample label indicates the estimated TiO₂ content, wt. %.

XRD patterns for pure TiO₂ and composite photocatalysts are presented in Figure 2. It can be seen that the XRD patterns of TiO₂/AC catalyst only have anatase peaks while broad activated carbon peaks are not detected due to the low AC content. Also a small amount of CaCO₃ admixture is observed in the AC sample.

In the TiO₂/SiO₂ samples, in addition to anatase peaks, there appears a broad silica peak at the 2θ value equal to 20–30° indicating its amorphous structure. This silica peak overlaps with the $2\theta = 25.3°$ peak of anatase. By and large the anatase peaks are similar for pure TiO₂ and composite TiO₂/AC and TiO₂/SiO₂ samples which indicates that TiO₂ crystallites have the same size in all cases because the value of the coherent-scattering domains size is about 11 nm for all samples.

Figure 3 shows the IR spectra of all synthesized photocatalysts, AC and SiO₂ supports measured by the FT-IR ATR technique. Since all measurements were carried out under ambient conditions the water $\delta_s(H_2O)$ absorption band at 1633 cm⁻¹ was recorded in all samples except for AC powder. A broad absorption band in the 2800–3750 cm⁻¹ range corresponds to the stretching vibration of the surface OH-groups and physically adsorbed H₂O molecules.

The 1055 and 1113 cm⁻¹ absorption bands in the spectrum of the pure TiO₂ sample correspond to the vibrations in sulfate complexes [35]. The presence of sulfur was additionally confirmed by the atomic emission spectroscopy (AES) results, which revealed about 1.3 wt. % of S. This means that bonded sulfate complexes remain on the catalyst surface even after thorough washing. The presence of sulfate groups on the photocatalyst surface was also observed for the TiO₂/AC sample. For the TiO₂/SiO₂ catalysts identification of surface sulfate group was difficult because their signals overlapped with the stretching vibration bands of Si-O-Si and Si-O-H bonds near the 1000 cm⁻¹ region.

Figure 1. Dependences of the specific surface area and pore volume on TiO_2 content for the TiO_2/AC and TiO_2/SiO_2 composite photocatalysts.

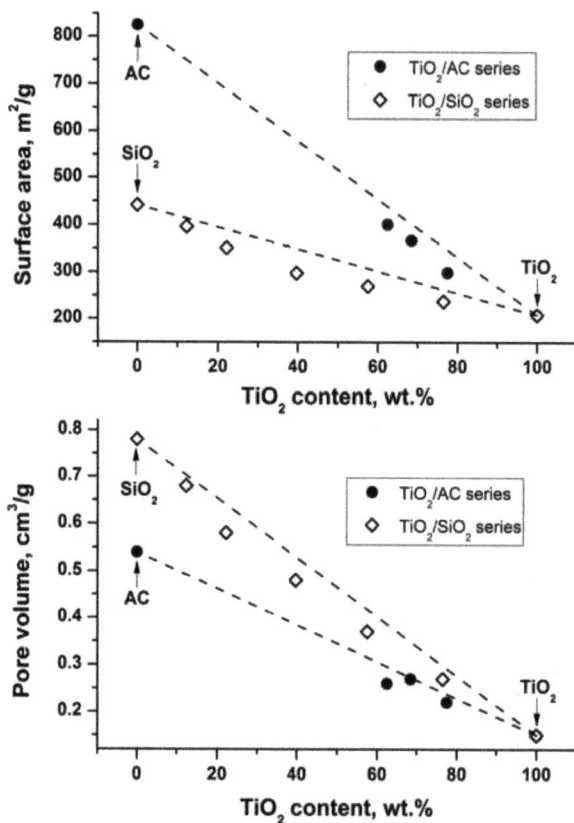

Figure 2. XRD patterns for the pure TiO_2, AC, SiO_2 and for the composite 70-TC and 40-TS photocatalysts.

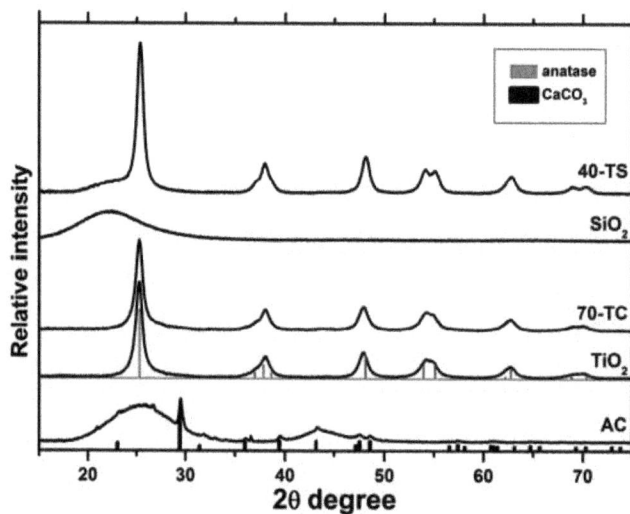

Figure 3. ATR FT-IR spectra for pure TiO₂, AC, SiO₂ and for the composite photocatalysts.

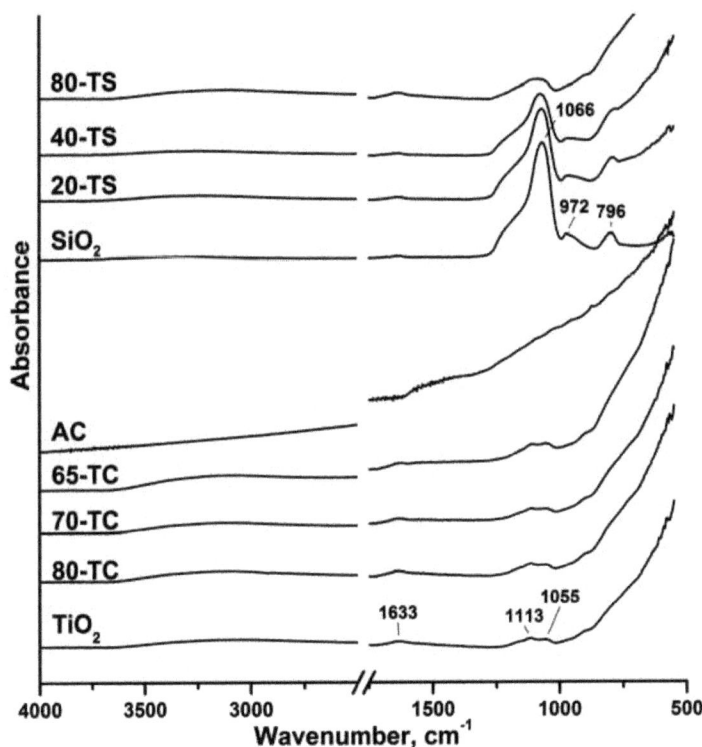

2.2. Kinetic Experiments

The main purpose of our work was to study the PCO of diethyl sulfide with the composite photocatalysts and to investigate their stability in long-term experiments. In this connection in the beginning we optimized the quantity of the photocatalyst. Then we chose the photocatalyst with adsorptivity and photocatalytic activity in good proportions and finally we investigated its stability in the DES PCO.

2.2.1. Effect of the Sample Quantity on the Photocatalytic Activity

The photocatalytic activities of pure TiO₂ and composite samples were measured in a continuous flow reactor in the reaction of acetone vapor PCO. The CO₂ formation rate was used as a measure of the photocatalytic activity. Photocatalysts were uniformly deposited onto a 3 × 3 cm glass support and then installed into the continuous flow reactor (see Experimental Section). The quantity of photocatalyst deposited was measured in mg/cm² units. For the quantity optimization experiments several glass supports with different quantities were prepared for all photocatalysts. For the pure TiO₂ the quantities were 0.25, 0.5, 1, 2 and 3 mg/cm². For the composite samples their quantities were adjusted in such a way that the quantities of contained TiO₂ were in the 0.2–3 mg/cm² range.

For example the 40-TS sample contains 39.7 wt. % of TiO_2 (Table 1) and to achieve the 0.25 mg/cm^2 value a quantity of 0.25/0.397 = 0.63 mg/cm^2 of 40-TS sample was deposited onto the glass support.

Figure 4 demonstrates the dependencies of the steady-state rate of CO_2 formation during acetone oxidation over TiO_2, TiO_2/AC and TiO_2/SiO_2 samples on the quantity of contained TiO_2. The higher is the contained TiO_2 on the glass support, the thicker the photocatalyst layer is.

Figure 4. Influence of the quantity of photocatalyst on its photocatalytic activity.

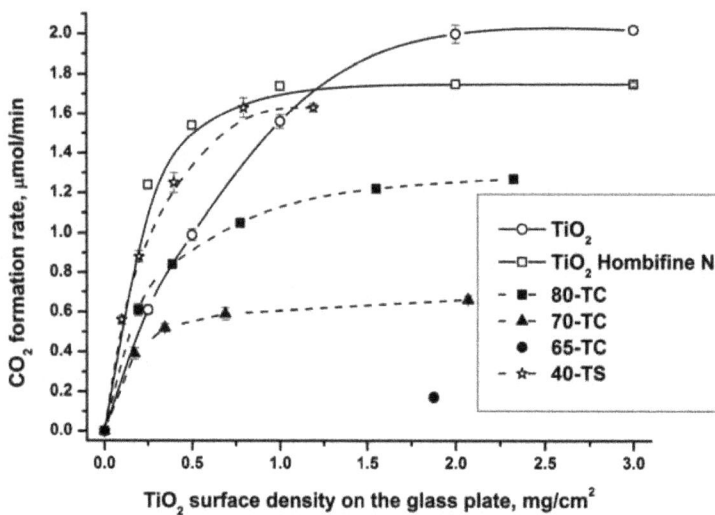

It could be seen that in all cases the PCO rate achieves the maximum value. It corresponds to the situation when the incident light is completely absorbed by the photocatalyst and any further increase in the quantity of photocatalyst leads to the formation of the bottom unirradiated photocatalyst layers which do not work.

The quantity of photocatalyst which corresponds to the maximum PCO rate depends on the TiO_2 content and its dispersion. For example for the commercial Hombifine N TiO_2 (Sachtleben Chemie GmbH, 100% anatase, S_{BET} = 350 m^2/g) the maximum rate quantity is about 1 mg/cm^2, whereas for the synthesized TiO_2 sample it is about 2 mg/cm^2.

It should be noted that at low sample quantity the 40-TS sample is more active than pure TiO_2. As it follows from the Table 1 both photocatalysts have the same size of TiO_2 crystallites—11 nm— therefore the difference of activities could be explained by a higher dispersion of the TiO_2 particles deposited onto the silica in the 40-TS sample than in the pure TiO_2.

On the other hand activities of the TiO_2/AC samples are lower than for pure TiO_2 because unlike silica, AC absorbs UV irradiation. Due to low photocatalytic activity of TiO_2/AC photocatalysts we used samples 80-TC and 70-TC with high TiO_2 content. Two conclusions could be reached from the above discussion:

(1) Comparison of the photocatalysts' activity should be done using high quantities when the thickness of the photocatalyst layer is sufficient for complete light absorption (e.g., 2–3 mg/cm^2). We used this approach when choosing a photocatalyst with good adsorptivity and photocatalytic activity (see Section 2.2.2);

(2) Studies of long-term photocatalyst use should be done using a relatively low TiO_2 quantity (e.g., 0.5 mg/cm^2) because in this case we can assume that the entire photocatalyst surface is irradiated and is involved in the reaction process. This is the reason why we investigated the diethyl sulfide oxidation with a 0.5 mg/cm^2 quantity of TiO_2.

2.2.2. Effect of TiO_2 Content on the Photocatalytic Activity of the TiO_2/SiO_2 Catalyst

Figure 5 demonstrates dependencies of the steady-state rate of CO_2 formation during acetone oxidation over the TiO_2/SiO_2 photocatalysts. The quantity of catalysts on the glass supports in these experiments was 3 mg/cm^2 in order to compare the highest possible photocatalytic activity of the samples.

Figure 5. Dependence of the CO_2 formation rate during acetone PCO on the TiO_2 content for TiO_2/SiO_2 series.

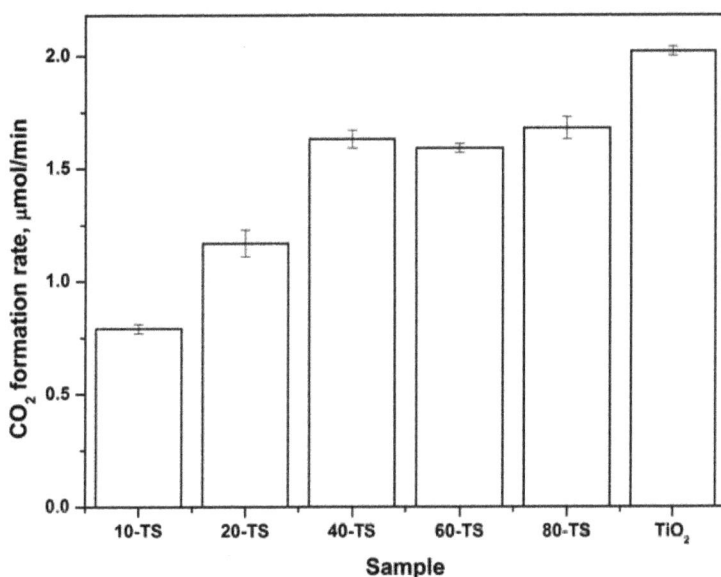

All TiO_2/SiO_2 samples demonstrate high activity, even at a low TiO_2 content, because silica does not absorb UV light. The CO_2 formation rate for the 10-TS sample which contains 12 wt. % of TiO_2 was 0.79 μmol/min and it was only 2.5 times lower than for a pure TiO_2 sample.

The oxidation rate increases with the increase of TiO_2 content and achieves almost the highest value for the 40-TS sample with 40 wt. % TiO_2 content. The 60-TS and 80-TS samples have slightly higher activity and it means that at 40 wt. % TiO_2 content SiO_2 particles are already completely covered with the TiO_2 particles. Therefore the following DES PCO experiments were conducted with the 40-TS sample which demonstrated high adsorption capacity due to its high content of porous support and at the same time high photocatalytic activity.

2.2.3. Kinetics of the DES PCO in a Static Reactor

The main objective of the experiments in the static reactor was to compare the kinetics of DES oxidation over pure TiO_2 and composite TiO_2/AC and TiO_2/SiO_2 photocatalysts and to compare the photocatalysts' deactivation during three consecutive DES PCO cycles.

Composite 80-TC, 70-TC and 40-TS samples as well as pure TiO_2 photocatalyst were chosen for these investigations. Sample quantities were correspondingly adjusted to 0.65, 0.73, 1.3, and 0.5 mg/cm^2 for 80-TC, 70-TC, 40-TS and pure TiO_2, so that the net TiO_2 quantity was equal to 0.5 mg/cm^2 in all cases. The same amount of active component (*i.e.*, TiO_2) placed in the reactor allowed us to carry out a valid comparison of photocatalyst deactivation for the samples with different TiO_2 content and to estimate the effect of the support.

H_2O, CO_2 and CO were detected as final gaseous oxidation products. The concentration of CO did not exceed the 55 ppm level and it was which much lower than the final CO_2 concentration which was equal to about 1400 ppm. The final surface products of DES PCO were sulfate complexes. The accumulation of sulfates on the photocatalysts surface was confirmed by FT-IR analysis and it was the reason of irreversible photocatalysts deactivation.

Acetaldehyde (CH_3CHO), formic acid ($HCOOH$), ethylene (C_2H_4) and SO_2 were detected in the gas phase as intermediates of DES PCO. All intermediates were completely oxidized to final products during the long-term irradiation. Noticeable concentrations were detected only for acetaldehyde and formic acid, therefore their kinetic curves were discussed along with DES removal and CO_2 accumulation kinetic curves.

SO_2 was detected in the gas phase only during the first PCO cycle and its concentration did not exceed the 30-50 ppm level. Quantitative analysis of ethylene was not performed due to its low concentration.

Kinetic curves of DES, acetaldehyde, formic acid and CO_2 during the first and the third cycle of 0.5 μL DES PCO in the static reactor over TiO_2, 80-TC, 70-TC and 40-TS samples are presented in the Figures 6 and 7, respectively.

Besides irreversible photocatalyst deactivation caused by attachment of sulfate ions to the photocatalyst surface, temporal deactivation was also observed. In the kinetic curves this is well illustrated by the intense accumulation of acetaldehyde in the gas phase and the increase of induction period for the CO_2 kinetic curves in the beginning of PCO run (compare the same samples in Figures 6 and 7).

The reason for this temporal deactivation is the formation of partial oxidation products like diethyl sulfoxide, diethyl sulfone and others [14,16]. These non-volatile compounds accumulate on the photocatalyst surface and hinder the PCO process. The continuous photocatalyst irradiation results in the gradual oxidation of surface non-volatile species making the catalyst surface available for further DES destruction. At this moment the fast removal of acetaldehyde from the gas phase and intensive accumulation of CO_2 begin. Formation of formic acid in the gas phase during DES PCO could be explained by its low adsorption on the photocatalyst surface due to its low molecular weight.

Figure 6. Kinetics of 0.5 μL DES PCO in the static reactor during the first cycle over the pure TiO_2, 80-TC, 70-TC and 40-TS samples.

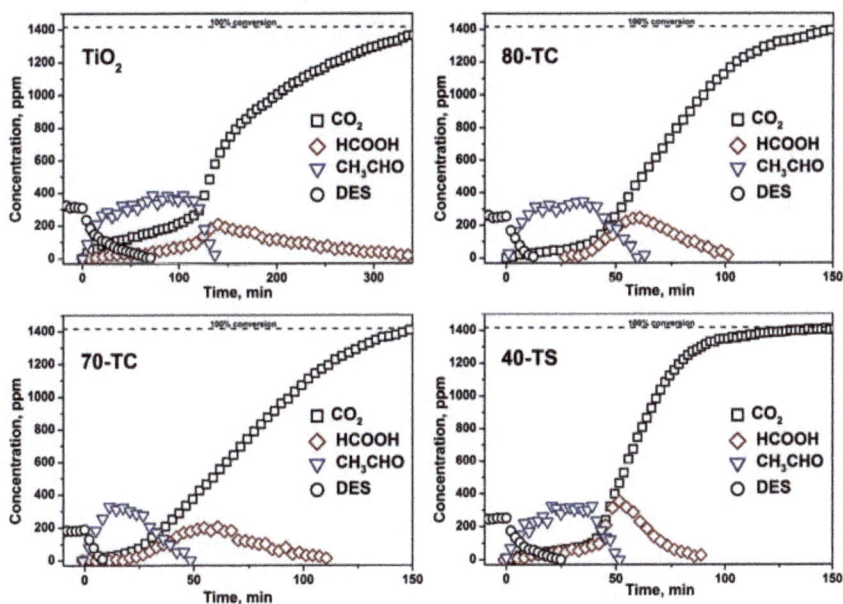

Figure 7. Kinetics of 0.5 μL DES PCO in the static reactor during the third cycle over the pure TiO_2, 80-TC, 70-TC and 40-TS samples.

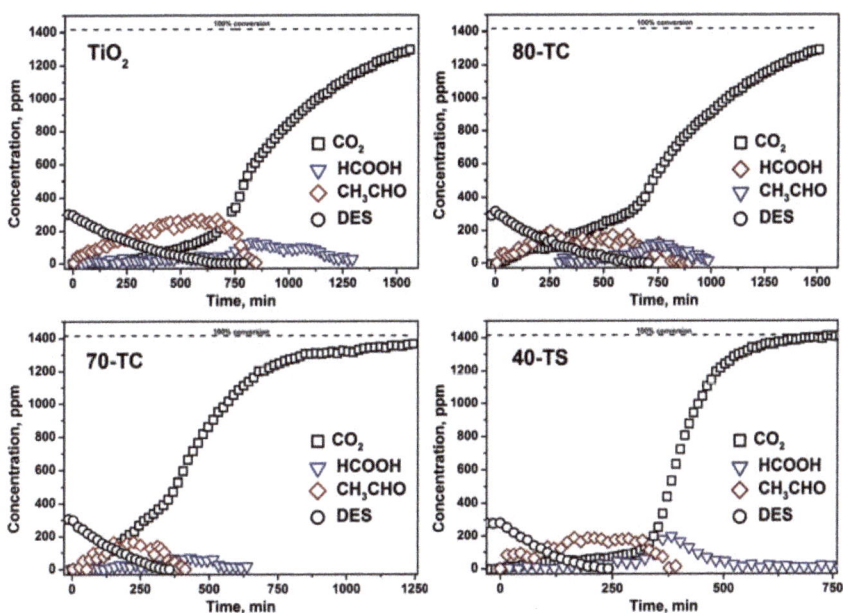

After the long-term irradiation complete oxidation of all intermediates was observed and the final CO_2 concentration reached the expected 1416 ppm value calculated from the mass balance.

The use of the composite photocatalyst increases the available surface. As a result the DES removal rate and kinetics of photooxidation change. A decrease of the time needed for complete removal of DES vapor from the gas phase was observed for the composite photocatalysts. For example, in the first oxidation cycle the time values of DES removal were 73, 14, 9 and 22 min for TiO_2, 80-TC, 70-TC and 40-TS samples, respectively. The increase of the rate of DES removal can be explained by reversible transfer of non-volatile intermediates from the TiO_2 surface onto the support surface (AC or SiO_2). As a result active sites on the TiO_2 surface remained free for further interaction with DES molecules. This effect was discussed in our previous work [20].

The fast DES removal in the case of composite photocatalysts led to a decrease of the induction period of CO_2 accumulation by about 2-fold (Figure 6). The initial rate of CO_2 accumulation after the induction period in the first oxidation cycle was 18.9, 18.7, 14.7 and 33.2 ppm/min for TiO_2, 80-TC, 70-TC and 40-TS samples, respectively. In contrast to the composite photocatalyst the CO_2 accumulation rate over pure TiO_2 sample was declining strongly with the increase of reaction time. As in the case of acetone PCO, the 40-TS demonstrated the highest activity in the DES PCO.

In each subsequent oxidation cycle over the same sample a decrease of DES PCO rate was observed. As a result in the third cycle the times of complete DES removal were 600, 550, 315 and 220 min, respectively, for the pure TiO_2, 80-TC, 70-TC and 40-TS samples (Figure 7). This means that a strong deactivation of the samples occurs. To estimate the extent of photocatalyst deactivation the time of 90% DES mineralization was calculated in each cycle. Calculated times for all samples in each oxidation cycle are presented in Figure 8.

Figure 8. Time values of 90% DES conversion into CO_2 in three oxidation cycles for pure TiO_2 and TiO_2/adsorbent composite photocatalysts.

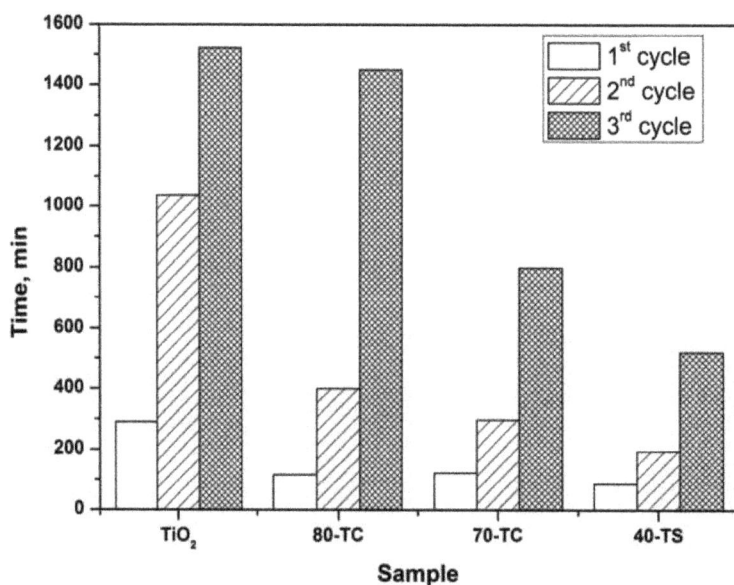

As it follows from Figure 8, the time of the complete mineralization of DES becomes higher for each subsequent PCO cycle. For example, for the pure TiO_2 sample in the first cycle the reaction was complete after 290 min, but in the third cycle it took 1522 min. Photocatalyst deactivation is decreasing in the following sequence: TiO_2 > 80-TC > 70-TC > 40-TS. For example, the sum of mineralization times in all three cycles for the pure TiO_2 sample is equal to 2848 min but for the most active and stable composite 40-TS photocatalyst it is only 800 min.

The excellent behavior of the 40-TS sample in the DES PCO can be explained by its high photocatalytic activity and large surface area which is available for the adsorption of intermediates. TiO_2/AC samples are less active than the 40-TS sample, but are still better than pure TiO_2 samples. It should be noted that the 70-TC sample demonstrated lower deactivation than the 80-TC sample due to its higher content of AC. In addition, the lowest concentration of gaseous intermediates, acetaldehyde and formic acid among all synthesized samples, was also observed for the 70-TC sample.

Decrease of deactivation in the case of composite photocatalyst as well as the increase of DES removal rates can be explained by reversible transfer of non-volatile intermediates, which are the reason of deactivation, from TiO_2 particles onto the support [20]. Another possible explanation for the increased activity of the composite photocatalysts is the possible transfer of OH radicals from TiO_2 onto the support surface. Such a possibility was shown by Carretero-Genevrier et al. [36], who demonstrated that OH radicals could migrate up to 10 nm distance from TiO_2 surface into the SiO_2 matrix.

Finally, turnover frequency (TOF) was calculated for all samples. The total amount of mineralized DES for three consecutive runs was $3 \times 0.5 = 1.5$ µL or 8×10^{18} molecules. For all samples the amount of active component (i.e., TiO_2) was the same—3.5 mg. To estimate the surface active sites concentration we used the value of 5×10^{14} a.s./cm^2 proposed by Ollis in 1980 [37].

Based on these data it is possible to calculate the total number of active sites:

$$5 \times 10^{14} \frac{a.s.}{10^{-4}m^2} \times 208 \frac{m^2}{g} \times 0.0035g = 3.6 \times 10^{18} \text{ a.s.} \quad (1)$$

The TiO_2 specific surface area from Table 1 was used for estimation of the active surface area because these experiments were performed at low sample quantity and we supposed that the entire photocatalyst surface was irradiated. The total time of complete DES mineralization was calculated as the sum of mineralization times in three consecutive PCO cycles presented in the Figure 8.

The estimated TON values are 1.3×10^{-5}, 1.9×10^{-5}, 3.0×10^{-5} and 4.6×10^{-5} s^{-1} for TiO_2, 80-TC, 70-TC and 40-TS samples, respectively. The TOF value for 40-TS composite photocatalyst is 3.5 times higher than for pure TiO_2 sample. Fast purification of air from the DES vapor over composite TiO_2/SiO_2 photocatalyst and its low deactivation during the long-term oxidation can thus be used for the development of purification methods against S-containing CWAs.

3. Experimental Section

3.1. Materials

The following chemical reagents were used for the catalyst preparation and oxidation experiments: titanyl sulfate ($TiOSO_4 \cdot 2H_2O$, >98%, Vekton, St. Petersburg, Russia), sulfuric acid (H_2SO_4, 93.5%–95.6%, PKF Ant, Russia), acetone (CH_3COCH_3, >99.8%, Mosreaktiv, Moscow,

Russia), DES (C$_2$H$_5$SC$_2$H$_5$, >98%, Fluka, Buchs, Switzerland). The reagents were used as supplied without further purification.

Activated carbon (AC) obtained by steam-gas activation of wood matter with S_{BET} = 825 m^2/g and silica with S_{BET} = 440 m^2/g and particle size of 10–40 µm were used for immobilization of TiO$_2$ particles. AC powder (Sorbent, Perm, Russia) was boiled before synthesis in distilled water during several hours to remove ionic impurities and finally washed out thoroughly by deionized water. Silica powder was supplied from Sigma-Aldrich (St. Louis, MO, USA) and used without any treatments. Titanyl sulfate water solution with a concentration of approximately 10 wt. % was used for TiO$_2$ deposition by the thermal hydrolysis method.

3.2. Synthesis of the Composite Photocatalyst

Composite photocatalysts were synthesized by thermal hydrolysis of TiOSO$_4$ according to the procedure described in details previously [22]. Typically, a certain amount of activated carbon or silica powder was suspended in a titanyl sulfate water solution (300 mL) and boiled for 5 hours under constant mixing. The calculated TiO$_2$ content in the sample was varied in the range of 65–80 wt. % for AC-containing samples and 10–80 wt. % for SiO$_2$-containing samples. The TiO$_2$ content was adjusted by adding a certain amount of support to 300 mL of TiOSO$_4$ solution. The samples containing AC or silica were marked as X-TC or X-TS correspondingly, where X was the TiO$_2$ content (wt. %). The reference TiO$_2$ sample was synthesized by the same procedure without addition of support (AC or silica) and marked as TiO$_2$.

3.3. Characterization Method

The Ti content in the synthesized samples was determined by atomic emission spectroscopy using an Optima 4300 DV spectrometer (PerkinElmer, Waltham, MA, USA). Content of TiO$_2$ was recalculated using these results according to the stoichiometric formula of oxides. The surface area and pore volume of the samples were measured by nitrogen adsorption at 77 K using the ASAP 2020 instrument (Micromeritics, Norcross, GA, USA). The specific surface area was calculated using the BET analysis and the pore volume was determined as total pore volume at P/Po~1. X-ray diffraction was applied to determine crystal phase composition and size of crystalline particles. XRD patterns were recorded using a D8 Advance (Bruker AXS GmbH, Karlsruhe, Germany) diffractometer with CuK$_\alpha$ radiation. The calculation of coherent-scattering domains size was performed using the Scherrer equation:

$$<D> = \frac{K\lambda}{\Delta(2\theta)cos\theta} \qquad (2)$$

with K equaled to 1.

The surface of samples and initial supports were investigated by FT-IR analysis using attenuated total reflectance technique. IR spectra of sample surface were registered using a Varian 640-IR FT-IR spectrometer (Varian Inc., Palo Alto, CA, USA) equipped by the ATR attachment. Samples were not treated in any way before analysis.

3.4. Kinetic Experiments

3.4.1. Acetone Oxidation

Acetone oxidation was investigated in the continuous flow unit described in details previously [38]. The continuous flow unit was equipped by an IR long-path gas cell (Infrared Analysis Inc., Anaheim, CA, USA) installed in a FT801 FT-IR spectrometer (Simex, Novosibirsk, Russia). Standard operational parameters were the following: acetone concentration—20 ± 4 μmol/L, temperature—40 °C, relative humidity—22% ± 2%, volumetric flow rate (U)—0.058 L/min. Detailed information about the effect of acetone concentration on the oxidation rate in the continuous flow unit is presented in Figure S1 in the Supplementary Information.

A certain amount of sample was uniformly deposited on a glass support of 9.1 cm² surface area and irradiated with UV light produced by a UV LED (Nichia, Tokushima, Japan) with λ_{max} ~373 nm. The sample irradiance in the 320–400 nm region was 9.7 mW/cm². The measurement of light intensity was performed using a Spectrilight spectroradiometer (International Light Technologies, Peabody, MA, USA). The emission spectrum of the UV LED is presented in Figure S2 in the Supplementary Information section.

The concentration of acetone and CO_2 in the reaction mixture was calculated from the FT-IR spectra using the integral form of the Beer-Lambert law:

$$\int_{\omega_1}^{\omega_2} A(\omega)\, d\omega = \varepsilon \times l \times C \tag{3}$$

where $A(\omega) = \lg\left(\dfrac{I_0(\omega)}{I(\omega)}\right)$ —absorbance, ω_1 and ω_2—limit of the corresponding absorption band

(cm⁻¹), ε —coefficient of extinction (L·μmol⁻¹·cm⁻²), l—optical path length (cm), C—gas phase concentration (μmol/L).

The rate of CO_2 formation was used to evaluate photocatalytic activity and was calculated according to the following formula:

$$W_{CO_2} = \Delta\, C_{CO_2} \times U \tag{4}$$

where $\Delta\, C_{CO_2}$ is the difference in CO_2 concentrations in the outlet and inlet air streams of the reactor and U is the volumetric flow rate.

3.4.2. DES Oxidation

Oxidation of the DES vapor was investigated in a 0.3 L static reactor installed in the cell compartment of a Nicolet 380 FT-IR spectrometer (Thermo Fisher Scientific Inc., Waltham, MA, USA). A detailed description of the experimental setup was presented in our previous work [22].

The sample was uniformly deposited onto a 7.0 cm² glass support which was placed in the reactor and irradiated with UV light produced by the UV LED described above during several hours in order to completely oxidize all organic species previously adsorbed on the catalyst surface during its storage. The sample irradiance in the 320–400 nm region was 10.2 mW/cm².

After sample training 0.5 μL of liquid DES was injected into the reactor and evaporated during 30 min to achieve adsorption-desorption equilibrium. Then the UV LED was turned on and IR spectra were taken periodically. Concentrations of DES and other oxidation products in the gas phase were calculated using the Beer-Lambert law described above. The details of the quantitative calculations using IR spectra can be found in [39]. The IR spectra of individual substances which were detected in the gas phase during the DES PCO and other information which was used for calculation of extinction coefficients for each substance are presented in the Supplementary Information section in Figure S3 and Table S1.

After complete mineralization of DES in the reactor (*i.e.*, when the amount of accumulated CO_2 reached the expected level calculated from the stoichiometric equations) the reactor was swept with fresh air and the next DES oxidation cycle was performed. Three oxidation cycles of the same amount of DES were performed for each sample.

4. Conclusions

Composite TiO_2/adsorbent photocatalysts were synthesized by the $TiOSO_4$ thermal hydrolysis method in the presence of activated carbon or SiO_2 and were tested in the photocatalytic oxidation of acetone and DES vapor. The following conclusions were made:

(1) The usage of composite photocatalyst results in up to an 8-fold decrease of DES removal time if compared with pure unmodified TiO_2. This could be explained by an increase of the available surface area in the case of composite photocatalyst and reversible transfer of non-volatile intermediates from TiO_2 surface to the support surface thus keeping the photocatalyst surface available for further interaction with substrate. Additionally the removal of intermediates—acetaldehyde and formic acid—occurs faster over composite photocatalyst;

(2) The long-term oxidation of DES leads to a strong deactivation of the photocatalyst. The deactivation decreases in the following sequence: $TiO_2 > TiO_2/AC > TiO_2/SiO_2$. The most active and stable catalyst is the TiO_2/SiO_2 one which contains 40 wt. % of TiO_2. The calculated TOF number for this sample is 3.5 times higher than for pure TiO_2.

Supplementary Materials

Supplementary materials can be accessed at: http://www.mdpi.com/1420-3049/19/12/21424/s1.

Acknowledgments

The work was performed with support of the Skolkovo Foundation (Grant Agreement for Russian educational organization №1 on 28.11.2013).

Author Contributions

Dmitry Selishchev synthesized the photocatalysts and performed experiments; Dmitry Selishchev and Denis Kozlov analyzed the data and wrote the manuscript. All authors read and approved the final version of the manuscript.

Conflicts of Interest

The authors declare no conflict of interest.

References

1. Watson, A.P.; Griffin, G.D. Toxicity of vesicant agents scheduled for destruction by the Chemical Stockpile Disposal Program. *Environ. Health Perspect.* **1992**, *98*, 259–280.
2. Munro, N.B.; Watson, A.P.; Ambrose, K.R.; Griffin, G.D. Treating exposure to chemical warfare agents: Implications for health care providers and community emergency planning. *Environ. Health Perspect.* **1990**, *89*, 205–215.
3. Munro, N. Toxicity of the organophosphate chemical warfare agents GA, GB, and VX: Implications for public protection. *Environ. Health Perspect.* **1994**, *102*, 18–38.
4. Perry Robinson, J.P., Ed. *Public Health Response to Biological and Chemical Weapons: WHO Guidance*, 2nd ed.; World Health Organization: Geneva, Switzerland, 2004; p. 340.
5. Ivarsson, U.; Nilsson, H.; Santesson, J., Eds. *A FOA Briefing book on Chemical Weapons: Threat, Effects, and Protection*; National Defence Research Establishment: Umea, Sweden, 1992.
6. Alexandrov, V.N.; Emel'yanov, V.I. *Poisonous Compounds (in Russian)*; Voenizdat: Moscow, Russia, 1990; p. 271.
7. Yang, Y.C.; Baker, J.A.; Ward, J.R. Decontamination of chemical warfare agents. *Chem. Rev.* **1992**, *92*, 1729–1743.
8. Zuo, G.M.; Cheng, Z.X.; Li, G.W.; Wang, L.Y.; Miao, T. Photoassisted Reaction of Sulfur Mustard under UV Light Irradiation. *Environ. Sci. Technol.* **2005**, *39*, 8742–8746.
9. Zuo, G.M.; Cheng, Z.X.; Li, G.W.; Shi, W.P.; Miao, T. Study on photolytic and photocatalytic decontamination of air polluted by chemical warfare agents (CWAs). *Chem. Eng. J.* **2007**, *128*, 135–140.
10. Bhatkhande, D.S.; Pangarkar, V.G.; Beenackers, A.A. Photocatalytic degradation for environmental applications—A review. *J. Chem. Technol. Biotechnol.* **2002**, *77*, 102–116.
11. Martyanov, I.N.; Klabunde, K.J. Photocatalytic Oxidation of Gaseous 2-Chloroethyl Ethyl Sulfide over TiO_2. *Environ. Sci. Technol.* **2003**, *37*, 3448–3453.
12. Fox, M.A.; Kim, Y.S.; Abdel-Wahab, A.A.; Dulay, M. Photocatalytic decontamination of sulfur-containing alkyl halides on irradiated semiconductor suspensions. *Catal. Lett.* **1990**, *5*, 369–376.
13. Vorontsov, A.V.; Panchenko, A.A.; Savinov, E.N.; Lion, C.; Smirniotis, P.G. Photocatalytic Degradation of 2-Phenethyl-2-chloroethyl Sulfide in Liquid and Gas Phases. *Environ. Sci. Technol.* **2002**, *36*, 5261–5269.
14. Kozlov, D.; Vorontsov, A.; Smirniotis, P.; Savinov, E. Gas-phase photocatalytic oxidation of diethyl sulfide over TiO_2: Kinetic investigations and catalyst deactivation. *Appl. Catal. B Environ.* **2003**, *42*, 77–87.
15. González-García, N.; Ayllón, J.A.; Doménech, X.; Peral, J. TiO_2 deactivation during the gas-phase photocatalytic oxidation of dimethyl sulfide. *Appl. Catal. B Environ.* **2004**, *52*, 69–77.

16. Vorontsov, A.V.; Savinov, E.V.; Davydov, L.; Smirniotis, P.G. Photocatalytic destruction of gaseous diethyl sulfide over TiO₂. *Appl. Catal. B Environ.* **2001**, *32*, 11–24.

17. Vorontsov, A. TiO₂ reactivation in photocatalytic destruction of gaseous diethyl sulfide in a coil reactor. *Appl. Catal. B Environ.* **2003**, *44*, 25–40.

18. Vorontsov, A.V. Photocatalytic transformations of organic sulfur compounds and H₂S. *Russ. Chem. Rev.* **2008**, *77*, 909–926.

19. Vorontsov, A.V.; Besov, A.S.; Parmon, V.N. Fast purification of air from diethyl sulfide with nanosized TiO₂ aerosol. *Appl. Catal. B Environ.* **2013**, *129*, 318–324.

20. Selishchev, D.; Kolinko, P.; Kozlov, D. Adsorbent as an essential participant in photocatalytic processes of water and air purification: Computer simulation study. *Appl. Catal. A Gen.* **2010**, *377*, 140–149.

21. Leary, R.; Westwood, A. Carbonaceous nanomaterials for the enhancement of TiO₂ photocatalysis. *Carbon* **2011**, *49*, 741–772.

22. Selishchev, D.S.; Kolinko, P.A.; Kozlov, D.V. Influence of adsorption on the photocatalytic properties of TiO₂/AC composite materials in the acetone and cyclohexane vapor photooxidation reactions. *J. Photochem. Photobiol. A Chem.* **2012**, *229*, 11–19.

23. Bouazza, N.; Lillo-Ródenas, M.A.; Linares-Solano, A. Photocatalytic activity of TiO₂-based materials for the oxidation of propene and benzene at low concentration in presence of humidity. *Appl. Catal. B Environ.* **2008**, *84*, 691–698.

24. Pucher, P.; Benmami, M.; Azouani, R.; Krammer, G.; Chhor, K.; Bocquet, J.F.; Kanaev, A.V. Nano-TiO₂ sols immobilized on porous silica as new efficient photocatalyst. *Appl. Catal. A Gen.* **2007**, *332*, 297–303.

25. Kitano, M.; Matsuoka, M.; Ueshima, M.; Anpo, M. Recent developments in titanium oxide-based photocatalysts. *Appl. Catal. A Gen.* **2007**, *325*, 1–14.

26. Xu, Y.; Langford, C.H. Photoactivity of titanium dioxide supported on MCM-41, zeolite X, and zeolite Y. *J. Phys. Chem. B* **1997**, *101*, 3115–3121.

27. Liu, S.X.; Chen, X.Y.; Chen, X. A TiO₂/AC composite photocatalyst with high activity and easy separation prepared by a hydrothermal method. *J. Hazard. Mater.* **2007**, *143*, 257–263.

28. Takeda, N.; Iwata, N.; Torimoto, T.; Yoneyama, H. Influence of carbon black as an adsorbent used in TiO₂ photocatalyst films on photodegradation behaviors of propyzamide. *J. Catal.* **1998**, *177*, 240–246.

29. Anderson, C.; Bard, A.J. An Improved Photocatalyst of TiO₂/SiO₂ Prepared by a Sol-Gel Synthesis. *J. Phys. Chem.* **1995**, *99*, 9882–9885.

30. Kolinko, P.A.; Smirniotis, P.G.; Kozlov, D.V.; Vorontsov, A.V. Cr modified TiO₂-loaded MCM-41 catalysts for UV-light driven photodegradation of diethyl sulfide and ethanol. *J. Photochem. Photobiol. A Chem.* **2012**, *232*, 1–7.

31. Panayotov, D.; Yates, J.T. Bifunctional Hydrogen Bonding of 2-Chloroethyl Ethyl Sulfide on TiO₂-SiO₂ Powders. *J. Phys. Chem. B* **2003**, *107*, 10560–10564.

32. Panayotov, D.A.; Paul, D.K.; Yates, J.T. Photocatalytic Oxidation of 2-Chloroethyl Ethyl Sulfide on TiO₂-SiO₂ Powders. *J. Phys. Chem. B* **2003**, *107*, 10571–10575.

33. Panayotov, D.; Kondratyuk, P.; Yates, J.T. Photooxidation of a Mustard Gas Simulant over TiO$_2$-SiO$_2$ Mixed-Oxide Photocatalyst: Site Poisoning by Oxidation Products and Reactivation. *Langmuir* **2004**, *20*, 3674–3678.

34. Ito, S.; Inoue, S.; Kawada, H.; Hara, M.; Iwasaki, M.; Tada, H. Low-Temperature Synthesis of Nanometer-Sized Crystalline TiO$_2$ Particles and Their Photoinduced Decomposition of Formic Acid. *J. Colloid Interface Sci.* **1999**, *216*, 59–64.

35. Saur, O. The structure and stability of sulfated alumina and titania. *J. Catal.* **1986**, *99*, 104–110.

36. Carretero-Genevrier, A.; Boissiere, C.; Nicole, L.; Grosso, D. Distance dependence of the photocatalytic efficiency of TiO$_2$ revealed by *in situ* ellipsometry. *J. Am. Chem. Soc.* **2012**, *134*, 10761–10764.

37. Childs, L.; Ollis, D. Is photocatalysis catalytic? *J. Catal.* **1980**, *66*, 383–390.

38. Korovin, E.; Selishchev, D.; Besov, A.; Kozlov, D. UV-LED TiO$_2$ photocatalytic oxidation of acetone vapor: Effect of high frequency controlled periodic illumination. *Appl. Catal. B Environ.* **2015**, *163*, 143–149.

39. Kozlov, D.; Besov, A. Method of Spectral Subtraction of Gas-Phase Fourier Transform Infrared (FT-IR) Spectra by Minimizing the Spectrum Length. *Appl. Spectrosc.* **2011**, *65*, 918–923.

Sample Availability: Samples are available from authors.

Preparation of a Titania/X-Zeolite/Porous Glass Composite Photocatalyst Using Hydrothermal and Drop Coating Processes

Atsuo Yasumori, Sayaka Yanagida and Jun Sawada

Abstract: Combinations of TiO$_2$ photocatalysts and various adsorbents have been widely studied for the adsorption and photocatalytic decomposition of gaseous pollutants such as volatile organic compounds (VOCs). Herein, a TiO$_2$-zeolite-porous glass composite was prepared using melt-quenching and partial sintering, hydrothermal treatment, and drop coating for preparation of the porous glass support and X-zeolite and their combination with TiO$_2$, respectively. The obtained composite comprised anatase phase TiO$_2$, X-zeolite, and the porous glass support, which were combined at the micro to nanometer scales. The composite had a relatively high specific surface area of approximately 25 m^2/g and exhibited a good adsorption capacity for 2-propanol. These data indicated that utilization of this particular phase-separated glass as the support was appropriate for the formation of the bulk photocatalyst-adsorbent composite. Importantly, the photocatalytic decomposition of adsorbed 2-propanol proceeded under UV light irradiation. The 2-propanol was oxidized to acetone and then trapped by the X-zeolite rather than being released to the atmosphere. Consequently, it was demonstrated that the micrometer-scaled combination of TiO$_2$ and zeolite in the bulk form is very useful for achieving both the removal of gaseous organic pollutants and decreasing the emission of harmful intermediates.

Reprinted from *Molecules*. Cite as: Yasumori, A.; Yanagida, S.; Sawada, J. Preparation of a Titania/X-Zeolite/Porous Glass Composite Photocatalyst Using Hydrothermal and Drop Coating Processes. *Molecules* **2015**, *20*, 2349-2363.

1. Introduction

Various atmospheric pollutants in the environment continue to cause considerable problems. In particular, volatile organic compounds (VOCs) such as formaldehyde, acetaldehyde, acetone, 2-propanol, and toluene, released mainly from paints and adhesives used in building and construction materials, give rise to sick building syndrome. Titanium dioxide (TiO$_2$) photocatalysts are well known to decompose organic compounds under ultraviolet (UV) light irradiation via oxidation reactions, and therefore, have been widely studied and are practically employed in combination with "black lights", such as fluorescent lamps, UV lamps, and sunlight to remove indoor VOCs [1–9].

The concentration of VOCs in living spaces is typically very low at several tens of ppm (by volume and used hereafter) to less than 1 ppm [10]. Under such conditions, a TiO$_2$ photocatalyst in the form of a coating material has two main disadvantages. First, the decomposition rate is not sufficiently high for gaseous pollutants at very low concentrations in the atmosphere, because the photocatalytic reaction mediated by TiO$_2$ (or any other solid semiconductor photocatalyst) occurs at or very close to the surface, and the diffusion process for the gaseous pollutants from the atmosphere to the surface of the TiO₂ is rate limiting. Second, the removal of VOCs is not effective under weak intensity radiation or during the nighttime.

One potential solution for overcoming such disadvantages of TiO₂ is the combination of a TiO₂ photocatalyst with an appropriate adsorbent. In such a composite, the pollutants are first concentrated in the pores of the adsorbent, and then photocatalytic degradation of the adsorbed or desorbed pollutants proceeds on the TiO₂ or in the pores due to the generation of oxygen radicals following light irradiation. Because the advantages of combining TiO₂ with an adsorbent are also observed for the degradation of dilute concentrations of pollutants in waste water, various adsorbents have been combined with TiO₂ for the rapid removal of low levels of pollutants in the environment [11]. For example, studies have been reported on the combination of TiO₂ photocatalysts with ceramic supports [12–16], porous glasses and glass fiber [17–21], activated carbon [22–26] and graphene [27–30], silica gel [31–35] and other mesoporous materials [36–40].

Zeolites are another important category of effective adsorbents that have been combined with TiO₂. Natural zeolite is an aluminosilicate mineral with micro pores. Various synthetic zeolites have been prepared with controllable pore sizes and surface properties and used as molecular sieves for the adsorption of molecules of particular sizes and polarities [41]. In addition, there are many studies of combinations of TiO₂ photocatalysts and zeolites for photochemical and photocatalytic science and applications [42,43]. For, example, studies have been reported on the adsorption and photocatalytic degradation of various organic pollutants [44–50] and nitrogen oxides (NOₓ) [51,52]. Among the synthesized zeolites, X- and Y-zeolites, which have faujasite (FAU) structure, are well known to have large pores (0.74 nm in diameter) due to their super cage structures and the ability to adsorb relatively large organic molecules [53]. Notably, the adsorption properties of those FAU-type zeolites are influenced by their compositions, and particularly the Si/Al ratio. The number of Na⁺ (or other alkaline or alkaline earth metal) ions accompanying the [AlO₄]⁻ tetrahedra on the surface of X-zeolite is relatively high because of its low Si/Al ratio (Si/Al ≈ 1.25) compared to that of other FAU-type zeolites, such as Y-zeolite (Si/Al ≈ 2.3) [53]. Because cations on the surfaces of zeolite pores exhibit strong electrostatic interactions with molecules having dipole or quadrupole moments, X-zeolite can adsorb molecules, such as H₂O and lower alcohols. Therefore, in the present study, X-zeolite with its high ability for the adsorption of polar VOCs was selected as the adsorbent for the model polar VOC pollutant 2-propanol.

For preparation of a composite comprising TiO₂ and X-zeolite, three important points were considered. (1) It is necessary for the X-zeolite to be adjacent to the TiO₂ to ensure that photo-generated oxygen radicals diffuse and attack the VOCs adsorbed in the X-zeolite before they are deactivated [54,55]; (2) TiO₂ and the X-zeolite should be co-loaded on appropriate supports to enable easy handling in practical applications [56]; (3) A transparent support for the TiO₂ and X-zeolite is required to allow passage of near-UV light for photoexcitation of the TiO₂. Previous studies mentioned above could achieve the first point that the zeolite was adjacent to the TiO₂ by use of mainly commercial particulate zeolite and appropriate synthesis procedures of TiO₂ fine particles. However, for the practical application, the above points (2) and (3) are essential. In addition, the support should also be porous to provide a highly dispersive condition of zeolite which can result in a large adsorption and reaction area. Therefore, porous silicate glass was selected because of its high specific surface area, transparency in the visible and UV range, and relatively high chemical durability, and zeolite was directly synthesized on the glass support. The disk-shaped silicate glass supports were prepared via the partial sintering of silicate glass frits comprising Na_2O-B_2O_3-CaO-ZrO_2-Al_2O_3-SiO_2 with ratios in the immiscibility region to achieve phase separation [57]. The porous glass was formed during subsequent hydrothermal treatment, which was used for the synthesis of the X-zeolite on the glass support. This process was expected to make the zeolite combine with the porous support much rigidly. The TiO_2 was then coated on the X-zeolite on the glass support by dropping a titanium alkoxide solution into the zeolite-porous glass composites. The crystalline phases and microstructures of the obtained TiO_2-coated zeolite-porous glass composites were characterized and their ability to adsorb and photocatalytically decompose 2-propanol was examined.

2. Results and Discussion

2.1. Crystalline Phases of the Composites

The XRD patterns of the partially sintered glass and composite samples are shown in Figure 1. The partially sintered glass exhibited only a halo pattern in the diffraction pattern (Figure 1a,b), indicating that the glass was not crystallized during the sintering process. The diffraction pattern of the zeolite-porous glass composite (Figure 1c) contained diffraction peaks for the X-zeolite of a FAU structure, while that for the TiO_2-coated composite sample (Figure 1d) exhibited the diffraction peak for anatase TiO_2 in addition to those for the X-zeolite, although the intensities of the latter peaks were slightly decreased. These results indicated that the X-zeolite precipitated following hydrothermal treatment and remained in the composite following the TiO_2 coating process.

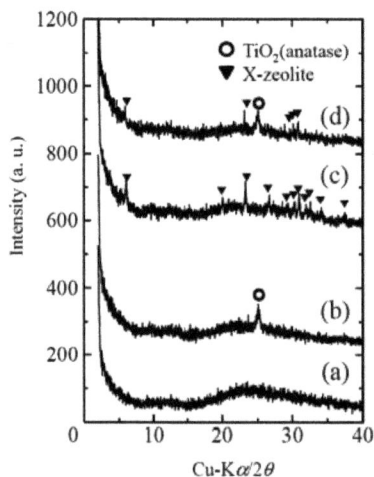

Figure 1. XRD patterns of the prepared samples: (a) partially sintered glass; (b) TiO_2-coated partially sintered glass; (c) zeolite-porous glass composite; and (d) TiO_2-coated zeolite-porous glass composite.

2.2. Micro Textures of the Composites

The micro textures of the TiO_2-coated partially sintered glass, the zeolite-porous glass composite, and the TiO_2-coated composite were observed via FE-SEM, and the results are shown in Figures 2–4, respectively. The TiO_2-coated partially sintered glass disk comprised glass particles with diameters of 100–200 μm (Figure 2a), which were similar to those of the raw glass frit particles. Sufficient spaces remained through which 2-propanol gas could diffuse to the interior of the disk. In addition, many small cracks were observed at the interfaces between the glass particles due to precipitation of the thick TiO_2 layer (Figure 2b). Interestingly, the zeolite-porous glass composite had a similar texture to that of the partially sintered glass on a macroscopic level, as can be seen in Figure 3a. However, in the microstructure near the surface that was directly exposed to the solution during hydrothermal treatment, many cubic precipitates were formed (Figure 3b), which were assigned to the X-zeolite based on their cubic shape [58] and the results of the XRD analysis. Micro pores formed by the selective etching of the phase separation texture during the hydrothermal treatment were also detected on the glass grains, as can be seen in Figure 3c. Finally, a thick TiO_2 layer covered the surface of the micro structure in the coated composite, but the texture of the sample was not changed significantly, as can be seen in Figure 4a–c. Figure 4d shows the coated X-zeolite as an example.

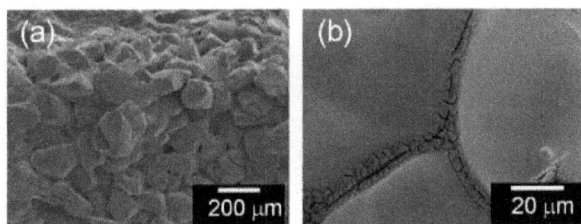

Figure 2. FE-SEM images of the TiO_2-coated partially sintered glass with (**a**) low and (**b**) high magnification.

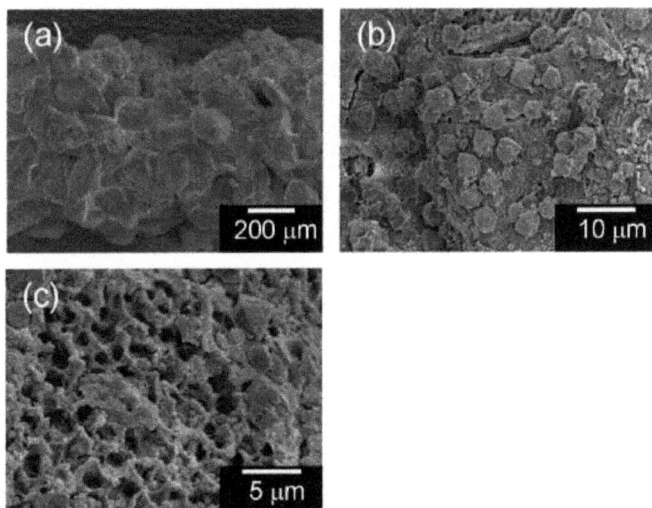

Figure 3. FE-SEM images of the zeolite-porous glass composite: (**a**) low magnification image; (**b**) precipitated zeolite particles; and (**c**) micro pores formed on the glass surface.

Figure 4. FE-SEM images of the TiO2-coated zeolite-porous glass composite: (**a**) low magnification image; (**b**) precipitated zeolite particles; (**c**) micro pores formed on the glass surface; and (**d**) TiO2 coating layer on the zeolite.

2.3. Adsorption Characteristics of the Composites

The adsorption and desorption isotherms of the zeolite-porous glass composite before and after TiO2 coating are shown in Figure 5. Both samples exhibited definite hysteresis in the high relative pressure region, and these isotherms corresponded to type IV isotherms as classified by International Union of Pure and Applied Chemistry (IUPAC) [59]. Typical FAU type zeolites exhibit type II isotherms with little or no adsorption and hysteresis in the same region [59,60]. In contrast, type IV isotherms are typically observed in silica gel or mesoporous silica [59]. Therefore, these isotherms indicated that the adsorption behavior of the composite could be attributed to the micropores of the zeolite and the mesopores of the porous glass in the low and the high relative pressure regions, respectively. Not surprisingly, the adsorption volume of the TiO2 coated sample was significantly decreased compared to that of the non-coated composite in the high relative pressure region. The multipoint Brunauer–Emmett–Teller (BET) specific surface area (Sg) of the zeolite-porous glass composite was approximately 30 m^2/g or approximately 4% that of a commercial X-zeolite (approximately 780 m^2/g, F9, Tosoh Co., Tokyo, Japan). After TiO2 coating, the samples retained a high Sg of approximately 25 m^2/g. These results indicated that the deposited TiO2 prevented the N2 gas from being adsorbed in the mesopores of the porous glass but allowed the gas to be adsorbed in the micropores of X-zeolite, and thus the combination of a TiO2 photocatalyst and X-zeolite adsorbent in a composite was successfully achieved. The Sg of the TiO2-coated partially sintered glass was less than the detection limit of the measurement system, suggesting that the coated TiO2 layer did not contribute to the adsorption of 2-propanol or acetone as an intermediate oxidative product of 2-propanol.

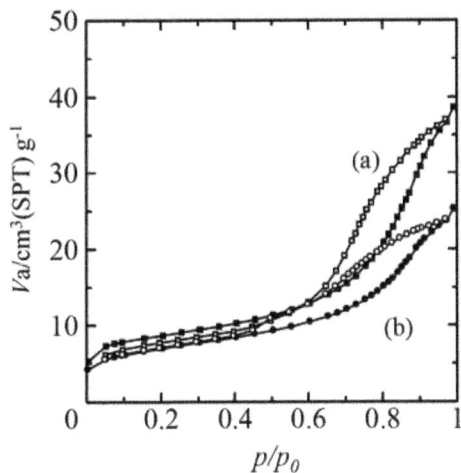

Figure 5. Adsorption and desorption isotherms of the zeolite-porous glass composite- (a) before and (b) after TiO₂ coating.

Next, the adsorption of 2-propanol by the prepared composites was examined. The changes in the concentration of 2-propanol in the glass container with time under dark conditions are shown in Figure 6 for the TiO₂-coated glass and the TiO₂ coated zeolite-porous glass composite. The TiO₂-coated glass adsorbed approximately 15% of the 2-propanol within the first 10 min, but then the concentration changed little up to 60 min. In contrast, the TiO₂-coated zeolite-porous glass composite adsorbed most of the 2-propanol within 10 min because it had a much higher specific surface area than the TiO₂-coated glass due to the presence of the X-zeolite.

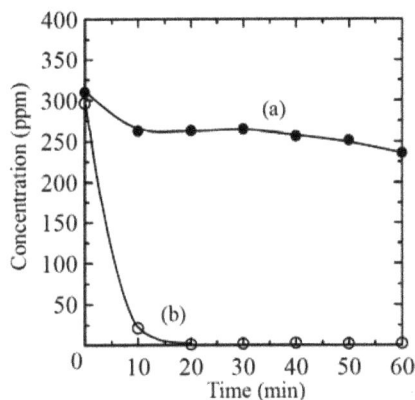

Figure 6. Change in the concentration of 2-propanol due to adsorption by the (a) TiO₂-coated partially sintered glass and (b) TiO₂-coated zeolite-porous glass composite.

2.4. Photocatalytic Properties of the Composites

The process of photocatalytic oxidation of 2-propanol has been well studied [54,61,62]. The reaction process is simply represented in Equations (1) and (2). Acetone is an intermediate product that is generated during the photocatalytic oxidative decomposition of 2-propanol:

$$C_3H_7OH + 1/2O_2 \rightarrow CH_3COCH_3 + H_2O \tag{1}$$

$$CH_3COCH_3 + 4O_2 \rightarrow 3CO_2 + 3H_2O \tag{2}$$

After adsorption of 2-propanol for 60 min, the samples were exposed to UV light irradiation, and the concentrations of 2-propanol, acetone, and CO_2 were determined by gas chromatography. The changes in the concentrations of these gaseous species during UV light irradiation are shown in Figure 7a,b for the TiO_2-coated partially sintered glass and the TiO_2-coated zeolite-porous glass composite, respectively. The concentration of CO_2 is presented as an increase over the initial concentration (ΔCO_2) as indicated by the values of the coordinates on the right vertical axis.

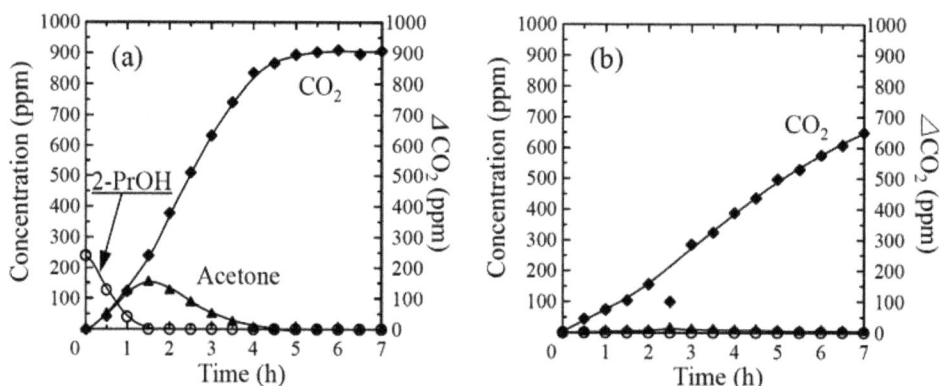

Figure 7. Changes in the concentrations of 2-propanol (open circles), acetone (filled triangles), and ΔCO_2 (filled diamonds, right axis) for the (**a**) TiO_2-coated partially sintered glass and (**b**) TiO_2-coated zeolite-porous glass composite.

In Figure 7a, the concentration of the residual 2-propanol gradually decreased and fell below the detection limit for the partially sintered TiO_2-coated glass after 1.5 h of UV light irradiation, while the concentration of acetone gradually increased to a maximum at 1.5 h, and then gradually decreased and fell below the detection limit after 4 h. The ΔCO_2 value also gradually increased to approximately 900 ppm and became saturated after approximately 5 h. This saturated value of approximately 900 ppm corresponded to the theoretical ΔCO_2 value expected for the complete conversion of 300 ppm of initial 2-propanol to H_2O and CO_2 via oxidation and decomposition as described in Equations (1) and (2).

With the TiO_2-coated partially sintered glass, a decrease in residual 2-propanol occurred simultaneously with an increase in the acetone and CO_2 concentrations, indicating that the 2-propanol was photocatalytically oxidized to acetone, a portion of which was immediately oxidized to CO_2 (Equations (1) and (2)), and a portion of which was released to the atmosphere. This behavior can be attributed to the lower adsorption capacity of the TiO_2-coated partially sintered glass.

In contrast, no desorption of 2-propanol to the atmosphere was detected for the TiO_2-coated zeolite-porous glass composite as shown in Figure 7b, and only a slight amount of acetone was observed. In addition, the ΔCO_2 value increased linearly with the UV light irradiation time. However, the generation rate for CO_2 was lower than that for the TiO_2-coated partially sintered glass, and the ΔCO_2 value was limited to approximately 650 ppm after 7 h. Considering the adsorption properties of X-zeolite as described in the introduction, the adsorption of 2-propanol and acetone on the X-zeolite in the present composite proceeded due to the electrostatic interactions between the Na^+ or K^+ ions on the surface of the X-zeolite and the 2-propanol and acetone. On the other hand, the adsorption of 2-propanol by the porous glass is considered to occur via hydrogen bonding, because Si-OH groups are likely to be present on the surface of the porous glass following hydrothermal treatment. However, based on the results of the FE-SEM analyses of the TiO_2-coated samples (Figures 2 and 4) and the decrease in the adsorption volume in the high relative pressure region (Figure 5), it is considered that the quantity of Si-OH groups decreased due to the subsequent repeated heat treatments at 400 °C and 500 °C necessary for fabrication of the TiO_2 coating and thick TiO_2 layer, respectively. Therefore, it can be concluded that the X-zeolite in the composite was the main adsorbent for the 2-propanol.

The schematic of the structure of the TiO_2-coated zeolite-porous glass composite are illustrated in Figure 8a,b in order to show its behaviors of adsorption and photocatalytic oxidation of 2-propanol and acetone, respectively. Even though CO_2 is not a polar molecule, it has a quadrupole moment and can thus also be adsorbed on X-zeolites via electrostatic interactions [63]. The adsorption behavior of the X-zeolite for 2-propanol, acetone, and CO_2 resulted in the slow generation rate of CO_2 from the zeolite-porous glass composite. In other words, although the X-zeolite strongly adsorbed 2-propanol and acetone, a certain amount of the CO_2 generated during irradiation with UV light was also adsorbed as shown in Figure 8b. These results suggest that the 2-propanol and acetone that desorbed from the X-zeolite (and the porous glass) were immediately oxidized by the TiO_2, which was adjacent to the X-zeolite. Alternatively, the 2-propanol and acetone adsorbed in the X-zeolite may have been attacked by oxygen radicals generated on the TiO_2 following irradiation with UV light. This oxidation process is possible to occur because the photocatalytic oxidation of 2-propanol sufficiently proceeded in the mechanically mixed zeolite and TiO_2 powder system [55], and in our composite system, the diffusion distance was enough short for the radicals generated from TiO_2 to reach the molecules adsorbed in the neighboring X-zeolite. Consequently, the TiO_2-coated zeolite-porous glass composite could sufficiently remove 2-propanol by the adsorption and the photocatalytic oxidation and also it could suppress the release of acetone to the atmosphere during the photocatalytic decomposition process owing to its high adsorption capacity.

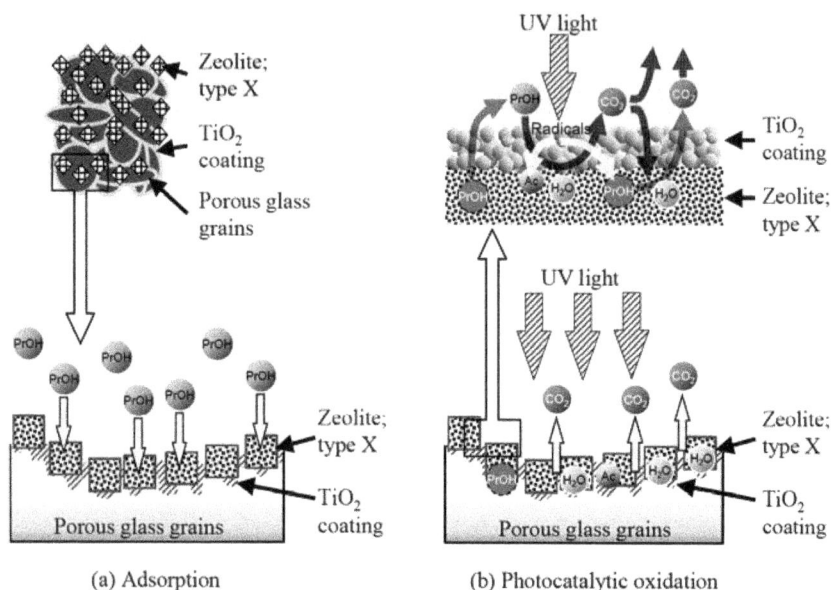

(a) Adsorption   (b) Photocatalytic oxidation

Figure 8. Schematic of (**a**) adsorption and (**b**) photocatalytic oxidation process by the TiO$_2$-coated zeolite-porous glass composite.

3. Experimental Section

3.1. Preparation of Partially Sintered Glass Supports

Porous silicate glass prepared using the phase separation phenomenon of oxide glasses was selected as the support. The sodium borosilicate glass system is considered to have a metastable immiscibility region and to separate into SiO$_2$-rich and Na$_2$O-B$_2$O$_3$ rich phases following appropriate thermal treatment above the glass transition temperature. In addition, SiO$_2$-rich porous glass can be obtained from such phase separated glasses with spinodal decomposition compositions upon further thermal treatment and subsequent selective leaching of the Na$_2$O-B$_2$O$_3$ rich phase [64]. Such a porous glass has been utilized as the support for TiO$_2$ photocatalysts [17,18]. However, for the synthesis of zeolites, it is necessary to use a hydrothermal treatment process in highly concentrated alkaline solution. Because typical SiO$_2$-rich glass has poor chemical resistance to alkaline solutions, it was necessary to use the mother glass, which has high durability in alkaline solutions, for synthesis of the glass-zeolite composite. Yazawa *et al.* reported that the Na$_2$O-B$_2$O$_3$-CaO-ZrO$_2$-Al$_2$O$_3$-SiO$_2$ glass system has a metastable immiscibility region, and the porous glass obtained from this system exhibited high durability in alkaline solutions [57]. Based on these results, a glass batch composition for the mother glass of 5.7Na$_2$O-9.2CaO-2.3Al$_2$O$_3$-3.2ZrO$_2$-22.7B$_2$O$_3$-56.9SiO$_2$ (mol %) was used in the present study.

Reagent-grade Na$_2$CO$_3$, H$_3$BO$_3$, SiO$_2$, CaCO$_3$, ZrO$_2$, and Al$_2$O$_3$ (Wako Pure Chemical, Osaka, Japan) were used for glass preparation without further purification. The glass batch was prepared by mixing the weighed raw materials and melting the mixture in an alumina crucible at 1500 °C for 1 h in air. The melt was then quenched and immediately heat-treated at 750 °C for 12 h to allow phase separation to occur. Next, the partially sintered glass support was prepared by first grinding the annealed, heat-treated (phase separated) glass into 106–150 μm diameter particles and then forming the particles into a rod-like shape. The rod was sintered at 700 °C for 1 h, and then disk shapes were cut, each with a diameter of approximately 11 mm, a thickness of 1 mm, and a weight of 0.1 g.

3.2. Synthesis of the Zeolite on the Partially Sintered Glass Support

To form micro pores on the partially sintered glass disks, hydrothermal treatment was used to simultaneously etch the phase separated glass particles that formed the disk and synthesize the zeolite.

FAU type zeolites are well known to be synthesized by use of a mixed NaOH-KOH solution [65]. Precise preparation conditions of X-zeolite on the phase separated glass were determined based on the results of the preliminary experiments and were described as followings. Two partially sintered glass disks, one on top of the other, were placed in an aqueous solution containing NaOH (3 mol/L, 25 mL) and KOH (4 mol/L, 5 mL). Next, silica gel (0.4 g) and $NaAlO_2$ (0.4 g), both purchased from Wako Pure Chemical and used without further purification, were added as the silica and alumina sources, respectively. The sample was tightly closed in a Teflon container (diameter: 35 mm; height: 55 cm in inner dimension; volume: approximately 50 mL) and was maintained at 75 °C for 24 h with stirring at 500 rpm. The hydrothermally treated sample was then washed with distilled water until the pH of the rinse water was approximately 8 and finally dried at 60 °C for 24 h.

3.3. TiO_2 Coating of the Samples

TiO_2 thin films were coated on the zeolite-glass composite and partially sintered glass using titanium diisopropoxide bis(acetylacetonate) $[(CH_3)_2CHO]_2Ti(C_5H_7O_2)_2$; TPA, Aldrich, St. Louis, MO, USA) as the TiO_2 source. A 2-propanol solution of TPA (1 mass% Ti) was stirred at approximately 2 °C for 1 h. Next, 200 μL of the TPA solution was dropped into the sample, dried at room temperature for 5 min, and subsequently heated at 400 °C for 10 min. This dropping and subsequent drying/heating cycle was repeated 6 times. Finally, the samples were heated at 500 °C for 2 h.

3.4. Characterizations of the Samples

The crystalline phases precipitated in the samples were examined using X-ray diffraction (XRD) analysis (LabX XRD-6100, Shimadzu, Kyoto, Japan) with a Cu-Kα radiation source. The micro textures of the surfaces and the cross sections of the samples were observed via field emission scanning electron microscopy (FE-SEM; S-4200, Hitachi, Tokyo, Japan) after platinum sputter coating. The adsorption and desorption isotherms and multipoint BET specific surface areas (Sg) of the samples were determined using the nitrogen adsorption technique (BELSORP-mini II, Nippon Bell, Osaka, Japan). The samples were heated at 200 °C for 24 h in a vacuum prior to measurement.

3.5. Gas Adsorption Ability and Photocatalytic Activity of the Samples

A calibrated gas generator (Permeater PD-1B, Gastech, Kanagawa, Japan) and dried air were used to produce air-diluted 2-propanol gas (approximately 300 ppm). A glass container (diameter: 4 cm; height: 6 cm; volume: approximately 65 mL) was placed in a glove bag filled with the air-diluted 2-propanol gas, and then the container filled with the same 2-propanol gas, immediately sealed and kept in the dark at 20 °C. The concentration of 2-propanol was determined using a gas chromatograph (GC, GC-8A, Shimadzu) with a thermal conductivity detector (TCD), a porous polymer beads column (Sunpak-A, 2 m, 160 °C, Shinwa Chemical, Kyoto, Japan), and a He carrier gas (20 mL/min). The gas in the sealed container was sampled (1 mL) every 10 min for 60 min using a micro syringe. After determination of the gas adsorption ability, the glass bottle was irradiation with UV light from a black light lamp (FL15BLB, peak wavelength = 0.30 mW/cm^2 at 365 nm). The concentrations of 2-propanol, acetone, and CO_2 in the glass container were determined every 30 min for 7 h using the same GC procedure.

4. Conclusions

A TiO_2-zeolite-porous glass composite was prepared using melt-quenching for the glass preparation, hydrothermal treatment for the synthesis of the X-zeolite, and drop coating for deposition of the TiO_2 thin film. The obtained composite comprised anatase phase TiO_2, X-zeolite, and the porous glass, which were combined at the micro to nanometer scale. Synthesis of the X-zeolite on the porous glass support was possible because a glass with high resistance to alkaline solutions was used. In addition, the TiO_2 and X-zeolite were solidified and formed into a disk shape with a relatively high specific surface area, and importantly, the TiO_2 as the photocatalyst was adjacent to the X-zeolite as the adsorbent. Furthermore, the X-zeolite in the composite was very effective for the adsorption of polar molecules, such as 2-propanol as a model pollutant and acetone and its oxidized intermediate. The suppression of the release of acetone due to the presence of the X-zeolite suggested that the combined use of TiO_2 and an adequate adsorbent is very useful for decreasing the emission of harmful intermediates generated during photocatalytic oxidative decomposition. Finally, the phase separated glass used as the support for the loading of the zeolite and TiO_2 made it possible to prepare a bulk photocatalyst-adsorbent composite with superior adsorption and photo degradation properties for 2-propanol, and it was considered to be a good candidate for a composite of photocatalyst and adsorbent for the practical applications of the removal of gaseous organic pollutants.

Author Contributions

A.Y. designed and organized the study; J.S. and S.Y. performed the experiments; J.S., S.Y., and A.Y. analyzed the data; S.Y. and A.Y. wrote the paper.

Conflicts of Interest

The authors declare no conflict of interest.

References

1. Peral, J.; Ollis, D.F. Heterogeneous photocatalytic oxidation of gas-phase organics for air purification Acetone, 1-butanol, butyraldehyde, formaldehyde, and *m*-xylene oxidation. *J. Catal.* **1992**, *136*, 554–565.

2. Fox, M.A.; Dulay, M.T. Heterogeneous photocatalysis. *Chem. Rev.* **1993**, *93*, 341–357.

3. Hoffmann, M.R.; Martin, S.T.; Choi, W.; Bahnemann, D.W. Environmental Applications of Semiconductor Photocatalysis. *Chem. Rev.* **1995**, *95*, 69–96.

4. Fujishima, A.; Rao, T.N.; Tryk, D.A. Titanium dioxide photocatalysis. *J. Photchem. Photobiol. C Photochem. Rev.* **2000**, *1*, 1–21.

5. Zhao, J.; Yang, X. Photocatalytic oxidation for indoor air purification: A literature review. *Build. Environ.* **2003**, *38*, 645–654.

6. Hashimoto, K.; Irie, H.; Fujishima, A. TiO$_2$ Photocatalysis: A Historical Overview and Future Prospects. *Jpn. J. Appl. Phys.* **2005**, *44*, 8269–8285.

7. Mo, J.; Zhang, Y.; Xu, Q.; Lamson, J.J.; Zhao, R. Photocatalytic purification of volatile organic compounds in indoor air: A literature review. *Atmos. Environ.* **2009**, *43*, 2229–2246.

8. Chen, H.; Nanayakkara, C.E.; Grassian, V.H. Titanium Dioxide Photocatalysis in Atmospheric Chemistry. *Chem. Rev.* **2012**, *112*, 5919–5948.

9. Yu, C.W.F.; Kim, J.T. Photocatalytic Oxidation for Maintenance of Indoor Environmental Quality. *Indoor Built Environ.* **2013**, *22*, 39–51.

10. Chapter 4: Summary of the guidelines. In *Air Quality Guidelines for Europe*, 2nd ed.; WHO Regional Office for Europe, WHO Regional Publications, European Series: Copenhagen, Demark, 2000; pp. 32–33.

11. Shaham-Waldmann, N.; Paz, Y. Modified Photocatalysts. In *Photocatalysis and Water Purification*; Pichat, P., Ed.; Wiley-VCH: Weinheim, Germany, 2013; pp. 107–111.

12. Kato, K. Photocatalytic Property of TiO$_2$ Anchored on Porous Alumina Ceramic Support by the Alkoxide Method. *J. Ceram. Soc. Jpn.* **1993**, *101*, 245–249.

13. Sauer, M.L.; Ollis, D.F. Acetone Oxidation in a Photocatalytic Monolith Reactor. *J. Catal.* **1994**, *149*, 81–91.

14. Obee, T.N.; Brown, R.T. TiO$_2$ Photocatalysis for Indoor Air Applications Effects of Humidity and Trace Contaminant Levels on the Oxidation Rates of Formaldehyde, Toluene, and 1,3-Butadiene. *Environ. Sci. Technol.* **1995**, *29*, 1223–1231.

15. Yasumori, A.; Ishizu, K.; Hayashi, S.; Okada, K. Preparation of a TiO$_2$ based multiple layer thin film photocatalyst. *J. Mater. Chem.* **1998**, *8*, 2521–2524.

16. Yasumori, A.; Shinoda, H.; Kameshima, Y.; Hayashi, S.; Okada, K. Photocatalytic and photoelectrochemical propertiesof TiO$_2$-based multiple layer thin film prepared by sol–gel and reactive-sputtering methods. *J. Mater. Chem.* **2001**, *11*, 1253–1257.

17. Anpo, M.; Aikawa, N.; Kubokawa, Y. Photoluminescence and photocatalytic activity of highly dispersed titanium oxide anchored onto porous Vycor glass. *J. Phys. Chem.* **1985**, *89*, 5017–5021.

18. Yamashita, H.; Honda, M.; Harada, M.; Ichihashi, Y.; Anpo, M.; Hirao, T.; Itoh, N.; Iwamoto, N. Preparation of Titanium Oxide Photocatalysts Anchored on Porous Silica Glass by a Metal Ion-Implantation Method. *J. Phys. Chem. B* **1998**, *102*, 10707–10711.

19. Ao, C.H.; Lee, S.C.; Yu, J.C. Photocatalyst TiO_2 supported on glass fiber for indoor air purification effect of NO on the photodegradation of CO and NO_2. *J. Photochem. Photobiol. A Chem*. **2003**, *156*, 171–177.

20. Ho, W.H.; Yu, J.C.; Yu, J. Photocatalytic TiO_2/Glass Nanoflake Array Films. *Langmuir* **2005**, *21*, 3486–3492.

21. Machida, F.; Daiko, Y.; Mineshige, A.; Kobune, M.; Toyoda, N.; Yamada, I.; Yazawa, T. Structures and Photocatalytic Properties of Crystalline Titanium Oxide-Dispersed Nanoporous Glass-Ceramics. *J. Am. Ceram. Soc.* **2010**, *93*, 461–464.

22. Takeda, N.; Torimoto, T.; Sampath, S.; Kuwabata, S.; Yoneyama, H. Effect of Inert Supports for Titanium Dioxide Loading on Enhancement of Photodecomposition Rate of Gaseous Propionaldehyde. *J. Phys. Chem.* **1995**, *99*, 9986–9991.

23. Torimoto, T.; Okawa, Y.; Takeda, N.; Yoneyama, H. Effect of Activated Carbon Content in TiO_2-loaded Activated Carbon on Photodegradation Behaviors of Dichloromethane. *J. Photochem. Photobiol. A Chem.* **1997**, *103*, 153–157.

24. Huang, B.; Saka, S. Photocatalytic activity of TiO_2 crystallite-activated carbon composites prepared in supercritical isopropanol for the decomposition of formaldehyde. *J. Wood Sci.* **2003**, *49*, 79–85.

25. Ao, C.H.; Lee, S.C. Combination effect of activated carbon with TiO_2 for the photodegradation of binary pollutants at typical indoor air level. *J. Photochem. Photobiol. A Chem.* **2004**, *161*, 131–140.

26. Ao, C.H.; Lee, S.C. Indoor air purification by photocatalyst TiO_2 immobilized on an activated carbon filter installed in an air cleaner. *Chem. Eng. Sci.* **2005**, *60*, 103–109.

27. Kamat, P.V. Graphene-Based Nanoarchitectures. Anchoring Semiconductor and Metal Nanoparticles on a Two-Dimensional Carbon Support. *J. Phys. Chem. Lett.* **2010**, *1*, 520–527.

28. Zhang, H.; Lv, X.; Li, Y.; Wang, Y.; Li, J. P25-Graphene Composite as a High Performance Photocatalyst. *ACS Nano* **2010**, *4*, 380–386.

29. Zhang, Y.; Tang, Z.; Fu, X.; Xu, Y. TiO_2–Graphene Nanocomposites for Gas-Phase Photocatalytic Degradation of Volatile Aromatic Pollutant: Is TiO_2–Graphene Truly Different from Other TiO_2–Carbon Composite Materials. *ACS Nano* **2010**, *4*, 7303–7314.

30. Kamegawa, T.; Yamahana, D.; Yamashita, H. Graphene Coating of TiO_2 Nanoparticles Loaded on Mesoporous Silica for Enhancement of Photocatalytic Activity. *J. Phys. Chem. C* **2011**, *114*, 15049–15053.

31. Domen, K.; Sakata, Y.; Kudo, A.; Maruya, K.; Onishi, T. The Photocatalytic Activity of a Platinized Titanium Dioxide Catalyst Supported over Silica. *Bull. Chem. Soc. Jpn.* **1988**, *61*, 359–362.

32. Yasumori, A.; Yamazaki, K.; Shibata, S.; Yamane, M. Preparation of TiO_2 Fine Particles Supported on Silica Gel as Photocatalyst. *J. Ceram. Soc. Jpn.* **1994**, *102*, 702–707.

33. Anderson, C.; Bard, A.J. An Improved Photocatalyst of TiO_2/SiO_2 Prepared by a Sol-Gel Synthesis. *J. Phys. Chem.* **1995**, *99*, 9882–9885.

34. Ding, Z.; Hu, X.; Lu, G.Q.; Yue, P.; Greenfield, P.F. Novel Silica Gel Supported TiO2 Photocatalyst Synthesized by CVD Method. *Langmuir* **2000**, *16*, 6216–6222.

35. Ismail, A.A.; Ibrahim, I.A.; Ahmed, M.S.; Mohamed, R.M.; El-Shall, H. Sol–gel synthesis of titania–silica photocatalyst for cyanide photodegradation. *J. Photochem. Photobiol. A Chem.* **2004**, *163*, 445–451.

36. Yu, J.C.; Wang, X.; Fu, X. Pore-Wall Chemistry and Photocatalytic Activity of Mesoporous Titania Molecular Sieve Films. *Chem. Mater.* **2004**, *16*, 1523–1530.

37. Li, H.; Bian, Z.; Zhu, J.; Huo, Y.; Li, H.; Lu, Y. Mesoporous $Au-TiO_2$ Nanocomposites with Enhanced Photocatalytic Activity. *J. Am. Chem. Soc.* **2007**, *129*, 4538–4539.

38. Tan, L.K.; Kumar, M.K.; An, W.W.; Gao, H. Transparent, Well-Aligned TiO_2 Nanotube Arrays with Controllable Dimensions on Glass Substrates for Photocatalytic Applications. *Appl. Mater. Interfaces* **2010**, *2*, 498–503.

39. Chu, S.; Inoue, S.; Wada, K.; Li, D.; Haneda, H.; Awatsu, S. Highly Porous (TiO_2-SiO_2-TeO_2)/Al_2O_3/TiO_2 Composite Nanostructures on Glass with Enhanced Photocatalysis Fabricated by Anodization and Sol-Gel Process. *J. Phys. Chem. B* **2003**, *107*, 6586–6589.

40. Qian, X.F.; Kamegawa, T.; Mori, K.; Li, H.X.; Yamashita, H. Calcium Phosphate Coatings Incorporated in Mesoporous $TiO_2/SBA-15$ by a Facile Inner-pore Sol-gel Process toward Enhanced Adsorption-photocatalysis Performances. *J. Phys. Chem. C* **2013**, *117*, 19544–19551.

41. Sircar, S.; Myers, A.L. 22 Gas Separation by Zeolites. Part V Applications. In *Handbook of Zeolite Science and Technology*; Auerbach, S.M., Carrado, K.A., Dutta, P.K., Eds.; CRC Press: New York, NY, USA, 2003; pp. 1063–1104.

42. Corma, A.; Garcia, H. Zeolite-based photocatalysts. *Chem. Commun.* **2004**, 1443–1459, doi:10.1039/B400147H.

43. Kuwahara, Y.; Yamashita, H. Efficient photocatalytic degradation of organics diluted in water and air using TiO_2 designed with zeolites and mesoporous silica materials. *J. Mater. Chem.* **2011**, *21*, 2407–2416.

44. Sampath, S.; Uchida, H.; Yoneyama, H. Photocatalytic Degradation of Gaseous Pyridine over Zeolite-Supported Titanium Dioxide. *J. Catal.* **1994**, *149*, 189–194.

45. Torimoto, T.; Ito, S.; Kuwabata, S.; Yoneyama, H. Effects of Adsorbents Used as Supports for Titanium Dioxide Loading on Photocatalytic Degradation of Propyzamide. *Environ. Sci. Technol.* **1996**, *30*, 1275–1281.

46. Xu, Y.; Langford, C.H. Photoactivity of Titanium Dioxide Supported on MCM41, Zeolite X, and Zeolite Y. *J. Phys. Chem. B* **1997**, *101*, 3115–3121.

47. Ichiura, H.; Kitaoka, T.; Tanaka, H. Preparation of composite TiO_2-zeolite sheets using a papermaking technique and their application to environmental improvement Part I Removal of acetaldehyde with and without UV irradiation. *J. Mater. Sci.* **2002**, *37*, 2937–2941.

48. Takeuchi, M.; Kimura, T.; Hidaka, M.; Rakhmawaty, D.; Anpo, M. Photocatalytic oxidation of acetaldehyde with oxygen on TiO$_2$/ZSM-5 photocatalyst: Effect of hydrophobicity of zeolites. *J. Catal.* **2007**, *246*, 235–240.

49. Donphai, W.; Kamegawa, T.; Careonpanich, M.; Nueangnoraj, K.; Mishihara, H.; Kyotani, T.; Yamashita, H. Photocatalytic performance of TiO$_2$–zeolite templated carbon composites in organic contaminant degradation. *Phys. Chem. Chem. Phys.* **2014**, *16*, 25004–25007.

50. Kamegawa, T.; Ishiguro, Y.; Kido, R.; Yamashita, H. Design of Composite Photocatalyst of TiO$_2$ and Y-Zeolite for Degradation of 2-Propanol in the Gas Phase under UV and Visible Light Irradiation. *Molecules* **2014**, *19*, 16477–16488.

51. Ichiura, H.; Kitaoka, T.; Tanaka, H. Preparation of composite TiO$_2$-zeolite sheets using a papermaking technique and their application to environmental improvement, Part II Effect of zeolite coexisting in the composite sheet on NO$_x$ removal. *J. Mater. Sci.* **2003**, *38*, 1611–1615.

52. Jan, Y.; Lin, L.; Karthik, M.; Bai, H. Titanium Dioxide/Zeolite Catalytic Adsorbent for the Removal of NO and Acetone Vapors. *J. Air Waste Manag. Assoc.* **2009**, *59*, 1186–1193.

53. Lobo, R.F. Introduction to the Structural Chemistry of Zeolites. Part II Synthesis and Structure. In *Handbook of Zeolite Science and Technology*; Auerbach, S.M., Carrado, K.A., Dutta, P.K., Eds.; CRC Press: New York, NY, USA, 2003; pp. 65–89.

54. Ohko, Y.; Kashimoto, K.; Fujishima, A. Kinetics of Photocatalytic Reactions under Extremely Low-Intensity UV Illumination on Titanium Dioxide Thin Films. *J. Phys. Chem. A* **1997**, *101*, 8057–8062.

55. Yamaguchi, K.; Inumaru, K.; Oumi, Y.; Sano, T.; Yamanaka, S. Photocatalytic decomposition of 2-propanol in air by mechanical mixtures of TiO$_2$ crystalline particles and silicalite adsorbent: The complete conversion of organic molecules strongly adsorbed within zeolitic channels. *Microporous Mesoporous Mater.* **2009**, *117*, 350–355.

56. Robert, D.; Keller, V.; Keller, N. Immobilization of a Semiconductor Photocatalyst on Solid Suppors: Methods, Materials, and Applications. In *Photocatalysis and Water Purification*; Pichat, P., Ed.; Wiley-VCH: Weinheim, Germany, 2013; pp. 145–178.

57. Yazawa, T.; Tanaka, H.; Eguchi, K.; Yokoyama, S. Novel alkali-resistance porous glass prepared from a mother glass based on the SiO$_2$-B$_2$O$_3$-RO-ZrO$_2$ (R=Mg, Ca, Sr, Ba and Zn) system. *J. Mater. Sci.* **1994**, *29*, 3433–3440.

58. Subotić, B.; Bronić, J. 5 Theoretical and Practical Aspects of Zeolite Crystal Growth. Part II Synthesis and Structure. In *Handbook of Zeolite Science and Technology*; Auerbach, S.M., Carrado, K.A., Dutta, P.K., Eds.; CRC Press: New York, NY, USA, 2003; pp. 129–203.

59. Sing, K.S.W.; Everett, D.H.; Haul, R.A.W.; Moscou, L.; Pierotti, R.A.; Rouquĕrol, J.; Siemieniewska, T. Reporting Physisorption Data for Gas/Solid Systems with Special Reference to the Determination of Surface Area and Porosity. *Pure Appl. Chem.* **1986**, *57*, 603–619.

60. Yang, Y.; Burke, N.; Zhang, J.; Huang, S.; Lim, S.; Zhu, Y. Influence of charge compensating cations on propane adsorption in X zeolites: Experimental measurement and mathematical modeling. *RSC Adv.* **2014**, *4*, 7279–7287.

61. Bickley, R.I.; Munuera, G.; Stone, F.S. Photoadsorption and Photocatalysis at Rutile Surfaces II. Photocatalytic Oxidation of Isopropanol. *J. Catal.* **1973**, *31*, 398–407.

62. Raillard, C.; Héquet, V.; Cloirec, P.L.; Legrand, J. Kinetic study of ketones photocatalytic oxidation in gas phase using TiO_2-containing paper: Effect of water vapor. *J. Photochem. Photobiol. A Chem.* **2004**, *163*, 425–431.
63. Cheung, O.; Hedin, N. Zeolites and related sorbents with narrow pores for CO_2 separation from flue gas. *RSC Adv.* **2014**, *4*, 14480–14494.
64. Doremus, R.H. 4 Phase Separation. Part One Formation and Structure of Glasses. In *Glass Science*; John Wiley & Sons: New York, NY, USA, 1973; pp. 44–73.
65. Kühl, G.H. Crystallization of low-silica faujasite ($SiO_2/Al_2O_3{\sim}2.0$). *Zeolites* **1987**, *7*, 451–457.

Sample Availability: Melt-quenched glass samples are available from the authors.

Some Observations on the Development of Superior Photocatalytic Systems for Application to Water Purification by the "Adsorb and Shuttle" or the Interphase Charge Transfer Mechanisms

Cooper Langford, Maryam Izadifard, Emad Radwan and Gopal Achari

Abstract: Adsorb and shuttle (A/S) and interfacial charge transfer are the two major strategies for overcoming recombination in photocatalysis in this era of nanoparticle composites. Their relationships are considered here. A review of key literature is accompanied by a presentation of three new experiments within the overall aim of assessing the relation of these strategies. The cases presented include: A/S by a high silica zeolite/TiO_2 composite, charge transfer (CT) between phases in a TiO_2/WO_3 composite and both A/S and CT by composites of TiO_2 with powered activated carbon (AC) and single-walled carbon nanotubes (SWCNT). The opportunities presented by the two strategies for moving toward photocatalysts that could support applications for the removal of contaminants from drinking water or that lead to a practical adsorbent for organics that could be regenerated photocatalytically link this discussion to ongoing research here.

Reprinted from *Molecules*. Cite as: Langford, C.; Izadifard, M.; Radwan, E.; Achari, G. Some Observations on the Development of Superior Photocatalytic Systems for Application to Water Purification by the "Adsorb and Shuttle" or the Interphase Charge Transfer Mechanisms. *Molecules* **2014**, *19*, 19557-19572.

1. Introduction

At present, the vast majority of the commercial application of photocatalysis depends on passive processes that may accomplish their photochemical goals slowly. This is a consequence of the small quantum yields typically reported and the limited UV energy available from low free or the low cost sources, either sunlight or fluorescent lamps installed for lighting purposes. Earlier in development of photocatalysis, major effort was expended on modification of the structure of TiO_2 to overcome one or both of the limitations. So far, radical improvement has eluded researchers. In recent years, improvement in manipulating the chemistry of the nanoscale has fueled a fresh campaign to achieve improvement by creating hybrids of the photocatalyst with other phases, such that the field of applications might expand. Two key strategies can be found in the literature. One is the combination of a phase with TiO_2 that can act as an acceptor for either the conduction band electron or the valence band hole to substantially inhibit recombination. The other is the "adsorb and shuttle" strategy that combines a phase that more extensively adsorbs the substrate and delivers it by surface diffusion to an adjacent TiO_2 site. Using an adsorbent, which could be photocatalytically generated, is advantageous for drinking water and waste water photocatalytic treatment systems. These two strategies emerged before the nanoparticle outburst, and some fundamental aspects from earlier studies will be discussed below along with more recent key references. Some new results will be reported. The transfer of carriers between phases with electron transfer from TiO_2 to WO_3 is an area related to the water program. The two oxides have little difference in adsorption, making this nearly a "pure" charge transfer case. Results for a pure adsorb and shuttle case will come from a study of zeolite ZSM-5 on TiO_2, a candidate in the effort to produce adsorbents that are photocatalytically regenerated. Finally, some results for the mixed case of carbon and TiO_2 will be discussed.

1.1. Adsorb and Shuttle (A/S)

The key issues with A/S were elucidated in a series of studies from Yoneyama's laboratories in the late 1990s. The approach involved loading a large excess of TiO_2 onto particles of well-known adsorbents. In the case of the substrate, 3,5-dichloro-N-(3-methyl-1-butyn-3-yl)benzamide (propyzamide), photodegradation was studied using photocatalyst coated onto clays, zeolites and activated carbon [1]. The catalysts were all 70 wt % TiO_2. Adsorption isotherms fitted the Langmuir model. The Langmuir maximum adsorption capacity parameter was critical, but so was the specific initial extent of adsorption after dark equilibration, a measure of adsorption "strength". Apparently, the decomposition rate (measured as the sum of solution and adsorbed loss) of propyzamide was large in the order of naked TiO_2 > 70 wt % TiO_2/mordenite > 70 wt % TiO_2/SiO_2 > 70 wt % TiO_2/AC (activated carbon), the order being opposite of the order of the Langmuir adsorption constant (strength), except for SiO_2. This appears to suggest that adsorb and shuttle was not useful. However, the rate of CO_2 production (mineralization, the goal of most work) was in the order: 70 wt % TiO_2/AC > 70 wt % TiO_2/SiO_2 > 70 wt % TiO_2/mordenite > naked TiO_2. This is exactly the order of the initial adsorbate loading of the photocatalysts at initiation of irradiation. The important following point is that tracking of intermediates showed that most were in solution in the case of bare TiO_2, but the overwhelming majority of intermediates was adsorbed in the case of TiO_2/AC. Thus, adsorb and shuttle can be a powerful tool for retaining intermediates and the completion of mineralization.

The key point of adsorb and shuttle, A/S, is that the concentration of the substrate close to the TiO_2 surface should greatly increase the probability of meeting between the substrate and a photoactivated site. An elementary illustration of this is the increase of a quantum yield when a favourable substrate concentration is raised, as is the case of an increase of quantum yield from 0.001 to 0.15 as the propanol concentration is raised from 0.001 M to 0.05 M [2]. Moreover, it is known that interfacial redox can compete well with recombination. For example, Colombo and Bowman [3] provide evidence that electron transfer can compete with recombination on a femtosecond time scale for particles encountering. Still, for the adsorbed substrate to reach the active site, it must migrate by surface diffusion. Consequently, surface diffusion must not be slow. Using propionaldehyde as a substrate, the relative surface diffusion was estimated on films containing TiO_2 by following the consumption of all of the substrate on a film illuminated over a part of the area. Takeda et al. [4] found relative surface diffusion to be in the order TiO_2/mordenite > TiO_2/silica > TiO_2/alumina > TiO_2/AC > TiO_2/zeolite A. An interesting related example of the problem of surface diffusion arose in an effort to produce a practical adsorbent based on a ZSM-5 hydrophobic zeolite absorber decorated with TiO_2 to allow photocatalytic regeneration, a project that reached the pilot scale. Vaisman *et al.* [5] reported that as the adsorption capacity increased, the photocatalytic kinetics decreased proportionately.

In summary, effective adsorb and shuttle, A/S, requires adsorbents with a delicate balance of adequate adsorption capacity, sufficient absorption strength and facile surface mobility. As the systems described above suggest, the best case seems to be a large excess of TiO_2 over the adsorbent.

1.2. Interphase Charge Transfer

The classic example of the reduction of the recombination rate is the deposition of small quantities of Pt on the TiO_2 surface to capture electrons in the Pt phase. The literature is extensive. There are also a number of semiconductor oxides that could inhibit recombination in TiO_2 if the matching of band gaps permits the transfer of an electron from the TiO_2 conduction band to a lower band edge or permits the hole in the TiO_2 conduction band to migrate to a higher energy conduction band edge. The examples are WO_3, which can play the role of electron acceptor and $Ni(OH)_2$, which can function as a hole acceptor. If the separation is to be long lived and build up in the acceptor phase, there must be electrochemical compensation. For example, if WO_3 accepts electrons, we look to W(V) chemical traps requiring cationic compensation. Tatsuma *et al.* [6] demonstrated that the presence of ionic conduction through water could support long-lived energy storage by supplying H^+ as the cation (note that the TiO_2 hole reacting with water produces a compensating H^+). Oxygen discharge of the "stored" electrons in WO_3 is slow, so energy storage providing reactivity in the dark is feasible. Thus, increased hole reactions from reduced recombination can be coupled to energy storage. Subsequently, Tatsuma *et al.* [7] demonstrated the bactericidal effect of the stored energy.

1.3. Carbon

Carbon has been mentioned above (along with a zeolite, silica or clay) as an adsorbent. It is, of course, the most widely used adsorbent for organics. However, there is also significant direct evidence of excited state electron transfer from TiO_2 to carbon. A striking example is the puzzle presented by Kedem *et al.* [8], who reported enhanced stability in polymer nanofibres containing TiO_2 and carbon nanotubes (CNTs). The CNTs reduced the rate of photocatalytic degradation of the (polyacrylonitrile) matrix. However, this did not interfere with photocatalytic reactions with several organic substrates and even enhanced the degradation of rhodamine 6G dye. This set of conflicting results can be understood as the CNTs functioning to transport charge carriers from the TiO_2 to CNT sites, which may induce a variety of effects (e.g., superoxide formation) depending on the degradation mechanism and reaction loci.

In combinations of carbon with photocatalysts, we must ask: what is the relative role of charge transfer *vs.* adsorb and shuttle?

The 6G dye measurements, made on aliquots of solution not accounting for the dye adsorbed on the CNTs, raises a final basic point. Xu and Langford [9] made an important observation about strongly adsorbed substrates using the dye XB3. For all the reactions induced by UV or visible light, the apparent initial rate of X3B loss in the aqueous phase increased with the initial equilibrated concentration of X3B. However, when the rate was determined by the decreased concentration both in the aqueous phase and on the catalyst surface, increase of real initial rate with the initial equilibrated concentration was observed only in the visible-light-induced reaction. That is, only when the reaction was initiated by dye sensitization that initiates dye oxidation directly upon excitation. It is important to measure total degradation, solution and surface, as done in the work in Yoneyama's group cited above.

2. Results and Discussion

2.1. TiO₂/ZSM-5 Adsorb and Shuttle

2.1. TiO$_2$/ZSM-5 Adsorb and Shuttle

The chosen adsorbent is the high silica zeolite ZSM-5 in a configuration similar to the system of Vaisman *et al.* [5] designed as an adsorbent for photocatalytic regeneration. ZSM-5, with a highly hydrophobic surface, presents an adsorption profile resembling activated carbon, but weaker for some highly hydrophobic molecules and stronger for more polar organics. It is used here in composite with Degussa P25 TiO$_2$ (PZS). The percentages of three different model substrates adsorbed on the surface of PZS and P25 after 30 min of stirring in the dark are shown in Table 1. It can be seen that there is an unequivocal increase in the amount of the 2,4,6-trichlorophenol (2,4,6-TCP) substrate adsorbed on the surface of PZS with a drastic increase observed in the case of sulfamethoxazole (SMX), while for atrazine, the adsorption was the least. The overall order of increase in percent adsorption is SMX > 2,4,6-TCP > atrazine.

Table 1. Adsorption percentage of the different model compounds at initial concentrations of 50.0 mg/L on the surface of PZS and P25 after the dark period. SMX, sulfamethoxazole; 2,4,6-TCP, 2,4,6-trichlorophenol.

Compound	Photocatalyst	
	Degussa P25%	PZS%
SMX	2.60	18.00
2,4,6-TCP	29.20	41.90
Atrazine	2.40	5.64

Figure 1 compares the photocatalytic performance of the PZS to that of commercial Degussa P25 as a reference. The first point corresponds to the normalized concentration after the dark adsorption period. Pseudo first-order rate constants (k) along with correlation coefficients (R^2) are listed in Table 2 (because of the rapid reaction, it was not a certain that the atrazine data do fit first order kinetics).

Figure 1. Photocatalytic degradation of the substrates with a UVA LED photoreactor. Solid lines, PZS (P25:ZSM-5:silica gel = 0.3:0.5:0.5 g/L); dashed lines, Degussa P25 (0.3 g/L).

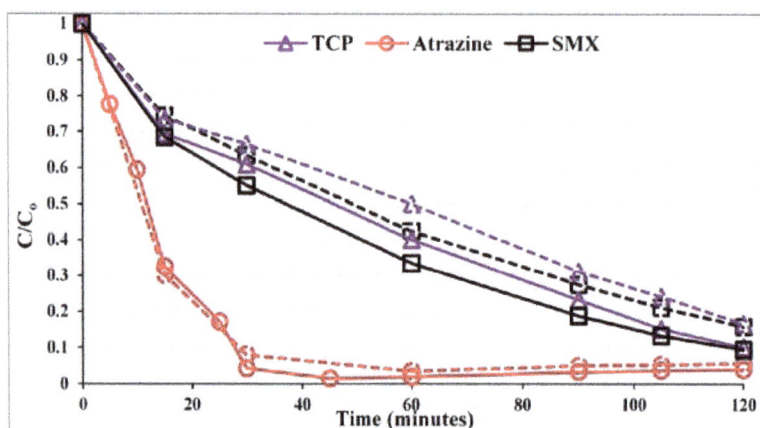

Table 2. Pseudo first order rate constants (min^{-1}) and R^2 fit measure for the substrates.

Compound	Degussa P25 min^{-1}		PZS min^{-1}	
SMX	k = 0.0147	R^2 = 0.996	k = 0.0189	R^2 = 0.9965
2,4,6-TCP	k = 0.0137	R^2 = 0.977	k = 0.0179	R^2 = 0.982
Atrazine	k = 0.055	R^2 = 0.923	k = 0.0976	R^2 = 0.966

As can be seen in Figure 1, for SMX and 2,4,6-TCP, the PZS shows noticeably higher photocatalytic activity than the Degussa P25. The enhancement in efficiency is not in proportion, as the adsorption strength of these compounds on the zeolite surface is different. While for atrazine, the photocatalytic activity was within the error, the same for both Degussa P25 and PZS. The order of photocatalytic activity improvement is SMX ~ 2,4,6-TCP > atrazine, which is consistent with the order of improved adsorption. This suggests that adsorb and shuttle may have been effective.

Substrates were extracted to track the extent of the retention of the initial substrate on the adsorbent surface during reaction. The amounts are reported in units of the solution concentration equivalent. Figure 2 shows data for 60 and 120 min.

2,4,6-TCP residue on the surface of both photocatalysts after 1 h of irradiation was small (less than 1 mg/L solution equivalent compared to an initial total of 50 mg/L). After 2 h of irradiation, no 2,4,6-TCP residues were detected on the surface of either photocatalyst. Furthermore, it can be seen from Figure 2 that, when the initial extent of dark adsorption is small, no residues were detected on the surface of either photocatalyst, as seen for atrazine on both photocatalysts and even for SMX on P25. The residue of SMX on the surface of PZS after 1 h of irradiation was 2.2 mg/L, which reduced to 1 mg/L after 2 h of irradiation. The comparison of the initial extent of dark adsorption results (Table 1) with these results suggests that 2,4,6-TCP surface diffusion [4] is faster than SMX.

Figure 2. Residue results of the substrates on the surface of PZS and Degussa P25 after one and two hour irradiation. The left bar shows 1-h data and the (clearly visible only for SMX) right bar 2-h data.

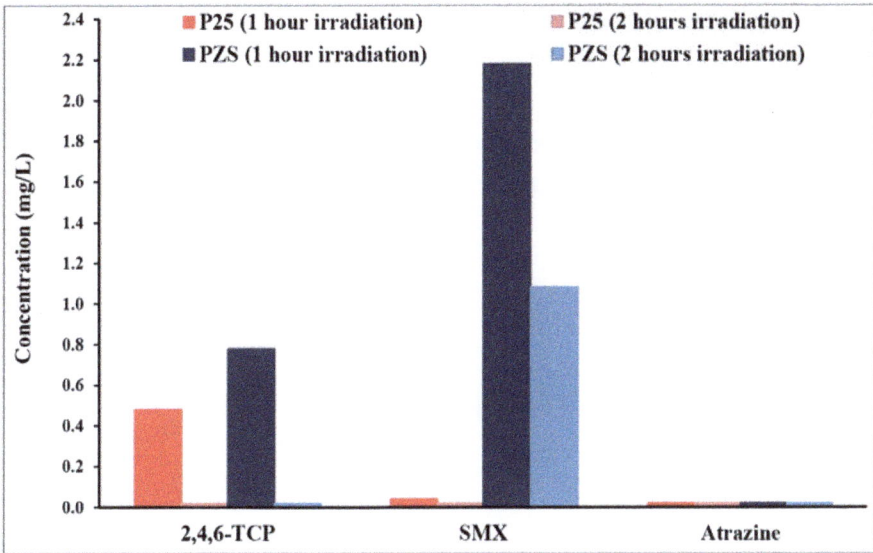

The key advantage of A/S is the promotion of mineralization [1]. The percentage of TOC removal of the substrates using PZS and Degussa P25 after 1 h of irradiation is shown in Figure 3. Using PZS improves the mineralization of SMX and shows that 2,4,6-TCP is ~45% mineralized after ~50% loss in the substrate. No mineralization of atrazine is seen, in agreement with McMurray *et al.* [10].

The results are a reminder that atrazine is one of the rare compounds not mineralized on TiO_2. The reaction stops at cyanuric acid [10].

Figure 3. Percentage TOC removal after 1 h of irradiation.

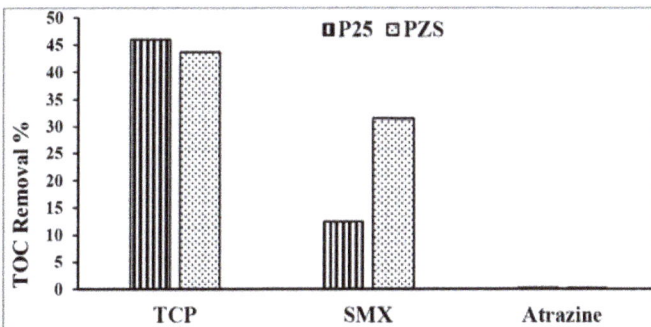

2.2. Interphase Charge Transfer: TiO_2/WO_3

One of the most striking demonstrations of the reduction of recombination rates arises from the persistence of reactivity after irradiation is terminated [11]. In this experiment, 4-chlorophenol

(4-CP) was chosen as a substrate to avoid any A/S contribution. Figure 4 shows the percent degradation of 4-CP after one hour of irradiation in a slurry of TiO_2/WO_3 in a 33-ppm solution of the 4-CP using the 365-nm LED reactor to allow precise pulsing. The duty cycle is varied. The figure shows the striking case of the efficiency increase in degradation reactions of 4-chlorophenol. A 1:1 duty cycle at a constant total dose with a pulse time of 10 minutes (10^4 ms) increases the efficiency of energy utilization by almost a factor of two. Titration of available electrons after termination of illumination by Fe(III) reduction is shown in Figure 5.

Figure 4. Pulsed *vs.* continuous irradiation in a 365-nm LED reactor degrading 4-chlorophenol (4-CP). Charge is 0.05 g WO_3/TiO_2 and 5 mL of 33 mg/L 4-CP receiving $(4.3 \pm 0.2) \times 10^{16}$ photons/s. Total irradiation: 60 min in each case, duty cycle (ratio minute light:minute dark) varied.

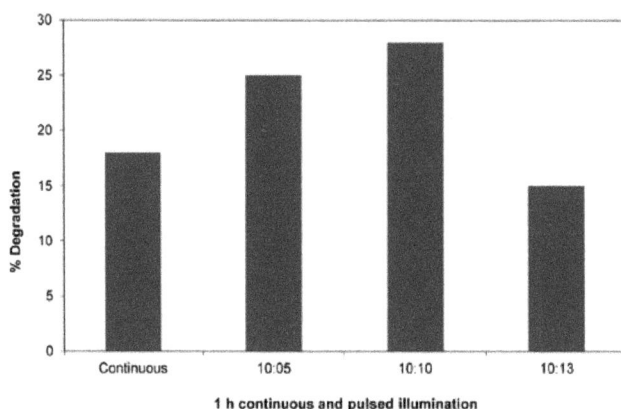

The extent of electron storage during the photodegradation of 4-CP was monitored by rapid titration with Fe(III) of a sample of the solution after successive times of illumination. Figure 5 shows the levels during the degradation of the 33 mg/L sample, as described above.

Figure 5. Stored electron levels during the photocatalytic degradation of 4-CP measured by Fe(III) titration with Fe(II) detection by phenanthroline complexation.

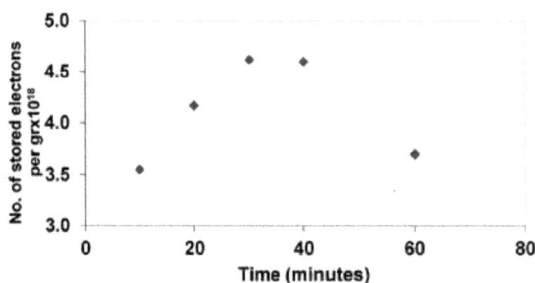

It has been reported [7] that charge storage with WO_3 allows inhibition of *E. coli* that are exposed in the dark to TiO_2/WO_3. This is confirmed by data in Table 3, where illumination was UVA (365 nm), and the sample was outflow water from a secondary wastewater treatment plant.

This water contains a mix of other organic molecules that can compete for reaction with excited TiO_2/WO_3. A 50-mL waste water sample with 10-ppm fulvic acid (FA) solution (as a supplemental hole scavenger) + 0.33 g WO_3-TiO_2 was used. In a first experiment, 50 mL of contaminated water was added to the TiO_2 slurry, and the mixture was incubated for 5 h. In the subsequent experiment, after 30 min in the dark, the slurry was irradiated for 2 h in the 365-nm LED reactor. Post irradiation, the 50-mL contaminated water sample was added and incubated for 5 h. The samples were measured for coliforms by the Colilert tray method, widely used for water monitoring. Note that the initial sample counts are not reproducible due to the biological activity in the stock.

Table 3. Coliform counts after exposure to TiO_2/WO_3 without and with pre-charging for 2 h. (MPN = most probable number of colonies).

Sample	Total Coliform (MPN)	*E coli* (MPN)	Decrease in *E. coli*
Stock solution	1,921.2	246.8	
WO_3/TiO_2 (dark blank)	872.0	161.0	35%
Stock solution	1,732.9	488.4	
WO_3/TiO_2 charged	108.4	26.6	95%

2.3. Carbon

Carbon is a good adsorbent and can act as an electron acceptor [12]. Consequently, both mechanisms may arise. An interesting potential application is the removal of emerging contaminants from water. For this reason, composites were made using the well-established TiO_2, Degussa P25 reference point. Figure 6 shows the adsorption isotherms for SMX on P25 loaded with 5% by weight of either AC or SWCNTs or 0.25% by weight of SWCNTs. These values are chosen, because preliminary results show good rates for photocatalysts with 0.25% SWCNTs and 5% activated carbon (AC).

The comparative amounts adsorbed as a function of the loading of carbon on TiO_2 is interesting and offers a possible explanation for the preliminary observation that 5.0% AC and 0.25% SWCNTs were both favourable loadings for the study of the photocatalytic reaction. The amounts of SMX adsorbed are similar.

A series of representative results for SMX degradation on the three photocatalysts are collected below for a photocatalyst loading of 1.0 g/L and initial solution SMX concentrations of 2.4×10^{-4} M. After the dark adsorption equilibration period, solution concentrations were 1.1×10^{-4} M for P25/AC, 2.3×10^{-4} M for P25 and 1.4×10^{-4} M for P25/SWCNT. Table 4 collects the rate constants. The rate constants for the total loss of substrate are compared in the table to apparent "rate constants" calculated from solution concentrations alone. Where adsorption is small (P25), the solution value is in agreement. However, differences are quite significant when carbon is added. The data shown here are a reminder that many rate constants in the literature have been reported on solution data where adsorption may have been significant. This can be quite misleading. It is especially true for studies involving dye molecules [9]. Figure 7 shows the kinetics of total SMX loss with P25, P25/AC and P25/SWCNT as photocatalysts plotted as first order reactions. Powdered activated carbon appears to act as an inhibitor in these circumstances.

SWCNTs may be marginally accelerating. This is consistent with reports that surface diffusion is faster on SWCNTs than powdered activated carbon [8].

Figure 6. Adsorption isotherms at 22 °C for SMX on activated carbon (AC) or single-walled carbon nanotubes (SWCNT). SWCNT: solid line, 5.0% by weight; dashed line, 0.25% by weight. C_e = solution concentration; q_e = adsorbed weight per unit mass (mg/g).

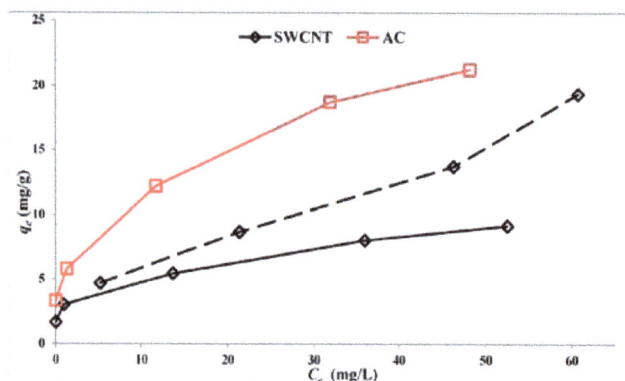

Table 4. First order rate constants for the total loss of SMX (4.2×10^{-4} M) with three photocatalysts.

Photocatalyst	Rate Constant (min^{-1})	R^2 Fitting Parameter	Apparent Solution "Rate Const." (min^{-1})
P25/AC	0.0103	0.988	0.0154
P25	0.0165	0.980	0.0166
P25/SWCNT	0. 0173	0. 990	0.0227

In an effort to detect possible electron storage from charge transfer in TiO$_2$/SWCNTs composites, reactions were carried out in the pulse irradiation mode with pulse cycle times down to 100 ms, with duty cycles of 50%, 30% and 10%. (with the relaxation time of TiO$_2$ alone near 70 ms. [13]l a shorter pulse time would not distinguish carbon effects). In no case was a significant difference between pulse irradiation and continuous irradiation observed for a constant energy dose. This could suggest that the SWCNTs function only in the A/S mode, but the alternative explanation, given the direct evidence for electron transfer, is that carbon mediates rapid electron transfer to O$_2$. There is evidence that nanocarbon can function as an efficient conductor, delivering the electrons [8] to an acceptor. Yao *et al.* [12] reported results for a physical mixture of carbon nanotubes with TiO$_2$ nanotubes. The solution conversion rates were superior to TiO$_2$ alone, but the physical mixture rate was only about half of the composite rate. The positive effect in a physical mixture was taken to be evidence of electron transfer. It does seem unlikely that the time of an encounter would allow efficient shuttle action, unless encounters are "sticky". Yao *et al.* used a dispersant to minimize this. An experiment with a physical mixture of P25 with SWCNTs (ratio as above) started from 95 mg/L SMX, which led to a surface concentration equivalent to 24 mg/L after dark adsorption (clearly carbon loading). Four hours of irradiation reduced the solution concentration and the surface residual to values comparable to those for the composite, tending to support a role for electron transfer.

Figure 7. Kinetics of the total loss of SMX on irradiation of slurries in a 365-nm batch reactor.

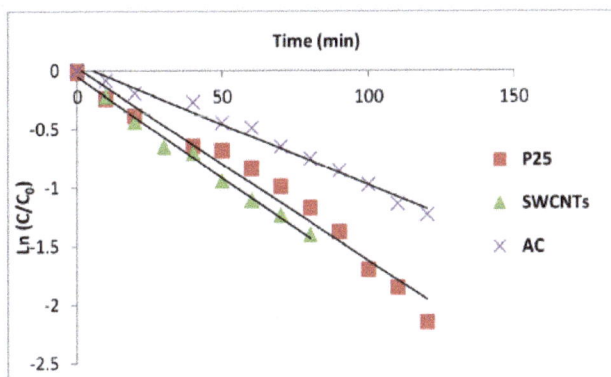

2.4. Discussion

2.4.1. Adsorb and Shuttle (A/S)

The new photocatalyst is assembled in a silica matrix. This resembles the systems described in Vaisman et al. [5] that were used to construct an adsorbent to be regenerated photocatalytically. Since good results are achieved with an excess of adsorbent over TiO_2, the adsorbent that is regenerated may be a good A/S target. The kinetic behaviour for the decrease of three substrates is reported in Figure 1. The TOC data in Figure 3 underlines the key advantage of A/S. Intermediates are retained on the photocatalyst assembly, and mineralization is favoured as observed by Torimoto et al. (1996). In the regeneration application, mineralization is important and may lead to the choice of a poorer catalyst for the loss of the initial target compound. The key challenge to the use of A/S is that adsorb and shuttle systems are highly dependent on the substrate.

2.4.2. Interphase Charge Transfer

A system to unambiguously demonstrate the transfer of carriers that inhibit recombination needs to minimize the potential for A/S. WO_3 may be a better adsorbent for some organic substrates than TiO_2, but the difference is not as great as with ZSM-5 or carbons. To further minimize A/S, a reaction with a weakly adsorbed substrate can be chosen. In this case, 4-chlorophenol is the chosen substrate. The pulse experiment is the most striking demonstration of the reduction of recombination, since it requires that carriers may remain to react after the light goes off. TiO_2 itself will have a short relaxation time for a reaction to occur after illumination stops, determined by the longest kinetic components of the recombination in TiO_2. This is on the order of ~70 ms [13]. With LED light sources, it could be feasible to examine a pulsing time below this limit, but the present system needed only minute-scale on-off sequences. A potentially practical application of a reduced recombination storing charge is shown, an increase in the light energy efficiency by optimizing the pulse widths and duty cycle in pulsed illumination. Of course there is a trade off in longer the overall residence time required to receive the dose.

Dark *E. coli* inhibition suggests that a system like TiO_2/WO_3 might overcome the limitation on a recycling design of a solar reactor for water disinfection plants, since the potential for organism regrowth in the dark phase of the recycle operation has been considered a barrier to the use of recycling.

2.4.3. Carbon

Carbon presents the problem of a co-catalyst that may play either of the two roles or both. For example, Torimoto *et al.* [1] included activated carbon as an effective adduct to TiO_2 among a list of adsorbents. In contrast, Yao *et al.* [12] report an elegant series of experiments with composites of carbon nanotubes and nano-TiO_2 that are interpreted entirely in terms of electron transfer to carbon. The transfer is verified by photoluminescence quenching. Twenty to one ratios of TiO_2 to SWCNTs gave their largest rate enhancement. The argument for electron transfer (*vs.* A/S) is further strengthened by observations on physical mixtures of C with TiO_2. However, only solution concentrations of the phenol substrate were reported, and surprisingly, measurements of remaining TOC after 4 h showed that P25 alone achieved the greatest carbon reduction (~80%). The most effective catalyst of phenol loss (>80% degradation in one hour) reduced TOC by only about ten percent in 4 h. The long lifetimes of intermediates suggests that A/S may have played a role. Data for surface residuals for [11] experiments would help to clarify.

If carbon nanotubes are good electron acceptors, why are efforts to measure energy storage more limited than TiO_2/WO_3? It was noted that W(V) can be stabilized by the capture of a cation and that the hole site reacts with water to provide a compensating H^+. In contrast, carbon is perhaps the better conductor to pass the electron on to, e.g., O_2. Good energy storage systems need a good way to stabilize the separated charges.

Here, experiments with two forms of carbon modeled after the literature, the common adsorbent activated carbon [1] and single-walled carbon nanotubes [12], are reported. Carbon has been reported to affect the band gap and absorbance limit of TiO_2. Irradiation with 365-nm LEDs assures comparison of the same energetic preparation of the excited state of TiO_2, which helps to focus on A/S *versus* charge transfer. The chosen substrate, SMX, is an agricultural antibiotic that is efficiently adsorbed by carbon, giving both pathways an opportunity. The results are mixed and suggest that both are involved. However, the pulse results imply that electron transfer must be accompanied by efficient transfer to oxygen. This then implicates the superoxide ion produced as a key player in the pathway.

3. Materials and Methods

3.1. Materials

TiO₂ (Degussa P25) powder (50 m²/g; 15%–30% rutile + 85%–70% anatase) was purchased from Degussa. Zeolite ZSM-5 (400 m²/g; Si/Al = 280) was purchased from Zeolyst International and calcined at 500 °C for an hour. Silica gel with particle sizes of 0.2–0.5 mm was purchased from Acros Organic. Single-walled carbon nanotubes (SWCNTs) and powdered activated carbon (AC) were obtained from Sigma-Aldrich and used as obtained without further purification. Sulfamethoxazole (SMX), 2,4,6-trichlorophenol (2,4,6-TCP), atrazine and 4-chlorophenol (4-CP) were selected as model substrates for organic pollutants for this report. They were purchased from Sigma-Aldrich and used as received. High performance liquid chromatography (HPLC)-grade acetonitrile and HPLC-grade water were used as a mobile phase in HPLC. Methyl alcohol, ACS grade, was purchased from Sigma-Aldrich. All of the solutions were prepared using "deionized" water (DI) from a Milli-Q system and characterized by its resistivity (18.2 MΩ).

3.1.1. Synthesis of ZSM-5 Containing Catalyst

A three-component (TiO₂, ZSM-5 and silica gel) composite photocatalyst was prepared as follows. First, ZSM-5 and TiO₂ were separately dispersed in 20 mL of methanol and sonicated for 30 min. Then, the ZSM-5 suspension was added to the TiO₂ suspension during stirring, and the ZSM-5/TiO₂ suspension was stirred for 15 min. This was followed by adding the silica gel powder during stirring, which continued for 15 more min. Finally, the methanol was evaporated while stirring; the composite catalyst was dried in an oven at 100 °C. then calcined in a furnace at 500 °C for 3 h. The composite catalyst components weights were maintained to obtain a ratio of TiO₂:ZSM-5:silica gel = 0.3:0.5:0.5 in the finished form of the catalyst. Above, PZS is used to refer to this ZSM-5-containing composite catalyst.

3.1.2. Synthesis of TiO₂/WO₃ Composite

WO₃/TiO₂ composite was prepared based on a sol-gel method [12] with titanium isopropoxide and phosphotungstic acid ($H_3PW_{12}O_{40}$) as precursors. Titanium isopropoxide (3.0 mL) and a specific amount of $H_3PW_{12}O_{40}$ were dissolved in 20 mL of isopropyl alcohol and deionized water, respectively. The alcohol solution was then added dropwise to the aqueous solution. After aging for 2 h, the white gel formed was dried at 100 °C and sintered at 500 °C for 5 h. The crystalline structure of this photocatalyst is reported to be anatase with an average particle size of 9.5–10 nm and a molar ratio of 0.04 for W/Ti [12].

3.1.3. Synthesis of Carbon/TiO₂ Composites

TiO₂/SWCNTs and TiO₂/AC composites with different mass ratios were prepared by a simple evaporation and drying process according to Yao et al. [12]. First, AC or SWCNTs were dispersed in 100 mL of water and sonicated for 10 min. TiO₂ powder was added to the suspension and sonicated for 20 more minutes. Then, the suspension containing AC or CNTs and TiO$_2$ particles was heated to 80 °C while stirring with air flowing across the suspension's surface to accelerate the evaporation of water. After the water evaporated, the composite was dried overnight in an oven at 104 °C. SWCNT composites were characterized by SEM and EDX. Carbon is found to be non-homogeneously distributed over the TiO$_2$. EDX measurements show that SWCNTs penetrate into the space between individual crystallites in the commercial TiO$_2$ aggregates.

3.2. Adsorption Isotherms

Adsorption studies were carried out in a batch mode as follows. Accurately weighed amounts of catalysts were added separately into glass vials containing an exact volume (20.0 mL) of different known initial concentrations of SMX. The glass vials were stirred in the dark for 45 min; then, the samples were filtered with a 0.20-μm syringe filter, and the residual concentration of SMX was measured. The amount of adsorbate uptake capacity at equilibrium, q_e, was calculated by mass balance as follows:

$$q_e = \frac{V(C_i - C_e)}{m}$$

where q_e is the equilibrium amount of solute adsorbed per unit mass of adsorbent (mg/g) and C_i and C_e are the initial and the equilibrium concentrations of the solute (mmol/L) in solution, respectively. V is the volume of the solution, and m is the weight of the adsorbent (g).

3.3. Photocatalytic Procedures

In all cases, the reactivity of the photocatalysts was measured as the disappearance of the substrate compounds (SMX, 2,4,6-TCP 4-Cl-phenol and atrazine). Accurately weighed amounts of different prepared photocatalyst composites were dispersed separately into a glass vial containing an exact volume (20.0 mL) of known initial concentration of the model substrates in a single component system. After 30 min of dark stirring, the suspension was irradiated. The reaction vessels were placed in a circular bench-scale 365-nm LED photoreactor, locally fabricated [14]. The inside diameter and depth of the reactor are 9 and 7 cm, respectively, and it is equipped with 90 LED 3-mW output lamps (NSHU5518), which are evenly distributed in 15 rows. The light intensity in the vessel was $(4.3 \pm 0.2) \times 10^{16}$ photons/s. A sample was withdrawn at pre-determined time intervals, centrifuged, filtered using a 0.2-μm syringe filter, and the residual concentration was measured by HPLC. Some surface concentrations were estimated from isotherms. In selected cases, compounds on the surface of composites and Degussa P25 were extracted with acetonitrile, centrifuged, filtered and the concentration measured by HPLC. Samples for total organic carbon (TOC) (irradiated for one to two hours) were centrifuged, passed through a 0.45-μm syringe filter and analyzed using an Apollo 9000 combustion TOC analyzer equipped with an autosampler.

3.4. Analytical Procedures

HPLC: A "Varian pro star 210" HPLC equipped with a PFP 100A column (Phenomenex kinetix™ 2.6 μm, LC Column 100 × 4.6 mm) with 20-μL injections and a 325 LC UV-Vis detector was used for the analysis of several substrates. Isocratic elution with a solvent mixture of 50% acetonitrile and 50% water at a flow rate of 1.00 mL·min^{-1} was used for the analysis of SMX. For phenols, a solvent mixture of 50% acetonitrile (0.1 formic) and 50% water (0.1 formic) at a flow rate of 1.25 mL·min^{-1} was used, and for atrazine a solvent mixture of 65% acetonitrile (0.1 formic) and 35% water (0.1 formic) at a flow rate of 1.00 mL·min^{-1} was used. The wavelength of detection was 270, 254 and 220 nm for SMX, phenols and atrazine, respectively. Coliform estimation: Total coliforms and *E. coli* counts were estimated by the Colilert tray method, which is widely employed in water monitoring. This process used trays and procedures from IDEXX Corporation for the Quanti Tray/2000® system. The most probable numbers (MPN) of colonies were estimated with the IDEXX MNP estimator.

3.5. Pulse and Stored Electron Studies

A series of experiments were performed to investigate the effect of the pulsed illumination. A locally fabricated programmable controller was used if a light pulse frequency from 1 to 999 ms was required; otherwise, manual off/on was used to achieve pulsing. For fast pulsing, illumination (on) and dark (off) periods in the 200–990-ms range and duty cycles of 10, 30 and 50% were used.

For stored electron measurements, a sample of 0.02 g (WO_3/TiO_2) plus 2 mL of 33 ppm 4-chlorophenol (4-CP) was irradiated for a chosen time. At that time, 2 mL iron(III) perchlorate (1×10^{-3} M) is added. Then, 1.0 mL to 1.0 mL acetate/acetic acid buffer (pH = 5.5), 0.5 mL ammonium fluoride (0.1 M) plus 0.5 mL 1,10-phenanthroline (0.1 M) are added. The number of electrons stored is calculated based on the absorbance of iron(II) phenanthroline at 510 nm [11].

4. Conclusions

The purpose of this study was to explore the relationship of "A/S" and "electron transfer" mechanisms to improve the efficiency of the degradation of organic compounds using three classes of photocatalysts: zeolite-TiO_2, WO_3-TiO_2 and carbon-TiO_2. The choice was dictated by the aim of comparing two cases, where only one of the mechanisms could be operative, to a case where both might compete. The specific mechanism for the zeolite is A/S, where WO_3-TiO_2 has identified electron transfer with little difference in surface properties between the two components of the composite. Both A/S and electron transfer can contribute to carbon-TiO_2 systems.

The A/S mechanism must balance adsorption with surface mobility to complete substrate delivery to TiO_2. When these two balance, gains can be disappointing, but the present results also show that it is important to monitor surface concentrations to avoid misleading evidence of "enhancement" from solution-only data.

Electron transfer has been confirmed to be contributed to by carbon. In the case of the carbon forms tested here, the gains are minimal. Only small improvement was observed in the case of SWCNTs, and the effect of carbon was detrimental in the case of using AC. This indicates slow surface diffusions, which can even overcome the effect of electron transfer and reduced recombination. However, some degree of limitation may reflect the absence of a relatively stable site for the transferred electron. Pulse experiments indicated short electron storage lifetimes. The long storage lifetime in WO_3 illustrates the desired condition, a stabilized electron site analogous to W(V).

Acknowledgments

The financial support of the Natural Sciences and Engineering Research Council of Canada is gratefully acknowledged. This support was part of a strategic research network, RES'EAU Waternet. (This project has terminated, and funds are no longer available to support open access publication). Emad Radwan Hafez thanks the government of Egypt for support.

Author Contributions

C.L. and G.A. developed the project and the outline of the research. E.R. and M.I. conducted the experimental studies reported. All authors contributed to interpretation and writing. C.L. was responsible for final editing of the manuscript with valuable help from E.R.

Conflicts of Interest

The authors declare no conflict of interest. The sponsors had no role in study design, data collection analysis or interpretation, writing of the manuscript or the decision to publish.

References

1. Torimoto, T.; Ito, S.; Kuwabata, S.; Yoneyama, H. Effects of adsorbents used as supports for titanium dioxide loading on photocatalytic degradation of propyzamide. *Environ. Sci. Technol.* **1996**, *30*, 1275–1281.
2. Lepore, G.P.; Pant, B.C.; Langford, C.H. Limiting quantum yield measurements for the disappearance of 1-propanol and propanal: An oxidative reaction study employing a TiO_2 based photoreactor. *Can. J. Chem.* **1993**, *71*, 2051–2059.
3. Colombo, D.P.; Bowman, R.M. Does interfacial charge transfer compete with charge carrier recombination? A femtosecond diffuse reflectance investigation of TiO_2 nanoparticles. *J. Phys. Chem.* **1996**, *100*, 18445–18449.
4. Takeda, N.; Ohtani, M.; Torimoto, T.; Kuwabata, S.; Yoneyama, H. Evaluation of diffusibility of adsorbed propionaldehyde on titanium dioxide-loaded adsorbent photocatalyst films from its photodecomposition rate. *J. Phys. Chem. B* **1997**, *101*, 2644–2649.
5. Vaisman, E.; Kabir, M.F.; Kantzas, A.; Langford, C.H. A fluidized bed photoreactor exploiting a supported photocatalyst with adsorption pre-concentration capacity. *J. Appl. Electrochem.* **2005**, *35*, 675–681.
6. Tatsuma, T.; Saitoh, S.; Ngaotrakanwiwat, P.; Ohko, Y.; Fujishima, A. Energy storage of TiO_2-WO_3 photocatalysis systems in the gas phase. *Langmuir* **2002**, *18*, 7777–7779.

7. Tatsuma, T.; Takeda, S.; Saitoh, S.; Ohko, Y.; Fujishima, A. Bactericidal effect of an energy storage tio2–wo3 photocatalyst in dark. *Electrochem. Commun.* **2003**, *5*, 793–796.

8. Kedem, S.; Rozen, D.; Cohen, Y.; Paz, Y. Enhanced stability effect in composite polymeric nanofibers containing titanium dioxide and carbon nanotubes. *J. Phys. Chem. C* **2009**, *113*, 14893–14899.

9. Xu, Y.; Langford, C.H. Uv- or visible-light-induced degradation of X3B on TiO2 nanoparticles: The influence of adsorption. *Langmuir* **2001**, *17*, 897–902.

10. McMurray, T.A.; Dunlop, P.S.M.; Byrne, J.A. The photocatalytic degradation of atrazine on nanoparticulate tio2 films. *J. Photochem. Photobiol. A Chem.* **2006**, *182*, 43–51.

11. Zhao, D.; Chen, C.; Yu, C.; Ma, W.; Zhao, J. Photoinduced electron storage in WO3/TiO2 nanohybrid material in the presence of oxygen and postirradiated reduction of heavy metal ions. *J. Phys. Chem. C* **2009**, *113*, 13160–13165.

12. Yao, Y.; Li, G.; Ciston, S.; Lueptow, R.M.; Gray, K.A. Photoreactive TiO2/carbon nanotube composites: Synthesis and reactivity. *Environ. Sci. Technol.* **2008**, *42*, 4952–4957.

13. Sczechowski, J.G.; Koval, C.A.; Noble, R.D. Evidence of critical illumination and dark recovery times for increasing the photoefficiency of aqueous heterogeneous photocatalysis. *J. Photochem. Photobiol. A Chem.* **1993**, *74*, 273–278.

14. Yu, L.; Achari, G.; Langford, C. Led-based photocatalytic treatment of pesticides and chlorophenols. *J. Environ. Eng.* **2013**, *139*, 1146–1151.

Sample Availability: Samples of the compounds are available from the authors.

Visible Light Induced Green Transformation of Primary Amines to Imines Using a Silicate Supported Anatase Photocatalyst

Sifani Zavahir and Huaiyong Zhu

Abstract: Catalytic oxidation of amine to imine is of intense present interest since imines are important intermediates for the synthesis of fine chemicals, pharmaceuticals, and agricultural chemicals. However, considerable efforts have been made to develop efficient methods for the oxidation of secondary amines to imines, while little attention has until recently been given to the oxidation of primary amines, presumably owing to the high reactivity of generated imines of primary amines that are easily dehydrogenated to nitriles. Herein, we report the oxidative coupling of a series of primary benzylic amines into corresponding imines with dioxygen as the benign oxidant over composite catalysts of TiO_2 (anatase)-silicate under visible light irradiation of $\lambda > 460$ nm. Visible light response of this system is believed to be as a result of high population of defects and contacts between silicate and anatase crystals in the composite and the strong interaction between benzylic amine and the catalyst. It is found that tuning the intensity and wavelength of the light irradiation and the reaction temperature can remarkably enhance the reaction activity. Water can also act as a green medium for the reaction with an excellent selectivity. This report contributes to the use of readily synthesized, environmentally benign, TiO_2 based composite photocatalyst and solar energy to realize the transformation of primary amines to imine compounds.

Reprinted from *Molecules*. Cite as: Zavahir, S.; Zhu, H. Visible Light Induced Green Transformation of Primary Amines to Imines Using a Silicate Supported Anatase Photocatalyst. *Molecules* **2015**, *20*, 1941-1954.

1. Introduction

Imines are a group of *N*-containing compounds, with a pivotal role as chemically and biologically useful intermediates in various cycloaddition, condensation and reduction reactions [1,2]. These compounds play a major role in pharmacophores, fragrances and numerous biologically active compounds [3]. For a long time, traditional condensation of amines with carbonyl compounds was regarded as the simplest way to prepare imines [4–6]. Highly reactive nature of aldehydes made handling difficult. This was overcome later by replacing aldehydes with alcohols and temporarily producing aldehydes *in-situ* within the reaction mixture and the subsequent reaction with an amine yield the imine compound [7,8]. Wang and co-workers have recently developed organosilicon supported TiO_2 catalyst for this reaction at 160 °C with added base to achieve a good imine yield [9]. However, these processes yield range of by-products and greatly affect the selectivity to the desirable product.

Amines are easily accessible compounds that can also be attractive precursors to synthesize imine by controlled oxidation. Until recently, several methods for oxidation of secondary amines to imines have been developed, while little attention has been devoted to the oxidation of primary amines. This is probably because of the generated imines, in which α-amino hydrogen is present, are generally intermediate products that are rapidly dehydrogenated to nitriles [10–12]. Subsequent research in the area has been dominated by the development of new catalytic processes that allow the aerobic oxidation of primary amines to imines under mild conditions. Angelici and co-workers reported aerobic oxidative homocoupling of primary amines to give imines, catalyzed by gold powder (50 μm size) and gold nanoparticles supported on alumina (5% Au/Al$_2$O$_3$) in toluene at 100 °C [13]. Following this study Au/C and CuCl catalysts have been employed in this reaction at 100 °C under molecular oxygen atmosphere [14,15]. It is becoming increasingly important to look for new materials that can catalyze reactions under moderate conditions (relatively low temperature and pressure). In this regard, the utilization of sunlight as an energy source to reduce the working temperature has recently attracted much attention [16–18]. Wang and co-workers have attained imines in excellent yields using mesoporous graphite carbon nitride photocatalyst at 80 °C [19]. In general, unavailability of structurally diverse amines has hampered the synthetic scope of oxidative coupling of benzylic amines to yield corresponding imines, yet this method is highly selective for imines.

Over the last few years, many efforts have been extended to organic redox-transformation reactions using TiO$_2$ photocatalysis [20–25]. However, to date, most of the reported reactions for the synthetic transformations using TiO$_2$ photocatalysts were carried out under UV irradiation and were usually associated with low selectivity [26,27]. Performing visible light induced selective transformations by photocatalysts is a challenge that has gained increasing attention. Recent discoveries demonstrated the surface modification of TiO$_2$ with noble metal complexes or nanoparticles rather than bulk doping might be a better strategy in light of new visible light responsive photocatalysts that could enhance the design of efficient redox reactions under visible light irradiation. Zhao and co-workers achieved this conversion of primary amine to imine with TiO$_2$ under UV light irradiation (>300 nm) [28] and later they also found it is also possible for this reaction to be initiated by visible light irradiation of λ > 420 nm [29]. According to them, amine molecules adsorbed onto TiO$_2$ forms a surface complex that could absorb visible light and so initiate electron transfer and ensuing reactions. Because the reaction takes place on the TiO$_2$ surface, we envisioned that ultrafine TiO$_2$ powders with large specific surface areas should exhibit better catalytic activity. A feasible approach to stabilizing TiO$_2$ nanocrystals is to disperse them in an inorganic medium, such as layered clays creating porous composite structures, while ensuring that most of the surface of TiO$_2$ crystals is accessible to various molecules [30]. Nevertheless, the structure of the composite solids has a profound impact on their catalytic performance [31]. The mesoporous composites of anatase nanocrystals and silicate are the catalyst materials of the optimal structure for high photocatalytic activity. Synthetic layered clay, laponite, can be used in the synthesis of the composite as silicate source [30].

Here in we report TiO$_2$ nanocrystal-silicate composite, prepared using laponite, as feasible photocatalyst for the selective oxidation of benzylamine to N-benzylidene benzylamine with excellent conversion and selectivity under the irradiation of visible light (λ > 460 nm). It is found that in the TiO$_2$-silicate composite catalyst TiO$_2$ is in anatase phase. Reference reactions with anatase show that under identical conditions TiO$_2$-silicate composite catalyst exhibited a much more superior catalytic activity to pure TiO$_2$ (anatase) powder. Nitrogen adsorption data confirms the large surface area of the composite catalyst. Furthermore, we also found that water can be used as the solvent. This catalyst could be employed for heterocoupling of two structurally diverse amines in the synthesis of imines as well as homocoupling of benzylic amines to imines, and the catalyst can be recycled up to five rounds without any significant loss of activity.

2. Results and Discussion

The aerobic photocatalytic oxidation of benzylamine to N-benzylidene benzylamine was chosen as the model reaction to optimize the reaction system. Reactions are carried out using 500 W halogen lamps where the light emitted is in 400–800 nm range. According to the data given in Table 1 it is apparent that TiO$_2$-silicate (abbreviated as TiO$_2$-S hereafter) is the most photocatalytically active photocatalyst towards this transformation. Catalyst samples were also prepared by loading Au and Pd nanoparticles (NPs) (3% by weight) and another sample with AuPd alloy NPs (1.5% weight of each metal) loaded on to TiO$_2$-S composite material (Characterization is provided in SI). We observed a lower imine product yield of 60% with Au@TiO$_2$-S, compared to 82% by TiO$_2$-S, despite the enhanced light absorption by Au NPs in the visible region due to localized surface plasmon resonance (LSPR) effect which is characterized by an intense band around 520 nm (Figure S1, Supplementary Information) [16–18]. This observation also serves as an example to support the fact, light absorption by a material is not the sole governing factor deciding catalysts ability to drive a particular chemical reaction under light irradiation. It appears that the reaction takes place on the surface of anatase, the loaded Au NPs lower the exposed surface area of TiO$_2$, the catalytically active sites of this system, lowering the accessibility to the reactants. Pd@TiO$_2$-S catalyst had similar activity to that of TiO$_2$-S, whereas AuPd@TiO$_2$-S was slightly sluggish. Results further indicate the unique potential of TiO$_2$ based materials towards oxidation reactions and importance of evaluation of surface modifications of TiO$_2$ for activity improvements. In control experiments, the reaction did not proceed without a photocatalyst or in the dark.

Table 1. Photocatalytic oxidation of benzylamine to N-benzylidene benzylamine over different catalyst materials and solvents. [a]

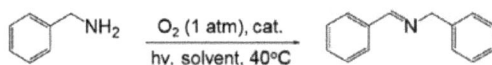

Entry	Catalyst	Solvent	Conv. (%) [b]	Sel. (%) [b]	Yield (%)
1	TiO$_2$-S	Acetonitrile	88	92	81
2	TiO$_2$-S	DMSO	18	100	18
3	TiO$_2$-S	THF	94	73	69
4	TiO$_2$-S	Toluene	74	97	72
5	Au@TiO$_2$-S	Acetonitrile	65	93	60
6	AuPd@TiO$_2$-S	Acetonitrile	88	90	79
7	Pd@TiO$_2$-S	Acetonitrile	89	96	85
8	Laponite	Acetonitrile	0	--	0
9	TiO$_2$(anatase)	Acetonitrile	51	100	51
10	H-titanate	Acetonitrile	73	97	71

[a] Reaction conditions: 50 mg catalyst, 0.5 mmol benzylamine, 5 mL solvent, 500 W halogen lamp (cut off wavelength below 400 nm) intensity 0.36 W/cm^2, 1 atm O$_2$, 24 h. [b] Determined by GC analysis. DMSO = dimethyl sulfoxide, THF = tetrahydrofuran.

As can be seen in Table 1, activity of TiO_2-S is superior to that of an equivalent amount of TiO_2 (anatase) as the photocatalyst material. In order to understand this change in behavior we closely studied the light absorption abilities of both TiO_2 (anatase) and TiO_2-S, in the presence and absence of benzylamine. UV-Visible diffuse reflectance spectra of benzylamine adsorbed TiO_2 (anatase) and TiO_2-S shows increased absorbance compared to solitary TiO_2 (anatase) and TiO_2-S, particularly in the visible region. This observation agrees well with previous reports, where electron rich molecules like amines make a charge transfer complex with TiO_2 and respond to visible light illumination [29]. It is also notable, the absorption of benzylamine adsorbed on TiO_2-S is significantly high compared to benzylamine adsorbed on TiO_2 (anatase) as shown in the Figure 1A. Even though TiO_2 is present in anatase phase in both TiO_2 and TiO_2-S photocatalysts used in the current study, the distribution of anatase particles is different in TiO_2-S. During TiO_2-S preparation, layered clay structure of precursor material laponite clay is lost as a result of the acidic titanium sol solution reacting with hydroxyl groups in the clay layers that are bound to magnesium ions within the layer [31]. Most of the magnesium in the clay was leached out in this way. Composition of the catalyst estimated by energy dispersive X-ray (EDX) confirms high weight percentage of silicate in the composite catalyst despite the leaching of Mg units. During preparation Si:Mg ratio (by weight) decreased from 1:0.58 to 1:0.25, this together with TEM image is a clear indication that ordered layer structure is damaged. Thus, TiO_2 in this TiO_2-S composite catalyst exists as discrete anatase crystals on fragmentized pieces of silicate. Correspondingly this composite structure restrains agglomeration of anatase particles leading to high exposed surface area of TiO_2. Brunauer-Emmett-Teller (BET) surface area of initial laponite clay changed from 330.6 m^2g^{-1} to 518.3 m^2g^{-1} in the final TiO_2-S catalyst material. The composite has porosity of about 0.4 cm^3/g and a mean pore size of 5 nm. Finally, in the obtained composite catalyst silica particles and anatase crystals exist as inter-dispersed phases in nanometer scale with a highly porous structure as can be seen in Figure 1B. X-ray diffraction (XRD) pattern of the catalyst only exhibit peaks responsible for the anatase phase of TiO_2 with no peaks related to silicate units or laponite clay, this indicates silica is present in the amorphous phase, and anatase particles of mean crystal size 4.22 nm (estimated by Debye-Scherrer equation using the broadening of the highly intense (101) XRD peak at $2\theta = 25.3°$) have homogeneously crystallized over amorphous silica moiety. This TiO_2-S structure obtained in the present study, offers high thermal and chemical stability, also provides ample opportunity for the reactant molecules to interact with energetic charge carriers. Smaller anatase particles reduce the possibility of charge recombination, since charge carriers are generated at the close proximity of surface and efficiently captured by benzylamine and oxygen molecules on the surface.

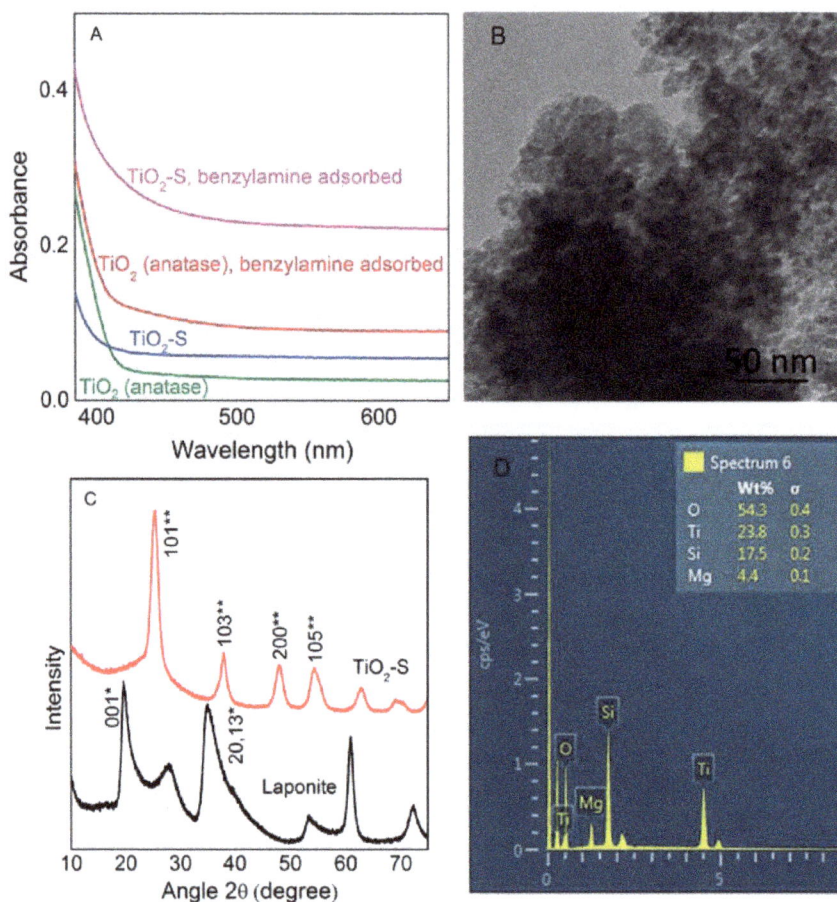

Figure 1. (A) UV-Visible diffuse reflectance spectra; (B) transmission electron microscopy (TEM) image; (C) X-ray diffraction (XRD) peak patterns indexed for *—silicate phase and **—anatase phase; (D) energy dispersive X-ray (EDX) spectra of TiO₂-S composite catalyst.

To further investigate the contribution from light in this reaction, we conducted a series of reactions at variable intensities (Figure 2A). The conversion rate of benzylamine on TiO₂-S catalyst increased gradually as the intensity increased, with the other reaction conditions unchanged (Experimental section). Selectivity to the product imine had a little influence on the intensity; however, overall imine yield (conversion rate x selectivity) increased with the intensity. Such a tendency reveals a strong dependence on the intensity for the light induced oxidative coupling of benzylamine, because in general light incident with a higher intensity is able to generate more energetic charge carriers (holes and electrons). Such conditions favor stronger interaction between benzylamine and the catalyst, and positively influence the reaction.

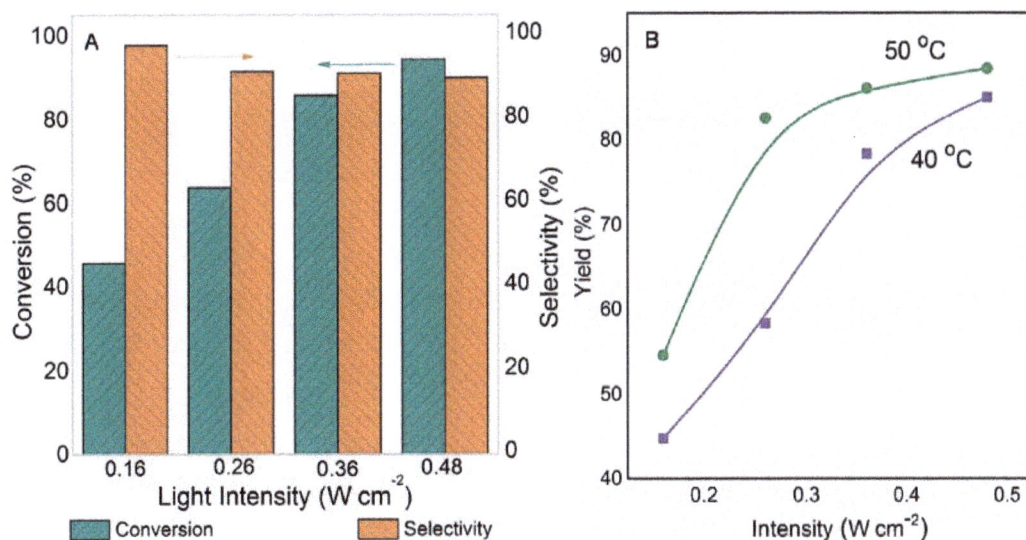

Figure 2. The effect of (**A**) light intensity and (**B**) temperature influence of the irradiation on the oxidative coupling of benzylamine.

Impact of the temperature on the yield of the reaction was studied by conducting the experiments at two different temperatures slightly above room temperature (40 °C and 50 °C). As shown in Figure 2B, observed yield was always high for the reaction under the study, oxidative coupling of benzylamine to imine at higher temperature for a given constant intensity within 40 to 50 °C temperature range. For instance, at 40 °C and 0.26 W·cm^{-2} intensity, 58% of benzylamine was converted to imine product whereas at 50 °C it was 83%. At high temperature however, the enhancement in the yield by increasing the intensity lessens since the selectivity to the imine product starts to decline, though reaction proceeds at a higher conversion rate. As the intensity was increased, the difference between the yields at 40 and 50 °C finally decreased, even though higher conversion rate was observed for 50 °C compared to that at 40 °C in all cases.

The dependence of yield on the irradiation wavelength was studied using five monochromatic light emitting diodes (LEDs) and it shows that higher photocatalytic yields are achieved under irradiation of short wavelengths (<460 nm). Anatase phase of TiO$_2$ exhibits a band gap of 3.2 eV (387.5 nm) where as in this case TiO$_2$-S is highly active up to 460 nm. Figure 3 demonstrates the apparent quantum yield (A.Q.Y) dependence on the incident wavelength; A.Q.Y. is a measure of imine yield per photon of energy absorbed per unit time. This finding indicates that composite TiO$_2$-S catalyst structure has a broad light response below 460 nm in the visible region due to the collective effects of benzylamine adsorbed TiO$_2$ (anatase) charge transfer surface complex and high population of defects in the composite photocatalyst structure. This reveals that TiO$_2$-S catalyst can function at a lower cut-off edge (460 nm) compared to 420 nm cut-off for solitary TiO$_2$ (anatase) system reported by Zhao and co-workers [29]. It is noteworthy that, in the composite structure of the catalyst there are contacts between silicate and anatase crystals. At these sites, the anatase surface is similar to the silica doped anatase surface that exhibits light absorption and visible light photocatalytic activity [32].

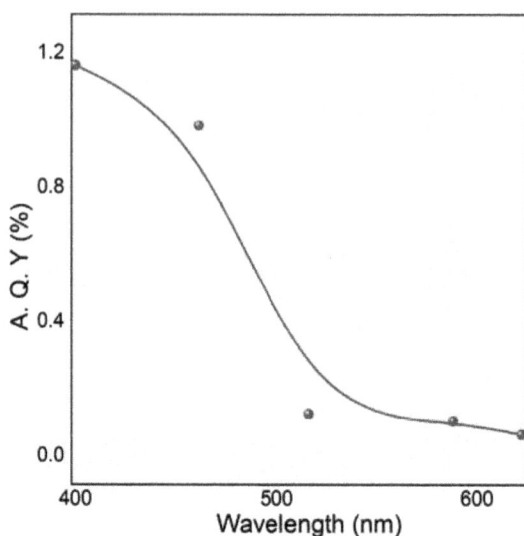

Figure 3. The effect of wavelength of the irradiation on the oxidative coupling of benzylamine.

According to the results summarized in Table 1, the solvent has important impact on the catalytic activity. Acetonitrile serves as the best solvent for this reaction while the poorest performance was observed in DSMO. Water is a viable solvent for organic reactions, and it is interesting to study the impact of water as the reaction medium. For some organic reactions, water exhibits special reactivity or selectivity due to its unique physical properties. In the present study, benzylamine oxidation reaction proceeded in water at a moderate conversion rate of 45.6% under the illumination of 500 W halogen lamp (400–800 nm) for 24 h, but with an excellent selectivity where the sole product being N-benzylidene benzylamine with an overall yield of 46%. Doubling the amount of catalyst from 50 mg to 100 mg of TiO_2-S enhanced the reaction yield to 62% without compromising the selectivity.

Motivated by this result, we expanded the scope of the substrates for the oxidation of amines to imines. Table 2 summarizes the photocatalytic oxidation of the benzylic amines to corresponding imines with their conversion rate and selectivity. Oxidation of primary benzylic amines substituted with an electron donating group (Table 2, entry 2–3) proceeded efficiently under visible light irradiation with good to high conversion rates and high selectivity for the imine product. Substituent group influences the conversion rate of the reaction than the selectivity to the imine product. The relatively low conversion rate for the oxidative coupling of 4-chlorobenzylamine (Table 2, entry 4) into its corresponding imine might be caused by inductive effects of C-Cl σ-bond polarity. No change was observed when aniline was subjected to the reaction, this is consistent with our hypothesis that the presence of a –H, bonded to the α-carbon is important for this transformation to take place. Furthermore, non-aromatic cyclic amines (Table 2, entry 7–8) did not yield the desired imine product. Control experiments carried out using cyclohexylamine (consist of a single α-hydrogen) produced the corresponding oxime (cyclohexanone oxime) instead of the imine. Benzaldehyde oxime was one of the products observed during the time course of the reaction of benzylamine.

Table 2. Aerobic oxidation of primary benzylic amines photocatalyzed by TiO₂-S under visible light irradiation. [a]

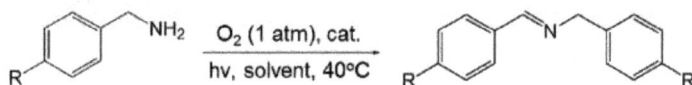

R—C₆H₄—CH₂—NH₂ → (O₂ (1 atm), cat.; hv, solvent, 40°C) → R—C₆H₄—CH=N—CH₂—C₆H₄—R

Entry	Substrate	Product	Con. (%) [b]	Select. (%) [b]	Yield (%)
1	benzylamine	N-benzylidenebenzylamine	88	92	81
2	4-methylbenzylamine	imine	96	92	88
3	4-MeO-benzylamine	imine (OMe)	100	96	96
4	4-Cl-benzylamine	imine (Cl)	57	100	57
			88 [c]	>99 [c]	87 [c]
5	phenethylamine (NH₂)	imine	91	35	32
7	piperidine-NH	bis-piperidine	51	20	10
8	cyclohexyl-CH₂NH₂	N-OH product	74	48	36

[a] Reaction condition: 50 mg catalyst, 0.5 mmol amine substrate, 5 mL acetonitrile, 500 W halogen lamp (cut-off wavelength below 400 nm) intensity 0.36 W/cm², 1 atm O₂, 24 h. [b] Determined by GC analysis. [c] Reaction time 36 h.

This reaction goes through the widely known intermediate benzaldehyde and a tentative mechanistic pathway is given in Scheme 1 based on the products observed. The photocatalysts contribution is mainly in the step of benzylamine oxidation to benzaldehyde, whereas the condensation of benzaldehyde with a benzylamine molecule leading to the imine product is faster. In the oxidation step, TiO₂-benzylamine surface complex absorb visible light (400–800 nm) and excite electrons. These excited electrons are then captured by oxygen molecules adsorbed on TiO₂ surface, then in the proceeding steps oxygen interacts with benzylamine and the substrate molecules lose the H bonded to the α-carbon atom, and oxidized to aldehyde. Thus, it is rational that under visible light irradiation, the oxygen molecules adsorbed on the catalyst capture the light excited electrons, and react with the H at the α-carbon. Role of oxygen is further confirmed, when the reaction was carried out in the air atmosphere benzylamine exhibit a relatively lower observed conversion rate of 51% and a selectivity of 96%, yielding 49% of imine after 24 h. This mechanism agrees well with the observed product selectivity results. At higher conversions of benzylamine, a decrease in the selectivity for the imine occurs and benzaldehyde appears in the products. This is due to the fact that, increased consumption of benzylamine in the solution could not ensure the complete condensation of aldehyde and the amine.

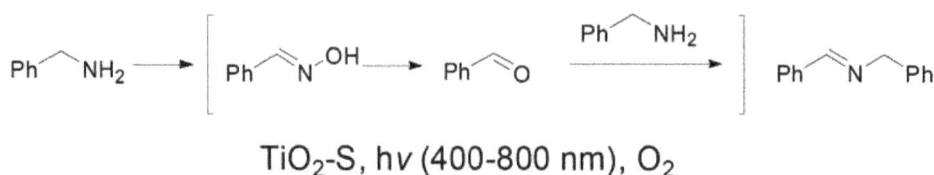

$$\text{TiO}_2\text{-S, h}\nu \text{ (400-800 nm), O}_2$$

Scheme 1. Tentative reaction pathway.

Ability of TiO$_2$-S photocatalyst to catalyze the oxidative cross-coupling of two benzylic amines with different substituent groups to yield a heterocoupled imine product was also studied using benzylamine, 4-methylbenzylamine and 4-methoxybenzylamine (two at a given reaction). Results demonstrated a poor selectivity since all four possible imines were observed in relatively similar yields after 24 h. Oxidative coupling of benzylamine with 4-methylbenzylamine had 95% of imine product yield. Self-coupling products of benzylamine (28%) and 4-methylbenzylamine (22%) were observed together with the two heterocoupled imines (50%), the product distribution of heterocoupling of benzylamine with 4-methylbenzylamine is as desired (~1:1:1:1) since difference in the nucleophilicities of "H" and methyl group is not significant. In order to evaluate the product distribution over the time span of this heterocoupled imine synthesis, we chose benzylamine and 4-methoxybenzylamine as the two benzylic amine substrates and the reaction profile is given in the Table 3, this reveals both the precursor imines produce the corresponding aldehydes as per the oxygenation step shown in Scheme 1, and then reacts with a free amine molecule to yield the final imine. Aldehyde of more electro deficient nucleus reacts faster with the more electron rich amine (P3) at early stages of the reaction and then with either amine as the reaction is progressing. Rate of aldehyde formation is slower in electron rich benzene nucleus, benzylamine in this system and it acts as the nucleophile (amine half), while 4 methoxybezylamine is easier to oxidize and preferentially be the aldehyde half. In the product distribution more P3 and P4 are observed during the whole cause of the reaction indicating high formation and reactivity of 4-methoxybenzaldehyde. Dual amine systems of benzylamine/aniline and 4-methoxybenzylamine/aniline yield only the self-coupled imines of benzylamine (92%) and 4-methoxybenzylamine (96%) respectively. Amount of aniline introduced in the reaction system remained unchanged even after the reaction, portraying its inert role in this photocatalyzed oxidative coupling reaction, anilne with a –NH$_2$ unit in its structure failed to participate in this heterocoupling reactions at least as the amine half.

Reusability of the catalyst is an important parameter in heterogeneous catalysis. The composite TiO$_2$-S photocatalyst studied in this system can be recovered readily from aqueous or organic solutions by simple filtration or sedimentation. The anatase nanocrystals in these composite samples are linked to silicate pieces such that grains in the μm scale are formed. Operational life of this catalyst examined over five consecutive rounds (Figure 4) revealed no apparent activity loss after five rounds. This further confirms the thermal and chemical stability of the catalyst. However selectivity towards the imine product was gradually decreased during each cycle lowering the overall product yield.

Table 3. Time conversion plot for oxidative coupling of benzylamine with 4-methoxybenzylamine [a].

Entry	Time (h)	Conversion (%)		Selectivity (%)			
		Benzylamine	4-Methoxybenzylamine	P1	P2	P3	P4
1	2	13	13	0	0	100	0
2	4	41	57	12	12	40	36
3	8	84	90	15	16	38	32
4	17	96	97	19	16	40	26
5	20	96	97	21	16	40	24

[a] Reaction Conditions: 25 mg catalyst, 0.25 mmol amine substrates, 2 mL acetonitrile, 500 W halogen lamp (cut-off wavelength below 400 nm) intensity 0.36 W/cm², 1 atm O_2.

P1   P2   P3   P4

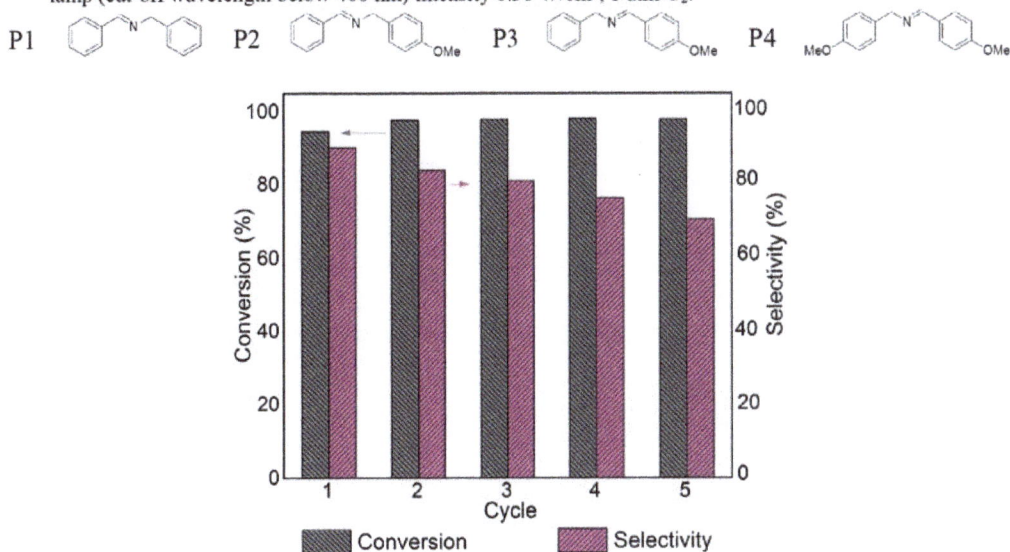

Figure 4. Reusability data of TiO_2-S catalyst for the oxidative coupling of benzylamine.

3. Experimental Section

3.1. General Information and Materials

The laponite clay was supplied from Fernz specialty chemicals Australia, all other chemicals were purchased from Sigma Aldrich (Castle Hill NSW, Australia) and used as received without further purification. Water used in all experiments was milli-Q water passing through an ultra-purification system.

3.2. General Procedure for the Synthesis of TiO₂-S Composite

TiO_2 precursor was prepared by hydrolyzing $Ti(OCH_3)_4$ in HCl for 3 h following a slightly modified method proposed by J. Sterte [30,31,33].

Initially 1.0 g of laponite was slowly dispersed in 50 mL of deionized water and kept stirring until it was transparent. Then 4.0 g of polyethylene glycol (FW 585) surfactant and the metal precursor solution was added drop wise with continuous stirring. Mixture was then transferred to teflon covered autoclaves and heated at 100 °C for 2 days. The solid was then recovered from centrifugation, followed by washing with water until no more chloride ions left (confirmed by a test with AgNO₃). Product was then dried in air and finally calcined at 500 °C for 20 h with the step being 2 °C·min⁻¹.

3.3. Characterization of TiO₂-S Composite

The diffuse reflectance UV/Vis (DR-UV/Vis) spectra were recorded on a Cary 5000 UV/Vis-NIR Spectrophotometer (Agilent, Santa Clara, CA, USA). X-ray diffraction (XRD) patterns of the samples were recorded on a Philips PANalytical X'Pert PRO diffractometer (PANalytical, Sydney, Australia) using CuKα radiation (l = 1.5418 Å) at 40 kV and 40 mA. Transmission electron microscopy (TEM) images were taken with a Philips CM200 Transmission electron microscope (Philips, Eindhoven, The Netherlands) employing an accelerating voltage of 200 kV. The specimens were fine powders deposited onto a copper micro grid coated with a holey carbon film. Nitrogen physisorption isotherms were measured on the Tristar II 3020 (Micromeritics, Norcross, GA, USA). Prior to the analysis, sample was degassed at 110 °C overnight under high vacuum. The specific surface area was calculated by the Brunauer-Emmett-Teller (BET) method from the data in a P/P° range between 0.05 and 0.2. The compositional data was determined by energy-dispersive X-ray spectroscopy (EDS) (EDAX, Mahwah, NJ, USA) attached to an FEI Quanta 200 scanning electron microscope (SEM, Quanta, OR, USA).

3.4. General Procedure for the Photocatalytic Reactions

Benzylic amine compound 0.5 mmol, 5 mL of solvent were measured to a clean dry reactor tube. 50 mg of the catalyst was added and finally the reactor was purged with oxygen gas. These reactors were kept magnetically stirring in front of a 500 W halogen lamp (except for dark and wavelength experiments) for 24 h at 40 °C. At the end of the reaction 1 mL samples were collected in to small glass vials after filtering out the solid catalyst using 0.2 μm milli pore filter. We tested for the products using a gas chromatograph (GC, Agilent, Santa Clara, CA, USA) equipped with a DB 5 column. For wavelength experiments, 5 monochromatic light emitting diodes (LEDs) of 390–410 nm, 460–462 nm, 515–517 nm, 587.5–560 nm or 620–625 nm was used.

4. Conclusions

We have successfully applied TiO₂-S composite photocatalyst in the oxidative coupling of benzylamine to imine under visible light irradiation. The numerous contacts between the anatase crystals and silicate and high population of defects in the composite photocatalyst are the possible reasons behind the enhanced visible light activity. The formation of imines proceed via an oxidation pathway: under visible light irradiation, the oxygen molecules adsorbed on the catalyst capture the light excited electrons, and react with the H bonded to the α-carbon of the substrate molecules, which is oxidized to aldehyde. The condensation of the aldehyde with amine yields the product imine. This photocatalyst has a very high activity in the region λ > 460 nm. This range is much broader compared to previously reported results for anatase materials ($\lambda > 420$ nm). Intensity, wavelength and reaction temperature can be tuned to optimize the reaction rate of TiO_2-S catalyzed oxidative coupling of benzylic amines. Water can be used as a solvent giving moderate conversion rate but sole product. These findings encourage us to further study the surface modified titania based materials for selective organic synthesis.

Supplementary Materials

Supplementary materials can be accessed at: http://www.mdpi.com/1420-3049/20/02/1941/s1.

Acknowledgments

This project was supported by Australian Research Council (ARC DP110104990).

Author Contributions

S.Z. and H.Z. designed the research; S.Z. performed the research, analysed the data and wrote the paper. Both authors read and approved the final manuscript.

Conflicts of Interest

The authors declare no conflict of interest.

References

1. Kobayashi, S.; Mori, Y.; Fossey, J.S.; Salter, M.M. Catalytic Enantioselective Formation of C-C Bonds by Addition to Imines and Hydrazones: A Ten-Year Update. *Chem. Rev.* **2011**, *111*, 2626–2704, doi:10.1021/cr100204f.
2. Adams, J.P. Imines, Enamines and Oximes. *J. Chem. Soc. Perkin Trans. 1* **2000**, 125–139, doi:10.1039/A808142E.
3. Largeron, M. Protocols for the Catalytic Oxidation of Primary Amines to Imines. *Eur. J. Org. Chem.* **2013**, *24*, 5225–5235.
4. Granzhan, A.; Riis-Johannessen, T.; Scopelliti, R.; Severin, K. Combining Metallasupramolecular Chemistry with Dynamic Covalent Chemistry: Synthesis of Large Molecular Cages. *Angew. Chem. Int. Ed.* **2010**, *49*, 5515–5518.
5. Belowich, M.E.; Stoddart, J.F. Dynamic Imine Chemistry. *Chem. Soc. Rev.* **2012**, *41*, 2003–2024.
6. Rasdi, F.R. M.; Phan, A.N.; Harvey, A.P. Rapid Determination of the Reaction Kinetics of an *N*-butylbenzaldimine Synthesis Using a Novel Mesoscale Oscillatory Baffled Reactor. *Procedia Eng.* **2012**, *42*, 1662–1675.
7. Alessandro, Z.; Jose, A.M.; Eduardo, P. One-Pot Preparation of Imines from Nitroarenes by a Tandem Process with an Ir-Pd Heterometallic Catalyst. *Chem. Eur. J.* **2010**, *16*, 10502–10506.
8. Kwon, M.S.; Kim, S.; Park, S.; Bosco, W.; Chidrala, R.K.; Park, J. One-Pot Synthesis of Imines and Secondary Amines by Pd-Catalyzed Coupling of Benzyl Alcohols and Primary Amines. *J. Org. Chem.* **2009**, *74*, 2877–2879.

9. Wang, H.; Zhang, J.; Cui, Y.M.; Yang, K.F.; Zheng, Z.J.; Xu. L.W. Dehydrogenation and Oxidative Coupling of Alcohol and Amines Catalysed by Organosilicon-Supported TiO$_2$@PMHSIPN. *RSC Adv.* **2014**, *4*, 34681–34686.

10. Yamaguchi, K.; Mizuno, N. Efficient Heterogeneous Aerobic Oxidation of Amines by a Supported Ruthenium Catalyst. *Angew. Chem. Int. Ed.* **2003**, *42*, 1480–1483.

11. Yamaguchi, K.; Mizuno, N. Scope, Kinetics, and Mechanistic Aspects of Aerobic Oxidations Catalysed by Ruthenium Supported on Alumina. *Chem. Eur. J.* **2003**, *9*, 4353–4361.

12. Mizuno, N.; Yamaguchi, K. Selective Aerobic Oxidations by Supported Ruthenium Hydroxide Catalysts. *Catal. Today* **2008**, *132*, 18–26.

13. Zhu, B.; Lazar, M.; Trewyn, B.G.; Angelici, R.J. Aerobic Oxidation of Amines to Imines Catalyzed by Bulk Gold Powder and by Alumina-Supported Gold. *J. Catal.* **2008**, *260*, 1–6.

14. Grirrane, A.; Corma, A.; Garcia, H. Highly Active and Selective Gold Catalysts for the Aerobic Oxidative Condensation of Benzylamines to Imines and One-Pot, Two-Step Synthesis of Secondary Benzylamines. *J. Catal.* **2009**, *264*, 138–144.

15. Patil, R.D.; Adimurthy, S. Copper-Catalyzed Aerobic Oxidation of Amines to Imines under Neat Conditions with Low Catalyst Loading. *Adv. Synth. Catal.* **2011**, *353*, 1695–1700.

16. Ke, X.; Zhang, X.; Zhao, J.; Sarina, S.; Barry, J.; Zhu, H. Selective Reductions using Visible Light Photocatalysts of Supported Gold Nanoparticles. *Green Chem.* **2013**, *15*, 236–244.

17. Zhu, H.; Ke, X.; Yang, X.; Sarina, S.; Liu, H. Reduction of Nitroaromatic Compounds on Supported Gold Nanoparticles by Visible and Ultraviolet Light. *Angew. Chem. Int. Ed.* **2010**, *122*, 9851–9855.

18. Zhang. X.; Ke, X.; Zhu, H. Zeolite-Supported Gold Nanoparticles for Selective Photooxidation of Aromatic Alcohols under Visible-Light Irradiation. *Chem. Eur. J.* **2012**, *18*, 8048–8056.

19. Su, F.; Mathews, S.C.; Mohlmann, L.; Antonietti, M.; Wang, X.; Blechert, S. Aerobic Oxidative Coupling of Amines by Carbon Nitride Photocatalysis with Visible Light. *Angew. Chem. Int. Ed.* **2011**, *50*, 657–660.

20. Shiraishi, Y.; Hirai, T.; Selective Organic Transformations on Titanium Oxide-Based Photocatalysts. *J. Photochem. Photobiol. C* **2008**, *9*, 157–170.

21. Palmisano, G.; Garcia-Lopez, E.; Marci, G.; Loddo, V.; Yurdakal, S.; Augugliaro, V.; Palmosano, L. Advances in Selective Conversions by Heterogeneous Photocatalysis. *Chem. Commun.* **2010**, *46*, 7074–7089.

22. Augugliaro, V.; Palmisano, L. Green Oxidation of Alcohols to Carbonyl Compounds by Heterogeneous Photocatalysis. *ChemSusChem* **2010**, *3*, 1135–1138.

23. Yurdakal, S.; Palmisano, G.; Loddo, V.; Augugliaro, V.; Palmisano, L. Nanostructured Rutile TiO$_2$ for Selective Photocatalytic Oxidation of Aromatic Alcohols to Aldehydes in Water. *J. Am. Chem. Soc.* **2008**, *130*, 1568–1569.

24. Tsukamoto, D.; Ikeda, M.; Shiraishi, Y.; Hara, T.; Ichikuni, N.; Tanaka, S.; Hirai, T. Selective Photocatalytic Oxidation of Alcohols to Aldehydes in Water by TiO$_2$ Partially Coated with WO$_3$. *Chem. Eur. J.* **2011**, *17*, 9816–9826.

25. Palmisano, G.; Augugliaro, V.; Pagliaro, M.; Palmisano, L. Photocatalysis: A Promising Route for 21st Century Organic Chemistry. *Chem. Commun.* **2007**, *33*, 3425–3437.

26. Fox, M.A.; Dulay, M.T. Heterogeneous Photocatalysis. *Chem. Rev.* **1993**, *93*, 341–357.

27. Maldotti, A.; Molinari, A.; Amadelli, R. Photocatalysis with Organized Systems for Oxofunctionalization of Hydrocarbons by O_2. *Chem. Rev.* **2002**, *102*, 3811–3836.

28. Li, N.; Lang, X.; Ma, W.; Ji, H.; Chen, C.; Zhao, J.C. Selective Aerobic Oxidation of Amines to Imines by TiO_2 Photocatalysis in Water. *Chem. Commun.* **2013**, *49*, 5034–5036.

29. Lang, X.; Ma, W.; Zhao, Y.; Chen, C.; Ji, H.; Zhao. J. Visible-Light-Induced Selective Photocatalytic Aerobic Oxidation of Amines into Imines on TiO_2. *Chem. Eur. J.* **2012**, *18*, 2624–2631.

30. Li, J.; Chen, C.; Zhao, J.; Zhu, H.; Orthman, J. Photodegradation of Dye Pollutants on TiO_2 Nanoparticles Dispersed in Silicate under UV-Vis Irradiation. *Appl. Catal. B: Environ.* **2002**, *37*, 331–338.

31. Zhu, H.Y.; Zhao, J.C.; Liu, J.W.; Yang, X.Z.; Shen, Y.N. General Synthesis of a Mesoporous Composite of Metal Oxide and Silicate Nanoparticles from a Metal Salt and Laponite Suspension for Catalysis. *Chem. Mater.* **2006**, *18*, 3993–4401.

32. Yang, D.; Cheng, C.; Zheng, Z.; Liu, H.; Waclawik, E.R.; Yan, Z.; Huang, Y.; Zhang, H.; Zhao, J.; Zhu, H. Grafting Silica Species on Anatase Surface for Visible Light Photocatalytic Activity. *Energy Environ. Sci.* **2011**, *4*, 2279–2287.

33. Sterte, J. Synthesis and Properties of Titanium Oxide Cross-Linked Montmorillonite. *Clays Clay Miner.* **1986**, *34*, 658–664.

Sample Availability: Not available.

A Review on Visible Light Active Perovskite-Based Photocatalysts

Pushkar Kanhere and Zhong Chen

Abstract: Perovskite-based photocatalysts are of significant interest in the field of photocatalysis. To date, several perovskite material systems have been developed and their applications in visible light photocatalysis studied. This article provides a review of the visible light ($\lambda > 400$ nm) active perovskite-based photocatalyst systems. The materials systems are classified by the B site cations and their crystal structure, optical properties, electronic structure, and photocatalytic performance are reviewed in detail. Titanates, tantalates, niobates, vanadates, and ferrites form important photocatalysts which show promise in visible light-driven photoreactions. Along with simple perovskite (ABO₃) structures, development of double/complex perovskites that are active under visible light is also reviewed. Various strategies employed for enhancing the photocatalytic performance have been discussed, emphasizing the specific advantages and challenges offered by perovskite-based photocatalysts. This review provides a broad overview of the perovskite photocatalysts, summarizing the current state of the work and offering useful insights for their future development.

Reprinted from *Molecules*. Cite as: Kanhere, P.; Chen, Z. A Review on Visible Light Active Perovskite-Based Photocatalysts. *Molecules* **2014**, *19*, 19995-20022.

1. Introduction

Photocatalysis has long been studied for clean energy and environmental applications. Over the past two decades, the number of applications based on photocatalysis has increased sharply, while a wide range of materials systems have been developed [1–4]. Photocatalysis has been of particular interest in the production of hydrogen from water using solar energy [5]. Further, conversion of CO_2 to hydrocarbons (fuels) is also of significant interest, as it is a solution to reduce CO_2 emissions across the globe [6,7]. Apart from the clean energy generation, photocatalysis has several promising applications in the environmental field. Some of the applications include degradation of volatile organic compounds (VOC) for water treatment [8], germicide and antimicrobial action [9–11], de-coloration of industrial dyes [12–14], nitrogen fixation in agriculture [15], and removal of NO_x/SO_x air pollutants [16–19]. These applications have driven the development of variety of materials systems which are suitable for specific applications. Although TiO_2-based materials are the most studied for photocatalytic applications, ternary and other complex oxide systems have been increasingly explored as photocatalysts. Among the various classes of materials studied, perovskites-based photocatalysts have unique photophysical properties and offer distinct advantages.

Perovskites are the class of compounds presenting the general formula ABO₃. Generally, in this crystal structure, the A site is occupied by the larger cation, while the B site is occupied by the smaller cation. Perovskites are one of the most important families of materials exhibiting properties suitable for numerous technological applications [20]. Perovskite compounds such as PbZrO₃,

BaTiO$_3$, PbTiO$_3$ are most commonly used piezoelectric compounds [21]. BiFeO$_3$ thin films show multiferroic behavior [22], while compounds such as SrTiO$_3$ have shown excellent photocatalytic properties [23,24]. The origin of such properties lies in the crystal structure of perovskites. The perovskite crystal structure has corner connected BO$_6$ octahedra and 12 oxygen coordinated A cations, located in between the eight BO$_6$ octahedra (Figure 1). The perfect structure of the octahedral connection results in a cubic lattice. However, depending on the ionic radii and electronegativity of the A and B site cations, tilting of the octahedra takes place, which gives rise to lower symmetry structures. As seen from the crystal structure, B site cations are strongly bonded with the oxygen (or any other anion) while, A site cations have relatively weaker interactions with oxygen. Depending on the type of the cations occupying the lattice sites, these interactions could be altered to yield the different perovskite crystal geometries.

Figure 1. Crystal structure of simple Perovskite, (**a**) BaTiO$_3$ and (**b**) double perovskite Na$_2$Ta$_2$O$_6$ (red: oxygen, green and purple: A site cation, grey and blue: BO$_6$ octahedra).

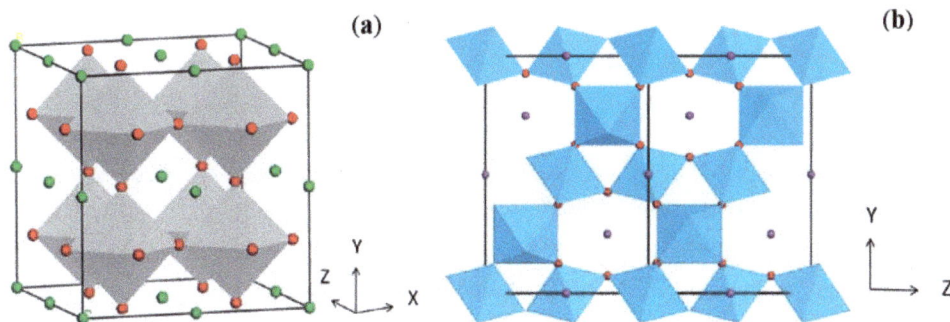

For example, different degrees of tilting of the octahedra give rise to different crystal fields, which result in different electronic and optical properties. The degrees of tilting may affect the band structure, electron and hole transport properties, photoluminescence, and dielectric behavior [25,26]. From the point of view of photocatalysis, perovskite structures may offer significant advantages over the corresponding binary oxides for several reasons. Firstly, perovskites could offer favorable band edge potentials which allow various photoinduced reactions. For example, as compared to the binary oxides, several perovskites have sufficiently cathodic conduction band (CB) energies for hydrogen evolution. Secondly, A and B site cations in the lattice give a broader scope to design and alter the band structure as well as other photophysical properties. In the case of double perovskites such as A$_2$B$_2$O$_6$, stoichiometric occupation of two cations at the B site is known to be beneficial for visible light photocatalysis. Thirdly, some studies have shown that it is possible to combine the effects such as ferroelectricity or piezoelectricity with the photocatalytic effect to benefit the photocatalytic activity.

Perovskite photocatalysts have been studied to a great extent because of their promise for being visible light active. A review of the present work on the visible light driven perovskite photocatalyts is essential to provide a broad overview and possible future directions. Shi *et al.* reported a general review of perovskite photocatalysts active under UV and visible light [27]. The current review article is focused on visible light active perovskite compounds. We emphasize the strategies used to develop or enhance the visible light absorption and subsequent photocatalytic activities. Further, we attempt to shed some light on the underlying principles specific to the perovskite crystal structure which play an important role in the photocatalytic activity, suggesting potential areas in the field where further work is needed. The first section of the article discusses the mechanism and thermodynamics of some of the most important photocatalytic reactions, while the later section reviews the material systems in detail. In the current review, perovskites are broadly divided into simple (ABO_3 type) perovskites and complex perovskites (double, layered, *etc.*).

2. Overview of Photocatalytic Reactions

Photocatalysis is a process that utilizes the energy input from incident radiation and the catalytic properties of the surface of a material to carry out and/or accelerate certain chemical reactions. To date, numerous chemical reactions have been studied, which are potentially useful in energy generation and environmental cleaning applications. Photocatalysis is known to be able to produce thermodynamically uphill reactions, which otherwise need intense energy inputs in terms of high temperature (or pressure). Understanding the mechanism of photocatalytic reactions is critically important to design and develop new photocatalytic materials. In this section, a brief review of mechanism and thermodynamics of most common photocatalytic reactions is presented. Figure 2 shows the reduction and oxidation levels of some of the common photocatalytic reactions with reference to vacuum and the normal hydrogen electrode (NHE). It is noted that these values provide an insight only on the thermodynamic feasibility of the reaction. It is seen that for the reduction reaction, the energy of the (photoexcited) electron should be higher (on the absolute vacuum scale) than the redox level. Therefore the CB potential of the photocatalyst should be located at a higher energy value than the reduction reaction of interest.

Figure 2. Energy levels of some of the important photocatalytic reactions with respect to NHE at pH = 0 [28].

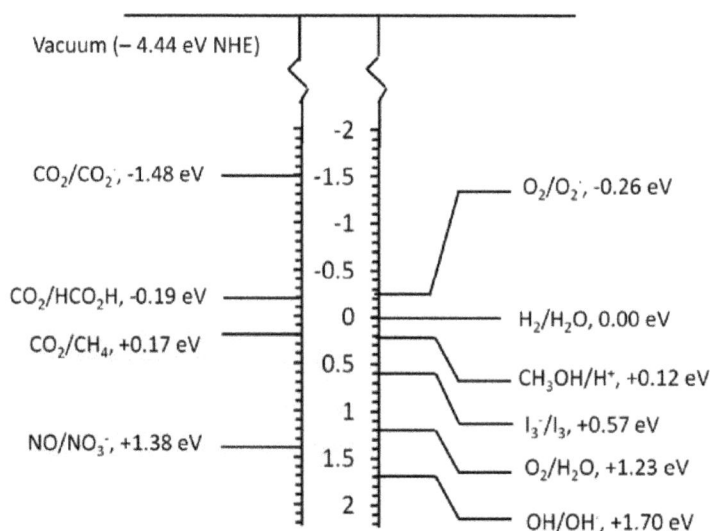

2.1. Photocatalytic Water Splitting

One of the most studied reactions is the direct splitting of water into hydrogen and oxygen. Figure 3 shows the schematics of the water splitting reaction according to the 4-photon model [29].

Figure 3. Band diagram and schematics of water splitting reaction over a photocatalyst surface [30].

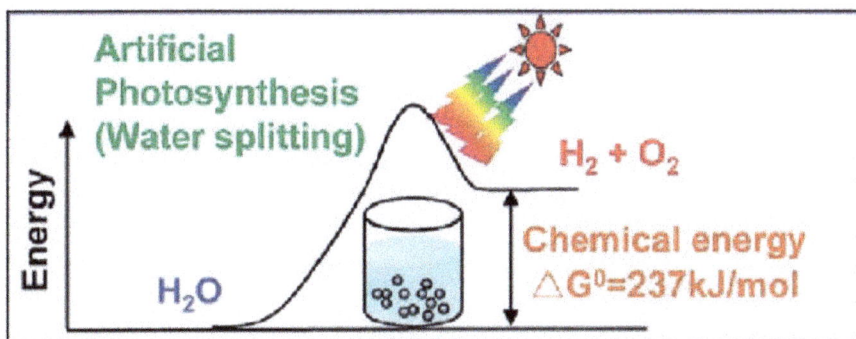

In the water splitting reaction, upon the radiation of photon with suitable wavelength, photoexcited pairs of electrons and holes are generated within a photocatalyst. Typically, electron-hole separation takes place, due to surface charge or co-catalyst loading. Direct oxidation of water molecules, chemisorbed on the surface of the photocatalyst (or co-catalyst) occurs, by the interaction of water molecule and hole in the valence band (VB) of the photocatalyst. This reaction results in liberation of an oxygen molecule and 4-protons. The protons then migrate to the sites of photoexcited electrons to form hydrogen molecules (Equations (1)–(3)).

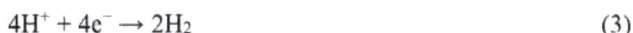

$$2H_2O \rightarrow 2H_2 + O_2 \; \Delta H = +234 \text{ kJ/mol} \tag{1}$$

$$2H_2O \rightarrow 4H^+ + O_2 \tag{2}$$

$$4H^+ + 4e^- \rightarrow 2H_2 \tag{3}$$

Evolution of hydrogen and oxygen using sunlight is considered as one of the most promising ways to generate hydrogen as a clean and renewable fuel. Like the water molecule, other molecules also undergo decomposition by the process of photocatalysis.

2.2. Photooxidation of Organic Molecules

Several organic compounds undergo photooxidation reactions, where a direct oxidation via photogenerated holes occurs or an indirect oxidation via hydroxyl ions takes place [31]. The degradation of organic molecules also takes place by reactive oxygen species (Figure 4). Organic dyes, aliphatics and aromatic hydrocarbons, and organic acids can be mineralized to CO_2 and H_2O by photocatalytic processes. Like organic compounds, hydroxyl ions and reactive oxygen species (ROS) are known to inactivate microorganisms by degrading their cell walls [10,32]. The photocatalytic inactivation of microbes is effective in antimicrobial, antifungal and antiviral applications. A later section reviews certain silver- and bismuth-based perovskites which display particularly efficient antimicrobial action under visible light.

Figure 4. Band diagram and schematics of degradation of organic compounds over a photocatalyst surface [12,33].

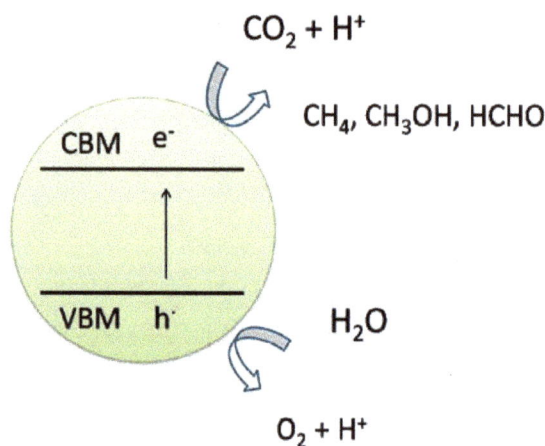

$$CO_2 + H^+$$

$$CH_4, CH_3OH, HCHO$$

CBM   e⁻

VBM   h·

$$H_2O$$

$$O_2 + H^+$$

2.3. Photocatalytic Conversion of CO₂ to Fuels

CO₂, with a standard enthalpy of formation of −393.5 kJ·mol⁻¹ at 298 K, is one of the most stable molecules. With appropriate adsorption and photocatalytic processes, reduction of CO₂ in presence of water could be performed to produce hydrocarbons (Figure 5). Possible chemical reactions of adsorbed CO₂ and protons are presented by the following equations (Equations (4)–(7)). It could be seen that different number of protons in the reactants, results in different hydrocarbons as products.

Figure 5. Schematics of CO₂ photoreduction reaction over a photocatalyst surface [34].

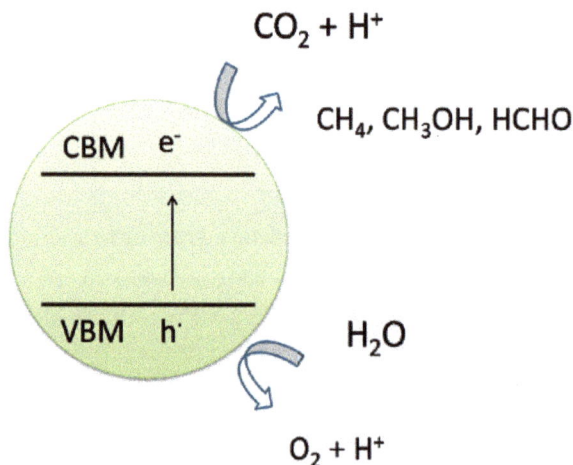

$$CO_2 + H^+$$

$$CH_4, CH_3OH, HCHO$$

CBM   e⁻

VBM   h·

$$H_2O$$

$$O_2 + H^+$$

Among these reactions, the reaction with eight protons converting CO_2 to methane is of significant interest. The photocatalytic reduction of CO_2 in the presence of water is a complex reaction and the photocatalyst must possess enough band potential for proton generation:

$$CO_2 + 2H^+ + 2e^- \rightarrow CO + H_2O \tag{4}$$

$$CO_2 + 6H^+ + 6e^- \rightarrow CH_3OH + H_2O \tag{5}$$

$$CO_2 + 8H^+ + 8e^- \rightarrow CH_4 + H_2O \tag{6}$$

$$2CO_2 + 12H^+ + 12e^- \rightarrow C_2H_5OH + H_2O \tag{7}$$

2.4. Photocatalytic Nitrogen Fixation

Like CO_2 reduction, atmospheric nitrogen could be reduced to ammonia or nitrates by the photocatalytic processes. The mechanism of nitrogen reduction is similar to that of CO_2, where chemically adsorbed nitrogen molecules react catalytically with protons and form compounds of nitrogen and hydrogen (Equations (8)–(10)):

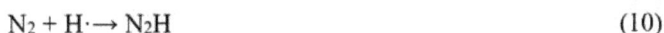

$$H_2O \ (hv/TiO_2) \rightarrow 2H^+ + _{1/2}O_2 + 2e^- \tag{8}$$

$$H^+ + e^- \rightarrow H \tag{9}$$

$$N_2 + H \cdot \rightarrow N_2H \tag{10}$$

The photocatalytic reduction of nitrogen is extremely useful in nitrogen photofixation processes for agricultural applications [35–37]. Although the process of photocatalytic nitrogen fixation is promising, efforts in this area have been severely limited. It is noted that the mechanism of the photocatalytic processes presented above is a simplified understanding, while the photocatalytic processes are complex in nature.

It is known that a given chemical reaction has a specific photooxidation or photoreduction level (potential) and thus the band potentials of the photocatalyst must satisfy the thermodynamic conditions. Intrinsic properties such as band gap (optical absorption) and band edge potentials determine the thermodynamic feasibility of photoinduced reactions under light irradiation. Apart from the basic conditions, there are several factors which affect the photocatalytic performance of the material system under consideration. Properties such as electron and hole effective mass, exciton lifetime and diffusion length, exciton binding energy affect the electron-hole separation and transport within the lattice. These properties are known to strongly influence the performance (kinetics/efficiency) of the photocatalytic reactions. Defects in the lattice, defect-induced energy states, localization of electrons on specific defect sites could determine the fate of the photoexited electron-hole pair. Finally, the electron transfer across semiconductor-electrolyte interface is significantly affected by surface states, surface band structure (depletion region induced electric field), and band bending. Such electronic properties of materials could be altered to suit specific photocatalytic applications. To date, numerous material systems have been evolved through systematic efforts of understanding and improving the electronic properties of materials. Among these materials perovskites have shown excellent promise for efficient photocatalysis under visible light irradiation, on account of their unique crystal structure and electronic properties. The perovskite crystal structure offers an excellent framework to tune the band gap values to enable visible light absorption and band edge potentials to suit the needs of specific photocatalytic reactions. Further, lattice distortion in perovskite compounds strongly influences the separation of photogenerated charge carriers. The following sections present some groups of materials that have shown visible light activity.

3. Simple Perovskites with Visible Light Response

3.1. Titanate Perovskites

Titanate perovskites have been studied for photocatalytic applications for a long time. Most of the titanate perovskites have band gap energy (E_g) value more than 3.0 eV, however they show excellent photocatalytic properties under UV radiation [1]. Using these titanates as host materials, doping is widely used to alter the optical properties and induce visible light absorption. TiO_2 (anatase) has a band gap of 3.2–3.4 eV and its CB potential is −0.3 to −0.6 eV above the water reduction level [38]. Certain perovskite titanates have CB energies more negative than TiO_2, making them more suitable candidates for hydrogen generation. Titanates also offer good photostability and corrosion resistance in aqueous solutions. In this section, a detailed review of $MTiO_3$ (M = Sr, Ba, Ca, Mn, Co, Fe, Pb, Cd, Ni) systems is presented. Figure 6 gives an overview of elements that form perovskite titanates.

Figure 6. Overview of elements forming perovskite titanates useful for visible light photocatalysis.

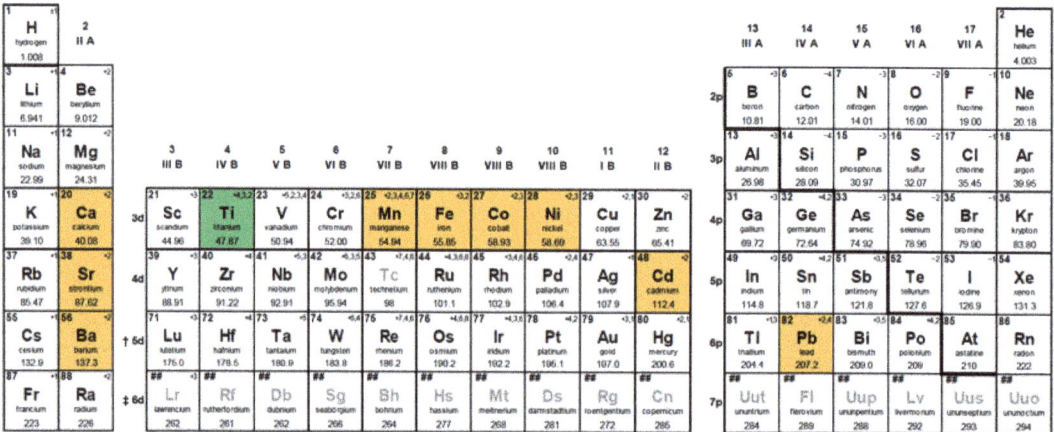

1 H hydrogen 1.008	2 IIA												13 IIIA	14 IVA	15 VA	16 VIA	17 VIIA	2 He helium 4.003
3 Li lithium 6.941	4 Be beryllium 9.012												5 B boron 10.81	6 C carbon 12.01	7 N nitrogen 14.01	8 O oxygen 16.00	9 F fluorine 19.00	10 Ne neon 20.18
11 Na sodium 22.99	12 Mg magnesium 24.31	3 IIIB	4 IVB	5 VB	6 VIB	7 VIIB	8 VIIIB	9 VIIIB	10 VIIIB	11 IB	12 IIB		13 Al aluminum 26.98	14 Si silicon 28.09	15 P phosphorus 30.97	16 S sulfur 32.07	17 Cl chlorine 35.45	18 Ar argon 39.95
19 K potassium 39.10	20 Ca calcium 40.08	21 Sc scandium 44.96	22 Ti titanium 47.87	23 V vanadium 50.94	24 Cr chromium 52.00	25 Mn manganese 54.94	26 Fe iron 55.85	27 Co cobalt 58.93	28 Ni nickel 58.69	29 Cu copper 63.55	30 Zn zinc 65.41		31 Ga gallium 69.72	32 Ge germanium 72.64	33 As arsenic 74.92	34 Se selenium 78.96	35 Br bromine 79.90	36 Kr krypton 83.80
37 Rb rubidium 85.47	38 Sr strontium 87.62	39 Y yttrium 88.91	40 Zr zirconium 91.22	41 Nb niobium 92.91	42 Mo molybdenum 95.94	43 Tc technetium 98	44 Ru ruthenium 101.1	45 Rh rhodium 102.9	46 Pd palladium 106.4	47 Ag silver 107.9	48 Cd cadmium 112.4		49 In indium 114.8	50 Sn tin 118.7	51 Sb antimony 121.8	52 Te tellurium 127.6	53 I iodine 126.9	54 Xe xenon 131.3
55 Cs cesium 132.9	56 Ba barium 137.3	71 Lu lutetium 176.0	72 Hf hafnium 178.5	73 Ta tantalum 180.9	74 W tungsten 183.8	75 Re rhenium 186.2	76 Os osmium 190.2	77 Ir iridium 192.2	78 Pt platinum 195.1	79 Au gold 197.0	80 Hg mercury 200.6		81 Tl thallium 204.4	82 Pb lead 207.2	83 Bi bismuth 209.0	84 Po polonium 209	85 At astatine 210	86 Rn radon 222
87 Fr francium 223	88 Ra radium 226	103 Lr lawrencium 262	104 Rf rutherfordium 261	105 Db dubnium 262	106 Sg seaborgium 266	107 Bh bohrium 264	108 Hs hassium 277	109 Mt meitnerium 268	110 Ds darmstadtium 281	111 Rg roentgenium 272	112 Cn copernicium 285		113 Uut ununtrium 284	114 Fl flerovium 289	115 Uup ununpentium 288	116 Lv livermorium 292	117 Uus ununseptium 293	118 Uuo ununoctium 294

3.1.1. SrTiO₃

$SrTiO_3$ is a simple cubic (Pm3m, a = 3.9 Å) perovskite with an indirect band gap of 3.25 eV [39]. When loaded with a co-catalyst such as Rh or NiO_x, $SrTiO_3$ shows stoichiometric water splitting under UV radiation [40] and has been studied extensively for developing visible light water splitting catalysts. Doping of the Ti site with Mn, Ru, Rh, and Ir was of significant interest in early days [41]. It is found that these dopants induce mid-gap states in the band gap allowing visible light absorption [42]. Mn and Ru doping are found useful for O_2 evolution, while dopants like Ru, Rh, and Ir are suitable for H_2 evolution [41]. Rh-doped $SrTiO_3$ thin films also shows cathodic photocurrent from overall water splitting under visible light, where 7% Rh doped $SrTiO_3$ showed 0.18% incident photo-to-electron conversion efficiency (IPEC) under 420 nm irradiation [43]. Using Rh-doped $SrTiO_3$ as a H_2 evolving photocatalyst, various Z scheme systems have been developed. In a significant demonstration, a novel electron mediator $[Co(bpy)3]^{3+/2+}$ was used for Rh-doped $SrTiO_3$ with $BiVO_4$ photocatalyst. Such a system showed a solar energy conversion efficiency of 0.06% under daylight [44]. Efforts in Z scheme photocatalysis have also been targeted towards eliminating the need for electron mediators by preparing composite photocatalysts. In such systems, electrons from an O_2 evolving photocatalyst recombine with holes from a H_2 evolving photocatalyst at the interface of the composite. The quality of the interface and the band alignment of the two semiconductors are important factors for the successful realization of mediator-free type Z schemes.

Rh-doped $SrTiO_3$ was combined with several O_2-evolving photocatalysts such as $BiVO_4$, $AgNbO_3$, Bi_2MoO_6, WO_3, or Cr/Sb-doped TiO_2 [45]. In those experiments the authors found that agglomeration of the photocatalyst particles occur under acidic conditions, which results in Z scheme photocatalysis. A combination of Rh-doped $SrTiO_3$ and $BiVO_4$ resulted in the best yield [45]. A schematic of microstructure and mechanism of water splitting of an agglomerated Z scheme photocatalyst is shown in Figure 7. In a recent effort, a composite of 1% Rh-doped $SrTiO_3$ loaded with 0.7% Ru and $BiVO_4$ was successfully prepared. Such a composite showed a stoichiometric water splitting reaction (pH 7) with quantum yield (QY) of 1.6% at 420 nm [23]. These studies successfully establish the feasibility of the "Z scheme" photocatalysis for candidates such as Rh-doped $SrTiO_3$ under visible light.

Figure 7. (a) Schematic microstructure and (b) band diagram of Z scheme photocatalysis using Rh-doped $SrTiO_3$ [45].

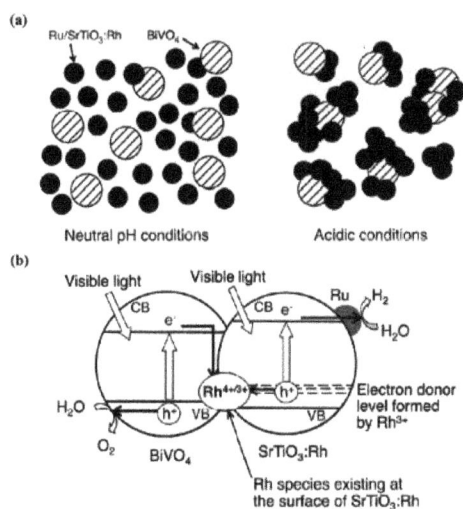

Further, the water splitting efficiency is dependent on the synthesis method used for Rh-doped $SrTiO_3$ [46]. The use of excess Sr in hydrothermal and complex polymerization methods proved useful for improving the apparent yield to 4.2% under 420 nm radiation [46]. Apart from mono-doping, co-doping has been employed in $SrTiO_3$ to pursue visible light driven photocatalysis. Co-doping of Sb (1%) and Rh (0.5%) was found useful for visible light photocatalysis and estimated H_2 and O_2 evolution rates for 1 m^2 surface area were 26 $mL \cdot h^{-1}$ and 13 $mL \cdot h^{-1}$, respectively [47]. Further, a composite system was prepared from co-doped La-Cr in $SrTiO_3$ and co-doped La-Cr Sr_2TiO_4 which showed visible light driven H_2 evolution (24 $\mu mol \cdot h^{-1} \cdot g^{-1}$) [48]. Composite preparation led to heterojunctions between doped phases and produced a synergistic effect for hydrogen evolution. Further, a solid solution of $AgNbO_3$ and $SrTiO_3$ was discovered to be a visible light active photocatalyst [49]. $(AgNbO_3)_{0.75}(SrTiO_3)_{0.25}$ showed promising performance for O_2 evolution and isopropanol (IPA) degradation under visible light.

Several efforts have been made to understand and design $SrTiO_3$-based photocatalysts. Particularly DFT-based band structure calculations provide useful insights into the electronic structure and its correlation with photocatalytic activity. It is shown that Rh doping in $SrTiO_3$ produces band-like states above the valence band maximum (VBM) which are responsible for the visible light absorption. The proximity of dopant-induced states to VBM helps efficient replenishment of electrons and suppresses electron trapping from CB [42]. Theoretical calculations indicate that a TiO_2-terminated $SrTiO_3$ surface with defects such as O and Sr vacancies would alter its electronic structure and induce visible light absorption peaking around 420 nm [50]. Such strategies could be useful in the development of low dimensional materials. Further, theoretical work on doped $SrTiO_3$ compounds show that certain dopants such as La strongly lower the effective mass of electrons and holes (near the valence band region), increasing the mobility of the photoexited carrier [51]. Along with the ground state band structure calculations, study of electron and hole masses, defect chemistry, photoexcited transport could be carried out to understand this system in detail. Understanding the excited state properties is useful in further development of new photocatalysts.

3.1.2. $BaTiO_3$

Like $SrTiO_3$, Rh doping in $BaTiO_3$ has been carried out and a quantum yield of 0.5% under 420 nm was reported [52]. Being a hydrogen evolving catalyst, this material has also been used as Z scheme with Pt/WO_3 for overall water splitting [52].

3.1.3. $CaTiO_3$

Calcium titanate is one of the common perovskite minerals with a band gap of 3.6 eV. Cu doping in $CaTiO_3$ is widely studied and visible light-driven photocatalytic water decomposition has been reported [53]. Cu doping not only induces visible light absorption, but also enhances hydrogen evolution under UV radiation when NiO_x co-catalyst is used. Studies of such doped systems, where dopants enhance the photocatalytic activity host materials are important to design efficient photocatalysts. Detrimental effects of doping on the properties such as electron-hole recombination, electron/hole effective mass, and reduced crystallinity should be studied and reported in detail. These studies are useful to gain insights on the photocatalytic activities of the doped systems. Co-doping of Ag and La at $CaTiO_3$ has been done to narrow the band gap and induce visible light absorption [54]. DFT studies also indicate that like $SrTiO_3$, TiO_2-terminated $CaTiO_3$ surfaces possess the capability of visible light absorption [55]. Along with alkali titanates, several transition metal titanates show promise for visible light photocatalysis. Figure 8 shows the empirically estimated band diagram of the $MTiO_3$ systems with respect to water oxidation and reduction levels.

Figure 8. Band edge potentials (*vs.* NHE; pH = 0) of $MTiO_3$ system [56].

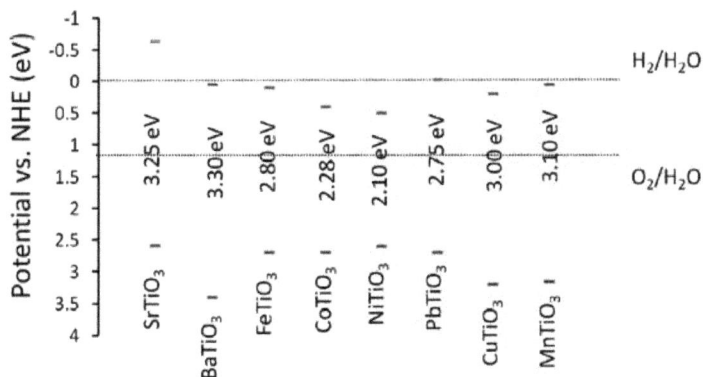

Alkali metal titanates such as Ba, Ca, and Sr have enough CB potential for hydrogen evolution. However, certain transition metal titanates do not possess the desired CB potential for water reduction, though they have band gap values in the visible region (such as Co, Ni, Fe *etc.*). These materials could be suitable for degradation of organics or other photooxidation reactions.

3.1.4. $CoTiO_3$

$CoTiO_3$ has a band gap in the visible region (E_g 2.28 eV). Recently this compound has been investigated for photocatalytic O_2 evolution reaction without any co-catalyst [57]. A yield of 64 $\mu mol \cdot g^{-1} \cdot h^{-1}$ was obtained under visible light. $CoTiO_3$ and TiO_2 composites have been studied for 2-propanol mineralization, however it is worth noting that the band positions of $CoTiO_3$ are not suitable for transferring photoexcited charges to TiO_2, offering no significant advantage [58].

3.1.5. $NiTiO_3$

$NiTiO_3$ has a reported band gap of around 2.16 eV and its light absorption spectra show peaks in visible region corresponding to crystal field splitting [59]. $NiTiO_3$ nanorods have been employed for degradation of nitrobenzene under visible light [59].

3.1.6. $FeTiO_3$

$FeTiO_3$ has a band gap of 2.8 eV and thus it absorbs visible light. Composites of $FeTiO_3$ and TiO_2 are studied for the degradation of 2-propenol under visible light. In such composites, TiO_2 acts as hole capturing phase, thereby separating the electron-hole recombination [60].

3.1.7. CdTiO₃

CdTiO$_3$ (E$_g$ 2.8 eV) nanofibers were synthesized and studied for the photodegradation of rhodamine 6G (R6G) dye [61].

3.1.8. PbTiO₃

PbTiO$_3$ has a band gap of 2.75 eV and has been investigated for visible light photocatalysis. Core-shell particles of nano-TiO$_2$ (shell) and micro-PbTiO$_3$ (core) were studied for effective charge separation and photocatalytic performance [62]. It is worth noting that PbTiO$_3$ is ferroelectric in nature while TiO$_2$ is dielectric. It is shown that the ferroelectric behavior helps electron-hole separation at the interface of two particles.

3.1.9. MnTiO₃

Further, F doped MnTiO$_3$ shows improved separation of charges and degrades rhodamine B under visible light [63]. Some of the promising titanates photocatalysts have been listed in Table 1, as a part of compilation of perovskite systems.

Table 1. Compilation of promising photocatalytic systems for hydrogen or oxygen evolution under visible light.

Material System	Irradiation (nm)	Photocatalytic Performance	Experimental Details	Ref.
1% Rh doped SrTiO$_3$ (0.5% Pt)	420–800	H$_2$ at 48.1 μmol·h^{-1} with sacrificial agent	20% methanol, 50 mg in 50 mL of solution	[64]
Rh: SrTiO$_3$: BiVO$_4$	>420	Z scheme Water splitting. H$_2$ at 128, O$_2$ at 61 μmol·h^{-1}	4.2% Efficiency, 50 mg 120 mL (FeCl$_3$ shuttle)	[46]
Cr-Sb co-doped SrTiO$_3$, (0.3% Pt)	>420	H$_2$ at 78, O$_2$ at 0.9 μmol·h^{-1} with sacrificial agents	in aqueous methanol and AgNO$_3$ solution	[65]
MCo$_{1/3}$Nb$_{2/3}$O$_3$ (0.2% Pt)	>420	H$_2$ at 1.4 μmol·h^{-1} with sacrificial agent	500 mg catalyst in 50 mL methanol, 220 mL water,	[66]
Sr$_{1-x}$NbO$_3$ (1% Pt)	>420	H$_2$ at 44.8 μmol·h^{-1} with sacrificial agent	0.025M oxalic acid, 0.1g catalyst in 200 mL,	[67]
AgNbO$_3$-SrTiO$_3$	>420	O$_2$ at 162 μmol·h^{-1} with sacrificial agent	0.5 g catalyst in 275 mL AgNO$_3$ solution,	[49]
LaFeO$_3$ (Pt co-catalyst)	400–700	H$_2$ at 3315 μmol·h^{-1} with sacrificial agent	H$_2$ = 3315, μmol·h^{-1}, 1 mg in 20 mL of ethanol	[68]
CaTi$_{1-x}$Cu$_x$O$_3$ (x = 0.02), NiO$_x$ co-catalyst	>400	H$_2$ at 22.7 μmol·h^{-1} with sacrificial agent	100 mg catalyst in 420 mL methanol solution	[53]
PrFeO$_3$, (Pt co-catalyst)	200W Tungsten source	H$_2$ at 2847 μmol·h^{-1} with sacrificial agent	1 mg in 20 mL ethanol solution	[69]
Bi doped NaTaO$_3$	>400	H$_2$ at 59.48 μmol·h^{-1} with sacrificial agent	100 mg catalyst in 210 mL of methanol solution	[70]
GdCrO$_3$—Gd$_2$Ti$_2$O$_7$ composite	>420	H$_2$ at 246.3 μmol·h^{-1} with sacrificial agent	4.1% apparent quantum efficiency, methanol solution	[71]
CoTiO$_3$	>420	O$_2$ at 64.6 μmol·h^{-1} with sacrificial agent	100 mg in 100 mL 0.04M AgNO$_3$ and La$_2$O$_3$ solution, 420 nm	[57]

3.2. Tantalate Perovskites

Alkali tantalates are particularly known for efficient overall water splitting reaction under UV irradiation as they possess both VB and CB potentials suitable for water splitting reaction [72–74]. To enable visible light photocatalysis, doping of various elements has been studied to achieve visible light activity.

3.2.1. NaTaO₃

Our group reported a detailed study on Bi-doped NaTaO₃ and showed that the bismuth doping site significantly affects the photocatalytic activity for hydrogen evolution [75,76]. Further, co-doping of La-Co, La-Cr, La-Ir, La-Fe in NaTaO₃ have shown successful visible light absorption and subsequent hydrogen evolution [77–81]. Co-doping of La-N in NaTaO₃ has been studied for hydrogen evolution by Zhao *et al.* [82]. These studies have indicated that both anion and cation doping in NaTaO₃ is useful for visible light photocatalytic applications. Among the doped NaTaO₃ systems, computational studies on the anionic (N, F, P, Cl, S) doping were reported by Han *et al.* which shows that certain anions like N and P may be useful for visible light absorption [83]. Additionally, doping of magnetic cations such as Mn, Fe, and Co in NaTaO₃ has also been studied using DFT-PBE [84]. Recently, our group studied DFT calculations of a number of doped NaTaO₃-based photocatalysts by PBE0 hybrid calculations (Figure 9) [85]. Further, anion doping was also explored in detail using (HSE06) hybrid DFT calculations, where N, P, C, and S doping at O sites were studied. The study also reports the thermodynamics and effect of coupling between N-N, C-S, and P-P on the optical and electronic properties [86]. DFT studies are useful in explaining the properties of existing materials systems and designing new materials. Particularly, use of hybrid functional such as PBE0 or HSE06, is able to accurately define the valence band structure and location of bands or energy states that are crucially important for visible light driven photocatalysis. Hybrid DFT calculations could be useful in predictive modeling, where, band gaps and band edge potential of useful doped photocatalysts are identified. An example of doped tantalate systems is shown in Figure 9.

Figure 9. Estimated band gaps and band edge potentials of doped and co-doped NaTaO₃ systems: DFT study to design novel photocatalyst [85]

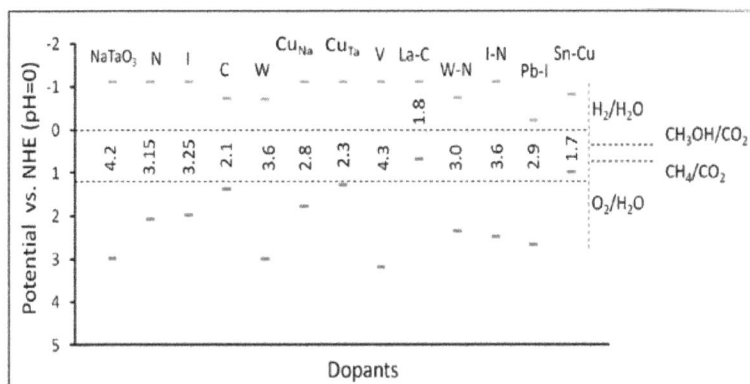

3.2.2. AgTaO₃

AgTaO₃ exhibits similar behavior to alkali tantalates, however, it has a smaller band gap value of 3.4 eV. AgTaO₃ doped with 30% Nb absorbs visible radiation and shows a stoichiometric overall water splitting reaction under visible light when loaded with NiO co-catalyst [87]. Co-doping of N-H and N-F in AgTaO₃ has been studied in detail. The study indicates that co-doping not only balances the charges but also improves the carrier mobility. N-F co-doped AgTaO₃ has an effective band gap value of 2.9 eV and shows H₂ generation under visible light [88].

3.2.3. KTaO₃

KTaO₃ (E_g 3.6 eV) photocatalysts have been studied for water splitting under UV radiation. However, work on development of visible light driven KTaO₃ based photocatalysts is limited.

3.3. Vanadium and Niobium Based Perovskites

Similar to tantalum (Ta)-based photocatalysts, Niobium (Nb)-based photocatalysts show good photocatalytic activity under UV irradiation.

3.3.1. KNbO₃ and NaNbO₃

Both KNbO₃ (E_g 3.14 eV) and NaNbO₃ (E_g 3.08 eV) have band gap values in the UV-responsive region, however suitable modifications of the band structure have resulted in visible light photocatalysis [89]. N-doped NaNbO₃ is a known visible light photocatalyst for the degradation of 2-propanol [90]. Nitrogen doping in KNbO₃ has been studied for water splitting as well as organic pollutant degradation [91]. First principles calculations predict that co-doping of La and Bi would induce visible light response in NaNbO₃ [92]. Recent work on ferroelectric perovskites of KNbO₃-BaNiNbO₃ shows that the solid solution of these compounds could absorb *six* times more light and shows *fifty* times more photocurrent than others [93]. Although photocatalytic properties are not known, this is an attractive candidate for visible light driven photocatalyst.

3.3.2. AgNbO₃

Replacing an A site alkali metal by silver reduces the band gap of the perovskite and AgNbO₃ has a band gap of around 2.7 eV. Studies have shown that the photocatalytic activity of AgNbO₃ strongly depends on the shape of the particles: polyhedron-shaped particles are favorable for O₂ evolution reactions [94]. Further, La doping was found to enhanced the hotocatalytic performance by 12-fold for gaseous 2-propenol degradation [95].

3.3.3. AgVO₃

AgVO₃ exists in two types of crystal structures, viz. α-AgVO₃ (E_g 2.5 eV) and β-AgVO₃ (E_g 2.3 eV) [96]. Both phases are photocatalytically active. However, β-AgVO₃ shows better photocatalytic performance than the α-phase. The CB potential of AgVO₃ is not sufficient for H₂ evolution, but it is suitable for the degradation of volatile organic compounds (VOCs) and O_2 evolution. β-$AgVO_3$ nanowires show excellent photocatalytic performance in the degradation of Rh B [97]. Composites of $AgBr$-$AgVO_3$ were reported to display respectable efficiency for Rh B degradation [98], while Ag-loaded $AgVO_3$ has shown good performance for degradation of bisphenol [99].

3.3.4. $CuNbO_3$

$CuNbO_3$ crystallizes in the monoclinic structure and has a band gap of 2.0 eV. It is an intrinsic p-type semiconductor and has shown 5% efficiency for photon to electron conversion when used as a photocathode. Being a stable material under irradiation, more investigations should be carried out on this material [100]. Tantalum, niobium, and vanadium belong to the same group in the periodic table. Perovskite compounds of these elements show decreasing band gap and CB potential values. This trend is attributed to the 3 d, 4 d and 5 d orbital energies in V, Nb and Ta, respectively.

3.4. Ferrite Perovskites

Most of the ferrite perovskites have their native band gaps in the visible region. Hematite and other iron oxide compounds have known shortcomings such as short exciton diffusion length, low electron conductivity, and lower conduction band edge potential [101]. However certain ferrite-based perovskites have shown good photocatalytic activities, circumventing the shortcomings seen in binary iron oxides.

3.4.1. $LaFeO_3$

$LaFeO_3$ (E_g 2.1 eV) has been explored for degradation of pollutants as well as hydrogen evolution under visible light. Sol-gel synthesized $LaFeO_3$ loaded with Pt co-catalyst showed high yield of hydrogen evolution (3,315 $\mu mol \cdot h^{-1} \cdot g^{-1}$, in the presence of ethanol) under 400 W tungsten light source [68]. Another study on this phase demonstrates high yield of H_2 and O_2 (1290 μmol and 640 μmol after three hours, respectively), without any co-catalyst loading [102]. Further, Thirumalairajan et al. showed shape dependent photocatalytic activity of $LaFeO_3$ for Rh B dye under visible light (>400 nm) [103]. Doping of Mn in $LaFeO_3$ has also been studied and it shows higher photocatalytic activity [104]. Lanthanum ferrite has demonstrated excellent photocatalytic activity under visible irradiation; however, studies on the fundamental photophysical properties are lacking the literature. Understanding the reasons why $LaFeO_3$ is a better photocatalyst that Fe_2O_3, in terms of comparative electronic properties such as electron-hole separation, mobility, photoexcited lifetimes etc. is important for further development in ferrite-based photocatalyst.

3.4.2. $BiFeO_3$

$BiFeO_3$ is a known ferroelectric material with a band gap of 2.3 eV. Recent studies have shown that $BiFeO_3$ could be used as a visible light photocatalyst [105]. The ferroelectric properties of $BiFeO_3$ could be utilized to enhance the electron-hole separation and improve the photocatalytic activity. Figure 10 shows the effect of polarization on the band bending of BiFeO₃ particles. Band bending affects the separation of electrons and holes in the space charge region and thus affects the photocatalytic activity. Gd-doped BiFeO₃ show enhanced photocatalytic degradation for rhodamine B degradation attributed to its ferromagnetic behavior [106]. Ca doping in BiFeO₃ leads to improved performance for photocatalytic degradation of Congo Red dye [107].

In another study, cations such as Y, Mg and Al were doped in BiFeO₃ and their photocatalytic performance was evaluated by degradation of Rh B under 400 nm radiation. The degradation performance was limited to C/C₀ of 0.8 within three hours [108]. The study of the effects of ferroelectric behavior on photocatalytic performance is a relatively new topic and it generally shows impressive activity for the degradation of organic pollutants, however, more experimental evidence and understanding are needed to establish a correlation between ferroelectric behavior and photocatalytic activity. Further studies on the stability and toxicity of bismuth-based materials should be carried out for realizing their practical applications. It is further noted that the tilting of octahedra in the perovskite crystal structure significantly affects its electronic properties. Nevertheless, the effect of tilting of octahedra on the photophysical properties such as electron-hole separation, electron transport, delocalization of charges has not been studied in detail. Such studies will prove useful in establishing the importance of perovskites in the field of photocatalysis.

Figure 10. Schematics of changes to band diagram upon polarization of Gd doped BiFeO₃ due to ferroelectricity [106].

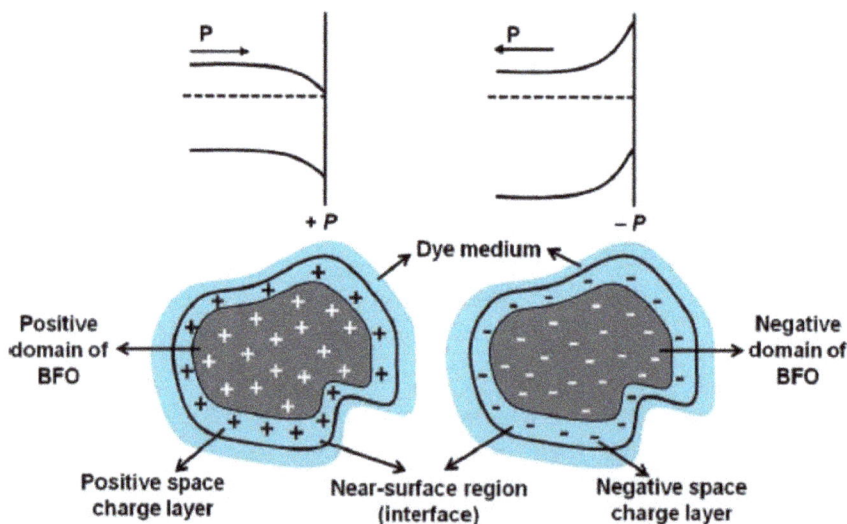

3.4.3. GaFeO₃

GaFeO₃ has been reported to show overall water splitting without any co-catalyst loading under visible light ($\lambda > 395$ nm) [109]. The authors also reported a yield of 0.10 and 0.04 $\mu\text{mol}\cdot\text{h}^{-1}$ under a 450 nm band pass filter. It is worth noting that the catalytic activity decreased due to deactivation of the catalyst.

3.4.4. YFeO₃

Among the other ferrites, YFeO₃ has a band gap value of 2.43 eV and showed photocatalytic activity *four* times that of TiO₂-P25 under > 400 nm visible light radiation (Rh B degradation) [110]

3.4.5. PrFeO₃

PrFeO₃ was evaluated for hydrogen evolution reaction from ethanol-water mixture and showed a yield of 2847 $\mu mol \cdot g^{-1} \cdot h^{-1}$ under 200W tungsten lamp irradiation [69].

3.4.6. AlFeO₃

Composites of AlFeO₃ and TiO₂ are reported to yield the sunlight driven photocatalytic degradation of methyl orange (MO) and eosin dye [111]. Ferrite based perovskites also offer advantage of magnetic recovery of the particles which is useful in practical applications.

3.5. Other Perovskite Systems

Among the other perovskite photocatalysts, compounds of bismuth, cobalt, nickel, and antimony (occupying B sites) have band gap values in the visible region. Pentavalent bismuth perovskites are known to be active photocatalysts under visible light radiation. Perovskites such as LiBiO₃ (E_g 1.63 eV), NaBiO₃ (E_g 2.53 eV), KBiO₃ (E_g 2.04 eV), and AgBiO₃ (E_g 2.5 eV) have all been investigated for degradation of organic pollutants [112]. NaBiO₃ shows better photocatalytic performance for phenol and methylene blue (MB) degradation as compared to other Bi^{5+} containing perovskites as well as P25-TiO₂ [112]. NaBiO₃ is also reported to have higher photocatalytic activity than N-doped TiO₂ photocatalyst [113]. NaBiO₃ and BiOCl composites have been studied for enhanced electron-hole separation and subsequent photocatalysis [114]. AgBiO₃ was shown to be effective in restricting the growth of *Microcystis aeruginosa* under simulated solar light [115]. This study shows that AgBiO₃ with band gap energy around 2.5 eV could be a potential algaecide under natural light. DFT studies show that NaBiO₃ has a strong band dispersion arising from Na 3s and O 2p hybridized orbitals, which contributes to the higher photocatalytic activity [112]. LaCoO₃ has a band gap value of 2.7 eV and oxygen deficient LaCoO₃ (LaCoO₃₋δ) has been studied for MO degradation (>400 nm) [116].

Other compounds such as LaNiO₃ has been studied for MO degradation using wavelengths greater than 400 nm [117]. Cu doping in LaNiO₃ has been studied for improved H₂ evolution [118]. Ilmenite type AgSbO₃ has an absorption edge onset at 480 nm, and it has been demonstrated for O₂ evolution under sacrificial agent as well as degradation of MB, RhB, and 4-chlorophenol [119]. Further, solid solution of LaCrO₃ and Na₀.₅La₀.₅TiO₃ was developed for hydrogen evolution [120]. Recently, Gd₂Ti₂O₇/GdCrO₃ composite was reported as photocatalyst p-n junction photocatalyst [71]. The study shows that GdCrO₃ has a band gap of 2.5 eV and is responsible for the visible light absorption [72].

4. Complex Perovskite Materials

4.1. Double Perovskites

Compounds with general formula $A_2B_2O_6$ belong to the double perovskites and they have similar crystal structures to simple perovskites. Double perovskites have the basic framework of corner connected BO_6 octahedra and A cations enclosed within, however, the connectivity of the octahedra may differ from structure to structure. Double perovskites could accommodate different cations at the A or/and B sites, taking a general form $AA'BB'O_6$. Accommodation of different cations at the A and B sites alters the photophysical properties of the compound significantly. Among the binary oxides, only a few compounds are known to have band gap values in the visible region (narrow gap), e.g., Fe_2O_3, WO_6, CuO, Bi_2O_3. However, these materials suffer from insufficient CB potential for hydrogen evolution. Some of the materials also suffer from poor stability and low mobility of photoexcited charges. On the other hand, many of the binary oxides are known to be efficient photocatalysts, however only activated by UV radiation (wide gap). Complex compounds offer a possibility of combining the elements from 'narrow gap' and 'wide gap' binary compounds to exploit the properties of both types of oxides, and thus are potentially useful as visible light photocatalysts. The following section reviews photocatalytic properties of double perovskites which are visible light active.

4.1.1. Sr_2FeNbO_6

Sr_2FeNbO_6 has a cubic crystal structure and a band gap of 2.06 eV. In its pristine form, it is a known photocatalyst, while 7% Ti doping at Fe site has shown two time enhancement of the hydrogen generation in methanol solution. A total of 28.5 $\mu mol \cdot h^{-1}$ and 650 $\mu mol \cdot h^{-1}$ were reported in presence of sacrificial agents and with 0.2% Pt as co-catalyst [121]. W doping in Sr_2FeNbO_6 has also been studied and demonstrated enhancement to hydrogen evolution under visible radiation [122].

4.1.2. La_2FeTiO_6

Hu et al. reported higher photocatalytic activity for La_2FeTiO_6 than for $LaFeO_3$ for degradation of p-chlorophenol under visible light irradiation [123].

4.1.3. Other Double Perovskites

Rare earth and bismuth-based double perovskites were studied for visible light photocatalysis. Compounds with general formula Ba_2RBiO_6, (R = La, Ce, Pr, Nd, Sm, Eu, Gd, Dy), were prepared and their photocatalytic activity was studied for MB degradation [124]. Authors found rare earth cation dependent photocatalytic performance, where compounds such as Ba_2EuBiO_6, Ba_2SmBiO_6, and Ba_2CeBiO_6 showed high photocatalytic activity. Complex perovskites such as $CaCu_3Ti_4O_{12}$ have also been studied for their photocatalytic performance. $CaCu_3Ti_4O_{12}$ was found to possess an indirect band gap of 1.27 eV and Pt-loaded $CaCu_3Ti_4O_{12}$ shows degradation of MO under radiation greater than 420 nm [125]. Photophysical properties of certain double perovskite compounds have been reported, however, more efforts are needed to investigate photocatalytic properties of these materials. Compounds such as Ba_2CoWO_6, Ba_2NiWO_6, Sr_2CoWO_6 and Sr_2NiWO_6 are stable perovskites compounds for O_2 evolution using sacrificial agents [126].

Certain tantalum-based compounds have been studied for degradation of organic compounds. N-doped $K_2Ta_2O_6$ is known to absorb visible light up to 600 nm and degrade formaldehyde under visible light [127]. Our group demonstrated that Bi doping in $Na_2Ta_2O_6$ causes visible light absorption and degradation of MB [128]. Although some work has been done with double perovskites as photocatalysts, understanding of their fundamental properties is limited. More work is needed to discover their advantages as visible light photocatalyst and develop novel materials systems.

4.2. Mixed Oxides

Mixture of oxides and nitrides or oxides and sulphides has been developed to engineer the band structure of the oxide photocatalysts suitable for the visible light absorption. Oxynitride and oxysulphide photocatalysts offer distinctive advantages over their doped counterparts. Earlier reports on oxynitrides have revealed that replacing oxygen by nitrogen at lattice sites, in a stoichiometric manner, narrows the band gap of the oxide, by pushing the VBM into the band gap [129]. Such a modification does not produce native point defects, which would otherwise be introduced in the case of N doping. Stoichiometric incorporation of nitrogen also avoids the localized states induced by N doping and reduces the possible electron-hole recombination. Similar composition containing mixtures of sulfur and oxygen *i.e.*, oxysulfides, has been developed for photocatalysts. Mixed oxysulfide perovskites of $Sm_2Ti_2S_2O_5$ (E_g 2.0 eV) is known for water oxidation and reduction reaction for oxygen and hydrogen evolution, respectively in presence of sacrificial agents under low photon energy wavelengths of 650 nm. The band structure of this phase reveals that the presence of sulphur narrows the band gap and enables visible light absorption [130].

Oxynitride compounds such as $CaNbO_2N$, $SrNbO_2N$, $BaNbO_2N$, and $LaNbON_2$ belong to the perovskite type crystal structures [131]. Photocatalytic hydrogen evolution has been reported under visible light from methanol solution. $SrNbO_2N$ (E_g 1.8 eV) has been investigated in detail, where the photoelectrode of $SrNbO_2N$ on a transparent conducting surface shows water oxidation reaction under no external bias [132]. Tantalum counterparts of these compounds were developed and utilized in the Z scheme photocatalysis. Compounds such as $CaTaO_2N$ and $BaTaO_2N$ were loaded with Pt co-catalyst and coupled with pt/WO_3 for Z scheme water splitting [133]. A solid solution of $BaTaO_2N$ and $BaZrO_3$ was formulated for hydrogen and oxygen evolution, which showed improved performance compared with the individual photocatalysts under visible light [134]. $LaTiO_xN_y$ is another perovskite type compound which shows high photocurrent density under visible light [135].

Apart from the double perovskites belonging to the general formula $AA'BB'O_6$, there are a several other compounds that show crystal structures close to the perovskite type structure, however such compounds are not included in the current review. Theoretically, double perovskites offer a wider scope to design photocatalysts by selecting suitable cations and AA' and BB' sites in the lattice. Work on design and development of double perovskite is currently limited and synthesis and characterization of new materials in this category are needed.

5. Summary and Outlook

A large number of perovskite-based compounds (over 80) have been studied for visible light driven photocatalytic applications. Perovskite structures offer abundant scope in designing novel compounds based on A and B site occupancy, which gives rise to a wide range of materials systems with unique properties. Among these compounds, photocatalysts with pristine band gaps in the visible region such as $LaFeO_3$, $PrFeO_3$, $NaBiO_3$, and $AgBiO_3$ (Table 2) show promising photocatalytic performance under visible radiation (>400 nm). Although there are reports on the photocatalytic activity of these compounds, detailed studies on these materials are limited. More efforts are needed to understand the structure-property relations in such ferrites and bismuth based compounds and improve their photocatalytic activity. Among the wide band gap semiconductors, $SrTiO_3$- and $NaTaO_3$-based photocatalysts are the most investigated systems. The strategy of doping foreign elements in wide band gap photocatalysts is widely used to induce visible light absorption and subsequently to enable photocatalytic activity. Nonetheless, very limited knowledge is available on the adverse effect of dopants on the photophysical properties of the compounds (e.g., if the benefit derived from dopant-induced visible light activity is overweighed by a loss of UV light activity). This should be properly investigated in the future. Appropriate dopants which retain the beneficial properties of the host materials while inducing visible light responses should be identified. Research work on complex perovskites show that these compounds offer distinct advantages as compared with simple perovskites. Designing complex perovskite compounds with suitable elements at A and B sites to yield desired photocatalytic properties is a challenge. We expect computational design would help shorten the selection process. Recent advances in computational tools such as DFT based band structure calculations are highly effective to design and understand novel materials systems.

Table 2. Compilation of promising photocatalytic systems for organic compounds degradation under visible light.

Materials System	Band Gap (eV)	Photocatalytic Tests Reported	Ref.
Ga doped $BiFeO_3$	2.18–2.50	Enhanced degradation of rhodamine B compared to pristine $BiFeO_3$	[106]
$LaFeO_3$	2.10	Nanospheres show higher rates of rhodamine B degradation than nanocubes and nanorods	[103]
$YFeO_3$	2.43	Rhodamine B degradation rate higher than P25 (>400 nm)	[110]
$NaBiO_3$	2.60	Bleaching rate of Methylene Blue higher than N doped TiO_2. (>400 nm)	[113]
$AgSbO_3$	2.58	Eddicient degradation of Rh B. MB, 4-chlorophenol (>420 nm)	[119,136]
$AgBiO_3$	2.50	Inhibition of *Microcystis*	[115]

As certain perovskite compounds exhibit ferroelectric, ferromagnetic, or piezoelectric effects, there is a need to understand the correlation between these effects and the photocatalytic activity to a greater depth. Such studies would certainly be helpful in the development of efficient visible light photocatalysts. On a final note, there has been significant progress in the development of visible light perovskites in the past years. This development has laid a good foundation for future work in this area. Further understanding of the crystal and electronic structural factors behind photocatalytic activity is needed for the future development of efficient visible light-driven perovskites.

Acknowledgments

Financial support through an SUG grant from the College of Engineering, Nanyang Technological University, Singapore is greatly appreciated.

Author Contributions

P.K. performed literature research, analysis, and drafted the paper. C.Z. initiated and supervised the work and provided insights. Both authors revised and approved the final manuscript.

Conflicts of Interest

The authors declare no conflict of interest.

References

1. Maeda, K. Photocatalytic water splitting using semiconductor particles: History and recent developments. *J. Photchem. Photobiol. C* **2011**, *12*, 237–268.
2. Qu, Y. Duan, X. Progress, challenge and perspective of heterogeneous photocatalysts. *Chem. Soc. Rev.* **2013**, *42*, 2568–2580.
3. Hou, W.; Cronin, S.B. A Review of Surface Plasmon Resonance-Enhanced Photocatalysis. *Adv. Funct. Mat.* **2013**, *23*, 1612–1619.
4. Osterloh, F.E. Inorganic nanostructures for photoelectrochemical and photocatalytic water splitting. *Chem. Soc. Rev.* **2013**, *42*, 2294–2320.
5. Cook, T.R.; Dogutan, D.K.; Reece, S.Y.; Surendranath, Y.; Teets, T.S.; Nocera, D.G. Solar energy supply and storage for the legacy and nonlegacy worlds. *Chem. Rev.* **2010**, *110*, 6474–6502.
6. Izumi, Y. Recent advances in the photocatalytic conversion of carbon dioxide to fuels with water and/or hydrogen using solar energy and beyond. *Coord. Chem. Rev.* **2013**, *257*, 171–186.
7. Tahir, M.; Amin, N.S. Recycling of carbon dioxide to renewable fuels by photocatalysis: Prospects and challenges. *Renew. Sust. Energy Rev.* **2013**, *25*, 560–579.
8. Tan, Y.N.; Wong, C.L.; Mohamed, A.R. An Overview on the Photocatalytic Activity of Nano-Doped-TiO2 in the Degradation of Organic Pollutants. *ISRN Mat. Sci.* **2011**, *2011*, 18.
9. Pelaez, M.; Nolan, N.T.; Pillai, S.C.; Seery, M.K.; Falaras, P.; Kontos, A.G.; Dunlop, P.S.M.; Hamilton, J.W.J.; Byrne, J.A.; O'shea, K.; *et al*. A review on the visible light active titanium dioxide photocatalysts for environmental applications. *Appl. Catal. B* **2012**, *125*, 331–349.
10. Dalrymple, O.K.; Stefanakos, E.; Trotz, M.A. Goswami, D.Y. A review of the mechanisms and modeling of photocatalytic disinfection. *Appl. Catal. B* **2010**, *98*, 27–38.

11. Zhang, Z.; Gamage, J. Applications of photocatalytic disinfection. *Int. J. Photoenergy* **2010**, doi:10.1155/2010/764870.

12. Konstantinou, I.K.; Albanis, T.A. TiO$_2$-assisted photocatalytic degradation of azo dyes in aqueous solution: Kinetic and mechanistic investigations: A review. *Appl. Catal. B* **2004**, *49*, 1–14.

13. Ahmed, S.; Rasul, M.G.; Martens, W.; Brown, R.; Hashib, M.A. Advances in Heterogeneous Photocatalytic Degradation of Phenols and Dyes in Wastewater: A Review. *Water Air Soil Pollut.* **2011**, *215*, 3–29.

14. Rajesh, J.T.; Praveen, K.S.; Ramchandra, G.K.; Raksh, V.J. Photocatalytic degradation of dyes and organic contaminants in water using nanocrystalline anatase and rutile TiO$_2$. *Sci. Technol. Adv. Mat.* **2007**, *8*, 455.

15. Bickley, R.I.; Vishwanathan, V. Photocatalytically induced fixation of molecular nitrogen by near UV radiation. *Nature* **1979**, *280*, 306–308.

16. Guo, C.; Wu, X.; Yan, M.; Dong, Q.; Yin, S.; Sato, T.; Liu, S. The visible-light driven photocatalytic destruction of NOx using mesoporous TiO2 spheres synthesized via a "water-controlled release process". *Nanoscale* **2013**, *5*, 8184–8191.

17. Zhao, J.; Yang, X. Photocatalytic oxidation for indoor air purification: A literature review. *Build. Environ.* **2003**, *38*, 645–654.

18. Lasek, J.; Yu, Y.-H.; Wu, J.C.S. Removal of NOx by photocatalytic processes. *J. Photchem. Photobiol. C* **2013**, *14*, 29–52.

19. Wang, H.; Wu, Z.; Zhao, W.; Guan, B. Photocatalytic oxidation of nitrogen oxides using TiO$_2$ loading on woven glass fabric. *Chemosphere* **2007**, *66*, 185–190.

20. Bhalla, A.S.; Guo, R.; Roy, R. The perovskite structure–a review of its role in ceramic science and technology. *Mat. Res. Innov.* **2000**, *4*, 3–26.

21. Damjanovic, D. Piezoelectric properties of perovskite ferroelectrics: unsolved problems and future research. *Ann. Chim.-Sci. Mat.* **2001**, *26*, 99–106.

22. Nuraje, N.; Su, K. Perovskite ferroelectric nanomaterials. *Nanoscale* **2013**, *5*, 8752–8780.

23. Jia, Q.; Iwase, A.; Kudo, A. BiVO$_4$-Ru/SrTiO$_3$:Rh composite Z-scheme photocatalyst for solar water splitting. *Chem. Sci.* **2014**, *5*, 1513–1519.

24. Sayama, K.; Mukasa, K.; Abe, R.; Abe, Y.; Arakawa, H. Stoichiometric water splitting into H$_2$ and O$_2$ using a mixture of two different photocatalysts and an IO$_3^-$/I$^-$ shuttle redox mediator under visible light irradiation. *Chem. Commun.* **2001**, *23*, 2416–2417.

25. Zhang, W.F.; Tang, J.; Ye, J. Photoluminescence and photocatalytic properties of SrSnO$_3$ perovskite. *Chem. Phys. Lett.* **2006**, *418*, 174–178.

26. Lin, W.H.; Cheng, C.; Hu, C.C.; Teng, H. NaTaO$_3$ photocatalysts of different crystalline structures for water splitting into H$_2$ and O$_2$. *Appl. Phys. Lett.* **2006**, *89*, 211904

27. Shi, J.; Guo, L. ABO$_3$-based photocatalysts for water splitting. *Prog. Nat. Sci. Mat. Int.* **2012**, *22*, 592–615.

28. Wang, W.N.; Soulis, J.; Jeffrey Yang, Y.; Biswas, P. Comparison of CO$_2$ photoreduction systems: A review. *Aerosol Air Qual. Res.* **2014**, *14*, 533–549.

29. Tang, J.; Durrant, J.R.; Klug, D.R. Mechanism of Photocatalytic Water Splitting in TiO_2. Reaction of Water with Photoholes, Importance of Charge Carrier Dynamics, and Evidence for Four-Hole Chemistry. *J. Am. Chem. Soc.* **2008**, *130*, 13885–13891.

30. Kudo, A.; Miseki, Y. Heterogeneous photocatalyst materials for water splitting. *Chem. Soc. Rev.* **2009**, *38*, 253–278.

31. Houas, A.; Lachheb, H.; Ksibi, M.; Elaloui, E.; Guillard, C.; Herrmann, J.-M. Photocatalytic degradation pathway of methylene blue in water. *Appl. Catal. B* **2001**, *31*, 145–157.

32. Cho, M.; Chung, H.; Choi, W.; Yoon, J. Linear correlation between inactivation of E. coli and OH radical concentration in TiO_2 photocatalytic disinfection. *Water Res.* **2004**, *38*, 1069–1077.

33. Turchi, C.S.; Ollis, D.F. Photocatalytic degradation of organic water contaminants: Mechanisms involving hydroxyl radical attack. *J. Catal.* **1990**, *122*, 178–192.

34. Habisreutinger, S.N.; Schmidt-Mende, L.; Stolarczyk, J.K. Photocatalytic reduction of CO_2 on TiO_2 and other semiconductors. *Angew. Chem. Int. Ed.* **2013**, *52*, 7372–7408.

35. Rusina, O.; Linnik, O.; Eremenko, A.; Kisch, H. Nitrogen Photofixation on Nanostructured Iron Titanate Films. *Chem. Eur. J.* **2003**, *9*, 561–565.

36. Rusina, O.; Macyk, W.; Kisch, H. Photoelectrochemical Properties of a Dinitrogen-Fixing Iron Titanate Thin Film. *J. Phys. Chem. B* **2005**, *109*, 10858–10862.

37. Zhu, D.; Zhang, L.; Ruther, R.E.; Hamers, R.J. Photo-illuminated diamond as a solid-state source of solvated electrons in water for nitrogen reduction. *Nat. Mater.* **2013**, *12*, 836–841.

38. Kavan, L.; Grätzel, M.; Gilbert, S.; Klemenz, C.; Scheel, H. Electrochemical and photoelectrochemical investigation of single-crystal anatase. *J. Am. Chem. Soc.* **1996**, *118*, 6716–6723.

39. Van Benthem, K.; Elsässer, C.; French, R.H. Bulk electronic structure of $SrTiO_3$: Experiment and theory. *J. Appl. Phys.* **2001**, *90*, 6156–6164.

40. Townsend, T.K.; Browning, N.D.; Osterloh, F.E. Overall photocatalytic water splitting with $NiOx$-$SrTiO_3$—A revised mechanism. *Energy Environ. Sci.* **2012**, *5*, 9543–9550.

41. Konta, R.; Ishii, T.; Kato, H.; Kudo, A. Photocatalytic Activities of Noble Metal Ion Doped $SrTiO_3$ under Visible Light Irradiation. *J. Phys. Chem. B* **2004**, *108*, 8992–8995.

42. Chen, H.-C.; Huang, C.-W.; Wu, J.C.S.; Lin, S.-T. Theoretical Investigation of the Metal-Doped $SrTiO_3$ Photocatalysts for Water Splitting. *J. Phys. Chem. C* **2012**, *116*, 7897–7903.

43. Iwashina, K.; Kudo, A. Rh-Doped $SrTiO_3$ Photocatalyst Electrode Showing Cathodic Photocurrent for Water Splitting under Visible-Light Irradiation. *J. Am. Chem. Soc.* **2011**, *133*, 13272–13275.

44. Sasaki, Y.; Kato, H.; Kudo, A. $[Co(bpy)_3]^{3+/2+}$ and $[Co(phen)_3]^{3+/2+}$ Electron Mediators for Overall Water Splitting under Sunlight Irradiation Using Z-Scheme Photocatalyst System. *J. Am. Chem. Soc.* **2013**, *135*, 5441–5449.

45. Sasaki, Y.; Nemoto, H.; Saito, K.; Kudo, A. Solar Water Splitting Using Powdered Photocatalysts Driven by Z-Schematic Interparticle Electron Transfer without an Electron Mediator. *J. Phys. Chem. C* **2009**, *113*, 17536–17542.

46. Kato, H.; Sasaki, Y.; Shirakura, N.; Kudo, A. Synthesis of highly active rhodium-doped SrTiO₃ powders in Z-scheme systems for visible-light-driven photocatalytic overall water splitting. *J. Mater. Chem. A* **2013**, *1*, 12327–12333.

47. Asai, R.; Nemoto, H.; Jia, Q.; Saito, K.; Iwase, A.; Kudo, A. A visible light responsive rhodium and antimony-codoped SrTiO₃ powdered photocatalyst loaded with an IrO₂ cocatalyst for solar water splitting. *Chem. Commun.* **2014**, *50*, 2543–2546.

48. Jia, Y.; Shen, S.; Wang, D.; Wang, X.; Shi, J.; Zhang, F.; Han, H.; Li, C. Composite Sr₂TiO₄/SrTiO₃(La,Cr) heterojunction based photocatalyst for hydrogen production under visible light irradiation. *J. Mater. Chem. A* **2013**, *1*, 7905–7912.

49. Wang, D.; Kako, T.; Ye, J. New Series of Solid-Solution Semiconductors (AgNbO₃)₁₋ₓ(SrTiO₃)ₓ with Modulated Band Structure and Enhanced Visible-Light Photocatalytic Activity. *J. Phys. Chem. C* **2009**, *113*, 3785–3792.

50. Fu, Q.; He, T.; Li, J.L.; Yang, G.W. Band-engineered SrTiO₃ nanowires for visible light photocatalysis. *J. Appl. Phys.* **2012**, *112*, 104322.

51. Wunderlich, W.; Ohta, H.; Koumoto, K. Enhanced effective mass in doped SrTiO₃ and related perovskites. *Physica B* **2009**, *404*, 2202–2212.

52. Maeda, K. Rhodium-Doped Barium Titanate Perovskite as a Stable p-Type Semiconductor Photocatalyst for Hydrogen Evolution under Visible Light. *ACS App. Mater. Interfaces* **2014**, *6*, 2167–2173.

53. Zhang, H.; Chen, G.; Li, Y.; Teng, Y. Electronic structure and photocatalytic properties of copper-doped CaTiO₃. *Int. J. Hydrog. Energy* **2010**, *35*, 2713–2716.

54. Zhang, H.; Chen, G.; He, X.; Xu, J. Electronic structure and photocatalytic properties of Ag–La codoped CaTiO₃. *J. Alloys Compd.* **2012**, *516*, 91–95.

55. Fu, Q.; Li, J.L.; He, T.; Yang, G.W. Band-engineered CaTiO₃ nanowires for visible light photocatalysis. *J. Appl. Phys.* **2013**, *113*, 104303.

56. Xu, Y.; Schoonen, M.A. The absolute energy positions of conduction and valence bands of selected semiconducting minerals. *Am. Miner.* **2000**, *85*, 543–556.

57. Qu, Y.; Zhou, W.; Fu, H. Porous Cobalt Titanate Nanorod: A New Candidate for Visible Light-Driven Photocatalytic Water Oxidation. *ChemCatChem* **2014**, *6*, 265–270.

58. Rawal, S.B.; Bera, S.; Lee, D.; Jang, D.-J.; Lee, W.I. Design of visible-light photocatalysts by coupling of narrow bandgap semiconductors and TiO₂: Effect of their relative energy band positions on the photocatalytic efficiency. *Catal. Sci. Technol.* **2013**, *3*, 1822–1830.

59. Qu, Y.; Zhou, W.; Ren, Z.; Du, S.; Meng, X.; Tian, G.; Pan, K.; Wang, G.; Fu, H. Facile preparation of porous NiTiO₃ nanorods with enhanced visible-light-driven photocatalytic performance. *J. Mater. Chem.* **2012**, *22*, 16471–16476.

60. Kim, Y.J.; Gao, B.; Han, S.Y.; Jung, M.H.; Chakraborty, A.K.; Ko, T.; Lee, C.; Lee, W.I. Heterojunction of FeTiO₃ Nanodisc and TiO₂ Nanoparticle for a Novel Visible Light Photocatalyst. *J. Phys. Chem. C* **2009**, *113*, 19179–19184.

61. Hassan, M.A.; Amna, T.; Khil, M.-S. Synthesis of High aspect ratio CdTiO₃ nanofibers via electrospinning: characterization and photocatalytic activity. *Ceram. Int.* **2014**, *40*, 423–427.

62. Li, L.; Zhang, Y.; Schultz, A.M.; Liu, X.; Salvador, P.A.; Rohrer, G.S. Visible light photochemical activity of heterostructured $PbTiO_3$-TiO_2 core-shell particles. *Catal. Sci. Technol.* **2012**, *2*, 1945–1952.

63. Dong, W.; Wang, D.; Jiang, L.; Zhu, H.; Huang, H.; Li, J.; Zhao, H.; Li, C.; Chen, B.; Deng, G. Synthesis of F doping $MnTiO_3$ nanodiscs and their photocatalytic property under visible light. *Mater. Lett.* **2013**, *98*, 265–268.

64. Shen, P.; Lofaro, J.C., Jr.; Woerner, W.R.; White, M.G.; Su, D.; Orlov, A. Photocatalytic activity of hydrogen evolution over Rh doped $SrTiO_3$ prepared by polymerizable complex method. *Chem. Eng. J.* **2013**, *223*, 200–208.

65. Kato, H.; Kudo, A. Visible-Light-Response and Photocatalytic Activities of TiO_2 and $SrTiO_3$ Photocatalysts Codoped with Antimony and Chromium. *J. Phys. Chem. B* **2002**, *106*, 5029–5034.

66. Yin, J.; Zou, Z.; Ye, J. A Novel Series of the New Visible-Light-Driven Photocatalysts $MCo_{1/3}Nb_{2/3}O_3$ (M = Ca, Sr, and Ba) with Special Electronic Structures. *J. Phys. Chem. B* **2003**, *107*, 4936–4941.

67. Xu, X.; Randorn, C.; Efstathiou, P.; Irvine, J.T.S. A red metallic oxide photocatalyst. *Nat. Mater.* **2012**, *11*, 595–598.

68. Tijare, S.N.; Joshi, M.V.; Padole, P.S.; Mangrulkar, P.A.; Rayalu, S.S.; Labhsetwar, N.K. Photocatalytic hydrogen generation through water splitting on nano-crystalline $LaFeO_3$ perovskite. *Int. J. Hydrog. Energy* **2012**, *37*, 10451–10456.

69. Tijare, S.N.; Bakardjieva, S.; Subrt, J.; Joshi, M.V.; Rayalu, S.S.; Hishita, S.; Labhsetwar, N. Synthesis and visible light photocatalytic activity of nanocrystalline $PrFeO_3$ perovskite for hydrogen generation in ethanol–water system. *J. Chem. Sci.* **2014**, *126*, 517–525.

70. Li, Z.; Wang, Y.; Liu, J.; Chen, G.; Li, Y.; Zhou, C. Photocatalytic hydrogen production from aqueous methanol solutions under visible light over $Na(Bi_xTa_{1-x})O_3$ solid-solution. *Int. J. Hydrog. Energy* **2009**, *34*, 147–152.

71. Parida, K.M.; Nashim, A.; Mahanta, S.K. Visible-light driven $Gd_2Ti_2O_7$/$GdCrO_3$ composite for hydrogen evolution. *J. Chem. Soc. Dalton Trans.* **2011**, *40*, 12839–12845.

72. Kato, H.; Asakura, K.; Kudo, A. Highly efficient water splitting into H_2 and O_2 over lanthanum-doped $NaTaO_3$ photocatalysts with high crystallinity and surface nanostructure. *J. Am. Chem. Soc.* **2003**, *125*, 3082–3089.

73. Yamakata, A.I.T.A.; Kato, H.; Kudo, A.; Onishi, H. Photodynamics of $NaTaO_3$ Catalysts for Efficient Water Splitting. *J. Phys. Chem. B* **2003**, *107*, 14383–14387.

74. Kato, H.; Kudo, A. Water splitting into H_2 and O_2 on alkali tantalate photocatalysts $ATaO_3$ (A = Li, Na, and K). *J. Phys. Chem. B* **2001**, *105*, 4285–4292.

75. Kanhere, P.D.; Zheng, J.; Chen, Z. Site Specific Optical and Photocatalytic Properties of Bi-Doped $NaTaO_3$. *J. Phys. Chem. C* **2011**, *115*, 11846–11853.

76. Kanhere, P.; Zheng, J.; Chen, Z. Visible light driven photocatalytic hydrogen evolution and photophysical properties of Bi^{3+} doped $NaTaO_3$. *Int. J. Hydrog. Energy* **2012**, *37*, 4889–4896.

77. Yi, Z.G.; Ye, J.H. Band gap tuning of $Na_{1-x}La_xTa_{1-x}Co_xO_3$ solid solutions for visible light photocatalysis. *App. Phys. Lett.* **2007**, *91*, 254108.

78. Yi, Z.G.; Ye, J.H. Band gap tuning of $Na_{1-x}La_xTa_{1-x}Cr_xO_3$ for H_2 generation from water under visible light irradiation. *J. App. Phys.* **2009**, *106*, 074910.

79. Yang, M.; Huang, X.; Yan, S.; Li, Z.; Yu, T.; Zou, Z. Improved hydrogen evolution activities under visible light irradiation over $NaTaO_3$ codoped with lanthanum and chromium. *Mater. Chem. Phys.* **2010**, *121*, 506–510.

80. Iwase, A.; Saito, K.; Kudo, A. Sensitization of $NaMO_3$ (M: Nb and Ta) photocatalysts with wide band gaps to visible light by Ir doping. *Bull. Chem. Soc. Jpn.* **2009**, *82*, 514–518.

81. Kanhere, P.; Nisar, J.; Tang, Y.; Pathak, B.; Ahuja, R.; Zheng, J.; Chen, Z. Electronic Structure, Optical Properties, and Photocatalytic Activities of $LaFeO_3$–$NaTaO_3$ Solid Solution. *J. Phys. Chem. C* **2012**, *116*, 22767–22773.

82. Zhao, Z.; Li, R.; Li, Z.; Zou, Z. Photocatalytic activity of La-N-codoped $NaTaO_3$ for H_2 evolution from water under visible-light irradiation. *J. Phys. D Appl. Phys.* **2011**, *44*, 165401.

83. Han, P.; Wang, X.; Zhao, Y.H.; Tang, C. Electronic structure and optical properties of non-metals (N, F, P, Cl,S)—Doped cubic $NaTaO_3$ by density functional theory. *Adv. Mater. Res.* **2009**, *79–82*, 1245–1248.

84. Zhou, X.; Shi, J.; Li, C. Effect of metal doping on electronic structure and visible light absorption of $SrTiO_3$ and $NaTaO_3$ (Metal = Mn, Fe, and Co). *J. Phys. Chem. C* **2011**, *115*, 8305–8311.

85. Kanhere, P.; Shenai, P.; Chakraborty, S.; Ahuja, R.; Zheng, J.; Chen, Z. Mono- and co-doped $NaTaO_3$ for visible light photocatalysis. *PCCP* **2014**, *16*, 16085–16094.

86. Wang, B.; Kanhere, P.; Chen, Z.; Nisar, J.; Pathak, B.; Ahuja, R. Anion-Doped $NaTaO_3$ for Visible Light Photocatalysis. *J. Phys. Chem. C* **2013**, *117*, 22518–22524.

87. Ni, L.; Tanabe, M.; Irie, H. A visible-light-induced overall water-splitting photocatalyst: conduction-band-controlled silver tantalate. *Chem. Commun.* **2013**, *49*, 10094–10096.

88. Li, M.; Zhang, J.; Dang, W.; Cushing, S.K.; Guo, D.; Wu, N.; Yin, P. Photocatalytic hydrogen generation enhanced by band gap narrowing and improved charge carrier mobility in $AgTaO_3$ by compensated co-doping. *PCCP* **2013**, *15*, 16220–16226.

89. Liu, J.W.; Chen, G.; Li, Z.H.; Zhang, Z.G. Hydrothermal synthesis and photocatalytic properties of $ATaO_3$ and $ANbO_3$ (A = Na and K). *Int. J. Hydrog. Energy* **2007**, *32*, 2269–2272.

90. Shi, H.; Li, X.; Iwai, H.; Zou, Z.; Ye, J. 2-Propanol photodegradation over nitrogen-doped $NaNbO_3$ powders under visible-light irradiation. *J. Phys. Chem. Solids* **2009**, *70*, 931–935.

91. Wang, R.; Zhu, Y.; Qiu, Y.; Leung, C.-F.; He, J.; Liu, G.; Lau, T.-C. Synthesis of nitrogen-doped $KNbO_3$ nanocubes with high photocatalytic activity for water splitting and degradation of organic pollutants under visible light. *Chem. Eng. J.* **2013**, *226*, 123–130.

92. Liu, G.; Ji, S.; Yin, L.; Xu, G.; Fei, G.; Ye, C. Visible-light-driven photocatalysts: (La/Bi + N)-codoped $NaNbO_3$ by first principles. *J. Appl. Phys.* **2011**, *109*, 063103.

93. Grinberg, I.; West, D.V.; Torres, M.; Gou, G.; Stein, D.M.; Wu, L.; Chen, G.; Gallo, E.M.; Akbashev, A.R.; Davies, P.K.; *et al.* Perovskite oxides for visible-light-absorbing ferroelectric and photovoltaic materials. *Nature* **2013**, *503*, 509–512.

94. Li, G.; Yan, S.; Wang, Z.; Wang, X.; Li, Z.; Ye, J.; Zou, Z. Synthesis and visible light photocatalytic property of polyhedron-shaped AgNbO₃. *J. Chem. Soc. Dalton Trans.* **2009**, *40*, 8519–8524.

95. Li, G.; Kako, T.; Wang, D.; Zou, Z.; Ye, J. Enhanced photocatalytic activity of La-doped AgNbO₃ under visible light irradiation. *J. Chem. Soc. Dalton Trans.* **2009**, *13*, 2423–2427.

96. Konta, R.; Kato, H.; Kobayashi, H.; Kudo, A. Photophysical properties and photocatalytic activities under visible light irradiation of silver vanadates. *PCCP* **2003**, *5*, 3061–3065.

97. Xu, J.; Hu, C.; Xi, Y.; Wan, B.; Zhang, C.; Zhang, Y. Synthesis and visible light photocatalytic activity of β-AgVO₃ nanowires. *Solid State Sci.* **2012**, *14*, 535–539.

98. Sang, Y.; Kuai, L.; Chen, C.; Fang, Z.; Geng, B. Fabrication of a Visible-Light-Driven Plasmonic Photocatalyst of AgVO₃@AgBr@Ag Nanobelt Heterostructures. *ACS App. Mater. Interfaces* **2014**, *6*, 5061–5068.

99. Ju, P.; Fan, H.; Zhang, B.; Shang, K.; Liu, T.; Ai, S.; Zhang, D. Enhanced photocatalytic activity of β-AgVO₃ nanowires loaded with Ag nanoparticles under visible light irradiation. *Sep. Purif. Technol.* **2013**, *109*, 107–110.

100. Joshi, U.A.; Palasyuk, A.M.; Maggard, P.A. Photoelectrochemical Investigation and Electronic Structure of a p-Type CuNbO₃ Photocathode. *J. Phys. Chem. C* **2011**, *115*, 13534–13539.

101. Lin, Y.; Yuan, G.; Sheehan, S.; Zhou, S.; Wang, D. Hematite-based solar water splitting: Challenges and opportunities. *Energy Environ. Sci.* **2011**, *4*, 4862–4869.

102. Parida, K.M.; Reddy, K.H.; Martha, S.; Das, D.P.; Biswal, N. Fabrication of nanocrystalline LaFeO₃: An efficient sol–gel auto-combustion assisted visible light responsive photocatalyst for water decomposition. *Int. J. Hydrog. Energy* **2010**, *35*, 12161–12168.

103. Thirumalairajan, S.; Girija, K.; Hebalkar, N.Y.; Mangalaraj, D.; Viswanathan, C.; Ponpandian, N. Shape evolution of perovskite LaFeO₃ nanostructures: A systematic investigation of growth mechanism, properties and morphology dependent photocatalytic activities. *RSC Adv.* **2013**, *3*, 7549–7561.

104. Wei, Z.-X.; Wang, Y.; Liu, J.-P.; Xiao, C.-M.; Zeng, W.-W.; Ye, S.-B. Synthesis, magnetization, and photocatalytic activity of LaFeO₃ and LaFe₀.₉Mn₀.₁O₃₋δ. *J. Mater. Sci.* **2013**, *48*, 1117–1126.

105. Gao, F.; Chen, X.Y.; Yin, K.B.; Dong, S.; Ren, Z.F.; Yuan, F.; Yu, T.; Zou, Z.G.; Liu, J.M. Visible-Light Photocatalytic Properties of Weak Magnetic BiFeO₃ Nanoparticles. *Adv. Mater.* **2007**, *19*, 2889–2892.

106. Mohan, S.; Subramanian, B.; Bhaumik, I.; Gupta, P.K.; Jaisankar, S.N. Nanostructured Bi₍₁₋ₓ₎Gd₍ₓ₎FeO₃—A multiferroic photocatalyst on its sunlight driven photocatalytic activity. *RSC Adv.* **2014**, *4*, 16871–16878.

107. Feng, Y.N.; Wang, H.C.; Luo, Y.D.; Shen, Y.; Lin, Y.H. Ferromagnetic and photocatalytic behaviors observed in Ca-doped BiFeO₃ nanofibres. *J. App. Phys.* **2013**, *113*, 146101.

108. Madhu, C.; Bellakki, M.B.; Manivannan, V. Synthesis and characterization of cation-doped BiFeO₃ materials for photocatalytic applications. *Indian J. Eng. Mater. Sci.* **2010**, *17*, 131–139.

109. Dhanasekaran, P.; Gupta, N.M. Factors affecting the production of H2 by water splitting over a novel visible-light-driven photocatalyst GaFeO$_3$. *Int. J. Hydrog. Energy* **2012**, *37*, 4897–4907.

110. Tang, P.; Chen, H.; Cao, F.; Pan, G. Magnetically recoverable and visible-light-driven nanocrystalline YFeO$_3$ photocatalysts. *Catal. Sci. Technol.* **2011**, *1*, 1145–1148.

111. Yuan, Z.; Wang, Y.; Sun, Y.; Wang, J.; Bie, L.; Duan, Y. Sunlight-activated AlFeO$_3$/TiO$_2$ photocatalyst. *Sci. China Ser. B* **2006**, *49*, 67–74.

112. Takei, T.; Haramoto, R.; Dong, Q.; Kumada, N.; Yonesaki, Y.; Kinomura, N.; Mano, T.; Nishimoto, S.; Kameshima, Y.; Miyake, M. Photocatalytic activities of various pentavalent bismuthates under visible light irradiation. *J. Solid State Chem.* **2011**, *184*, 2017–2022.

113. Kako, T.; Zou, Z.; Katagiri, M.; Ye, J. Decomposition of Organic Compounds over NaBiO$_3$ under Visible Light Irradiation. *Chem. Mater.* **2006**, *19*, 198–202.

114. Chang, X.; Yu, G.; Huang, J.; Li, Z.; Zhu, S.; Yu, P.; Cheng, C.; Deng, S.; Ji, G. Enhancement of photocatalytic activity over NaBiO$_3$/BiOCl composite prepared by an in situ formation strategy. *Catal. Today* **2010**, *153*, 193–199.

115. Yu, X.; Zhou, J.; Wang, Z.; Cai, W. Preparation of visible light-responsive AgBiO$_3$ bactericide and its control effect on the Microcystis aeruginosa. *J. Photochem. Photobiol.* **2010**, *101*, 265–270.

116. Sun, M.; Jiang, Y.; Li, F.; Xia, M.; Xue, B.; Liu, D. Dye degradation activity and stability of perovskite-type LaCoO$_{3-x}$ (x = 0~0.075). *Mater. Trans.* **2010**, *51*, 2208–2214.

117. Tang, P.; Sun, H.; Cao, F.; Yang, J.; Ni, S.; Chen, H. Visible-light driven LaNiO$_3$ nanosized photocatalysts prepared by a sol-gel process. *Adv. Mater. Res.* **2011**, *279*, 83–87.

118. Li, J.; Zeng, J.; Jia, L.; Fang, W. Investigations on the effect of Cu^{2+}/Cu^{1+} redox couples and oxygen vacancies on photocatalytic activity of treated LaNi$_{1-x}$Cu$_x$O$_3$ (x=0.1, 0.4, 0.5). *Int. J. Hydrog. Energy* **2010**, *35*, 12733–12740.

119. Singh, J.; Uma, S. Efficient Photocatalytic Degradation of Organic Compounds by Ilmenite AgSbO$_3$ under Visible and UV Light Irradiation. *J. Phys. Chem. C* **2009**, *113*, 12483–12488.

120. Shi, J.; Ye, J.; Zhou, Z.; Li, M.; Guo, L. Hydrothermal Synthesis of Na$_{0.5}$La$_{0.5}$TiO$_3$-LaCrO$_3$ Solid-Solution Single-Crystal Nanocubes for Visible-Light-Driven Photocatalytic H$_2$ Evolution. *Chem. Eur. J.* **2011**, *17*, 7858–7867.

121. Borse, P.H.; Cho, C.R.; Yu, S.M.; Yoon, J.H.; Hong, T.E.; Bae, J.S.; Jeong, E.D.; Kim, H.G. Improved photolysis of water from ti incorporated double perovskite Sr$_2$FeNbO$_6$ lattice. *Bull. Korean Chem. Soc.* **2012**, *33*, 3407–3412.

122. Borse, P.H.; Lim, K.T.; Yoon, J.H.; Bae, J.S.; Ha, M.G.; Chung, E.H.; Jeong, E.D.; Kim, H.G. Investigation of the physico-chemical properties of Sr$_2$FeNb$_{1-x}$W$_x$O$_6$ (0.0 ≤ x ≤ 0.1) for visible-light photocatalytic water-splitting applications. *J. Korean Phys. Soc.* **2014**, *64*, 295–300.

123. Hu, R.; Li, C.; Wang, X.; Sun, Y.; Jia, H.; Su, H.; Zhang, Y. Photocatalytic activities of LaFeO$_3$ and La$_2$FeTiO$_6$ in p-chlorophenol degradation under visible light. *Catal. Commun.* **2012**, *29*, 35–39.

124. Hatakeyama, T.; Takeda, S.; Ishikawa, F.; Ohmura, A.; Nakayama, A.; Yamada, Y.; Matsushita, A.; Yea, J. Photocatalytic activities of Ba_2RBiO_6 (R = La, Ce, Nd, Sm, Eu, Gd, Dy) under visible light irradiation. *J. Ceram. Soc. Jpn.* **2010**, *118*, 91–95.

125. Clark, J.H.; Dyer, M.S.; Palgrave, R.G.; Ireland, C.P.; Darwent, J.R.; Claridge, J.B.; Rosseinsky, M.J. Visible light photo-oxidation of model pollutants using $CaCu_3Ti_4O_{12}$: An experimental and theoretical study of optical properties, electronic structure, and selectivity. *J. Am. Chem. Soc.* **2011**, *133*, 1016–1032.

126. Iwakura, H.; Einaga, H.; Teraoka, Y. Photocatalytic Properties of Ordered Double Perovskite Oxides. *J. Novel Carbon Resour. Sci.* **2011**, *3*, 1–5.

127. Zhu, S.; Fu, H.; Zhang, S.; Zhang, L.; Zhu, Y. Two-step synthesis of a novel visible-light-driven $K_2Ta_2O_{6-x}N_x$ catalyst for the pollutant decomposition. *J. Photochem. Photobiol. A* **2008**, *193*, 33–41.

128. Kanhere, P.; Tang, Y.; Zheng, J.; Chen, Z. Synthesis, photophysical properties, and photocatalytic applications of Bi doped $NaTaO_3$ and Bi doped $Na_2Ta_2O_6$ nanoparticles. *J. Phys. Chem. Solids* **2013**, *74*, 1708–1713.

129. Hara, M.T.T.; Kondo, J.N.; Domen, K. Photocatalytic reduction of water by TaON under visible light irradiation. *Catal. Today* **2004**, *90*, 313–317.

130. Ishikawa, A.; Takata, T.; Kondo, J.N.; Hara, M.; Kobayashi, H.; Domen, K. Oxysulfide $Sm_2Ti_2S_2O_5$ as a Stable Photocatalyst for Water Oxidation and Reduction under Visible Light Irradiation ($\lambda \le 650$ nm). *J. Am. Chem. Soc.* **2002**, *124*, 13547–13553.

131. Siritanaratkul, B.; Maeda, K.; Hisatomi, T.; Domen, K. Synthesis and Photocatalytic Activity of Perovskite Niobium Oxynitrides with Wide Visible-Light Absorption Bands. *ChemSusChem* **2011**, *4*, 74–78.

132. Maeda, K.; Higashi, M.; Siritanaratkul, B.; Abe, R.; Domen, K. $SrNbO_2N$ as a Water-Splitting Photoanode with a Wide Visible-Light Absorption Band. *J. Am. Chem. Soc.* **2011**, *133*, 12334–12337.

133. Higashi, M.; Abe, R.; Takata, T.; Domen, K. Photocatalytic Overall Water Splitting under Visible Light Using $ATaO_2N$ (A = Ca, Sr, Ba) and WO_3 in a IO_3^-/I^- Shuttle Redox Mediated System. *Chem. Mater.* **2009**, *21*, 1543–1549.

134. Maeda, K.; Domen, K. Preparation of $BaZrO_3$–$BaTaO_2N$ solid solutions and the photocatalytic activities for water reduction and oxidation under visible light. *J. Catal.* **2014**, *310*, 67–74.

135. Le Paven-Thivet, C.; Ishikawa, A.; Ziani, A.; Le Gendre, L.; Yoshida, M.; Kubota, J.; Tessier, F.; Domen, K. Photoelectrochemical Properties of Crystalline Perovskite Lanthanum Titanium Oxynitride Films under Visible Light. *J. Phys. Chem. C* **2009**, *113*, 6156–6162.

136. Kako, T.; Kikugawa, N.; Ye, J. Photocatalytic activities of $AgSbO_3$ under visible light irradiation. *Catal. Today* **2008**, *131*, 197–202.

Tungsten Trioxide as a Visible Light Photocatalyst for Volatile Organic Carbon Removal

Yossy Wicaksana, Sanly Liu, Jason Scott and Rose Amal

Abstract: Tungsten trioxide (WO_3) has been demonstrated to possess visible light photoactivity and presents a means of overcoming the UV-light dependence of photocatalysts, such as titanium dioxide. In this study, WO_3 nanostructures have been synthesised by a hydrothermal method using sodium tungstate ($Na_2WO_4·2H_2O$), sulphate precursors and pH as structure-directing agents and parameters, respectively. By altering the concentration of the sulphate precursors and pH, it was shown that different morphologies and phases of WO_3 can be achieved. The effect of the morphology of the final WO_3 product on the visible light photoactivity of ethylene degradation in the gas phase was investigated. In addition, platinum (Pt) was photodeposited on the WO_3 structures with various morphologies to enhance the photocatalytic properties. It was found that the photocatalytic properties of the WO_3 samples greatly depend on their morphology, chemical composition and surface modification. WO_3 with a cuboid morphology exhibited the highest visible light photoactivity compared to other morphologies, while adding Pt to the surface improved the performance of certain WO_3 structures.

Reprinted from *Molecules*. Cite as: Wicaksana, Y.; Liu, S.; Scott, J.; Amal, R. Tungsten Trioxide as a Visible Light Photocatalyst for Volatile Organic Carbon Removal. *Molecules* **2014**, *19*, 17747-17762.

1. Introduction

Volatile organic compounds (VOCs) are the group of airborne organic compounds capable of damaging human health and the environment. Indoor environments have been reported to contain up to ten times greater VOC pollutant levels than outdoor environments [1]. Indoor VOCs can originate from building materials, paints, glues, lacquer, carpet, office furnishings, cleaning compounds, cigarette smoke and other items [2,3]. Increasingly stringent regulations regarding acceptable VOC concentrations necessitate the implementation of technologies capable of meeting these requirements. Technologies based on adsorption or scrubbing the gas streams are capable of removing the VOCs, but generate secondary waste streams, which require disposal. Photocatalysis is an alternative technology for treating VOCs, converting them into comparatively benign water and carbon dioxide [4].

Titanium dioxide (TiO_2) is generally the semiconductor of choice due to its high photoactivity, low cost, ready availability and non-toxic properties. However, TiO_2 can only be activated by ultra-violet (UV) light ($\lambda \leq 380$ nm), limiting the use of sunlight as the light source and rendering it virtually unusable in indoor environments without the presence of an external UV light source [5]. Tungsten trioxide (WO_3) is a comparatively less studied semiconductor, which is capable of being activated by visible light ($\lambda \leq 450$ nm) [6] and, consequently, may be a more suitable semiconductor for degrading VOCs in an indoor environment.

Several methods, including chemical vapour deposition (CVD) [7], thermal evaporation [8], electrochemical techniques [9], a spray pyrolysis approach [10], template-mediated synthesis [11], the sol-gel process [12] and hydrothermal reactions [13], have been reported for WO₃ nanostructure synthesis. Of the listed methods, hydrothermal synthesis is a facile technique, which is well-suited to producing a range of nanostructured morphologies by simple variations to the precursor solution. For instance, controlled synthesis of WO₃ nanostructures by the hydrothermal method has been performed with the help of structure-directing chemicals, like Na₂SO₄, Rb₂SO₄, K₂SO₄, Li₂SO₄, FeSO₄ and Na₂S [14–18]. This work focuses on fabricating tungsten oxide photocatalysts using a hydrothermal technique with varying pH, as well as the amount and type of sulphate precursor. Emphasis is placed on the influence of preparation conditions on the characteristics of WO₃ nanostructures and, subsequently, on its capacity to photodegrade gas-phase ethylene using visible light as the energy source.

The relative energy of the electrons in the WO₃ conduction band restricts their capacity to reduce oxygen, which results in a build-up of these electrons, followed by an increased incidence of recombination with holes and, ultimately, a decrease in photocatalytic performance. Improvements in WO₃ photoactivity, so as to counteract this limitation, may be achieved by closely controlling particle morphology [19] or by loading platinum (Pt) deposits on the surface. The noble metal Pt is believed to greatly assist in electron transfer during the oxidation-reduction reactions in the photocatalysis process [20]. Pt loaded onto WO₃ nanoparticles has been shown to enhance aqueous acetic acid mineralisation compared to bare WO₃ photocatalysts under visible light illumination [21]. Similarly, WO₃ photocatalysts with different platinum loadings have been proven to be more efficient for phenol oxidation [22], as well as tetracycline oxidation [23] in an aqueous suspension. Even at a 0.1% Pt loading on WO₃ nanoparticles, notable improvement of the water splitting reaction has been demonstrated [24]. A beneficial impact of Pt deposits on WO₃ was also observed in the gas phase for acetaldehyde photodegradation [25]. In this work, the influence of loading the surface with nanosized Pt deposits on visible light photoactivity is also considered.

2. Results and Discussion

The alkaline solution precursor of Na₂WO₄·2H₂O essentially contains stable WO_4^{2-} ions. With the supply of H⁺ ions from the resin, the WO_4^{2-} ions gradually underwent condensation reactions in the column to form paratungstate ions, such as $[W_{12}O_{41}]^{10-}$ and $[H_2W_{12}O_{40}]^{6-}$ (at pH ~ 4.0–7.0), and the metatungstic acid, $(WO_3)_n·xH_2O$, (at pH ~ 1–4), according to the equations below [26].

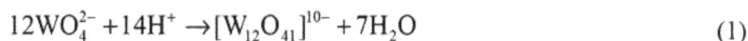

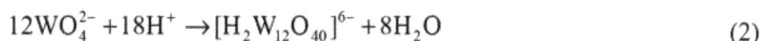

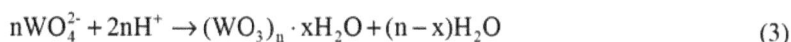

$$12WO_4^{2-} + 14H^+ \rightarrow [W_{12}O_{41}]^{10-} + 7H_2O \tag{1}$$

$$12WO_4^{2-} + 18H^+ \rightarrow [H_2W_{12}O_{40}]^{6-} + 8H_2O \tag{2}$$

$$nWO_4^{2-} + 2nH^+ \rightarrow (WO_3)_n·xH_2O + (n-x)H_2O \tag{3}$$

The condensation reaction involves protonation of tungsten oxyanions, formation of oxygen bridging between the protonated monomeric tungsten oxyanions and the release of water molecules. The rates of condensation are strictly controlled by the flow rate of solution and the amount of resin used.

WO_3 particles prepared from the neat tungstic acid precursor (*i.e.*, W0) were found from ICP-AES analysis to contain a sodium atomic content of 2.78×10^{-4} per atom of W. Since the starting solution contained two atoms of Na per atom of W, the significant reduction from the original amount indicates high removal efficiency by the ion-exchange process. The W0 particles were yellow in colour and identified by XRD analysis (spectra are located in the Supporting Information, Figure S1) to be a mixture of the monoclinic WO_3 phase (cell constants: a = 7.2970 Å, b =7.5390 Å, c = 7.6880 Å; JCPDS 01-071-2141) and the orthorhombic $WO_3 \cdot \frac{1}{3}H_2O$ phase (cell constants: a = 7.3590 Å, b = 12.5130 Å, c = 7.7040 Å; JCPDS 00-035-0270). The mixture of phases in W0 conforms to the tungstic acid dehydration evolution profile as described by Livage and Guzman [27]. SEM imaging indicates that the W0 morphology comprises a mixture of large slabs with clusters of columnar crystals, as observed from Figure 1A.

Figure 1F shows the SEM image for the commercial Sigma Aldrich WO_3 (WSA). WSA consists of nanoparticles with a size distribution from 30–100 nm and, from XRD (spectra are located in the Supporting Information, Figure S1) was found to possess a pure monoclinic crystalline phase.

2.1. Effect of Sulphate Anions and pH on WO₃ Characteristics

The impact of sulphate (originating from Na_2SO_4) as the shape directing agent on WO_3 morphology is illustrated in Figure 1B,C for W0.3NaS and W7.6NaS (see Experimental Section 3.2 and Table 1 for an explanation of the sample abbreviations), respectively. The two samples show more elongated structures when compared to W0. The W0.3NaS particles consist mainly of randomly orientated columnar crystals and individual particles, while the W7.6NaS particles exhibit a bundled structure consisting of aligned nanorods. The difference in the morphology can be attributed predominantly to the addition of the SO_4^{2-} anions, as the pH values of the final solutions were similar (Table 1). It is readily apparent that the SO_4^{2-} anions promote anisotropic growth. One manner by which shape-controlling additives can behave is to "cap" the growth of particles along a particular crystal plane. That is, they preferentially interact with one or two crystal faces, hindering growth on those planes, which then favours growth on the "uncapped" face, resulting in elongated structures [28]. To explain the tendency of the nanorods to align parallel to each other for W7.6NaS, it is speculated that the lateral capillary forces along the length of the nanorod are higher compared to its width, causing side-by-side alignment rather than end-to-end. The aggregation of the pre-formed nanorods by oriented attachment [29] may be energetically favoured to reduce the surface energy of the system, and consequently, the nanorod WO_3 bundles are obtained for W7.6NaS.

Figure 1. SEM images of: (**A**) hydrothermally synthesised WO₃ with no additives (W0); (**B**) hydrothermally synthesised WO₃ with Na₂SO₄ added at a $SO_4^{2-}:WO_4^{2-}$ ratio of 0.3 (W0.3NaS); (**C**) hydrothermally synthesised WO₃ with Na₂SO₄ added at a $SO_4^{2-}:WO_4^{2-}$ ratio of 7.6 (W7.6NaS); (**D**) hydrothermally synthesised WO₃ with H₂SO₄ added at a $SO_4^{2-}:WO_4^{2-}$ ratio of 0.3 (W0.3HS); (**E**) hydrothermally synthesised WO₃ with H₂SO₄ added at a $SO_4^{2-}:WO_4^{2-}$ ratio of 7.6 (W7.6HS); (**F**) commercial Sigma Aldrich WO₃ (WSA).

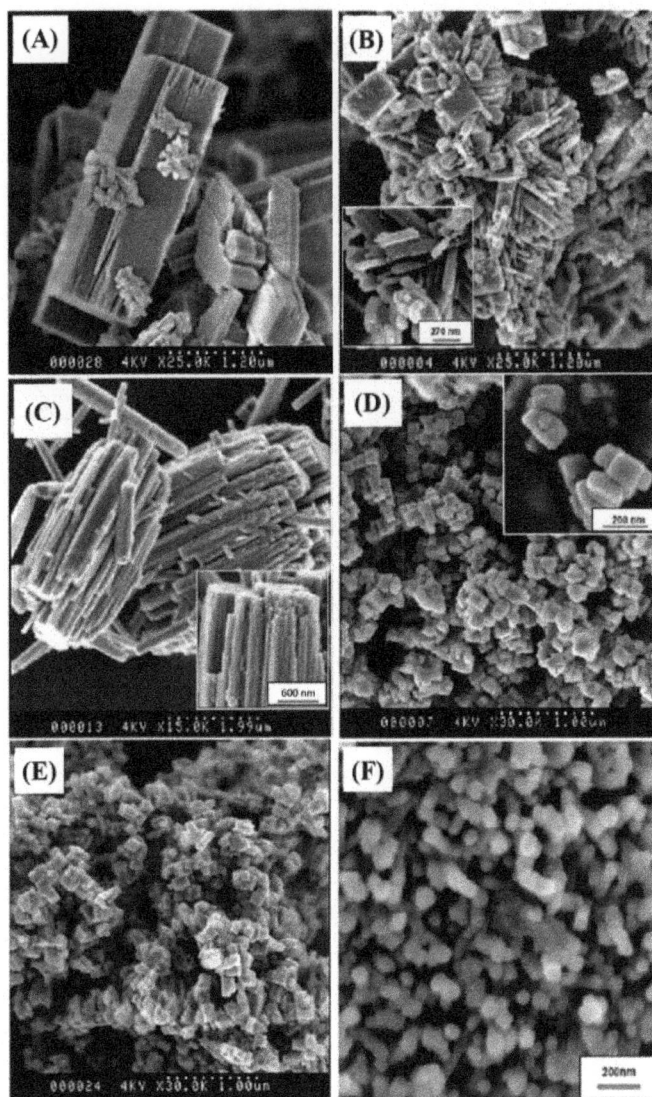

Using different amounts of the sulphate additive in hydrothermal synthesis can produce different WO₃ crystalline phases. At the lower SO_4^{2-}/WO_4^{2-} ratio (W0.3NaS), XRD indicated that the crystalline phase was the same as W0 (Table 1 and Supporting Information, Figure S1). The addition of excess sulphate additive induces a change in the crystalline phase from a mixture of monoclinic and orthorhombic in W0.3NaS to hexagonal-phase WO_3 (cell constants: a = 7.3244 Å, b =7.3244 Å, c = 7.6628 Å; JCPDS 01-085-2459) in W7.6NaS. This observation suggests that not only does the sulphate additive play a crucial role in altering the morphology, but also in governing the crystalline phase of the product. Studies by Gu *et al.* [30] and Huang *et al.* [17] have also demonstrated the role of sulphate salt addition in determining crystalline phase of the WO_3 product. Upon adding an increased amount of sulphate alkali metal salts, it was reported that the hexagonal phase becomes gradually dominant, which agrees with our result. Here, although the sodium ions in the Na_2SO_4 additive were removed during the ion exchange process, it is possible that a certain amount of Na^+ ions still exist in the precursor solution (especially in W7.6NaS with a much higher Na_2SO_4 concentration) and therefore act as stabilising ions for the hexagonal and triangular tunnels in the formation of metastable hexagonal WO_3.

Table 1. Selected synthesis parameters and characteristics for the different WO_3 photocatalysts. Included are the characteristics for commercial Sigma Aldrich WO_3 (WSA).

Sample	Precursor Solution	Molar Ratio $\frac{SO_4^{2-}}{WO_4^{2-}}$	Final pH	Morphology	Crystalline Phase (Chemical Formula)	Surface Area (m^2/g)	Band-Gap (eV)
W0	20 mL 0.1 M H_2WO_4	-	1.61	slabs with columnar clusters	monoclinic (WO_3) + orthorhombic ($WO_3 \cdot \frac{1}{3}H_2O$)	n.a.	n.a.
W0.3NaS	20 mL 0.1 M H_2WO_4 + SO_4^{2-} (from Na_2SO_4)	0.3	1.3	columnar crystals	monoclinic (WO_3) + orthorhombic ($WO_3 \cdot \frac{1}{3}H_2O$)	n.a.	n.a.
W7.6NaS	20 mL 0.1 M H_2WO_4 + SO_4^{2-} (from Na_2SO_4)	7.6	1.15	nanobundles	hexagonal (WO_3)	44.4	2.68
W0.3HS	20 mL 0.1 M H_2WO_4 + H_2SO_4	0.3	<0.3	nanocubes	monoclinic (WO_3) + orthorhombic ($WO_3 \cdot \frac{1}{3}H_2O$)	n.a.	n.a.
W7.6HS	20 mL 0.1 M H_2WO_4 + H_2SO_4	7.6	<0.3	nanocubes	monoclinic (WO_3) + orthorhombic ($WO_3 \cdot \frac{1}{3}H_2O$)	7.0	2.75
WSA	Commercial WO_3 (Sigma Aldrich)	-	-	nanoparticles	monoclinic (WO_3)	8.3	2.61

To further investigate the unique behaviour of sulphate ions, H_2SO_4 was also used as a capping agent. The amount of H_2SO_4 added was adjusted, such that the SO_4^{2-}/WO_4^{2-} ratios were equivalent to those used during W0.3NaS and W7.6NaS synthesis. A consequence of using H_2SO_4 was a simultaneous increase in the free H^+ ions present, whereby the pH dropped to below 0.3. At this low pH, the resulting WO_3 crystal morphology did not resemble any of the previous systems. Aggregated nanocubes (edge length of approximately 100–200 nm) were obtained for both W0.3HS and W7.6HS (Figure 1D,E, respectively), with the size of the WO_3 particles being noticeably smaller than for the Na_2SO_4 additive. The diffraction lines for both powder samples

(XRD spectra are located in Supporting Information, Figure S1) were designated as containing mainly monoclinic WO_3 (cell constants: a = 7.2970 Å, b = 7.5390 Å, c = 7.6880 Å; JCPDS 01-071-2141) and partially orthorhombic $WO_3 \cdot \frac{1}{3}H_2O$ (cell constants: a = 7.3590 Å, b = 12.5130 Å, c = 7.7040 Å; JCPDS 00-035-0270). It appears that the addition of more H_2SO_4 has little influence in terms of the overall morphology of the product or the crystalline phase, at least for SO_4^{2-}/WO_4^{2-} ratios over the range 0.3 to 7.6.

It is suspected that the presence of the H^+ ions is mainly responsible for both the cuboid morphology and the monoclinic-orthorhombic crystalline phase. pH has been demonstrated by Reis et al. [31] to have a marked influence on the crystalline phase of structures formed during hydrothermal acid hydrolysis of sodium tungstate ($Na_2WO_4.2H_2O$), while Gu et al. [28] established pH variations over the range <0.5, 1.0–1.4 and 1.5–2.0 gave nanoparticle, nanobundle and nanowire WO_3 structures, respectively. A possible explanation for the particle size shrinkage observed for the hydrothermally produced W0.3HS and W7.6HS samples might be that at a low pH value, the tungstic acid solution is so supersaturated that nucleation occurs homogeneously, rapidly generating a large number of small WO_3 nuclei [32]. This nucleation reduces the supersaturation and consumes most of the resources for growth, which then results in a briefer ripening process.

Band-gaps derived from UV-Vis spectra (see Supporting Information, Figure S2), which are an indicator of the wavelength of light capable of photoactivating the material, ranged between 2.61 for WSA to 2.75 for W7.6HS (Table 1), which is typical for WO_3 [24]. Table 1 also indicates that the specific surface area of the WO_3 nanobundles is around five- to six-times greater than both the nanocubes and the nanoparticles.

2.2. Pt/WO₃ Characteristics

TEM micrographs showing Pt deposits on the various particle morphologies are provided in Figure 2. Figure 2A suggests that for W7.6NaS, there are Pt deposits (with diameters over the range 5–10 nm) loaded onto the nanorods when using UV-A light, although similarities in contrast between the Pt and WO_3 make them difficult to discern. Identifying Pt deposits on W7.6NaS following visible light photodeposition (Figure 2B) was challenging with no clear evidence of Pt deposits being present in the TEM image. The presence of Pt being photodeposited on the WO_3 nanocubes (W7.6HS) using both UV-A and visible light was more apparent, as illustrated in Figure 2C,D, respectively. In both instances, the Pt deposits are hemispherical and dispersed in nature, with photodeposition under UV-A light appearing to give a greater prevalence of smaller (2–3 nm) deposits. Visible light photodeposition appears to favour Pt deposits in the range of 5–10 nm, although photodeposition using UV light also provided some larger Pt deposits (~10 nm). The WSA sample exhibited the most distinct difference in Pt photodeposit characteristics arising from the different light sources. Figure 2E indicates that Pt photodeposits from UV-A light are present as individual, roughly hemispherical deposits, although, again, the low contrast makes them difficult to identify. Visible light Pt photodeposition on WSA (Figure 2F) results in clusters of small Pt deposits accumulating on the WO_3 particles. Pt nanoparticles within the clusters appear to have diameters of 2–3 nm, while the clusters themselves are up to 20 nm in diameter. It also appears that individual Pt deposits are present on the WO_3 particles. From the TEM images, we can infer that the different WO₃ morphologies and the light wavelengths impact the Pt photodeposition process. Particle morphology (e.g., nanocube *versus* irregular) will govern the number of potential sites (e.g., steps, edges, defects) where Pt deposition is favoured, as too will the crystalline phase (e.g. monoclinic *versus* hexagonal).

Figure 2. TEM micrographs of Pt photodeposited on WO₃ nanorod bundles W7.6NaS using (**A**) UV-A light and (**B**) visible light; WO₃ nanocubes W7.6HS using (**C**) UV-A light and (**D**) visible light; Sigma Aldrich WO₃ WSA using (**E**) UV-A light and (**F**) visible light. Arrows indicate the Pt deposits.

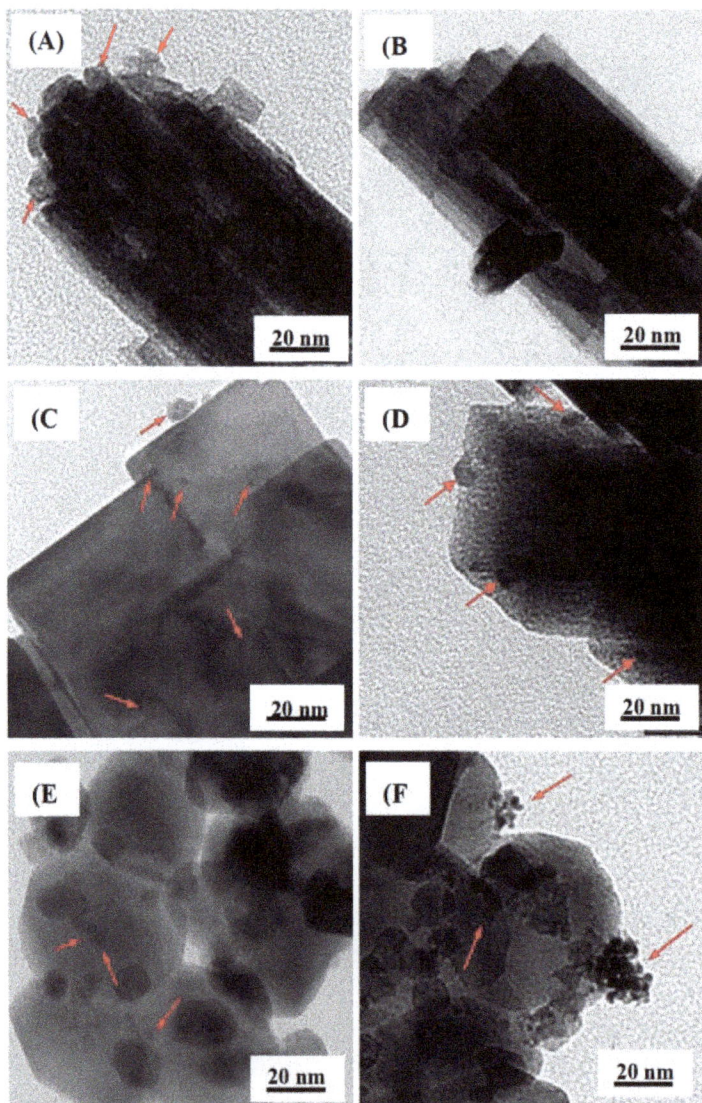

The valence state of the deposited platinum nanoparticles plays an important role during charge trapping and interfacial charge transfer, and subsequently, the photocatalytic efficiency of the resulting platinised metal oxide. Figure 3 shows the XPS spectra of the platinised WO₃ nanocuboids (W7.6HS) in the platinum (Pt4f) region (70–79 eV) arising from both UV-A and visible light photodeposition. In terms of the platinum oxidation state, the spectrum could be deconvoluted into two pairs of doublets. The main doublets located *ca.* 71.2 and 74.8 eV were attributed to atomic state Pt⁰, while the second pair of doublets at *ca.* 73.1 eV and 76.2 eV corresponded to the oxidation states of Pt^{II}. By comparing the normalised peak areas, Pt⁰ was found to be the dominant species from both UV-A and visible light photodeposition. However, the relative percentage of Pt^{II} in the deposits was greater for the visible light particles (44%) than the UV-A particles (21%). The difference in the extent of reduced Pt between the two samples may arise from the higher tendency of Pt^{IV} to be reduced (Equation (5)) compared to Pt^{II} (Equation (4)) [33]. Although we have employed a longer irradiation time during the photodeposition of platinum on WO₃ supports using visible light (3 h) compared to UV (1 h), we cannot exclude the possibility that the total amount of photons from the visible light illumination were not enough to reduce the Pt precursor into Pt metal.

$$[PtCl_6]^{2-} + 2e^- \rightarrow [PtCl_4]^{2-} + 2Cl^- \qquad\qquad E^0 = 0.68 \text{ V} \qquad (4)$$

$$[PtCl_4]^{2-} + 2e^- \rightarrow Pt + 4Cl^- \qquad\qquad E^0 = 0.755 \text{ V} \qquad (5)$$

Figure 3. XPS spectra over the platinum Pt4f region for platinum photodeposited on WO₃ nanocubes (W7.6HS) using: **(A)** UV-A light for 1 h; **(B)** visible light for 3 h.

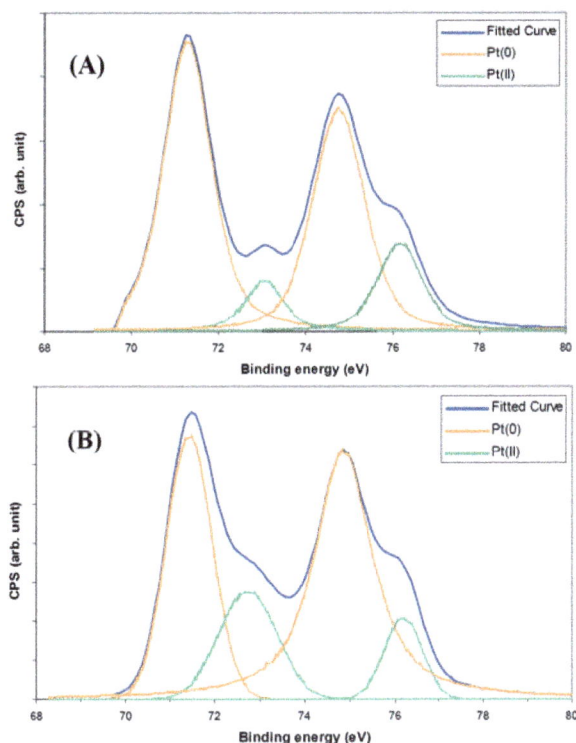

2.3. Ethylene Photodegradation

Ethylene photodegradation profiles for neat WO_3 nanocubes (W7.6HS) and WO_3 nanocubes loaded with Pt using either visible light (3 h) or UV-A (1 h) are provided in Figure 4A. Upon illumination, the ethylene concentration drops sharply within the first 2.5 min for all samples, after which it increases slowly, eventually reaching a stable conversion level. The gradual increase in the ethylene concentration after the initial drop was postulated to be due to the formation of intermediate products, which compete with ethylene during the photocatalytic degradation process. Unfortunately, these intermediate products were not able to be detected by the GC/flame ionisation detector (FID) instrument, such that no identification of these products was available. Once stabilised, ethylene conversion was in the order neat $WO_3 \approx Pt/WO_3$ (visible light 3 h) < Pt/WO_3 (UV-A light 1 h). Despite displaying similar ethylene conversion levels after 30 min of illumination, it appears that, at least initially, the WO_3 nanocubes platinised using visible light are more active than the neat WO_3 nanocubes. The findings indicate that loading platinum deposits on the WO_3 nanocubes improves their capacity for photodegrading ethylene. The greater photoactivity exhibited by the nanocubes loaded with Pt using UV-A light may arise from the greater portion of Pt^0 within the deposits (Figure 3). Pt^0/TiO_2 has been reported to be more active than Pt^{II}/Pt^{IV} species on TiO_2 for photocatalytic degradation of various organic contaminants [34,35]. The Pt^{II} or Pt^{IV} can undergo consecutive oxidation/reduction cycles, which consume excited charge carriers and lowers performance. A control system comprising only the washed silica beads is included in Figure 4 for comparison showing that the ethylene photodegradation derives solely from the neat and platinised WO_3 nanostructures.

Figure 4. Photodegradation profiles displaying: **(A)** ethylene conversion with time for neat WO_3 nanocubes (W7.6HS) and WO_3 nanocubes platinised using visible light for 3 h or UV-A light for 1 h. Included is a silica control profile. **(B)** Steady-state ethylene conversion (as a percentage of initial ethylene concentration) for neat and platinised (visible light for 3 h or UV-A light for 1 h) WO_3 nanobundles (W7.6NaS), nanoparticles (WSA) and nanocubes (W7.6HS).

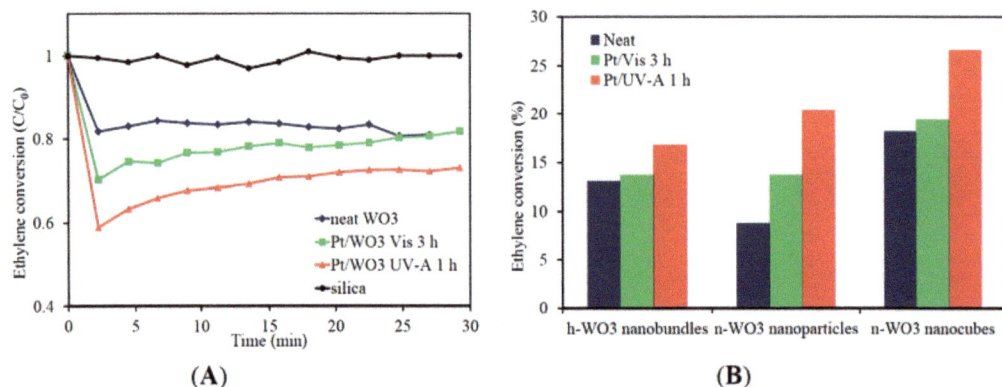

(A)

(B)

Figure 4B demonstrates that particle morphology has an influence on photocatalytic performance. In the case of neat WO_3, the activity for ethylene photodegradation was in the order of nanocubes > nanobundles > nanoparticles. Despite having similar surface areas (Table 1), the WO_3 nanocubes exhibit approximately 10% greater conversion of ethylene over the nanoparticles once the photoactivities had stabilised. This may be due to the more "edged" nature of the cuboid morphology, reducing the photogenerated electron-hole recombination, similar to that described by Kato et al. [36] in their study on photocatalytic water splitting by $NiO/NaTaO_3$. Alternately, the relative proportion of different exposed crystal facets for the different morphologies may have influenced the photoactivity, as was demonstrated by Xie et al. for WO_3 [37]. They observed that WO_3 in a sheet-like form possessed a greater percentage of the (002) facet compared with WO_3 cuboids. The dominance of the (002) facet blue-shifted the band-gap of the sheet-like structure, as well as increased the reduction potential of the conduction (and valence) band(s), relative to the cuboid structure. The deeper valence band maximum of the cuboid structure was used to account for it being the more effective structure when evolving oxygen during photocatalytic water oxidation in the presence of an electron acceptor. Interestingly, the nanobundles have a comparably higher surface area than the other nanostructures and a partially edged morphology, but do not display a correspondingly larger photoactivity. This may result from the hexagonal crystalline phase not being as active as the monoclinic-orthorhombic phase, with this postulation requiring further investigation.

Platinising the WO_3 nanoparticles and nanocubes appears to have a more pronounced effect on ethylene photodegradation than for the nanobundles, especially when UV-A is used as the light source. In this instance, the photoactivity order becomes Pt/nanocubes > Pt/nanoparticles > Pt/nanobundles. This could derive from the hexagonal crystalline phase of the nanobundles being less active than the monoclinic phase and/or the comparative lack of Pt deposits on the nanobundles. Further study is needed to identify the parameter responsible for the result.

In general, the presence of metal deposits on the surface of a photocatalyst acts as an electron sink, trapping electrons and allowing for greater photogenerated hole availability for photodegradation reactions. However, the Pt deposits can also facilitate the transfer of electrons to electron scavengers in the system, most often O_2 (Equation (6)). Wang et al. [38] suggested that the consumption of electrons by O_2 is the rate limiting step in photocatalysis, and the presence of metal deposits can help alleviate this. In the instance of WO_3, the reduction potential for the photogenerated electron (present in the conduction band) is not large enough for the single electron reduction of O_2. That is, it does not possess enough energy to be picked up by O_2 in accordance with Equation (6). However, the photogenerated electron does have enough energy to participate in the multi-electron reduction of O_2 (Equations (7) and (8)), but the need for more than one electron in close proximity to invoke this reduction is not as favourable. This is depicted schematically in Figure 5. The valence and conduction band potentials of TiO_2 are also provided for comparison.

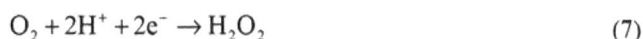

$$O_2 + e^- \rightarrow O_2^-$$
(6)

$$O_2 + 2H^+ + 2e^- \rightarrow H_2O_2$$
(7)

$$O_2 + 4H^+ + 2e^- \rightarrow 2H_2O \tag{8}$$

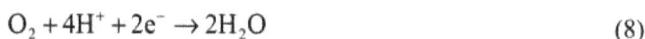

The presence of Pt facilitates the multi-electron reduction of O_2 due to the ready availability of electrons in close proximity to one another on the Pt deposit. Consequently, the Pt reduces the limitation of the smaller conduction band electron potential of WO_3, in turn promoting the photocatalytic activity. Similar to the result from this study, Abe et $al.$ [21] also found that the decomposition of organic compounds under visible light irradiation with platinum loaded WO_3 sample was enhanced significantly and postulated this to be due to the promotion of multi-electron O_2 reduction on the Pt co-catalyst.

Figure 5. Valence and conduction band potentials for WO_3 and single and multi-electron potentials for oxygen. The valence and conduction band potentials for TiO_2 are included for comparison.

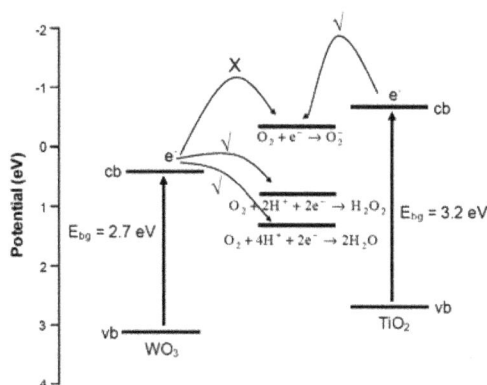

3. Experimental Section

3.1. Materials

Commercial tungsten (VI) oxide nanopowder, WO_3 (Sigma-Aldrich, St. Louis, MO, USA, denoted as "WSA"), with a mean particle size of less than 100 nm was used as the benchmark photocatalyst. For the hydrothermal synthesis of WO_3 nanostructures, the following chemicals were used as received: $Na_2WO_4 \cdot 2H_2O$ (99.0%, A.C.S. reagent, Sigma-Aldrich); Na_2SO_4, (99.0%, Ajax Finechem Pty. Ltd., Auburn, NSW, Australia) and "Amberlite" IR-120 (H) ion-exchange resin (BDH Laboratory Supplies).

Platinum (Pt) was loaded onto the WO_3 nanostructures by a photodeposition method. H_2PtCl_6 (Sigma-Aldrich) was employed as the platinum precursor, and methanol, CH_3OH (99.7%, Ajax Finechem Pty. Ltd.), acted as the hole-scavenger in the system. Nitrogen gas, N_2 (>99%, Coregas Pty. Ltd., Yennora, NSW, Australia), was used to purge the solution during the photodeposition reaction.

Visible light photodegradation focused on ethylene (508 ppm, Coregas Pty. Ltd.) as the VOC model pollutant. Compressed air (79% N_2, 21% O_2, Coregas, Pty. Ltd.) acted as the diluent of the gas stream, while compressed nitrogen, N_2 (>99%, Coregas Pty. Ltd.), was used as the carrier gas in the gas chromatograph. The flame ionisation detector (FID) was fuelled by compressed hydrogen, H_2 (Coregas, Pty. Ltd.), blended with compressed air.

3.2. Hydrothermal Synthesis of WO₃ Nanostructures

Five variants of WO_3 particles were hydrothermally synthesised: one without any shape-directing agent, denoted as "W0" (where "W" represents tungsten trioxide and "0" represents no shape directing agents); two with different concentrations of Na_2SO_4, denoted as "WₓNaS" (where "x" represents the $SO_4^{2-}:WO_4^{2-}$ molar ratio during hydrothermal synthesis and "NaS" indicates that Na_2SO_4 was the shape directing agent); and two with different concentrations of H_2SO_4, denoted as "WₓHS" (where "HS" indicates that H_2SO_4 was the shape directing agent). An ion-exchange column containing approximately 35 g of "Amberlite" IR-120 (H) was set up to remove sodium ions (Na^+) from the NaWO₄ solution. When preparing the precursor for W0 and WₓHS synthesis, a solution of $Na_2WO_4·2H_2O$ (0.1 M) was passed through the column at 1 mL/min, and the eluent was collected. The resulting tungstic acid ($H_2WO_4·2H_2O$) solution displayed a pale yellow colour and possessed a pH between 1.61 and 1.67. In the case of the WₓNaS precursor, a specified amount of Na_2SO_4 (see Table 1) was added to the $H_2WO_4·2H_2O$ (0.1 M) solution and then passed through the ion-exchange column to remove the Na ions from the Na_2SO_4.

To prepare the W0 particles, 20 mL of the H_2WO_4 precursor solution was initially placed in a 50-mL Teflon tube. In the case of WₓHS particle synthesis, a small amount of H_2SO_4 with a specified concentration (see Table 1) was added to the H_2WO_4 precursor solution in the Teflon tube. For WₓNaS particle synthesis, 20 mL of the mixed tungstic-sulphate precursor was placed in the 50-mL Teflon tube. Following precursor addition, the Teflon tube was then sealed in a stainless steel autoclave and heated in an oven at 200 °C for 10 h. The resulting precipitate was recovered by centrifuging (Beckmann Coulter Allegra 25R, 10,000 rpm), washed at least five times with deionised water to remove any unreacted precursors and air-dried in a 60 °C oven for approximately 15 h.

3.3. Photodeposition of Platinum on WO₃ Nanostructured Supports

Platinum was loaded onto the WO_3 nanostructure surface using photodeposition in a 500-mL Pyrex glass annular reactor surrounding a 20-W NEC T10 black light blue lamp (λ_{max} = 360 nm). A 1 g/L WO_3 slurry was dispersed ultrasonically for 15 min, with 550 mL of this slurry transferred to the photoreactor and circulated for 30 min under illuminated conditions. Light pre-treatment was designed to remove adsorbed organic impurities from the particle surface. The UV lamp was then turned off, and 2000 µg carbon, in the form of methanol, were added to the system as a hole scavenger. The metal precursor (H_2PtCl_6) was added to give the required Pt loading (1.0 at. %); the pH was adjusted to 3 using dilute perchloric acid, and the system was purged with nitrogen gas at a flow rate of 50 mL/min for 20 min. The conditioned slurry was illuminated for 60 min, recovered by centrifuging and washed with deionised water for a minimum of five times. The washed particles were dried in an oven at 60 °C for 12 h and then ground and stored in a desiccator prior to use.

As a comparison, the effect of using visible light as the source of illumination during Pt photodeposition was investigated. In this instance, an 18-W fluorescent light (Sylvania Luxline Plus F18/860, Erlangen, Germany) with a cut-off filter ($\lambda > 420$ nm) was employed as the light source. The photodeposition procedure was the same as for the UV-lamp apart from the illumination period (during Pt photodeposition) being extended to 3 h, due to the weaker irradiance of the fluorescent lamp.

3.4. Photocatalytic Oxidation of Ethylene

Neat and Pt-loaded WO_3 nanostructures were assessed for gas-phase ethylene photodegradation using an annular-type packed-bed photoreactor, as described previously [39]. The photoreactor consisted of a 400 mm-long Pyrex glass tube (15 mm outside diameter (o.d.), 1.2 mm wall thickness) containing a glass filler tube (12 mm o.d.). The packed-bed reactor was prepared by filling the annular gap with a catalyst/silica bead mixture in a 1:10 ratio (total weight of 0.66 g) to give a catalyst bed approximately 50 mm in length. The bed was supported at each end by washed silica beads and held in position by quartz wool. Prior to use, the silica beads were cleaned with 1 wt % HCl solution and then rinsed with deionised water until no pH change was observed. The washed beads were then dried in an oven at 100 °C. Illumination was provided by four 6-W Sylvania fluorescent lamps spaced equidistantly around the reactor. Inlet ethylene concentration was maintained at 50 ppm in air at a flow rate of 20 mL/min. Ethylene concentration was measured using a Shimadzu GC-8A (Shimadzu Co., Tokyo, Japan) equipped with an FID. Gas component separation was achieved with an Alltech HAYESEP Q 80/100 column (Alltech Associates, Inc., Deerfield, IL, USA).

Photocatalysis experiments involved passing 10 mL/min air through the packed bed for 10 min, whereby the lamps were switched on for 60 min to remove impurities from the photocatalyst surface. The lamps were then switched off, and the ethylene:air mix passed through the bed (20 mL/min) until ethylene concentration in the reactor effluent was the same as that in the reactor inlet. At this point, the lamps were switched on and the reactor effluent analysed for ethylene concentration. Samples were taken every 2 min.

3.5. Characterisation

Morphology characterisation of the samples was obtained by a scanning electron microscope (Hitachi S900 SEM, Tokyo, Japan) at an applied voltage of 4 kV and a high resolution transmission electron microscope (CM200 TEM, Philips Co., Amsterdam, The Netherlands) operated at 200 kV. X-ray diffraction (XRD) spectroscopy with a Philips X'pert Pro MPD (Philips Co., Eindhoven, The Netherlands), Cu Kα1 radiation $\lambda = 1.54060$ Å, 45 kV, 40 mA, was used to identify the crystalline phase of the product. N_2 physisorption on a Micromeritics TriStar 3000 (Micromeritics, Norcross, GA, USA) was employed to evaluate the specific surface area of the product. Prior to analysis, the sample was degassed at 150 °C under vacuum overnight. The Brunauer–Emmett–Teller (BET) model (5-points) was used to determine the specific surface area. The photoresponses and band gap properties of all tungsten oxide catalysts were measured using UV-Vis spectroscopy (Varian Cary

300) from 200–800 nm, with barium sulphate (BaSO₄) as the reference material. The oxidation states of platinum nanodeposits were determined by X-ray photoelectron spectroscopy (XPS-EscaLab 220-iXL, Al Kα radiation (1486.6 eV), Thermo VG Scientific Ltd., East Grinstead, UK). Inductively coupled plasma-atomic emission spectroscopy (ICP-AES, Varian Vista AX, Varian, Palo Alto, CA, USA) was used to determine the Na content within the synthesised particles.

4. Conclusions

The presence of SO_4^{2-} anions and pH control has been demonstrated to play an important role in controlling the final morphology and crystalline phase of hydrothermally synthesised WO_3 nanostructures. SO_4^{2-} anions promoted the formation of hexagonal nanobundles, while at pH values below 0.3, nanocube formation with a monoclinic-orthorhombic crystalline structure was favoured. The influence of WO_3 nanostructure morphology and crystalline phase on its capacity to photodegrade ethylene using visible light was investigated. The WO_3 nanocubes provide the best photodegradation performance due to their unique geometric configuration. The presence of Pt deposits improved the photoactivity of the nanoparticles and nanocubes, which was assigned to the Pt deposits ability to facilitate the multi-electron reduction of O_2. Utilising different light sources (*i.e.*, visible light or UV-A) to photodeposit the Pt on the WO_3 nanostructures altered the Pt deposit morphology, size and oxidation state, which also influenced the photocatalytic performance. The smallest photoactivity improvement was apparent for Pt loaded on the nanobundles, with this tentatively attributed to their hexagonal crystalline phase.

Supplementary Materials

Supplementary materials can be accessed at: http://www.mdpi.com/1420-3049/19/11/17747/s1.

Acknowledgments

The authors would like to acknowledge the assistance of personnel at the Mark Wainwright Analytical Centre (UNSW) with the XPS analysis, ICP-AES analysis and XRD training, as well as the Electron Microscopy Unit (UNSW) for help with the SEM/TEM imaging.

Author Contributions

Yossy Wicaksana was responsible for the experimental component of the study, including WO_3 particle synthesis, particle characterisation, photocatalytic experiments, compilation and interpretation of the results and manuscript proofing. Sanly Liu was responsible for preparation and proofing of the manuscript. Jason Scott was responsible for research direction and editing the manuscript. Rose Amal was responsible for research direction and proofing the manuscript.

Conflicts of Interest

The authors declare no conflict of interest.

310

References

1. Kotzias, D. Indoor air and human exposure assessment-needs and approaches. *Exp. Toxicol. Pathol.* **2005**, *57*, 5–7.

2. Kostiainen, R. Volatile organic compounds in the indoor air of normal and sick houses. *Atmos. Environ.* **1995**, *29*, 693–702.

3. Jo, W.K.; Park, K.H. Heterogeneous photocatalysis of aromatic and chlorinated volatile organic compounds (vocs) for non-occupational indoor air application. *Chemosphere* **2005**, *57*, 555–565.

4. Mo, J.; Zhang, Y.; Xu, Q.; Lamson, J.J.; Zhao, R. Photocatalytic purification of volatile organic compounds in indoor air: A literature review. *Atmos. Environ.* **2009**, *43*, 2229–2246.

5. Wang, S.; Ang, H.M.; Tade, M.O. Volatile organic compounds in indoor environment and photocatalytic oxidation: State of the art. *Environ. Int.* **2007**, *33*, 694–705.

6. Yu, J.; Qi, L. Template-free fabrication of hierarchically flower-like tungsten trioxide assemblies with enhanced visible-light-driven photocatalytic activity. *J. Hazard. Mater.* **2009**, *169*, 221–227.

7. Wang, X.P.; Yang, B.Q.; Zhang, H.X.; Feng, P.X. Tungsten oxide nanorods array and nanobundle prepared by using chemical vapor deposition technique. *Nanoscale Res. Lett.* **2007b**, *2*, 405–409.

8. Ponzoni, A.; Comini. E.; Ferroni, M.; Sberveglieri, G. Nanostructured WO_3 deposited by modified thermal evaporation for gas-sensing applications. *Thin Solid Films* **2005**, *490*, 81–85.

9. Baeck, S.H.; Choi, K.S.; Jaramillo, T.F.; Stucky, G.D.; McFarland, E.W. Enhancement of photocatalytic and electrochromic properties of electrochemically fabricated mesoporous WO_3 thin films. *Adv. Mater.* **2003**, *15*, 1269–1273.

10. Arutanti, O.; Ogi, T.; Nandiyanto, A.B.D.; Iskandar, F.; Okuyama, K. Controllable crystallite and particle sizes of WO_3 particles prepared by a spray-pyrolysis method and their photocatalytic activity. *Am. Inst. Chem. Eng. J.* **2014**, *60*, 41–49.

11. Satishkumar, B.C.; Govindaraj, A.; Nath, M.; Rao, C.N.R. Synthesis of metal oxide nanorods using carbon nanotubes as templates. *J. Mater. Chem.* **2000**, *10*, 2115–2119.

12. Badilescu, S.; Ashrit, P.V. Study of sol-gel prepared nanostructured WO_3 thin films and composites for electrochromic applications. *Solid State Ion.* **2003**, *158*, 187–197.

13. Song, X.C.; Zheng, Y.F.; Yang, E.; Wang, Y. Large-scale hydrothermal synthesis of WO_3 nanowires in the presence of K_2SO_4. *Mater. Lett.* **2007**, *61*, 3904–3908.

14. Salmaoui, S.; Sediri, F.; Gharbi, N. Characterization of h-WO_3 nanorods synthesized by hydrothermal process. *Polyhedron* **2010**, *29*, 1771–1775.

15. Gu, Z.; Zhai, T.; Gao, B.; Sheng, X.; Wang, Y.; Fu, H.; Ma, Y.; Yao, J. Controllable assembly of WO_3 nanorods/nanowires into hierarchical nanostructures. *J. Phys. Chem. B* **2006**, *110*, 23829–23836.

16. Rajagopal, S.; Nataraj, D.; Mangalaraj, D.; Djaoued, Y.; Robichaud, J.; Khyzhun, O.Y. Controlled growth of WO_3 nanostructures with three different morphologies and their structural, optical, and photodecomposition studies. *Nanoscale Res. Lett.* **2009**, *4*, 1335–1342.

17. Huang, K.; Pan, Q.; Yang, F.; Ni, S.; Wei, X.; He, D. Controllable synthesis of hexagonal WO₃ nanostructures and their application in lithium batteries. *J. Phys. D: Appl. Phys.* **2008**, *41*, 155417.

18. Xu, Z.; Tabata, I.; Hirogaki, K.; Hisada, K.; Wang, T.; Wang, S.; Hori, T. Preparation of platinum-loaded cubic tungsten oxide: A highly efficient visible light-driven photocatalyst. *Mater. Lett.* **2011**, *65*, 1252–1256.

19. Amano, F.; Ishinaga, E.; Yamakata, A. Effect of particle size on the photocatalytic activity of WO₃ particles for water oxidation. *J. Phys. Chem. C* **2013**, *117*, 22584–22590.

20. Denny, F.; Scott, J.; Chiang, K.; Teoh, W.Y.; Amal, R. Insight towards the role of platinum in the photocatalytic mineralisation of organic compounds. *J. Mol. Catal. A: Chem.* **2007**, *263*, 93–102.

21. Abe, R.; Takami, H.; Murakami, N.; Ohtani, B. Pristine simple oxides as visible light driven photocatalysts: Highly efficient decomposition of organic compounds over platinum-loaded tungsten oxide. *J. Am. Chem. Soc.* **2008**, *130*, 7780–7781.

22. Sclafani, A.; Palmisano, L.; Marc, G.; Venezia, A.M. Influence of platinum on catalytic activity of polycrystalline WO₃ employed for phenol photodegradation in aqueous suspension. *Sol. Energy Mater. Sol. Cells* **1998**, *51*, 203–219.

23. Zhang, G.; Guan, W.; Shen, H.; Zhang, X.; Fan, W.; Lu, C.; Bai, H.; Xiao, L.; Gu, W.; Shi, W. Organic additives-free hydrothermal synthesis and visible-light-driven photodegradation of tetracyline of WO₃ nanosheets. *Ind. Eng. Chem. Res.* **2014**, *53*, 5443–5450.

24. Bamwenda, G.R.; Arakawa, H. The visible light induced photocatalytic activity of tungsten trioxide powders. *Appl. Catal. A: Gen.* **2001**, *210*, 181–191.

25. Murata, A.; Oka, N.; Nakamura, S.; Shigesato, Y. Visible-light active photocatalytic WO₃ films loaded with Pt nanoparticles deposited by sputtering. *J. Nanosci. Nanotechnol.* **2012**, *12*, 5082–5086.

26. Choi, Y.G.; Sakai, G.; Shimanoe, K.; Miura, N.; Yamazoe, N. Preparation of aqueous sols of tungsten oxide dihydrate from sodium tungstate by an ion-exchange method. *Sens. Actuator B-Chem.* **2002**, *87*, 63–72.

27. Livage, J.; Guzman, G. Aqueous precursors for electrochromic tungsten oxide hydrates. *Solid State Ion.* **1996**, *84*, 205–211.

28. Gu, Z.; Li, H.; Zhai, T.; Yang, W.; Xia, Y.; Ma, Y.; Yao, J. Large-scale synthesis of single-crystal hexagonal tungsten trioxide nanowires and electrochemical lithium intercalation into the nanocrystals. *J. Solid State Chem.* **2007**, *180*, 98–105.

29. Wang, H.L.; Ma, X.D.; Qian, X.F.; Yin, J.; Zhu, Z.K. Selective synthesis of CdWO₄ short nanorods and nanofibers and their self-assembly. *J. Solid State Chem.* **2004**, *177*, 4588–4596.

30. Gu, Z.; Ma, Y.; Yang, W.; Zhang, G.; Yao, J. Self-assembly of highly oriented one-dimensional h-WO₃ nanostructures. *Chem. Commun.* **2005**, *28*, 3597–3599.

31. Reis, K.P.; Ramanan, A.; Whittingham, M.S. Hydrothermal synthesis of sodium tungstates. *Chem. Mater.* **1990**, *2*, 219–221.

32. Wang, J.; Khoo, E.; Lee, P.S.; Ma, J. Controlled synthesis of WO₃ nanorods and their electrochromic properties in H₂SO₄ electrolyte. *J. Phys. Chem. C* **2009**, *113*, 9655–9658.

33. Lide, D.R. *CRC Handbook of Chemistry and Physics*, 88th Ed.; CRC Press: Boca Raton, FL, USA, 2007–2008; pp. 8-20–8-29.

34. Lee, J.; Choi, W. Photocatalytic reactivity of surface platinized TiO_2: Substrate specificity and the effect of Pt oxidation state. *J. Phys. Chem. B* **2005**, *109*, 7399–7406.

35. Vorontsov, A.; Savinov, E.; Zhensheng, J. Influence of the form of photodeposited platinum on titania upon its photocatalytic activity in CO and acetone oxidation. *J. Photochem. Photobiol. A: Chem.* **1999**, *125*, 113–117.

36. Kato, H.; Asakura, K.; Kudo, A. Highly efficient water splitting into H_2 and O_2 over lanthanum-doped $NaTaO_3$ photocatalysts with high crystallinity and surface nanostructure. *J. Am. Chem. Soc.* **2003**, *125*, 3082–3089.

37. Xie, Y.P.; Liu, G.; Yin, L.; Cheng, H.-M. Crystal facet-dependent photocatalytic oxidation and reduction reactivity of monoclinic WO_3 for solar energy conversion. *J. Mater. Chem.* **2012**, *22*, 6746–6751.

38. Wang, C.M.; Heller, A.; Gerischer, H. Palladium catalysis of O_2 reduction by electrons accumulated on TiO_2 particles during photoassisted oxidation of organic compounds. *J. Am. Chem. Soc.* **1992**, *114*, 5230–5234.

39. Lee, S.L.; Scott, J.; Chiang, K.; Amal, R. Nanosized metal deposits on titanium dioxide for augmenting gas-phase toluene photooxidation. *J. Nanopart. Res.* **2009**, *11*, 209–219.

Sample Availability: Samples of the compounds are not available from the authors.

Surface Properties and Photocatalytic Activity of KTaO₃, CdS, MoS₂ Semiconductors and Their Binary and Ternary Semiconductor Composites

Beata Bajorowicz, Anna Cybula, Michał J. Winiarski, Tomasz Klimczuk and
Adriana Zaleska

Abstract: Single semiconductors such as KTaO₃, CdS MoS₂ or their precursor solutions were combined to form novel binary and ternary semiconductor nanocomposites by the calcination or by the hydro/solvothermal mixed solutions methods, respectively. The aim of this work was to study the influence of preparation method as well as type and amount of the composite components on the surface properties and photocatalytic activity of the new semiconducting photoactive materials. We presented different binary and ternary combinations of the above semiconductors for phenol and toluene photocatalytic degradation and characterized by X-ray powder diffraction (XRD), UV-Vis diffuse reflectance spectroscopy (DRS), scanning electron microscopy (SEM), Brunauer–Emmett– Teller (BET) specific surface area and porosity. The results showed that loading MoS₂ onto CdS as well as loading CdS onto KTaO₃ significantly enhanced absorption properties as compared with single semiconductors. The highest photocatalytic activity in phenol degradation reaction under both UV-Vis and visible light irradiation and very good stability in toluene removal was observed for ternary hybrid obtained by calcination of KTaO₃, CdS, MoS₂ powders at the 10:5:1 molar ratio. Enhanced photoactivity could be related to the two-photon excitation in KTaO₃-CdS-MoS₂ composite under UV-Vis and/or to additional presence of CdMoO₄ working as co-catalyst.

Reprinted from *Molecules*. Cite as: Bajorowicz, B.; Cybula, A.; Winiarski, M.J.; Klimczuk, T.; Zaleska, A. Surface Properties and Photocatalytic Activity of KTaO₃, CdS, MoS₂ Semiconductors and Their Binary and Ternary Semiconductor Composites. *Molecules* **2014**, *19*, 15339-15360.

1. Introduction

Since the discovery of the photocatalytic splitting of water on TiO₂ photochemical electrodes by Fujishima and Honda in 1972 [1], semiconductor-based photocatalysts have attracted increasing interest due to their potential applications in solar energy conversion, hydrogen evolution and photodegradation of organic pollutants [2–11]. Among different photocatalysts, titanium dioxide has become the most studied and widely used semiconductor material, however, it can only be excited by UV light, leading to low efficiency for utilizing solar energy [12]. Therefore, it is still of great importance to develop new types of photocatalysts which should be highly active, photostable and activated by low powered and low cost irradiation sources. Furthermore, one of the promising approaches is combining some semiconductors to form composites which can improve the efficiency of a photocatalytic system because of novel or enhanced properties that do not exist in individual components [13–16]. It was demonstrated that the Bi₂WO₆-based photocatalysts systems could be efficiently used in chemical synthesis and fuel production [7], whilst the graphene based composites seem to be a promising material for solar energy conversion and selective transformations of organic compounds [11]. Zhang *et al.* showed that the nanocomposites of TiO_2-graphene exhibits much higher photocatalytic activity and stability than bare titanium dioxide toward the gas-phase degradation of benzene [17].

Among semiconductors, potassium tantalate with a wide band gap could be an interesting alternative to the well-known titanium dioxide. Tantalates possess conduction bands consisting of Ta5d orbitals located at a more negative position than titanates (Ti3d) or niobates (Nb4d). Therefore, the high potential of the conduction band of tantalates could lead to being more advantageous in the photocatalytic reaction [18]. It has been reported by Liu *et al.* that sodium and potassium tantalates showed higher activities for water splitting than those of niobates synthesized by hydrothermal route. Moreover, the highest photocatalytic activity was exhibited by $NaTaO_3$ powder with cubic crystalline structure [19]. It was also observed that when NiO co-catalysts were loaded on the tantalate semiconductors, the photocatalytic activities were drastically increased [20]. Furthermore, there could be correlation between the photocatalytic activity of ABO_3 tantalates and preparation method as well as their crystal structure. Lin *et al.* showed that sol gel sodium tantalate with monoclinic crystalline structures exhibited higher photoactivity for water splitting than that with orthorhombic structure and obtained by solid-state method [21]. In other study Torres-Martínez *et al.* prepared sodium tantalate samples doped with Sm and La using sol-gel or solid state reaction methods. The best half-life time ($t_{1/2}$ = 65 min) of photocatalytic degradation of methylene blue under UV light was shown by $NaTaO_3$ doped with Sm obtained by sol-gel technique and heat treated at 600 °C [22]. However, in the literature there is still not enough information about photocatalytic activity of tantalate for degradation of organic pollutants, especially potassium tantalate has received very little research attention in this area.

In contrast to potassium tantalate, cadmium sulfide is among the most exhaustively investigated semiconductors. It has a wide range of applications, including solar cells, optoelectronic devices, fluorescence probes, sensors, laser light-emitting diodes and photocatalysis [23–25]. Unfortunately, bare CdS photocatalyst has very low separation efficiency of photogenerated electron-hole pairs and undergoes photocorrosion which limit its practical application [26]. In order to improve photoactivity and to inhibit the photocorrosion, cadmium sulfide is usually coupled with other semiconductors, including CdS/TiO_2 [27–29], CdS/ZnO [30,31], CdS/ZnS [32], CdS/WO_3 [33] as well as CdS/MoS_2. In particular, MoS_2 semiconductor is an efficient co-catalyst. Moreover, the unique layered structure of MoS_2 endows many important properties, such as anisotropy, chemical stability and anti-photo corrosion. [34]. There were few studies on the MoS_2-CdS system obtained by impregnation [35], ball-milling combined calcination [36], electrodeposition and chemical bath deposition [37] or hydrothermal [38] methods for hydrogen production under visible light. Zong *et al.* showed that individual CdS and MoS_2 particles were almost inactive in hydrogen evolution compared to the MoS_2 particles deposited onto CdS. Moreover, researchers observed that the rate of hydrogen evolution on MoS_2/CdS was higher than on CdS particles loaded with other catalysts such as Pt, Ru, Rh, Pd and Au. This result was explained by the better electron transfer between MoS_2 and CdS [39].

Binary and ternary composites based on potassium tantalate, cadmium sulfide and molybdenum disulfide could be new examples of photoactive materials. In this study we obtained various combinations of KTaO₃, CdS, MoS₂ single semiconductors as well as their precursor solutions using calcination and hydro/solvothermal routes. The effect of the preparation method, type and amount of the composite components on the surface properties and photocatalytic activity in phenol degradation in the aqueous phase as well as activity and photostability in toluene degradation in the gas phase was investigated. Light emitting diodes (LEDs) were used in the gas phase measurements as a promising irradiation source, which allow to reduce power consumption and costs of photocatalytic processes.

2. Results and Discussion

2.1. BET Surface Area

The surface area and the pore volume of $KTaO_3$, CdS, MoS_2 semiconductors and their binary and ternary composites prepared by hydro/solvothermal as well as hydro/solvothermal mixed solutions and calcination methods are summarized in Table 1. The surface area of as-prepared samples fluctuated from less than 0.1 to 17.5 m^2/g and was dependent on type and amounts of semiconductors as well as preparation methods. The pure CdS and MoS_2 semiconductors had BET surface area about 1.2 and 1.8 m^2/g, respectively. The CdS-MoS_2 composites containing more or the same amounts of CdS as MoS_2 caused an increase in surface area whereas in the presence of the excess MoS_2 in the composite a drop in the surface area was observed as compared with pure semiconductors. The sample CdS-MoS_2 4-1 presented the highest surface area (about 11.1 m^2/g) from among CdS-MoS_2 composites. The $KTaO_3$-based composites prepared by hydro/solvothermal methods had higher BET surface areas than composites which were calcined. Extremely small surface area and nearly zero pore volume were observed for the $KTaO_3$-MoS_2 10-1_C and $KTaO_3$-CdS-MoS_2 10-1-1_C samples. In the case of binary $KTaO_3$-based composites it could be seen that potassium tantalate containing a small amount of MoS_2 had lower surface area than potassium tantalate containing a small amount of CdS. Moreover, in the case of ternary composites it could be also observed that the high loading of CdS resulted in a significant increase in surface area. The pore volumes of obtained photocatalysts were very low and fluctuated from zero to 0.0088 cm^3/g. It was observed that the pore volumes increased with increasing the surface area.

2.2. XRD Analysis

Figure 1a shows XRD patterns for pure CdS and MoS_2 as well as CdS-MoS_2 samples with different CdS:MoS_2 molar ratio. Low observed intensity of the XRD peaks for MoS_2 and for CdS-MoS_2 with 1:5 ratio is likely caused by low crystallinity of the MoS_2 compound. As can be seen, adding less than 20% of CdS to MoS_2 causes a dramatic change in XRD pattern. CdS is a dominant phase and MoS_2 is hardly present. No traces of MoS_2 have been found for higher CdS concentration (CdS-MoS_2 1:1 and CdS-MoS_2 5:1 samples). Lattice parameters for CdS ($P6_3mc$, s.g. #186) were refined by using the LeBail method.

Table 1. Sample label, preparation method, BET surface area and pore volume of obtained single semiconductors and their composites.

Sample Label	KTaO₃:CdS:MoS₂ / Molar Ratio	Preparation Method	BET Surface Area/ [m²/g]	Pore Volume/ [cm³/g]
KTaO₃	1:0:0	hydrothermal	0.1	0.0001
CdS	0:1:0	solvothermal	1.2	0.0006
MoS₂	0:0:1	hydrothermal	1.8	0.0009
CdS-MoS₂ 5-1	0:5:1	solvothermal mixed solutions	5.0	0.0026
CdS-MoS₂ 4-1	0:4:1	solvothermal mixed solutions	11.1	0.0056
CdS-MoS₂ 1-1	0:1:1	solvothermal mixed solutions	10.4	0.0052
CdS-MoS₂ 1-5	0:1:5	solvothermal mixed solutions	1.0	0.0006
KTaO₃-CdS 10-1_MS	10:1:0	solvothermal mixed solutions	17.5	0.0088
KTaO₃-CdS 10-1_C	10:1:0	hydro/solvothermal and calcination	0.4	0.0002
KTaO₃-MoS₂ 10-1_MS	10:0:1	hydrothermal mixed solutions	2.8	0.0014
KTaO₃-MoS₂ 10-1_C	10:0:1	hydrothermal and calcination	<0.1	~0
KTaO₃-CdS-MoS₂ 10-1-1_MS	10:1:1	solvothermal mixed solutions	4.0	0.0019
KTaO₃-CdS-MoS₂ 10-1-1_C	10:1:1	hydro/solvothermal and calcination	<0.1	~0
KTaO₃-CdS-MoS₂ 10-5-1_MS	10:5:1	solvothermal mixed solutions	10.3	0.0051
KTaO₃-CdS-MoS₂ 10-5-1_C	10:5:1	hydro/solvothermal and calcination	0.5	0.0003

For CdS we obtained a = 4.1341(8) Å and c = 6.710(1) Å in very good agreement with reported by Wiedemeire *et al.* [40]. Adding MoS₂ does not change *a* lattice parameter, whereas *c* increases with increasing MoS₂ and for the sample which contains the highest amount of MoS₂ (sample b) the refined *c* is 6.733(3) Å. This clear change of the unit cell size and relative change of the intensity of three most dominant XRD peaks between 24 and 29 degrees, allow us to conclude that Mo atoms are incorporated into CdS hexagonal crystal structure. Figure 1b displays powder X-ray diffraction patterns for pure KTaO₃ prepared by the hydrothermal method. Two different methods were used to obtain KTaO₃-CdS and KTaO₃-MoS₂ composites. As can be seen from the XRD patterns for KTaO₃-CdS 10:1_C and KTaO₃-MoS₂ 10:1_C samples, the calcination method does not produce new phases and KTaO₃ together with CdS are observed. For the latest compound the most intense peaks are hardly visible between 25 and 30 deg. Contrary to the calcination method, the hydro/solvothermal mixed solutions route produces more complex compounds. For the system with CdS, the majority phase is Ta₂O₅ with small amount of TaS₂ and K₂Ta₁₅O₃₂. For the system with MoS₂ almost pure pyrochlore-like K₂Ta₂O₆ was found. This defect pyrochlore structure was described in details and also tested as a photocatalyst by Hu *et al.* [41].

Figure 1. Powder XRD patterns for **(a)** single CdS, MoS_2 and binary CdS-MoS_2 nanocomposites, **(b)** single $KTaO_3$ and binary $KTaO_3$-based nanocomposites, **(c)** ternary $KTaO_3$-based nanocomposites.

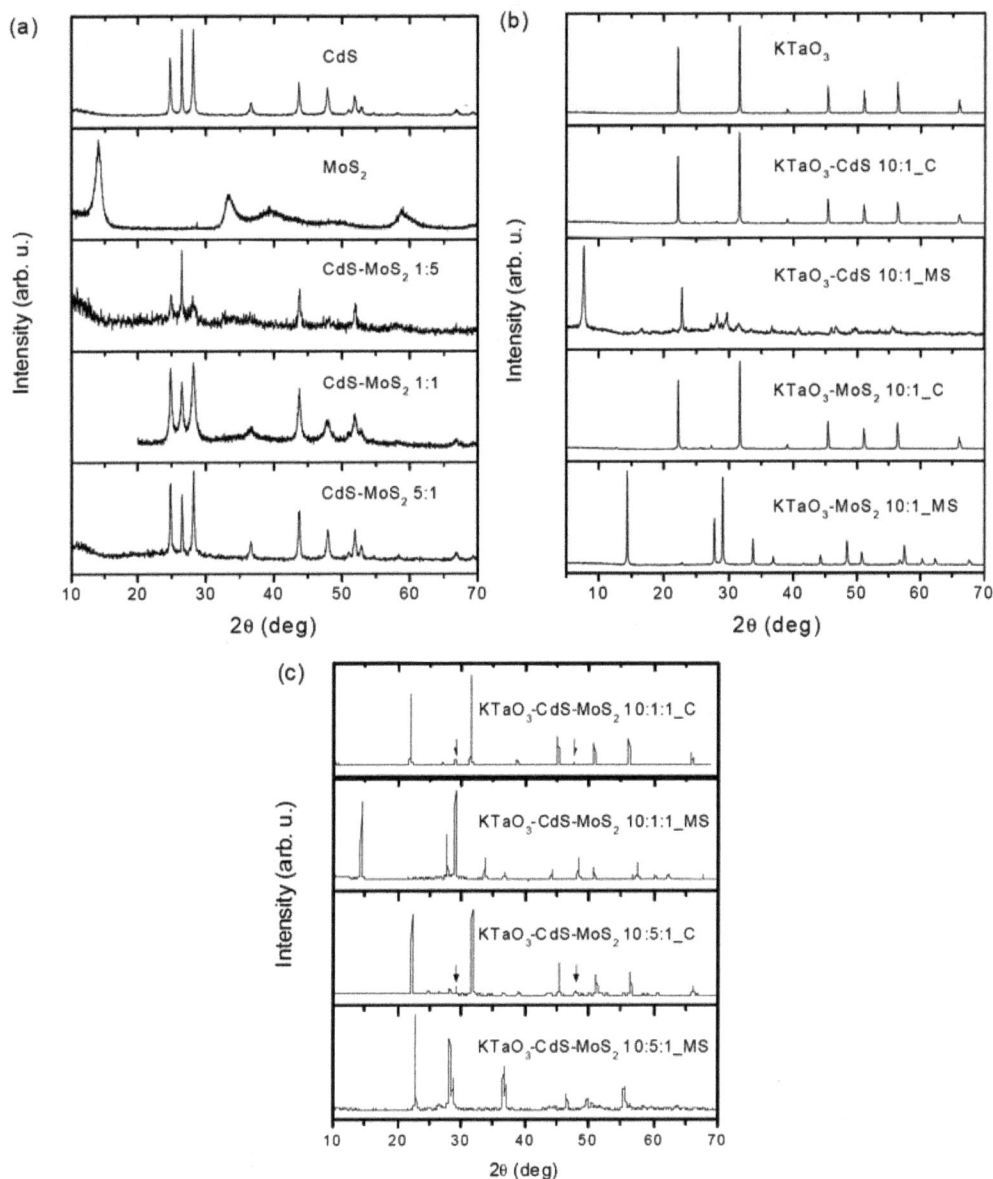

Figure 1c shows XRD patterns for ternary $KTaO_3$-CdS-MoS_2 composites with two different molar ratios and two various preparation routes. Similarly to what was observed for binary composites produced by calcination method, the majority phase is $KTaO_3$, however, a small amount of $CdMoO_4$ emerges with the two most intense peaks marked by arrow. While the solvothermal mixed solutions method causes decomposition of $KTaO_3$, the main products detected are $K_2Ta_2O_6$ and Ta_2O_5 together with Mo-doped CdS for the sample 10:1:1 and 10:5:1, respectively.

2.3. Morphology

The morphology of the products was observed by SEM and is presented in the Figures 2–5. Figure 2 shows an SEM image of single semiconductors prepared by hydro/solvothermal methods. As clearly shown, the KTaO₃ have a cube-like shape with various sizes. Using higher magnification image of these particles, we notice that small KTaO₃ nanocubes with an average size of about 0.2–1 μm grow at the surface of bigger cubes having widths of 5–8 μm. Pure CdS clearly indicates nanoleaf structures having the length of about 3 μm. In the case of MoS₂ we observed that the product has the shape of flower-like microspheres with an average diameter of 5 μm.

Figure 2. SEM images of single semiconductors obtained by hydrothermal method: (**a**,**b**) KTaO₃; (**c**) CdS; (**d**) MoS₂.

(a) (b)

(c) (d)

Figure 3 shows SEM images of binary CdS-MoS₂ composites with different CdS and MoS₂ molar ratios. In the case of CdS-MoS₂ 5-1 sample containing the highest molar ratio of CdS, we still observed nanoleaf structure indicating the presence of CdS. XRD analysis also showed that no traces of MoS₂ have been found for higher CdS concentration. The increase of MoS₂ ratio resulted in the formation of hexagonal shaped nanostructures with average edge size of about 100–125 nm (Figure 3b).

As can be seen from Figure 3c,d, further increase in molar ratio of MoS_2 to CdS causes a large change in microstructures. We can observe bonded structures of microspheres with diameters ranging from 0.08 to 1 µm.

Figure 3. SEM images of binary $CdS-MoS_2$ composites obtained by solvothermal mixed solution methods with different molar ratio of CdS: (**a**) $CdS:MoS_2 = 5:1$; (sample $CdS-MoS_2$ 5-1); (**b**) $CdS:MoS_2 = 4:1$ (sample $CdS-MoS_2$ 4-1); (**c**) $CdS:MoS_2 = 1:1$ (sample $CdS-MoS_2$ 1-1); and (**d**) $CdS:MoS_2 = 1:5$ (sample $CdS-MoS_2$ 1-5).

(a)

(b)

(c)

(d)

Figure 4 displays SEM images of binary $KTaO_3-CdS$ and $KTaO_3-MoS_2$ composites prepared by two different routes. In the case of the $KTaO_3-CdS$ and $KTaO_3-MoS_2$ samples, which were obtained by calcination, we can see mainly the structure of cubes, which indicates the presence of $KTaO_3$. In contrast, the solvothermal mixed solutions route leads to the formation of other nanostructures, suggesting the presence of more complex compounds: $K_2Ta_{15}O_{32}$, Ta_2O_5, TaS_2 and $K_2Ta_2O_6$ for the system with CdS and MoS_2 which were confirmed by XRD analysis. The transition of the $KTaO_3$ structure from cubic to octahedral was observed for the samples $KTaO_3-MoS_2$ 10-1 and $KTaO_3-CdS-MoS_2$ 10-1-1 obtained by solvothermal mixed solutions. In the case of calcinated $KTaO_3-CdS-MoS_2$ 10-5-1 composite we can see nanoleaves of CdS deposited on the surface of large cubes of $KTaO_3$ (Figure 5d). Contrary to calcination, solvothermal mixed solutions method causes dramatic changes in the structure—neither cubes nor nanoleaf are observed (Figure 5c).

Figure 4. SEM images of binary KTaO₃-CdS and KTaO₃-MoS₂ composites obtained with different molar ratio and using two preparation routes: (**a**) KTaO₃-CdS (10:1) obtained by solvothermal mixed solutions (sample KTaO₃-CdS 10-1_MS); (**b**) KTaO₃-CdS (10:1) obtained by calcination of single previously synthesized semiconductors (sample KTaO₃-CdS 10-1_C); (**c**) KTaO₃-MoS₂ (10:1) obtained by solvothermal mixed solutions (sample KTaO₃-MoS₂ 10-1_MS); and (**d**) KTaO₃-MoS₂ (10:1) obtained by calcination of single previously synthesized semiconductors (sample KTaO₃-MoS₂ 10-1_C).

(a)

(b)

(c)

(d)

2.4. UV-Vis Properties

DRS UV-Vis absorption spectra in the wavelength range of 200–800 nm of as-prepared samples were investigated and the results are shown in Figure 6a–c. Figure 6a depicts the spectra for pure CdS and MoS₂ semiconductors and their binary composites with varying molar ratio between CdS and MoS₂. As it can be seen, all the samples could absorb both UV and visible light. The absorption edge of single CdS is at about 510 nm, which coincides with the literature. Moreover, it was previously reported that the absorption properties of CdS are strongly shape-dependent [42]. It could also be seen that the loading of MoS₂ on CdS improved the light absorption as compared with pure CdS. Liu *et al.* also presented enhanced absorption properties for CdS-MoS₂ composites to photocatalytic H₂ production [38]. In comparison with single CdS, composites with excess of CdS exhibited a red-shift and a less steep absorption edge.

Figure 5. SEM images of ternary KTaO₃-CdS-MoS₂ composites obtained with different molar ratio and using two different preparation route: (a) KTaO₃-CdS-MoS₂ (10:1:1) obtained by solvothermal mixed solutions (sample KTaO₃-CdS-MoS₂ 10-1-1_MS); (b) KTaO₃-CdS-MoS₂ (10:1:1) obtained by calcination of single previously synthesized semiconductors (sample KTaO₃-CdS-MoS₂ 10-1-1_C); (c) KTaO₃-CdS-MoS₂ (10:5:1) obtained by solvothermal mixed solutions (sample KTaO₃-CdS-MoS₂ 10-5-1_MS); and (d) KTaO₃-CdS-MoS₂ (10:5:1) obtained by calcination of single previously synthesized semiconductors (sample KTaO₃-CdS-MoS₂ 10-5-1_C).

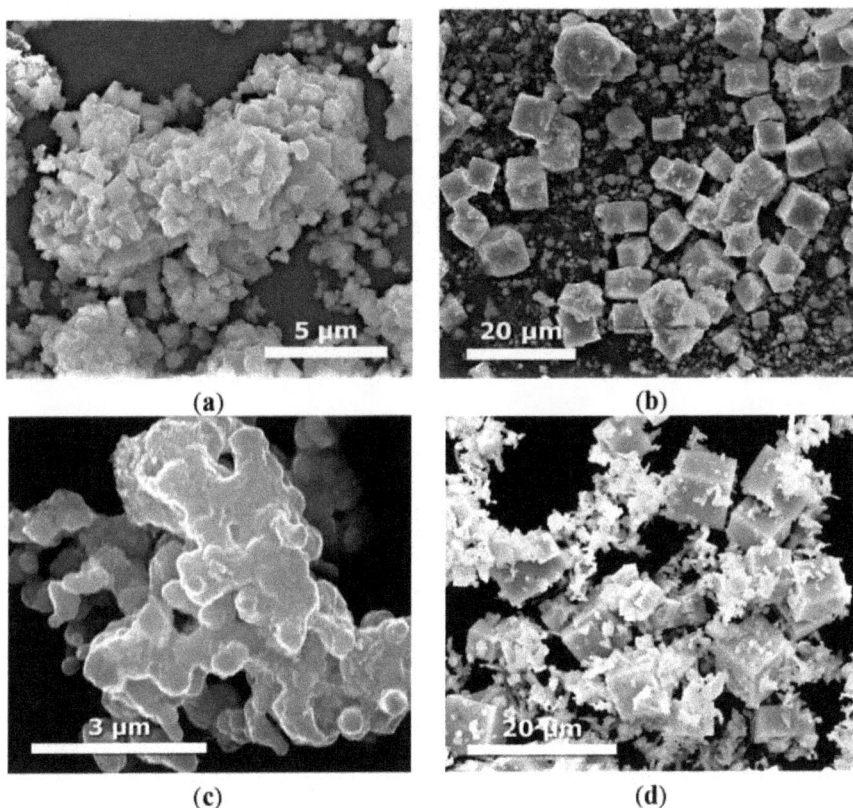

(a) (b) (c) (d)

The absorbance was very similar for pure MoS₂ and composite with 1:1 molar ratio between components. However, for composites containing excess MoS₂, absorption was more intense. Figure 6b displays absorption spectra of single KTaO₃ and binary KTaO₃-based composites containing small amounts of CdS or MoS₂ and prepared by two different procedures. In the case of single KTaO₃, there is an obvious absorption band centered at about 310 nm and no absorption peak is detected above 310 nm. For octahedral KTaO₃ nanocrystalline obtained by Zou et al. using hydrothermal method, the absorption peak was detected at 265 nm [43]. It was also observed that binary KTaO₃-based composites loaded with small amount of CdS or MoS₂ had steeper absorption edges and maximum absorption shifted to shorter wavelengths as compared with single potassium tantalate. Moreover, KTaO₃-CdS composites showed second steep absorption edges in the visible light region. Figure 6c shows the spectra for ternary KTaO₃-CdS-MoS₂ composites containing different amount of CdS and prepared by various methods. Loading of larger amounts of CdS onto KTaO₃-CdS-MoS₂ composites significantly improved the spectral absorption. For ternary composites containing greater amount of cadmium sulfide and obtained by both methods, there were widest absorption ranges among all obtained samples but only for a calcined composite a steep absorption edge (at about 510 nm) was observed. In general, loading MoS₂ onto CdS shifted absorbance peaks to longer wavelengths, just as loading CdS onto KTaO₃ enhanced absorption features. It may be suggested that the best adsorption properties could probably be achieved for the ternary KTaO₃-CdS-MoS₂ composites containing appropriate molar ratio between semiconductors.

Figure 6. The UV-Vis diffuse reflectance spectra of (**a**) single CdS, MoS₂, and binary CdS-MoS₂ nanocomposites, (**b**) single KTaO₃ and binary KTaO₃-based nanocomposites, (**c**) ternary KTaO₃-based nanocomposites.

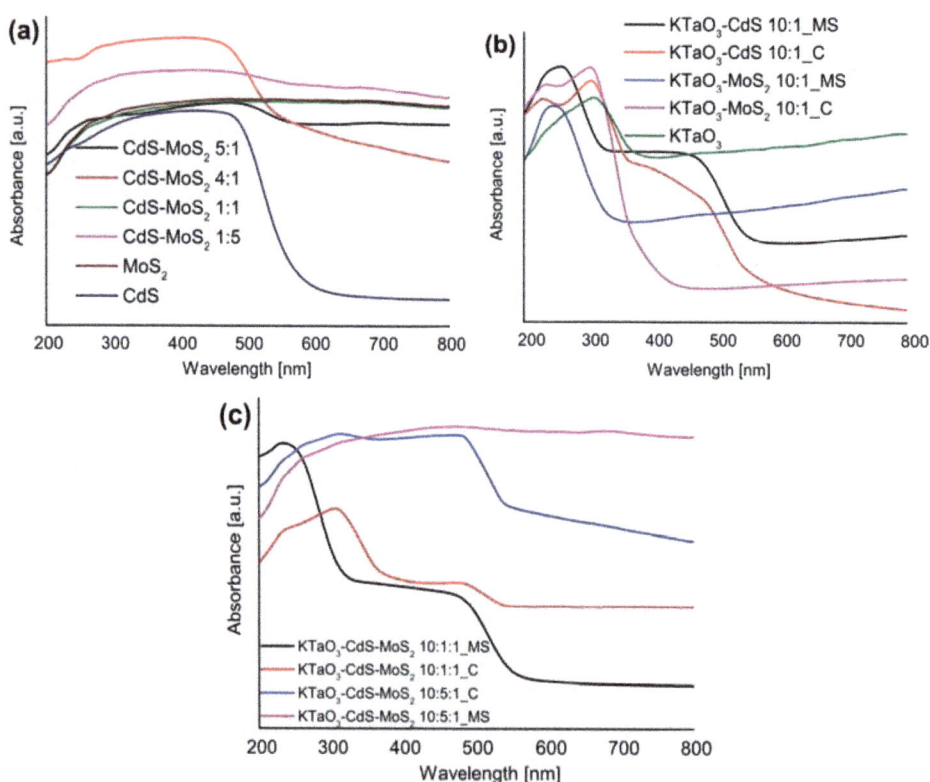

2.5. The Photocatalytic Degradation of Phenol in the Aqueous Phase

Photocatalytic activity of as-prepared semiconductors and their nanocomposites was estimated by examining the reaction of phenol degradation in the presence of UV-Vis light irradiation and for selected samples photoactivity under visible light was also analyzed. No degradation of phenol was observed in the absence of photocatalyst or illumination. Photocatalytic activity under UV-Vis and visible light is presented as phenol degradation rate (Table 2) and as efficiency of phenol removal after 60 min of irradiation (Figures 7 and 8). Commercially available P25 TiO$_2$ was used as a standard reference material in photocatalytic activity measurements.

Figure 7. The percentage degradation of phenol at various time intervals under UV-Vis light in the presence: (**a**) single CdS, MoS$_2$, and CdS-MoS$_2$ nanocomposites; (**b**) single KTaO$_3$ and binary KTaO$_3$-based nanocomposites; (**c**) ternary KTaO$_3$-based nanocomposites.

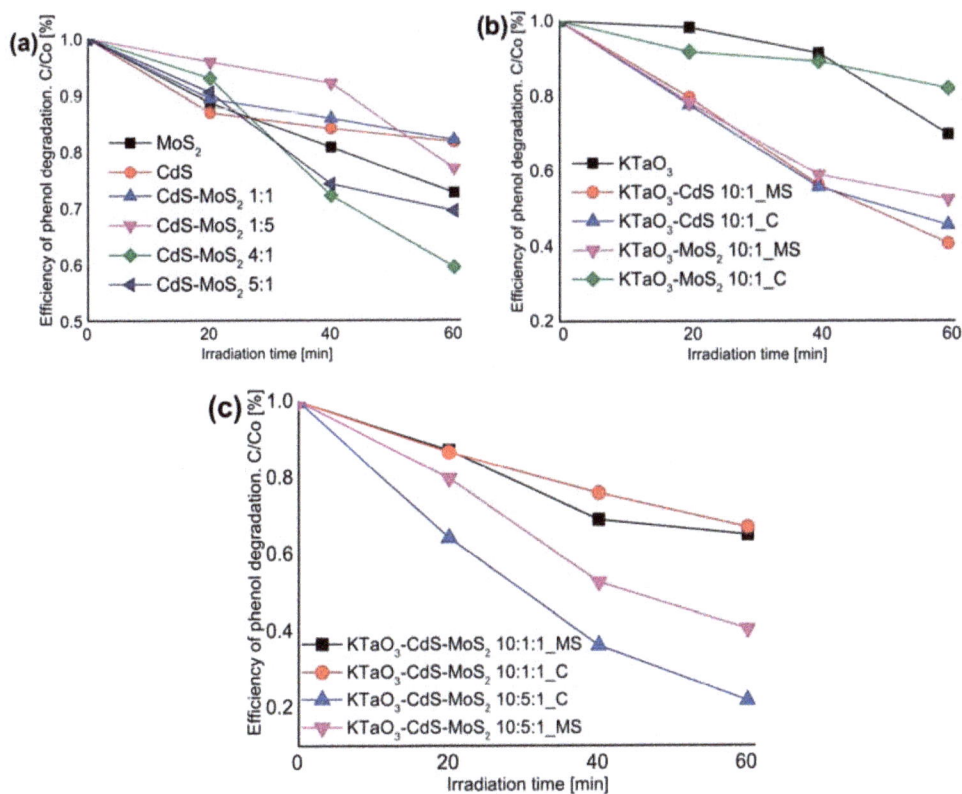

Figure 7a shows the results of phenol degradation under UV-Vis light in the presence of pure CdS and MoS$_2$, and their binary composites synthesized by hydro/solvothermal methods. In comparison with pure CdS and MoS$_2$, photocatalytic activity for degradation of phenol increased in the presence of CdS-MoS$_2$ nanocomposites containing excess of CdS. In the presence of CdS-MoS$_2$ 4:1 and CdS-MoS$_2$ 5:1, the percentage degradation of phenol after 60 min was about 41% and 31%, respectively, which indicates that with decreasing excess of CdS, the photocatalytic activity is relatively higher. For composites containing more or the same amount of MoS$_2$ as CdS, photoactivity was comparable with pure semiconductors. The highest phenol degradation rate, about 1.41 µmol·dm^{-3}·min^{-1}, was observed in the case of the CdS-MoS$_2$ 4:1 composite. For the other samples degradation rates were low and fluctuated from 0.62 to 0.90 µmol·dm^{-3}·min^{-1}. Zong and co-workers reported that the photocatalytic H$_2$ production achieved a maximum when the loading amount of MoS$_2$ on CdS was about 0.2 wt % [39]. Figure 7b displays the results of phenol degradation under UV-Vis light in the presence of single KTaO$_3$ and binary potassium tantalate-based composites containing a small amount of CdS or MoS_2 and prepared by different routes. The phenol degradation after 60 min of irradiation was only 18% in the case of the calcined $KTaO_3$-MoS_2 sample, which is lower than in the presence of single $KTaO_3$ (about 30%). While for the $KTaO_3$-MoS_2 sample obtained by hydrothermal method phenol degradation was more efficient (about 48%). It could be also observed that photoactivities were similar and increased to 59% and 55% for $KTaO_3$-CdS 10-1 composites prepared by hydrothermal and calcination method, respectively. It may be suggested that loading cadmium sulfide onto $KTaO_3$ enhanced photoactivity, while the influence of introduction of molybdenum disulfide on the photoactivity was dependent on preparation technique. Figure 7c depicts the results of phenol degradation under UV-Vis light for ternary $KTaO_3$-CdS-MoS_2 composites containing various amounts of CdS as well as prepared by different methods. In general, more addition of CdS significantly enhanced the photocatalytic activity of ternary composites. In the presence of the $KTaO_3$-CdS-MoS_2 10-1-1 sample the percentage degradation of phenol was low (about 35%) and quite comparable for both methods. In the case of the $KTaO_3$-CdS-MoS_2 10-5-1 composites photoactivities were much more efficient. After 60 min of irradiation, for the sample prepared by hydro/solvothermal method phenol degradation reached 60%. At the same time in the presence of the $KTaO_3$-CdS-MoS_2 10-5-1 sample prepared by the calcination method about 80% of phenol was degraded and it was the highest efficiency among all obtained samples. Therefore, appropriate ratio between $KTaO_3$, CdS and MoS_2 semiconductors in the composites is supposed to be responsible for the highest efficiency of phenol photodegradation. This deduction coincides with the UV-Vis properties of obtained samples described in Section 3.3.

Figure 8. The percentage degradation of phenol at various time intervals under visible light in the presence of selected composites.

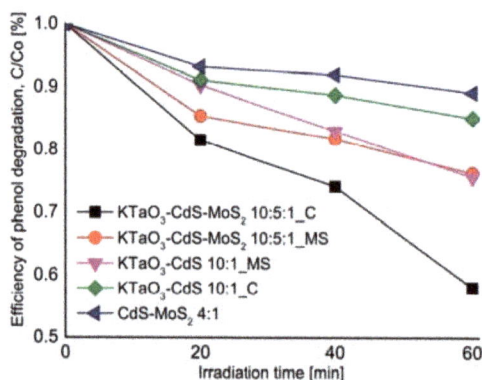

Figure 8 shows the results of phenol degradation under visible light for selected samples which exhibited the highest efficiency of phenol degradation under UV-Vis light irradiation. Among these, the highest activity (about 42%) after 60 min of visible irradiation was also exhibited by calcined ternary $KTaO_3$-CdS-MoS_2 10-5-1 composites. In the case of the binary calcined composite containing small amount of CdS, photoactivity reached about 15%. In the presence of samples obtained by the solvothermal mixed solutions route activity was more efficient (24%). The lowest activity (about 11%) was observed for the composite containing CdS and MoS$_2$ at the 4:1 molar ratio.

2.6. The Photocatalytic Degradation of Toluene in the Gas Phase

The photocatalytic activity and stability of as-prepared photocatalysts were also evaluated by degradation of the toluene in the gas phase under UV light irradiation using low powered irradiation source, e.g., LEDs (λ_{max} = 375 nm) in four subsequent measurement cycles. The efficiency of toluene photodegradation after 60 min in the presence of as-prepared semiconductor samples is given in Table 2.

It could be seen that CdS, KTaO$_3$ and MoS$_2$ semiconductors and their composites were photoactive in toluene degradation. In the case of single semiconductors, removal efficiency in the presence of CdS reached about 57% after a 60-min process and activity was stable in four subsequent cycles of irradiation. While in the presence of KTaO$_3$ and MoS$_2$ photocatalytic activity decreased from 64% and 46% (1st cycle) to 37% and 22% (4th cycle), respectively. The highest activity and stability under UV light irradiation from all among CdS-MoS$_2$ composites was exhibited by samples at the molar ratio 4:1 and 1:1.

Table 2. Photocatalytic activities of obtained samples for toluene photodegradation in four subsequent measurement cycles and for phenol degradation presented as phenol degradation rate.

Sample Label	Phenol Degradation Rate under UV-Vis/(μmol·dm^{-3}·min^{-1})	Toluene Degradation after 1 h Irradiation (LEDs, λ_{max} = 375 nm) [%]			
		1st Cycle	2nd Cycle	3rd Cycle	4th Cycle
KTaO$_3$	0.79	64	63	42	37
CdS	0.61	57	57	57	52
MoS$_2$	0.90	46	23	22	22
CdS-MoS$_2$ 1-5	0.77	57	53	44	27
CdS-MoS$_2$ 1-1	0.62	61	53	62	52
CdS-MoS$_2$ 5-1	0.81	70	60	49	48
CdS-MoS$_2$ 4-1	1.41	53	56	60	62
KTaO$_3$-CdS 10-1_MS	2.08	47	45	41	40
KTaO$_3$-CdS 10-1_C	1.75	53	48	52	50
KTaO$_3$-MoS$_2$ 10-1_MS	1.69	55	51	49	51
KTaO$_3$-MoS$_2$ 10-1_C	0.55	46	34	37	35
KTaO$_3$-CdS-MoS$_2$ 10-1-1_MS	1.15	50	52	48	39
KTaO$_3$-CdS-MoS$_2$ 10-1-1_C	1.11	53	54	49	40
KTaO$_3$-CdS-MoS$_2$ 10-5-1_MS	1.99	50	41	43	41
KTaO$_3$-CdS-MoS$_2$ 10-5-1_C	2.81	48	48	50	46
TiO$_2$ (P25)	2.87	98	96	95	95

The lowest photostability presented the CdS-MoS$_2$ sample when the molar ratio was 1:5. It suggests that an excess of MoS$_2$ caused the decrease in photostability. The CdS-MoS$_2$ composite at the molar ratio 5:1 exhibited the highest activity in the 1st cycle (about 70%) but this activity decreased to 48% in the 4th cycle of photoirradiation. In the case of binary $KTaO_3$-based composites it was observed that the photoactivity about 50% and the best stability during four cycles of irradiation was exhibited by potassium tantalate with a small amount of cadmium sulfide ($KTaO_3$-CdS 10-1 sample) prepared by the calcination method and potassium tantalate with molybdenum disulfide ($KTaO_3$-MoS_2 10-1 sample) obtained by the hydrothermal mixed solutions route. While the photocatalytic activity of the $KTaO_3$-MoS_2 10-1 sample prepared by calcination method after four cycles of irradiation was less by 11% as compared with the first cycle. It indicates that the activity is dependent not only on the type and content of the composite components but also on the preparation method. All of prepared ternary $KTaO_3$-CdS-MoS_2 composites exhibited about 50% efficiency degradation of toluene after the 1st cycle of irradiation, but activity didn't change after four cycles only in the case of the sample $KTaO_3$-CdS-MoS_2 10-5-1 obtained by the calcination method.

2.7. The Origin of Photocatalytic Activity

UV and visible light induced photoactivity presented as efficiency of phenol removal after 60-min irradiation for the most active binary and ternary composites is shown in Figure 9. Highest activity under UV-Vis light of $KTaO_3$-CdS-MoS_2 (10:5:1) composite, obtained by solvothermal preparation of single semiconductors followed by calcination, could be related to the presence of heterojunctions between semiconductors. One pot reaction used for preparation of the $KTaO_3$-CdS-MoS_2 composite did not allow the formation of well-developed crystal structure of all semiconductors, as was shown in XRD analysis. However, the structure obtained by the sintering of three components could favors photogenerated charge carriers transfer between phase boundary of semiconductors resulted in higher photoactivity. The $KTaO_3$-CdS-MoS_2 nanocomposite could be excited by two photons due to utilization of a heterojunction between $KTaO_3$ with two low band gap semiconductors (*i.e.*, CdS and MoS_2). As shown in Figure 10, UV light could induce excitation in solid $KTaO_3$ (E_g = 3.4 eV) and generate free electrons (e^-) and holes (h^+). Since the VB position of $KTaO_3$ (+3.1 V) is lower than the VB band of MoS_2 (+1.8 V), h^+ transfer would then occur from $KTaO_3$ to MoS_2. At the same time UV or visible light could be absorbed by CdS (E_g = 2.5 eV) and generate e^- and h^+ in CdS. As the CB position of CdS (−0.6 V) is higher in energy than the CB of solid MoS_2 (+0.05 V), electron transfer could occur from CdS to MoS_2, thereby causing charge separation at the CdS-MoS_2 junction. Finally positioning the CB of $KTaO_3$ (−0.3 V) higher in energy than the VB of CdS (+1.8 V) causes e^- transfer from $KTaO_3$ to CdS and recombination of e^- in $KTaO_3$ with h^+ generated in CdS, thus completing the photoexcitation cycle in which the excited state of MoS_2 has been achieved via a two-photon process. Thus, organic compounds (such as phenol or toluene) could be degraded via active oxygen species generated at the surface of excited MoS_2 in two photons process.

Figure 9. Efficiency of phenol removal under UV-Vis and visible light (λ > 420 nm) for selected binary and ternary composites.

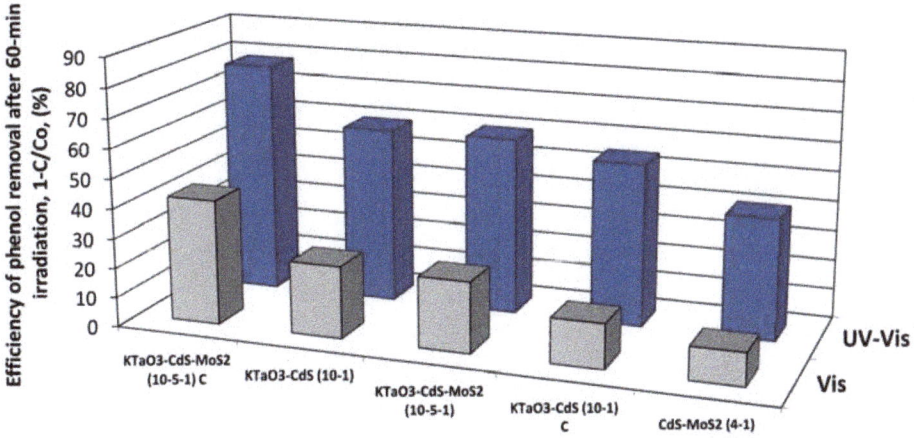

Figure 10. Probable UV-Vis two-photon excitation cycle of KTaO₃-CdS-MoS₂ composite: (1) UV induced excitation in solid KTaO₃ (E_g = 3.4 eV) to generate free electrons (e^-) and holes (h^+), and since the VB position of KTaO₃ is lower than the VB band of MoS₂, h^+ transfer would then occur from KTaO₃ to MoS₂; (2) UV or visible light excitation of solid CdS (E_g = 2.5 eV) also generates e^- and h^+ in CdS, and as the CB position of CdS is higher in energy than the CB of solid MoS₂, electron transfer could occur from CdS to MoS₂, thereby causing a charge separation at the CdS-MoS₂ junction; (3) positioning the CB of KTaO₃ higher in energy than the VB of CdS causes e^- transfer from KTaO₃ to CdS and recombination of e^- in KTaO₃ with h^+ generated in CdS, thus completing the photoexcitation cycle in which the excited state of MoS₂ has been achieved via a two-photon process.

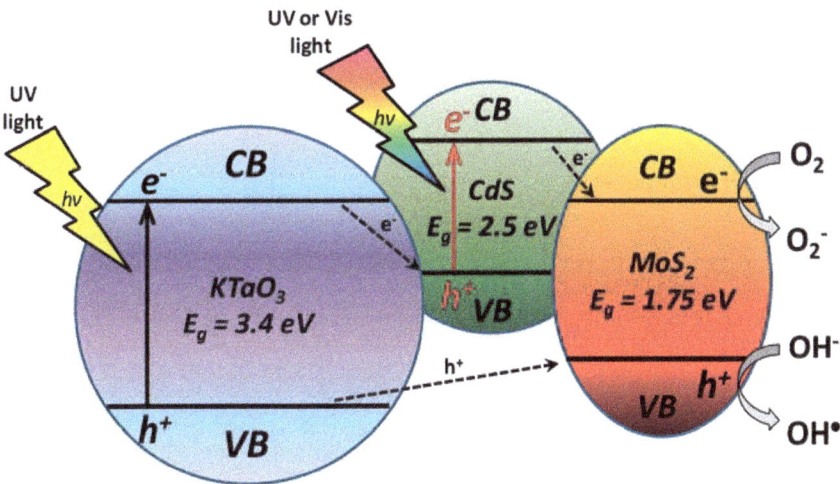

Additionally, higher activity under UV-Vis light of the KTaO₃-CdS-MoS₂ 10-5-1_C sample could be also related to the presence of $CdMoO_4$ detected by XRD analysis (see Figure 1c). Photocatalytic activity of pure and Ag-modified $CdMoO_4$ under simulate solar irradiation was reported by others [44,45]. The valence band of $CdMoO_4$ consists of the hybrid orbitals of O 2p as well as Cd 6s and the conduction band consists of Mo 4d orbital and the band gap energies between them is about 3.4 eV. Thus, $CdMoO_4$ presented at the surface of composite could also adsorb UV light and enhance activity in observed photodegradation reaction. Lower UV and Vis-induced photoactivity of KTaO₃-CdS-MoS₂ (10:5:1) composite obtained by mixed solvothermal reaction, comparing to that one prepared by calcination step, could be related the decomposition of $KTaO_3$ and formation of $K_2Ta_2O_6$ and Ta_2O_5 as main component of as-prepared samples.

3. Experimental

3.1. Materials and Instruments

Tantalum (V) oxide (>99% Aldrich, Poznan, Poland) and potassium hydroxide (Chempur, pure p.a.) were used as precursors for the preparation of $KTaO_3$. Thiourea (Aldrich, 99%) has been chosen as a sulfur source to synthesize sulfides. $CdCl_2 \cdot 2H_2O$ (Sigma-Aldrich, ≥99%) and $Na_2MoO_4 \cdot 2H_2O$ (Sigma-Aldrich, ≥99.5%) were used as precursors of Cd and Mo, respectively. Ethanol, polyethylene glycol 400 (PEG-400), ethylene glycol (EG) were purchased from POCH S.A. (Gliwice, Poland) and used without further purification. Deionized water was used for all reactions and treatment processes. A commercial form of TiO_2 (P25, crystalline composition: 80% anatase, 20% rutile, surface area 50 g/m²) from Evonik (Essen, Germany) was used for the comparison of the photocatalytic activity.

Nitrogen adsorption–desorption isotherms at 77 K were measured using a Micromeritics Gemini V (model 2365) physisorption analyzer (Micromeritics Instrument, Norcross, GA, USA). Specific surface areas were calculated following typical Brunauer–Emmett–Teller (BET) method using the adsorption data in the relative pressure (p/p₀) range from 0.05 to 0.3. Prior to adsorption measurements the samples were degassed under vacuum at 200 °C for 2 h.

Diffuse reflectance spectra (DRS) of the synthesized materials were characterized using the Thermo Scientific Evolution 220 UV-Visible spectrophotometer (Thermo Scientific, Waltham, MA, USA) equipped with ISA-220 integrating sphere accessory. The UV-Vis DRS spectra were recorded in the range of 200–800 nm using a barium sulfate reference.

Powder X-ray diffraction (PXRD, Philips/PANalytical X'Pert Pro MPD diffractometer, (PANalytical, Almelo, The Netherlands) with Cu Kα radiation λ = 1.5418 Å) was used to determine the phase composition and calculate lattice parameters of polycrystalline samples.

The morphology of the semiconductor composites was investigated with FEI Quanta 250 FEG scanning electron microscope (SEM; FEI, Hillsboro, OR, USA) working in high vacuum mode. Energy-dispersive X-ray spectroscopy (EDS) measurements were carried out using SEM-integrated EDAX Apollo-SDD detector (EDAX Inc., Mahwah, NJ, USA). Accelerating voltage was set to 30 kV. Standardless analysis was conducted with the EDAX TEAM software with *eZAF* quantization method.

3.2. Synthesis of Single KTaO₃, CdS and MoS₂ Semiconductors

KTaO₃, CdS and MoS₂ semiconductors were prepared by hydro/solvothermal methods using an autoclave with a capacity of 500 mL as the reactor. In a typical procedure for the preparation of potassium tantalate, KOH (30 g) was dissolved in deionized water (60 mL), then Ta₂O₅ (11 g) and PEG-400 (3 mL) were added. This mixture (marked as *solution A*) was stirred for 1 h before it was transferred into a Teflon-lined stainless steel autoclave. The autoclave was sealed and got heated at 200 °C for 24 h. After cooling naturally to room temperature, the resulting powder was washed several times by centrifugation with distilled water and ethanol respectively and dried in an oven at 70 °C for 8 h. Finally, some white powder was obtained.

Cadmium sulfide was prepared based on the methodology presented by Zhong *et al.* [42]. In a typical procedure, CdCl₂ (18.3 g) was added into deionized water (160 mL) and EG (160 mL) under continuous stirring for 15 min. After that, thiourea (15.2 g) was added to the solution and stirring was continued for an additional 20 min. This mixture (marked as *solution B*) was transferred into the autoclave which was sealed and maintained at 190 °C for 24 h and allowed to cool down to room temperature naturally. The crystalline powder product was washed several times by centrifugation with distilled water and ethanol, respectively and dried in the oven at 70 °C for 8 h. Finally, some yellow powder was obtained.

Molybdenum disulfide was synthesized using a method similar to reported by Wang *et al.* [46]. To prepare MoS₂, Na₂MoO₄·2H₂O (12.1 g), thiourea (15.6 g) and PEG-400 (3.5 g) were dissolved in deionized water (300 mL). The solution was stirred continuously for 15 min. The resulting mixture (marked as *solution C*) was transferred into the teflon-lined stainless autoclave which was maintained at 200 °C for 24 h, then cooled to room temperature. The precipitate was washed with deionized water and dried in the oven at 80 °C for 6 h. Finally, some black powder was obtained.

3.3. Synthesis of Binary and Ternary Semiconductor Composites

The KTaO₃, CdS and MoS₂ powders or solutions A, B, C prepared according to the methods described in Section 2.2 were combined to form semiconductor nanocomposites by the calcination or by the hydro/solvothermal mixed solutions methods, respectively. All prepared samples with different molar ratios between semiconductors are presented in Table 1.

To prepare KTaO₃-CdS, KTaO₃-MoS₂ and KTaO₃-CdS-MoS₂ composites by the calcination method, KTaO₃ and CdS or/and MoS₂ powders, respectively were mixed together sufficiently with various molar ratios. The powder mixture was calcined in the oven at 500 °C for 3 h with a heating rate of 3 °C/min. Then the heated mixture was removed from the oven and allowed to cool to room temperature naturally.

A typical hydro/solvothermal mixed solutions process for the preparation of CdS-MoS₂; KTaO₃-CdS; KTaO₃-MoS₂; KTaO₃-CdS-MoS₂ composites with various molar ratios between semiconductors was as follows: the required amounts of solutions B and C; A and B; A and C; A and B and C, respectively were mixed together and stirred continuously for 45 min using a magnetic stirrer. Then the as-prepared mixture was placed in the autoclave and heated at 200 °C for

24 h. The resulting precipitate was washed with distilled water and ethanol, respectively and dried in the oven at 70 °C for 8 h.

3.4. Measurement of Photocatalytic Activity in the Aqueous Phase

The photocatalytic activity of $KTaO_3$, CdS and MoS_2 single semiconductors and their nanocomposite powder in ultraviolet (UV) and visible light (Vis) was estimated by monitoring the decomposition rate of 0.21 mM phenol in the aqueous solution. Phenol was selected as a model contaminant because it is a non-volatile and common organic pollutant found in various types of industrial wastewater. Photocatalytic degradation runs were preceded by blind tests in the absence of a photocatalyst or illumination. The aqueous phase containing the photocatalyst (125 mg), deionized water (24 mL) and phenol (1 mL, C = 500 mg/dm^3) was placed in a photocatalytic reactor (V = 25 cm^3) equipped with a 30 mm-thick quartz window. The temperature of the aqueous phase during the experiments was maintained at 10 °C by an external circulating water bath. The prepared suspension was stirred using magnetic stirrer and aerated (V = 5 dm^3/h) for 30 min in the dark to reach the adsorption equilibrium and then the content of the reactor was photoirradiated with a 1000 W Xenon lamp (Oriel Instruments, Stratford, CT, USA) which emitted both UV and Vis irradiation. The optical path included a water filter and glass filters (GG420, Optel, Opole, Poland) to cut off IR and/or UV, respectively. GG glass filter transmitted light of wavelength greater than 420 nm. During the irradiation, 1 cm^3 of suspension sample was collected at regular time periods and filtered through syringe filters (Ø = 0.2 μm) to remove the photocatalyst particles. Phenol concentration was estimated by colorimetric method after derivatization with diazo-p-nitroaniline using UV-Vis spectrophotometer (DU-7, Beckman, Warsaw, Poland).

3.5. Measurement of Photocatalytic Activity in the Gas Phase

The photocatalytic activity of the prepared semiconductors and their nanocomposite powders was also determined in the toluene degradation process. Toluene, an important volatile organic compound (VOC), was used as a model air contaminant. The photocatalysts activity tests were carried out in the flat stainless steel reactor (V = 30 cm^3) equipped with a quartz window, two valves and a septa. As an irradiation source there was used an array of 25 LEDs (λ_{max} = 375 nm, 63 mW per diode) which was described in our previous study [47]. In a typical measurement the semiconductor powder (about 0.1 g) was suspended in a small amount of water and loaded as a thick film on a glass plate (3 cm × 3 cm) using painting technique. The obtained semiconductors coated support was dried and then placed at the bottom side of the photoreactor followed by closing the reactor with a quartz window. The gaseous mixture from a cylinder was passed through the reactor space for 1 min. The concentration of toluene in a gas mixture was about 150 ppm. After closing the valves, the reactor was kept in the dark for 15 min to reach adsorption equilibrium. A reference sample was taken just before starting irradiation. To estimate toluene concentration the samples were taken every 10 min during 60 min of irradiation. The photocatalytic stability was estimated in four subsequent cycles of toluene degradation. The analysis of toluene concentration in the gas phase was carried out using a Perkin Elmer Clarus 500 GC (Perkin Elmer, Waltham,

MA, USA) equipped with a 30 m × 0.25 mm Elite-5 MS capillary column (0.25 µm film thickness) and a flame ionization detector (FID). The samples (200 µL) were injected by using a gas-tight syringe. Helium was used as a carrier gas at a flow rate of 1 mL/min.

4. Conclusions

A series of novel binary and ternary composite photocatalysts was obtained based on the combination of potassium tantalate, cadmium sulfide and molybdenum disulfide powders as well as their hydro/solvothermal precursor solutions. It was observed that calcination is an effective method to prepare various $KTaO_3$, CdS and MoS_2 semiconductor combinations, the while solvothermal mixed precursor solutions route led to obtaining other composite components. The UV-Vis DRS spectra showed that the loading MoS_2 onto CdS shifted absorbance peaks to longer wavelengths as well as loading CdS onto $KTaO_3$ enhanced absorption properties. On the other hand, an excess of MoS_2 in composite caused a decrease in activity both for phenol and toluene degradation. It may be suggested that the best absorption properties and the highest activity could probably be achieved for the ternary composites containing appropriate molar ratio between above semiconductors but it needs further investigation. We found that the ternary semiconductor hybrid prepared by calcination of $KTaO_3$, CdS and MoS_2 powders at the 10:5:1 molar ratio exhibited very good stability during four measurement cycles in toluene degradation and excellent photocatalytic performance in phenol degradation among all obtained photocatalysts–the activity reached 80% under UV-Vis and 42% under Vis light 60-min irradiation. This relatively high photoactivity under visible light could be related to the presence of heterojunction between semiconductors. Due to the band structure of component semiconductors, two-photon excitation together with charge carrier transfer between $KTaO_3$, CdS and MoS_2 could be expected. However, ternary composite with 10:1:1 molar ratio between semiconductors exhibited relatively low activity which proved that the photocatalytic efficiency in aqueous phase reaction strictly depends on amount of composite components and need further investigation. In general, this work demonstrates that novel composites based on $KTaO_3$, CdS and MoS_2 are promising as photocatalysts for the degradation of organic pollutants in both the gas and aqueous phases.

Acknowledgments

This work was supported by National Science Centre, Poland (research grant Preparation and characteristics of novel three-dimensional semiconductor-based nanostructures using a template-free methods, contract No. 2011/03/B/ST5/03243).

Author Contributions

AZ designed research; BB, AC, MJW, TK and AZ performed research and analyzed the data; BB, AC, TK and AZ wrote the paper. All authors read and approved the final manuscript.

Conflicts of Interest

The authors declare no conflict of interest.

References

1. Fujishima, A.; Honda, K. Electrochemical photolysis of water at a semiconductor electrode. *Nature* **1972**, *238*, 37–38.
2. Chen, X.B.; Mao, S.S. Titanium dioxide nanomaterials: Synthesis, properties, modifications, and applications. *Chem. Rev.* **2007**, *107*, 2891–2959.
3. Tong, H.; Ouyang, S.X.; Bi, Y.P.; Umezawa, N.; Oshikiri, M.; Ye, J.H. Nanophotocatalytic materials: Possibilities and challenges. *Adv. Mater.* **2012**, *24*, 229–251.
4. Bahnemann, D. Photocatalytic water treatment: Solar energy applications. *Sol. Energy* **2004**, *77*, 445–459.
5. Ahmad, M.; Ahmed, E.; Zhang, Y.W.; Khalid, N.R.; Xu, J.F.; Ullah, M.; Hong, Z.L. Preparation of highly efficient *Al*-doped ZnO photocatalyst by combustion synthesis. *Curr. Appl. Phys.* **2013**, *4*, 697–704.
6. Liao, C.H.; Huang, C.W.; Wu, J.C.S. Hydrogen production from semiconductor-based photocatalysis via water splitting. *Catalysts* **2012**, *2*, 490–516.
7. Zhang, N.; Ciriminna, R.; Pagliaro, M.; Xu, Y.-J. Nanochemistry-derived Bi_2WO_6 nanostructures: Towards production of sustainable chemicals and fuels induced by visible light. *Chem. Soc. Rev.* **2014**, *43*, 5276–5287.
8. Zhang, N.; Zhang, Y.; Xu, Y.-J. Recent progress on graphene-based photocatalysts: Current status and future perspectives. *Nanoscale* **2012**, *4*, 5792–5813.
9. Zhang, N.; Liu, S.; Xu, Y.-J. Recent progress on metal core@semiconductor shell nanocomposites as a promising type of photocatalyst. *Nanoscale* **2012**, *4*, 2227–2238.
10. Pan, X.; Yang, M.-Q.; Fu, X.; Zhang, N.; Xu, Y.-J. Defective TiO_2 with oxygen vacancies: Synthesis, properties and photocatalytic applications. *Nanoscale* **2013**, *5*, 3601–3614.
11. Yang, M.-Q.; Xu, Y.-J. Selective photoredox using graphene-based composite photocatalysts. *Phys. Chem. Chem. Phys.* **2013**, *15*, 19102–19118.
12. Fujishima, A.; Zhang, X.; Tryk, D.A. TiO_2 photocatalysis and related surface phenomena. *Surf. Sci. Rep.* **2008**, *63*, 515–582.
13. Marschall, R. Semiconductor composites: Strategies for enhancing charge carrier separation to improve photocatalytic activity. *Adv. Funct. Mater.* **2014**, *17*, 2420–2440.
14. Yang, G.; Yan, Z.; Xiao, T. Preparation and characterization of SnO/ZnO/TiO_2 composite semiconductor with enhanced photocatalytic activity. *Appl. Surf. Sci.* **2012**, *258*, 8704–8712.
15. Emeline, A.V.; Kuznetsov, V.N.; Ryabchuk, V.K.; Serpone, N. On the way to the creation of next generation photoactive materials. *Environ. Sci. Pollut. Res.* **2012**, *19*, 3666–3675.
16. Serpone, N.; Emeline, A.V. Semiconductor photocatalysis—Past, present, and future outlook. *J. Phys. Chem. Lett.* **2012**, *3*, 673–677.

17. Zhang, Y.; Tang, Z.-R.; Fu, H.; Xu, Y.-J. TiO2-graphene nanocomposites for gas-phase photocatalytic degradation of volatile aromatic pollutant: Is TiO2-graphene truly different from other TiO2-carbon composite materials? *ACS Nano* **2010**, *4*, 7303–7314.

18. Kato, H.; Kudo, A. Photocatalytic water splitting into H2 and O2 over various tantalate photocatalysts. *Catal. Today* **2003**, *78*, 561–569.

19. Liu, J.W.; Chen, G.; Li, Z.H.; Zhang, Z.G. Hydrothermal synthesis and photocatalytic properties of ATaO3 and AnbO3 (A = Na and K). *Int. J. Hydrogen Energy* **2007**, *32*, 2269–2272.

20. Kudo, A. Development of photocatalyst materials for water splitting. *Int. J. Hydrog. Energy* **2006**, *31*, 197–202.

21. Lin, W.; Chen, C.; Hu, C.; Teng, H. NaTaO3 photocatalysts of different crystalline structures for water splitting into H2 and O2. *Appl. Phys. Lett.* **2006**, *89*, 211904.

22. Torres-Martínez, L.M.; Cruz-López, A.; Juárez-Ramírez, I.; Meza-de la Rosa, M.E. Methylene blue degradation by NaTaO3 sol-gel doped with Sm and La. *J. Hazard. Mater.* **2009**, *165*, 774–779.

23. Zhu, H.; Jiang, R.; Xiao, L.; Chang, Y.; Guan, Y.; Li, X.; Zeng, G. Photocatalytic decolorization and degradation of Congo Red on innovative crosslinked chitosan/nano-CdS composite catalyst under visible light irradiation. *J. Hazard. Mater.* **2009**, *169*, 933–940.

24. Zhai, T.Y.; Fang, X.S.; Bando, Y.S.; Liao, Q.; Xu, X.J.; Zeng, H.B.; Ma, Y.; Yao, J.N.; Golberg, D. Morphology-dependent stimulated emission and field emission of ordered CdS nanostructure arrays, *ACS Nano* **2009**, *3*, 949–959.

25. Dongre, J.K.; Nogriya, V.; Ramrakhiani, M. Structural, optical and photoelectrochemical characterization of CdS nanowire synthesized by chemical bath deposition and wet chemical etching. *Appl. Surf. Sci.* **2009**, *255*, 6115–6120.

26. Zhang, H.; Zhu, Y. Significant visible photoactivity and antiphotocorrosion performance of CdS photocatalysts after monolayer polyaniline hybridization. *J. Phys. Chem. C* **2010**, *114*, 5822–5826.

27. Zhong, M.; Shi, J.; Xiong, F.; Zhang, W.; Li, C. Enhancement of photoelectrochemical activity of nanocrystalline CdS photoanode by surface modification with TiO2 for hydrogen production and electricity generation. *Sol. Energy* **2012**, *86*, 756–763.

28. Li, X.; Xia, T.; Xu, C.; Murowchick, J.; Chen, X. Synthesis and photoactivity of nanostructured CdS–TiO2 composite catalysts. *Catal. Today* **2014**, *225*, 64–73.

29. He, D.; Chen, M.; Teng, F.; Li, G.; Shi, H.; Wang, J.; Xu, M.; Lu, T.; Ji, X.; Lv, Y. Enhanced cyclability of CdS/TiO2 photocatalyst by stable interface structure. *Superlattices Microstruct.* **2012**, *51*, 799–808.

30. Panigrahi, S.; Basak, D. Morphology driven ultraviolet photosensitivity in ZnO–CdS composite. *J. Colloid Interface Sci.* **2011**, *364*, 10–17.

31. Jana, T.K.; Pal, A.; Chatterjee, K. Self assembled flower like CdS–ZnO nanocomposite and its photo catalytic activity. *J. Alloys Compd.* **2014**, *583*, 510–515.

32. Liu, S.; Li, H.; Yan, L. Synthesis and photocatalytic activity of three-dimensional ZnS/CdS composites. *Mater. Res. Bull.* **2013**, *48*, 3328–3334.

33. Liu, X.; Yan, Y.; Da, Z.; Shi, W.; Ma, C.; Lv, P.; Tang, Y.; Yao, G.; Wu, Y.; Huo, P.; *et al.* Significantly enhanced photocatalytic performance of CdS coupled WO_3 nanosheets and the mechanism study. *Chem. Eng. J.* **2014**, *241*, 243–250.

34. Zhao, Y.; Zhang, Y.; Yang, Z.; Yan, Y.; Sun, K. Synthesis of MoS_2 and MoO_2 for their applications in H_2 generation and lithium ion batteries: A review. *Sci. Technol. Adv. Mater.* **2013**, *14*, 043501.

35. Zong, X.; Wu, G.; Yan, H.; Ma, G.; Shi, J.; Wen, F.; Wang, L.; Li, C. Photocatalytic H_2 Evolution on MoS_2/CdS Catalysts under Visible Light Irradiation. *J. Phys. Chem. C* **2010**, *114*, 1963–1968.

36. Chen, G.; Li, D.; Li, F.; Fan, Y.; Zhao, H.; Luo, Y.; Yu, R.; Meng, O. Ball-milling combined calcination synthesis of MoS_2/CdS photocatalysts for high photocatalytic H_2 evolution activity under visible light irradiation. *Appl. Cat. A* **2012**, *443–444*, 138–144.

37. Liu, Y.; Yu, Y.X.; Zhang, W.D. MoS_2/CdS Heterojunction with high photoelectrochemical activity for H_2 evolution under visible light: The Role of MoS_2. *J. Phys. Chem. C* **2013**, *117*, 12949–12957.

38. Liu, Y.; Yu, H.; Quan, X.; Chen, S. Green synthesis of feather-shaped MoS_2/CdS photocatalyst for effective hydrogen production. *Int. J. Photoenergy* **2013**, *2013*, 247516–247520.

39. Zong, X.; Yan, H.; Wu, G.; Ma, G.; Wen, F.; Wang, L.; Li, C. Enhancement of photocatalytic H_2 evolution on CdS by loading MoS_2 as cocatalyst under visible light irradiation. *J. Am. Chem. Soc.* **2008**, *130*, 7176–7177.

40. Wiedemeier, H.; Khan, A.A. Phase studies in the system managanese sulfioe–cadmium sulfide. *Trans. Metall. Soc. AIME* **1968**, *242*, 1969–1972.

41. Hu, C.-C.; Yeh, T.-F.; Teng, H. Pyrochlore-like $K_2Ta_2O_8$ synthesized from different methods as efficient photocatalysts for water splitting. *Catal. Sci. Technol.* **2013**, *3*, 1798–1804.

42. Zhong, S.; Zhang, L.; Huang, Z.; Wang, S. Mixed-solvothermal synthesis of CdS micro/nanostructures and their optical properties. *Appl. Surf. Sci.* **2011**, *257*, 2599–2603.

43. Zou, Y.; Hu, Y.; Gu, H.; Wang, Y. Optical properties of octahedral $KTaO_3$ nanocrystalline. *Mater. Chem. Phys.* **2009**, *115*, 151–153.

44. Adhikari, R.; Malla, S.; Gyawali, G.; Sekino, T.; Lee, S.W. Synthesis, characterization and evaluation of the photocatalytic performance of Ag-$CdMoO_4$ solar light driven plasmonic photocatalysts. *Mat. Res. Bull.* **2013**, *48*, 3367–3373.

45. Wang, W.-S.; Zhen, L.; Xu, C.-Y.; Shao, W.-Z.; Chen, Z.-L. Formation of $CdMO_4$ porous hollow nanospheres via self-assembly accompanied with Ostwald ripening process and their photocatalytic performance. *CrystEngComm* **2013**, *15*, 8014–8021.

46. Wang, S.; Li, G.; Du, G.; Jiang, X.; Feng, C.; Guo, Z.; Kim, S. Hydrothermal synthesis of molybdenum disulfide for lithium ion battery applications. *Chin. J. Chem. Eng.* **2010**, *18*, 910–913.

47. Nischk, M.; Mazierski, P.; Gazda, M.; Zaleska, A. Ordered TiO_2 nanotubes: The effect of preparation parameters on the photocatalytic activity in air purification process. *Appl. Catal. B* **2014**, *144*, 674–685.

Sample Availability: Samples of the compounds are available from the authors.

Chapter 3:
Air, Water and Surface Decontamination

The Viability of Photocatalysis for Air Purification

Stephen O. Hay, Timothy Obee, Zhu Luo, Ting Jiang, Yongtao Meng, Junkai He, Steven C. Murphy and Steven Suib

Abstract: Photocatalytic oxidation (PCO) air purification technology is reviewed based on the decades of research conducted by the United Technologies Research Center (UTRC) and their external colleagues. UTRC conducted basic research on the reaction rates of various volatile organic compounds (VOCs). The knowledge gained allowed validation of 1D and 3D prototype reactor models that guided further purifier development. Colleagues worldwide validated purifier prototypes in simulated realistic indoor environments. Prototype products were deployed in office environments both in the United States and France. As a result of these validation studies, it was discovered that both catalyst lifetime and byproduct formation are barriers to implementing this technology. Research is ongoing at the University of Connecticut that is applicable to extending catalyst lifetime, increasing catalyst efficiency and extending activation wavelength from the ultraviolet to the visible wavelengths. It is critical that catalyst lifetime is extended to realize cost effective implementation of PCO air purification.

Reprinted from *Molecules*. Cite as: Hay, S.O.; Obee, T.; Luo, Z.; Jiang, T.; Meng, Y.; He, J.; Murphy, S.C.; Suib, S. The Viability of Photocatalysis for Air Purification. *Molecules* **2015**, *20*, 1319-1356.

1. Introduction

Using light to achieve clean air and water resources through photocatalytic oxidation is a goal of scientists worldwide [1–3]. Success depends on the air or water stream to be purified [4–7]. Our focus is on the challenges of purifying an air medium, primarily indoor air. United Technologies Research Center (UTRC) devoted significant resources towards this goal over the last two decades. Air itself is a mixed media that may contain a variety of both particulate and gaseous components. Photocatalysis is a widely generic term that applies to chemical change enabled by photon activated catalysis. The chemical change is usually oxidation, but in some cases reduction can be effected. The catalyst is generally a metal oxide semiconductor, usually titania, with an appropriate band gap energy that allows adsorption of an ultra-violet photon to generate electron hole pairs which initiate the chemical change. Generically:

$$hv + TiO_2(s) \longrightarrow TiO_2(s) + h^+ + e^- \tag{1}$$

For titania the band gap is centered near 360 nm. In air saturated with water vapor and under ambient light, water vapor has chemically adsorbed creating a partially hydroxylated surface. With this in mind we sometimes express Equation (1) as:

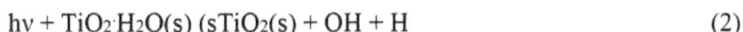

$$hv + TiO_2 \cdot H_2O(s) \, (sTiO_2(s) + OH + H \tag{2}$$

This allows us to think about the chemical change in terms of hydroxyl or proton attack on an adsorbed species and this can be useful in understanding and discussing the chemical changes initiated by photocatalysis. The surface chemistry is extremely rich and complex and depends on morphology of the bulk and surface, and specifically on the termination of the semiconductor surface bond and the myriad species that can adsorb on the surface.

In evaluating the effect of a photocatalytic oxidation (PCO) based air purifier we only need understand what goes in and what comes out in relation to our goal. It is critical that one completely understands the medium to be purified, the resultant fluid, and the desired outcome. If we are talking about polluted ground water the desired outcome is simply to remove the pollutant by chemical change to benign products. In this case we are working with a well-defined system and we need only understand the surface adsorption phenomenon and photocatalytic reactions of a few species. For example, if we look at ground water contaminated with chlorinated solvents from a dormant degreasing pit, then the solvents most commonly used were TCE and PCE. The gas over the contaminated ground water would consist of water saturated air, contaminated with TCE and/or PCE, and we look to construct an air purifier that removes the contaminants [7] or chemically changes them to benign or easily removable products. This system is well defined, has a fixed set of species to oxidize with a slowly varying source rate, and serves to define one limit of the variety of uses for an air purifier.

Another extreme occurs when evaluating the effect of a catalytic air purifier on indoor air. In this case the challenge fluid is complex and the goal is either to effect a change to a healthier environment, without impacting comfort, or to reduce outdoor air intake while maintaining air quality and comfort. The latter allows energy savings to be realized by minimizing conditioning (heating and cooling). In some cases, the size of the Heating, Ventilation and Air Conditioning (HVAC) equipment may be reduced resulting in capital savings on equipment. The components of indoor air that affect the human condition are myriad and both particulate and gaseous. Within the set of all particles, ultra-fine particles have been directly linked to heart health [8]. Bioaerosols can be allergens, asthmatic triggers, or mold spores [9], and some particles are benign. Within the set of gaseous products some are carcinogens; some cause respiratory distress; some are toxic; some are odiferous and some are benign. If we wish to treat indoor air to make it "healthy", one technology alone will not suffice to treat the wide range of particulates that may be encountered, as well as the wide range of gaseous components.

In fact, the stated goal of creating a healthy environment is itself nebulous. Individuals respond differently to different exposures [8]. In order to comment on either the efficiency or validity of an air purifier in this case, we first need to understand the challenge. Therefore, we need to understand indoor air and its components, we need to understand how the mixture of species adsorbs on the catalyst surface. We need to understand how this mixture reacts in an Ultra-Violet Photocatalytic Oxidation (UVPCO) air purifier and what is contained in the resultant mixture of effluents. In addition we need to understand this both initially and on a steady-state basis since UVPCO air purifiers will not have total mineralization capacity for all species and may produce hazardous by-products. In other words, transient effects may be important both as the air purifier reaches steady-state operation and as the building environment changes diurnally and seasonally.

To accomplish this we will look closely at indoor air and its components. We will restrict the photocatalyst to titania and modified titania. We will examine what occurs to those components when passing through a photocatalytic reactor and review the approach taken by UTRC in the construction of an UVPCO air purifier and validation of its effect in a real indoor environment.

2. Results and Discussion

The challenge in defining indoor air is that it is a mixed medium with variable components [10]. Indoor air is a mixture of outdoor air and recycled indoor air that may have been conditioned. Outdoor air comes from either or both ventilation and infiltration and is highly dependent on the quality of the outdoor air. Recycled air has been contaminated with a plethora of sources including construction materials, furnishings and occupants. In reality, each building whether an individual home or high-rise office building presents a unique and time-varying challenge, but if we limit ourselves to certain typical types of buildings in typical places, we can begin to define indoor air sufficiently to understand the operation of a photocatalytic air purifier.

2.1. Indoor Air

As stated, indoor air consists of both aerosols and gaseous components. Aerosols can be mineral, liquid or biological. In general, aerosols are either removed from indoor air by filtration or fine particles remain suspended in air. In a typical office building environment, large particles are filtered by low pressure drop filtration on the air intake and in the recycled flow during conditioning. Some bioaerosols have been demonstrated to be oxidized by photocatalysts, but, in general, this causes either reversible or irreversible deactivation of the photocatalyst and is undesirable. For this reason a higher quality filter is placed in the air stream prior to the catalyst, at the expense of an increase in pressure drop. The exact specifications of the filter are dependent on the design of the HVAC system in which the purification unit is located, specifically the pressure drop tolerance, the air flow rate and the dimensions of the photocatalytic unit. The goal is simply to insure minimal contact of aerosols with the surface of the photocatalyst. Particles will either be trapped by the filter or fly over the surface of the photocatalyst without impingement. A useful target is the highest efficiency filter allowed by pressure drop constraint. This implies that we can largely ignore the aerosols in indoor air and focus on the effect on the photocatalyst on the gaseous components.

The US EPA has commissioned the Building Assessment Survey and Evaluation (BASE) study [11] to determine the gaseous components in indoor air in 100 randomly selected office buildings in the United States. Using standardized protocols they collected extensive data in 37 cities in 25 states. The preliminary results of this study are partially summarized in the first two columns of Table 1.

In normal indoor air, there are ca. 200 individual gaseous components, most in the 10 ppb range or lower, and most are volatile organic compounds (VOCs). The average tolerance index of the air found in office buildings by the BASE study is 0.884. In problem indoor air, air that has generated complaints and or illness, there may be a considerably higher total or higher concentrations of individual components, resulting in a significantly higher tolerance index.

Table 1. Selected VOCs found in the US EPA BASE study and their contribution to UTRC's proposed tolerance metric.

Compound	Average Concentration (ppbv)	Contribution to Tolerance Index	Origin
Formaldehyde	11.6	0.387	TLV
Acetaldehyde	4.2	0.0627	OT
Benzene	2.1	0.0339	SMAC
Toluene	8.3	0.00519	OT
1,4-Dichlorobebzene	1.4	0.0117	OT
Carbon disulfide	1.1	0.0115	OT
Styrene	1.6	0.0114	OT
Butyl acetate	3.3	0.0106	OT
m,p-Xylenes	3.3	0.00243	OT
Decane	1.4	0.00190	OT
Undecane	2.0	0.00167	OT
Acetone	27.1	0.00131	SMAC
Dodecane	2.0	0.0010	OT
Naphthalene	13.0	0.342	OT

If our goal is to change air quality, we can simply rate the effect of an air purifier based on its efficiency. However, in treating indoor air our goal is to create cleaner or healthier air. This goal is somewhat nebulous as there are different effects that can be exhibited by different VOCs. Some VOCs such as formaldehyde and benzene are carcinogens, some are toxic, some are odorous and some are benign. In order to quantify the effect of a photocatalytic air purifier one needs a metric to describe the ability of humans to tolerate an environment with any given mixture of contaminants. One such metric is proposed by Hollick and Sangiovanni [12]. Based on existing metrics for allowed VOCs in Spacecraft, odor thresholds, toxicity levels or exposure limits, the proposed tolerance metric defines a yardstick for measuring the tolerance of human beings to a given indoor air mixture. This metric also allows us to model the effect of a photocatalytic air purifier on the building environment. They define a Tolerance Index, T_i, for each VOC component in indoor air as follows:

$$T_i = C_i/A_i \qquad (3)$$

where C_i is the actual concentration of contaminant i and A_i is the maximum acceptable concentration for each contaminant. The metrics used were derived from three factors:

1. NASA's Spacecraft Maximum Allowed Concentration (SMAC) levels
2. One tenth of ACGIH's Threshold Limit Values (TLV) levels
3. Odor Thresholds (OT)

Multiple limits are used since no consensus exists in the indoor air scientific community as to what constitutes a healthy indoor environment and what measurable metric can be used to assess indoor air. Most species are limited by their odor threshold. TLVs are eight hour exposure limits over a worker's lifetime. One tenth of the TLV value was taken to insure a conservative metric for air exposure which can occur on a twenty four hour basis.

Air quality is assessed by summing the individual contributions to the tolerance matrix:

$$\Sigma Ti < 1.0 \text{ indicates acceptable VOC concentrations} \qquad (4)$$

The top fifteen contributors to the tolerance metric culled from the EPA Base Study are shown in the latter two columns of Table 1. The dominant effect on air quality, based on this metric, is due to formaldehyde (limited by TLV) and napthalene (limited by odor). The use of a metric such as this allows validation of the operation of a UVPCO air purifier on complex media, such as indoor air. Other similar metrics have been proposed, but, so far, none have gained wide acceptance and use.

2.2. Design of an Air Purifier

A photocatalytic air purifier uses a catalyst, a substrate to support the catalyst in the air flow, and a light source. In addition, the purifier must be housed in an accessible and safe fashion, while integrating the device into the air supply it was designed to act upon.

The design [13–17] of a PCO air-purifier balances efficiency and cost. The capital cost, alone, is due to the housing, the light source and the catalyst/substrate. Cost is spread about equally across these three categories. For indoor air purification the cost of operation consists of the electricity used in the lamps, and ballasts, if fluorescent lamps are used. Maintenance costs include lamp, ballasts and catalyst replacement. Catalyst lifetime dictates its replacement schedule. If the device is placed in a flowing airstream provided by an existing ventilation system, the design must also consider the allowable pressure drop. If a fan is used, that cost is added to the initial and maintenance costs. If a filter is used to protect the lamps and or the catalyst, filter replacement cost is added to the maintenance cost, and if the filter is high efficiency, the filterdrop must be considered.

The standard catalyst for many photocatalytic applications is Degussa P-25. P-25 is cost effective, readily available and exhibits high photocatalytic activity for many species of interest in indoor air. The catalyst support needs to provide a large surface area in contact with the flowing air stream, a low pressure drop and structural integrity. The support should also lend itself to ease of manufacture and have sufficient integrity to survive shipping and handling. Ease of prototyping and the ability to easily modify the design for size are also considerations. Based on these considerations, UTRC uses a modular approach to the purifier design utilizing aluminum honeycomb as a catalyst substrate. The basic module is a light source and a titania catalyst supported on the honeycomb. A PCO air purifier of variable efficiency (see Figure 1) can be constructed by using multiples of this module and this can be designed to fit pre-existing spatial constraints.

Titania is activated by photons with energy greater than the band gap (ca. 360 nm.) Light sources (see Table 2) may be fluorescent specialty lamps, LEDs or any other photon emitter having the required wavelength. The Sun is free, but light is hard to deliver where needed, and is only available during daylight hours. The cheapest, longest lifetime and most readily available light sources are UV fluorescent lamps. UTRC uses fluorescent lamps in their modular design. LED sources need lower wavelengths and longer lives to be a viable alternative. UV fluorescent lamps are based on the mercury vapor spectra. Germicidal lamps emit principally at 254 nm. Black light fluorescent lamps are coated with a manufacturer-dependent phosphor. This causes slight variations in emission spectra, but are generally centered near 360 nm and have a *ca.* 50 nm FWHM bandwidth. Lifetimes are approximate and both manufacturer and mode-of-operation dependent.

Side View

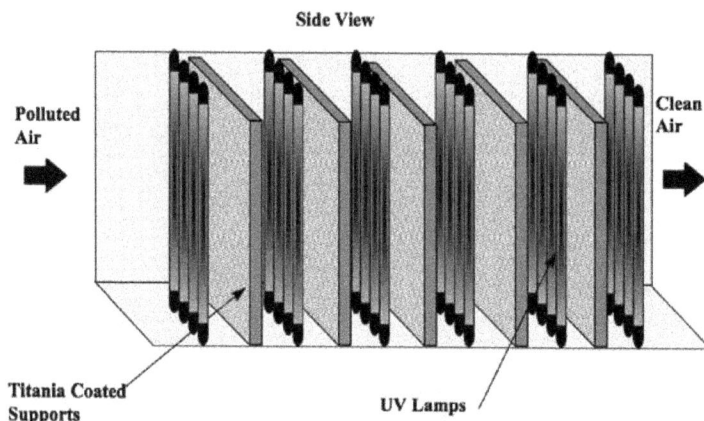

Polluted Air

Clean Air

Titania Coated Supports

UV Lamps

Figure 1. Generic Multi-stage, Honeycomb-Monolith Photocatalytic Reactor.

In rate measurements performed in a flat plate reactor [18] which will be described later in greater detail, we see no measurable difference in photocatalytic (precursor disappearance) rates obtained with germicidal lamps and those obtained employing black light lamps. This is attributed to the tradeoff that exists between the black light source where the emission band overlaps the titania band gap adsorbing *ca.* 70% of the emitted photons and the *ca.* 70% fewer photons per watt generated at 254 nm. Photocatalysis is a photon initiated process, and the small differences between the number of photons absorbed per Watt at these two wavelengths is ameliorated by the *ca.* 0.6 power intensity dependence observed at the intensities used for rate measurements and purifier design. Therefore, the sole considerations for fluorescent lamp choice are cost and the outcome desired. Indoor air contains bioaerosols both viable, such as mold spores, bacteria and airborne viruses and nonviable, such as allergens. Germicidal lamps can inactivate viable bioaerosols as they fly through the irradiant field. If this effect is desired, then germicidal lamps are the lamp of choice and the housing must be designed to be resistant to damage by UV light. Black light wavelengths are not strongly germicidal and are more material friendly. In either case, bioaerosols can settle on the photocatalyst and cause deactivation, blocking the surface until they are mineralized. If bioaerosols are mineralized to non-volatile compounds, the deactivating effect may be permanent. Most bioaerosols are either captured by a filter or fly past the catalyst surface entrained in the airstream.

The housing should have easy access to the UV bulb, catalyst/substrate and filter replacement and should fit in a building airstream, preferably downstream of the HVAC unit. The housing also must be sized appropriately for the space available. All interior surfaces should reflect the wavelengths used to excite the photocatalyst. If UVC excitation is used, all parts exposed to UVC radiation should be resistant to UV degradation.

Table 2. Common UV light sources for a UVPCO air purifier.

Photon Emitter	UV Wavelength Range	Approximate Lifetime
Sun	UVA and UVB; most UVC is adsorbed in the atmosphere	exists with daylight
Black Light Fluorescent	UVA (260 nm ± 50 nm FWHM)	5000–12,000 h, usually limited by phosphor degradation
Germicidal Fluorescent	UVC (254 nm)	10,000–20,000 h
LEDs	Various, 190 nm to 1100 nm	wavelength dependent, a few thousand hours at short wavelengths

It is cost prohibitive to design an air purifier to be 100% efficient. In most cases a design goal of *ca.* 10%–20% single-pass removal efficiency for formaldehyde is achievable and will result in cleaner air through recycling. An effective clean air delivery rate (CADR$_{eff}$), which is the preferred design parameter rather than single-pass efficiency, is calculated based on the mole fraction (X) of individual contaminants in the air, the reactors single pass efficiency (SPE) and the air flow rate:

$$CADR_{eff} = (SPE)(Flow\ rate) = X_1CADR_1 + X_2CADR_2 + ... \qquad (5)$$

The effective CADR increases with increasing number of modules, and with flow rate, as shown in Figure 2. Using an effective CADR allows us to compare the effect of a photocatalytic air purifier with ventilation. ASHRE requires outdoor ventilation of 15 CFM per person unless air purification is used. If used, verification is required that indoor air quality is maintained. Using an effective CADR allows comparison of the cost of ventilation (heating and cooling) with the cost of air purification.

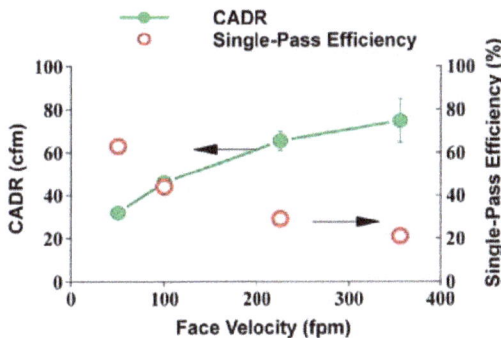

Figure 2. The effect of flow velocity on the effective CADR is opposite to the effect on SPE and gives a more accurate picture of the effect of air purification to HVAC professionals. A generic HVAC design is assumed.

This is the basic modular design used in a series of prototypes and products deployed by United Technologies. UTRCs steps for the design and validation of a UVPCO air purifier are:

1. Measure reaction rates as a function of humidity and contaminant concentration.
2. Understand the effect on rates due to mixture of contaminants

3. Model and validate the effect of prototype air purifiers
4. Validate prototypes in indoor air
5. Design and validate products

The model was used to design the prototypes which, in turn, were used to validate the model. As will be shown, external validation was also effected through cooperation with the University of Arizona, Harvard University, Danish Technical University, the University of Wisconsin, the University of Connecticut, the University of Nottingham, Lawrence Berkeley National Laboratory and others. Some results remain unpublished and unavailable for review.

2.3. Reaction Rate Studies

A complete set of intrinsic rate data, assembled as outlined above, serves as essential input to a design procedure for a photocatalytic air-purifier.

The rate (R) for photocatalytic oxidation of a contaminant species (X) over TiO_2 can be expressed as:

$$R = k_{obs} I^n [X_{(s)}]^m [H_2O_{(s)}] [O_{2(s)}] = k'_{obs} [X_{(s)}]^m, \text{ where} \tag{6}$$

$$k'_{obs} = k_{obs} I^n [H_2O_{(s)}] [O_{2(s)}] \tag{7}$$

In other words, at constant UV intensity and constant water vapor and oxygen concentrations the rate is proportional to the surface coverage of the species X. At low concentrations the rate (R) is linear with respect to the contaminant X, so we may express the relation as:

$$R_X \alpha [X_{(s)}] = S_X \tag{8}$$

where the rate of disappearance of species X is proportional to the surface concentration of X. The photocatalysis of gaseous species can be viewed as a multi-step process where adsorption of gaseous species onto the catalyst surface occurs first. All the interesting chemistry in this process occurs at the gas-solid interface between the photocatalyst, for example, solid titanium dioxide (TiO_2), and a contaminated airstream. A basic description of the process is accomplished by separating the varied chemical and physical processes that occur into four different categories:

1. Coadsorption of the gas phase species on the semiconductor surface. This includes water, oxygen molecules, the species to be oxidized, and any other species present in the gas phase that compete for surface sites.
2. Activation of the semiconductor surface by a UV photon, generation of electron-hole pairs, followed by the competing processes of recombination and trapping. The trapping species are generally believed to be surface oxygen and water respectively resulting in hydroxyl (OH) and superoxide (O_2^-) radicles
3. Initiation, where the free radicals produced by trapping the electron-hole pair initiate attack on the species to be oxidized. This step removes the precursor and the rate of removal is the rate generally measured.

4. Propagation, where sequential free radical attack causes degradation of the reactant species to products and, in some cases, stable by-products. Deactivation of the catalyst, either reversible or irreversible can occur during this step. Intermediate free radicals can bond to the catalyst surface or non-volatile products can form.

The solid titania surface, in air and ambient light, is an active surface, in which water has chemisorbed forming ca. one third hydroxyl terminal groups. Molecular physical adsorption from the gas phase is dominated by the strongest force, *i.e.*, by the largest molecule-to-surface-site binding energy. For small molecules the dominant intermolecular forces are hydrogen bonding, dipole-dipole interactions and London forces. One of the earliest published studies of this effect is Obee and Hay [19] and their results show marked dependence on surface binding energy. In brief, they demonstrate that organic molecular functionality and the resultant hierarchy of intermolecular forces (IMFs) dictated the relative reaction rate. 1-Butanol is shown to have a larger rate of photocatalytic removal than 2-butanone, which is larger than 1-butene, which is larger than *n*-butane.

One way that the adsorption of molecules on a surface can be expressed mathematically is to relate the surface concentration S_i to the collision frequency of the molecules with the surface, pr and the retention time, ncy of the molecules with the surface:

$$S_i = Z\tau = <v> n_i \sigma\tau \tag{9}$$

The collision frequency can be expressed as a product of the average molecular velocity $<v>$, the gas phase number density n_i, and the collision cross-section σ. The average molecular velocity can be expressed in turn as:

$$<v> = \{8kT/\pi m\}^{1/2} \tag{10}$$

where T = temperature and m = mass of the species. The average time spent on the surface, τ, can be expressed as:

$$\tau = \tau_0 e^{Q/kT} \tag{11}$$

where q_0 is a constant, and Q is the binding energy to surface. If we insert these expressions into Equation (11) we obtain:

$$S_i = \{8kT/\pi m\}^{1/2} n\sigma \tau_0 e^{Q/kT} \tag{12}$$

Equation (12) tells us that surface coverage S_i depends directly on the gas phase number density or concentration and on the molecular mass, the binding energy to the surface and the bulk temperature. What we expect is a photocatalytic removal rate that depends on light intensity and surface species coverage. If light intensity, concentration, and temperature are kept constant, and the variation in molecular mass is small, surface coverage will depend on binding energy as shown by 1-butanol, 2-butanone, 1-butene and *n*-butane. The inverse square root dependence of the surface coverage to the molecular mass is slightly misleading since the binding energy to the surface also depends on molecular mass. This is also illustrated by Obee and Hay [19], who performed a series of rate measurements with the straight chain alkanes, *n*-butane, *n*-hexane and *n*-decane. In the limit of other binding energies being approximately equal the molecular size effect dominates. The larger molecule exhibits the largest Van der Walls effect and the largest observed removal rate. The rate of mineralization is distinctly different. Acting conversely to the above effect, mineralization reaction sequences require additional radical attack or addition steps to mineralize larger molecules.

Indoor air is predominantly composed of N_2, O_2, H_2O and CO_2 with trace contaminants as discussed above. Of these major components only water binds strongly to the hydroxylated titania surface. It is water therefore that is the major adsorbent on the titania surface. All trace contaminants that we wish to oxidize must compete with water adsorption. This competition affects their disappearance rate.

2.3.1. The Effect of Concentration on Rate and the Extent of Mineralization

In a well-conditioned building (20% to 60% RH at 20 °C) water concentration is in the 6000 ppmv to 16,000 ppmv range. In order to observe the effect of contaminant concentration on removal rate we must fix water concentration and light intensity and wavelength to achieve the relation given in Equation (10). At the drier end of building air (6000 ppmv of water) we minimize the effect of multiple water layers covering the surface of the titania. Figure 3 shows the removal rate for 14 common air contaminates over a range of concentrations from 0.10 to 100 ppmv. Light Intensity was fixed at 1 W/cm^2 using standard UVC germicidal fluorescent lamps and water concentration was kept standard at 6000 ppmv. At lower concentrations (<0.50 ppmv) the curve of oxidation rate *vs.* gas-phase concentration is near linear, at higher concentrations the curve appears to roll off or stabilize, while at still higher concentrations some species rates decrease while other species increase. Various competing effects contribute to these observations; TCE, for example, can dissociate under UV irradiation.

As concentration increases, some species saturate the surface, some form additional layers of secondarily adsorbed species, some intermediates chemisorb, blocking active sites, and some form gas-phase free radicals by radical metathesis. Each contaminant behaves uniquely in its competition with water and other contaminant molecules to adsorb (and react) on titania. Hay and Obee [7] showed a map of the product space for TCE as a function of concentration. Light intensity, water vapor concentration and oxygen content are held constant, and they look at the products formed when TCE is photo oxidized over titania (Degussa P-25). The carbon in TCE mineralizes completely to CO_2 at concentrations less than 1 ppmv. Over this limit the carbon fraction of CO_2 decreases, the CO carbon fraction increases from *ca.* 0.0 at 1.0 ppmv TCE to *ca.* 0.70 at *ca.* 20 ppmv TCE. Phosgene ($COCl_2$) appears as a product over 1.1 ppmv TCE and increases with TCE concentration to ca. 11 ppmv TCE, then decreases with increasing TCE concentration. Dichloroacetyl chloride begins to appear at *ca.* 10 ppmv TCE and its carbon fraction rises with increasing TCE concentration (see Figure 4). In the limit of low concentration, when the number of active sites dominates the number of adsorbed contaminant molecules, complete mineralization will occur. The exact range of concentrations for this limit to be observed depends on light intensity, the nature of the photocatalyst and the contaminant molecule. At higher concentrations incomplete mineralization results in byproducts. Hay and Obee observed stable molecular byproducts that desorb from the surface reentering the gas stream. Other more transient byproducts exist but do not survive long enough to be detectible in the gas phase. This is a critical observation for treating indoor air quality, where most contaminants are in the ≤ 10 ppbv range. We expect complete mineralization of contaminant molecules unless the contaminant exists at high concentration due to a spill or specific high contaminant emission source. They also observe gas phase oxidation of TCE and PCE in the absence of a photocatalyst. Photodisociation can occur at the wavelengths in use. This effect dominates at high concentrations.

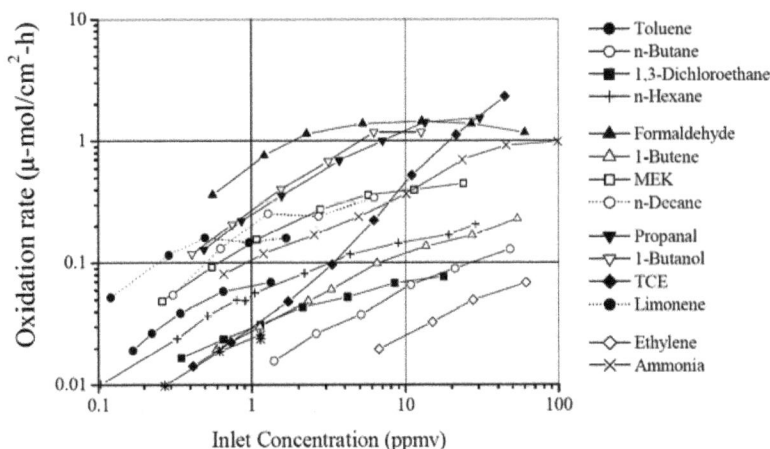

Figure 3. Measured PCO reaction rates for VOCs of interest in indoor air; UV 1-mW/cm^2, 6000 ppm water level.

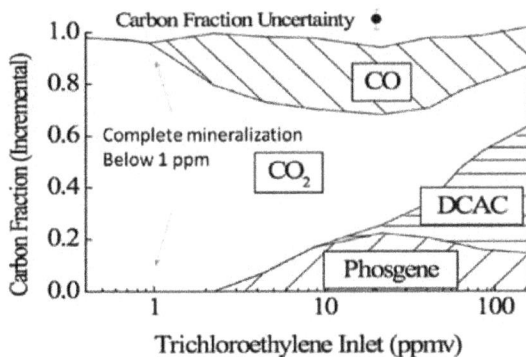

Figure 4. Measured PCO products vary with increasing concentration. Complete mineralization occurs at low concentrations. A different version of this diagram is shown in ref. [7].

2.3.2. The Effect of Humidity on Rate

As previously discussed, water is the major adsorbent on the hydroxylated titania surface and, as such, all other adsorbed species compete with water to adsorb. Species with a strong affinity to the surface compete with a higher degree of success. Obee [18] showed this effect of humidity on the photooxidation rate of toluene and formaldehyde, which possess a weak and a strong molecular dipole moment respectively. One expects formaldehyde with its strong molecular dipole to compete successfully, and this is illustrated by the data. Formaldehyde oxidizes at a faster rate. At

2.2 ppmv the reaction rate of formaldehyde is faster and increases with increasing humidity to *ca.* 2500 ppmv water, the rate then gradually declines with higher water concentration. If we equate the rates observed to surface coverage this behavior is only partially explained by the competition between water and formaldehyde for surface adsorption sites. Above *ca.* 2500 ppmv water, formaldehyde's diminishing disappearance rate is explained by increased competition from water due to an increase in the gas phase water concentration. However, in photocatalysis, water has a dual function. H_2O adsorbs on available sites blocking formaldehyde from adsorbing, and is split by the photo-activated catalyst to produce radicals. In the limit of low concentrations of water and formaldehyde, there is no coadsorption affect and the rate of formaldehyde disappearance increases with increasing formaldehyde concentration. However, here the formaldehyde concentration is kept constant, so the increase in formaldehyde disappearance rate correlates to increasing water concentration. This can be attributed to increased radical production from water and resultant larger turnover rate for formaldehyde on the surface. If this description is accurate, changing the contaminant molecule from formaldehyde to a larger, but more weakly bound molecule, such as toluene, results in a lower surface concentration of contaminant and a resultant lower oxidation rate. At similar gas phase concentrations (2.2 ppmv formaldehyde and 5.4 ppmv toluene) the disappearance rate of toluene is depressed by a factor of *ca.* 10 over the formaldehyde rate.

As water vapor concentration increases to those representing high humidity in a building, both reaction rates reduce, the weaker bound species affected to a greater degree. A bimolecular Langmuir-Hinschelwood rate expression seems to fit the data well, especially at lower water concentration. Obee and Brown [20] studied the effect of water vapor concentration on the same two species and butadiene. In all cases the effect of humidity was to increase the rate as humidity decreases until a limit of increased oxidation is reached. This effect is observed when water is less than 1000 ppmv (approximately 30% RH). These are very dry conditions for indoor air and this effect can be neglected in most case. In a well-conditioned building humidity is maintained between 20% and 60% RH at *ca.* 20 °C, oxidizable contaminants in indoor air will exhibit higher reaction rates at the lower end of the humidity range.

2.3.3. The Effect of Temperature on Rate

Temperature can affect a semiconductor such as titania by promoting thermal catalysis. In controlled buildings the temperature range is small, and this effect, while shown to exist, is negligible. Surface coverage does depend on temperature, which affects the collision frequency of the gas phase molecules with the titania surface, but this is a small effect in the tight range of temperatures manifested by conditioned indoor air. Ordinarily, the expectation is that surface concentration decreases as temperature increases. In some cases, when partitioning across the surface occurs with large differences in binding energies, a different phenomenon dominates and the reaction rate for the more weakly bound species can increase with temperature.

If we restate Equation (12) and hold concentration constant while allowing temperature to vary, surface coverage is proportional to the binding energy, the temperature and the mass of the species. We can express the result as:

$$S \, \alpha \, e^{Q/kT}/m^{1/2} \tag{13}$$

If we form a ratio of the rate of surface coverages of water to ethylene we find:

$$S_w/S_{et} \, \alpha \, e^{(Q_w - Q_{et})/kT} \text{ or } e^{Q_w/kT}$$
$$\text{when } Q_e << Q_w \tag{14}$$

This shows that the ratio of relative surface coverage depends only on the binding energy difference and the temperature. When the binding energy of ethylene is much less than the binding energy of water, this ratio will decrease with increasing temperature, and so as the ratio of surface coverage of water to that of ethylene increases, so will the ratio of measured rates. This effect was seen by Obee and Hay [21], over the temperature range 2 °C to 48 °C. They modeled this effect using a temperature dependent form of the Lanqmuir-Hinshelwood rate expression. The photocatalytic removal rate of ethylene is observed to increase with increasing temperature. This effect will be observed for all species with small binding energies. Increasing temperature does not occur to any great extent in buildings, so this effect can be neglected when treating indoor air.

2.3.4. The Effect of Mixtures on Rates

The effects of coadsorption on the titania surface also dictates how a mixture of oxidizable gaseous contaminants will react photocatalytically. In the limit of low humidity and low concentrations of contaminants all species will adsorb on the catalyst surface partitioned solely by the effects of relative concentration and binding energy. In this limit, all oxidizable species react simultaneously. As humidity or total contaminant concentration increases, increasing competition develops for adsorption sites, and as concentrations increase the species with the strongest adsorption binding energy dominates the photocatalytic process. In the extreme of high contaminant concentration, Zorn, et al. [22] observed this effect in a recirculating photoreactor system. Sequential reaction occurs based on the strength of bonding to the surface. They studied the compounds ethanol, ethanal, propanone, propene and propane both singly and in mixtures at the 100s of ppmv concentration level. All five individual reaction rates were measured and fell in a sequence dictated by relative strength of bonding to the surface. In mixture of propene and propanone, a degradation of the propene oxidation rate was observed until all propanone was oxidized. In a separate experiment, where ethanol was injected into a photoreactor after propanone is ca. 80% removed by PCO, the reaction of the ketone was halted by the alcohol and further disrupted by the aldehyde intermediate observed in ethanal PCO. Only when the alcohol and aldehyde intermediate disappear does the photooxidation of the ketone resume. In indoor air we expect the contaminant mixture to be near the limit of low concentration. Normal indoor air has a total contaminant load that is less than 1 ppmv, usually much less. In sick buildings, or in special circumstances the contaminant load may be higher and in these cases, the effects of coadsorption and byproducts become more significant. This effect is correlated with some success to Henry's Law both in Zorn, et al., and Hodgson, et al. [23].

As we have seen with the example of TCE, byproduct formation can depend on parent concentration. This could be related to the ratio of activated surface site or radicals formed to adsorbed species and the number of steps in the mineralization reaction sequence. Byproducts form when mineralization proceeds through sequential free radical attack and a stable molecule forms as an intermediate. If a stable intermediate forms it competes with all other species to be retained on the surface or further oxidized. If this analogy holds, one expects more byproducts at higher total contaminant concentration in a mixture such as indoor air. In indoor air this means that we expect by-product formation to be important if any volatile VOC is introduced into the environment in high quantities. This could be the consequence of a spill or use of a highly volatile solvent during maintenance. As in the experiment conducted by Zorn, *et al.*, a byproduct will be oxidized further in time by a recycling system, but understanding the potential for such events is critical.

If the surface morphology of the catalyst is altered, the coadsorption phenomenon may be altered. Wei *et al.*, patented [24,25] a method of creating a 3% WO_3 coating on Degussa P-25. This modifies the surface from the hydrated titania surface to one that is partially hydrated titania and partially WO_3. The photochemical removal rate is enhanced for most VOCs, formaldehyde excluded, by this modification. Enhanced photocatalytic efficiency is shown for propanal, toluene and butene. High efficiency also ameliorates the effect of humidity by lessoning the effect of water coadsorption (see Figure 5). This manifests in higher removal rates for these species at high water vapor concentrations. This is attributed to the surface modification where water does not hydrogen-bond to WO_3 as it would to the hydroxylated titania. This allows most VOCs more efficient competition for adsorbent site. In indoor air applications, surface modification of titania can enhance overall purifier efficiency.

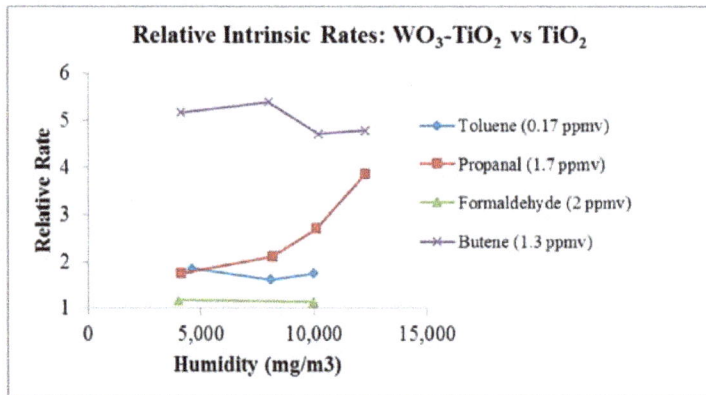

Figure 5. A partial surface layer of WO_3 modifies the catalyst surface, this changes the binding energies in such a manner that VOC adsorption can compete with water adsorption on the catalyst surface. The relative rate is the PCO removal rate observed for (WO_3-TiO_2) minus the PCO removal rate for (TiO_2) divided by the removal rate for (TiO_2).

2.3.5. Deactivation

Deactivation of the titania photocatalyst surface can occur by either reversible or irreversible means [26]. When a VOC containing carbon, oxygen and hydrogen is oxidized by the hydroxyl or superoxide radical to complete mineralization, the products are carbon dioxide and water. This regenerates the partially hydroxylated titania surface. Radicals formed during the mineralization reaction sequence may chemisorb on the surface. This can occur in a reversible fashion. If the chemisorbed species can oxidize further by radical attack to carbon dioxide and water the original surface regenerates itself. This regeneration may also be effected by calcining. In practice a photocatalytic air purifier may suffer reversible deactivation due to periodic fluctuations in VOC concentration. Operation during times when the building is devoid of occupants, which is associated with low total contamination rates, can serve to regenerate the catalyst. Irreversible deactivation can also occur if a non oxidizable species forms on the surface permanently blocking previously active sites or denying access for adsorption. Cao, *et al.* [27] studied toluene oxidation at high (relative to indoor air) concentration (10 ppm) concentrations using various forms of titania and platinized titania. The study is performed using UTRCs flat plate reactor, as described above, to measure rates. They contrast the performance of Degussa P-25 with three nanoscale titania materials, one partially platinized. The three nanoscale materials were prepared by sol-gel methodology and differed either in calking temperature (350 °C or 420 °C) or the addition of 0.5% Pt (350 °C). The nanoscale titania was found to be more reactive to toluene that P-25, but rapid deactivation was observed due to blocking of active sites by the partially oxidized intermediates, benzaldehyde and benzoic acid. The platinized nanoscale titania exhibits a lower reaction rate but deactivates slower. The deactivation was reversed by heating at temperatures above 420 °C. The more active the surface, the faster deactivation can occur.

In the same reactor, Huang, *et al.* [28] studied the photooxidation of tetraethylamine (TEA) over Degussa P-25. The disappearance rate was found to vary with concentration, humidity and light intensity in a manner consistent with that discussed above. The study was conducted with TEA in the concentration range 5.0 ppmv to 20.5 ppmv. Again deactivation is observed. Fourier transform infrared (FTIR) and temperature programmed desorption with a mass spectroscopic detector (TPD-MS) were used to probe the deactivation. They observed a marked concentration dependence on the degradation. At 20.5 ppmv TEA deactivation occurs rapidly and is near total after 90 min., while at 5.0 ppmv some activity remains after ten hours. They also observe that incomplete oxidation (or byproduct formation) increases as deactivation proceeds. If this behavior is universal, then as deactivation proceeds by-product formation could increase. Intermediates are either chemisorbed on the surface, as with TEA, released as stable molecules into the air stream, or adsorbed on the surface and further mineralized.

These varied results yield a snapshot of how an air purifier affects indoor air. We expect the effects to be humidity and temperature dependent. The effect of temperature is minimal in a conditioned environment. Increasing humidity will depress reaction rates but tailoring the surface morphology to lessen water adsorption can ameliorate these effects. At indoor air concentrations, we expect all species to photo-oxidize independently unless there is a spike in the concentration of one or more contaminants. Recycling air will minimize the effect of a momentary increase in VOC concentration, but the effect of by-products (both transient and steady-state) on occupants needs more study.

3. Experimental Section

3.1. Photocatalytic Reaction Rate Reactor

The main purpose of our photocatalytic flat-plate reactor is to generate intrinsic oxidation rates for selected gaseous species of importance to indoor air quality. Design features include uniform irradiation of the photocatalyst surface, radiation of appropriate range of wavelength, and elimination of mass-transport influence (on the reactants and reaction byproducts) between the flow field and photocatalyst surface. Later these data are fed into an air purifier design procedure that explicitly accounts for non-uniform distribution of radiation on the photocatalyst surface and mass-transfer effects between the flow field and photocatalyst surface.

Our photocatalytic reactor design is shown in Figure 6. Details of the reactor, detection apparatus, and experimental procedures used to obtain rate data are given elsewhere [18,20]. The reactor design allows for measurements of intrinsic destruction rates free from mass-transport (diffusion) effects, and for study of the effects of contaminant concentration, humidity, and UV intensity-dependencies on their rates of disappearance.

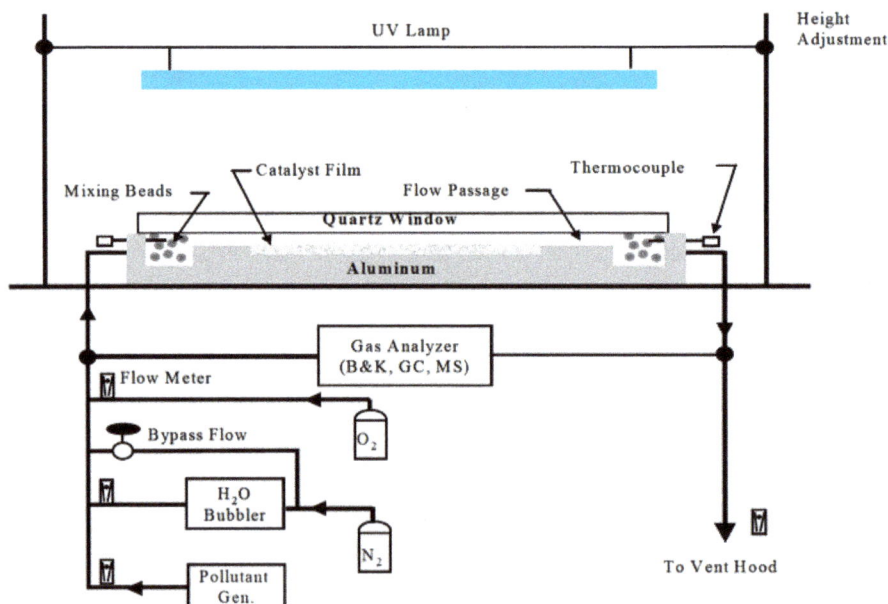

Figure 6. Photo-oxidation Reactor.

A flow by-pass valve allows complete adjustment of the delivered humidity level. An approximate atmospheric level of oxygen (15%) and nitrogen (85%) closely mimics the targeted application environment (residential and building air). The oxygen level is not critical as long as the level is maintained above about ~1% [2,29], (reaction rates follows Langmuir-Hinshelwood kinetics and is zero order for oxygen levels exceeding ~1%).

Titania-coated glass (or aluminum) plate are placed in a well (25 mm by 46 cm) milled from an aluminum block, and covered by an appropriate quartz window (UV transparent). Gaskets between the quartz window and aluminum block creat a flow passage of 25 mm (width) by 1–2 mm (height) above the titania-coated glass-plates.

In this reactor an opaque film of the photocatalyst, Degussa P-25 titania, is deposited on flat 25 mm wide microscope slides using a wash-coat process. The wash-coat solution is prepared by suspending the titania (5% by weight) in distilled water. The slides are dipped in the wash-coat solution several times, air dried between dipping, and then oven dried at 70 °C. This process is repeated until a sufficient loading ($\geq$0.74 mg/cm^2 film per side) is achieved.

In a study of film loading by Jacoby [30] the oxidation rate of trichloroethylene increased with film loading up to a P-25 titania loading of 0.5 mg/cm^2 and remained constant for all higher loadings. This finding suggests that the oxidation rate maximizes at a film loading of 0.5 mg/cm^2 and that additional film loading adds nothing to the oxidation rate. This conclusion should not depend on the specific contaminant used. The titania film of 0.74 mg/cm^2 film loading was determined to be opaque to UVA by placing a coated plate between UV black-light lamps and a UV power meter. This finding coupled with the conclusion drawn from Jacoby's thesis finding suggests that the UV radiation is being maximally utilized in the oxidation process. In some tests UVC radiation is used. In this case, irradiation occurs deeper into the adsorption profile for titania and similar opaqueness is anticipated due to a higher density of adsorption-allowed states.

Variation of UV intensity is achieved simply by adjusting the distance between the photocatalyst surface and lamp. Although the reactor can accommodate a photocatalyst up to 12 inches length along the flow direction, a UV opaque mask is used to select a section of the photocatalyst for irradiation. Commercial lamps of a bi-axial design are used. The lamp specifications typically includes the lamp power and flux at a selected distance from the lamp. To check for UV flux uniformity a UV meter is used to scan the exposed photocatalyst.

For the data generated by the glass-plate reactor, the oxidation rate is defined as:

$$r = 2.45(X_{in} - X_{out})Q/A \qquad (15)$$

where r (m-mole/cm^2-h) is the oxidation rate, and X_{in} (ppmv) and X_{out} (ppmv), are the inlet and outlet ethylene concentrations, respectively, Q (lpm) is the volumetric flow rate, and A (cm^2) is the area of the titania-coated glass-plate; the numerical coefficient accounts for the units change.

The absence of mass-transport effects between the photocatalyst and the flow field is demonstrated by measuring the oxidation rate as a function of the approach velocity (or volumetric flow rate) all the while keeping the residence time (length of irradiated catalyst in the flow direction divided by the approach velocity) through the reactor constant [30], Such a data plot will exhibit typically two distinct regions: a low velocity region in which the reaction rate increases with increase in the approach velocity, and a higher velocity region in which the reaction rate is constant, that is, has reached a plateau. The lowest approach velocity at which the plateau first appears established the minimum approach velocity in which mass transfer effects are negligible.

This so determined minimum approach velocity is found for a specified humidity level and UV intensity. For any subsequent reaction rate that is lower than the rate at the plateau, this reaction rate is also mass-transport free. Conversely, if a change in UV flux or humidity resulted in a reaction rate greater than the one found at the plateau, then one would need to generate a new plot of reaction rate *versus* approach velocity to determine a new minimum approach velocity.

For the photocatalyst titanium dioxide, reaction rates follow Langmuir-Hinshelwood kinetics [18,20,31,32]. And for sub-ppm concentration levels the kinetics are linear dependent (see Figure 3) on the contaminant level. A large change in contaminant concentration across the photoreactor would result in a significant change in reaction rates between inlet and outlet catalyst sub-elements and the overall reaction rate is considered integral. The difference in contaminant concentration between the inlet and outlet of the photo-reactor should be kept small, a fractional change less than ~0.15 is desirable. In doing so the reaction chemistry at the reactor inlet and outlet is essentially the same and reaction rate can be considered differential. Differential reaction rates are more highly valued over integral rates as they are easier to implement in an air-purifier design code.

Humidity level has a significant impact on the reaction rate for a given contaminant [18,20,32], In these studies the reaction rate is shown to follow a bimolecular expression of the Langmuir-Hinshelwood (L-H) kinetic rate form:

$$r(X_c, X_w, I_G) = K_0 \frac{K_1 X_c}{(1 + K_1 X_c + K_2 X_w)} \frac{K_4 X_w}{(1 + K_3 X_c + K_4 X_w)} \tag{16}$$

where K_0 (m-mole/cm²-h) is the rate constant for a given UV intensity (I_G), K_1, K_2, K_3, K_4 (ppmv^{-1}) are the Langmuir adsorption equilibrium constants (ratio of adsorption to desorption rates), and X_c (ppmv) and X_w (ppmv) are the gas-phase concentrations of the contaminant and water vapor, respectively. The two fractional terms on the right-hand-side represent competitive adsorption between the contaminant and water for the same adsorption site [33].

As rate varies as I_a, the value of the exponent depends on the UV level and ranges between 0 and 1 [20]. At low UV levels an exponent of 1 is expected, while at intermediate UV levels an exponent of 0.5 is expected, and at high UV levels an exponent of 0 is expected. For example, at the UV flux of 0.71 mW/cm² the exponent was 0.64 for toluene, while at 33 mW/cm² a value of 0.55 was found [20]. This range represents typical UV power in UTRCs PCO air purifier.

3.2. Modeling

As discussed, a generic photocatalytic air-purifier can be thought of as consisting, as a minimum, of three basic elements: a source of radiation, a photocatalyst and its support, and a container to support these latter two elements and as a means for channeling contaminated air from and returning treated air back to an occupied space. Commercially available cylindrical UV lamps offer an easily implemented, economical source of UV radiation. Reticulated foam and honeycomb monoliths, which have been used as supports for thermal catalysts in catalytic converters, have a long studied history, and provide suitable supports for the photocatalyst. An example diagram of a generic modular photocatalytic air purifier is displayed in Figure 1. The reactor consists of six banks of UV lamps with five photocatalytic coated honeycomb monoliths, each monolith being positioned between a banks of four lamps. Although reticulated foam and honeycomb monoliths offer similar photocatalytic performance, the honeycomb offers much lower pressure drop. Commercial and residential HVAC systems are all very sensitive to pressure drops from any ancillary environmental equipment such as air-purifiers, and so the honeycomb becomes the preferred choice for the photocatalytic reactor [32,34].

3.2.1. 1-D Model

A one dimensional (1–D) model of the generic multi-stage, honeycomb-monolith photocatalytic reactor has been described in detail [13,32]. This model assumes a monochromatic radiation field at the entrance of each honeycomb channel, but allows for a non-uniform (2-D) radiation field across the honeycomb face. Within each honeycomb passage a 1-D treatment is used that incorporates gas-solid mass-transport correlations for circular flow cross-section, averaged over the channel length. Within each channel the radiation field model rigorously accounts for diffuse reflectance of the catalyst film coating on the channel walls. Reactions on the photocatalyst incorporate Langmuir-Hinshelwood kinetics to accurately account for the influence of UV flux, contaminant concentration, and humidity.

A schematic diagram of the workings of the computer coded version of the model is shown in Figure 7. Because the model is essentially one-dimensional, it is fast in execution and thus makes for a highly efficient tool for engineering design.

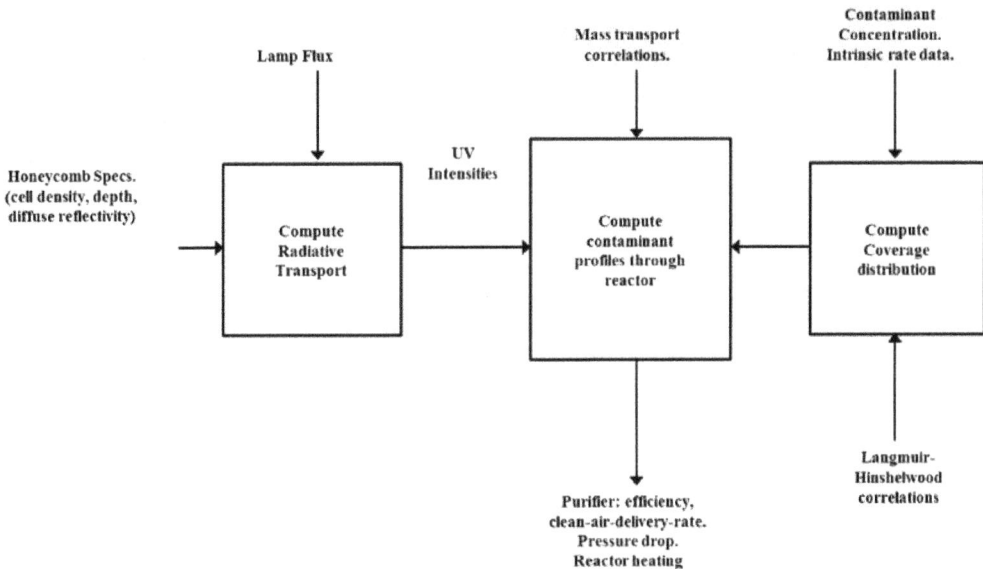

Figure 7. Photocatalytic reactor design flow diagram.

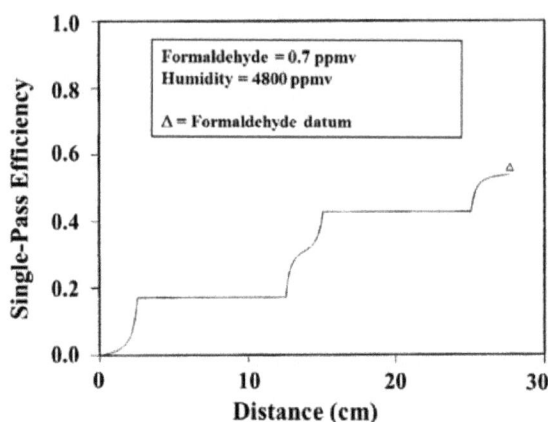

Figure 8. Comparison of model and experiment for formaldehyde in a 2-UV banks, 3-honeycomb monolith photoreactor. A similar version of this figure is published in Ref. [13].

A sample output of the model is given in Figure 8, showing the removal efficiency for the contaminant formaldehyde. In this reactor two banks of UV lamps are sandwiched between three honeycombs monoliths. A single measurement datum, taken at the reactor outlet, is included for comparison. The predicted changes in efficiency shown in the figure occur along the axial length of the honeycomb passage way. The plateaus correspond to the space between the monoliths wherein no reactions can occur. The agreement between model and the single datum is excellent.

3.2.2. 3D Model

A first-principles mathematical model was developed by Hussain, *et al.*, and describes the performance of a photocatalyst honeycomb monolith photocatalytic reactor for air purification [35], The single-channel, 3-D advection-diffusion-reaction model assumes steady-state operation, negligible axial dispersion, and negligible heat-of-reaction. The reactor model accounts rigorously for entrance effects arising from the developing fluid-flow field and uses a first-principles radiation-field sub-model for the UV flux profile down the monolith length. The model requires specification of intrinsic photocatalytic reaction rates that include the influence on local UV light intensity and local reactant concentration, and uses reaction-rate expressions and kinetic parameters determined independently using UTRCs flat-plate reactor [36]. Model output predictions are similar to that shown in Figure 8, and validate the 1-D model.

3.2.3. 1-D and 3-D Model Validation

Both the 1-D and 3-D model predictions were favorably compared to experimental demonstration-scale formaldehyde and toluene conversion measurements for a range of inlet contaminant concentrations, air humidity levels, monolith lengths, and for various monolith/lamp-bank configurations [13,32,34]. In addition to these contaminants the 1-D model included successful comparisons with other contaminant species, such as, ammonia and hexane [32]. In the 3-D model this agreement was realized without benefit of any adjustable photocatalytic reactor model parameters, radiation-field sub-model parameters, or kinetic sub-model parameters. For formaldehyde the agreement between both models and data was excellent. Both models tended to systematically over predict toluene conversion data by about 30%, but which falls within the accepted limits of experimental kinetic parameter accuracy. The explanation for this disparity follows from the expected stronger photocatalyst bonding (affinity) for formaldehyde than for toluene [20]. As a direct consequence, photo-oxidation of toluene is more likely to be interfered by background co-contaminants. The empirical data for formaldehyde were collected at relatively high concentrations, greater than 1-ppm, while the concentration of toluene was set at a low level of ~0.3-ppm (this is necessitated by the fact that a reversible deactivation of the photocatalyst occurs for toluene levels above about 1-ppm [20], The data developed using the demonstration reactor used unfiltered building air seeded with either formaldehyde or toluene. Since background contaminant levels in the sub-ppm range would be expected, the observed systematic disparity with toluene is likely entirely or in part due to background contaminant interference. The generally satisfactory agreement between experimental and modeling data indicates that computer modeling can be used to guide the design of scaled-up reactors for practical applications. The 1-D model offers the advantage of ease of code implementation, but also is fast in execution which makes it conducive for parametric studies and as a sub-element in an optimization host code.

3.3. Validation

The modular design we describe has been tested in various locations, with a variety of results. Real world testing is complex, time consuming and expensive. There is no substitute for deployment of PCO based air purifiers in building environments. Laboratory conditions are well controlled environments, whereas building conditions are defined by variable source rates. Source rates vary with occupancy, usage, the age of the building and its furnishings, and with outdoor ventilation air quality. The composition of indoor air contaminates is rarely static and can vary seasonally. Given unlimited resources, a test environment (a building) could be well characterized with respect to indoor air contaminant species (both aerosol and gaseous), and with respect to hourly variations, due to occupancy and usage, and seasonal variations. A UVPCO air purifier would then be designed to act on this variable challenge and to satisfy the building HVAC design constraints. Lastly, the indoor air would be characterized again during testing with and without the use of the air purifier. In most cases compromises in the testing regimen allow validation in these environments with the resources available.

3.3.1. Perceived Air Quality

Kolarick and Wargocki [37] of DTU placed a prototype air purifier, provided by UTRC, in an environment with poor air quality due to off-gassing from building materials. They showed that the air purifier had a significant effect on perceived air quality. The effect of the air purifier was to simulate an increase in e the outdoor ventilation rate by factor of 0.02 to 3. The improvement depended on the specific pollutants encountered. When the indoor air was polluted by human

358

bio-effluents the perceived air quality worsened. This effect is attributed to an increase in byproducts from the incomplete oxidation of alcohols. Alcohols such as methanol, ethanol, amyl alcohol, *etc.* are known human bio-effluents. This effect is similar to that discussed above.

Kolarick, *et al.* [38] performed a similar experiment but included Proton-Transfer-Reaction Mass Spectroscopy (PTR-MS), Gas Chromatograhic Mass Spectroscopy and High-Pressure Liquid Chromotography measurements of trace pollutants. Analytic measurements are difficult on indoor air contaminants due to interferences between compounds and the low concentration levels and the results vary with the analytical technique used. The operation of a photocatalytic air purifier improved air quality in an office polluted by typical building materials. Of the methods used, subjective assessments and PTR-MS measurements were the most effective in demonstrating the effect. An increase in byproducts is observed by GC-MS, these byproducts are predominantly acetic acid, acetone and 2-butanone and are far below odor thresholds. These by-products do not contribute to UTRCs proposed tolerance metric as they are limited by odor threshold. PTR-MS did not show the same increase, but detected a few ppb increase in formaldehyde, acetaldehyde and acetone during the first hour after start up. These species will contribute to the tolerance metric and these byproducts must be well understood to determine any impact on human health.

Both the above studies were performed with a high efficiency (seven stages) UVPCO prototype air purifier, provided by UTRC, and the conclusions pertain to perceived air quality, not occupant health.

3.3.2. Analytical Air Quality

In a similar study, a smaller prototype (two stages) air purifier was supplied to LBNL, who challenged the reactor with a mixture of 27 VOCs commonly found in office buildings. Hodgson, *et al.* [22] reported that operation of the air purifier compensated for reducing the outdoor air intake by 50%. They assumed the unit was installed in the building air flow after the outdoor and recirculated air are combined and the recirculated flow was three times the outdoor air intake. In addition they observed formaldehyde, acetaldehyde, acetone formic acid and acetic acid byproducts. Of these, formaldehyde would contribute the most to an air quality metric such as the tolerance index.

3.3.3. Building Air Quality

Lemcoff, *et al.* [39] performed a building study at UTRC's Technical Education Center where a specifically designed UVPCO air purifier is installed in a rooftop unit that feeds a *ca.* 100 person capacity educational room. VOC testing is carried out using US EPA protocol techniques both prior to and during air purifier operation. Summa canisters are used to sample hydrocarbons and chlorinated solvents and adsorbent cartridges to analyze for aldehydes. The application of the URC proposed Tolerance Index is used to determine the level of air quality. The purifier is installed in the recycled air (nominally 15,000 ft^2/min, 55 °C) subsequent to the 35 ton Rooftop Unit and prior to variable air ventilation (VAV) unit at the room exhausts. An economizer was installed to lower the rate of outdoor ventilation in good weather and humidity was controlled through use of a desiccant wheel. A schematic is shown in Figure 9. Sensors are used to continuously monitor temperature, relative humidity, particle concentration, total VOC and CO_2. Comparable pollutant levels to those found in the EPA BASE study are found. When UTRCCo proposed tolerance metric is applied to the levels found for one study performed with an occupied room (50 occupants), operation of the air purifier is predicted to reduce the tolerance metric from a steady state of ca. 0.64 without air purification to a steady state of 0.29 with the air purifier. The outdoor air flow was reduced during operation to show potential energy savings. Application of the tolerance metric to the indoor air with purification and with reduced air show the Index varied over the range 0.48 to 0.62, indicating that air quality was maintained with reduced air flow and 15% energy savings. The reduced flow would also allow for a 17% capacity reduction in the roof top unit. These results are consistent with those of LBNL. Unfortunately, when the catalyst is examined after three months of operation the catalyst was found to be partially deactivated. Some catalyst activity was regenerated by operation in a clean humid environment for several days.

UVPCO air purifiers deployed in both the US and France [6] were tested for catalyst efficiency after varied periods of operation in different office environments. Catalyst deactivation is observed in all cases, some deactivation was reversed when catalyst monoliths are exposed to UVC operation in a clean environment. The catalyst used was either 100% Degussa P-25 or a mixture of 50% Degussa P-25 and 50% Millenium 50 supported on aircraft grade aluminum honeycomb. Examination of catalyst surface by XPS shows a surface contaminated with silicon, carbon and to a lesser extent nitrogen. Electron microprobe data of a catalyst taken from an air purifier deployed in France indicated a strong 230–260 nm Si rich layer on the surface of the catalyst. Silcones are the most prevalent silicon containing VOC or SVOC components of indoor air. Hay, et al. [6] studied siloxane deactivation of titania in the flat plate reactor described above. A strong correlation is observed between surface Si as seen by XPS and the extent of deactivation. They also show a strong correlation of the extent of deactivation observed in catalysts (3% WO_3 on Degussa P-25) deployed in prototypes operating in office environments, with the laboratory results. They attribute purifier deactivation to ambient siloxanes oxidized by PCO action to amorphous SiO_2. This is observed by TEM in catalysts 80% deactivated by PCO in the presence of tetramethyl siloxane (TMS or D4 siloxane). Ambient air testing in the locations where the prototypes were deployed indicated an average siloxane concentration of 0.2 ppmv was present with DMCPS (decamethylcyclopentasiloxane) the most prevalent. Hay, et al., also demonstrate that UVPCO air purifier lifetime can be extended with an adsorbent prefilter selected to adsorb high molecular weight compounds such as siloxanes.

Validation of UVPCO air purifiers in laboratory situations simulating indoor air and in buildings, where prototypes operated in an actual office environment yield results that are both promising and disconcerting. These air purifiers can impact the quality of air positively but questions remain about the potential impact of byproducts and the lifetime of catalysts. In order to provide impetus for the deployment of this technology as product, consumers must see a tangible benefit. Improved air quality has been linked to improved productivity, but no metric exists correlating measurable quantities directly to improved air quality. Building owners, concerned with the cost of day-to-day operation of building HVAC systems see no payback in nebulous increased productivity but look to the hard numbers available for energy use (operating cost) and equipment capacity (capital cost).

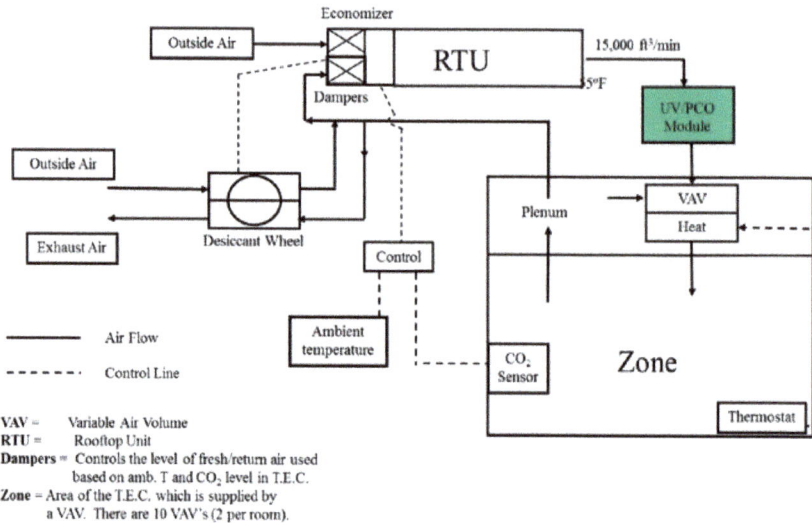

Figure 9. HVAC Schematics of air purifier installation; 35 ton rooftop with VAV, economizer and desiccant wheel.

The demonstration conducted in UTRC's Technical Education Center indicates that both capacity reduction and energy use reduction can be obtained utilizing UVPCO air purification to reduce outdoor air intake. This is validated by the work done at DTU and LBNL. This indicates that a business case can exist to replace outdoor air with purified air. UTRC's proposed tolerance metric, if valid, is a strong indicator that this can be done while simultaneously improving total air quality. At this stage of development, however, each case must be considered independently. This business case depends strongly on ambient air intake, if little cooling or heating is required, there are little energy savings to be gained by reducing the extent of conditioning. In addition, the effects of byproduct formation must be better understood in terms of their impact on human health.

3.4. Current and Future Research

As we have discussed, ambient air itself is an elusive target, chemical off-gassing changes as new materials are developed and incorporated into building materials, furnishings, clothing, and personal care products. A century ago no concerns over siloxanes existed, now silicon oxide whiskers grow on electrostatic particle removers, interfering with operation [40], and causing UVPCO deactivation. Siloxanes appear to be ubiquitous and have even been identified in Artic ice [41].

The action of a photocatalytic air purifier can change ambient air. A purifier can mineralize VOCs if designed and operated appropriately, but is cost prohibitive to do so. One hundred percent mineralization is achievable only in a laboratory. Spatial requirements in buildings dictate the use of small air purifiers. These have various single pass efficiencies which depend, as we have discussed, on species, concentration, temperature and coadsorption phenomenon. In addition, they depend strongly on the choice of catalyst. BET surface area, porosity, crystallinity and surface morphology can affect adsorption. In addition the photocatalyst surface changes over time as various types of deactivation occur. As deactivation reduces the number of active sites available, incomplete mineralization becomes more prevalent, promoting the production of byproducts. Not only is by-product formation a concern when the catalyst surface is pristine, but the effect of deactivation on byproduct formation must be evaluated. As active surface sites are removed from photocatalytic action, the numbers of radicals available for mineralization decrease. This decrease may simulate the effect of increasing concentration. In other words, as the ratio of available radicals to parent molecules is decreased, complete mineralization is decreased.

To facilitate commercialization of this technology, a critical need exists to minimize the effect of deactivation and to increase photoreactivity. Extending the wavelength region for efficiency photon adsorption further into the visible can also increase the ease of commercialization. The business case in buildings is based on the capital cost of the air purifier and the cost of operation. Is there a return on investment? The efficiency of contaminant removal relates directly to the size of the air purifier and therefore the capital cost. The catalyst lifetime relates directly to the operating cost. If the lifetime is too short the return on investment time increases. Extension of excitation wavelength into the visible could allow the use of LEDs, which are commercially available at longer lifetimes in longer wavelengths. This in turn could lower both capital and operational costs due to the choice of light source. Current activity ongoing at the University of Connecticut is focused on efficient siloxane trapping, deactivation resistant photocatalysts, high efficiency catalysts and visible light activated catalysts.

As was described previously, deactivation of metal-based catalyst through either coking or poisoning by particular species has been a continuing problem. One such example is that of the titanium dioxide system, in which TiO_2 is deactivated via siloxane groups that deposit amorphous SiO_2 onto the surface of the catalyst. The main effect of this is partial to complete deactivation of the catalyst, dependent on the amount of siloxane present [6]. This occurs through the blocking of active sites (and the surface) by the silica deposited, eventually leading to almost total deactivation.

To counteract deactivation, a variety of methods have been employed, including the development of a high surface area material and the tuning of the surface area and pore sizes of the material. The objective is to create an active material where deposits of amorphous silica do not encapsulate the surface. UTRC patented one such a technique [42], and another was pioneered at UConn by Horvath [43]. Through heat treatments, these parameters can be tuned. Additional research has shown that pore size and surface area can be maintained at high calcination temperatures through the use of silane coatings followed by the removal of the silica layers post heat treatment. A porous silica overlayer can extend catalyst lifetime [44,45].

Figure 10 shows the dependence of surface area and pore volume on temperature for doped titania systems. Increases in temperature lead to decreases for both. From these images, these parameters may be modulated via different dopants, though studies revolving around these materials are ongoing. Increases in overall surface area and pore volume are theorized to increase the lifetime of the catalyst system as the loading of silica required to deactivate would theoretically increase.

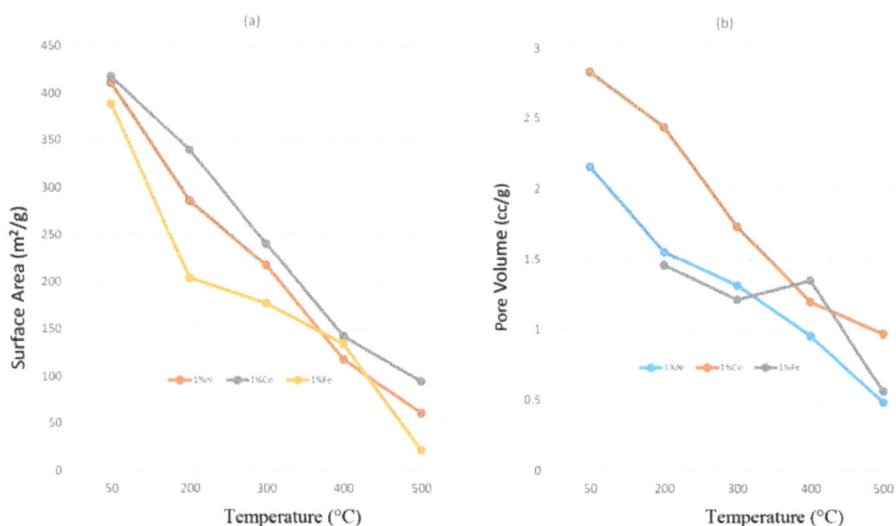

Figure 10. (a) Surface area change and **(b)** pore volume change in doped titania systems as a function of calcination temperature (T = 50 °C, 200 °C, 300 °C, 400 °C and 500 °C).

One attractive technique is to simulate the high activity of Degussa P25 in a high surface area material. P25 is a benchmark material for titanium based photocatalyst studies, employed by UTRC and many other laboratories. Since 1990, significant research has been focused on P25 due to its unique composition and high relative activity. This high activity has been shown in different photocatalytic reaction systems but few are better photocatalytically under similar conditions. The reason why P25 has such excellent activity is attributed to a synergetic effect between anatase and rutile [46]. P25 is a mixed phase material that is composed of anatase and rutile phases, the specific ratio has been reported between 80:20 and 75:25 [47]. While P25 is attractive photocatalytically, its disadvantages are not negligible. The surface area of P25 is relatively low (around 50 m²/g). P25 is a nonporous material and is visible light inactive. After intense studies on P25, researchers are trying to develop new photocatalysts that have P25hat haveactivity combined with a high surface area, a porous structure and are visible light active.

To achieve high surface area with a porous structure, mesoporous TiO₂ materials are a prime candidate. Mesoporous TiO₂ was first synthesized by Antonelli et al. in 1995 [48]. The reported surface area of mesoporous TiO₂ material is as high as 1200 m²/g [49]. This high surface area of offers more active sites during photocatalytic reactions. Moreover, the nanocrystalline walls and the pore structure will allow photogenerated electrons and holes to reach the surface easier and further favor the surface photocatalytic oxidation/reduction reactions. The main drawback of mesoporous TiO₂ materials is their low thermal stability. Thus, most of the mesoporous TiO₂ materials can only possess the anatase phase. Mixed phase or rutile phase titania materials are difficult to form while maintaining the mesopore structure.

To achieve visible light photocatalytic activity, metal and non-metal dopants are commonly used to stretch the absorption band of photocatalysts to the visible light region. Different metals have been doped into TiO$_2$ nanomaterials; W, V, Ce, Zr, Fe, Cu, Ag are all shown to effectively stretch the absorption band to the visible light range (>400 nm) [50,51]. Not only metal dopants, but also various nonmetal elements such as B, C, N, F, S, Cl and Br have been successfully doped into TiO$_2$ nanomaterials. After dopants have been introduced into the TiO$_2$ structure, the electronic, optical and photoelectrical properties of TiO$_2$ photocatalyst are modified [52].

One of the important properties of dopant modified TiO$_2$ structures is the phase. The most photocatalytic active phase is the anatase phase, which has a band gap energy at 3.2 eV (387 nm). The thermally stable phase is the rutile phase, which has a band gap energy of 3.0 eV (412 nm). The anatase phase is not visible light active. The photon energy from visible light (>400 nm) is not strong enough to separate electron/hole pairs. Theoretically, the rutile phase should be photocatalytically active under visible light with an appropriate band gap. However, rutile shows a high recombination rate which leads to poor photocatalytic activity. Combining to make a mixed phase material should lead to photocatalytic activity that should be significantly improved. Commercially available P25 is the best example of such mixed phase materials.

If one takes advantage of a mesoporous structure and the mixed phase simultaneously, the photocatalytic activity of this material should be significantly improved. Achieving two phases (anatase and rutile) and maintaining a mesoporous structure at the same time is the biggest challenge. Phase transformation promoters could significantly decrease the anatase-to-rutile phase transition temperature. Al, Co, Cu, Fe, Cr, V and Zn are all reported as efficient phase transformation promoters [53–58]. If the transition temperature can be brought to lower than lower than 450 °C, a mesostructure can be maintained at the same time.

A mesoporous mixed phase TiO$_2$ material is expected to be an efficient visible-light-activated photocatalyst. The mesopores facilitate the diffusion of photogenerated electrons and holes to the particle surface. More importantly, the synergetic effect between the anatase phase and the rutile phase causes an efficient charge separation across phase junctions. The mesoporous mixed phase TiO$_2$ will not only have enhanced photocatalytic activity but also higher adsorption ability, improved thermal stability, longer life and be suitable for both liquid phase and gas phase reactions.

Recently developed University of Connecticut (UCT) mesoporous materials offer the opportunity to prepare mesoporous mixed phase TiO$_2$ materials [59]. Through controlling the sol-gel chemistry of inorganic sols in inverse micelles and NO$_x$ chemistry, highly thermal stable highly thermal stable (450 °C) mesoporous TiO$_2$ materials can be prepared. Further, by modifying the synthesis method with certain metal dopants (phase transition promoters), thermally stable mesoporous TiO$_2$ with mixed phase can be prepared.

The mesoporous mixed phase TiO$_2$ (UCT-TiO$_2$) was prepared based on the method discussed above. This material is composed of 60% anatase and 40% rutile based on calculations from X-ray diffraction patterns. The adsorption ability of UCT-TiO$_2$ was compared with Degussa P25 in the dark to adsorb methylene blue (MB) dye in 2 h. Figure 11 shows that the UCT-TiO$_2$ shows much higher adsorption ability than P25.

In addition, the visible light (>400 nm) photocatalytic activity of UCT-TiO$_2$ was tested by degrading methylene blue dye. As shown in Figure 10, MB dye can be completely removed by UCT-TiO$_2$ in 2 h. P25 displays poor degradation performance under visible light.

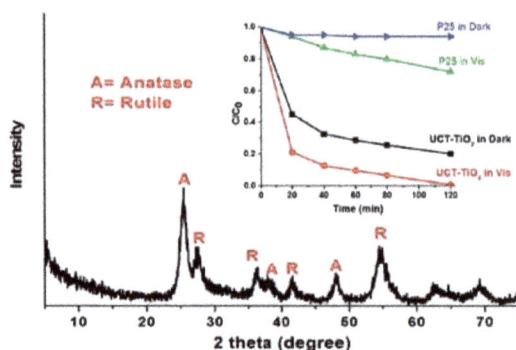

Figure 11. X-ray diffraction pattern for UCT-TiO₂ materials. Inset image: adsorption ability and photocatalytic ability of UCT-TiO₂ compared to P25 tested in dark and visible light conditions, respectively. The efficiency was calculated by C/C₀ (dye concentration in different time/initial dye concentration).

The mesoporous mixed phase TiO₂ (UCT-TiO₂) is a potential new photocatalyst that can be used in liquid phase reactions such as dye degradation, organic compound decomposition, and also for gas phase reactions such as VOC degradation and CO oxidation reactions. The unique mesostructure combined with the synergetic effect between mixed phase junctions offers the opportunity for more potential photocatalytic applications. This type of catalyst is promising for increasing reactor performance for indoor air, and potentially allowing energy efficient LED activation.

Siloxanes are a class of anthropogenic chemicals having a multitude of applications in the production of household, automotive, construction, and personal care products, as well as acting as intermediates in the production of silicon polymers. Siloxanes are found to be ubiquitous in the air, water, sediment, sludge, and biota. Due to their widespread use, siloxanes have received notable attention as emerging organic environmental contaminants over the past two decades. Most low molecular weight siloxane compounds volatize quickly into atmosphere to pollute the air and high molecular weight siloxane compounds remain in the water and soil. Siloxanes need to be cleaned in the air because of their potential for long range transport and bioaccumulation. They also, of course, deactivate photocatalysts. Table 3 lists common types of linear and cyclic siloxanes.

Table 3. Siloxanes and their physical properties.

Siloxane Type	Formula	Abbreviation	Molecular Weight	Vapor Pressure (Torr, 25 °C)
Hexamethyldisiloxane	$C_6H_{18}OSi_2$	L2, MM	162	31
Octamethyltrisiloxane	$C_8H_{24}O_2Si_3$	L3, MDM	237	3.9
Decamethyltetrasiloxane	$C_{10}H_{30}O_3Si_4$	L4, MD₂M	311	0.43
Dodecamethylpentasiloxane	$C_{12}H_{36}O_4Si_5$	L5, MD₃M	385	0. 1022
hexamethylcyclotrisiloxane	$C_6H_{18}O_3Si_3$	D3	222	10
Octamethycyclotetrasiloxane	$C_8H_{24}O_4Si_4$	D4	296	1.3
Decamethylcyclopentasiloxane	$C_{10}H_{30}O_5Si_5$	D5	371	0.4
Dodecamethylcyclohexasiloxane	$C_{12}H_{36}O_6Si_6$	D6	445	0. 0494

As we have seen, siloxanes in ambient air can disrupt the operation of a PCO air purifier causing rapid deactivation through conversion to amorphous silica on the catalyst surface. Hay, *et al.* [6] demonstrated that approximate lifetime doubling occurs when an air purifier is protected by an adsorbent filter. Lifetimes can be extended further as more efficient siloxane traps are developed. There are various methods to remove siloxane including biological methods, cooling, absorption, catalysts and adsorption. Among these techniques, solid adsorbent is the simplest way to remove siloxane for which various types of active materials have been applied such as silica gel, alumina, activated carbon and so on (Table 4). The pollutant is adsorbed by physical interaction with the surface.

Table 4. Siloxane adsorbents in the literature.

Type of Materials	Materials Details	Adsorption Capacity (g Siloxane/g Adsorbent)	Surface Area (m²/g)	Type of Siloxane	Ref
Molecular sieve	13× molecular sieve 45/60 mesh	0.01	-	D5	[59]
Zeolite	Faujasite NaX	0.276	500	D3	[60]
activated carbon (ACs)	NORIT RB4	0.41	>1000	D3	[61]
MgO	Commercial, periclase phase	N/A	31	D3	[60]
CaO	Ex-CaCO₃ calcination, cubic phase	0.003	31–40	D3	[60]
Silica gel	Fluka (particle size: 1–3 mm)	0.1	-	D5	[59]
Alumina	Duksan Co. (65.7% porosity)	0.1775	-	D4,D5	[62]

At present, active carbon is the most efficient and cheapest solid adsorbent which is used to clean the siloxanes in the air.

However, active carbon will have a dramatic decrease of performance in the presence of humidity. This is attributed to blocking the adsorption sites by the formation of hydrogen bonds [63]. Mesoporous aluminosilicate (UCT-15) belongs to the family of UCT mesoporous materials, which have high surface area, crystalline walls and monomodal pore size distributions. The UCT materials have been used for H_2S adsorption and showed remarkably high adsorption capacities.

The mesoporous aluminosilicate was synthesized by dissolving TEOS and aluminum nitrate (Si:Al molar ratio = 5) in a solution containing HNO_3, 1-butanol, and P123 in the beaker at room temperature and under magnetic stirring. The obtained clear gel was placed in oven for 4 h. Then samples were grained and calcined under air at 450 °C for 4 h (2 °C/min heating rate).

Adsorbent performance of D4 adsorption tests were run by passing adsorbents with carrier gas (N_2) which contains siloxane and water moisture. The concentration of water moisture in the carrier gas was 1.7% (molar). After passing through the adsorbents, the gas wash bottle was used to adsorb the residue siloxane in the carrier gas. GC/MS with a DB-5 column was used determine the siloxane concentration in the trap solvent.

Figure 12a shows adsorbed amount of siloxane over time by mesoporous aluminosilicate compared with active carbon. On the adsorption figure, the more siloxane the adsorbent adsorbed, the better the adsorbent. Figure 12b shows the pore size distribution of mesoporous aluminosilicate adsorbent and the surface area of aluminosilicate is 229 m^2/g. The pore size distribution figure confirms the mesoporous structure of mesoporous carbon adsorbents. We can see from Figure 12a that mesoporous aluminosilicate works much better than actived carbon under the moisture condition. This is because the hydrophobicity [64] of the aluminosilicate surface could weaken the blocking effect of active adsorption sites.

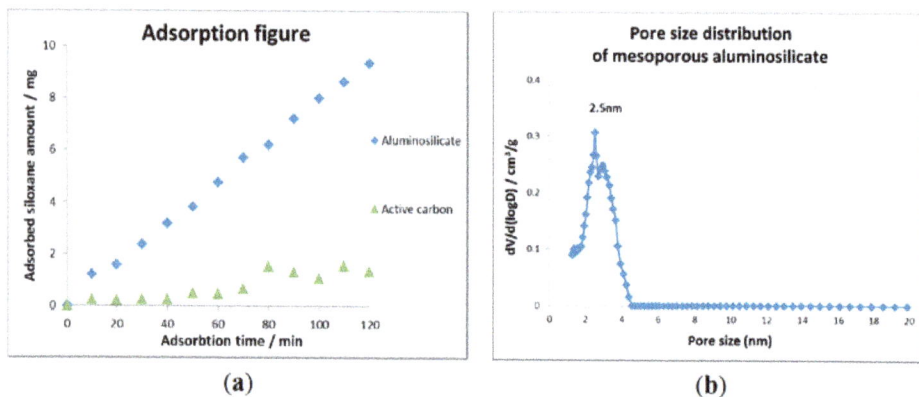

(a)   (b)

Figure 12. (a) Adsorption of siloxane on the adsorbents over time, UCT material compared to activated carbon; (b) Pore size distribution of mesoporous aluminosilicate.

Manganese oxides comprise a large group of compounds due to the multi-valent nature of such oxides and the complexity of the forms in how the octahedral units connect with each other. The application of these compounds has been pursued widely in adsorption, catalysis, energy storage and environmental pollutant removal [65]. In particular, due to the strong oxidative properties, manganese oxide was successfully used for abatement of a large category of environmentally hazardous materials. These include carbon monoxide [66,67], VOCs [68], dyes [69], hydrocarbons [70], halogenated hydrocarbons [71], and organophosphates [72]. Both thermal and photocatalytic processes were studied to achieve sizable decomposition of the pollutants [70,73]. The photocatalytic approach is beneficial for consuming less energy when conducted under ambient conditions as compared to the harsh conditions of thermal degradation. Amorphous manganese oxide (AMO) shows remarkable activities and stabilities in oxidative photocatalysis [70] and is a promising candidate for titania replacement in a PCO air purifier. A typical synthesis of AMO is done by reducing $KMnO_4$ with oxalic acid under ambient conditions [74].

Figure 13a,b show the morphology of AMO. The amorphous nature can be seen from both the weak X-ray diffraction (Figure 12c inset) and high resolution transmission electron microcopy (HRTEM) (Figure 13b) data. Figure 13c displays the O_2-TPD profile of AMO, which exhibits two desorption peaks. The first broad peak that spans over 350–550 °C represents the large amount of adsorbed oxygen; the second one centered at 600 °C indicates lattice oxygen [74].

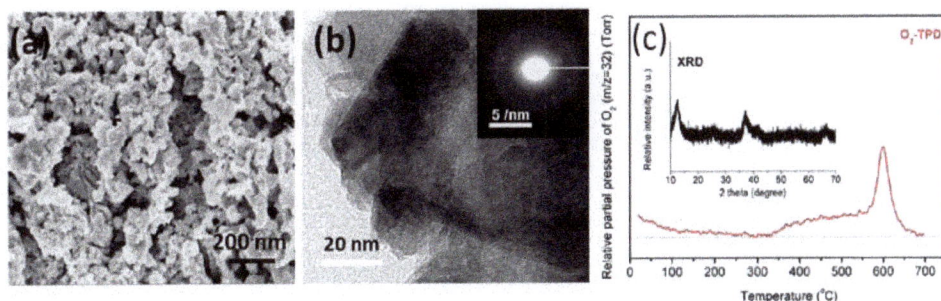

Figure 13. (a) SEM images; **(b)** HRTEM, (in set, selected area diffraction (SAD));
(c) O_2-temperature programmed desorption (inset, XRD pattern) of AMO.

The uniqueness of AMO is due to the composition of randomly oriented nanosize domains (*ca.* 10 nm), mixed valency, and large surface area (*ca.* 180 cm^2/g). More importantly, the good photocatalytic activity was correlated with the high mobility of lattice oxygen and ample surface adsorbed oxygen, and their migration to and from atmospheric oxygen, which assures high stabilities under continuous irradiation [75]. The transformation of surface adsorbed oxygen upon irradiation can be described by the following equation [75]:

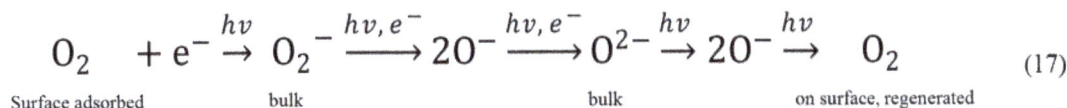

$$O_2 + e^- \xrightarrow{hv} O_2^- \xrightarrow{hv,\,e^-} 2O^- \xrightarrow{hv,\,e^-} O^{2-} \xrightarrow{hv} 2O^- \xrightarrow{hv} O_2 \qquad (17)$$

Surface adsorbed   bulk   bulk   on surface, regenerated

A plausible explanation of the photo activity is that surface adsorbed O_2 can accept photogenerated electrons, and become O_2^-. Upton further excitation under light, O_2^- reacts with electrons and forms oxygen radicals (O^-), which continue to react with electrons to form O^{2-}(bulk). Next, O^{2-}(bulk) could migrate to the surface as O_2 (regenerated). In the whole process, radicals such as ˙OH and O_2^- (superoxide anion radicals) could be formed and play reactive roles. Both species were probed in a recent study utilizing AMO as a photocatalyst to degrade *N*- nitrosodimethalamine (NDMA) to form NO_3^- and HCOH. The efficiency of the degradation is comparable to that of TiO_2 (Degussa P25) as a photocatalyst [75].

Considerable work has been done toward liquid phase dye degradation with different manganese oxide catalysts in our group. Segal *et al.* [76] use octahedral molecular sieves (OMS), octahedral layered (OL), and amorphous manganese oxide (AMO) materials to decompose pinacyanol chloride (PC) dye. The result shows that metal-doped OMS-2 materials have the highest activities compared with OMS-1, OL-1, and commercial MnO_2 in the rate of decomposition of PC dye. The presence of H_2O_2 can inhibit the dye degradation ability of the manganese oxide materials. Other parameters such as catalyst concentration, pH, and structural changes that occur in the catalysts are also studied in the literature. The manganese oxide catalyzed dye decomposition follows an adsorption/oxidation/desorption process. Sriskandakumar *et al.* [69] reported decomposition of methylene blue (MB) dye with several doped and undoped OMS manganese oxide materials by using tertiary-butyl hydrogen peroxide (TBHP) as oxidant which could enhance the degree of MB dye decomposition.

Figure 14 shows the free-standing membrane from the self-assembly of ultralong MnO₂ nanowires recently made by our group for an *in-situ* dye degradation study. The smooth MnO₂ nanowire membrane (Figure 14a) was made according to Yuan's method [77]. The nanowires (Figure 14b were uniformly distributed with lengths of tens of micrometers. With the *in-situ* membrane reaction system (Figure 14c) the dye degradation reaction can easily be controlled. Methyl orange dye was partly decomposed after 120 min at room temperature by MnO₂ nanowire membrane (Figure 14d). Further studies are still going on with different metal doped MnO₂ nanowire membranes and reduced graphene oxide (RGO)/MnO₂ nanowire membranes for different organic dye decompositions.

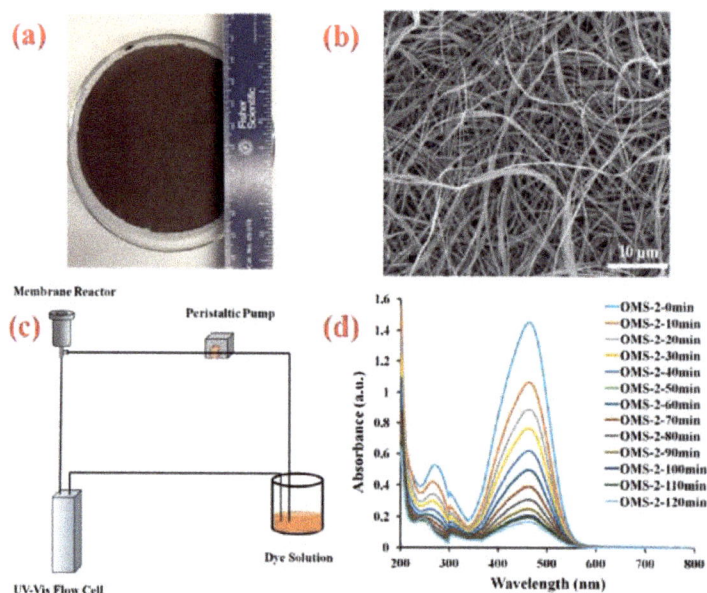

Figure 14. MnO₂ nanowire membrane for dye degradation: (**a**) overall view of MnO₂ nanowire membrane; (**b**) SEM image of the MnO₂ nanowire membrane; (**c**) *in-situ* membrane reaction system; (**d**) the UV-Vis spectra of the catalytic degradation of methyl orange by MnO₂ nanowire membrane.

Developing these various materials can enable cost effective, efficient deployment of photocatalytic air purifiers.

4. Conclusions

Understanding the effect of a photocatalytic purifier on air is best accomplished by thorough study of the coadsorption phenomenon. In a simple system with limited quantity of contaminants the study is straightforward. The PCO reactor can be designed to be cost effective and practical. In indoor air the system is complex and varying, the design of the air purifier is constrained by cost and the assumptions made about the environment. Practical validation experiments have demonstrated that two significant barriers remain, catalyst lifetime and by-product formation.

Photocatalyst deactivation has been demonstrated to occur rapidly in ambient air containing siloxanes. Sequential radical attack on the parent molecule creates an amorphous silica cap on the catalyst surface preventing further oxidation of air entrained contaminates.

Mineralization by-products occur due to incomplete oxidation of parent VOCs. When the number of active sites is large compared to the adsorbed species, mineralization is encouraged and by-products are minimized. This is seen in product space maps such as the example shown with TCE. Conversely, when surface coverage of the parent molecule is high, this ratio decreases and incomplete mineralization occurs. This is seen both at high TCE concentration and in PCO purifier validation studies when alcohols are present in the ambient airstream. If this ratio impacts complete mineralization and byproduct formation, then the effect of an ageing catalyst also could be significant. As the PCO air purifier operates in ambient air, deactivation occurs and byproduct formation could increase with the decreasing number of active sites.

Of these effects, rapid deactivation is the most significant impediment to implementation of this technology. Pre-removal of siloxane or other deactivating agents and/or the development of deactivation resistant catalysts is critical. Higher surface area catalyst with an increased number of active sites could decrease the formation of byproducts. Visible light photocatalysts could impact operational costs. All these areas are fruitful avenues for further study, providing ample grist for the academic mill.

PCO air purifiers are viable now for simple systems, such as remediation or industrial waste streams. They are not viable for indoor air until the effect by-product formation is understood and minimized and catalyst lifetime is extended.

Other research groups have contributed significantly to the application of photocatalytic technology to indoor air. Other reactor designs will affect the parameters we have discussed, some modifying the effect of byproduct formation through longer dwelling time. Each modification to ameliorate one problem may exacerbate another. Longer dwelling time may increase pressure drop increasing cost of implementation and may promote deactivation by siloxanes. The authors have attempted to describe through reviews the methodology appropriate to photocatalytic product development; and how to design, prototype, and validate those products in the real world. There are of course a wide variety of reactor designs that offer a different set of barriers to product development. Many of these designs have been proposed and tested in academic environments but few, if any, have been developed into products and tested in buildings. It is the same methodology described in this review that is appropriate for product development based on current and future reactor innovations. It is the manner in which UTRC approached this very complex and intriguing problem and the knowledge gained during this journey that is instructive for future product development.

Acknowledgments

This review is dedicated to all the researchers across the globe who contributed to our understanding of the use of PCO technology in treating indoor air. The authors specifically wish to acknowledge Susan Brandes for dedication and leadership in this area. Other contributors include Joe Sangiovanni, Heidi Hollick, Robert Hall, Jim Friehaut, Bernie Woody, Mary Saroka, James Davies, Norberto Lemcoff, Catherine Thibaox-Erkey, Treese Campbell, Greg Dobbs, Suzanne

Opalka, and Tom Vanderspurt. Funding was supplied by United Technologies and their divisions Carrier Corporation and Hamilton Sundstrand and SLS acknowledges support for this work from the Fraunhofer CEI Center at UCONN and the US Department of Energy Office of Basic Energy Sciences Division of Chemical Geological Sciences and Biological Sciences under grant DE-FG02-86ER13622-A000.

Author Contributions

S. O. Hay contributed to the entire paper; T. Obee to Sections 1, 2, and 3; Z. Luo, T. Jiang, Y. Meng, J. He and S. Murphy contributed to Section 3.4; and S. Suib to Sections 2 and 3.

Conflicts of Interest

The authors declare no conflict of interest.

References

1. Peral, J.; Ollis, D.F. Heterogeneous photocatalysis oxidation of gas-phase organics for air purification: Acetone, 1-butanol, butyraldehyde, formaldehyde, and *m*-xylene oxidation. *J. Catal.* **1992**, *136*, 554–565.
2. Dibble, L.; Raupp, G. Kinetics of the gas-solid heterogeneous photocatalytic oxidation of trichloroethylene by near UV illuminated titanium dioxide. *Catal. Lett.* **1990**, *4*, 345–354.
3. Pichat, P.; Disdier, J.; Hoang-Van, C.; Mas, D.; Goutallier, G.; Gaysee, C. Purification/deodorization of indoor air and gaseous effluents by TiO_2 photocatalysis. *Catal. Today* **2000**, *63*, 363–369.
4. Zhao, J.; Yang, X. Photocatalytic oxidation for indoor air purification: A literature review. *Build. Environ.* **2003**, *38*, 645–654.
5. Mo, J.; Zhang, Y.; Xu, Q.; Lamson, J.; Zhao, R. Photocatalytic purification of volatile organic compounds in indoor air: A literature review. *Atmos. Environ.* **2009**, *43*, 2229–2246.
6. Hay, S.O.; Obee, T.N.; Thibaud-Erkay, C. The deactivation of photocatalytic based air purifiers by ambient siloxanes. *Appl. Catal. B Environ.* **2010**, *99*, 445–431.
7. Hay, S.O.; Obee, T.N. The augmentation of UV photocatalysis oxidation with trace quantities of ozone. *J. Adv. Oxid. Technol.* **1999**, *4*, 209–212.
8. Kampa, M.; Castanas, E. Human Health effects of air pollution. *Environ. Pollut.* **2008**, *151*, 362–367.
9. Spengler, J.D.; Samet, J.M.; McCarthy, J.F. *Indoor Air Quality Handbook*; McGraw-Hill Co.: New York, NY, USA, 2001.
10. Sundell, J. On the history of indoor air quality and health. *Indoor Air* **2004**, *14*, 51–58.
11. Building Assessment Survey and Evaluation (BASE) Study. Available online: http://www.epa.gov/iaq/base/index.html (accessed on 8 September 2014).

12. Hollick, H.H.; Sangiovanni, J.J. A proposed indoor air quality metric for estimation of the combined effects of gaseous contaminates on human health and comfort. In *Air Quality and Comfort in Airliner Cabins, ASTM STP 1393*; Nagda, N.L., Ed.; American Society for Testing and Materials: West Conshohocken, PA, USA, 2000; pp. 76–98.

13. Hall, R.; Hall, R.J.; Sangiovanni, J.J.; Hollick, H.H.; Obee, T.N.; Hay, S.O. Design of Air Purifiers for Aircraft Passenger Cabins Based on Photocatalytic Oxidation Technology. In *Air Quality and Comfort in Airliner Cabins*; Nagda, N.L., Ed.; American Society for Testing and Material: New Orleans, LA, USA, 2000.

14. Benfeld, P.; Hall, R.J.; Obee, T.N.; Hay, S.O.; Sangiovanni, J.J. A Feasibility Study of Photocatalytic Air Purification for Commercial Passenger Airlines; UTRC Report R97–1.300.9702; UTRC: East Hartford, CT, USA, 1997.

15. Reisfeld, B.; Chaing, R.H.L.; Josserand, O.; Dunshee, K.; Jomard, T.; Drago, T.E.; Hay, S.O.; Obee, T.N.; Sangiovanni, J.J.; Hall, R.J.; *et al.* Photocatalytic Air Purifier for a Fan Coil Unit. US 12/512,584, 30 July 2001.

16. Vanderspurt, T.H.; Davies, J.A.; Hay, S.O.; Obee, T.N.; Opalka, S.M.; Wei, D. Air Purification System. US Patent Application US 12/302,615 A1, 31 May 2007.

17. Dobbs, G.M.; Obee, T.N.; Sheehan, D.S.; Freihaut, J.D.; Hay, S.O.; Lemcoff, N.O.; Sangiovanni, J.J.; Saroka, M.; Hall, R.C. Gas Phase Contaminant Removal with Low Pressure Drop. US 10/857,301 29 May 2003.

18. Obee, T.N. Photooxidation of Sub-PPM Toluene and Formaldehyde Levels on Titania Using a Glass-Plate Reactor. *Environ. Sci. Technol.* **1996**, *30*, 3578–3584.

19. Obee, T.N.; Hay, S.O. The estimation of photocatalytic rate constraints based on molecular structure: Extending to multi-component systems. *J. Adv. Oxid. Technol.* **1999**, *4*, 147–152.

20. Obee, T.N.; Brown, R.T. TiO_2 Photocatalysis for Indoor Air Applications: Effects of Humidity and Trace Contaminant Levels on the Oxidation Rates of Formaldehyde, Toluene, and 1,3-Butadiene. *Environ. Sci. Technol.* **1995**, *29*, 1223–1231.

21. Obee, T.N.; Hay, S.O. The effect of moisture and temperature on the photooxidation of ethylene on titania. *Environ. Sci. Technol.* **1997**, *31*, 2032–2038.

22. Zorn, M.E.; Hay, S.O.; Anderson, M.A. Effect of molecular functionality on the photocatalytic oxidation of gas-phase mixtures. *Appl. Catal. B Environ.* **2010**, *99*, 420–427.

23. Hodgson, A.T.; Destaillats, H.; Sullivan, D.P.; Fisk, W.J. Performance of ultraviolet photocatalytic oxidation for indoor air cleaning applications. *Indoor Air* **2007**, *4*, 305–316.

24. Wei, D.; Obee, T.; Hay, S.; Vanderspurt, T.; Schmidt, W.; Samgiovanni, J. Tungsten Oxide/Titanium Dioxide Photocatalyst for Improving Indoor Air Quality. U.S. Patent 20040241040 A1, 30 May 2003.

25. Obee, T.N.; Hay, S.O.; Vanderspurt, T. A kinetic study of photocatalytic oxidation of VOCs on WO_3 coated titanium dioxide. In Proceedings of the Eleventh International Conference on TiO_2 Photocatalysis Fundamentals and Applications, Pittsburg, PA, USA, 22–25 September 2006.

26. Peral, J.; Ollis, D.F. TiO_2 photocatalyst deactivation by gas-phase oxidation of heteroatom organics. *J. Mol. Catal. A Chem.* **1997**, *115*, 347–354.

27. Cao, L.; Gao, Z.; Suib, S.L.; Obee, T.N.; Hay, S.O. Photocatalytic oxidation of toluene on nanoscale TiO_2 catalysis: Studies of deactivation and regeneration. *J. Catal.* **2000**, *196*, 253–261.

28. Huang, A.; Cao, L.; Chen, J.; Spiess, F.J.; Suib, S.L.; Obee, T.N.; Hay, S.O. Photocatalytic degradation of triethylamine on titanium oxide thin films. *J. Catal.* **1999**, *188*, 40–47.

29. Jacoby, W.A.; Blake D.M.; Noble R.D.; Koval C.A. Kinetics of the oxidation of trichloroethylene in air via heterogeneous photocatalysis. *J. Catal.* **1995**, *157*, 87–96.

30. Jacoby, W.A. Destruction of Trichloroethylene in Air via Gas-Solid Heterogeneous Photocatalysis. Ph.D. Dissertation, University of Colorado, Boulder, CO, USA, 1993.

31. Hill, C.G., Jr. *An Introduction to Chemical Engineering Kinetics & Reactor Design*; John Wiley & Sons: New York, NY, USA, 1977; Chapter 6.

32. Hall, R.T.; Bendfeldt, P.; Obee, T.N.; Sangiovanni, J.J. Computational and Experimental Studies of UV/Titania Photocatalytic Oxidation of VOCs in Honeycomb Monoliths. *J. Adv. Oxidat. Technol.* **1998**, *3*, 243–251.

33. Adamson, A.W. *Physical Chemistry of Surfaces*, 3rd ed.; John Wiley & Sons: New York, NY, USA, 1976; Chapter 15.

34. Hall, R.T.; Obee, T.N.; Hay, S.O.; Sangiovanni, J.J.; Bonczyk, P.A.; Freihaut, J.D.; Genovese, J.E.; Sribnik, F. Photocatalytic Oxidation Technology for Trace Contaminant Control in Aircraft and Spacecraft. In Proceedings of the 26th International Conference on Environmental Systems, Monterey, CA, USA, 8–11 July 1996.

35. Hossain, M.M.; Raupp, G.B.; Hay, S.O.; Obee, T.N. Three Dimensional Developing Flow Model of a Photocatalytic Honeycomb Monolith Reactor. *AIChE J.* **1999**, *45*, 1309–1321.

36. Li Puma, G.; Slavado Estiville, I.; Obee, T.N.; Hay, S.O. Kinetics Rate Model of the Photocatalytic Oxidation of Trichloroethelene in Air over $TiO2$ Thin-Films. *Sep. Purif. Technol.* **2009**, *67*, 226–232.

37. Kolarick, J.; Wargocki, P. Can a photocatalytic air purifier be used to improve the percieved air quality indoors? *Indoor Air* **2010**, *20*, 255–262.

38. Kolarick, J.; Wargocki, P.; Skorek-Osikowska, A.; Wisthaler, A. The effect of a photocatalytic air purifier on indoor air quality quantified usuing different measuring methods. *Build. Environ.* **2010**, *3*, 255–262.

39. Lemcof, N.O.; Hollick, H.H.; Obee, T.N.; Davies, J.A.; Hay, S.O. Energy efficient alternative ventilation using ultra-violet photocatalytic oxidation technology. In Proceedings of the Eighth International Conference on TiO_2 Photocatalysis, Montreal, ON, Canada, 26–29 October 2003.

40. Chen, J.; Davidson, J.H. Chemical Vapor Deposition of Silicon Dioxide by Direct-Current Corona Discharge in Dry Air Containing Octamethylcyclotetrasiloxane Vapor: Measurement of the Deposition Rate. *Plasma Chem. Plasma Process.* **2004**, *24*, 169–188.

41. Warner, N.A.; Evenset, A.; Christensen, G.; Gabrielsen, G.W.; Borga, K.; Leknes, H. Volatile siloxanes in the european artic: Assessment of sources and spatial distribution. *Environ. Sci. Technol.* **2010**, *44*, 7705–7710.

42. Hugener-Campbell, T.; Ollis, D.F.; Vanderspurt, T.H.; Schmidt, Wayde R.; Obee, T.N.; Hay, S.O.; Kryzman, M.A. Photocatalytic Device with Mixed Photocatalyst/Silica Structure. U.S. Patent 8617478, 31 December 2013.

43. Horvath, D.T. Synthesis, Characterization, and Photocatalytic Testing of Titania-Based Aerogels for the Degradation of Volatile Organic Compounds. Honors Scholar Thesis, University of Connecticut, Storrs, CT, USA, 2012.

44. Hugener-Campbell, T.; Vanderspurt, T.H.; Ollis, D.F.; Hay, S.O.; Obee, T.N.; Schmidt, W.R.; Kryzman, M.A. Preparation and Manufacture of an Overlayer for Deactivation. U.S. Patent 8309484, 31 May 2007.

45. Hay, S.O.; Brandes, S.D.; Lemcoff, N.O.; Obee, T.N.; Schmidt, W.R. Photocatalyst Protection. US 8263012 B2, 13 April 2010.

46. Su, R.; Bechstein, R.; Sø, L.; Vang, R.T.; Sillassen, M.; Esbjörnsson, B.; Palmqvist, A.; Besenbacher, F. How the anatase-to-rutile ratio influences the photoreactivity of TiO_2. *J. Phys. Chem. C* **2011**, *115*, 24287–24292.

47. Ohtani, B.; Prieto-Mahaney, O.O.; Li, D.; Abe, R. What is Degussa (Evonik) P25? Crystalline composition analysis, reconstruction from isolated pure particles and photocatalytic activity test. *J. Photochem. Photobiol. A Chem.* **2010**, *216*, 179–182.

48. Antonelli, D.M.; Ying, J.Y. Sythesis of hexagonally packed mesoporous TiO_2 by a modified sol-gel method. *Angew. Chem. Int. Ed. Engl.* **1995**, *34*, 2014–2017.

49. Yoshitake, H.; Sugihara, T.; Tatsumi, T. Preparation of wormhole-like mesoporous TiO_2 with an extremely large surface area and stabilization of its surface by chemical vapor deposition. *Chem. Mater.* **2002**, *14*, 1023–1029.

50. Liu, H.; Wu, Y.; Zhang, J. A new approach toward carbon-modified vanadium-doped titanium dioxide photocatalyst. *ACS Appl. Mater. Interfaces* **2011**, *3*, 1757–1764.

51. Xing, M.; Wu, Y.; Zhang, J.; Chen, F. Effect of synergy on the visible light activity of B, N and Fe co-doped TiO_2 for the degradation of MO. *Nanoscale* **2010**, *2*, 1233–1239.

52. Chen, X.; Mao, S.S. Titanium dioxide nanomaterials: Synthesis, properties, modifications and applications. *Chem. Rev.* **2007**, *107*, 2891–2959.

53. Li, C.; Shi, L.; Xie, D.; Du, H. Morphology and crystal structure of Al-doped TiO_2 nanoparticles synthesized by vapor phase oxidation of titanium tetrachloride. *J. Non-Cryst. Solids* **2006**, *352*, 4128–4135.

54. Hanaor, D.A.H.; Sorrell, C.C. Review of the anatase to rutile phase transformation. *J. Mater. Sci.* **2010**, *46*, 855–874.

55. Shannon, R.D.; Pask, J.A. Kinetics of the anatase-rutile transformation. *J. Am. Ceram. Soc.* **1965**, *48*, 391–398.

56. Carneiro, J.O.; Teixeira, V.; Portinha, A.; Magalhães, A.; Coutinho, P.; Tavares, C.J.; Newton, R. Iron-doped photocatalytic TiO_2 sputtered coatings on plastics for self-cleaning applications. *Mater. Sci. Eng. B* **2007**, *138*, 144–150.

57. Venezia, A.M.; Palmisano, L.; Schiavello, M. Structural changes of titanium oxide induced by chromium addition as determined by an X-Ray diffraction study. *J. Solid State Chem.* **1995**, *114*, 364–368.

58. Poyraz, A.S.; Kuo, C.-H.; Biswas, S.; King'ondu, C.K.; Suib, S.L. A general approach to crystalline and monomodal pore zide mesoporous materials. *Nat. Commun.* **2013**, *4*, doi:10.1038/ncomms3952.

59. Schweigkofler, M.; Niessner, N. Removal of siloxanes in biogases. *J. Hazard. Mater.* **2001**, *83*, 183–196.

60. Montanari, T.; Finocchio, E.; Bozzano, I.; Garuti, G.; Giordano, A.; Pistarino, C.; Busca, G. Purification of landfill biogases from siloxanes by adsorption: A study of silica and 13X zeolite adsorbents on hexamethylcyclotrisiloxane separation. *Chem. Eng. J.* **2010**, *165*, 859–863.

61. Finocchio, E.; Garuti, G.; Baldi, M.; Busca, G. Decomposition of hexamethylcyclotrisiloxane over solid oxides. *Chemosphere* **2008**, *72*, 1659–1663.

62. Park, J.K.; Lee, G.M.; Lee, C.Y.; Hur, K.B.; Lee, N.H. Analysis of Siloxane Adsorption Characteristics Using Response Surface Methodology. *Environ. Eng. Res.* **2012**, *17*, 117–122.

63. Cabrera-Codony, A.; Montes-Morán, M.; Sánchez-Polo, M.; Martín, M.J.; Gonzalez-Olmos, R. Biogas upgrading: Optimal activated carbon properties for siloxane removal. *Environ. Sci. Technol.* **2014**, *48*, 7187–7195.

64. Chen, C.Y.; Li, H.X.; Davis, M.E. Studies on mesoporous materials: I. Synthesis and characterization of MCM-4. *Microporous Mater.* **1993**, *2*, 17–26.

65. Suib, S.L. Porous manganese oxide octahedral molecular sieves and octahedral layered materials. *Acc. Chem. Res.* **2008**, *41*, 479–487.

66. Genuino, H.C.; Meng, Y.; Horvath, D.T.; Kuo, C.H.; Seraji, M.S.; Morey, A.M.; Joesten, R.L.; Suib, S.L. Enhancement of Catalytic Activities of Octahedral Molecular Sieve Manganese Oxide for Total and Preferential CO Oxidation through Vanadium Ion Framework Substitution. *ChemCatChem* **2013**, *5*, 2306–2317.

67. Özacar, M.; Poyraz, A.S.; Genuino, H.C.; Kuo, C.-H.; Meng, Y.; Suib, S.L. Influence of silver on the catalytic properties of the cryptomelane and Ag-hollandite types manganese oxides OMS-2 in the low-temperature CO oxidation. *Appl. Catal. A* **2013**, *462*, 64–74.

68. Genuino, H.C.; Dharmarathna, S.; Njagi, E.C.; Mei, M.C.; Suib, S.L. Gas-phase total oxidation of benzene, toluene, ethylbenzene, and xylenes using shape-selective manganese oxide and copper manganese oxide catalysts. *J. Phys. Chem. C* **2012**, *116*, 12066–12078.

69. Sriskandakumar, T.; Opembe, N.; Chen, C.-H.; Morey, A.; King'ondu, C.; Suib, S.L. Green decomposition of organic dyes using octahedral molecular sieve manganese oxide catalysts. *J. Phys. Chem. A* **2009**, *113*, 1523–1530.

70. Cao, H.; Suib, S.L. Highly efficient heterogeneous photooxidation of 2-propanol to acetone with amorphous manganese oxide catalysts. *J. Am. Chem. Soc.* **1994**, *116*, 5334–5342.

71. Chen, J.; Lin, J.C.; Purohit, V.; Cutlip, M.B.; Suib, S.L. Photoassisted catalytic oxidation of alcohols and halogenated hydrocarbons with amorphous manganese oxides. *Catal. Today* **1997**, *33*, 205–214.

72. Segal, S.R.; Suib, S.L.; Tang, X.; Satyapal, S. Photoassisted decomposition of dimethyl methylphosphonate over amorphous manganese oxide catalysts. *Chem. Mater.* **1999**, *11*, 1687–1695.

73. Meng, Y.; Genuino, H.C.; Kuo, C.-H.; Huang, H.; Chen, S.-Y.; Zhang, L.; Rossi, A.; Suib, S.L. One-Step Hydrothermal Synthesis of Manganese-Containing MFI-Type Zeolite, Mn–ZSM-5, Characterization, and Catalytic Oxidation of Hydrocarbons. *J. Am. Chem. Soc.* **2013**, *135*, 8594–8605.

74. Meng, Y.; Song, W.; Huang, H.; Ren, Z.; Chen, S.-Y.; Suib, S.L. Structure-Property Relationship of Bifunctional MnO_2 Nanostructures: Highly Efficient, Ultra-Stable Electrochemical Water Oxidation and Oxygen Reduction Reaction Catalysts Identified in Alkaline Media. *J. Am. Chem. Soc.* **2014**, *136*, 11452–11464.

75. Genuino, H.C.; Njagi, E.C.; Benbow, E.M.; Hoag, G.E.; Collins, J.B.; Suib, S.L. Enhancement of the photodegradation of *N*-nitrosodimethylamine in water using amorphous and platinum manganese oxide catalysts. *J. Photochem. Photobiol. A* **2011**, *217*, 284–292.

76. Segal, S.R.; Suib, S.L.; Foland, L. Decomposition of pinacyanol chloride dye using several manganese oxide catalysts. *Chem. Mater.* **1997**, *9*, 2526–2532.

77. Yuan, J.; Laubernds, K.; Villegas, J.; Gomez, S.; Suib, S.L. Spontaneous Formation of Inorganic Paper-Like Materials. *Adv. Mater.* **2004**, *16*, 1729–1732.

A Comparison of the Environmental Impact of Different AOPs: Risk Indexes

Jaime Giménez, Bernardí Bayarri, Óscar González, Sixto Malato, José Peral and Santiago Esplugas

Abstract: Today, environmental impact associated with pollution treatment is a matter of great concern. A method is proposed for evaluating environmental risk associated with Advanced Oxidation Processes (AOPs) applied to wastewater treatment. The method is based on the type of pollution (wastewater, solids, air or soil) and on materials and energy consumption. An Environmental Risk Index (E), constructed from numerical criteria provided, is presented for environmental comparison of processes and/or operations. The Operation Environmental Risk Index (E_{Oi}) for each of the unit operations involved in the process and the Aspects Environmental Risk Index (E_{Aj}) for process conditions were also estimated. Relative indexes were calculated to evaluate the risk of each operation (E/NOP) or aspect (E/NAS) involved in the process, and the percentage of the maximum achievable for each operation and aspect was found. A practical application of the method is presented for two AOPs: photo-Fenton and heterogeneous photocatalysis with suspended TiO_2 in Solarbox. The results report the environmental risks associated with each process, so that AOPs tested and the operations involved with them can be compared.

Reprinted from *Molecules*. Cite as: Giménez, J.; Bayarri, B.; González, Ó.; Malato, S.; Peral, J.; Esplugas, S. A Comparison of the Environmental Impact of Different AOPs: Risk Indexes. *Molecules* **2015**, *20*, 503-518.

1. Introduction

Sustainability is one of today's most important goals in any human activity, and decreased environmental impacts must be a basic objective for its achievement. Therefore, many businesses are implementing environment management systems following such standards as the ISO 14001 [1] or EMAS [2], which are the starting point for pollution control.

There are several different methods for evaluating environmental impact. Life Cycle Assessment (LCA) is one of the most widely used. LCA includes four steps: goal and scope definition, inventory analysis, Life Cycle Impact Assessment (LCIA) and interpretation of results. In the first step (goal and scope), the limits of the system and the focus of the analysis are defined. Inventory analysis involves compiling an inventory of raw materials, energy consumption and pollutants production. In the LCIA step, the data from the inventory are assigned to different impact categories and characterized according to the corresponding factors. These impact categories depend on whether midpoint methods [3,4], defining only impacts in categories such as ozone depletion, climate change, *etc.*, or endpoint methods [5,6], which analyze final impacts on human health, environment or resources consumption, are used.

Environmental consequences can also be analysed using an environmental risk index, as provided for in Directive 96/82/EC [7], which is to be replaced as of June 1, 2015, by Directive 2012/18/EU [8]. It is based on the analysis of four parameters: risk sources (analysis of possible accidents including hazardous substances involved), primary control systems (preventive and protective measures to reduce the consequences of an accident, for example, retention ponds), transport systems (pollutant concentration profile in the area affected by a spill), and sensitive receptors (agricultural areas, protected species, monuments, *etc.*, present in the area affected by a spill). With these factors, the environmental consequences index can be calculated, which when combined with the probability of an accident, results in an environmental risk index.

A third method of environmental analysis is based on Directive 2004/35/CE [9]. This Directive proposes a methodology very similar to the one used for risk evaluation in industrial safety. The first step is the identification of hazards and possible environmental accidents by techniques such as the event tree. The second step is modelling the accidents and evaluating their consequences, considering the setting where they occur. Then risk is evaluated by including their probability, estimated, for instance, by tree fault analysis. Finally, an economic assessment is made of damage caused by the accidents and the cost of recovery of affected areas.

Finally, the methodology (Arteche *et al.* [10]) selected for the estimation of environmental impact in this study is based on application of risk indexes [11]. The first stage is identification of the steps in the processes under study in three different situations: normal, abnormal and emergency operations. The environmental impact is evaluated for each considering two aspects, pollution (wastewater, solids, air or soil) and consumption (raw materials, water, energy). Risk assessment is based on frequency (probability), hazard and amount (volume). Numerical criteria are provided to assign a value to the environmental risk, thus enabling comparison of the environmental impact of processes and/or operations. The numerical criteria proposed by Arteche *et al.* [10] were updated and adapted to the particular cases studied, which were laboratory or small pilot plant operations, and current legislation. The values found in this paper are therefore noticeably different from those originally proposed by Arteche *et al.* [10]. The final result is the Global Environmental Risk Index (E), which gives an overview of the environmental priority of each activity related to other analogous activities. Its evaluation is the main purpose of this paper. Other related indexes are also evaluated for comparison of the various process operations and aspects involved. This method is amply explained in the following sections, and a practical application is developed for two AOPs, photo-Fenton and heterogeneous photocatalysis with TiO_2 in suspension, both carried out in a Solarbox. This application enables evaluation of such processes according to their risk.

2. Results and Discussion

The environmental impact indexes are calculated as shown in Table 1, where the operations involved in the evaluation are given on the left, and the points evaluated are at the top. As shown in the second column, three types of operations are considered: normal, abnormal and emergency. The operations included in each group are described in the third column. Photocatalytic treatment of metoprolol tartrate salt (MET), a common β-blocker found in wastewater [12], in aqueous solution with TiO_2 in suspension in the Solarbox, is presented as an example. In this case, the operations included are (see third column in Table 1):

- Normal: Preparation of MET solution, Weighting TiO_2, Experiment Development, Sampling, Analysis, Reactor emptying and cleaning.
- Abnormal: Disassembly of experimental equipment for maintenance.
- Emergency: Tank or reactor breakdown and/or solvent spillage.

Two aspects are evaluated—pollution and consumption (second row in Table 1)—and each includes several specific parameters. Four parameters are considered for pollution: wastewater, solid waste, air emission and soil pollution (third row in Table 1). Consumption includes raw materials (in our case, MET), water and energy (third row in Table 1). For each of these (fourth row in Table 1), frequency (A), hazard (B) and amount (C) are considered. The final result, which is calculated as the product of these values, as shown in Column R_{ij}, gives an idea of the risk of that aspect. Numerical values assigned to evaluate each item are from the criteria in Tables 2–4. The values proposed by Arteche [10] are taken as a first approximation. However, they have been modified and updated to current legislation and adapted to laboratory, pilot plant or industrial scale for this study.

Table 1. Environmental Risk Identification and Evaluation. Process tested: MET treatment by photocatalysis with TiO2 in the Solarbox. Parameters A–C are defined in Tables 2–4, respectively.

		Environmental Aspects																																			
		Pollution																			Consumption																
		1. Waste Water				2. Solid Waste				3. Air Emissions				4. Soil Pollution				5. Raw Materials				6. Water				7. Energy					
			A	B	C	Rij		A	B	C	Rij		A	B	C	Rij	A	B	C	Rij		A	B	C	Rij		A	B	C	Rij		A	B	C	Rij	E_{oi}	%$E_{oi,max}$
Normal Operation	1. Preparation of MET solution											MET powders	6	1	1	6					MET	6	1	1	6	Water	6	2	1	12						24	0.42
	2. Weighting of TiO2											TiO2 powders	6	1	1	6					TiO2	6	1	1	6											12	0.21
	3. Experiment Development						Gloves	6	1	1	6																				Electricity for lamp, pumps, stirrer	6	5	2	60	66	1.16
	4. Sampling						Filters	6	1	2	12																									12	0.21
Operation	5. Analysis	Aqueous solution with organics	6	10	3	180						Air TOC	6	5	2	60					Reagents	6	10	2	120	Water for preparations	6	2	1	12	Electricity for equipment	6	5	3	90	462	8.11
	6. Reactor emptying and cleaning	Aqueous solution with organics	6	3	4	72	Adsorbent paper	6	1	1	6															Water for cleaning	6	2	2	24	Electricity for pump, stirrer	6	5	2	60	162	2.84
Abnormal Operation	7. Disassembly of experimental equipment for maintenance						Crashing of reactor	1	1	1	1																									1	0.02
Emergency Operation	8. Tank or reactor breakdown and/or solvent spillage	Aqueous solution with organics	1	3	2	6	Adsorbent material	1	1	1	1																									7	0.12
	E_{Ai}	258					26					72					0				132					48					210					$E =$ 746	-
	%$E_{Ai,max}$	3.23					0.33					0.90					0				1.65					3.00					5.25					-	-

Table 2. Parameter A (Frequency): Numerical values for Parameter A, assigned by the frequency (probability) of occurrence of each event.

		Aspect to Be Evaluated						
		Waste-water	Solid Wastes	Air Emission	Soil Pollution	Raw Material	Water	Energy
Frequency	*Accidental*	1	1	1	1	1	1	1
	Yearly	2	2	2	2	2	2	2
	Monthly	4	4	4	4	4	4	4
	Every two weeks	5	5	5	5	5	5	5
	Weekly	6	6	6	6	6	6	6
	Three times a week	8	8	8	8	8	8	8
	Daily	10	10	10	10	10	10	10

Table 3. Parameter B (Hazard): Numerical values for Parameter B, assigned by how hazardous each event is.

Emission or Consumption Hazard	Rate	Notes
Wastewater		
No contact with raw materials or products	1	
Contact Non-hazardous	3	(1),(3)
Contact with Products with signal word Warning	6	(1),(3)
Contact with Products with signal word Danger	10	(1),(3)
Solid Wastes		
Not in Decision 2000/532/EC	1	(2),(3)
Not hazardous in Decision 2000/532/EC	5	(2),(3)
Hazardous in Decision 2000/532/EC	10	(2),(3)
Air Emissions		
Non-hazardous	1	(1),(3)
Products with signal word Warning	5	(1),(3)
Products with signal word Danger	10	(1),(3)
Soil Pollution		
Non-hazardous	1	(1),(3)
Products with signal word Warning	5	(1),(3)
Products with signal word Danger	10	(1),(3)
Raw Materials		
Non-hazardous	1	(1),(3)
Products with signal word Warning	5	(1),(3)
Products with signal word Danger	10	(1),(3)
Water		
Any water consumption	2	
Energy		
Any energy consumption	5	

(1) Classification according to CLP Regulation Signal Word. Regulation (EC) No 1272/2008 aligns existing EU legislation to the United Nations Globally Harmonised System (GHS). This new regulation on classification, labelling and packaging ("CLP Regulation") contributes to the GHS aim for the same hazards to be described and labelled in the same way worldwide [13]; (2) Decision 2000/532/EC establishes a list of hazardous wastes [14,15]; (3) When there is a mix of products, if one of them has the signal word Danger, the score assigned is 10. If more than two products have one hazard level, the value

assigned is the one corresponding to the next level. For instance, if more than two products have the signal word Warning, the value assigned is also 10 (corresponding to the signal word Danger).

Table 4. Parameter C (Amount): Numerical values for Parameter C, assigned by the amount of product or energy consumption involved.

Discharge/Emissions or Consumption	Rate
Wastewater	
<0.001 m³/year	1
0.001–0.01 m³/year	2
0.01–0.1 m³/year	3
0.1–1 m³/year	4
1–10 m³/year	5
10–100 m³/year	6
100–1000 m³/year	7
1000–10,000 m³/year	8
10,000–100,000 m³/year	9
>100,000 m³/year	10
Solid Wastes	
<0.001 tn/year	1
0.001–0.01 tn/year	2
0.01–0.1 tn/year	3
0.1–1 tn/year	4
1–10 tn/year	5
10–100 tn/year	6
100–1000 tn/year	7
1000–10,000 tn/year	8
10,000–100,000 tn/year	9
>100,000 tn/year	10
Air Emission (Polluted Air Flow-Rate)	
<1 Nm³/year	1
1–10 Nm³/year	2
10–100 1Nm³/year	3
100–1000 Nm³/year	4
10^3–10^4 Nm³/year	5
10^4–10^5 Nm³/year	6
10^5–10^6 Nm³/year	7
10^6–10^7 Nm³/year	8
10^7–10^8 Nm³/year	9
>10^8 Nm³/year	10
Soil Pollution (Amount of Product Discharged)	
<0.001 tn/year	1
0.001–0.01 tn/year	2
0.01–0.1 tn/year	3
0.1–1 tn/year	4
1–10 tn/year	5
10–100 tn/year	6
100–1000 tn/year	7
1000–10,000 tn/year	8
10,000–100,000 tn/year	9
>100,000 tn/year	10

Table 4. *Cont.*

Discharge/Emissions or Consumption	Rate
Raw Materials	
<0.001 tn/year	1
0.001–0.01 tn/year	2
0.01–0.1 tn/year	3
0.1–1 tn/year	4
1–10 tn/year	5
10–100 tn/year	6
100–1000 tn/year	7
1000–10,000 tn/year	8
10,000–100,000 tn/year	9
>100,000 tn/year	10
Water	
<0.1 m^3/year	1
0.1–1 m^3/year	2
1–10 m^3/year	3
10–100 m^3/year	4
100–1000 m^3/year	5
1000–10,000 m^3/year	6
10,000–100,000 m^3/year	8
>100,000 m^3/year	10
Energy	
<50 kW.h/year	1
50–500 kW.h/year	2
500–2500 kW.h/year	3
2500–15,000 kW.h/year	5
15,000–75,000 kW.h/year	7
75,000–150,000 kW.h/year	9
>150,000 kW.h/year	10

As an example, the proposed method has been applied to the photocatalytic treatment of MET with TiO_2 in suspension in a Solarbox (Table 1). For each operation and each environmental aspect involved in the process, a R_{ij} index can be evaluated, where subscript i indicates the operation and subscript j indicates the environmental aspect. In our case, i varies from 1–8, because eight operations are involved in the process (see third column in Table 1). On the other hand, subscript j varies from 1–7, because there are seven environmental aspects (see third row in Table 1). The specific agent is shown in the cell corresponding to each operation and for each environmental aspect. For instance, in Environmental Aspect 1 (wastewater), the specific wastewater generated is shown in the corresponding cells (see Table 1): *aqueous solution with organics* for Operation 6 (Reactor emptying and cleaning) and *aqueous solution with organics* for Operation 8 (Tank or reactor breakdown and/or solvent spillage).

Data related to energy or raw materials and reagent consumption are required for any environmental impact assessment. Electricity consumption data for photocatalytic treatment of MET in the Solarbox are shown in Table 5. Obviously, the consumption of raw materials must be known.

This is summarized in Table 6. The consumption of reagents for analysis must also be included in raw materials consumption. These data are presented in Table 7.

Table 5. Electricity consumption in MET photocatalytic treatment experiments in a Solarbox for 50 experiments/year.

Equipment	Experiment			
	Time Use Equipment (h/exp)	*Power (kW)*	*Consumption (kWh)*	*Yearly Consumption (kWh/y)*
Lamp (1000 W)	5.5	1.00	5.50	275
Pump (250–500 W)	5.5	0.40	2.20	110
Thermostatic bath (240 W at 20 °C)	5.5	0.24	1.32	66
Stirrer (1–5 W)	5.5	0.003	0.02	0.8
Total consumption/year (kWh/y)				452
Equipment	Analysis			
	Time Use Equipment (h/exp)	*Power (kW)*	*Consumption (kWh)*	*Yearly Consumption (kWh/y)*
HPLC (12 samples × 15 min/sample)	3	2.50	7.50	375
TOC (12 samples × 15 min/sample)	3	2.20	6.60	330
Spectrophotometer (DQO+Fe+H_2O_2+SUVA)	0.2	0.25	0.05	2.5
Water deionization device (preparation 1 L water)	0.1	0.10	0.01	0.5
Total consumption/year (kWh/y)				708

Table 6. Reagent consumption in the photocatalytic experiments with a 1 L reaction volume for 50 experiments/year.

Reagents	Concentration—Amount/Exp	Total Amount/Exp	Yearly Consumption (kg)
Metoprolol tartrate salt	50 mg/L	50 mg	0.003
TiO_2	0.4 g/L	0.4 g	0.02
Millipore Water		1.00 L	50

An explanation for one of the lines in Table 1 is given below to clarify its use (see shaded cells). For instance, for *Reactor emptying and cleaning* (Operation 6), subscript i (1–8) in the R_{ij} index is 6. Wastewater is generated in this operation, which implies that subscript j (1–7) in R_{ij} is 1. Wastewater frequency is weekly, and according to Table 2 criteria, a value of 6 can be assigned to Parameter A. Considering that it is an aqueous solution in contact with non-hazardous products, a value of 3 is assigned to parameter B, according to Table 3 criteria. Finally, the amount discharged is from 100 to 1000 L/year and a value of 4 is assigned to Parameter C, according to the criteria in Table 4. Thus, risk R_{ij} (R_{61}) associated with Operation 6 and wastewater discharged has a value of 72 (product of AxBxC). Similarly, for the same operation (Operation 6), the risk for water consumption (Aspect 6) is R_{66}, and the value for Parameter A is 6 (weekly frequency). For Parameter B, the value is 2 (water consumption hazard). Finally, Parameter C is 2 (between 100 and 1000 L/year). Thus, $R_{66} = 24$. The rest of the risk indexes associated with *reactor emptying and cleaning* (Operation 6) can be calculated in a like manner.

Table 7. Reagent consumption in the analysis of photocatalytic experiments for 50 experiments/year.

Parameter	Amount/Sample	Samples/Exp	Total Amount/Exp	Yearly Consumption
MET analysis				
Acetonitrile HPLC	2.55 mL	15	38.25 mL	1.5 kg
Acidified water HPLC	10.2 mL	15	153.00 mL	7.7 kg
TOC analysis				
Synthetic air TOC (150 mL/min × 15/min/sample)	2250 mL	15	33750 mL	1.5 Nm^3
COD analysis				
Dichromate COD	1.5 mL	4	6 mL	0.3 kg
H$_2$SO$_4$ COD	3.5 mL	4	14 mL	0.7 kg
BOD				
Reagents	Negligible	0		0
Lyophilized capsules	0.0625 capsules	4	0.25	12.5 capsules
Toxicity				
Osmotic adjuster	0.25 mL	6	1.5 mL	0.1 kg
Dilution water	7.5 mL	6	45 mL	2.3 kg
Bacteria restorative (1 mL/container)	0.056 mL	6	0.333 mL	0.02 kg
Bacteria	0.056 containers	6	0.333 cont.	17 containers
Sampling				
Filters for samples	1 filter	15	15 filters	750 filters

Except for acetonitrile (density = 786 kg/m^3), solution density was considered to be 1000 kg/m^3 because they are very diluted aqueous solutions.

All the R_{ij} values in Table 1 were calculated as described in the examples above. When they have all been determined, the environmental risk index E_{Oi} associated with each operation, or the environmental risk index E_{Aj} associated with each environmental aspect, can be calculated. Thus, the second-last column in Table 1 shows the environmental risk associated with each operation E_{Oi}, calculated as the sum of the risks:

$$\sum_{j=1}^{n} R_{ij} = E_{Oi} \tag{1}$$

As an example, total risk E_{O6} associated with Operation 6 can be calculated as the sum of the risks associated with each aspect:

$$E_{Oi} = E_{O6} = \sum_{j=1}^{n} R_{6j}$$

$$= R_{61} + R_{62} + R_{63} + R_{64} + R_{65} + R_{66} + R_{67} = 72 + 6 + 0 + 0 + 0 + 24 + 60 = 162 \tag{2}$$

This process is summarized schematically in Figure 1.

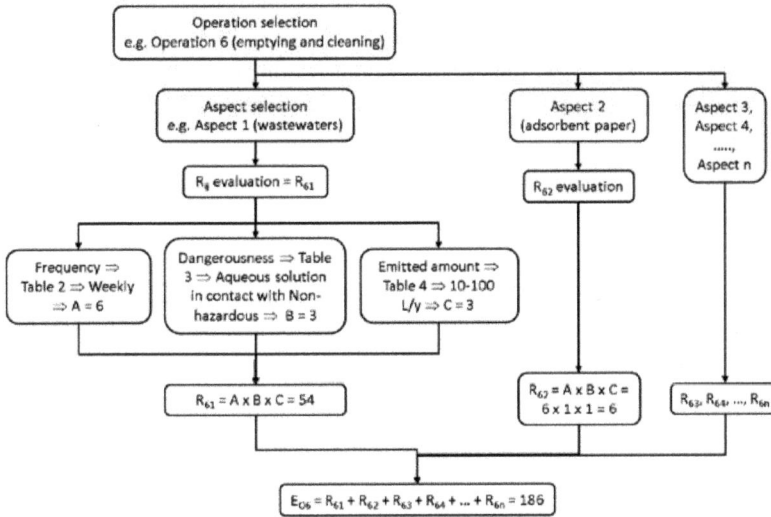

Figure 1. Evaluation of Environmental Risk Index: Schematic view.

The last row in Table 1 shows the environmental risk associated with each environmental aspect E_{Ai}, calculated as the sum of the risks for all the operations included in the process:

$$\sum_{i=1}^{n} R_{ij} = E_{Aj} \tag{3}$$

For example, for wastewater (see Environmental Aspect 1), the E_{Ai} is:

$$E_{Aj} = E_{A1} = \sum_{i=1}^{n} R_{i1} \tag{4}$$
$$= R_{11} + R_{21} + R_{31} + R_{41} + R_{51} + R_{61} + R_{71} + R_{81} = 0+0+0+0+180+72+0+6 = 258$$

The Global Environmental Risk Index (E) for this process is the sum of the E_{Oi} or E_{Aj} values. This index appears in the second-to-the-last cell (bottom-right) in Table 1:

$$E = \sum_{i=1}^{m} E_{Oi} = \sum_{j=1}^{n} E_{Aj} \tag{5}$$

Applying all of this to the values in Table 1 results in:

$$E = \sum_{i=1}^{m} E_{Oi} = E_{O1} + E_{O2} + E_{O3} + E_{O4} + E_{O5} + E_{O6} + E_{O7} + E_{O8} \tag{6}$$
$$= 24+12+66+12+462+162+1+7 = 746$$

Or

$$E = \sum_{j=1}^{n} E_{Aj} = E_{A1} + E_{A2} + E_{A3} + E_{A4} + E_{A5} + E_{A6} + E_{A7} \tag{7}$$
$$= 258 + 26 + 72 + 0 + 132 + 48 + 210 = 746$$

An interesting index is the percentage of risk for the maximum possible ($\%E_{Oi,max}$). The maximum R_{ij} possible for each of the first five aspects (wastewater, solid wastes, air emission, soil pollution, raw materials consumption), is 1000, which is the product of the maximums possible for A, B and C, according to Tables 2–4. For water consumption, the maximum possible is 200 and for energy consumption the maximum is 500. Thus, the E_{Oi} maximum for any operation is 5700. Dividing 5700 by the E_{Oi} for each operation and multiplying by 100, $\%E_{Oi,max}$ can be found for each operation (see the last column of Table 1).

Similar reasoning follows for $E_{Ai,max}$. Thus, the maximum E_{Ai} for each aspect is 8000 for the first five aspects (wastewater, solid waste, air emission, soil pollution, raw materials consumption), that is, 1000 for each aspect (10 for frequency, 10 for hazard and 10 for the amount) multiplied by the eight operations in our case. $E_{Ai,max}$ is $E_{Ai} \times (100/8000)$. The maximum of E_{Ai} for water consumption is 1600. Thus, $E_{Ai,max}$ for water consumption is 3.0%. For energy consumption, the maximum is 4000 and $E_{Ai,max}$ is 5.25.

The environmental risk indexes can be found the same way for photo-Fenton MET treatment in a Solarbox. As in photocatalysis, first you need to know the electricity and raw materials consumed. The electricity consumption data are the same as for the photocatalytic experiments and the only change is the time needed for each experiment, which in this case is 3.5 h. Thus, the total consumption of electricity by year, corresponding to the experiments performed, is 288 kWh/y. The rest of the data are the same as in Table 5.

Data corresponding to reagent consumption appears in Table 8. Note that changes from Table 6 (photocatalysis) are TiO_2 disappearance and $FeSO_4$ and H_2O_2 appearance.

Table 8. Reagent consumption in photo-Fenton experiments with a 1-L reaction volume for 50 experiments/year.

Reagents	Concentration— Amount/Exp	Total Amount Reagent/Exp.	Yearly Consumption (Kg)
Metoprolol tartrate salt	50 mg/L	50 mg	0.003
$FeSO_4.7H_2O$ (7 mg/L Fe)	7 mg/L	34.75 mg	0.002
H_2O_2 (100 mg/L)	100 mg/L	333.3 mg	0.017
Millipore Water		1.00 L	50

Neither reagents for analysis, nor filters for samples (Table 7) are needed in this case, while Fe and H_2O_2 analyses are. These changes from Table 7 (photocatalysis) are observed in Table 9 (photo-Fenton).

The risk indexes for photo-Fenton experiments can be evaluated using the amounts shown in the tables above, similar to what was explained for photocatalysis. Results are presented in Table 10.

Comparisons can be made by taking the value found for each operation in each process. For the two processes studied (photocatalysis and photo-Fenton), the operations with the most risk are those related to analysis (Operation 5 in Table 1 and Operation 6 in Table 10) due to the use of highly hazardous reagents. In both cases, Operation 7, related to reactor emptying and cleaning, is the second most risky, which seems logical because wastewater generated may contain hazardous products. Finally, the third position in this virtual ranking of environmental risk of operations is running the experiment (Operation 3 in Table 1 and Operation 4 in Table 10). For the seven aspects analysed in each operation, photo-Fenton values are usually higher. There are two reasons for that, the number of operations involved is higher and the products used in photo-Fenton are more hazardous than those used in photocatalysis.

Table 9. Reagent consumption in photo-Fenton experiment analyses for 50 experiments/year.

Parameter	Amount/Sample	Samples/Exp	Total Amount/Exp	Yearly Consumption
MET analysis				
Acetonitrile HPLC	2.55 mL	15	38.25 mL	1.5 kg
Acidified water HPLC	10.2 mL	15	153 mL	7.7 kg
TOC analysis				
Synthetic air TOC (150 mL/min × 15/min/sample)	2250 mL	15	33750 mL	1.5 Nm^3
COD analysis				
Dichromate COD	1.5 mL	4	6 mL	0.3 kg
H_2SO_4 COD	3.5 mL	4	14 mL	0.7 kg
H_2O_2 analysis				
Reagents for H_2O_2 determination	1.5 mL	2	3 mL	0.15 kg
Fe analysis				
Phenanthroline for Fe determination	1.0 mL	2	2 mL	0.1 kg
BOD				
Reagents	Negligible	0	0	0
Lyophilized capsules	0.0625 capsules	4	0.25 capsules	12.5 capsules
Toxicity				
Osmotic adjuster	0.25 mL	6	1.5 mL	0.1 kg
Dilution water	7.5 mL	6	45 mL	2.3 kg
Bacteria restorative (1 mL/container)	0.056 mL	6	0.333 mL	0.02 kg
Bacteria	0.056 containers	6	0.333 cont.	17 containers

Except for acetonitrile (density = 786 kg/m^3), solution density is considered to be 1000 kg/m^3 because they are very diluted aqueous solutions.

Of course, when comparisons are made, it has to be taken into account that the absolute values of E depend on the number of operations involved in the process, and it seems logical for E to increase when the number of operations is higher. Thus, from the results of Tables 1 and 10, photo-Fenton is more dangerous than photocatalysis because their risk indexes (E = E_{Oi} = E_{Aj}) are 994 and 746, respectively.

However, the ratio of E to the number of operations (E/NOP) can be instructive with respect to process risk and a good parameter for processes comparison, because it represents an average. In addition, this ratio gives information on how hazardous process operations are. Of course, high E/NOP ratios imply more risky operations. The same may be said of the ratio of E to the number of aspects in the process (E/NAS). There are always seven aspects analysed in the process, as seen in Tables 1 and 10. Thus, an increase in the E/NAS ratio means that the process environmental hazard increases. The E/NOP and E/NAS ratios are the highest for photo-Fenton (see Table 11).

Table 10. Environmental Risk Identification and Evaluation. Process tested: photo-Fenton MET treatment in the solarbox.

Each environmental aspect cell lists the label followed by its A, B, C and R_{ij} values.

	Environmental Aspects — Pollution				Consumption			E_{oi}	$\%E_{oi,\,max}$
	1. Waste water (A B C R_{ij})	2. Solid Wastes (A B C R_{ij})	3. Air Emission (A B C R_{ij})	4. Soil Pollution (A B C R_{ij})	5. Raw Materials (A B C R_{ij})	6. Water (A B C R_{ij})	7. Energy (A B C R_{ij})		
Normal Operation									
1. Preparation of MET solution			MET powders 6 1 1 6		MET 6 1 1 6	Water 6 2 1 12		24	0.42
2. FeSO$_4$ addition					FeSO$_4$ 6 5 1 30			30	0.53
3. H$_2$O$_2$ addition					H$_2$O$_2$ 6 10 1 60			60	1.05
4. Experiment Development		Gloves 6 1 1 6					Electricity for lamp, pumps, stirrer 6 5 2 60	66	1.16
5. Sampling								0	0
Operation									
6. Analysis	Aqueous solution with organics 6 10 3 180		Air TOC 6 5 2 60		Reagents 6 10 2 120	Water for preparation 6 2 1 12	Electricity for equipment 6 5 3 90	462	8.11
7. Reactor emptying and cleaning	Aqueous solution with organics 6 10 4 240	Adsorbent paper 6 1 1 6				Water for cleaning 6 2 2 24	Electricity for pump, stirrer 6 5 2 60	330	5.79
Abnormal Operation									
8. Disassembly of experimental equipment for maintenance		Crashing of reactor 1 1 1 1						1	0.02
Emergency Operation									
9. Tank or reactor breakdown and/or solvent spillage	Aqueous solution with organics 1 10 2 20	Adsorbent material 1 1 1 1						21	0.37
E_{Ai}	440	14	66	0	216	48	210	$E = 994$	-
$\%E_{AL,\,max}$	4.89	0.16	0.73	0	2.40	2.67	4.67	-	-

Table 11. Summarized final basic parameters for the environmental comparison of the processes.

Process	E	E/NOP	E/NAS	Av. %E_{OI} max	Av. %E_{AI} max
Photocatalysis	746	93	107	1.64	2.05
Photo-Fenton	994	110	142	1.94	2.22

Another parameter that can give an idea of the risk involved in the tested process is the percentage with respect to the maximum possible to be achieved for each operation and aspect (see last row and last column in Tables 1 and 10). The averages are presented in Table 11, for both items and processes. The first comment to be made is that in all cases the values are very low. This means that the hazard levels of the processes analysed are very low.

3. Experimental Section

The processes in the study are well known and have been previously described elsewhere [16–18]. However, a short description is presented here to make it easier for the reader to understand. Metoprolol tartrate salt (MET) (a model pollutant) was treated with photo-Fenton or photocatalysis with TiO_2 in a solarbox, as described previously [17].

Experiments were carried out in a solarbox (from CO.FO.ME.GRA) with an Xe lamp (1000 W). The solution to be treated was prepared in a feed tank (1 L) and pumped into a tubular reactor located at the bottom of the solarbox in the axis of a parabolic mirror. From there, the suspension was continuously recirculated to the feed tank. The feed tank was continuously stirred and the temperature was kept constant using a thermostatic bath. The experimental devices used are described at length elsewhere [16–18].

The parameters and variables monitored and analysed during the experiments were: MET concentration, TOC, COD, BOD, toxicity, *etc.* (see Tables 7 and 9).

MET removal was the same in both processes, but was degraded much faster by photo-Fenton than by photocatalysis (70 min and 180 min, respectively, for 90% MET removal), for experiments done with the same experimental equipment and under the same conditions. However, mineralisation was very similar in both processes. On the other hand, photocatalysis proved to be more energy efficient (mg of MET removal/kJ of useful light) than photo-Fenton. A detailed comparison of these two technologies for MET removal is available in a previous paper [17].

4. Conclusions

Several indexes have been estimated: absolute Environmental Risk Index (E_{Oi}) associated with each of the unit operations involved in the process, absolute Environmental Risk Index (E_{Aj}) related to the conditions of process running, and Global Environmental Risk Index (E) of the process studied. In addition, E/NOP and E/NAS relative indexes for evaluating the risk of each operation or aspect involved in the processes have been proposed. With these indexes, operations and processes, *etc.*, may be compared to find the one that is environmentally best. This procedure was developed for lab scale, but its results could be used for plant design or operation, simply by entering data from large

AOP wastewater treatment plants. It is important to highlight that the key point of the procedure is the correct selection of operations involved in each process to be analysed.

It should also be mentioned that the lab-scale operations with the highest risk are the analytical protocols (reinforcing the idea of further R&D on these protocols, especially when applied to evaluate processes for improving environmental protection). In our case, photocatalysis seems to involve less environmental risk than photo-Fenton.

Acknowledgments

The authors thank the Ministry of Science and Innovation of Spain (projects CTQ2011-26258 and NOVEDAR 2010 CSD2007-00055) and AGAUR—Generalitat de Catalunya (project 2009SGR 1466) for the financial support.

Author Contributions

All the authors discussed and planned the paper. Jaime Giménez wrote the first draft version, and Jaime Giménez, Bernardí Bayarri, Óscar González, Sixto Malato, José Peral and Santiago Esplugas made their comments, modifications, and suggestions to this first draft. Thus work on this paper was really shared by all authors.

Conflicts of Interest

The authors declare that there is no conflict of interest.

References and Notes

1. ISO 14001:2004: Environmental Management Systems—Requirements with Guidance for Use.
2. EMAS: Regulation (EC) No 1221/2009 of the European Parliament and of the Council of 25 November 2009 on the voluntary participation by organisations in a Community eco-management and audit scheme (EMAS), repealing Regulation (EC) No 761/2001 and Commission Decisions 2001/681/EC and 2006/193/EC.
3. Guinée, J.B.; Gorrée, M; Heijungs, R. *LCA—An Operational Guide to the ISO-Standards—Part 2a: Guide*; CML: Leiden, The Netherlands, 2001.
4. Guinée, J.B. Gorrée, M; Heijungs, R. LCA—An Operational Guide to the ISO-Standards—Part 2b: Operational Annex; CML: Leiden, The Netherlands, 2001.
5. Goedkoop, M.; Spriensma, R. The Eco-Indicator 99: A damage Oriented Method for Life Cycle Impact Assessment. Methodology Report; PRé Consultants: Amersfoort, The Netherlands, 2001.
6. Goedkoop, M.; Spriensma, R. The Eco-Indicator 99: A Damage Oriented Method for Life Cycle Impact Assessment. Methodology Annex; PRé Consultants: Amersfoort, The Netherlands, 2001.
7. Directive 96/82/EC, of 9 December 1996, on the control of major-accident hazards involving dangerous substances.

8. Directive 2012/18/EU of the European Parliament and of the Council, of 4 July 2012, on the control of major-accident hazards involving dangerous substances, amending and subsequently repealing Council Directive 96/82/EC.
9. Directive 2004/35/CE, of 21 April 2004, on environmental liability with regard to the prevention and remedying of environmental damage.
10. Arteche, F.; Avellaneda, A.; Ferrer-Vidal, V.; Hernández, L.; Masip, J.M. *Guía Para la Implantación y Desarrollo de un Sistema de Gestión Medioambiental*; Generalitat de Catalunya; Departament de Medi Ambient: Barcelona, Spain, 1997.
11. Giménez, J.; Bayarri, B.; Latorre, M.; Illana, E.; Carreira, E.; Contreras, S.; Esplugas, S. Integration of quality, environment, safety and health management systems in AOPs laboratories. In Proceedings of the 4th European Meeting on Solar Chemistry and Photocatalysis: Environmental Applications, Las Palmas de Gran Canaria, Spain, 8–10 November 2006.
12. Corcoran, J.; Winter, M.J.; Tyler, C.R. Pharmaceuticals in the aquatic environment: A critical review of the evidence for health effects in fish. *Crit. Rev. Toxicol.* **2010**, *40*, 287–304.
13. Regulation (EC) No 1272/2008 of the European Parliament and of the Council of 16 December 2008 on classification, labelling and packaging of substances and mixtures, amending and repealing Directives 67/548/EEC and 1999/45/EC, and amending Regulation (EC) No 1907/2006.
14. 2000/532/EC: Commission Decision of 3 May 2000 replacing Decision 94/3/EC establishing a list of wastes pursuant to Article 1(a) of Council Directive 75/442/EEC on waste and Council Decision 94/904/EC establishing a list of hazardous waste pursuant to Article 1(4) of Council Directive 91/689/EEC on hazardous waste (notified under document number C(2000) 1147).
15. Ley 22/2011, de 28 de julio, de residuos y suelos contaminados. Transposition of Directive 2008/98/EC of the European Parliament and of the Council of 19 November 2008 on waste and repealing certain Directives.
16. De la Cruz, N.; Dantas, R.F.; Giménez, J.; Esplugas, S. Photolysis and TiO_2 photocatalysis of the pharmaceutical propranolol: Solar and artificial light. *Appl. Catal. B—Environ.* **2013**, *130–131*, 249–256.
17. Romero, V.; González, O.; Bayarri, B.; Marco, P.; Giménez, J.; Esplugas, S. Performance of different advanced oxidation Technologies for the abatement of the beta-blocker metoprolol. *Catal. Today* **2015**, *240*, 86–92.
18. Romero, V.; Méndez-Arriaga, F.; Marco, P.; Giménez, J.; Esplugas, S. Comparing the photocatalytic oxidation of Metoprolol in a solarbox and a solar pilot plant reactor. *Chem. Eng. J.* **2014**, *254*, 17–29.

Sample Availability: Not available.

A Review of Heterogeneous Photocatalysis for Water and Surface Disinfection

John Anthony Byrne, Patrick Stuart Morris Dunlop, Jeremy William John Hamilton, Pilar Fernández-Ibáñez, Inmaculada Polo-López, Preetam Kumar Sharma and Ashlene Sarah Margaret Vennard

Abstract: Photo-excitation of certain semiconductors can lead to the production of reactive oxygen species that can inactivate microorganisms. The mechanisms involved are reviewed, along with two important applications. The first is the use of photocatalysis to enhance the solar disinfection of water. It is estimated that 750 million people do not have accessed to an improved source for drinking and many more rely on sources that are not safe. If one can utilize photocatalysis to enhance the solar disinfection of water and provide an inexpensive, simple method of water disinfection, then it could help reduce the risk of waterborne disease. The second application is the use of photocatalytic coatings to combat healthcare associated infections. Two challenges are considered, *i.e.*, the use of photocatalytic coatings to give "self-disinfecting" surfaces to reduce the risk of transmission of infection via environmental surfaces, and the use of photocatalytic coatings for the decontamination and disinfection of medical devices. In the final section, the development of novel photocatalytic materials for use in disinfection applications is reviewed, taking account of materials, developed for other photocatalytic applications, but which may be transferable for disinfection purposes.

Reprinted from *Molecules*. Cite as: Byrne, J.A.; Dunlop, P.S.M.; Hamilton, J.W.J.; Fernández-Ibáñez, P.; Polo-López, I.; Sharma, P.K.; Vennard, A.S.M. A Review of Heterogeneous Photocatalysis for Water and Surface Disinfection. *Molecules* **2015**, *20*, 5574-5615.

1. Heterogeneous Photocatalysis

According to the IUPAC Gold Book, photocatalysis is the "*change in the rate of a chemical reaction or its initiation under the action of ultraviolet, visible or infrared radiation in the presence of a substance—the photocatalyst—that absorbs light and is involved in the chemical transformation of the reaction partners.*" A photocatalyst is defined as a "*catalyst able to produce, upon absorption of light, chemical transformations of the reaction partners. The excited state of the photocatalyst repeatedly interacts with the reaction partners forming reaction intermediates and regenerates itself after each cycle of such interactions*" [1]. In heterogeneous photocatalysis, the photocatalyst is present as a solid with the reactions taking place at the interface between phases, *i.e.*, solid-liquid or solid-gas. For the purposes of this review we will be concerned with heterogeneous photocatalysis taking place at the surface of solid semiconductor materials, which act to absorb the photon energy and provide active sites for the adsorption of reactants. In semiconductor photocatalysis, the primary reactions are electrochemical oxidation or reduction reactions involving hole and electron transfer from the photo-excited semiconductor.

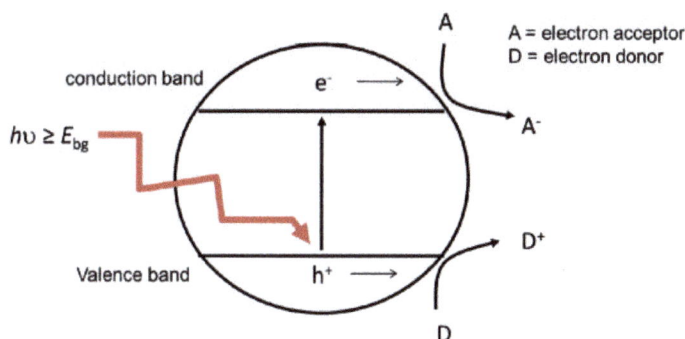

Figure 1. Schematic showing the basic mechanism of heterogeneous photocatalysis.

Figure 1 shows a schematic of the basic mechanism involved. The semiconductor is excited by the absorption of electromagnetic radiation with energy equal to or greater than the band gap energy (E_{bg}). This results in the promotion of an electron from the valence band to the conduction band, leaving a positive hole in the valence band. The electron/hole pairs may recombine with the energy being re-emitted as heat or light, or the charge carriers can migrate to the particle surface. The conduction band electron can be passed on to an electron acceptor with a more positive electrochemical reduction potential than the conduction band edge potential. The valence band hole may accept electrons from donor species with a less positive electrochemical reduction potential than the valence band edge potential. Overall, these processes result in the reduction of an acceptor species and the oxidation of a donor species, where both reactions are driven by the potential difference generated by the absorption of electromagnetic radiation. The potential difference generated is close to the band gap energy of the semiconductor.

Reactions that are thermodynamically downhill (−ve ΔG) are photocatalytic, and reactions that are thermodynamically uphill (+ve ΔG) are photosynthetic. Nevertheless, it is common to use the term photocatalytic to describe up-hill reactions e.g., "photocatalytic" water splitting. For the purposes of photocatalytic disinfection, we are primarily concerned with surface redox reactions, which lead to the generation of reactive oxygen species (ROS). Figure 2 shows a general scheme for the production of reactive oxygen species where oxygen is acting as the electron acceptor and water or hydroxyl ions is acting as the electron donor. In this example, titanium dioxide (TiO_2) is the semiconductor as the band edge potentials must be in suitable positions to drive the reactions of interest. Other semiconductor materials may be used—*vide infra*. The valence band hole may have an electrochemical reduction potential positive enough to oxidize water to yield hydroxyl radical and the conduction band should be negative enough to reduce molecular oxygen to yield superoxide radical anion, and via subsequent electron transfer, yield peroxide and hydroxyl radical. Overall, in the presence of oxygen and water, the photocatalytic mechanism can generate a mixture of ROS, which inactivate microorganisms or degrade organic chemical contaminants. For more detailed information on the physicochemical mechanisms of heterogeneous photocatalysis the reader should consult previously published reviews [2–8].

In this review, the mechanism of inactivation of microorganisms is discussed and two specific applications of photocatalytic disinfection are reviewed, *i.e.*, the solar photocatalytic disinfection of water and photocatalytic coatings to combat healthcare associated infections. In the last section we review the development of novel photocatalytic materials with respect to the inactivation of microorganisms.

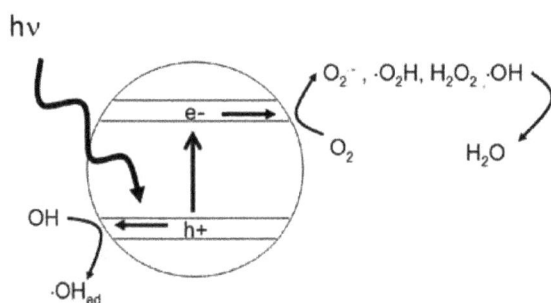

Figure 2. Schematic of photocatalytic mechanism on a titanium dioxide particle leading to the production of reactive oxygen species.

2. Mechanism of Photocatalytic Inactivation of Microorganisms

We are surrounded by microorganisms that colonize our skin, intestines and the environment we live in. In most cases we live in harmony with these microscopic entities, in some cases symbiotically, and we utilize them to produce food, degrade waste, generate essential chemicals and clean water. However, some microorganisms are the cause of disease and death. These pathogenic microorganisms cause diseases, such as Ebola hemorrhagic fever, typhoid, and cholera, and illnesses, such as flu and the common cold. They can be transmitted from person to person and sometimes from species to species, and transmission can be via direct contact, via body fluids, airborne in aerosol, infected surfaces, food, and water. Over the last century, humans have developed an armory of defenses against pathogenic microorganisms, including vaccines, antibiotics, disinfectants, food hygiene practices, water disinfection (chlorination, ozonation, UVC), sterilization, *etc.* Unfortunately pathogenic microorganisms rapidly evolve and adapt to resist our defense systems and we must utilize science and engineering to deal with existing and emerging pathogens. Heterogeneous photocatalysis has been shown to be effective for the inactivation of a wide range of pathogenic microorganisms, including some which are resistant to other methods of disinfection.

Since Matsunaga *et al.* first reported the inactivation of bacteria using TiO_2 photocatalysis in 1985 [9] there have been more than 1000 research papers published in the area. The effectiveness of photocatalysis against microorganisms, including bacteria (cells [10,11], spores [12] and biofilms [11]), viruses [13], protozoa [14], fungi [15] and algae [16] has been investigated, and this work has been reviewed by McCullagh *et al.* [17], Malato *et al.* [8], and Robertson *et al.* [18]. In general photocatalytic disinfection in water requires minutes or tens of minutes of direct UVA exposure (using TiO_2 as the photocatalyst) and it is considered to be quite a slow microbial inactivation process, as compared to e.g., UVC disinfection (seconds of direct exposure). The mechanism of photocatalytic inactivation is different from that of UVC disinfection. Whilst the majority of papers published in the area focus on the assessment of novel materials, new reactor systems or the effect of experimental parameters on the rate of inactivation, a significant number of studies have specifically investigated ROS interaction with the biological structures within microorganisms in an attempt to elucidate the mechanism resulting in the loss of organism viability. A review of the mechanisms involved in photocatalytic disinfection was conducted by Dalrymple *et al.* [19], however, the exact sequence of events leading to loss of viability is not completely understood. Continued insight into the mechanisms of attack of ROS on microorganisms will allow researchers to optimize materials and reactor design to improve the rate and efficacy of photocatalytic disinfection [20].

Upon excitation, a range of ROS can be generated at the semiconductor particle-solution interface. Of these the hydroxyl radical (HO$^{\bullet}$) has been suggested to be the primary species responsible for microorganism inactivation, however superoxide radical anion (O$_2^{\bullet-}$), hydroperoxyl radical (HO$_2^{\bullet}$) and hydrogen peroxide (H$_2$O$_2$) have been shown to contribute to the biocidal process [21]. Within biology the toxicity of "free radicals" is well known. Biological systems have enzymatic process to convert ROS into less toxic species (e.g., catalase and superoxide dismutase) and hydroxyl radicals play a pivotal role in the reaction of white blood cells with pathogens and apoptosis (programmed cell death). The prevalence of ROS within the environment in which microorganisms thrive and the evolution of defense systems against these active species are of particular relevance to photocatalysis. Unlike antibiotics, which target a specific biological process within the lifecycle of bacterial organisms (only), ROS attack is not specific to one site or an individual pathway. This permits the use of ROS generated by photocatalysis against a wide range of pathogens, and as important, the development of bacterial resistance to photocatalysis is considered to be almost impossible [22]. In photocatalytic disinfection, the ROS must be produced at concentrations greater than that which the microorganism can protect against in order to result in complete inactivation, thereby preventing bacterial recovery and re-growth.

ROS interaction with microbes typically occurs from the outside of the organism, in towards the sensitive metabolic processes and genetic materials within organisms. The "strength" of the outer layers of the organism effectively dictates the ability of the organism to survive, the thick protein, carbohydrate and lipid structures surrounding protozoa and bacterial spores present a more challenging target than viruses, fungi and bacteria—in that order [8]. Due to the relative ease of culture and detection, bacteria have been the most widely studied organisms, with *Escherichia coli* (*E. coli*) the primary species used.

Typically bacterial organisms are classified based upon the content of their outer cell layers, which surround an internal liquid based cytoplasmic matrix, comprising genetic material and the biochemical systems used for energy production, cell regulation and reproduction. The cytoplasm is bound by the cell membrane—a phospholipid bilayer, containing cross membrane proteins structures, which regulate transmission of chemicals into and out of the cytoplasm—and maintaining the integrity of this structure is of particular importance for bacterial viability. The cell membrane in Gram-positive bacteria is surrounded by a cell wall comprising a thick layer (20–80 nm) of porous peptidoglycan, responsible for structural integrity and the retention of Gram stain. Chains of lipoteichoic acid extend from the cell membrane through the cell wall and play roles in cell binding reactions. The cell wall in Gram-negative bacteria is much more complex. A thin layer of peptidoglycan (7–11 nm) is encapsulated by a second phospholipid bilayer, termed the outer membrane. This is populated with long extending chains of lipopolysaccharide, which elicit strong immune responses in animals, and as with the inner cell membrane contains cross membrane protein channels.

Given the complexity of microorganisms, it is perhaps understandable why the full mechanism of photocatalytic inactivation is still unknown, however, the accepted sequence of events taking place during photocatalytic inactivation of microorganisms is that prolonged ROS attack results in damage of the cell wall, followed by compromise of the cytoplasmic membrane and direct attack of intracellular components (Figure 3). To this end, microscopy based studies have revealed the formation of pores within cell wall and cell membrane structures [23], the degradation of peptidoglycan has been reported [24], lipid peroxidation within phospholipids membranes demonstrated [24–27], degradation of porin proteins within cell membranes [28], the detection of intercellular compounds and genetic material exterior to the cell confirmed [9,29], and direct DNA damage reported via genetic analysis [30]. More specific modes of action discussed include membrane damage leading to inactivation of respiratory pathway chemistry [24] and loss of fluidity and increased ion permeability [27]. The general consensus relates to hydroxyl radical mediated lipid peroxidation of the outer cell wall components, and perhaps the most conclusive evidence for this was recently reported by Kubacka et al. [31]. Following a detailed and systematic investigation into the levels of a range of genetic and protein markers during of TiO_2 photocatalysis the authors concluded that extensive radical induced cell wall modifications are the main factor responsible for the high biocidal performance of TiO_2-based nanomaterials. Some workers have also proposed additional mechanistic action via "inside-out" processes, with Malato et al., describing that UVA irradiation of intracellular chromophores in the presence of oxygen could produce single stranded DNA breaks and nucleic acid modifications [8] and Gogniat and Dukan reporting DNA damage via Fenton reaction-generated hydroxyl radicals during bacterial recovery [32].

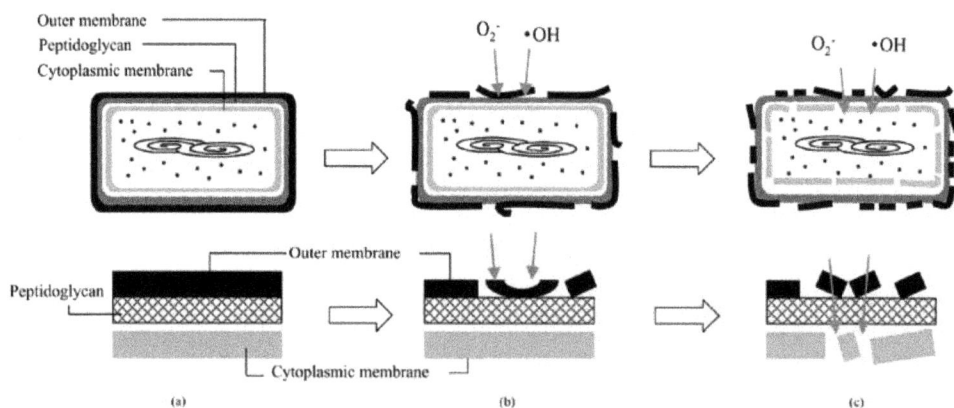

Figure 3. (a–c) Schematic illustration of the process of *E. coli* inactivation on photo-excited TiO_2. In the lower row, the part of cell envelope is magnified. (Reproduced from Sunada, K.; Watanabe, T.; Hashimoto, K. Studies on photokilling of bacteria on TiO_2 thin film. *J. Photoch. Photobio A* **2003**, *156*, 227–233 [27]).

Whilst photocatalysis has great potential to be used as a biocidal technology, caution should to be exercised when conducting photocatalytic disinfection assays as organisms have the potential to recover from sub-lethal ROS exhibited stress and re-grow. Several authors have reported the need for complete inactivation of organisms and specific analysis to confirm that the treatment prevents subsequent bacterial re-growth [11,33]. The transfer of antibiotic resistance following sub-lethal photocatalytic treatment has recently been reported [34].

3. Solar Photocatalytic Disinfection of Water

Water is an important natural resource and safe drinking water is vital for human existence and good quality of life. Clean water resources are becoming depleted due to population growth, over-use of resources and climate change. Since the adoption of the Millennium Development Goals (MDG), the WHO/UNICEF Joint Monitoring Programme for Water Supply and Sanitation has reported on progress towards achieving Target 7c: "reducing by half the proportion of people without sustainable access to safe drinking water and basic sanitation" [35]. Although the world met the MDG drinking water target, 748 million people, mostly the poor and marginalized, still lack access to an improved drinking water source. Almost a quarter of these (173 million) rely on untreated surface water, and over 90% live in rural areas. It is estimated that there will be 547 million people without an improved drinking water supply in 2015. Many more are forced to rely on sources that are microbiologically unsafe, leading to a higher risk of contracting waterborne diseases, including typhoid, hepatitis A and E,polio and cholera [35–38]. The WHO estimated that in 2008 diarrheal disease claimed the lives of 2.5 million people [39]. For children under five, this burden is greater than the combined burden of HIV/AIDS and malaria [40]. A total of 58 countries from all continents reported a cumulative total of 589,854 cholera cases in 2011, representing an increase of 85% from 2010. The greatest proportion of cases was reported in Latin America and Africa [41].

Although the MDG drinking-water target refers to sustainable access to safe drinking water, the MDG indicator—"use of an improved drinking water source"—does not include a measurement of either drinking water safety or sustainable access. This means that accurate estimates of the proportion of the global population with sustainable access to safe drinking water are likely to be significantly lower than estimates of those using improved drinking water sources.

Piped-in water supplies are a long-term goal and interventions to improve water supplies at the source (point of distribution) have long been recognized as effective in preventing waterborne disease. Recent reviews have shown household-based (point-of-use) interventions to be significantly more effective than those at the source for the reduction of diarrheal diseases in developing regions (possibly due to contamination of water between collection and use). As a result, there is increasing interest in such household-based interventions that can deliver the health gains of safe drinking water at lower cost. Household water treatment and safe storage (HWTS) is one option for improving the quality of water for consumption within the home, especially where water handling and storage is necessary and recontamination is a real risk between the point of collection and point of use. Living conditions in many humanitarian crises also call for effective HWTS. The practice of household water treatment and safe storage can help improve water quality at the point of consumption, especially were drinking water sources are distant, unreliable or unsafe. HWTS is a stop-gap measure only and does not replace the obligation of a service provider to supply access to safe drinking water. Household water treatment (HWT) methods include boiling, filtration, adding chlorine or bleach, and solar disinfection.

In 2008, Clasen and Haller reported on the cost and cost effectiveness of household based interventions to prevent diarrhea [42]. They compared: chlorination using sodium hypochlorite following the "Safe Water System" (SWS) developed and promoted by the US Centers for Disease Control and Prevention (CDC); gravity filtration using either commercial "candle" style gravity filters or locally fabricated pot-style filters developed by Potters for Peace; solar disinfection following the "SODIS" method in which clear 2 L PET bottles are filled with raw water and then exposed to sunlight for 6–48 h; and flocculation disinfection using Procter and Gamble PUR® sachets, which combine an iron-based flocculent with a chlorine-based disinfectant to treat water in 10 L batches. They concluded that household-based chlorination was the most cost-effective. Solar disinfection was only slightly less cost-effective, owing to its almost identical cost but lower overall effectiveness. Given that household-based chlorination requires the distribution of sodium hypochlorite, solar disinfection has a major advantage in terms of non-reliance on chemical distribution.

Sunlight is freely available on Earth and the combined effects of heat and UV from the sun can inactivate pathogenic organisms present in water. Of course, there are a number of parameters that affect the efficacy of the solar disinfection (SODIS) process, including the solar intensity and temperature (weather conditions), and the level and nature of the contamination (some pathogens are more resistant to SODIS than others). One approach to SODIS enhancement is the use of heterogeneous photocatalysis.

Furthermore, the presence of some water pathogens is also an issue in developed regions. Urban wastewater treatment plants (WWTP) are among the main sources of antibiotics' release into the aquatic environment worldwide, and therefore these are considered as hotspots of antibiotics and promote the genetic selection and generation of antibiotic resistant bacteria in water [43]. Therefore, the occurrence of antimicrobial resistant pathogens in the effluents of urban WWTP is evidence that conventional secondary water treatment cannot effectively control them. A number of bacterial species have been identified which exhibit acquired resistance to one or more types of antibiotics [44]. It is also recognized that chlorination of drinking water as tertiary treatment affects the fraction of antibiotic-resistant bacteria by potentially assisting in microbial selection, and regrowth or reactivation of antibiotic-resistant bacteria after chlorination in wastewater [45]. Nowadays, one of the major scientific challenges is to reduce the incidence of waterborne disease globally and also address the increasingly virulent and antibiotic resistant pathogens such as vanomycin-resistant enterococci (VRE) and methicillin-resistant *Staphylococcus aureus* [46]. Therefore alternative new technologies for water disinfection are needed to mitigate these problems at all levels.

The photocatalytic enhancement of solar water disinfection has been successfully proven in many occasions using solar light, either natural or artificial, to photo-excite TiO_2 nanoparticles suspended or immobilized against a wide number of model water pathogens, which go from bacteria (*E. coli*, *E. faecalis*, total coliforms, *Salmonella*, *Pseudomonas*, etc.), virus (MS2 phage, RNA bacteriophage, phi164, *etc.*), spores of bacteria as *Bacillus subtillis* and fungi like *Fusarium, Candida albicans, Aspergilllus niger, Phytophthora, etc.*, and parasites like *Cryptosporidium parvum* [10,12,14,47–50].

The configuration of the catalyst in the reactor can significantly alter in the disinfection results. Typically there are two ways to use the TiO_2 for water treatment purposes, (i) as aqueous suspensions of TiO_2 particles; and (ii) as immobilized TiO_2 over an inert and robust support which should be resistant to the photocatalytic process and also to the hydrodynamic conditions (flow and pressure) in the photo-reactor during the process. The configuration of the catalyst in the photo-reactor depends, among others factors, on the final application of this water treatment. If the system is designed for drinking water purification for human consumption, the use of suspended TiO_2 particles, as a part of a routine intervention for improving the potability of water at the house-hold level (point-of-use water treatment) may not be acceptable due to concerns over toxicity of nanomaterials [51] and the photocatalyst particles would have to be removed before consumption; However, one should point out that TiO_2 is used as a food additive (E171) and is found in a wide range of food products, including coffee creamer, and the use of a few mg L^{-1} for the disinfection of water would result in a lower intake of TiO_2 than the average US citizen. To avoid any complication of using dispersed titania, one can utilize the photocatalyst as an immobilized system, but this will result in lower rates of disinfection or more complex systems with higher cost [8,52].

Different types of supported TiO_2 for improvement of the efficiency of solar water disinfection like cylinders, pills, balls, mesh, *etc.*, have been investigated, for example TiO_2 deposited on glass rashing rings inside a tubular reactor of a CPC solar prototype was compared with slurry systems [53]. The maximum efficiency was found with the suspended TiO_2 particles reactor, due to the better surface contact between bacteria and catalyst. However, fixed-bed reactors lead to inactivation rate quite close to that of the slurry. It is not only the high titania surface area of this configuration which is responsible for the bacteria inactivation but the important contribution of the mechanical and osmotic stress have to be considered. The main advantage of the fixed-bed TiO_2 catalyst is the non-requirement for recovery of the catalyst following treatment; however the longevity of the catalyst must be considered. In the study by Sordo *et al.* [53], the catalyst was tested for ten reaction cycles without deactivation being observed. Most of these immobilized systems can be re-used for several cycles, as long as the catalyst is not fouled by inorganic species such as phosphates and sulfates. More work is needed on photocatalyst longevity under operational conditions. Alrousan *et al.*, tested solar photocatalytic (SPC-DIS) and solar disinfection (SODIS) of water at pilot scale using different reactor configurations with and without immobilized TiO_2 (Evonik Aeroxide P25) (Figure 4). The model organism used was *E. coli*. The use of compound parabolic collectors improved the SODIS and SPC-DIS of water; however, the improvement was less significant compared to the improvements reported previously for SODIS in static batch reactors. Kinetic fitting yields a log-linear component (first order rate constant). The following order was found for *k* where coated refers to TiO_2 coating and the equals sign indicates no significant difference; uncoated external = coated internal ≥ double coated tube ≥ uncoated double tube. It is known that *E. coli* is inactivated by SODIS and it may be a "soft" target for comparing the effectiveness of SODIS *vs.* SPC-DIS. Nevertheless, photocatalysis presents advantages in terms of the non-recovery of inactivated organisms and the inactivation of SODIS resistance organisms [54].

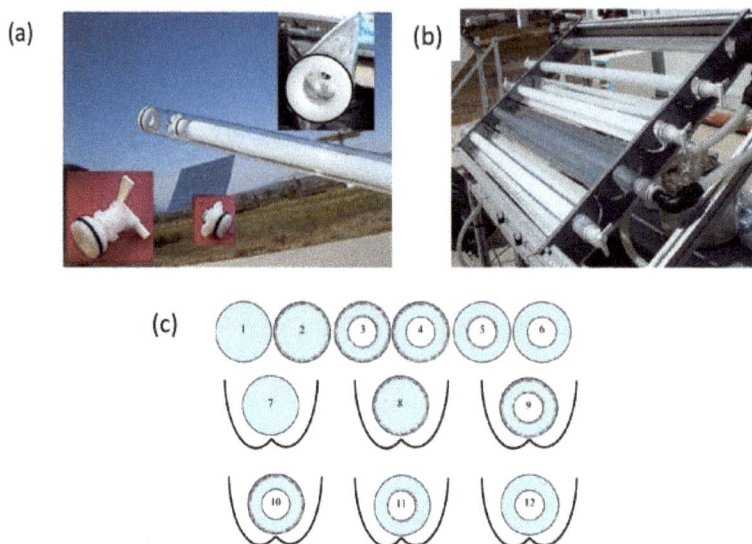

Figure 4. Photographs showing the double tube configuration with internal tube cap and the valve for external tube (**a**) and the solar photocatalytic reactor with and without CPC during disinfection tests (**b**). Schematic cross-section representation of the different reactor configurations tested in the solar reactor (**c**), (1)/(7) uncoated single tube without/with CPC; (2)/(8) coated single tube without/with CPC; (3)/(9) coated double tube without/with CPC; (4)/(10) coated external–uncoated internal without/with CPC; (5)/(11) coated internal–uncoated external without/with CPC; (6)/(12) uncoated double tube without/with CPC (reproduced from Alrousan, D.M.A.; Polo-López, M.I.; Dunlop, P.S.M.; Fernández-Ibáñez, P.; Byrne, J.A. Solar photocatalytic disinfection of water with immobilized titanium dioxide in re-circulating flow CPC reactors. *Appl. Catal. B* **2012**, *128*, 126–134 [54]).

3.1. Photocatalytic Materials Tested under Solar or Solar Simulated Conditions

Most research studies on photocatalytic disinfection have been carried out with the commercial TiO_2 *Evonik Aeroxide P25* (formerly *Degussa P25*) photocatalyst. Nevertheless, some research has been carried out with pure anatase and doped titanium dioxide. For solar applications, visible light active materials are desirable, to increase the photon absorption beyond the solar UVA spectrum, which is only *ca.* 4% of the global solar component reaching the Earth surface. However, the smaller band gap, while absorbing a greater number of solar photons, gives a narrower voltage window to drive the redox reactions at the interface. Furthermore, metal sulfide semiconductors, which absorb in the visible region of the spectrum, tend to undergo photo-anodic corrosion [52]. Unfortunately, the number of publications concerning the photocatalytic activity of these materials for the inactivation of microorganisms is limited. The UV activity of undoped TiO_2 may be greater than the visible light activity of a doped material. Therefore, for solar applications, the efficiency should be tested under simulated solar irradiation or under real sun conditions. Rengifo-Herrera and Pulgarin reported on the photocatalytic activity of N, S co-doped and N-doped commercial anatase (Tayca

TKP 102) TiO$_2$ powders towards phenol oxidation and *E. coli* inactivation [55]. However, these novel materials did not present any enhancement as compared to *Evonik Aeroxide P25* under simulated solar irradiation. They suggest that while the N or N, S co-doped TiO$_2$ may show a visible light response, the localized states responsible for the visible light absorption do not play an important role in the photocatalytic activity. More research is required to determine if visible-light active materials can deliver an increase in the efficiency of photocatalysis under solar irradiation. Nowadays, considering cost, chemical and photochemical stability, availability, and lack of toxicity, the most suitable catalyst reported to date for the disinfection of water is TiO$_2$.

Some researchers have been investigating the use of titanium dioxide—reduced graphene oxide (TiO$_2$-RGO) composites to improve the photocatalytic efficiency. This material may be easily synthesized by the photocatalytic reduction of exfoliated graphene oxide (GO) by TiO$_2$ (*Evonik Aeroxide P25*) under UV irradiation in the presence of methanol as a hole acceptor. TiO$_2$-RGO composites were compared to suspended TiO$_2$ (P25) for the disinfection of water contaminated with *E. coli* cells and *F. solani* spores under natural sunlight (Figure 5). Rapid water disinfection was observed with both *E. coli* and *F. solani*. An enhanced rate in the *E. coli* inactivation efficiency was observed with the TiO$_2$-RGO composite compared to P25 alone. In this work, the materials were evaluated with filtered sunlight; the major part of the solar UVA was cut-off ($\lambda > 380$ nm). In this case, a much greater time was required for inactivation of *E. coli* with TiO$_2$ P25 but the same inactivation rate was observed for the TiO$_2$-RGO indicating visible light activity, which could be attributed to singlet oxygen production by TiO$_2$-RGO composites, which would lead to *E. coli* inactivation [20].

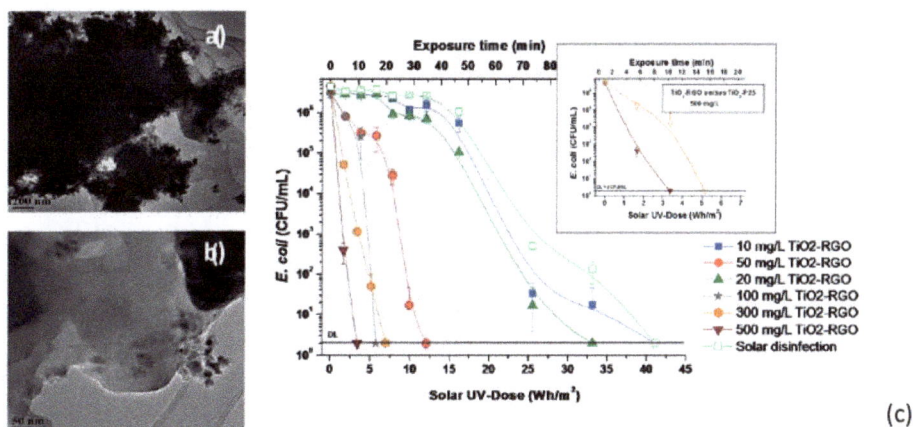

Figure 5. (a) TiO$_2$-GO aggregate before photoreduction; (b) TiO$_2$-RGO after UV assisted photoreduction and (c) *E. coli* inactivation at several TiO$_2$-RGO concentrations. Figure inserts shows efficiency of TiO$_2$-RGO and TiO$_2$-P25 on the *E. coli* inactivation (reproduced from Fernández-Ibáñez, P.; Polo-López, M.I.; Malato, S.; Wadhwa, S.; Hamilton, J.W.J.; Dunlop, P.S.M.; D'Sa, R.; Magee, E.; O'Shea, K.; Dionysiou, D.D.; Byrne, J.A. Solar Photocatalytic Disinfection of Water using Titanium Dioxide Graphene Composites. *Chem. Eng. J.* **2015**, *261*, 36–44 [20]).

Other researchers have investigated the efficacy of several nanomaterials under natural sunlight for water disinfection. The materials were employed as suspended nanoparticles in stirred batch reactors using mainly anatase phase (P25 and PC500), or mainly rutile phase (Ruana), and samples of Bi_2WO_6; these two last materials having the advantage of absorbing more solar photons. The photocatalyst activities were compared with solar disinfection (without catalyst) under natural sunlight using *E. coli* as model microorganism. This work proved that the adding of any kind of photo-catalyst to the water accelerated the bactericidal action of solar irradiation and led to complete disinfection during treatment times that varied from 60 to 150 min. The photocatalytic disinfection efficiency was not enhanced by the increase of catalyst concentration above 0.5 g/L for P-25, PC500 and Bi_2WO_6, where 10^6 CFU/mL were completely inactivated within 5 min, 30 min and more than 150 min of solar exposure under clear sky, respectively. An increase of the concentration to 1 g/L slightly decreased the total inactivation time. Rutile (Ruana) catalyst behaves differently where the optimal concentration was lower than for the other titania materials and agglomeration of the particles occurred as the concentration of catalyst was increased. Anions and cations released during *E. coli* inactivation was monitored *i.e.*, sodium, ammonium, potassium, magnesium, calcium, chlorine, sulfate and nitrate, as indicators of damage to bacterial cells. Only ammonium and potassium were formed with varied quantities depending on the photocatalyst used. Ammonium can be formed by the photocatalytic oxidation of amino acids, which composed the protein present in cells membranes. The appearance of K^+ during bacterial inactivation is in agreement with other studies that demonstrated that the loss of K^+ was followed by the loss of cell viability, while other contributions suggested that damage in cell membrane occurs after or with the K^+ loss and this can also cause cell death [56].

The morphology of photocatalyst nanoparticles has been also investigated, as well as the relationship between well-tuned TiO_2 photocatalyst and their exposed facet with the photocatalytic activity against *Fusarium* spores in water. Four TiO_2 morphologies were tested: nanotubes (NT), nanoplates (NPL), nanorods (NR) and nanospheres (NS). The solar photocatalytic properties were compared with the disinfection of spores of *Fusarium solani* in water. The solar inactivation of the resistant spores of *Fusarium* in water was demonstrated to be related to the exposed TiO_2 facets. At very low concentrations of photocatalyst, the inactivation of *F. solani* over TiO_2 nanospheres showed the best disinfection efficiency with respect to the others morphologies. The simultaneous presence of formic acid during photocatalytic disinfection and decontamination showed that the presence of this organic acid strongly retards the disinfection reaction in the case of TiO_2 nanospheres while formic acid degradation occurred simultaneously with *F. solani* inactivation in the case of TiO_2 NT [57].

Other new materials have been developed and investigated for improvement of the solar photocatalytic efficiency. As an example, $Ag-BiVO_4$ composites were synthesized and their photocatalytic disinfection activity was tested against *E. coli* under visible light ($\lambda > 420$ nm) [58]. This work reported that the deposition of silver nanoparticles on the surface of $BiVO_4$ led to a significant improvement of the photocatalytic activity. These composites had a red shift edge on the UV-Vis absorption, which increased with the amount of deposited silver. Photocatalytic inactivation of *E. coli* in the presence of $Ag/BiVO_4$ resulted in the total disinfection of the cells (10^7 to less than 1 CFU/mL within 3 h). This photocatalytic activity was stable in repeated runs under natural sunlight.

They attributed this significant photocatalytic enhancement of Ag/BiVO₄ to the effect of metallic silver nanoparticles. Ag particles can act as an electron sink on the surface of semiconductors (BiVO₄), which prevents the recombination of e^-/h^+ pairs. When the metal and semiconductor are in contact, a Schottky barrier is formed. The Schottky barrier height is equal to the energy barrier that electrons need to overpass to migrate from the metal to the semiconductor. Electrons will naturally migrate from the semiconductor to the metal towards reaching the equilibrium chemical potential. This results in the accumulation of electrons on the metal (negative charges) and excessive positive charges on the semiconductor interface, therefore in an efficient separation of e^-/h^+ pairs during the photocatalytic process. This may explain that Ag nanoparticles on BiVO₄ form a Schottky barrier at the interface and enhance the photoactivity of the semiconductor composite, which promotes the separation of photo-induced e^-/h^+ pairs for the generation of reactive oxygen species. The development of novel photocatalytic materials will be discussed in more detail later in Section 5.

3.2. Solar Reactor Design and Applications

There are several approaches to improving the efficacy of solar disinfection and solar photocatalysis disinfection of water. These enhancements should consider the following aspects:

(i)   Increasing the appropriate solar photons entering the photo-reactor system.
(ii) Improving the efficacy of the treatment against certain resistant water pathogens.
(iii) Increasing the total volume of treated water for a certain treatment time (solar exposure).
(iv) Reducing complexity of the water treatment system and decreasing the user dependency of the process.
(v)   Using low-cost systems based on simple designs and local and cheap materials to construct the water disinfection systems for application in developing countries.
(vi) Design optimized photo-reactor systems that avoid post-treatment recovery or re-growth of microorganisms and prevent recontamination of treated water due to inappropriate handling or storage.

The first research on reactors using compound parabolic collectors (CPC) with TiO₂ solar photocatalysis for water disinfection was reported by Vidal *et al.* [59]. This solar photo-reactor had a 4.5 m² CPC solar collector that maximized the collected solar radiation for long exposure periods (several hours). They showed a 5-log decrease of *E. coli* and *Enterococcus faecalis* in 30 min of solar treatment with suspended TiO₂ (P25; 0.5 g/L) and natural solar irradiation (average UVA irradiance: 25 W/m²). They also reported an economic analysis of this technology for future application not only to solar photocatalytic disinfection, but also to decontamination of organic pollutants. Mcloughlin *et al.* [60] evaluated three geometrical configurations of solar mirrors at lab-scale, solar photo-reactors with aluminum reflectors consisting of compound parabolic, parabolic and V-groove profiles and showed that they enhanced the bactericidal effect of the natural solar radiation using *E. coli* as the model microorganism. The CPC was found to be more efficient than the parabolic or V-groove profiles. They also demonstrated that bacterial photo-inactivation using sunlight alone can be enhanced by low loadings of TiO₂ (P25 at few mg/L) suspended in the water.

Up to now, there are few commercial systems based on solar disinfection or solar photocatalytic disinfection. One example of a commercial system is the Solar Bag available from Puralytics (see Figure 6) [61]. Prototypes and bench scale reactors are continuously being developed by different groups with the aim of developing effective, reliable and low-cost solar reactors for water purification. These contributions are investigating the main factors affecting the disinfection process and performance under real conditions of solar radiation, real contaminated water (drinking water, wastewater, *etc.*), naturally occurring microorganisms, and in the presence of naturally organic and inorganic matter. One of the major challenges in this field is to create a system that disinfect and decontaminate water at the same time. A large part of the research on solar reactors for water disinfection has been carried out using the mere action of solar radiation, *i.e.*, the synergistic effect of solar thermal heating and the detrimental effects of solar UVA (and some small percentage of UVB). Other research is mainly focused in the photocatalytic acceleration of the inactivation of microorganisms. It is important to put together both fields, the concepts of both processes to better accomplish the task of efficient solar reactor systems for water disinfection and decontamination. One of the most critical criteria for good solar photo-reactor performance is the increase of the volume of treated water output in a reasonable solar treatment time. To address this objective one must take into account the following limiting technical aspects.

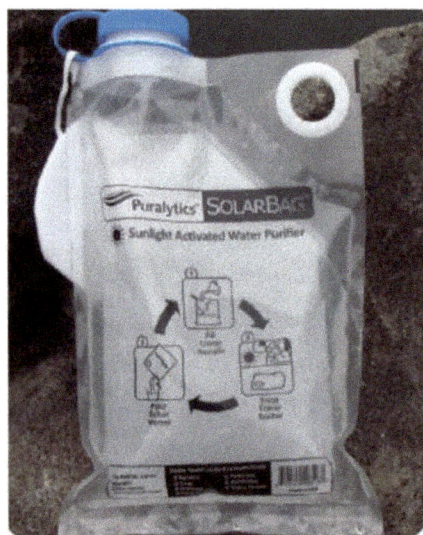

Figure 6. Solar bag commercially available from Puralytics, which utilizes photocatalysis (from Puralytics website [61]).

(i)   The incoming photon-flux inside solar photo-reactor must be as high as possible (using either CPC solar mirrors or other low-cost reflectors, which increase the solar light collection). It should be efficiently and homogeneously distributed in the photo-reactor water, even for large volumes of water. Therefore light scattering and absorption of water samples is a key factor, thus there must be a compromise between the optical properties of the water and the physical path length of the photo-reactor (diameter if it is a tube, or depth if it is a pond or raceway). If the UV-Vis transmission of the water is good enough (turbidity < 10 NTU), the optical reactor path length (*i.e.*, diameter of the tubular photo-reactor) can be increased until 10–20 cm (Figure 7b) [62]. The use of solar mirrors accelerates the disinfection rate, the increase in solar irradiance decreases the treatment time and accelerates the disinfection rate, but not necessarily in a linear proportionality manner [63].

(ii)   When the water does not have good UV-Vis transmission (*i.e.*, when suspended TiO_2 is used or turbid waters are treated), then the reactor tube must be reduced to few centimeters and the large volume requirement may be accomplished by connection of several photo-reactor modules [64].

(iii) The use of immobilized photocatalyst is desirable, mainly for drinking water applications. In this case, the water flow rate must be low enough to increase the residence time required to achieve the desired disinfection in a single pass (the contact time will depend on the nature of the pathogen and the immobilized photocatalyst used) so that the recovery mechanisms of some microorganisms cannot effectively work after the treatment within the residence time. This important limitation has been widely investigated [53,65,66] with the conclusion that the flow rate must be low enough (only a few L/min) to ensure that a minimum lethal dose of energy is delivered before the water circulates to dark regions of the reactor. In the dark regions of the reactor, bacteria may recover and re-grow when the damage produced by the disinfecting agent (either by direct action of solar photons or by hydroxyl radicals) is not strong enough to achieve total inactivation (*i.e.*, complete disinfection) and desired disinfection efficiencies may not be achieved [63,65,66]. Therefore, the design of the reactor must be done with prior knowledge of the lethal solar energy dose required so that complete disinfection is achieved in one run. Obviously, this energy requirement for complete killing depends strongly on several parameters like water composition, turbidity and nature of the biological contamination. A number of examples of the influence of the presence of ions, organic matter and natural occurring bacteria present in wastewater are available in the literature [65,67,68].

(iv)   There are limitations of solar disinfection (without any photocatalyst) when it is scaled-up through the use of large batch volumes or continuous flow recirculation reactors [63]. Increasing the flow rate has a negative effect on inactivation of bacteria, as at a given time point there needs to be maximum exposure of bacteria to UV to ensure inactivation as compared to having bacteria exposed to sub-lethal doses over a long period of time. When the water is kept static under solar light it is constantly illuminated and hence the required uninterrupted UV dose can be easily achieved. With continuous flow systems, the lethal dose can be delivered but in an intermittent manner and complete inactivation is not observed. This statement has important implications for those attempting to scale-up solar systems through the use of pumped, re-circulatory, continuous flow reactors. If the operational parameters are set such that the microbial pathogens are repeatedly exposed to sub-lethal doses of solar radiation followed by a period within which the cells have an opportunity to recover or repair, complete inactivation may not be achieved.

(v) In photocatalytic disinfection, the electron acceptor is normally dissolved oxygen, which is easily available from the air. In the case of static batch systems, the concentration of dissolved oxygen will be rapidly depleted and must be replenished to maintain photocatalytic activity. Furthermore, the solubility of oxygen in water is reduced by temperature. This must be taken into consideration as the temperature within solar irradiated reactors can reach 55 °C. New designs of reactor must address the need for replenishment of dissolved oxygen in photocatalytic disinfection systems. This was investigated in the recent contribution of Garcia-Fernandez *et al.* [67]. These authors investigated the influence of temperature and dissolved oxygen using a solar 60 L-CPC reactor with suspended TiO_2 (0.1 g/L) (Figure 7a). They injected air in the reactor pipeline (160 L/h) and evaluated several controlled (fixed) temperatures (15, 25, 35 and 45 °C) for the photocatalytic inactivation of two models of water pathogen, *E. coli* (vegetative cell model of fecal contamination) and *Fusarium solani* (a spore model of resistant fungi). In this work, they also assessed the effect of the chemical composition of the water and compared real urban wastewater with synthetic ones. They observed that increasing the water temperature, from 15 to 45 °C, had clear benefits on the disinfection rate for both pathogens. They also observed that air injection led to an important enhancement on the inactivation efficiency, which was even stronger for *F. solani* spores, the most resistant pathogen evaluated. The composition of the water matrix significantly affected negatively the efficiency of the photocatalytic treatment, showing a better inactivation rate in simulated urban wastewater effluent than for real urban wastewater effluent.

(vi) An alternative to enhance photocatalytic performance may be to introduce other oxidants e.g., H_2O_2, persulfate radical, *etc.* however, this would give rise to a dependence on consumable chemicals which may be affordable or undesirable depending on the final application of the system.

(vii) Other aspects like reducing the user dependence of the process or making the system as cheap and robust as possible are worthy to consider so that reactors can be used worldwide for solving a number of issues related with water safety, mainly for human consumption purposes in developing countries. The automated SODIS reactor developed by Polo-Lopez *et al.* [66] addressed some practical problems associated with increasing the treated water output using a continuous flow concept. This novel sequential batch photo-reactor was designed and constructed with the aim of decreasing the treatment time, increasing the total volume of water treated per day and reducing user-dependency. The photoreactor incorporated a CPC concentration factor of 1.89 and the treatment time was automatically controlled by an electronic UVA sensor (Figure 7c). The feedback sensor system controlled the gravity-filling of the reactor from an untreated water reservoir, and controlled the discharge of the treated water into a clean reservoir tank following receipt of the pre-defined UVA dose. The reactor was tested using *E. coli* in well water under real sun conditions in Southern Spain. They found that this system would permit processing of six sequential batches of 2.5 L each day (*i.e.*, 15 L of solar purified water each per day). The system is modular, therefore it may be scaled up to allow several CPC photoreactors to be used under the control of a single UVA sensor. For example, six systems like this could produce around 90 L of potable water per day (for several households), and it could produce approximately 31,500 L during a typical year. Of course static systems do not present a good option when using immobilized photocatalyst, as mass transfer limitations will predominate unless a mechanism for forced convection and mixing is introduced.

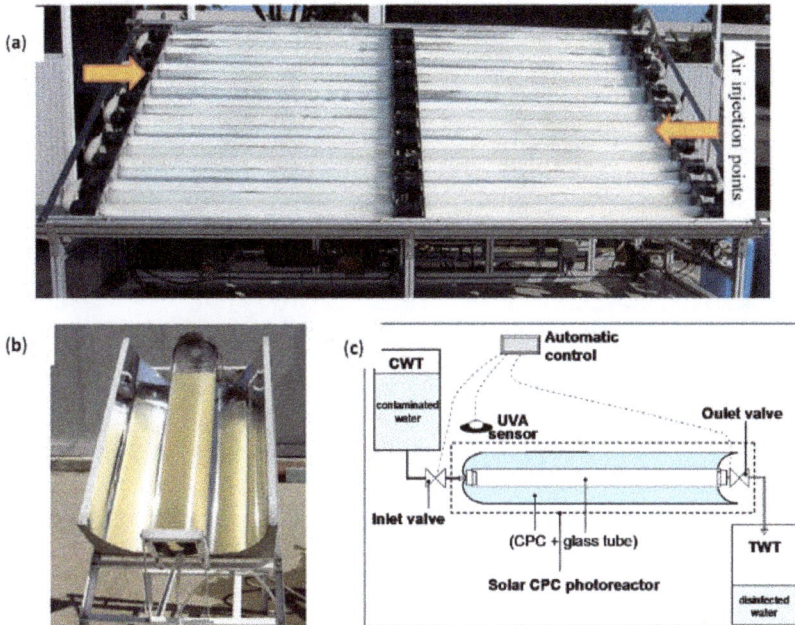

Figure 7. (a) Images of front view of the solar 60 L-CPC reactor at PSA facilities (4.5 m² of collector mirrors) with air injection points indicated; (b) Enhanced SODIS batch reactor filled with 100 NTU turbid water; (c) Schematic of the sequential batch system.

4. Photocatalytic Coatings for Healthcare Applications

There is a need for new and/or complimentary methods of disinfection to combat the on-going problem of health care associated infections (HCAI). A healthcare associated infection is defined as an infection contracted during the hospital stay [69]. The estimated cost of HCAI to the National Health Service (NHS) in the UK is at least £1 billion per year. HCAI's cause added stress and inconvenience to patients, and it has been estimated that around 5000 deaths per year can be attributed to HCAI's [70]. Recent reports suggest that the incidence of certain strains has decreased but infection from other strains is increasing [71]. Healthcare patients are more susceptible to infection as they may be already ill with low immunity, their own endogenous flora can present as opportunistic pathogens, and they are at risk from infection from other pathogens in the healthcare environment which may be transmitted via the air, water or from contact with contaminated persons or surfaces [11]. Although the extent of this risk has not been clearly established, it is known that hospital surfaces and/or medical devices can become contaminated with infectious pathogens, which can potentially play a role in the spread of HCAI. This is further substantiated by reviews, which recommend increased cleaning to reduce incidence of HCAI [72,73].

One of the key factors in HCAI transmission is the ability of some pathogens to remain viable from days to months on surfaces and medical devices, therefore acting as a perpetual source of infection [74] and becoming a reservoir of infection [75]. Routine decontamination and cleaning is essential in the healthcare environments to reduce the risk of transmission of infection, and deep cleaning is essential following an outbreak of infection [76]. It has been shown that shared medical equipment can become contaminated with bio-burden (a population of viable infectious agents contaminating a medical device [77] and this increases the risk of transmission of infection to patients [70]. For example, stethoscopes (which are commonly used) can be a vector if not decontaminated properly [78]. An additional source of HCAI's is transmission from other medical implements, such as contaminated surgical equipment, or invasive medical devices, such as catheters [79,80]. There are a number of factors which determine the role surfaces play in the spread of infectious agents: the longevity of the organisms; the frequency of which the site is touched; and the concentration of pathogens on the surface, *i.e.*, if it is high enough to result in spreading to patients [72]. The most common hospital pathogens are Methicillin-resistant *Staphylococcus aureus* (MRSA), *Clostridium difficile*, *Escherichia coli* (*E. coli*) (and its resistant form extended beta lactamase or ESBL *E. coli*) and Vancomycin-resistant *enterococci* (VRE). Near patient sites should be cleaned and disinfected. These include bed rails, equipment stands and environmental surfaces, *etc.*, common hand touch sites, such as door handles, and all medical appliances. Photocatalytic coatings may provide an additional means of surface disinfection and decontamination to help reduce the incidence of HCAI's.

Photocatalytic TiO_2 coatings and composites have been utilized for commercial application in tiles, paving slabs and self-cleaning glass [81,82]; however, there are few commercial medical applications, although self-cleaning tiles by TOTO are reported to be effective for use in hospitals, yet published data and case studies are not readily found [83]. There are a range of approaches to the formation of TiO_2 coatings on surfaces including sol-gel, chemical vapor deposition (CVD) and physical vapor deposition (PVD). Anatase is usually reported to be the most active crystal phase of titania and formation normally requires elevated temperatures between 225 °C and 550 °C. Therefore coating temperature labile materials with photocatalytically active titania is a challenge.

Some researchers have investigated photocatalytic coatings for disinfection and decontamination of devices and materials including catheters [79], lancets [84], dental adhesives [85] and implants [86]. Other researchers have investigated the use of photocatalytic coatings for environmental surface disinfection [87,88].

4.1. Self-Disinfecting Coatings for Environmental Surfaces

In 2003, Kühn *et al.* [87] reported on the use of UV activated photocatalytic coatings for surface disinfection. As UV activation was required for the TiO_2 coatings, and levels are low under ambient conditions, they used UV sources with light guides and a light-guiding sheet to ensure UV irradiation of the surface. They investigated the inactivation of *E. coli*, *Pseudomonas aeruginosa*, *Staphylococcus aureus* (*S. aureus*), *Enterococcus faecium* and *Candida albicans*. The photocatalytic coating was a thin film of TiO$_2$ deposited on Plexiglass and the surface was irradiated from above and below. Their results indicated a 6-log$_{10}$ reduction in colony forming units (CFU) after 60 min for *P. aeruginosa, S. aureus* and *E. faecium*. For *C. albicans* a 2-log$_{10}$ reduction in CFU over 60 min was observed. *E. coli* and *P. aeruginosa* showed the best responses in this experiment as they had the greatest reduction efficiencies out of all the bacteria. The controls without a photocatalyst (which were subject to UVA light only) showed a bacterial reduction after 60 min for *E. coli* and 10 min for *P. aeruginosa*. Analysis by light microscopy indicated that photocatalytic bacterial inactivation results from direct damage to the cell walls. For surface disinfection, a 3–5-log$_{10}$ reduction in the amount of infectious agents is normally adequate. It is not clear how the UV irradiation could be delivered in a real healthcare environment without risk to staff and patients.

Page *et al.*, carried out research on the antibacterial activity of titania and silver-titania composite films prepared using a sol-gel dip coating method on glass slides [88]. Clinically relevant strains of bacteria were used as the model organisms, including *S. aureus, E. coli* and *B. cereus*. The titania and silver-doped titania nanoparticulate thin films were created using a dip-coating procedure followed by heat treatment at 500 °C. The photocatalytic activity of the films was determined using stearic acid degradation monitored by FTIR. The samples and their controls were initially illuminated using a 254 nm germicidal UV lamp for 30 min to allow time for activation and disinfection of the films. The disinfection experiments were undertaken with irradiation from black-light tubes with main emission at $\lambda = 365$ nm (1.4 mW/cm^2). Irradiation times ranged from 2 to 6 h. The films showed photocatalytic activity for the inactivation of all strains studied, with the Ag-TiO$_2$ films showing better inactivation rates as compared to the TiO$_2$.

The photocatalytic disinfection properties of sputtered TiO$_2$ films were tested by Miron *et al.* [89]. They tested the bactericidal ability of their films against *Diplococcus pneumoniae, Staphylococcus aureus* and *E. coli*. The bacteria were grown and then transferred to the films. After 1 h UV light irradiation (light intensity 1 mW/cm^2) initial bacterial membrane destruction was observed on the sputtered TiO$_2$ films. After 6 h irradiation the bacteria were completely destroyed; however, 6 h irradiation is a significantly long time for UV photocatalytic disinfection. When compared to the uncoated glass substrate, there was no noticeable destruction of bacteria after 6 h.

Dunlop *et al.* [11] investigated the inactivation of clinically relevant pathogens on photocatalytic coatings. A method was developed to assess the disinfection efficiency of photocatalytic surfaces which allowed the determination of pathogen viability as a function of treatment time, the assessment of the surface for viable surface bound organisms following disinfection and measurement of the re-growth potential of inactivated organisms. The developed method was used to investigate the inactivation of extended-spectrum beta-lactamase (ESBL) *Escherichia coli*, methicillin resistant *Staphylococcus aureus* (MRSA), *Pseudomonas aeruginosa* and *Clostridium difficile* spores using immobilized films of titania nanoparticles (*Evonik Aeroxide P25*). A 99.9% reduction in viability (3-log kill) was observed for all bacterial cells within 80 min of photocatalytic treatment (under UVA irradiation). Complete surface inactivation was demonstrated and bacterial re-growth following photocatalytic treatment was not observed. More than 99% inactivation (2.6-log reduction) was observed when the photocatalytic surfaces were contaminated with *C. difficile* spores.

As TiO₂ photocatalysis is only activated by UV photons, this reduces its potential for use in the disinfecting of environmental surfaces under solar or ambient lighting conditions. Therefore, attempts have been made to create visible light active (VLA) photocatalysts by, for example, doping TiO₂ with nonmetals. Wong *et al.* tested the efficiency of carbon doped (C-TiO₂) and nitrogen doped TiO₂ (N- TiO₂) created by ion-assisted electron beam evaporation [90]. It was found that the N-TiO₂ had better bactericidal ability than the C-TiO₂. Using the N-TiO₂ for the bactericidal experiments inactivation of human pathogens *Shigella flexneri*, *Listeria monocytogenes*, *Vibrio parahaemolyticus*, *Staphylococcus aureus*, *Streptococcus pyogenes*, and *Acinetobacter baumannii* as well as laboratory strain *E. coli* were tested. They achieved less than a 1-log₁₀ reduction (90%) in the *E. coli* population using an incandescent light and N-TiO₂, however this was after only 25 min illumination (light intensity 3×10^4 lux) therefore this is a relatively good inactivation time for VLA disinfection. Prior to this, it was found that when testing a range of visible light intensities, this higher value (3×10^4 lux) showed the best inactivation. Inactivation of the human pathogens was less successful as they were only reduced by 50% and the authors conclude they were possibly more resistant to N-TiO₂ mediated treatment due to presence of enzyme systems.

Mitoraj *et al.*, tested visible light activated photocatalysts for use on surfaces in areas that require clean and sterile conditions e.g., a hospital [91]. A variety of pathogens (*Escherichia coli*, *Staphylococcus aureus*, *Enterococcus faecalis*, *Candida albicans*, *Aspergillus niger*) were tested in suspension using C-TiO₂ and TiO₂ modified with platinum (IV) chloride, irradiated with a high-pressure mercury lamp (of varying light intensities 1.8 W/cm² and 1.0 W/cm²). The inactivation of the pathogens follows the same order as listed above. According to the author, this is because the density and complexity of the cell wall increases from the gram-negative *E. coli*, to gram positive *S. aureus* and *E. faecalis* to the *C. albicans* and *A. niger* fungi. The authors also tested the antimicrobial activity of immobilized catalysts, which is more relevant to a self-disinfecting surface. The bactericidal efficiency of an immobilized catalyst is influenced by several factors including the ease with which light and oxygen can access the photocatalyst surface, the distance between pathogens and surface and the ability of the radicals to penetrate the microbial cells. When the C-TiO₂ was immobilized, it was not capable of inactivating either *E. coli* or *S. aureus*. The TiO₂-platinum (IV) chloride suspension was immobilized onto plastic plates and showed fast inactivation of *E. coli*, when irradiated with a halogen lamp of light intensity 0.2 W/cm². After 30 min, 98% of the bacteria were inactivated. In the dark the TiO₂-platinum (IV) chloride was also capable of inactivating the bacteria meaning it has some bactericidal ability of its own, however this was at a slower rate. The reason for the bactericidal ability of the catalyst under irradiation is thought to be due to platinum species becoming desorbed from the catalyst surface and subsequently becoming partially reduced to platinum (II) in the presence reducing agents (such as photogenerated e⁻). Both platinum (II) and platinum (IV) complexes can inhibit bacterial DNA, RNA and protein synthesis. Therefore it was proposed that the fast activity of the photocatalyst under light irradiation was mainly due to the toxicity of the catalyst when irradiated, however the authors do not state if the disinfection process was further enhanced by the photogeneration of ROS (˙OH, *etc.*). The ability of the catalyst to inactivate bacteria in the dark shows it also has inherent toxic qualities. The other microorganisms were not tested using this method; however the results show that this immobilized photocatalyst has potential to become a self-disinfecting surface, however the disinfection affect was probably not due to VLA photocatalysis. It is somewhat surprising that the authors did not test the efficiency of TiO_2 loaded with Pt nano-clusters, as the latter has been reported to show high photocatalytic efficiency in many studies.

Dunnill *et al.* tested the efficiency of $S-TiO_2$ and $N-TiO_2$ for the inactivation of *E. coli* under white light illumination. The films were prepared using atmospheric pressure chemical vapor deposition (APCVD) [92]. The light source used was one common in UK hospitals, a fluorescent lamp which had a luminosity of 5965 lux when measured at 20 cm from the lamp. The TiO_2 thin films were prepared using APCVD using a custom CVD reactor, with a nitrogen or sulfur source being added for the doping. The sulfur or nitrogen precursor was added during the vapor stage to the deposition chamber. The authors found that good interstitial doping of nitrogen was achieved when low quantities were added, therefore giving good photocatalytic activity. Nitrogen incorporation at 0.13 at.% gave the best photocatalytic activity and sulfur incorporation of 0.1 at.% gave the best activity. The absorption edge for the *N*- and *S*-doped samples were found to be 2.9 eV and 3.0 eV, respectively, while the undoped TiO_2 had an optical band gap of 3.2 eV (normal for anatase). The samples were tested for their photocatalytic disinfection properties using *E. coli*. The best efficiency was found on samples pre-irradiated for 24 h as this provides a clean surface before application of bacteria. After bacterial inoculation onto the samples, they were further illuminated with white light for 24 h and the nitrogen-doped sample appeared to perform slightly better than the sulfur-doped sample with kill rates of 99.9% and 99.5% respectively. The undoped TiO_2 and glass only substrates showed little or no kill, showing that the visible light was not able to activate the TiO_2. Although this study reports VLA, it is a very slow process. The indoor lighting with doped catalysts was able to inactivate the bacteria, with the N-doped sample slightly outperforming the *S*-doped sample; however the time taken for disinfection was very long (24 h). Nevertheless, if photocatalytic coatings can provide a residual disinfection mechanism, this may reduce the risk of transmission of HAI via contaminated surfaces.

More recent work by Dunnill *et al.*, investigated the ability of nanoparticulate Ag loaded titania thin films to kill bacteria under hospital lighting conditions [93]. It is well known that Ag has some antimicrobial activity and nano-Ag is already used in existing products such as catheters, ventilator tubing and surfaces. Silver ions are toxic to microbes and since they are able to move from the antimicrobial material into the microorganisms the cells become damaged by the ions. The purpose of the coatings made in their study was to determine if they were visible light active and to exploit the antibacterial effects of both the silver and the photocatalysis using a common hospital light source. The films were fabricated using a sol–gel method to coat glass slides with TiO_2 in the form of anatase. Ag nanoparticles were then loaded onto the surface by UV photo-reduction of Ag onto the TiO_2 from $AgNO_3$ solution to give $Ag-TiO_2$ substrates. The substrates were then annealed at 500 °C. Optical absorbance gave an optical band gap of 2.8 eV. Experiments were carried out to determine photo-induced superhydrophilicity, degradation of stearic acid and the antibacterial effect of the films. *E. coli* and epidemic MRSA (EMRSA-16) were used as the model microorganisms. With the MRSA a significant enhancement in the inactivation was observed with the $Ag-TiO_2$ film under white light irradiation as compared to the TiO_2 film. For *E. coli*, a 99.996% (4.4 log_{10} CFU) decrease in the number of viable bacteria was observed on the Ag-TiO$_2$ films incubated under white light for 6 h, compared with the TiO$_2$ films incubated under the same light conditions for the same time period; However, a similar decrease in the bacterial number was demonstrated with the Ag-TiO$_2$ film in the absence of light indicating that the observed inactivation of *E. coli* was due to the Ag in the dark.

An important consideration is that the rate of photocatalytic disinfection is rather slow and the presence of organic contaminants will compete for ROS. Therefore, photocatalytic coatings for the disinfection of environmental surfaces under ambient lighting conditions should be additional to normal cleaning routines and may provide some residual surface disinfection to reduce the risk of transmission of infection.

4.2. Photocatalytic Coatings for the Disinfection and Decontamination of Medical Devices

Catheters and other medical tubes are extensively used in hospitals to administer medicine or nutrients into arteries and drain fluids or urine from the urethra or digestive organs. Indwelling catheters have the potential to cause infection and around 20% of healthcare acquired infection occurs to the urinary tract, potentially by the use of catheters. Catheters are also an attractive reservoir for microorganisms. To combat this some catheters with antibacterial action have been developed, for example those with silver impregnated in them. However, this is not a long-term solution as silver ions are released from the wall of the material and will eventually wear out. Ohko *et al.* developed TiO$_2$ coated silicone catheters in an effort to exploit the antibacterial and self-cleaning properties of the coating when exposed to UV light [79]. Silicone is one of the most common materials for catheter manufacture; therefore this type of catheter was used. However, because this material has low wettability, it is difficult to coat the TiO$_2$ directly onto it. Therefore the catheter was first dipped in sulfuric acid to improve the overall wettability then dipped into a sol containing titanium dioxide. The existence of the coating on the catheter still allowed it to remain adequately flexible and hard for practical use. The authors found that the coating on the inside of the catheter could be irradiated by low-intensity UV light when transmitted from the outside. Therefore it is expected that the photocatalytic disinfecting qualities could be experienced on the inside as well as the outside of the catheter. The degradation of methylene blue was tested and it was found that on the inside of the tube the methylene blue was bleached, meaning that the UV light was transmitted to the inside of the tube and photocatalysis took place. The bactericidal efficiency was also tested using *E. coli* and it was found after 60 min irradiation the survival ratio had decreased to a negligible amount. The authors state this kind of catheter has potential for use in intermittent catheterization. Dunlop *et al.* studied the efficacy of photocatalytic disinfection to inactivate *Staphyloccocus epidermidis* cells within a biofilm [11]. Following 3 h of UVA irradiation, 96.5% of the biofilm cells on the TiO$_2$ film were determined to be non-viable. Importantly, inactivation of cells throughout the 3–4 µm thick biofilm was observed.

Lancets, used to prick the finger for self-monitoring of blood glucose levels, and other needles must be completely sterile as they enter the body. The use of needles and in particular lancets has increased in recent years due to the rise of diabetes. Two important factors in the development of needles are to create a low piercing resistance and also the need to sterilize the needles by safe and low cost methods. There are some problems with current sterilization methods of lancets, for example the need for specialized equipment for gamma-ray sterilization. Therefore Nakamura *et al.*, assessed whether TiO$_2$ photocatalysis could be a low cost alternative sterilization method and studied the benefits of adding a TiO$_2$ coating to the lancets [84]. In this study the medical grade lancets were coated with a uniform nano-layer of TiO$_2$ by sputter deposition. The antimicrobial activity was assessed using *E. coli* K12. After 45 min of UV irradiation 83% of the bacteria were inactivated. The uncoated control lancet showed 33% inactivation and lancets in the dark showed no inactivation. Un-annealed lancets showed similar results to the uncoated control lancet when subject to UV light, meaning that the un-annealed TiO$_2$ coating was not photoactive and that any killing was probably due to the UV light. The lancets must have a low lancing resistance this to reduce the pain induced when someone pricks their finger. This is important, as diabetics must carry out this task several times a day. The annealed TiO$_2$ film showed low lancing resistance.

Oka *et al.* studied the ability of a titanium dioxide coating on percutaneous implants to inhibit bacterial colonization [86]. Metal pins are widely used for the application of skeletal traction or for external fixation devices in the repair of orthopedic fractures. These pins cannot be completely sterile as they are not isolated from the environment but rather are a link from outside of the skin, which is usually colonized with bacteria, to the bone, which is not normally colonized [94]. A considerable number of fracture fixations, either external of percutaneous, can become infected. Pin tract infection is a considerable complication and is normally combated with antibiotics. However, due to the increasing resistance of certain pathogens, like MRSA, antibiotics may not prove effective for inactivating bacteria. If bacteria is not inhibited and infection around the pin occurs this can cause pin loosening, pin removal or chronic osteomyelitis. If the fixation is changed from external to internal there will be a risk of deep infection, because the pin will penetrate the skin barrier. Another problem of implant-related infection is the risk of biofilm formation. Biofilm is a complex build-up of many layers of bacteria, which surround themselves in a protective exopolymeric fluid. Biofilms have an advantage over normal free-living bacteria in that they are extremely resistant to treatment from antibiotics [75]. To avoid the formation of biofilm and pin tract infection in general, bacteria must not be allowed to adhere and colonize on the implant. The authors therefore created a TiO$_2$ photocatalyst, which was processed from a pure titanium plate via direct oxidation. MRSA was used as the test pathogen, as it is one frequently associated with percutaneous tract infection. After 60 min UVA irradiation of the photocatalyst the survival rate was negligible. The authors also acknowledge that there were several limitations to this study including the way they applied a large inoculation of bacteria directly to the implant is not how bacterial colonization occurs if the infection occurred clinically. Also, UVA irradiation cannot reach the subcutaneous part of the pin. An animal model was also used to test colonization of bacteria onto the implant and it was found that fewer bacteria colonized the photocatalyst surface than the pure titanium surface. Hydrophilicity is another important factor of TiO$_2$ photocatalysis. Oxygen vacancies are created when the photogenerated holes react with superoxide anion and oxygen atoms are ejected. The oxygen vacancies then become occupied with water molecules, which make the surface hydrophilic. It is thought a hydrophilic surface deters bacteria from adhering to it therefore the hydrophilicity combined with the self-cleaning effect of the TiO$_2$ photocatalyst, which would inactivate and remove bacteria attached, would provide an effective and functional coating. Tsuang *et al.* also studied the use of TiO_2 photocatalytic coatings on percutaneous implants [94]. The coating was applied to stainless steel plate by a sol-gel dip coating method. The authors found a significant reduction in the amount of *E. coli* colonies above the TiO_2 coated metal plate. Other pathogens were tested in suspension, however they would not provide useful results whenever the TiO_2 coating would be immobilized onto a metal pin.

Shiraishi *et al.* researched the bactericidal ability of TiO_2 coated metal implants against *S. aureus*, which is associated with many surgical site infections (SSI) [95]. SSI are normally a result of contamination during surgery. Consequently, there is a need to develop methods to reduce bacterial infection linked to implants. Pure titanium and medical grade stainless steel (SU316) sample disks were synthesized with TiO_2 using a plasma source ion implantation (PSII) system. The authors state that this method is very feasible for creating TiO_2 coated implants as it can deposit the films completely and uniformly over three dimensional surfaces and there is little other existing evidence of the bacterial ability of PSII deposited-TiO_2 metal implants. The photocatalytic activity of the TiO_2 coated stainless steel and titanium as well as the control was assessed via the degradation of methylene blue. It was found there was no degradation on the control disk irradiated with UVA (peak wavelength 352 nm/light intensity 2.0 mW/cm^2) or the dark control TiO_2 coated disks that were not subject to UVA. Degradation was only observed on disks with UVA irradiated TiO_2. The light control (UVA irradiation with no photocatalyst) showed a gradual decline in the viability of the bacteria. The most appreciable results were seen on the TiO_2 coated discs under UVA irradiation. TiO_2 films on titanium showed a clear reduction in cell viability and complete kill after 90 min irradiation. TiO_2 on stainless steel showed a quicker reduction in cell viability than that of the titanium supporting TiO_2 and complete kill was achieved after 60 min. The authors conclude that TiO_2 deposited by PSI could have the potential to reduce the occurrence of SSIs with regards to medical implants and the photocatalytic bactericidal ability to inactive *S. aureus* has many potential benefits for sterilizing contaminated surfaces of bio-implants.

Medical polymers are also at risk from contamination. Therefore improving the antimicrobial properties of polymers is important. Polymethyl methacrylate (PMMA) is commonly used in the manufacture of medical implants such as in the fabrication of ophthalmic intraocular lenses (IOL) as well as dentures and bone cement. Because it is optically transparent it can be used for replacement intraocular lenses in the eye when the original lens is removed in the treatment of cataracts [96] and also for tonometer tips, which are used to measure intraocular eye pressure [97]. This material is useful because of its good mechanical properties, mouldability and use for ophthalmic rehabilitation. The existing properties of PMMA could further be enhanced by the addition of antibacterial TiO_2 coating that had high transmittance in the visible region and prove useful for ophthalmic applications. Polymers have proven difficult to coat with TiO_2 however, as evaporated inorganic coatings do not adhere successfully to polymers. Another obstacle is the need for polymer surfaces to be kept at a relatively low temperature, as they are thermally sensitive. Su *et al.*, created translucent TiO_2 films onto PMMA using a sol-gel dip coating method [98]. Prior to this the PMMA surfaces were activated by pre-treating with low pressure DC glow discharge plasma. After coating, the PMMA was dried in an oven for 30 min at 60 °C. XRD analysis found that anatase TiO_2 was present. The bactericidal ability properties of the TiO_2 coated PMMA was tested using *E. coli* BL21 and *S. aureus* with what the authors describe as "indoor natural light" (average intensity at 365 nm of 143.9 $\mu W/cm^2$ and at 297 nm of 6.7 $\mu W/cm^2$ during daytime antibacterial test hours). After 15 min of illumination, 50% of the *E. coli* BL21 had been inactivated. After 2 h, none of the bacteria were detected on the agar plate. The controls: PMMA in the dark, PMMA with light and TiO_2/PMMA in the dark did not show any reduction in the number of bacteria. The authors also found that after bacterial adhesion tests, the amount of adherent *S. aureus* was decreased by 89%–92% and adherent *E. coli* was decreased by 96%–98%. The PMMA alone was not capable of decreasing the amount of adherent bacteria meaning that the induced surface hydrophilicity of the TiO_2 may have given it better anti-adhesion properties. The authors conclude that their research created successful translucent TiO_2 coatings onto PMMA that had efficient antibacterial and anti-adhesion properties. The translucence may limit the applications of these types of coatings.

Suketa *et al.* were one of the first groups to report on the use of photocatalytic coatings for dental implants [99]. Oral implants normally have better clinical success when the surface of the implant is altered to improve the integration between bone and implant. One of these alterations is to increase the surface roughness; however an increased surface roughness is thought to increase the risk of a bacterial infection, which can lead to a build-up of plaque. The presence of plaque on a surface for a long period of time can result in peri-implantitis which affects both soft and hard tissues around the osseointegrated implants resulting in the development of implant pockets and loss of supporting bone. Therefore to compromise between the need for a rough surface and fact that rough surfaces potentially harbor more bacteria than smooth ones, an alternative sterilizing method is needed to clean the rough surface and/or remove the formation of plaque biofilm. The authors suggest that TiO_2 photocatalyst coating would be advantageous for two reasons, the photocatalyst normally is more efficient with a high surface area and an implant with a roughened surface will have a large surface area. *Actinobacillus actinomycetemcomitans* and *Fusobacterium nucleatum* were the test pathogens, which are thought to play a key role in the causation of peri-implantitis. The photocatalyst coatings were created using PSII. It was found that the surface roughness of the anatase and control aluminum oxide were not significantly different. The bacterial tests showed that the TiO_2 disk subject to UVA illumination inactivated 100% of the bacteria after 120 min. There was also a reduction in colony forming units with light control sample; however it was much less than that of the photocatalytic test. A TiO_2 surface should not prevent the osseointegration of implant and bone, and it has been shown that TiO_2 surfaces created by PSII did not prevent osseointegration in a rabbit. The photocatalytic surface prepared in this study was shown to have good surface roughness therefore it should be able to attach successfully to bone and not prevent osseointegration. After implantation of the implant, flap elevation would be needed to sterilize the contaminated surface because the UVA radiation would not be able penetrate thick human tissues. The authors highlight some challenges that need to be addressed including; 120 min of UVA is too long for clinical use, and that more intense UV would damage human tissue. Furthermore, bacterial tests would be needed on the other pathogens associated with peri-implantitis. Nonetheless their research has shown the potential for TiO_2 photocatalysis in this kind of application.

Another type of dental application investigated was the use of TiO_2 in dental adhesives for restoration procedures and other purposes [85]. Many dental restorations are unsuccessful and a replacement is needed due to the onset of dental caries. The adhesives may provide a site that allows bacteria to attach onto and colonize; therefore the development of materials that have bactericidal ability at the tooth-composite boundary may be beneficial in combating dental caries. Hence there is a potential for the use of TiO_2 coatings. TiO_2 can increase bioactivity, in this case the interfacial bonding of a material to bone tissue by means of formation of biologically active hydroxyapatite (HA) [85]. Bioactivity is important as it can improve the bond between tooth and material. Many dental adhesives are not bioactive therefore adding a photocatalytically active TiO_2 may also prove useful as a bioactive layer. The dental adhesives in this study were fabricated to include TiO_2 nanoparticles and the bactericidal ability and bioactivity was studied. The commercially available dental adhesive material was mixed with the P25 nanoparticles containing different weights. *Staphylococcus epidermidis*, a gram-positive species, was used as the test pathogen as it is normally part of the skin flora, forms biofilms and is also resistant to antibiotics. The different light sources were used: a low intensity UV light of 1.2 mW/cm^2 and a strong UV light, used with a filter to block wavelengths below 320 nm, to produce a UVA intensity of 7.5 mW/cm^2. The low intensity light was used for 30 or 120 min whereas the high intensity light was used for 7 or 10 min. The nanoparticle adhesives (NP adhesives) containing 10% and 20% P25 subject to the low intensity irradiation for 120 min both showed significant reduction in the amount of colony forming units (CFU) after 18 h incubation. After 30 min low intensity irradiation and 12 h of incubation there was no regrowth of bacteria. However for the same conditions and 18 h incubation the bacterial regrowth was extensive showing that the bacteria must have experienced non-fatal damage during the 30 min, which only delayed or slowed the regrowth. The control glass plate subject to low intensity light showed no bactericidal effect. Using high intensity UV light, efficient bacterial elimination was observed on 20% NP adhesive after 10 min irradiation. Due to the negative impact of UV radiation on tissues within the body and in particular the gums, the intensity should be kept as low as possible and also localized to just the tooth area possibly by using a light guiding tube. It is likely that the adhesive would be sterilized once during or after the curing process by administering the UV light. As the adhesive would then be covered by a composite it would be difficult for the light to reach the photocatalytic coating therefore reducing the antimicrobial effect. For a clinically feasible irradiation time, which is less than 60 s [85], a high intensity would be needed. This could be achieved by using a regular dental curing light and therefore result in a superior antibacterial effect. The coatings also proved to be bioactive, after a week in simulated body fluid the NP adhesives showed build-up of HA crystals, which is typical of bioactive behavior.

Creutzfeldt-Jakob disease (CJD) is a rare and fatal condition that affects the brain. CJD appears to be caused by an abnormal infectious protein called a prion. These prions accumulate at high levels in the brain and cause irreversible damage to nerve cells. While the abnormal prions are technically infectious, they are very different to viruses and bacteria. Prions are not destroyed by the extremes of heat and radiation used to kill bacteria and viruses, and antibiotics or antiviral medicines have no effect on them [100]. Sporadic CJD is the most common type of CJD although it is still very rare, affecting only one to two in every million people each year in the UK. There were 104 recorded deaths from sporadic CJD in the UK during 2013. Variant CJD is likely to be caused by consuming meat from a cow that has been infected with a similar prion disease called bovine spongiform encephalopathy (BSE, "mad cow disease"). There have been 177 recorded cases of variant CJD in the UK to date and there was one recorded death from the condition in the UK during 2013. There were eight deaths from familial CJD and similar inherited prion diseases in the UK during 2013. Iatrogenic CJD can occur if instruments used during brain surgery on a person with CJD are not properly cleaned between each surgical procedure before re-use on another person. There was one death from iatrogenic CJD in the UK during 2013 [101]. Paspaltsis *et al.*, reported on the inactivation of prion protein using titanium dioxide photocatalysis in 2006 [102]. They used P25 in suspension under UVA irradiation to treat solutions containing recombinant prion proteins PrP (normal isoform) and PrP^{SC} (scrapie, abnormal isoform), including suspensions of inoculated brain homogenate. The analysis of the protein was undertaken using electrophoresis and immunoblotting. Experiments were also carried out to determine infectivity using an animal model. TiO$_2$ photocatalysis under UVA irradiation and the addition of H$_2$O$_2$ showed complete degradation of PrP^{SC} after 12 h, as determined by Western Blots. The *in vivo* assessment of infectivity of inoculated brain homogenates indicated that the infectivity titre was substantially reduced following photocatalytic treatment.

Ahmed, Byrne and Keyes reported on the degradation of β-amyloid peptides on the surface of Ag-TiO$_2$ films [103]. The degradation of β-sheet peptides is relevant as the abnormal conformation of the prion protein is mainly β-sheet. TiO$_2$ films were prepared on stainless steel using rf magnetron sputtering and the films were then modified by the photocatalytic reduction of Ag$^+$ from solution. The adsorption of β-amyloid (1–42), photolytic and photocatalytic degradation of the peptide was studied using Raman spectroscopy, X-ray photoelectron spectroscopy (XPS), and atomic force microscopy (AFM). The Ag-TiO$_2$ films were mainly anatase and the Ag was predominantly Ag (>0%). Ag loading of the TiO$_2$ markedly enhanced the Raman signal (*ca.* 15-fold), but caused significant changes to the protein spectrum indicating non-specific binding of β-amyloid side chain residues to the silver. The amide modes remained well-resolved and were used to estimate the conformational change induced in the protein by the silver. Raman analysis showed an increase in the intensity of the band at ~1665 cm^{-1} assigned to the disordered conformation of the β-amyloid, suggesting that the adsorption at the silver sites induced conformational changes in the peptide. Contaminated surfaces were exposed to UVB irradiation and further conformational changes in the β-amyloid were observed which mildly inhibited amyloid fibril formation.

Table 1 summarizes some of the papers reviewed, the photocatalyst used, the model microorganism used, and the proposed application.

5. Novel Photocatalytic Materials

Although the bulk of photocatalytic disinfection concerns TiO$_2$ as the photocatalyst, other materials have been investigated [104]. ZnO is the second most commonly investigated photocatalytic material after TiO$_2$ and a number of ZnO based compounds have been reported for photocatalytic disinfection. For instance, ZnO nanoparticles [105] and nanorod films [106] have been investigated for the inactivation of *E. coli* disinfection under UV irradiation (ZnO is also a wide band semiconductor). The main goal of novel material development has been to produce visible light active materials. Two slightly different strategies have been employed; the development of narrower band gap materials and modification/sensitizing of existing wider band gap materials.

Principally materials with visible light activity should benefit from increased efficiency under solar irradiation due to the greater number of photons and improved activity under indoor or ambient lighting conditions.

Table 1. Showing application, type of photocatalyst, microorganism tested and advantages/disadvantages.

Proposed Application	Photocatalyst	Microorganisms Tested	Reference
Environmental surfaces and medical devices	Evonik Aeroxide P25 immobilized on glass slides	extended-spectrum beta-lactamase (ESBL) *Escherichia coli*, methicillin resistant *Staphylococcus aureus* (MRSA), *Pseudomonas aeruginosa* and *Clostridium difficile* spores	[11]
General environmental cleaning	TiO$_2$ (P25) on glass	*E. coli, P. aeruginosa, S. aureus, E. faecium, C.albicans*	[87]
Coating for hard surfaces in hospital environment	TiO$_2$ and Ag-TiO$_2$ on glass slides via sol-gel dip coating method	*E. coli, S. aureus, B. cereus*	[88]
Self-disinfecting catheters	TiO$_2$ dip coated onto silicone catheters	*E. coli*	[79]
Lancet	TiO$_2$ layer created by sputter depositions	*E. coli*	[84]
Percutaneous implant	Direct oxidation of pure Titanium plates	MRSA	[86]
Metal pin for external/percutaneous fixation	TiO$_2$ coated on stainless steel via sol-gel dip coating	*E. coli*	[93]
Metal implant	Stainless steel and titanium coated with TiO$_2$ via PSII	*S. aureus*	[94]
Intraocular lenses	PMMA plasma pre-treated and dip-coated with TiO$_2$	*E. coli* *S. aureus*	[97]
Dental implant	Titanium disc coated via PSII	*A. actinomyce-temcomitans F. nucleatum*	[98]
Dental adhesives	TiO$_2$ nanoparticles mixed with commercially available dental adhesives	*S. epidermidis*	[85]
Antibacterial surface	TiO$_2$ films sputter deposited onto silicon wafers	*Diplococcus pneumoniae, S. aureus, E. coli*	[89]
VLA coatings for environmental surfaces	N-doped TiO$_2$ created by ion-assisted electron beam evaporation	*S. flexneri, L. monocytogenes, V. para-haemolyticus, S. aureus, S. pyogenes, A. baumannii, E. coli*	[90]
Self-disinfecting hospital surfaces	C-TiO$_2$ and TiO$_2$ modified with platinum (IV) chloride	*E. coli, S. aureus, E. faecalis, C. albicans, A. niger*	[91]
Antimicrobial healthcare surfaces	Sulfur and nitrogen doped titanium dioxide composites create via APCVD	*E. coli*	[92]
Antimicrobial healthcare surfaces	Ag loaded TiO$_2$ films created via sol-gel	*E. coli, EMRSA*	[93]

5.1. Narrow Band Gap Materials for Visible Light Activity

The majority of new visible light active photocatalysis research is focused on water splitting/hydrogen evolution reactions. However the reaction path from water to hydrogen is complex involving a number of ROS intermediates, making many of these materials capable of disinfection through ROS attack and other mechanisms. Therefore, the literature concerned with water splitting literature is rich with possible new photocatalysts for disinfection applications.

The path to development of visible light active photocatalyst materials started with testing of simple binary materials consisting of two elements such as WO_3 Fe_2O_3, *etc.* These materials can produce reactive oxygen species under visible irradiation [107] and have been reported to elicit disinfection [108,109]. However, many of the narrow band gap photocorrode under irradiation [110,111]. Narrow band gap (1.8–2.5 eV) materials include Fe_2O_3, SiC CdS, and Cu_2O. Wider band gap (2.6–2.8 eV) materials such as WO_3 and MoO_6 are more stable the small change in band gap for these materials, compared to TiO_2 (3.0–3.2 eV), results in limited improvements in efficiency for visible light activity [112]. Figure 8 shows the band gap and band edge potentials for different semiconductor materials, which have been investigated for water splitting.

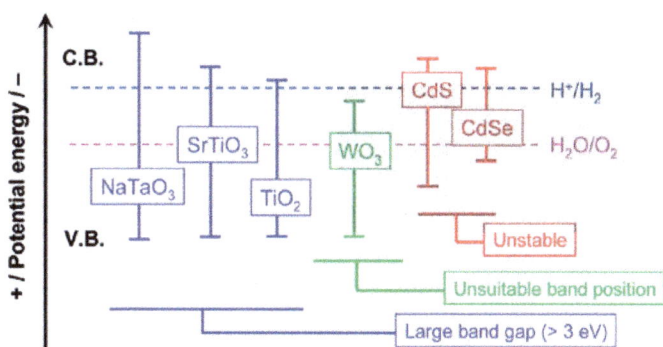

Figure 8. Band gap energies and band edge potentials for different photocatalytic materials with respect to the water splitting couples (with permission from Maeda, K.; Domen, K. New Non-Oxide Photocatalysts Designed for Overall Water Splitting under Visible Light. *J. Phys. Chem. C* **2007**, *111*, 7851–7861 [113]).

Metal carbide, nitride and sulfide materials generally have narrower band gaps than oxide materials but they tend to photocorrode in aqueous media. Some non-metal oxide materials have been investigated e.g., $ZnIn_2S_4$ was reported to show visible light photocatalytic activity for the inactivation of *E. coli* inactivation (under electrochemical bias) [114].

The search for stable visible light active materials photocatalyst materials have been pursued from both a design inspired and combinatorial chemistry approaches. Using combinational chemistry, a large number of potential materials with varying ratios of elements can be rapidly generated and, although assessment of their photocatalytic activity may require a lot of effort [115]. Design-based material developments concentrating on specific photocatalyst properties have been useful in guiding strategies for further material development.

Some visible light active photocatalyst materials have been designed using transition metal elements known to absorb strongly in the visible region due to d-d transitions. Iron centered catalysts prepared using combinatorial routes have been particularly popular [115]. New visible light active materials including $CuFeO_2$ [116], $CuY_xFe_{2-x}O_4$ [117], $LaFeO_3$ [118] have been reported as photocatalysts. Strontium titanate ferrite has been reported to have antibacterial activity under visible light irradiation [119]. Other transition metal ion centered photocatalyst materials reported include: manganese centered $CuMn_2O_4$, and $ZnMn_2O_4$ [120], Chromium centered $BaCr_2O_4$ [121] and $SrCrO_4$ [122], Cobalt centered $NiCo_2O_4$ [123], $CuCo_2O_4$ [117], $LaCoO_3$ and $La0.9Sr0.1CoO_3$ the latter of which demonstrated photocatalytic disinfection against *E. coli* [124].

Novel photocatalysts based on vanadium centers are an interesting group of photocatalyst materials as vanadium oxides generally have narrow band gaps < 1 eV. However, when nanostructured to isolate a body centered cubic (BCC) phase this material shows an optical band-gap ~2.7 eV, resulting in photocatalyst, with high quantum efficiency [125]. Other vanadate photocatalysts combining additional elements have been reported like $InVO_4$ [126], Vanadium $BiVO_4$ [127,128] wherein the BCC structure is maintained. Interestingly $BiVO_4$ supports a number of crystal phases but only the monoclinic phase shows photocatalytic activity [129]. Bismuth vanadate photocataysts have also been reported for disinfection [130,131] with further enhancements in rate of disinfection reported with silver as a co-catalyst [58].

Other design led photocatalyst developments have concentrated on photocatalyst stability. Within the binary materials discussed above there is a general rule that occupancy of the d orbital determines stability with d^0 and d^{10} materials remaining stable under reaction whilst all other materials exhibit photocorrosion [113]. This explains the difference in stability between Fe_2O_3 or ZnO (d^5 and d^8) and TiO_2 or WO_3 (d^0) photocatalysts. However, the consequence of no d-orbital electron transitions often means photon excitation energies are generally large and in the UV domain and oxides of d^0 or d^{10} elements have band gaps too large for efficient use of the solar spectrum [113]. To overcome this limitation oxynitride compounds of d^0 materials, titanium oxynitrides and niobium oxynitrides, were developed. These materials have shown activity for disinfection with PdO nanoparticle modified TiON nanofibers reported for *E. coli* inactivation under visible light [132].

Two metal-layered oxides in perovskite structures have also been investigated [133]. Initial studies using d^0 parents titanate and tantalite perovskites with a site occupancy of alkali metals produced stable catalysts, like $SrTiO_3$, albeit with wider band gaps [134]. For a full review of the crystal structures (double, ternary, layered, *etc.* perovskites) band gaps produced from d^0 metals with alkali metals and other light metals, the reader is referred to the paper by Eng *et al.* [135]. In the latter work the band gaps were measured in d^0 compounds and it was found that in all crystal structures the band gap was largely determined by the central d^0 metal ion. This considered, some researchers are utilizing d^0 perovskite as photocatalysts under visible irradiation. Solid solution perovskites of d^0 metals such as $CeCo_{0.05}Ti_{0.95}O_{3.97}$ [136] or d^0 metal compounds incorporating heavier ions such as Bi, Ag, In or Sn to produce $AgNbO_3$ [137] $NiNb_2O_6$ $NiTa_2O_6$ [138] and other analogues have been reported as visible light activated photocatalysts. With the niobate materials such as $Bi_2O_2CO_3/Bi_3NbO_7$ composites, $K_4Nb_6O_{17}$ and Ag/Cu modified $K_4Nb_6O_{17}$ reported for the inactivation of *E. coli* [139,140].

Organic semiconductors unbound to a surface represent homogeneous photocatalysts and are not the topic of this review and are not be used for disinfection due to issues associated with their recovery. However low dimensional organic semiconducutors like graphene oxide represent a new class of photocatalyst that can be separated from solution due to their large size in two dimensions.

Graphene oxide, C_3N_4 and low dimensional photocatalysts have been reported for disinfection [141]. Due to the narrow band gaps they are normally supported on other materials as sensitizers to form semiconductor stacks a further new class of materials, produced by combining two existing photocatalysts.

Oxide semiconductor sensitized materials will be further discussed below. For example, tri-layer metal-free heterojunction photocatalysts based on RGO, α-S$_8$ and CN sheets have been reported. Layered in different orders, namely CN-RGO-S$_8$ and RGO-CN-S$_8$, have been fabricated. Both of the photocatalytic structures have demonstrated bactericidal effect towards *E. coli* under visible-light irradiation with CN-RGO-S$_8$ having higher inactivation rate than RGO-CN-S8 in aerobic conditions and *vice versa* [142]. Anatase TiO_2 coated multiwalled carbon nanotubes (MWNT) demonstrated higher photocatalytic activity as compared to P25. TiO_2 coated MWNT can inactivate *B. cereus* two times faster than P25. However, the same efficiency was not observed for *E. coli*, which could be due to the steric hindrance provided by the appendages of *E. coli* towards the 1-D nanotubes [143].

5.2. Modification and Sensitization of Existing Materials

Photocatalyst materials have been adapted for visible light activated disinfection by dye sensitized [144], plasmon sensitized [145] and semiconductor sensitized routes—*vide infra*. The mechanism of ROS generation in these materials is generally driven through oxygen reduction reactions by electron injected into the conduction band of the host semiconductor.

The addition of a second semiconductor material to an existing photocatalyst can provide a number of functions some simultaneously including; promoting charge carrier separation [146], acting as a charge carrier sink [147], acting as a co-catalyst to promote a specific reaction [148], and sensitization.

When a second semiconductor is used as a surface sensitizer the second material can be considered as an independent photocenter. In the example of *E. coli* disinfection [149] with CdS loaded TiO_2, CdS can be excited independently of the parent photocatalyst injecting electrons in a manner similar to that of a dye sensitization. Surface sensitizers themselves are often nanostructured to alter their band-gap with the aim of increasing activity [150]. Some groups have extended this philosophy to consider molecular clusters of a few atoms as sensitizers [151]. These surface sensitizers often act in a manner similar to the action of the parent semiconductor with narrower band gap materials producing states above the valence band that can be observed via XPS while larger band gap materials did not produce VLA [152]. Another consideration when joining semiconductor materials and creating semiconductor junctions is the effect on the parent material. There is evidence that strongly polarizing species can affect the lattice structure raising the valence band in some materials [153].

5.3. Modification by Doping (Including Surface Loading)

Doping of photocatalysts to produce a visible light response is one of the most commonly reported approaches to improving the visible light activity. Matsunaga *et al.*, in 1985, provided one of the earliest accounts of doped TiO_2 for photoelectrochemical sterilization of microbial cells. Pt loaded

TiO$_2$ powders were demonstrated to inactivate *L. acidophilus*, *S. cerevisiae* and *E. coli* within 2 h [9]. Doping of ZnO nanostructures have been demonstrated to enhance the inactivation process either by shifting the band towards visible side or by dopant acting as co-catalyst. A number of dopants, including Pd [154], Ce [155], Co [156] and Ag [157], have been demonstrated to enhance the bacterial inactivation. C. Karunakaran *et al.*, have investigated the effect of preparation method on photocatalytic inactivation activity of ZnO and Ag doped ZnO materials towards *E. coli*. The materials have been synthesized by three methods *i.e.*, sol-gel [157] combustion [158] and microwave synthesis [159] out of which sol-gel synthesized materials have demonstrated the highest photocatalytic disinfection efficiency.

Metal ion dopants although the most studied for visible light activity there is a lack of consensus over efficacy, with as many reports claiming enhanced activity as a reduction in activity [160]. TiO$_2$ doped with a variety of elements have been investigated for photocatalytic disinfection. Vohra *et al.*, have explored Ag$^+$ doped P25 TiO$_2$ for disinfection of indoor air. *B. cereus*, *S. aureus*, *E.coli*, *A. niger*, and MS2 bacteriophage have been successfully inactivated, demonstrating high disinfection efficiency [161]. Ag doped TiO$_2$ has also been effective in water disinfection [162]. Cu and S doped TiO$_2$ nanoparticles have been effectively utilized for inactivation of *E. coli* and *M. lylae*, respectively [163,164]. Non-metals C, N, S, B, and the halogens as dopants for TiO$_2$ and ZnO have been reviewed by Rehman *et al.* [165] and also by Im *et al.* [166]. Often co-doped materials are reported to have higher rates than the single dopant regimes as demonstrated by Li and co-workers, comparing nitrogen and carbon nitrogen co-doping on the inactivation rate for *E. coli* [167]. Second generation photocatalysts like WO$_3$ [168] can also be doped to improve visible light activity, and present a further approach to developing new materials for disinfection [169].

6. Conclusions

In semiconductor photocatalysis, the primary reactions are electrochemical oxidation or reduction reactions involving hole and electron transfer from the photo-excited semiconductor. These redox reactions, in the presence of water and oxygen, can result in the production of reactive oxygen species (ROS), which can attack and inactivate microorganisms. The ROS include the hydroxyl radical (HO$^{\bullet}$), which has been suggested to be the primary species responsible for microorganism inactivation, however superoxide radical anion (O$_2^{\bullet-}$), hydroperoxyl radical (HO$_2^{\bullet}$) and hydrogen peroxide (H$_2$O$_2$) have been shown to contribute to the biocidal process. The ROS attack indiscriminately and therefore, emergence of antimicrobial resistance to photocatalysis is unlikely; however, a photocatalytic treatment must be adequate to avoid repair and regrowth of target organisms.

Around 750 million people are without access to an improved source for drinking and many more rely on sources that are unsafe. The development of a simple, yet inexpensive, water disinfection technology might help address the risk of waterborne disease in developing regions. Solar disinfection is recognized as an appropriate house-hold based treatment intervention and photocatalysis may be applied to enhance the solar disinfection efficiency; however, there are several challenges to be addressed before photocatalysis can be cheaply and efficiently deployed in developing regions.

In developing regions there is much concern about the risks of healthcare associated infections, which can result in death, increased bed stay, increased patient stress, and increased costs to the health service providers. One approach to reduce the risk of transmission of HCAI's is to use "self-disinfecting" coatings on environmental surfaces within the healthcare setting e.g., bed rails, table tops, door handles. Photocatalytic coatings may be suitable for some surfaces; however, the coatings must be active for the inactivation of pathogenic microorganisms under ambient or indoor lighting conditions, and this poses a major challenge for the photocatalysis community. An important consideration is that the rate of photocatalytic disinfection is rather slow and the presence of organic contaminants will compete for ROS. Therefore, photocatalytic coatings for the disinfection of environmental surfaces under ambient lighting conditions should be additional to normal cleaning routines and may provide some residual surface disinfection to reduce the risk of transmission of infection. Another approach to reduce the incidence of HCAIs is to use photocatalytic coatings in the decontamination of medical devices including catheters, diagnostic tools and surgical instruments. The number of publications in the area of photocatalysis to address HCAIs is small but growing, and this is an important opportunity for researchers in the field. To assist in the development of efficient photocatalytic technology for solar water disinfection and the disinfection of surfaces in healthcare environments, novel materials are being explored which may be able to utilize visible light. Many of these materials have been investigated for other applications, mainly solar energy harvesting, but only a small number of these novel materials have been tested for their disinfection properties. Photocatalytic disinfection is a rapidly growing, challenging, and multi-disciplinary field, requiring the collaboration of researchers in microbiology and the physical sciences.

Acknowledgments

The authors wish to acknowledge financial support from the following; Department of Employment and Learning Northern Ireland (DELNI), Science Foundation Ireland (SFI) and the National Science Foundation (NSF) for funding under the US-Ireland R&D Partnership (NSF-CBET award No.1438721); financial support provided by the Transnational Access to Research Infrastructures within the European project SFERA II (Grant Agreement No. 312.643) under the 7th Framework Program, and Spanish Ministry of Economy and Competitiveness for the financial support within the AQUASUN project (ref. CTM2011-29143-C03-03); Ulster University for funding student PKS, and DELNI for funding student ASMV.

Conflicts of Interest

The authors declare no conflict of interest.

References

1. IUPAC. Compendium of Chemical Terminology. In *The Gold Book*, 2nd ed.; McNaught, A.D., Wilkinson, A., Eds.; Blackwell Scientific Publications: Oxford, UK, 1997.
2. Mills, A.; Le Hunte, S. An overview of semiconductor photocatalysis. *J. Photochem. Photobiol. A* **1997**, *108*, 1–35.

3. Tryk, D.A.; Fujishima, A.; Honda, K. Recent topics in photoelectrochemistry: Achievements and future prospects. *Electrochim. Acta* **2000**, *45*, 2363–2376.

4. Bahnemann, D. Photocatalytic water treatment: Solar energy applications. *Sol. Energy* **2004**, *77*, 445–459.

5. Agrios, A.G.; Pichat, P. State of the art and perspectives on materials and applications of photocatalysis over TiO2. *J. Appl. Electrochem.* **2005**, *35*, 655–663.

6. Augugliaro, V.; Litter, M.; Palmisano, L.; Soria, J. The combination of heterogeneous photocatalysis with chemical and physical operations: A tool for improving the photoprocess performance. *J. Photochem. Photobiol. C* **2006**, 7, 127–144.

7. Fujishima, A.; Zhang, X.; Tryk, D.A. TiO2 photocatalysis and related surface phenomena. *Surf. Sci. Rep.* **2008**, *63*, 515–582.

8. Malato, S.; Fernandez-Ibanez, P.; Maldonado, M.I.; Blanco, J.; Gernjak, W. Decontamination and disinfection of water by solar photocatalysis: Recent overview and trends. *Catal. Today* **2009**, *147*, 1–59.

9. Matsunaga, T.; Tomoda, R.; Nakajima, T.; Wake, H. Photoelectrochemical sterilization of microbial cells by semiconductor powders. *FEMS Microbiol. Lett.* **1985**, *29*, 211–214.

10. Dunlop, P.S.M.; Byrne, J.A.; Manga, N.; Eggins, B.R. The photocatalytic removal of bacterial pollutants from drinking water. *J. Photochem. Photobiol. A* **2002**, *148*, 355–363.

11. Dunlop, P.S.M.; Sheeran, C.P.; Byrne, J.A.; McMahon, M.A.S.; Boyle, M.A.; McGuigan, K.G. Inactivation of clinically relevant pathogens by photocatalytic coatings. *Photochem. Photobiol. A* **2010**, *216*, 303–310.

12. Dunlop, P.S.M.; McMurray, T.A.; Hamilton, J.W.J.; Byrne, J.A. Photocatalytic inactivation of Clostridium perfringens spores on TiO2 electrodes. *J. Photochem. Photobiol. A* **2008**, *196*, 113–119.

13. Watts, R.J.; Kong, S.; Orr, M.P.; Miller, G.C.; Henry, B.E. Photocatalytic inactivation of coliform bacteria and viruses in secondary wastewater effluent. *Water Res.* **1995**, *29*, 95–100.

14. Sunnotel, O.; Verdoold, R.; Dunlop, P.S.M.; Snelling, W.J.; Lowery, C.J.; Dooley, J.S.G.; Moore, J.E.; Byrne, J.A. Photocatalytic inactivation of Cryptosporidium parvum on nanostructured titanium dioxide films. *J. Water Health* **2010**, *8*, 83–91.

15. Sichel, C.; de Cara, M.; Tello, J.; Blanco, J.; Fernández-Ibánez, P. Solar photocatalytic disinfection of agricultural pathogenic fungi: *Fusarium* species. *Appl. Catal. B-Environ.* **2007**, *74*, 152–160.

16. Linkous, C.A.; Carter, G.J.; Locuson, D.B.; Ouellette, A.J.; Slattery, D.K.; Smith, L.A. Photocatalytic Inhibition of Algae Growth Using TiO2, WO3, and Cocatalyst Modifications. *Environ. Sci. Technol.* **2000**, *34*, 4754–4758.

17. McCullagh, C.; Robertson, J.M.C.; Bahnemann, D.W.; Robertson, P.J.K. The application of TiO2 photocatalysis for disinfection of water contaminated with pathogenic micro-organisms: A review. *Res. Chem. Intermed.* **2007**, *33*, 359–375.

18. Robertson, P.K.J.; Robertson, J.M.C.; Bahnemann, D.W. Removal of microorganisms and their chemical metabolites from water using semiconductor photocatalysis. *J. Hazard. Mater.* **2012**, *211*, 161–171.

19. Dalrymple, O.K.; Stefanakos, E.; Trotz, M.A.; Goswami, D.Y. A review of the mechanisms and modeling of photocatalytic disinfection. *Appl. Catal. B-Environ.* **2010**, *98*, 27–38.

20. Fernandez-Ibanez, P.; Polo-Lopez, M.I.; Malato, S.; Wadhwa, S.; Hamilton, J.W.J.; Dunlop, P.S.M.; D'Sa, R.; Magee, E.; O'Shea, K.; Dionysiou, D.D.; *et al.* Solar photocatalytic disinfection of water using titanium dioxide graphene composites. *Chem. Eng. J.* **2015**, *261*, 36–44.

21. Cho, M.; Chung, H.; Choi, W.; Yoon, J. Linear correlation between inactivation of *E. coli* and OH radical concentration in TiO₂ photocatalytic disinfection. *Water Res.* **2004**, *38*, 1069–1077.

22. Goulhen-Chollet, F.; Josset, S.; Keller, N.; Keller, V.; Lett, M.C. Monitoring the bactericidal effect of UV-A photocatalysis: A first approach through 1D and 2D protein electrophoresis. *Catal. Today* **2009**, *147*, 169–172.

23. Wu, P.; Xie, R.; Imlay, J.A.; Shang, J.K. Visible-light-induced photocatalytic inactivation of bacteria by composite photocatalysts of palladium oxide and nitrogen-doped titanium oxide. *Appl. Catal. B-Environ.* **2009**, *88*, 576–581.

24. Kiwi, J.; Nadtochenko, V. Evidence for the mechanism of photocatalytic degradation of the bacterial wall membrane at the TiO₂ interface by ATR-FTIR and laser kinetic spectroscopy. *Langmuir* **2005**, *21*, 4631–41.

25. Huang, Z.; Maness, P.; Blake, D.M.; Wolfrum, E.J.; Smolinski, S.L.; Jacoby, W.A. Bactericidal mode of titanium dioxide photocatalysis. *J. Photochem. Photobiol. A* **2000**, *130*, 163–170.

26. Wainwright, M. Methylene blue derivatives—Suitable photoantimicrobials for blood product disinfection? *Int. J. Antimicrob. Agents* **2000**, *16*, 381–394.

27. Sunada, K.; Watanabe, T.; Hashimoto, K. Studies on photokilling of bacteria on TiO₂ thin film. *J. Photochem. Photobiol. A* **2003**, *156*, 227–233.

28. Carré, G.; Hamon, E.; Ennahar, S.; Estner, M.; Lett, M.C.; Horvatovich, P.; Gies, J.P.; Keller, V.; Keller, N.; Andre, P. TiO₂ Photocatalysis Damages Lipids and Proteins in *Escherichia coli*. *Appl. Environ. Microbiol.* **2014**, *80*, 2573–2581.

29. Rincon, A.G.; Pulgarin, C. Field solar *E. coli* inactivation in the absence and presence of TiO₂: Is UV solar dose an appropriate parameter for standardization of water solar disinfection? *Sol. Energy* **2004**, *77*, 635–648.

30. Pigeot-Rémy, S.; Simonet, F.; Errazuriz-Cerda, E.; Lazzaroni, J.C.; Atlan, D.; Guillard, C. Photocatalysis and disinfection of water: Identification of potential bacterial targets. *Appl. Catal. B-Environ.* **2011**, *104*, 390–398.

31. Kubacka, A.; Diez, M.S.; Rojo, D.; Bargiela, R.; Ciordia, S.; Zapico, I.; Albar, J.P.; Barbas, C.; Martins dos Santos, V.A.P.; Fernandez-Garcia, M.; *et al.* Understanding the antimicrobial mechanism of TiO₂-based nanocomposite films in a pathogenic bacterium. *Sci. Rep.* **2014**, *4*, doi:10.1038/srep04134.

32. Gogniat, G.; Dukan, S. TiO₂ Photocatalysis Causes DNA Damage via Fenton Reaction-Generated Hydroxyl Radicals during the Recovery Period. *Appl. Environ. Microbiol.* **2007**, *73*, 7740–7743.

33. Rincon, A.-G.; Pulgarin, C. Bactericidal action of illuminated TiO₂ on pure Escherichia coli and natural bacterial consortia: Post-irradiation events in the dark and assessment of the effective disinfection time. *Appl. Catal. B-Environ.* **2004**, *49*, 99–112.

426

34. Dunlop, P.S.M.; Ciavola, M.; Rizzo, L.; McDowell, D.A.; Byrne, J.A. Effect of photocatalysis on the transfer of antibiotic resistance genes in urban wastewater. *Catal. Today* **2014**, in press.

35. WHO; UNICEF. *Progress on Drinking Water and Sanitation: 2014 Update*; World Health Organization; UNICEF: Geneva, Switzerland, 2014.

36. Clasen, T.; Edmondson, P. Sodium dichloroisocyanurate (NaDCC) tablets as an alternative to sodium hypochlorite for the routine treatment of drinking water at a household level. *Int. J. Hyg. Environ. Health* **2006**, *209*, 173–181.

37. WHO. *Economic and Health Effects of Increasing Coverage of Low Cost Household Drinking Water Supply and Sanitation Interventions to Countries off-Track To Meet MDG Target 10*; World Health Organization: Geneva, Switzerland, 2007.

38. Burch, J.D.; Thomas, K.E. Water disinfection for developing countries and potential for solar thermal pasteurization. *Sol. Energy* **1998**, *64*, 87–97.

39. WHO. *Cause-Specific Mortality: Regional Estimates for 2008*; World Health Organization: Geneva, Switzerland, 2011. Available online: http://www.who.int/healthinfo/global_burden_disease/estimates_regional/en/index.html (accessed on 20 March 2015).

40. Liu, L.; Johnson, H.L.; Cousens, S.; Perin, J.; Scott, S.; Lawn, J.E.; Rudan, I.; Campbell, H.; Cibulskis, R.; Li, M.; *et al.* Child Health Epidemiology, Global, regional, and national causes of child mortality: An updated systematic analysis for 2010 with time trends since 2000. *Lancet* **2012**, *379*, 2151–2161.

41. WHO. *Cholera*; Weekly Epidemiological Report, 2011. Available online: http://www.who.int/wer/2012/wer873132/en/index.html (accessed on 20 March 2015).

42. Clasen, T.F.; Haller, L. *Water Quality Interventions to Prevent Diarrhoea: Cost and Cost-Effectiveness*; World Health Organization: Geneva, Switzerland, 2008.

43. Michael, I.; Rizzo, L.; McArdell, C.S.; Manaia, C.M.; Merlin, C.; Schwartz, T.; Dagot, C.; Fatta-Kassinos, D. Urban wastewater treatment plants as hotspots for the release of antibiotics in the environment: A review. *Water Res.* **2013**, *47*, 957–995.

44. Bouki, C.; Venieri, D.; Diamadopoulos, E. Detection and fate of antibiotic resistant bacteria in wastewater treatment plants: A review. *Ecotox. Environ. Safe* **2013**, *91*, 1–9.

45. Huang, J.J.; Hu, H.Y.; Tang, F.; Li, Y.; Lu, S.Q.; Lu, Y. Inactivation and reactivation of antibiotic-resistant bacteria by chlorination in secondary effluents of a municipal wastewater treatment plant. *Water Res.* **2011**, *45*, 2775–2781.

46. Hedin, G.; Rynbäckm, J.; Loré, B. Reduction of bacterial surface contamination in the hospital environment by application of a new product with persistent effect. *J. Hosp. Infect.* **2010**, *75*, 112–115.

47. Markowska-Szczupak, A.; Ulfig, K.; Morawski, A.W. The application of titanium dioxide for deactivation of bioparticulates: An overview. *Catal. Today* **2011**, *169*, 249–257.

48. Alrousan, D.M.A.; Dunlop, P.S.M.; McMurray, T.A.; Byrne, J.A. Photocatalytic inactivation of *E. coli* in surface water using immobilised nanoparticle TiO_2 films. *Water Res.* **2009**, *43*, 47–54.

49. Amezaga-Madrid, P.; Silveyra-Morales, R.; Cordoba-Fierro, L.; Nevarez-Moorillon, G.V.; Miki-Yoshida, M.; Orrantia-Borunda, E.; Solí, F.J. TEM evidence of ultrastructural alteration on Pseudomonas aeruginosa by photocatalytic TiO_2 thin films. *J. Photochem. Photobiol. B* **2003**, *70*, 45–50.

50. Méndez-Hermida, F.; Ares-Mazás, E.; McGuigan, K.G.; Boyle, M.; Sichel, C.; Fernández-Ibáñez, P. Disinfection of drinking water contaminated with Cryptosporidium parvum oocysts under natural sunlight and using the photocatalyst TiO_2. *J. Photochem. Photobiol. B* **2007**, *88*, 105–111.

51. Weir, A.; Westerhoff, P.; Fabricius, L.; von Goetz, N. Titanium Dioxide Nanoparticles in Food and Personal Care Products. *Environ. Sci. Technol.* **2012**, *46*, 2242–2250.

52. Byrne, J.A.; Fernandez-Ibanez, P.A.; Dunlop, P.S.M.; Alrousan, D.M.A.; Hamilton, J.W.J. Photocatalytic Enhancement for Solar Disinfection of Water: A Review. *Int. J. Photoenergy* **2011**, *12*, doi:10.1155/2011/798051.

53. Sordo, C.; van Grieken, R.; Marugán, J.; Fernández-Ibáñez, P. Solar photocatalytic disinfection with immobilised TiO_2 at pilot-plant scale. *Water Sci. Technol.* **2010**, *61*, 507–512.

54. Alrousan, D.M.A.; Polo-López, M.I.; Dunlop, P.S.M.; Fernández-Ibáñez, P.; Byrne, J.A. Solar photocatalytic disinfection of water with immobilised titanium dioxide in re-circulating flow CPC reactors. *Appl. Catal. B-Environ.* **2012**, *128*, 126–134.

55. Rengifo-Herrera, J.A.; Pulgarin, C. Photocatalytic activity of N, S co-doped and N-doped commercial anatase TiO_2 powders towards phenol oxidation and *E. coli* inactivation under simulated solar light irradiation. *Sol. Energy* **2010**, *84*, 37–43.

56. Helali, S.; Polo-López, M.I.; Fernández-Ibáñez, P.; Ohtani, B.; Amano, F.; Malato, S.; Guillard, C. Solar photocatalysis: A green technology for *E. coli* contaminated water disinfection. Effect of concentration and different types of suspended catalyst. *J. Photochem. Photobiol. A* **2013**, *276*, 31–40.

57. Turki, A.; Kochkara, H.; García-Fernández, I.; Polo-López, M.I.; Ghorbel, A.; Guillard, C.; Berhault, G.; Fernández-Ibáñez, P. Solar photocatalytic inactivation of *Fusarium solani* over TiO_2 nanomaterials with controlled morphology—Formic acid effect. *Catal. Today* **2013**, *209*, 147–152.

58. Booshehri, A.Y.; Goh, S.C.K.; Hong, J.; Jiang, R.; Xu, R. Effect of depositing silver nanoparticles on $BiVO_4$ in enhancing visible light photocatalytic inactivation of bacteria in water. *J. Mater. Chem. A* **2014**, *2*, 6209–6217.

59. Vidal, A.; Díaz, A.I.; el Hraiki, A.; Romero, M.; Muguruza, I.; Senhaji, F.; González, J. Solar photocatalysis for detoxification and disinfection of contaminated water: Pilot plant studies. *Catal. Today* **1999**, *54*, 283–290.

60. McLoughlin, O.A.; Fernández Ibáñez, P.; Gernjak, W.; Malato Rodríguez, S.; Gill, L.W. Photocatalytic disinfection of water using low cost compound parabolic collectors. *Sol. Energy* **2004**, *77*, 625–633.

61. Puralytics. Available online: http://www.puralytics.com/html/home.php (accessed on 22 December 2014).

62. Ubomba-Jaswa, E.; Fernández-Ibáñez, P.; Navntoft, C.; Polo-López, M.I.; McGuigan, K.G. Investigating the microbial inactivation efficiency of a 25 L batch solar disinfection (SODIS) reactor enhanced with a compound parabolic collector (CPC) for household use. *J. Chem. Technol. Biotechnol.* **2010**, *85*, 1028–1037.

63. Ubomba-Jaswa, E.; Navntoft, C.; Polo-López, M.I.; Fernandez-Ibáñez, P.; McGuigan, K.G. Solar disinfection of drinking water (SODIS): An investigation of the effect of UV-A dose on inactivation efficiency. *Photochem. Photobiol. Sci.* **2009**, *8*, 587–595.

64. Polo-López, M.I.; Fernández-Ibáñez, P.; García-Fernández, I.; Oller, I.; Salgado-Tránsito, I.; Sichel, C. Resistance of Fusarium sp spores to solar TiO$_2$ photocatalysis: Influence of spore type and water (scaling-up results). *J. Chem. Technol. Biotechnol.* **2010**, *85*, 1038–1048.

65. Sichel, C.; Tello, J.C.; de Cara, M.; Fernández-Ibáñez, P. Effect of UV solar intensity and dose on the photocatalytic disinfection of bacteria and fungi. *Catal. Today* **2007**, *129*, 152–160.

66. Polo-López, M.I.; Fernández-Ibáñez, P.; Ubomba-Jaswa, E.; Navntoft, C.; Garcia-Fernandez, I.; Dunlop, P.S.M.; Schmidt, M.; Byrne, J.A.; McGuigan. K.G. Elimination of water pathogens with solar radiation using and automated sequential batch CPC Reactor. *J. Hazard. Mater.* **2011**, *196*, 16–21.

67. García-Fernández, I.; Fernández-Calderero, I.; Polo-López, M.I.; Fernández-Ibáñez, P. Disinfection of urban effluents using solar TiO$_2$ photocatalysis: A study of significance of dissolved oxygen, temperature, type of microorganism and water matrix. *Catal. Today* **2014**, *240*, 30–38.

68. Polo-López, M.I.; Oller, I.; Castro-Alferez, M.; Fernández-Ibáñez, P. Assessment of solar photo-Fenton, photocatalysis, and H$_2$O$_2$ for removal of phytopathogenic fungi spores in synthetic and real effluents of urban wastewater. *Chem. Eng. J.* **2014**, *257*, 122–130.

69. Ducel, G.; Fabry, J.; Nicolle, L. *Prevention of Hospital Acquired Infections: A Practical Guide*, 2nd ed.; World Health Organisation: Geneva, Switzerland, 2002.

70. Anon. Improving Patient Care by Reducing the Risk of Hospital Acquired Infection: A Progress Report, 2004. Available online: https://www.nao.org.uk/report/improving-patient-care-by-reducing-the-risk-of-hospital-acquired-infection-a-progress-report/ (accessed on 30 April 2014).

71. Anon. MRSA Rates Slashed, but other Bugs a Threat. NHS Choices 2012. Available online: http://www.nhs.uk/news/2012/05may/Pages/mrsa-hospital-acquired-infection-rates.aspx (accessed on 25 April 2014).

72. Boyce, J.M. Environmental contamination makes an important contribution to hospital infection. *J. Hosp. Infect.* **2007**, 65, 50–54.

73. Dancer, S.J. The role of environmental cleaning in the control of hospital-acquired infection. *J. Hosp. Infect.* **2009**, *73*, 378–385.

74. Kramer, A.; Schwebke, I.; Kampf, G. How long do nosocomial pathogens persist on inanimate surfaces? A systematic review. *BMC Infect. Dis.* **2006**, *6*, doi:10.1186/1471-2334-6-130.

75. Pratt, R.J.; Pellowe, C.M.; Wilson, J.A.; Loveday, H.P.; Harper, P.J.; Jones, S.R.; McDougall, C.; Wilcox, M.H. National Evidence-Based Guidelines for Preventing Healthcare-Associated Infections in NHS Hospitals in England. *J. Hosp. Infect.* **2007**, *65*, S1–S64.

76. Weber, D.J.; Rutala, W.A.; Miller, M.B.; Huslage, K.; Sickbert-Bennett, E. Role of hospital surfaces in the transmission of emerging health care-associated pathogens: Norovirus, Clostridium difficile, and Acinetobacter species. *Am. J. Infect. Control* **2010**, *38*, S25–S33.

77. Lee, C.C. *Environmental Engineering Dictionary*, 4th ed.; Government Institutes: Lanham, MD, USA, 2005.

78. Schabrun, S.; Chipchase, L. Healthcare equipment as a source of nosocomial infection: A systematic review. *J. Hosp. Infect.* **2006**, *63*, 239–245.

79. Ohko, Y.; Utsumi, Y.; Niwa, C.; Tatsuma, T.; Kobayakawa, K.; Satoh, Y.; Kubota, Y.; Fujishima, A. Self-sterilizing and self-cleaning of silicone catheters coated with TiO$_2$ photocatalyst thin films: A preclinical work. *J. Biomed. Mater. Res.* **2001**, *58*, 97–101.

80. Brun-Buisson, C. New Technologies and Infection Control Practices to Prevent Intravascular Catheter-related Infections. *Am. J. Respir. Crit. Care Med.* **2001**, *164*, 1557–1558.

81. Pichat, P. Self-cleaning materials based on solar photocatalysis in "New and future developments in catalysis". In *Solar Catalysis*; Suib, S.L., Ed.; Elsevier: Amsterdam, The Netherlands, 2013; Volume 7, pp. 167–190.

82. Pilkington, Self-cleaning glass. Available online: https://www.pilkington.com/en-gb/uk/householders/types-of-glass/self-cleaning-glass (accessed on 27 March 2015).

83. TOTO. Amazing New Tiles that Keep Their Beauty and Cleanliness by the Power of Light. 2011 Available online: http://www.toto.com.hk/tech/hydrotect.html (accessed on 17 April 2014).

84. Nakamura, H.; Tanaka, M.; Shinohara, S.; Gotoh, M.; Karube, I. Development of a self-sterilizing lancet coated with a titanium dioxide photocatalytic nano-layer for self-monitoring of blood glucose. *Biosens. Bioelectron.* **2007**, *22*, 1920–1925.

85. Welch, K.; Cai, Y.; Engqvist, H.; Strømme, M. Dental adhesives with bioactive and on-demand bactericidal properties. *Dent. Mater.* **2010**, *26*, 491–499.

86. Oka, Y.; Kim, W.-C.; Yoshida, T.; Hirashima, T.; Mouri, H.; Urade, H.; Itoh, Y.; Kubo, T. Efficacy of titanium dioxide photocatalyst for inhibition of bacterial colonization on percutaneous implants. *J. Biomed. Mater. Res. B Appl. Biomater.* **2008**, *86*, 530–540.

87. Kühn, K.P.; Chaberny, I.F.; Massholder, K.; Stickler, M.; Benz, V.W.; Sonntag, H.-G.; Erdinger, L. Disinfection of surfaces by photocatalytic oxidation with titanium dioxide and UVA light. *Chemosphere* **2003**, *53*, 71–77.

88. Page, K.; Palgrave, R.G.; Parkin, I.P.; Wilson, M.; Savin, S.L.P.; Chadwick, A.V. Titania and silver-titania composite films on glass-potent antimicrobial coatings. *J. Mater. Chem.* **2007**, *17*, 95–104.

89. Miron, C.; Roca, A.; Hoisie, S.; Cozorici, P.; Sirghi, L. Photoinduced bactericidal activity of TiO$_2$ thin films obtained by radiofrequency magnetron sputtering deposition. *J. Optoelectron. Adv. Mater.* **2004**, *7*, 915–919.

90. Wong, M.-S.; Chu, W.-C.; Sun, D.-S.; Huang, H.-S.; Chen, J.-H.; Tsai, P.-J.; Lin, N.-T.; Yu, M.-S.; Hsu, S.-F.; Wang, S.-L.; *et al.* Visible-Light-Induced Bactericidal Activity of a Nitrogen-Doped Titanium Photocatalyst against Human Pathogens. *Appl. Environ. Microbiol.* **2006**, *72*, 6111–6116.

91. Mitoraj, D.; Janczyk, A.; Strus, M.; Kisch, H.; Stochel, G.; Heczko, P.B.; Macyk, W. Visible light inactivation of bacteria and fungi by modified titanium dioxide. *Photochem. Photobiol. Sci.* **2007**, *6*, 642–648.

92. Dunnill, C.W.; Aiken, Z.A.; Pratten, J.; Wilson, M.; Parkin, I.P. Sulfur- and Nitrogen-Doped Titania Biomaterials via APCVD. *Chem. Vapor Depos.* **2010**, *16*, 50–54.

93. Dunnill, C.W.; Page, K.; Aiken, Z.A.; Noimark, S.; Hyett, G.; Kafizas, A.; Pratten, J.; Wilson, M.; Parkin, I.P. Nanoparticulate silver coated-titania thin films—Photo-oxidative destruction of stearic acid under different light sources and antimicrobial effects under hospital lighting conditions. *J. Photochem. Photobiol. A* **2011**, *220*, 113–123.

94. Tsuang, Y.-H.; Sun, J.-S.; Huang, Y.-C.; Lu, C.-H.; Chang, W.H.-S.; Wang, C.-C. Studies of Photokilling of Bacteria Using Titanium Dioxide Nanoparticles. *Artif. Organs* **2008**, *32*, 167–174.

95. Shiraishi, K.; Koseki, H.; Tsurumoto, T.; Baba, K.; Naito, M.; Nakayama, K.; Shindo, H. Antibacterial metal implant with a TiO$_2$-conferred photocatalytic bactericidal effect against Staphylococcus aureus. *Surf. Interface Anal.* **2009**, *41*, 17–22.

96. Bombac, D.; Brojan, M.; Fajfar, P.; Kosel, F.; Turk, R. Review of materials in medical applications. *RMZ-Mater. Geoenviron.* **2007**, *54*, 471.

97. Amin, S.Z.; Smith, L.; Luthert, P.J.; Cheetham, M.E.; Buckley, R.J. Minimising the risk of prion transmission by contact tonometry. *Br. J. Ophthalmol.* **2003**, *87*, 1360–1362.

98. Su, W.; Wang, S.; Wang, X.; Fu, X.; Weng, J. Plasma pre-treatment and TiO$_2$ coating of PMMA for the improvement of antibacterial properties. *Surf. Coat. Technol.* **2010**, *205*, 465–469.

99. Suketa, N.; Sawase, T.; Kitaura, H.; Naito, M.; Baba, K.; Nakayama, K.; Wennerberg, A.; Atsuta, M. An Antibacterial Surface on Dental Implants, Based on the Photocatalytic Bactericidal Effect. *Clin. Implant Dent. Relat. Res.* **2005**, *7*, 105–111.

100. National Health Service UK. Creutzfeldt-Jakob Disease. Available online: http://www.nhs.uk/conditions/Creutzfeldt-Jakob-disease/Pages/Introduction.aspx (accessed on 28 December 2014).

101. National CJD Research and Surveillance Unit. CJD in the UK by calendar year, 2014. Available online: http://www.cjd.ed.ac.uk (accessed on 28 December 2014).

102. Paspaltsis, I.; Kotta, K.; Lagoudaki, R.; Grigoriadis, N.; Poulios, I.; Sklaviadis, T. Titanium dioxide photocatalytic inactivation of prions. *J. Gen. Virol.* **2006**, *87*, 3125–3130.

103. Ahmed, M.H.; Keyes, T.E.; Byrne, J.A. The photocatalytic inactivation effect of Ag–TiO$_2$ on β-amyloid peptide (1–42). *J. Photochem. Photobiol. A* **2013**, *254*, 1–11.

104. Zhang, D.; Li, G.; Yu, J.C. Inorganic materials for photocatalytic water disinfection. *J. Mater. Chem.* **2010**, *20*, doi:10.1039/B925342D.

105. Gondal, M.A.; Dastageer, M.A.; Khalil, A.; Hayat, K.; Yamani, Z.H. Nanostructured ZnO synthesis and its application for effective disinfection of *Escherichia coli* microorganism in water. *J. Nanopart. Res.* **2011**, *13*, 3423–3430.

106. Rodríguez, J.; Paraguay-Delgado, F.; López, A.; Alarcón, J.; Estrada, W. Synthesis and characterization of ZnO nanorod films for photocatalytic disinfection of contaminated water. *Thin Solid Films* **2010**, *519*, 729–735.

107. Hill, J.C.; Choi, K.-S. Effect of Electrolytes on the Selectivity and Stability of n-type WO_3 Photoelectrodes for Use in Solar Water Oxidation. *J. Phys. Chem. C* **2012**, *116*, 7612–7620.

108. Wang, P.; Huang, B.; Qin, X.; Zhang, X.; Dai, Y.; Whangbo, M-H. $Ag/AgBr/WO_3 \cdot H_2O$: Visible-Light Photocatalyst for Bacteria Destruction. *Inorg. Chem.* **2009**, *48*, 10697–10702.

109. Basnet, P.; Larsen, G.K.; Jadeja, R.P.; Hung, Y-C.; Zhao, Y. α-Fe_2O_3 Nanocolumns and Nanorods Fabricated by Electron Beam Evaporation for Visible Light Photocatalytic and Antimicrobial Applications. *Appl. Mater. Interfaces* **2013**, *5*, 2085–2095.

110. Meissner, D.; Memming, R.; Kastening, B. Photoelectrochemistry of Cadmium Sulfide 1. Reanalysis of Photocorrosion and Flat-Band Potential. *J. Phys. Chem.* **1988**, *92*, 3476–3483.

111. Huang, L.; Peng, F.; Yu, H.; Wang, H. Preparation of cuprous oxides with different sizes and their behaviors of adsorption, visible-light driven photocatalysis and photocorrosion. *Solid State Sci.* **2009**, *11*, 129–138.

112. Hamilton, J.W.J.; Byrne, J.A.; Dunlop, P.S.M.; Brown, N.M.D. Photo-Oxidation of Water Using Nanocrystalline Tungsten Oxide under Visible Light. *Int. J. Photoenergy* **2008**, doi:10.1155/2008/185479.

113. Maeda, K.; Domen, K. New Non-Oxide Photocatalysts Designed for Overall Water Splitting under Visible Light. *J. Phys. Chem. C* **2007**, *111*, 7851–7861.

114. Yu, H.; Quan, X.; Zhang, Y.; Ma, N.; Chen, S.; Zhao, H. Electrochemically Assisted Photocatalytic Inactivation of *Escherichia coli* under Visible Light Using a $ZnIn_2S_4$ Film Electrode. *Langmuir* **2008**, *24*, 7599–7604.

115. Woodhouse, M.; Herman, G.S.; Parkinson, B.A. Combinatorial Approach to Identification of Catalysts for the Photoelectrolysis of Water. *Chem. Mater.* **2005**, *17*, 4318–4324.

116. Takata, T.; Tanaka, A.; Hara, M.; Kondo, J.N.; Domen, K. Recent progress of photocatalysts for overall water splitting. *Catal. Today* **1998**, *44*, 17–26.

117. Yan, J.; Yang, H.; Tang, Y.; Lu, Z.; Zheng, S.; Yao, M.; Han, Y. Synthesis and photocatalytic activity of $CuYyFe_{2-y}O_4$–$CuCo_2O_4$ nanocomposites for H_2 evolution under visible light irradiation. *Renew. Energy* **2009**, *34*, 2399–2403.

118. Yang, J.; Zhong, H.; Li, M.; Zhang, L.; Zhang, Y. Markedly enhancing the visible-light photocatalytic activity of $LaFeO_3$ by post-treatment in molten salt React. *Kinet. Catal. Lett.* **2009**, *97*, 269–274.

119. Zhang, L.; Tan, P.Y.; Chow, C.L; Lim, C.K.; Tan, O.K.; Tse, M.S.; Sze, C.C. Antibacterial activities of mechanochemically synthesized perovskite strontium titanate ferrite metal oxide. *Colloids Surf. A* **2014**, *456*, 169–175.

120. Bessekhouad, Y.; Trari, M. Photocatalytic hydrogen production from suspension of spinel powders AMn_2O_4 (A = Cu and Zn). *Int. J. Hydrog. Energy* **2002**, *27*, doi:10.1016/s0360-3199(01)00159-8.

121. Wang, D.; Zou, Z.; Ye, J. A new spinel-type photocatalyst $BaCr_2O_4$ for H_2 evolution under UV and visible light irradiation. *Chem. Phys. Lett.* **2003**, *373*, 191–196.

122. Yin, J.; Zou, Z.; Ye, J. Photophysical and photocatalytic properties of new photocatalysts $MCrO_4$ (M=Sr, Ba). *Chem. Phys. Lett.* **2003**, *378*, 24–28.

123. Cui, B.; Lin, H.; Li, Y.Z.; Li, J.B.; Sun, P.; Zhao, X.C.; Liu, C.J. Photophysical and Photocatalytic Properties of Core-Ring Structured NiCo$_2$O$_4$ Nanoplatelets. *J. Phys. Chem. C* **2009**, *113*, 14083–14087.

124. Wu, W.; Zhang, H.; Chang, S.; Gao, J.; Jia, L.J. Study on the antibacterial performance of perovskite LaCoO$_3$ under visible light illumination. *Chem. Ind. Times* **2009**, *7*, 25–28.

125. Wang, Y.; Zhang, Z.; Zhu, Y.; Li, Z.; Vajtai, R.; Ci, L.; Ajayan, P.M. Nanostructured VO$_2$ Photocatalysts for Hydrogen Production. *ACS Nano* **2008**, *2*, 1492–1496.

126. Ye, J.; Zou, Z.; Oshikiri, M.; Matsushita, A.; Shimoda, M.; Imai, M.; Shishido, T. A novel hydrogen-evolving photocatalyst InVO$_4$ active under visible light irradiation. *Chem. Phys. Lett.* **2002**, *356*, 221–226.

127. Kudo, A.; Ueda, K.; Kato, H. Photocatalytic O$_2$ evolution under visible light irradiation on BiVO$_4$ in aqueous AgNO$_3$ solution. *Catal. Lett.* **1998**, *53*, 229–230.

128. Kohtani, S.; Makino, S.; Kudo, A.; Tokumura, K.; Ishigaki, Y.; Matsunaga, T.; Nikaido, O.; Hayakawa, K.; Nakagaki, R. Photocatalytic degradation of 4-n-nonylphenol under irradiation from solar simulator: Comparison between BiVO$_4$ and TiO$_2$ photocatalysts. *Chem. Lett.* **2002**, *7*, 660–661.

129. Tokunaga, S.; Kato, H.; Kudo, A. Selective Preparation of Monoclinic and Tetragonal BiVO$_4$ with Scheelite Structure and Their Photocatalytic Properties. *Chem. Mater.* **2001**, *13*, 4624–4628.

130. Xie, B. Bactericidal Activity of monoclinic BiVO$_4$ Under Visible Light Irradiation. *Chin. J. Disinfect.* **2010**, *1*, 14–16.

131. Wang, W.; Yu, Y.; An, T.; Li, G.; Yip, H.Y.; Yu, J.C.; Wong, P.K. Visible-Light-Driven Photocatalytic Inactivation of *E. coli* K-12 by Bismuth Vanadate Nanotubes: Bactericidal Performance and Mechanism. *Environ. Sci. Technol.* **2012**, *46*, 4599–4606.

132. Li, Q.; Li, Y.W.; Wu, P.; Xie, R.; Shang, J.K. Palladium Oxide Nanoparticles on Nitrogen-Doped Titanium Oxide: Accelerated Photocatalytic Disinfection and Post-Illumination Catalytic "Memory". *Adv. Mater.* **2008**, *20*, 3717–3723.

133. Pena, M.A.; Fie, J.L.G. Chemical Structures and Performance of Perovskite Oxides. *Chem. Rev.* **2001**, *101*, 1981–2017.

134. Tanaka, H.; Misono, M. Advances in designing perovskite catalysts. *Curr. Opin. Solid State Mater. Sci.* **2001**, *5*, 381–387.

135. Eng, H.W.; Barnes, P.W.; Auer, B.M.; Woodward, P.M. Investigations of the electronic structure of d0 transition metal oxides belonging to the perovskite family. *J. Solid State Chem.* **2003**, *175*, 94–109.

136. Fei, D.Q.; Hudaya, T.; Adesina, A.A. Visible-light activated titania perovskite photocatalysts: Characterisation and initial activity studies. *Catal. Commun.* **2005**, *6*, 253–258.

137. Kato, H.; Kobayashi, H.; Kudo, A. Role of Ag+ in the Band Structures and Photocatalytic Properties of AgMO$_3$ (M: Ta and Nb) with the Perovskite Structure. *J. Phys. Chem. B* **2002**, *106*, 12441–12447.

138. Yea, J.; Zou, Z.; Matsushita, A. A novel series of water splitting photocatalysts NiM$_2$O$_6$ (M = Nb; Ta) active under visible light. *Int. J. Hydrog. Energy* **2003**, *28*, 651–655.

139. Gan, H.; Zhang, G.; Huang, H. Enhanced visible-light-driven photocatalytic inactivation of *Escherichia coli* by $Bi_2O_2CO_3/Bi_3NbO_7$ composites. *J. Hazard. Mater.* **2013**, *250*, 131–137.

140. Lin, H.; Lin, H. Visible-light photocatalytic inactivation of *Escherichia coli* by $K_4Nb_6O_{17}$ and Ag/Cu modified $K_4Nb_6O_{17}$. *J. Hazard. Mater.* **2012**, *217*, 231–237.

141. Gao, P.; Liu, J.; Sun, D.D.; Ng, W. Graphene oxide–CdS composite with high photocatalytic degradation and disinfection activities under visible light irradiation. *J. Hazard. Mater.* **2013**, *250–251*, 412–420.

142. Wang, W.; Yu, J.C.; Xia, D.; Wong, P.K.; Li, Y. Graphene and g-C_3N_4 Nanosheets Cowrapped Elemental α-Sulfur As a Novel Metal-Free Heterojunction Photocatalyst for Bacterial Inactivation under Visible-Light. *Environ. Sci. Technol.* **2013**, *47*, 8724–8732.

143. Krishna, V.; Pumprueg, S.; Lee, S.-H.; Zhao, J.; Sigmund, W.; Koopman, B.; Moudgi, B.M. Photocatalytic Disinfection with Titanium Dioxide Coated Multi-Wall Carbon Nanotubes. *Proc. Saf. Environ. Prot.* **2005**, *83*, 393–397.

144. Yao, K.S.; Wang, D.Y.; Chang, C.Y.; Weng, K.W.; Yang, L.Y.; Lee, S.J.; Cheng T.C.; Hwang, C.C. Photocatalytic disinfection of phytopathogenic bacteria by dye-sensitized TiO_2 thin film activated by visible light. *Surf. Coat. Technol.* **2007**, *202*, 1329–1332.

145. Hou, Y.; Li, X.; Zhao, Q.; Chen, G.; Raston, C.L. Role of Hydroxyl Radicals and Mechanism of Escherichia coli Inactivation on Ag/AgBr/TiO_2 Nanotube Array Electrode under Visible Light Irradiation. *Environ. Sci. Technol.* **2012**, *46*, 4042–4050.

146. Scanlon, D.O.; Dunnill, C.W.; Buckeridge, J.; Shevlin, S.A.; Logsdail, A.J.; Woodley, S.M.; Catlow, C.R.A.; Powell, M.J.; Palgrave, R.G.; Parkin, I.P.; *et al.* Band alignment of rutile and anatase TiO_2. *Nat. Mater.* **2013**, *12*, 798–801.

147. Chen, H.M.; Chen, C.K.; Chen, C.-J.; Cheng, L.-C.; Wu, P.C.; Cheng, B.H.; Ho, Y.Z.; Tseng, M.L.; Hsu, Y.-Y.; Chan, T.-S.; *et al.* Plasmon Inducing Effects for Enhanced Photoelectrochemical Water Splitting: X-ray Absorption Approach to Electronic Structures. *ACS Nano* **2012**, *6*, 7362–7372.

148. Tributsch, H. Multi-electron transfer catalysis for energy conversion based on abundant transition metals. *Electrochim. Acta* **2007**, *52*, 2302–2316.

149. Hayden, S.C.; Allam, N.K.; el-Sayed, M.A. TiO_2 Nanotube/CdS Hybrid Electrodes: Extraordinary Enhancement in the Inactivation of *Escherichia coli*. *J. Am. Chem. Soc.* **2010**, *132*, 14406–14408.

150. Vogel, R.; Hoyer, P.; Weller, H. Quantum-Sized PbS, CdS, Ag2S, Sb2S3, and Bi2S3 Particles as Sensitizers for Various Nanoporous Wide-Bandgap Semiconductors. *J. Phys. Chem.* **1994**, *98*, 3183–3188.

151. Tada, H.; Jin, Q.; Iwaszuk, A.; Nolan, M. Molecular-Scale Transition Metal Oxide Nanocluster Surface-Modified Titanium Dioxide as Solar-Activated Environmental Catalysts. *J. Phys. Chem. C* **2014**, *118*, 12077–12086.

152. Nolan, M. Surface modification of TiO_2 with metal oxide nanoclusters: A route to composite photocatalytic materials. *Chem. Commun.* **2011**, *47*, 8617–8619.

153. Iwaszuka, A.; Nolan, N. SnO-nanocluster modified anatase TiO_2 photocatalyst: Exploiting the Sn(II) lone pair for a new photocatalyst material with visible light absorption and charge carrier separation. *J. Mater. Chem. A* **2013**, *1*, 6670–6677.

154. Khalil, A.; Gondal, M.A.; Dastageer, M.A. Augmented photocatalytic activity of palladium incorporated ZnO nanoparticles in the disinfection of Escherichia coli microorganism from water. *Appl. Catal. A* **2011**, *402*, 162–167.

155. Karunakaran, C.; Gomathisankar, P.; Manikandan, G. Preparation and characterization of antimicrobial Ce-doped ZnO nanoparticles for photocatalytic detoxification of cyanide. *Mater. Chem. Phys.* **2010**, *123*, 585–594.

156. Xu, C.; Cao, L.; Su, G.; Liu, W.; Qu, X.; Yu, Y. Preparation, characterization and photocatalytic activity of Co-doped ZnO powders. *J. Alloys Compd.* **2010**, *497*, 373–376.

157. Karunakaran, C.; Rajeswari, V.; Gomathisankar, P. Enhanced photocatalytic and antibacterial activities of sol–gel synthesized ZnO and Ag-ZnO. *Mater. Sci. Semicond. Proc.* **2011**, *14*, 133–138.

158. Karunakaran, C.; Rajeswari, V.; Gomathisankar, P. Combustion synthesis of ZnO and Ag-doped ZnO and their bactericidal and photocatalytic activities. *Superlattice Microst.* **2011**, *50*, 234–241.

159. Karunakaran, C.; Rajeswari, V.; Gomathisankar, P. Optical, electrical, photocatalytic, and bactericidal properties of microwave synthesized nanocrystalline Ag-ZnO and ZnO. *Solid State Sci.* **2011**, *13*, 923–928.

160. Hamilton, J.W.J.; Byrne, J.A.; McCullagh, C.; Dunlop, P.S.M. Electrochemical Investigation of Doped Titanium Dioxide. *Int. J. Photoenergy* **2008**, *2008*, doi:10.1155/2008/631597.

161. Vohra, A.; Goswami, D.Y.; Deshpande, D.A.; Block, S.S. Enhanced photocatalytic disinfection of indoor air. *Appl. Catal. B-Environ.* **2006**, *65*, 57–65.

162. Sökmena, M.; Candana, F.; Sümer, Z. Disinfection of *E. coli* by the Ag-TiO_2/UV system: Lipid peroxidation. *J. Photochem. Photobiol. A* **2001**, *143*, 241–244.

163. Karunakarana, C.; Abiramasundaria, G.; Gomathisankara, P.; Manikandana, G.; Anandib, V. Cu-doped TiO_2 nanoparticles for photocatalytic disinfection of bacteria under visible light. *J. Colloid Interface Sci.* **2010**, *352*, 68–74.

164. Yu, J.C.; Ho, W.; Yu, J.; Hoyin Yip, H.; Wong, P.K.; Zhao, J. Efficient Visible-Light-Induced Photocatalytic Disinfection on Sulfur-Doped Nanocrystalline Titania. *Environ. Sci. Technol.* **2005**, *39*, 1175–1179.

165. Rehman, S.; Ullah, R.; Butt, A.M.; Gohar, N.D. Strategies of making TiO_2 and ZnO visible light active. *J. Hazard. Mater.* **2009**, *170*, 560–569.

166. Im, J.S.; Yun, S.; Lee, Y. Investigation of multielemental catalysts based on decreasing the band gap of titania for enhanced visible light photocatalysis. *J. Colloid Interface Sci.* **2009**, *336*, 183–188.

167. Li, Q.; Xie, R.; Li, Y.W.; Mintz, E.A.; Shang, J.K. Enhanced Visible-Light-Induced Photocatalytic disinfection of *E. coli* by Carbon-Sensitized Nitrogen-Doped Titanium Oxide. *Environ. Sci. Technol.* **2007**, *41*, 5050–5056.

168. Hameed, A.; Gondal, M.A.; Yamani, Z.H. Effect of transition metal doping on photocatalytic activity of WO_3 for water splitting under laser illumination: role of 3d-orbitals. *Cat. Commun.* **2004**, *5*, 715–719.

169. Bagabas, A.; Gondal, M.; Khalil, A.; Dastageer, A.; Yamani, Z.; Ashameria, M. Laser-induced photocatalytic inactivation of coliform bacteria from water using Pd-loaded nano-WO_3. *Stud. Surf. Sci. Catal.* **2010**, *175*, 279–282.

Coupled Microwave/Photoassisted Methods for Environmental Remediation

Satoshi Horikoshi and Nick Serpone

Abstract: The microwave-induced acceleration of photocatalytic reactions was discovered serendipitously in the late 1990s. The activity of photocatalysts is enhanced significantly by both microwave radiation and UV light. Particularly relevant, other than as a heat source, was the enigmatic phenomenon of the non-thermal effect(s) of the microwave radiation that facilitated photocatalyzed reactions, as evidenced when examining various model contaminants in aqueous media. Results led to an examination of the possible mechanism(s) of the microwave effect(s). In the present article we contend that the microwaves' non-thermal effect(s) is an important factor in the enhancement of TiO_2-photoassisted reactions involving the decomposition of organic pollutants in model wastewaters by an integrated (coupled) microwave-/UV-illumination method (UV/MW). Moreover, such coupling of no less than two irradiation methods led to the fabrication and ultimate investigation of microwave discharged electrodeless lamps (MDELs) as optimal light sources; their use is also described. The review focuses on the enhanced activity of photocatalytic reactions when subjected to microwave radiation and concentrates on the authors' research of the past few years.

Reprinted from *Molecules*. Cite as: Horikoshi, S.; Serpone, N. Coupled Microwave/Photoassisted Methods for Environmental Remediation. *Molecules* **2014**, *19*, 18102-18128.

1. Introduction

1.1. Microwave Radiation in Chemistry

Microwave radiation has become one of the more popular technologies, both domestically and industrially. It describes the low-energy electromagnetic radiation that spans the frequency range from 30 GHz to 300 MHz; that is, the wavelengths from 100 cm to 1 cm. Two rather familiar devices that make extensive use of this low-energy radiation are the domestic microwave oven and the cellular phone. Early (since 1949) industrial use of microwave radiation involved the thermal molding of wood and plastics, and the drying of medicinal products, fibers, teas, and cigarettes. In recent years, microwaves have been used in, among other applications, the sintering of ceramics, in cancer treatment (hyperthermia), in the drying and sterilization of foodstuffs, and in the vulcanization of rubber.

In the inorganic chemistry area, active research in the use of microwave radiation focused at the microwave sintering of ceramics (in the early 1980s) [1], in which the principal feature was the formation of compact crystal grains in a short time at relatively low temperatures. In the organic chemistry field, the use of microwave radiation to drive organic syntheses was not explored until the mid-1980s, at least not until the first two studies reported in 1986 by Gedye and coworkers [2] and by Giguere *et al.* [3] on microwave-enhanced organic processes using domestic microwave ovens. Since then, organic chemists have discovered the benefits of microwaves to drive synthetic reactions, as a consequence of which industries began to manufacture microwave ovens specifically designed for research laboratories. The number of reports on the use of microwaves as an energy source to drive chemical reactions has witnessed an astronomical growth since the early 1990s [4].

1.2. The Microwave-/Photo-Assisted Methodology

The photoassisted oxidative (and reductive) decomposition of pollutants by means of TiO_2 semiconductor nanoparticulates is an effective and attractive oxidation (reduction) method in the general area of Advanced Oxidation Technologies. Several review articles have appeared that summarize environmental protection using TiO_2 materials as the photomediators, if not as photocatalysts [5–8]. Applications of photoassisted treatments to air pollution have been developed by TiO_2 fixation on such suitable substrate supports as filters in air conditioners, for instance [9]. However, this photoassisted degradation methodology is not suitable for large-scale wastewater treatment because the degradation rates of organic compounds dissolved in wastewaters tend to be rather slow. In this regard, relatively little has been done in this area in the last decade as large-scale treatments of organic pollutants in aquatic environments have not been without some problems, not least of which is the low photodegradation efficiency, a result of several factors, most notably: (i) the poor adsorption of wastewater organic pollutants on the TiO_2 surface; (ii) the penetration of UV light tends to be shallow in turbid wastewaters; (iii) the need for dissolved oxygen in the photoassisted degradations; and (iv) the need to immobilize TiO_2 nanoparticles in their use in aquatic ecosystems. Also relevant, the processing time has been the principal problem in actual wastewater treatments that have used TiO_2 nanomaterials. Many of the above problems could be resolved if the activity of the photocatalysts were improved. To achieve such an objective we proposed some time ago [10] the coupling of both microwave radiation and ultraviolet radiation so as to enhance the activity of photocatalysts. With the latter coupled (*i.e.*, integrated) methodology, it was possible to enhance the photoassisted degradation processes in TiO_2 dispersions by the added assistance of microwave radiation in the remediation of wastewaters contaminated with such pollutants as dyes, polymers, surfactants, herbicides, and endocrine disruptors, among others.

An integrated microwave-/photoassisted methodology presents certain advantages in wastewater treatment. In this technique, a feature of the reaction on the TiO_2 surface involves thermal and specific effects (e.g., non-thermal effects) originating from the absorption of microwave radiation by the metal-oxide nanoparticulates. As a case in point, a microwave specific effect was inferred for P-25 TiO_2 that exhibited lattice distortion and oxygen vacancies when used in UV-driven/microwave-assisted photocatalyzed reactions [11]. Differences between various TiO_2 batches with regard to microwave-specific effect(s) were examined using microwaves of different frequencies [12], and by examining the effects of the microwaves' magnetic and electric fields [13] in photoassisted processes involving TiO_2 and ZnO nanomaterials. Increased formation of •OH radicals on the surface of various TiO_2 specimens that were exposed to microwave radiation correlated with increased photoactivity [11]. The microwave specific effect, not encountered when the photocatalyzed reaction is subjected to conventional heating, inferred that the composition and the electronic characteristics of the TiO_2 nanomaterials were important factors.

In the present article, we review some of our research studies carried out during the past few years that have focused on improving the photocatalytic activity of TiO_2 by exposing it to microwave radiation. The enhancement of the photoactivity of TiO_2, in particular, and metal-oxide nanomaterials, in general, by microwave non-thermal effect(s) is discoursed in this review paper. We further describe the utility of microwave discharged electrodeless lamps (MDELs) as an optimal light source that embeds the integrated UV/MW technique.

2. Experimental Setup of an Integrated Microwave/Photoreactor System

Continuous microwave irradiation of a wastewater sample can typically be achieved in a single-mode applicator using a 2.45-GHz microwave generator, a power monitor (to assess incident and reflected microwave power), a three stub tuner and an isolator (an air cooling device) such as the one fabricated by the Hitachi Kyowa Engineering Co. Ltd. (Hitachi, Ibaraki Prefecture, Japan; Figure 1a). A typical reactor setup might contain a model wastewater sample (30 mL) containing TiO_2 particles (Evonik P25; 60 mg loading) introduced into a closed high-pressure 150-mL Pyrex glass cylindrical reactor. Subsequent irradiation of the reactor with UV light from a super high-pressure Hg lamp can be achieved through the means of a light guide. The solution temperature is normally measured with an optical fiber thermometer. A pressure gauge and a release bulb are also connected to the cover of the reactor. The reaction mixture is then stirred continually using a magnetic bar during the UV light irradiation or coupled UV/microwave (MW) irradiation.

Three different methodologies have typically been used to examine the photodecomposition of aqueous samples of pollutants in aqueous TiO_2 dispersions. The first is the photo-/microwave-assisted method using UV light and microwave irradiation in the presence of TiO_2 (UV/MW). The second method entails UV irradiation alone (UV), whereas the third method involves a thermally-assisted photodegradation of the TiO_2 dispersions using UV light and externally applied conventional heat (UV/CH; typically from an oil bath or an electric heat mantle). In the present context, the external heat can be supplied by coating one part of the cylindrical photoreactor with a thin metallic film on one side at the bottom of the reactor, whereas the uncoated side permits the UV radiation to reach the dispersion. The pressure and the rate of increase of temperature (error typically $\leq \pm 1$ °C) in the UV/CH method are maintained at levels otherwise identical to those used in the UV/MW methodology. As such, no differences in temperature profiles are observed when using either microwave dielectric heating or conventional heating.

Figure 1. (a) Photograph of an integrated microwave/photoreactor system having a single-mode applicator, and **(b)** schematic of the system and a typical plot (inset) of the change of temperature with irradiation time for an aqueous TiO₂ dispersion under microwave irradiation.

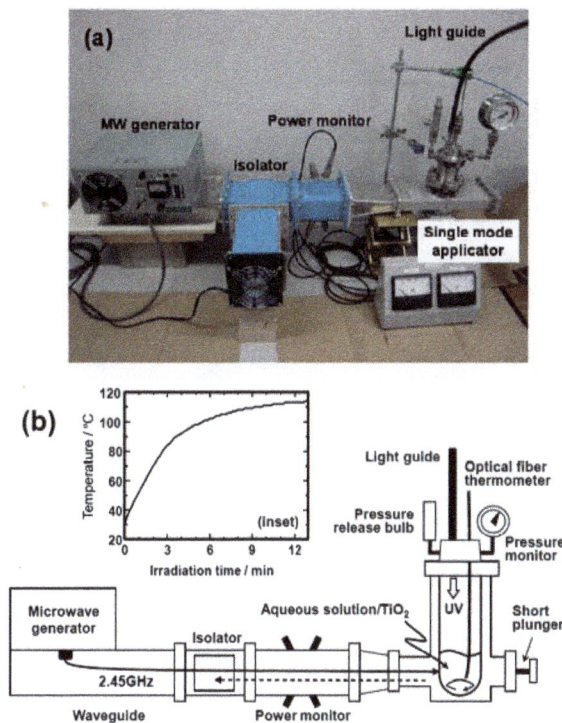

3. Description and Discussion of Results from the Various Studies

3.1. Degradation of Substances Incompatible with TiO₂ Photocatalysts

The photooxidative remediation of wastewaters with TiO₂ as the photocatalyst has been reported in many international journals (see for example the references in [14]). In the degradation of a dye substrate, the photoassisted reaction can be accomplished by means of electron transfer from the excited dye that is exposed to visible light. In this regard, it is noteworthy that the Japanese Industrial Standards (JIS) includes the methylene blue dye as a standard wastewater substance to ascertain the activity of photocatalysts [15]. Some dyes are poorly photodecomposed and thus are not useful in ascertaining the photoactivity of a metal-oxide photocatalyst. For example, the rate of photodegradation of the cationic dye rhodamine-B (RhB) is slow in acidic aqueous media because the surface of TiO₂ particles is positively charged (Ti-OH₂$^+$; pI = 6.3). However, RhB has proven as an interesting model compound to examine the microwave effect. In earlier studies, the major focus of our studies was on the degradation of organic pollutants, as exemplified by the degradation of the rhodamine-B (RhB) dye catalyzed by TiO₂ semiconductor particles under both UV and microwave irradiation [10,16]. Changes in color intensity of the RhB dye solutions occurring under various conditions are illustrated in Figure 2. The photodegradation of RhB is clearly evident on using the TiO$_2$-assisted UV/MW method. These observations demonstrate that a method that can treat large quantities of pollutants in wastewaters by a hybrid combination of microwaves and TiO$_2$ photoassisted technologies is conceivable. The photodegradation by this metal oxide is unaffected by conventional heating (CH)—compare, for example, the results from the UV and the UV/CH methods in the presence of TiO$_2$ (Figure 2) [17].

Figure 2. Visual comparison of color fading in the degradation of RhB solutions (0.05 mM) subsequent to being subjected to various degradation methods for 150 min. From left to right: initial RhB solution; RhB subjected to photoassisted degradation (UV); EhB subjected to integrated microwave-/photo-assisted degradation (UV/MW); RhN subjected to thermal- and photo-assisted degradation (UV/CH). Reproduced from [17]. Copyright 2009 by Elsevier B.V.

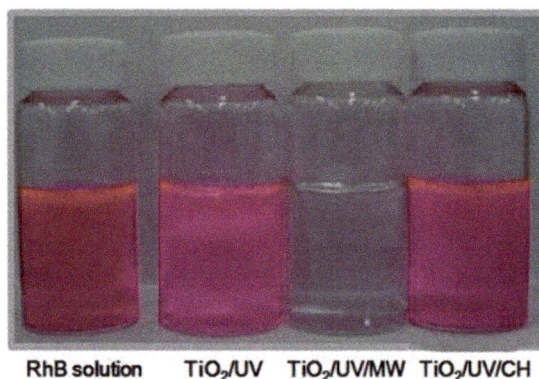

RhB solution   TiO$_2$/UV   TiO$_2$/UV/MW   TiO$_2$/UV/CH

Microwave effects were examined by the temporal decay of total organic carbon (TOC) in the degradation of aqueous RhB solutions under four different methodologies. Results displayed in Figure 3a show that for a TiO$_2$ loading of 60 mg (volume 30 mL) there is no distinction between the efficacy of the UV and UV/CH methods. In the absence of the metal oxide TiO$_2$, no changes in TOC occurred when the RhB dye solution was irradiated only by the microwaves, even after 3 h. At a TiO$_2$ loading of 30 mg (volume 30 mL) the UV/MW method proved very efficient in decreasing the TOC by nearly a factor of six from 18.6 mg·L^{-1} (ppm) to about 3 mg·L^{-1} after 3 h.

The degradation of RhB was also examined at two different UV-light irradiances (0.3 and 2.0 mW·cm^{-2}) to assess the microwave effects through the loss of TOC as depicted in Figure 3b. Clearly, the degradation of RhB by the UV/ MW was faster even at the lower irradiance of 0.3 mW·cm^{-2} than occurred by the UV method at the higher irradiance of 2.0 mW·cm^{-2}. Evidently, the situation at the lower UV-light irradiance accentuated the effect of the microwave radiation.

Figure 3. (a) Decrease of total organic carbon (TOC) in the decomposition of RhB solution (initial TOC concentration, $18.6 \ mg \cdot L^{-1}$; 30 mL) by MW (without TiO_2), UV (60 mg), UV/CH (60 mg) and UV/MW (30 mg); **(b)** Temporal evolution of the decrease of TOC during the degradation of RhB solution (0.050 mM, 30 mL) at a radiance of 0.3 and 2.0 $mW \cdot cm^{-2}$; **(c)** Decrease of TOC values for the influence of different added gases on the degradation of RhB (0.050 mM); **(d)** Decrease of TOC for RhB solutions (0.050 mM, 30 mL) with TiO_2 loading (30 mg) by the UV/MW method (microwave applied power at 150 W, 225 W and 300 W); **(e)** Temporal evolution of the formation of NH_4^+ ions in the decomposition of RhB (0.050 mM) using the UV, UV/CH and UV/MW methods; the radiance was 0.3 $mW \cdot cm^{-2}$. Reproduced from [10]. Copyright 2002 by the American Chemical Society.

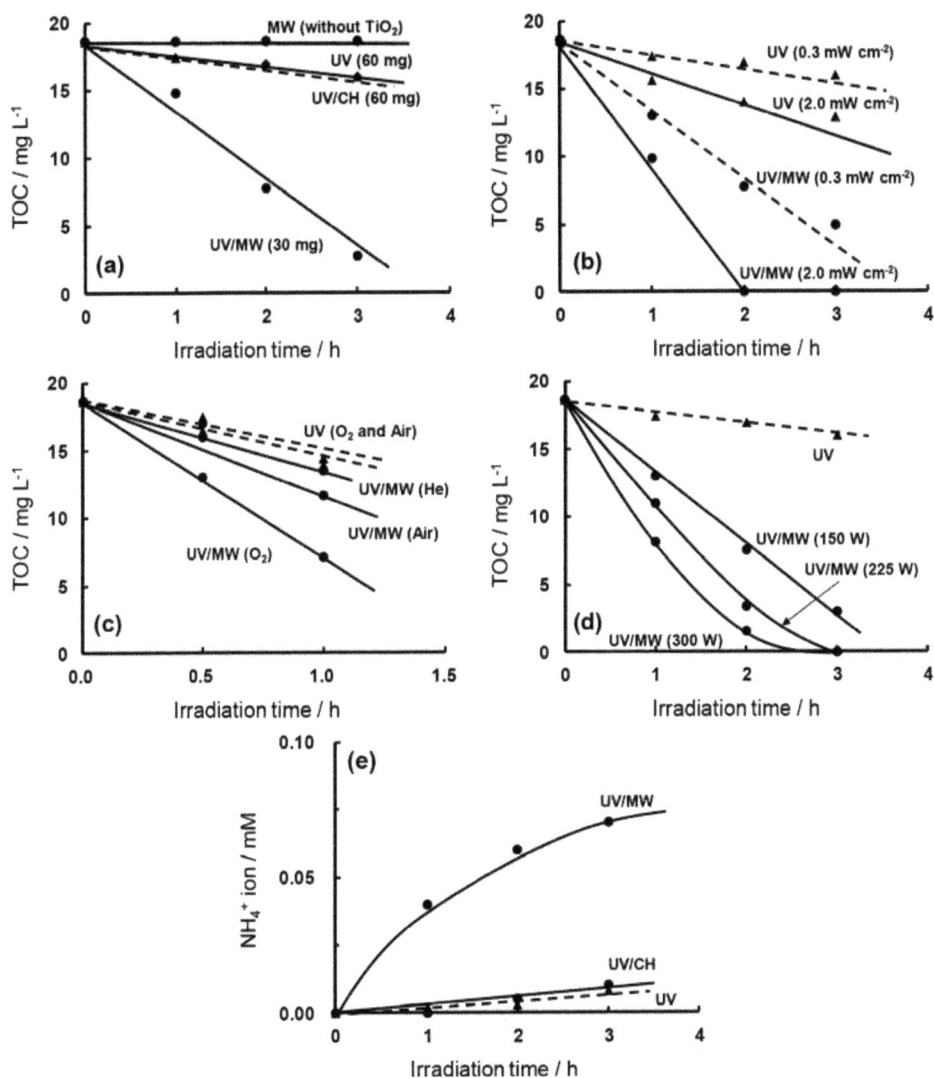

The disappearance of TOC in aqueous RhB solutions under various atmospheric conditions (air, oxygen gas, and helium gas) in the presence of TiO_2 is illustrated in Figure 3c. Recombination of photogenerated valence band holes with conduction band electrons after UV irradiation of TiO_2 particles is known to compete with formation of reactive oxygen species ($^\bullet OH$ and $^\bullet OOH$ radicals) that typically lead to the degradative processes. The degradation of an organic pollutant via a TiO_2 photoassisted reaction typically follows the order oxygen gas > air > inert gas (e.g., He). In the present instance, integrating microwave irradiation to UV irradiation for an oxygen-saturated solution led to twofold enhancement in the decrease of TOC [10]. Moreover, the decrease of TOC in the air-equilibrated RhB solution by the UV/MW method was enhanced relative to the degradation of oxygen-saturated RhB solutions by the UV method alone. Within experimental error, the decomposition rate of the helium-purged RhB solution by the UV/MW method was nearly the same as the rate for an oxygen-purged dye solution by the UV method alone, again indicating the influence of microwave factors in the degradation process.

The impact of microwave power on the degradation of RhB solutions is illustrated in Figure 3d. The temperature rise from ambient (25 °C) at power levels of 150, 225 and 300 W was 62, 75 and 138 °C, respectively, within the first 20 min of irradiation. The increase in microwave power enhanced the degradation of RhB solutions, as witnessed by the decrease of TOC for an irradiation period of about 2 h: the decrease in TOC by the UV/MW was 61% at a microwave power of 150 W, 82% for 225 W, and 92% at 300 W. In the absence of MW irradiation, the extent of TOC decrease by the UV method alone was only 14%. Clearly, under the conditions used, microwave power output enhanced the decomposition dynamics.

Formation of NH_4^+ (and NO_3^-) ions in the decomposition of RhB was also investigated by the UV, UV/CH and UV/MW methods in the presence of TiO_2 nanoparticulates (only data of NH_4^+ are shown in Figure 3e) [10]. Again, the UV/MW method led to more significant changes than either the UV or UV/CH methods alone in converting the two nitrogen atoms in the RhB structure. In this regard, the mineralization yields of the two nitrogen atoms, given as the sum of the yields of NH_4^+ and NO_3^- ions, respectively, after 3 h of irradiation were 77% (70% + 7%) for the UV/MW procedure, 12.8% (10% + 2.8%) for the UV method, and 8% (8% + 0%) for the UV/CH method. The increased formation yield of NH_4^+ ions by the UV/MW method was significant by comparison with the UV and UV/CH methods. These features inferred that the mechanistic details in the decomposition of RhB in aqueous media in the presence of TiO_2 differed under UV/MW irradiation relative to the UV-induced degradation.

The initially-formed intermediates from the degraded RhB were identified by electrospray ESI ionization mass spectra (positive ion mode) and subsequently confirmed by HPLC/absorption spectroscopy [16]. An initial adsorption model of RhB molecule on the TiO_2 surface was proposed by computer simulations that led to estimates of frontier electron densities of all atoms in the RhB structure, which afforded inferences as to the position of $^\bullet OH$ (or $HOO^\bullet$) radical attack on the RhB structure. Results from these simulations led to the proposed degradation mechanisms summarized in Scheme 1 [16]. For the UV method, RhB approaches the positively charged TiO_2 surface through the two oxygen atoms in the carboxylate function bearing the greater negative charge, with further assistance provided by the repulsion between the positively charged nitrogen atoms and the positive TiO$_2$ surface. Accordingly, de-ethylation of RhB by the UV method was rather inefficient as evidenced by the formation of intermediate (**I**) only (LC/MSD analysis).

Scheme 1. Proposed initial mechanistic steps in the degradation of RhB dye by the UV and UV/MW methods. Reproduced from [16]. Copyright 2002 by the American Chemical Society.

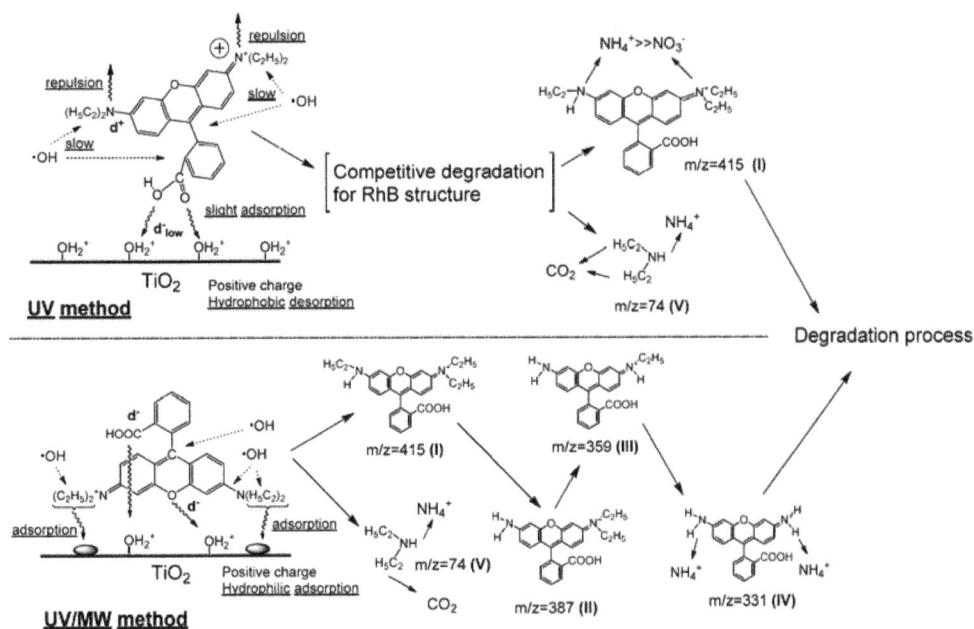

Diethylamine was formed through the cleavage of the C-N bond in the C-N(C$_2$H$_5$)$_2$ fragment of RhB. For the UV/MW method, the increase of the hydrophobic nature of TiO$_2$ through microwave irradiation facilitated adsorption of RhB through the aromatic rings aided by the three oxygen atoms in RhB; that is, RhB lay flat on the particle surface. The principal intermediates formed in the degradation of RhB by the UV/MW method were the N-de-ethylated species **I** to **IV**. Ultimately, the amino groups of intermediate **IV** were converted predominantly into NH$_4^+$ ions. The hydrophobic methyl component of the ethyl group could not adsorb onto the hydrophilic TiO$_2$ surface, so that transformation of the nitrogen atoms was not a priority event in the photoassisted process.

The surface electric charge of TiO$_2$ particles is another important factor that impinges on the adsorption of substrates on the TiO$_2$ surface. Accordingly, the electric charge on the TiO$_2$ surface was ascertained by a zeta potential analysis using the coupled microwave/UV irradiation system [13]. Under UV light irradiation alone, the zeta-potential was positive in acidic media and decreased with increase in pH of the aqueous TiO$_2$ dispersions. The point of zero charge (pzc) of the TiO$_2$ particles under UV irradiation was attained at pH ~ 6.7. On the other hand, the typical zeta-potential curve was not evident under simultaneous UV/MW irradiation conditions. In the latter case, the zeta-potential remained somewhat positive in the 0–20 mV range for dispersions throughout the pH range 4–9. A change of the TiO_2 surface charge by the microwave radiation changes the adsorbed state of RhB in a manner that facilitates the cationic nitrogen group to approach the TiO_2 surface.

3.2. Degradation of 2,4-Dichlorophenoxyacetic Acid (2,4-D) in Aqueous TiO₂ Dispersion

Many studies have reported the photocatalyzed decomposition of such chlorinated pollutants as the agrochemicals containing the triazine skeleton (atrazine) [18], dibromochloropropane [19], hexachlorocyclohexane [20], *p,p'*-DDT [21], *p,p'*-DDE [22], 2,4-dichlorophenol [23], kelthane [24], polychlorinated biphenyls (PCB) [25], polychlorinated dibenzo-*p*-dioxins [26], pentachlorophenol (PCP) [27], and 2,4,5-trichlorophenoxyacetic acid [28]. Results indicate that complete dechlorination (reduction reaction) of chlorinated compounds by the photocatalytic degradative method tends to be rather slow toward the mineralization of organic carbon atoms (oxidation reaction) to carbon dioxide. As an example, the TiO_2-catalyzed photodecomposition of the herbicide 2,4-dichlorophenoxyacetic acid (2,4-D) in aqueous dispersions has been investigated extensively for several years [29–32]. This agricultural chemical is a widely used and highly toxic synthetic phytohormone (toxin) that the United States Environmental Protection Agency has classified as a suspected endocrine disruptor. Not surprisingly then that 2,4-D was chosen as a model pollutant in in examining the dismantling of chlorine-containing compounds [33].

The degradation of 2,4-D was monitored by the loss of UV absorption at 285 nm and loss of TOC using the UV and UV/MW methods for oxygen-purged 2,4-D/TiO_2 dispersions. It was shown [33] that the decomposition rate of 2,4-D by the UV/MW method is accelerated when compared with the UV method (Figure 4a). Loss of TOC (Figure 4b) is consistent with the UV spectral observation. Detection and identification of the intermediates produced during the degradation of the 2,4-D substrate were carried out by electrospray mass spectral techniques (LC-MSD, Agilent Technologies Inc., Alpharetta, GA, USA).

Figure 4. Disappearance of (**a**) UV absorption and (**b**) TOC in the photodegradation of 2,4-D by the UV and UV/MW methods. Reproduced from [33]. Copyright 2007 by Elsevier B.V.

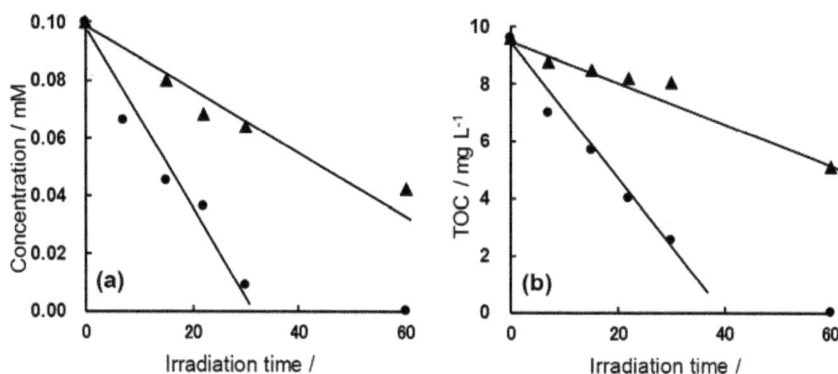

Results from LC-MSD suggested the initial mechanistic steps (Scheme 2) for the photodegradation of 2,4-D by the UV and UV/MW methods [33]. The initial step in the UV method was taken as the cleavage of the C-C bond in the -CH₂-COOH group in 2,4-D to yield formic acid and 2,4-dichlorophenoxymethanol (**II**) through addition of an •OH radical to the -•CH₂ fragment (step A). Degradation of intermediate (**II**) then produced 2,4-dichlorophenoxyformaldehyde (**VI**), which on further irradiation yielded 2,4-dichlorophenol (**III**), followed by formation of 3,5-dichloro-1,2-benzenediol (**IV**) and 4-chloro-1,2-benzenediol (*i.e.*, 4-chlorocatechol; **V**). In step B, the 6-hydroxy-2,4-dichlorophenoxyacetic acid (**I**) was formed by attack of •OH radicals on the benzene ring, after which it led to the 3,5-dichloro-1,2-benzenediol species (**IV**) and to 4-chlorocatechol (**V**). For the UV/MW method, degradation of 2,4-D gave initially 6-hydroxy-2,4-dichlorophenoxyacetic acid (**I**) by the same mechanism as the UV method (step C). The acetic acid and formic acid intermediates were preceded by cleavage of the O-C bond in the O-CH₂COOH fragment of 6-hydroxy-2,4-dichlorophenoxy acetic acid (**I**) and of 2,4-D (step D) to give the 2,4-dichlorophenol intermediate (**III**) and subsequently the 3,5-dichloro-1,2-benzenediol species (**IV**). The differences in these initial degradations are due to the initial cleavage position of the 2,4-D molecular structure. In other words, the C-C bond (83 kcal·mol⁻¹) fragment was cleaved in the UV method, whereas the O-C bond (84 kcal·mol⁻¹) was cleaved in the UV/MW method.

Scheme 2. Proposed initial degradation mechanism of the photodegradation of 2,4-D by the UV and UV/MW methods. Reproduced from [33]. Copyright 2007 by Elsevier B.V.

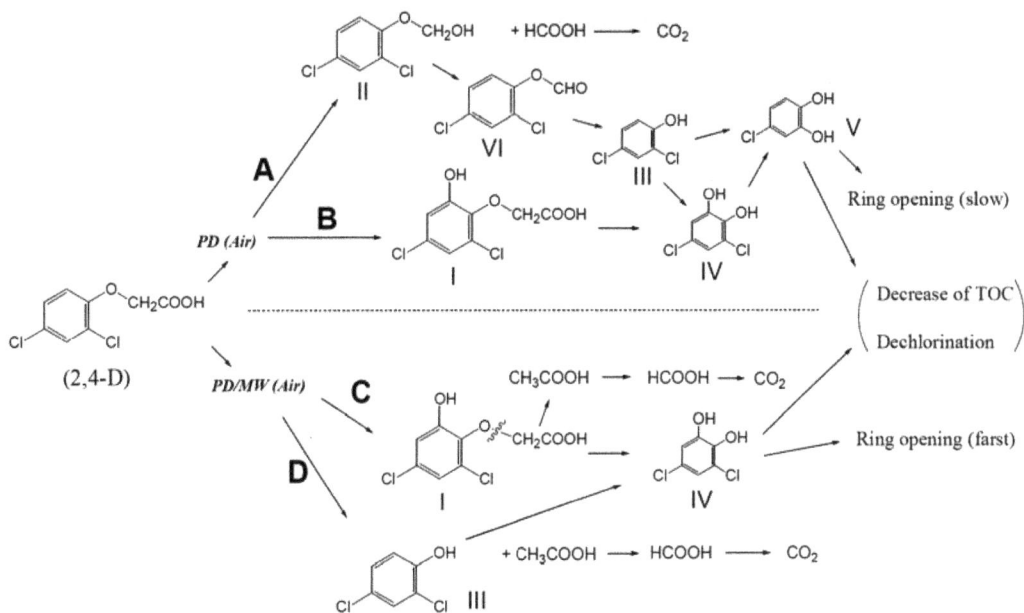

3.3. Degradation of 1,4-Dioxane in Aqueous TiO₂ Dispersions

The degradation of an aqueous 1,4-dioxane solution with TiO₂ was monitored by the decrease of TOC (Figure 5) [34]. The extent of decrease of TOC was 86% when the degradation was carried out by the UV/MW method for a 2-h irradiation period. By contrast, the decrease of TOC was only

13% and 21% by the UV and UV/CH methods, respectively. Under microwave irradiation alone, no decomposition of 1,4-dioxane occurred. The intermediates produced in the degradation of 1,4-dioxane were ascertained by LC-MS and HPLC-UV/Vis analyses [34]. The two oxygen atoms in the dioxane adsorbed on the TiO_2 surface, following which peroxidation and/or hydroxylation events occurred with subsequent opening of the ring and ultimate formation of CO_2 gas. The 1,2-diformyloxyethane (*i.e.*, ethylene glycol diformate) and ethylene glycol intermediates were observed early in the decomposition by the UV/MW method, whereas they were generated gradually under UV and UV/CH irradiations. The degradation of aqueous 1,2-diformyloxyethane solution (0.10 mM) and aqueous ethylene glycol solution (0.10 mM) was also performed using the UV, UV/CH and UV/MW methods. The extent of TOC decrease for 1,2-diformyloxyethane was 2% (UV), 17% (UV/CH) and 79% (UV/MW), while the decrease of TOC for ethylene glycol was 17% (UV), 16.3% (UV/CH) and 76% (UV/MW). Clearly, the decomposition events of two of the intermediates identified in the degradation of 1,4-dioxane were also significant when carried out by the UV/MW method.

Figure 5. Decrease of TOC for the degradation of 1,4-dioxane. Reproduced from [34]. Copyright 2002 by the Japan Society of Color Material.

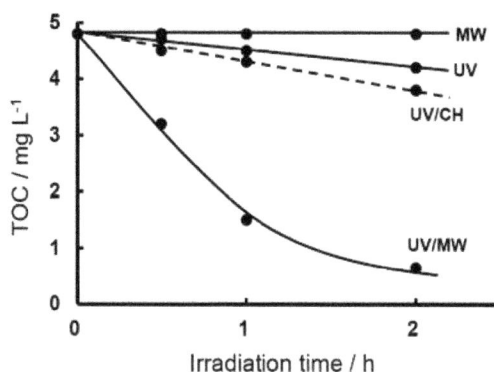

3.4. Non-Thermal Microwave Effect(s) in TiO₂ Photoassisted Reactions

3.4.1. Low Molecular Weight Organics

The degradation of many water contaminants by the UV/MW method is improved significantly when compared with the UV method. In addition to the decomposition of the initial contaminant, the decomposition of the resulting intermediates is also enhanced by the UV/MW method. To ascertain these inferences, studies examined the microwave specific (*i.e.*, non-thermal) effects in relation to the chemical structure of low molecular weight compounds, whose photoassisted mineralization in the presence of titanium dioxide typically leads to formation of CO_2 by an oxidation process via alcohol, aldehyde, and carboxylic acid intermediates (Reaction (1)) [35]. Delineating the effect of microwave radiation on these compounds provided an important base for clarifying the mechanistic features of the degradation and mineralization:

$$R\text{-}CH_3 \rightarrow R\text{-}CH_2OH \rightarrow R\text{-}CHO \rightarrow R\text{-}COOH \rightarrow RH + CO_2 \qquad (1)$$

Accordingly, the tendency of water-soluble organics to be photomineralized under microwave and UV irradiation was examined to establish the mechanistic features using model substrates bearing alcoholic (methanol, ethanol 1-propanol, ethylene glycol and glycerin) and carbonyl functions (acetone, formic acid and acetic acid). Trends were seen from the decrease of the kinetics of loss of TOC with irradiation time [36]. The dynamics of the temporal losses of TOC on irradiation of the various dispersions by the UV, UV/MW and UV/CH methods are reported in Table 1.

Table 1. Kinetics (k) in the loss of TOC for 0–120 min of irradiation by the UV, UV/CH and UV/MW methods, except for methanol, formic acid and acetic acid that were irradiated only for 0–90 min. TH factor: increased rate by UV/CH method relative to UV; TH-MW factor: increased rate by UV/MW method relative to UV; and MW factor: increase in rate by the UV/MW method with respect to UV/CH. Reproduced from ref. [36]. Copyright 2007 by Elsevier B.V.

Compound	Mineralization Rate Constants, k_{TOC} (10^{-3} min^{-1})			Increase in Rates by Different Factors		
	UV	UV/CH	UV/MW	TH [a]	TH-MW [b]	MW [c]
Methanol	1.0	9.1	25.3	9.1	25.3	2.8
Ethanol	0.9	2.7	3.4	3.0	3.8	1.3
1-Propanol	2.0	1.8	2.1	0.9	1.0	1.2
Ethylene glycol	1.6	2.2	9.0	1.4	5.6	4.1
Glycerin	1.8	6.3	11.9	3.5	6.6	1.9
Acetone	2.7	3.6	6.0	1.3	2.2	1.7
Formic acid	8.4	11.3	23.1	1.3	2.7	2.0
Acetic acid	3.9	6.1	20.5	1.6	5.3	3.4

[a] Thermal factor by comparing the k_{TOC} between the UV/CH relative to UV; [b] Thermal factor caused by the microwave irradiation on comparing the k_{TOC} obtained from the UV/MW method relative to UV; [c] Microwave factor from the comparison of k_{TOC}'s from the UV/MW method relative to the UV/CH method.

Methanol: The rate of mineralization of methanol by the UV/MW method was nearly three times faster than by the UV/CH method, even though the temperatures at which the process occurred were identical. As well, the thermal effect was also not insignificant—the photoassisted mineralization of this substrate under conventional heating was nearly an order of magnitude faster; e.g., compare the UV/CH method with the UV method (the TH factor).

Ethanol: Adding one C atom to methanol caused no changes in the kinetics of the photoassisted mineralization of this substrate by the UV method. By contrast, the rates were significantly attenuated for the UV/CH and UV/MW by factors of ~3 and 3.8, respectively. The thermal factor also caused a threefold increase in the rate of mineralization, whereas the MW effect was somewhat subdued (factor of 1.3) indicating that the MW effect was somewhat limited relative to the TH factor.

1-Propanol: With the addition of two C atoms to methanol, both the TH and the MW factors no longer seemed to play a role in the mineralization of this substrate. The kinetics of mineralization were practically identical.

Ethylene glycol: Relative to methanol, this substrate contains an additional C atom as well as an additional OH function. The TH effect was considerably attenuated; *i.e.*, the rate of mineralization by the UV/CH method relative to the photoassisted UV method was just slightly faster. By contrast, the MW factor caused a significant fourfold increase in the rate of mineralization, whereas the combined TH–MW effect led to a nearly six-fold increase in the kinetics.

Glycerin: The structure of glycerin contains three C atoms and three OH functions, *i.e.*, one additional -CH_2-OH unit with respect to ethylene glycol. The dynamics of the photoassisted mineralization of these two highly OH-loaded substrates by the UV method were experimentally similar. However, the TH factor increased the rates substantially by nearly a factor of 3.5 as evidenced by comparing the UV method with the UV/CH method. The MW factor also impacted on the mineralization of this substrate with the rate of mineralization by the UV/MW method being twofold faster than by the UV/CH method. Just as evidenced for ethylene glycol, the mineralization of glycerin and the corresponding TH and MW factors were somewhat attenuated relative to methanol, yet the photoassisted process (UV method) was enhanced, albeit to a small extent relative to methanol. Evidently, both the number of -CH_2- units and -OH functions seem to play a role in either promoting or attenuating the degradation of the substrates with the assistance of the TH and MW effects.

Acetone: The TH effect and the MW effect had only a small influence on the dynamics of mineralization of acetone, whereas the rate of the photoassisted mineralization by the UV method was not very different from that of 1-propanol.

Formic acid: Of all the substrates examined, formic acid was mineralized faster by UV irradiation of the aqueous TiO_2 suspension. Compared to methanol, the dynamics of mineralization by the UV/CH and UV/MW methods were rather similar. Regardless, while the TH factor was significantly different from that of methanol (1.3 *versus* 9.1; see column 5 of Table 2), the MW effect was only slightly smaller (2.0 *versus* 2.8).

Acetic acid: The rates of mineralization of acetic acid were about two times slower for both the UV and UV/CH methods relative to formic acid, whereas the rates for the UV/MW method were rather similar. However, even though the influence of the TH factors were similar for both formic and acetic acids, the MW effect was somewhat greater for the latter.

3.4.2. Degradation of the Endocrine Disruptor BPA

In the degradation of various pollutants in wastewaters, the effect of microwaves manifested itself in the enhancement of the reaction dynamics of some aromatics that otherwise could not be obtained by conventional heating. In this regard, we examined the microwave-assisted photodegradation of bisphenol-A (BPA) in the presence of TiO_2 at near-ambient temperature so as to unravel some details on the importance of the microwave non-thermal factor(s) [37]. BPA is one of many listed endocrine disruptors that tend to accumulate in the natural environment with serious consequential damage to the reproductive cycle in various animal species. To further delineate between the microwaves' thermal and non-thermal factors, the degradation of BPA was later revisited [38] using an integrated microwave/ photoreactor (MPR) system that was equipped with a cooling system (see Figure 6a). The Pyrex reactor consisted of a double-layered structure with internal and external diameters of 75 and 100 mm, respectively. The silicone oil refrigerant was circulated through the inner part of this double layered structure using a circulation cooling apparatus. The silicone oil did not absorb the microwave radiation, and the temperature was maintained at −20 °C by the recirculation apparatus operating at the maximal flow rate possible. The course of the BPA degradation was followed using three experimental methods: (a) the UV/MW; (b) the UV/MW method under controlled ambient temperature with the cooling system (UV/MW/Cool); and (c) the photoassisted method alone (UV) mediated by TiO_2. The temperatures of the solution by the UV and UV/MW/Cool methods were kept at 21 °C, whereas the temperature in the UV/MW method reached 85 °C after 1 h of irradiation.

Figure 6. (a) Schematic illustration of the microwave-assisted photoreactor (MPR) coupled to a cooling system in the microwave multimode applicator; **(b)** Temporal decrease of the concentration of bisphenol-A (BPA: 0.050 mM) during its decomposition in aqueous media by photoassisted oxidation (UV), by the microwave-/photo-assisted oxidation (UV/MW) method, and by the integrated microwave-/photo-assisted degradation under cooling conditions (UV/MW/Cool). Reproduced from [38]. Copyright 2007 by Elsevier B.V.

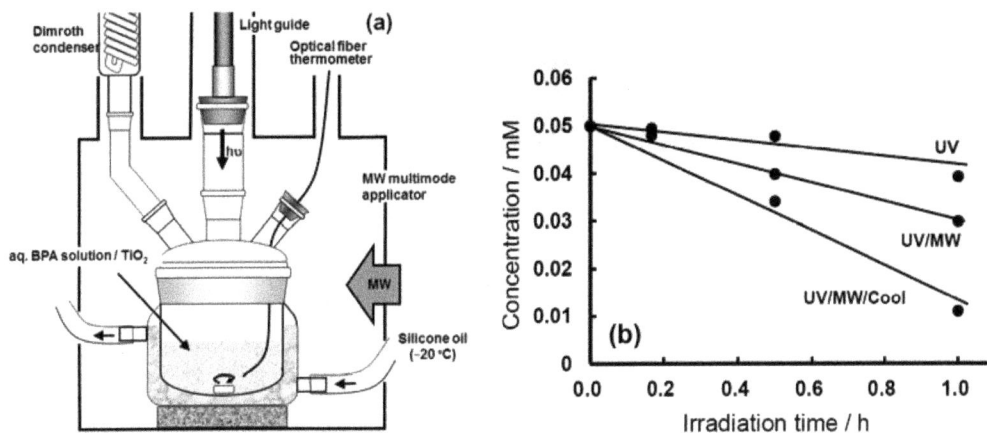

The time course of the degradation of bisphenol-A by the UV, UV/MW and UV/MW/Cool methods is reported in Figure 6b. The relevant degradation dynamics subsequent to UV/MW irradiation were *ca.* twofold faster than for the UV method alone. The UV/MW/Cool method (*ca.* 21 °C) was twofold faster than for the degradation of BPA by the UV/MW method taking place at the higher temperature. The extent of degradation of BPA followed the order: UV/MW/Cool (78%) > UV/MW (42%) > UV (21%) after microwave and/or UV irradiation for 1 h. Clearly, the microwave-/photo- assisted degradation of BPA was most efficient when carried out at near-ambient temperature, under which the microwave-assisted photodegradation of BPA was not only due to a microwave thermal effect, but also to a significant non-thermal effect that might implicate the formation of hot spots on the TiO_2 particle surface. The photodegradation was enhanced under constant ambient temperature. The non-thermal effect of the microwave radiation was also shown in the selective radical synthesis of 3-cyclohexyl-1-phenyl-1-butanone [39].

3.5. Increase in Radical Species on TiO_2 under Microwave Irradiation

When a semiconductor particle is illuminated at wavelengths corresponding to photon energies equal to or larger than the bandgap energy, valence band electrons are excited to the conduction band and holes are produced in the valence band. Most of the electron-hole pairs produced recombine after some tens of picoseconds with the energy being released as emission of photons, phonons or both [40]. Coordination defects at the surface of the particle and defects within the particle lattice trap the remaining charges [41]. The holes trapped at the surface have a highly reactive oxidation potential and the electrons have a highly reactive reduction potential. Thus, photogenerated holes and electrons can induce reactions at the surface that, for historical reasons, are often referred to as photocatalytic reactions [42]. Titanium dioxide has a bandgap of 3.20 eV (anatase crystal), which corresponds to a wavelength of 387 nm. This means that electron-hole pairs are created when TiO_2 is radiated with UV-light with wavelengths shorter than 387 nm (Reaction (2)). To the extent that not all the photogenerated electrons and holes recombine, some of the holes can migrate to the surface and react with surface-bound -OH groups and/or water molecules surrounding the particles that lead ultimately to the formation of hydroxyl radicals (Reaction (3)). Dissolved oxygen molecules react with conduction band electrons (e^-) to yield superoxide radical anions ($O_2^{-\bullet}$; Reaction (4)), which on protonation generate the hydroperoxy radicals $^\bullet OOH$ (Reaction (5)). Accordingly, the photooxidation of organic substrates with the UV/TiO_2-driven photoassisted process (Reaction (6)) depends on the concentration of $^\bullet OH$ (and/or $^\bullet OOH$) radicals produced by the photooxidation of surface hydroxyl groups and/or chemisorbed H_2O.

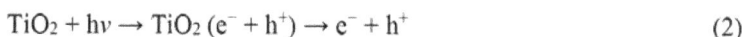

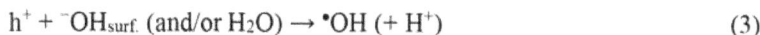

$$TiO_2 + h\nu \rightarrow TiO_2 \ (e^- + h^+) \rightarrow e^- + h^+ \tag{2}$$

$$h^+ + {}^-OH_{surf.} \ (and/or \ H_2O) \rightarrow {}^\bullet OH \ (+ \ H^+) \tag{3}$$

$$e^-_{cb} + O_2 \rightarrow O_2^{-\bullet} \tag{4}$$

$$O_2^{-\bullet} + H^+ \rightarrow {}^\bullet OOH \tag{5}$$

$${}^\bullet OH \ (or \ {}^\bullet OOH) + organic \ pollutant \rightarrow Oxidative \ products \tag{6}$$

The possible enhancement of the photoactivity of metal-oxide specimens subsequent to being exposed to microwave radiation from the viewpoint of the amount of $^\bullet OH$ radicals generated was also investigated [43]. Formation of $^\bullet OH$ radicals during TiO_2-assisted photooxidations that were driven simultaneously by UV light and microwave radiation was probed by electron spin resonance spectroscopy employing a novel setup in which the ESR sample (contained the DMPO spin-trap agent and TiO_2 particles in aqueous media) was irradiated by both UV light and microwave radiation [11,43]. In this case, microwave radiation was produced using a magnetron microwave generator (frequency, 2.45 GHz), a three-stub tuner, a power monitor, and an isolator (Figure 7).

The UV irradiation source was an Ushio 250-W mercury lamp; the emitted UV light irradiated the sample at an angle to the horizontal plane using a fiber optic light guide.

Figure 7. Setup used to generate $^{\bullet}$OH radicals in water alone under MW irradiation, in an aqueous TiO_2 dispersion by MW irradiation alone, and by the UV and UV/MW methods. Reproduced from [43]. Copyright 2003 by Elsevier B.V.

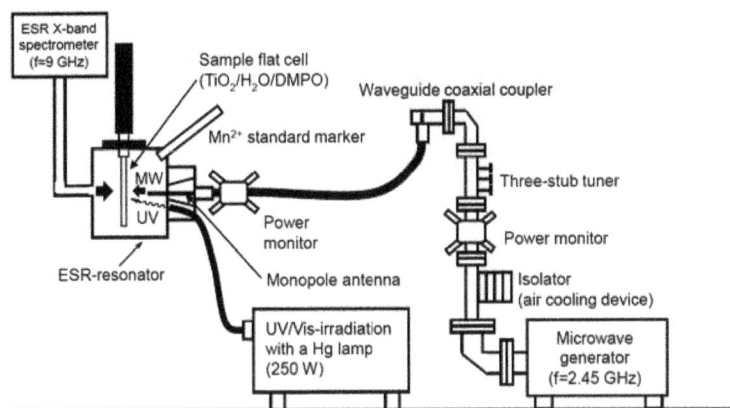

The number of $^{\bullet}$OH radicals generated under various experimental conditions is summarized in Table 2. For P25 titania, the number of $^{\bullet}$OH radicals produced by the UV/MW method was nearly 30% greater than the quantity generated by the UV method alone [43]. A fivefold increase in incident microwave power from 3 to 16 W led to a significant increase (*ca.* 40%) in the number of $^{\bullet}$OH radicals. Such an increase was sufficient to increase the efficiency of the photooxidation of the organic pollutant in water.

Table 2. Number of DMPO-$^{\bullet}$OH spin adducts produced in the various heterogeneous systems under microwave irradiation, UV irradiation, and MW/UV irradiation relative to those formed in the rutile TiO_2 specimen for the $TiO_2/H_2O/MW$ heterogeneous system. Reproduced from [43]. Copyright 2003 by Elsevier B.V.

Methodology	P25	UV100	Anatase	Rutile
UV	182	45	110	110
UV/MW (3 W)	259	51	92	76
UV/MW (16 W)	369	-	-	-

For the UV100 sample, the increase in the number of $^{\bullet}$OH radicals produced was only 10% greater on increasing the MW power five times. On the other hand, the number of $^{\bullet}$OH radicals generated for the pristine anatase and rutile TiO_2 samples decreased under microwave irradiation. The P25 specimen was clearly influenced by the microwaves and generated $^{\bullet}$OH radicals efficiently under the influence of microwave effects. Therefore, the rate of decomposition was enhanced when P25 was used to decompose the wastewater sample by the UV/MW method. On the other hand, to the extent that the quantity of $\cdot$OH radicals produced by the other TiO_2 does not increase even when irradiated with microwaves, the rate of decomposition is not enhanced. To test this assertion, the photodegradation of aqueous 4-chlorophenol (4-CP) solution was examined using four different kinds of TiO$_2$ particles [14,15]: Evonik P25 titania, Hombikat UV100 TiO$_2$, Ishihara ST01 titanium dioxide (Ishihara Sangyo Kaisha Ltd., Nishi-ku, Osaka, Japan), and pristine anatase and rutile titania products (Wako Pure Chemicals Co. Ltd., Chuo-ku, Osaka, Japan). The physical properties of these various TiO$_2$ specimens and the corresponding zero-order rates for the degradation of 0.025 mM 4-CP in aqueous media by the UV, UV/MW and UV/CH methods are reported in Table 3 [11].

Table 3. Physical properties of various TiO$_2$ specimens and zero-order rates for the degradation of 4-chlorophenol (4-CP: 0.025 mM) using several methods: namely, the UV, UV/MW and UV/CH methods. Reproduced from [11]. Copyright 2009 by Copyright 2002 by the American Chemical Society.

Sample	Anatase (%)	Particle Size (nm)	Surface Area (m$^2 \cdot$g^{-1})	Zero-Order Rates (10^{-4} mM$\cdot$min^{-1})		
				UV	UV/MW [a]	UV/CH [a]
P25	82	33	52	1.0	2.5	1.7
UV100	100	10	323	0.6	1.4	1.5
Anatase	100	369	10	1.5	1.7	1.8
Rutile	0	263	16	0.02	0.05	0.17

[a] Reactions carried out under otherwise identical temperature conditions.

The microwave non-thermal effect(s) (UV/MW *versus* UV/CH) is clearly indicated in the Evonik P25 sample, which consists of well interwoven anatase (*ca.* 80%) and rutile (*ca.* 20%) crystallite forms; this sample is a well-known material that has manifested significant photoactivity in several photoreductive and photooxidative processes. The microwave thermal effect was observed in other TiO$_2$ specimens. It is curious that the microwaves affect the P25 system and enhance its activity as deduced from our studies, from which we inferred to be due to some structural features (see below) inherent in this TiO$_2$ system. Results are in agreement with the increase in the number of •OH radicals produced by the microwave radiation.

Heat treatment of P25 titania and ST01 TiO$_2$ in the presence of H$_2$ caused the former to change color from white to light blue, while the latter changed from white to pale yellow with the colors being stable even after exposure of the samples to air oxygen. The change in the extent of lattice distortion of the heat/H$_2$-treated ST01 specimen was two-times greater than for the P25 sample under otherwise identical conditions. The UV-visible absorption spectra of both sets of specimens revealed a broad unresolved band envelope above 400 nm that was attributed [44] to the formation of *F*-type color centers originating from oxygen vacancies [45]. The kinetics of photodegradation of 4-CP by these heat/H$_2$-treated specimens were enhanced significantly under UV/MW irradiation (*ca.* three times) relative to irradiation by the UV/CH and UV methods. As well, the treated P25 sample was nearly 25% more efficient than the untreated sample, while the corresponding ST01 specimen was 85% more efficient. Such increased efficiency under UV/MW irradiation for the heat/H$_2$-treated ST-01 specimen relative to the pristine ST01 sample was not due to a microwave thermal effect, but to a non-thermal effect of the microwaves impacting on the nanostructure of the metal-oxide samples [44].

4. Microwave Discharge Electrodeless Lamp

4.1. The Need for More Efficient UV Light Sources

It was shown in the preceding section that microwave radiation is effective in enhancing the photocatalytic activity of metal-oxide specimens. Accordingly, a treatment method that can treat larger quantities of pollutants in wastewaters is conceivable by a hybrid combination of the microwave technology and the photocatalytic technology. However, the light source can be a problem in the scale-up of microwave-assisted TiO_2-photoassisted processes. The bandgap energy of anatase TiO_2 is 3.20 eV, so that UV light less than 387 nm is necessary to activate this metal oxide. Generally, an electrode Hg lamp is used as the UV light source in Advanced Oxidation Processes. However, it is difficult to set up a typical mercury lamp in a microwave field from the viewpoint of the electrical discharge. In addition, it is difficult to open the window for the light beam to the microwave applicator because of possible leakages of the microwave radiation. Moreover, a small hole can only irradiate a small fraction of the dispersions. Accordingly, novel microwave-driven electrodeless lamps (MDELs) were developed by us (also photoreactors) and others as UV light sources to overcome many of the problems noted (see e.g., [46]).

4.2. Preparation of the MDEL Light Sources

Optimized conditions to obtain the best gas mixture ratios and internal gas pressures in the MDEL devices were examined using the setup illustrated in Figure 8 [46]. A quartz ampoule connected to the vacuum system was positioned in the microwave waveguide. The size of the quartz ampoule (MDEL) was 145 mm (length) by 18 mm (diameter). The initial internal pressure in the ampoule was set at 10^{-3} Torr (or $ca.$ 0.13 Pa) using a turbo molecular pump assisted by a rotary pump. Subsequently, mercury and argon gases (target gases) were introduced into the ampoule with the amount adequately adjusted by a mass controller. The pressure inside the ampoule was monitored with a capacitance manometer. The gas-mixture ratios were calculated from the volume ratio of each gas.

Figure 8. **(a)** Experimental setup for the examination of optimized conditions for a microwave discharge electrodeless lamp (MDEL); **(b)** photograph of nitrogen/argon mixed plasma light with the source of the MW radiation. Reproduced from [46]. Copyright 2009 by the Royal Society of Chemistry.

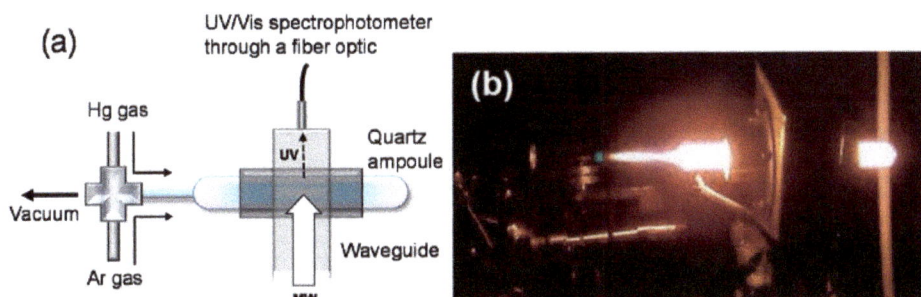

The UV-visible spectra of the emitted light plasma and the corresponding light intensities for each gas (and mixture) subjected to microwave irradiation were monitored through a fiber optic connected to a UV-Visible spectrophotometer (Figure 8a). The most suitable gas, gas-mixture ratio, and gas pressure for the MDEL device was determined using three criteria: (1) light intensity, (2) spectral pattern, and (3) self-ignition of the gases by MW irradiation alone. In one case, subsequent to evacuating the MDEL quartz envelope to 133×10^{-7} Pa, the system was purged with argon gas (133×10^{-3} Pa) after which a small quantity of liquid mercury was added. Figure 8b displays a photograph of the light emitted by the MDEL subsequent to MW irradiation.

A 100-W microwave source was used to irradiate the MDEL device positioned in the multimode microwave applicator. The emitted vacuum UV light (VUV) was attenuated by the ozone generated from the oxidation of atmospheric oxygen. However, this decrease in irradiance could be minimized by purging the multimode applicator with nitrogen gas. The main peaks of the VUV and UV range were 185 nm and 254 nm, respectively (Figure 9) [47]. These peaks are similar to the wavelengths emitted by a traditional low-pressure mercury lamp.

Figure 9. Vacuum-UV, UV, Visible and near-IR wavelengths emitted by the MDEL under microwave irradiation. Reproduced from [47]. Copyright 2009 by the Royal Society of Chemistry.

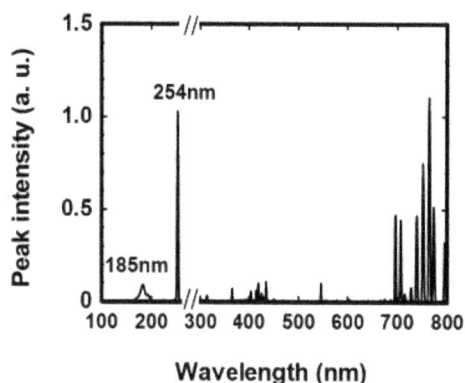

4.3. Traditional Hg Lamp versus an MDEL Light Source in the Photodegradation of 2,4-D in Aqueous TiO₂ Dispersions

An MDEL device of the same size as a domestic low-pressure mercury electrode lamp was used to examine the effect of the MDEL light source against a conventional Hg light source in the purification of a model wastewater sample containing the 2,4-D under otherwise identical power consumption [46]. The two UV electrode Hg lamps [dimensions: 10 cm (length) × 3 cm (diameter)] were located on top and bottom of a quartz pipe reactor. The TiO₂/2,4-D aqueous dispersion was circulated using a peristaltic pump coupled to a cooling device to maintain the solution temperature at *ca.* 27 °C.

The degradation yields for 2,4-D using the electrode Hg lamp and the electrodeless Hg-filled lamp (MDEL) were 33% and 50%, respectively, after a 30-min irradiation period under otherwise identical conditions (Figure 10a) [46]. The corresponding levels of dechlorination (Figure 10b) of 2,4-D were 20% and 33%, respectively. The total electric power used to power the electrode and electrodeless Hg lamp systems was in both cases 150 W. However, the average light irradiance for the electrode Hg lamp was three times greater (*ca.* 12 mW·cm^{-2}) than that of the electrodeless lamp (*ca.* 4 mW·cm^{-2}). Clearly, the degradation of 2,4-D by the TiO$_2$/MDEL/MW method, even at a smaller light irradiance, was faster than by the TiO$_2$/electrode Hg-lamp method without microwaves but at higher light irradiance. Hence, microwave irradiation enhanced the overall degradation of 2,4-D and made up for the poor transparency of the dispersion and for the lower irradiance used.

4.4. Degradation of Perfluoroalkoxy Acids in Aqueous Media Using Small MDELs

Perfluoroalkylated pollutants have been detected globally in the wildlife [47], including in such remote areas as the Arctic [48]. Stable and chemically inert (C-F bond energy: 568 kJ·mol^{-1} or 5.882 eV, making it one of the thermodynamically strongest bond known), these perfluoroalkylated chemicals repel water and oil, reduce surface tension better than other surfactants, and work well under harsh conditions [49]. Perfluorooctanoic acid (PFOA, C$_7$F$_{15}$COOH) is one such pollutant that is used to make fluoropolymers that can release the PFOA precursor by transformation of some fluorinated telomeres. In 2006, eight major companies and the Environmental Protection Agency (USA) launched a stewardship program in which the industry was committed: (i) to reduce global manufacturing emissions and product content of perfluorooctanoic acid and related chemicals by 95 percent by 2010; and (ii) to work toward total elimination of emissions and product contents within the 2010–2015 period.

Figure 10. (a) Temporal variations in the concentration of 2,4-D during its degradation in aqueous solutions using MDEL and electrode Hg lamps; **(b)** formation of Cl$^-$ ions during the dechlorination of the 2,4-D solution. Reproduced from [46]. Copyright 2009 by the Royal Society of Chemistry.

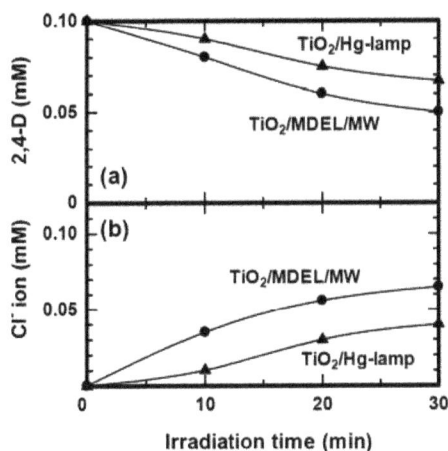

PFOA does not degrade naturally, and even with the use of advanced oxidation processes, its decomposition is rather difficult; as well, pyrolysis of PFOA necessitates relatively high temperatures [50]. Accordingly, a methodology that operates under milder conditions is most desirable for the decomposition and defluorination of PFOA and other perfluoroalkyls. Germane to this effort, Hori and coworkers reported the photodegradation of PFOA in aqueous media with and without a heteropolyacid photocatalyst ($H_3PW_{12}O_{40} \cdot H_2O$) [51] or in the presence of persulfate ($S_2O_8{}^{2-}$) [52]. In addition, the photoassisted defluorination and degradation of PFOA in aqueous media was also achieved by direct photolysis with 185-nm and/or 254-nm light from a low-pressure mercury lamp [53]. The extent of defluorination of PFOA attained by the latter method was 17% with the 185-nm VUV light after a 2-h irradiation period. In order to achieve the photodecomposition of perfluoroalkyls by the UV/MW method, we began to fabricate MDELs of a particular nature and design in our laboratory in Japan.

Small size microwave discharge granulated electrodeless lamps (MDELs) were constructed using vacuum-UV transparent synthetic quartz as the envelope and a mixture of Hg and Ar as the gas-fills [54]. Dimensions of the devices were 10 mm (length) by 5 mm (external diameter)—see Figure 11a. Subsequent to evacuating the MDEL quartz envelope to 133×10^{-7} Pa, the system was purged with argon gas (133×10^{-3} Pa) after which a small quantity of liquid mercury was added. Continuous microwave radiation was produced using a Hitachi Kyowa Engineering System microwave generator (frequency, 2.45 GHz; maximal power, 800 W), an isolator, a power monitor and a short-circuit plunger. The 300-mL air-equilibrated aqueous PFOA solutions were circulated with a peristaltic pump through the multipass MW/UV reactor (Figure 11b) containing the MDELs (20 pieces) at a flow rate of 600 mL·min^{-1}. Note that no TiO$_2$ was used in this experiment.

Figure 11. (a) Photograph of small MDELs and **(b)** the experimental setup of small MDEL device in a single mode microwave apparatus. Reproduced from [54]. Copyright 2011 by Elsevier B.V.

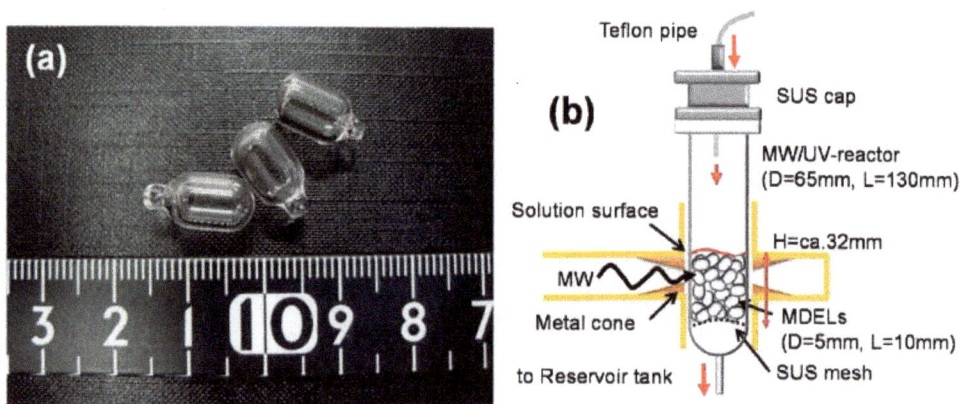

Figure 12 illustrates the defluorination of perfluorooctanoic acid in aqueous media using the 20 MDELs as the light sources [54]. The extent of defluorination of PFOA was *ca.* 80% after 200 min, but reached 100% upon further irradiation for another 200 min.

Figure 12. Time profiles of the extent of the photoassisted defluorination of perfluorooctanoic acid (PFOA) in aqueous solutions in the photoreactor containing the 20 MDELs systems. Reproduced from [54]. Copyright 2011 by Elsevier B.V.

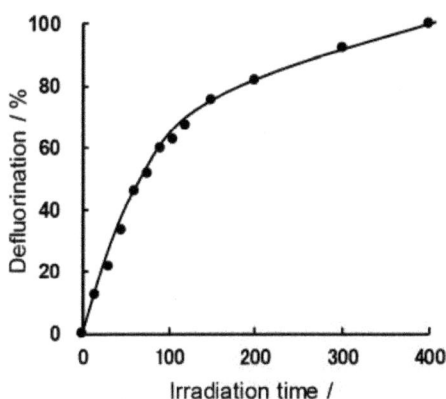

4.5. Purification of Water Using TiO₂-Coated MDEL Systems in Natural Disasters

The great earthquake that occurred in eastern Japan on 11 March 2011 was soon followed by a huge tsunami; hundreds of houses collapsed either directly by the ground movement or else were destroyed during the subsequent tsunami. The demand for drinking water by sterilization of rain water increased immediately after the earthquake disaster. In connection with it, the needs of water for other uses (e.g., water to rinse off mud, water for toilets, *etc.*) also grew exponentially. Accordingly, we recently examined the possible sterilization of resurgent water in the earthquake-stricken area using MDELs and rain water collected in a pond (used as model rain water) using the setup illustrated in Figure 13 [55].

To the extent that an electric supply is typically unavailable in such areas after the earthquake and ensuing tsunami, the experimental setup also included a system of solar cells that were connected to the equipment so that water purification could be achieved. The sterilization equipment consisted of a reaction vessel that contained 150 pieces of bead-shaped MDELs (10 mm long by 5 mm in external diameter) positioned in the single mode microwave applicator. Sterilization was carried out by continuous introduction of the rain water sample from the reaction container located in the upper part of the setup. The equipment was so designed as to process natural water continuously on site. However, if used for long periods of time the surface of the various MDELs tends to become dirty. To overcome this issue, a thin layer of TiO₂ nanoparticles was coated on the surface of the MDELs, such that the surface of the MDELs would remain clean by the well-known self-cleaning properties of titanium dioxide.

Figure 13. Water sterilization equipment used to sterilize rain water samples using the solar cells located on the right hand side of the photograph and the TiO₂-coated MDELs (150 pieces) [55].

Preliminary experiments showed that a single pass is sufficient to sterilize more than 95% of the rain water (rate: 0.4 mL·min^{-1}) through the reactor containing the TiO₂-coated MDELs. Moreover, 100% killing of *Escherichia coli* (*E. coli*) was observed by a single pass through the reactor. In fact, sterilization of rain water was complete after only two passes through the reactor aided by an appropriate pump. Importantly, this equipment can also be used to process waters continuously that may have been contaminated with agricultural chemicals such as insecticides.

5. Summary Remarks

Some 15 years have passed since the discovery that led to the improvement in photocatalyst activity by the microwave radiation. The photocatalyst is a material that absorbs the energy of electromagnetic waves (UV light) and changes it into chemical energy. In this regard, microwave radiation also consists of electromagnetic waves. The notion of irradiating TiO₂ with microwaves may appear strange at first because the photon energy (1×10^{-5} eV) of the microwaves of frequency 2.45 GHz is several orders of magnitude lower than the bandgap energy required (3.0–3.2 eV) to activate the TiO₂ semiconductor. Moreover, microwave radiation brings other effects to bear to a photocatalyst other than heat. Microwave non-thermal effects have been deduced to contribute significantly to the enhancement of a TiO₂-photoassisted reaction, as it may affect both the surface and the crystalline structure of the metal oxide toward reactions taking place at the surface. The mechanism of the effect of microwave radiation on photocatalyzed reactions has been resolved gradually. The enhanced treatment of wastewaters through improvement in photocatalyst activity when exposed to microwave radiation is now a clear possibility. Coupling the microwave radiation with UV light in TiO_2-photoassisted processes can contribute significantly to the treatment of wastewaters as a novel advanced oxidation technology (AOT).

Acknowledgments

This article would not have been possible without the fruitful collaboration of many University and industrial researchers, and not least without the cooperation of many students whose names appear in many of the earlier publications. One of us (SH) is grateful to the Japan Society for the Promotion of Science for financial support (JSPS; Grant-in-aid for Scientific Research No. C-25420820). A grant from the Sophia University-wide Collaborative Research Fund to S.H is also appreciated. One of us (NS) thanks Albini of the University of Pavia (Italy) for his continued hospitality during the many winter semesters in his PhotoGreen Laboratory.

Author Contributions

SH wrote the first draft of this article that was then followed by modifications by NS. Both authors have read and approved the final manuscript.

Conflicts of Interest

The authors declare no conflict of interest.

References

1. Kingston, H.M.; Haswell, S.J. (Eds.) *Microwave-Enhanced Chemistry*; American Chemical Society: Washington, DC, USA, 1997.
2. Gedye, R.; Smith, F.; Westaway, K.; Ali, H.; Baldisera, L.; Laberge, L.; Rousell, J.R. The use of microwave ovens for rapid organic synthesis. *Tetrahedron Lett.* **1986**, *27*, 279–282.
3. Giguere, R.J.; Bray, T.L.; Duncan, S.M. Application of commercial microwave ovens to organic synthesis. *Tetrahedron Lett.* **1986**, *27*, 4945–4948.
4. Loupy, A.; Varma, R.S. Microwave effects in organic synthesis, Mechanistic and reaction medium considerations. *Chem. Today* **2006**, *24*, 36–39.
5. Hashimoto, K.; Irie, H.; Fujishima, A. TiO_2 photocatalysis: A historical overview and future prospects. *Jpn. J. Appl. Phys.* **2005**, *44*, 8269–8285.
6. Kabra, K.; Chaudhary, R.; Sawhney, R.L. Treatment of hazardous organic and inorganic compounds through aqueous-phase photocatalysis: A review. *Ind. Eng. Chem. Res.* **2004**, *43*, 7683–7696.

7. Konstantinou, K.; Albanis, T.A. Photocatalytic transformation of pesticides in aqueous titanium dioxide suspensions using artificial and solar light: intermediates and degradation pathways. *Appl. Catal. B: Environ.* **2003**, *42*, 319–335.

8. Bhatkhande, D.S.; Pangarkar, V.G.; Beenackers, A.A. Photocatalytic degradation for environmental applications—A review. *J. Chem. Technol. Biotechnol.* **2002**, *77*, 102–116.

9. Fujishima, A.; Hashimoto, K.; Watanabe, T. *TiO₂ Photocatalysis: Fundamentals and Applications*; BKC Incorporated: Tokyo, Japan, 1999.

10. Horikoshi, S.; Hidaka, H.; Serpone, N. Environmental remediation by an integrated microwave/UV illumination method. 1. Microwave-assisted degradation of rhodamine-B dye in aqueous TiO₂ dispersions. *Environ. Sci. Technol.* **2002**, *36*, 1357–1366.

11. Horikoshi, S.; Sakai, F.; Kajitani, M.; Abe, M.; Emeline, A.V.; Serpone, N. Microwave-Specific Effects in Various TiO₂ Specimens. Dielectric Properties and Degradation of 4-Chlorophenol. *J. Phys. Chem. C* **2009**, *113*, 5649–5657.

12. Horikoshi, S.; Sakai, F.; Kajitani, M.; Abe, M.; Serpone, N. Microwave frequency effects on the photoactivity of TiO₂: Dielectric properties and the degradation of 4-chlorophenol, bisphenol A and methylene blue. *Chem. Phys. Lett.* **2009**, *470*, 304–307.

13. Horikoshi, S.; Matsubara, A.; Takayama, S.; Sato, M.; Sakai, F.; Kajitani, M.; Abe, M.; Serpone, N. Characterization of microwave effects on metal-oxide materials: Zinc oxide and titanium dioxide. *Appl. Catal. B: Environ.* **2009**, *91*, 362–367.

14. Konstantinou, I.K.; Albanis, T.A. TiO₂-assisted photocatalytic degradation of azo dyes in aqueous solution: Kinetic and mechanistic investigations. A review. *Appl. Catal. B: Environ.* **2004**, *49*, 1–14.

15. *Japanese Industrial Standards R 1703–2:2007*; Japanese Standards Association: Tokyo, Japan, 2007.

16. Horikoshi, S.; Hidaka, H.; Serpone, N. Environmental remediation by an integrated microwave/UV illumination method. V. Thermal and non-thermal effects of microwave radiation on the photocatalyst and on the photodegradation of rhodamine-B under UV/Vis radiation. *Environ. Sci. Technol.* **2002**, *37*, 5813–5822.

17. Horikoshi, S.; Serpone, N. Photochemistry with microwaves. Catalysts and environmental applications. *J. Photochem. Photobiol. C: Photochem. Rev.* **2009**, *10*, 96–110.

18. Pelizzetti, E.; Minero, C.; Piccinini, P.; Vincenti, M. Phototransformations of nitrogen containing organic compounds over irradiated semiconductor metal oxides: Nitrobenzene and atrazine over TiO₂ and ZnO. *Coord. Chem. Rev.* **1993**, *125*, 183–193.

19. Bahnemann, D.W.; Moning, J.; Chapman, R. Efficient Photocatalysis of the irreversible one-electron and two-electron reduction of halothane on platinized colloidal titanium dioxide in aqueous suspension. *J. Phys. Chem.* **1987**, *91*, 3782–3788.

20. Barbeni, M.; Morello, M.; Pramauro, E.; Pelizzetti, E.; Vincenti, M.; Borgarello, E.; Serpone, N.; Jamieson, M.A. Sunlight photodegradation of 2,4,5-trichlorophenoxy acetic acid and 2,4,5,trichlorophenol on TiO₂. Identification of intermediates and degradation pathway. *Chemosphere* **1987**, *16*, 1165–1179.

21. Sabin, F.; Turk, T.; Vogler, A. Photo-oxidation of organic compound in the presence of titanium dioxide: Determination of the efficiency. *J. Photochem. Photobiol. A: Chem.* **1992**, *63*, 99–106.

22. Zhou, Q.; Ding, Y.; Xiao, J.; Liu, G.; Guo, X. Investigation of the feasibility of TiO_2 nanotubes for the enrichment of DDT and its metabolites at trace levels in environmental water samples. *J. Chromatogr. A* **2007**, *1147*, 10–16.

23. D'Oliveira, J.-C.; Minero, C.; Pelizzetti, E.; Pichat, P. Photodegradation of dichlorophenols and trichlorophenols in TiO_2 aqueous suspensions: kinetic effects of the positions of the Cl atoms and identification of the intermediates. *J. Photochem. Photobiol. A: Chem.* **1993**, *72*, 261–267.

24. Nome, F.; Rubira, A.F.; Franco, C.; Ionescu, L.G. Limitations of the pseudophase model of micellar catalysis. The dehydrochlorination of 1,1,1-trichloro-2,2-bis(p-chlorophenyl)ethane and some of its derivatives. *J. Phys. Chem.* **1982**, *86*, 1881–1885.

25. Tunesi, S.; Anderson, M.A. Photocatalysis of 3,4-DCB in TiO_2 aqueous suspensions; effects of temperature and light intensity; CIR-FTIR interfacial analysis. *Chemosphere* **1987**, *16*, 1447–1456.

26. Pelizzetti, E.; Carlin, V.; Minero, C.; Gratzel, M. Enhancement of the rate of photocatalytic degradation on TiO_2 of 2-chlorophenol, 2,7-dichlorodibenzodioxin and Atrazine by inorganic oxidizing species. *New J. Chem.* **1991**, *15*, 351–359.

27. Minero, C.; Pelizzetti, E.; Malato, S.; Blanco, J., Large solar plant photocatalytic water decontamination: Degradation of pentachlorophenol. *Chemosphere* **1993**, *26*, 2103–2119.

28. Mailhot, G.; Astruc, M.; Bolte, M. Degradation of tributyltin chloride in water photoinduced by iron(III). *Appl. Organometal. Chem.* **1999**, *13*, 53–61.

29. Chamarro, E.; Esplugas, S. Photodecomposition of 2,4-dichlorophenoxy acetic acid: Influence of pH. *J. Chem. Technol. Biotechnol.* **1993**, *57*, 273–279.

30. Muller, T.S.; Sun, Z.; Kumer, G.; Itoh, K.; Murabayashi, M. The combination of photocatalysis and ozonolysis as a new approach for cleaning 2,4-dichlorophenoxy acetic acid polluted water. *Chemosphere* **1998**, *36*, 2043–2055.

31. Terashima, Y.; Ozaki, H.; Giri, R.R.; Tano, T.; Nakatsuji, S.; Takanami, R.; Taniguchi, S. Photocatalytic oxidation of low concentration 2,4-D solution with new TiO_2 fiber catalyst in a continuous flow reactor. *Water Sci. Technol.* **2006**, *54*, 55–63.

32. Watanabe, N.; Horikoshi, S.; Suzuki, K.; Hidaka, H.; Serpone, N. Mechanistic inferences of the photocatalyzed oxidation of chlorinated phenoxy acetic acids by electrospray mass spectral techniques and from calculated point charges and electron densities on all atoms. *New J. Chem.* **2003**, *27*, 836–843.

33. Horikoshi, S.; Hidaka, H.; Serpone, N. Environmental remediation by an integrated microwave/UV-illumination technique: IV. Non-thermal effects in the microwave-assisted degradation of 2,4-dichlorophenoxy acetic acid in UV-irradiated TiO_2/H_2O dispersions. *J. Photochem. Photobiol. A: Chem.* **2003**, *159*, 289–300.

34. Horikoshi, S.; Kajitani, M.; Serpone, N.; Abe, M. Investigation of the promoted degradation mechanism of 1,4-dioxane using a novel microwave-assisted photocatalytic method. *J. Jpn. Soc. Colour Mater.* **2008**, *82*, 51–55.

35. Hashimoto, K.; Kawai, T.; Sakata, T. Photocatalytic Reactions of Hydrocarbons and Fossil Fuels with Water. Hydrogen Production and Oxidation. *J. Phys. Chem.* **1984**, *88*, 4083–4088.

36. Horikoshi, S.; Abe, M.; Serpone, N. Influence of alcoholic and carbonyl functions in microwave-assisted and photo-assisted oxidative mineralization. *Appl. Catal. B: Environ.* **2009**, *89*, 284–287.

37. Atkinson A.; Roy, D. *In vitro* conversion of environmental estrogenic chemical bisphenol-A to DNA binding metabolite(s). *Biochem. Biophys. Res. Commun.* **1995**, *210*, 424–433.

38. Horikoshi, S.; Kajitani, M.; Serpone, N. The microwave-/photo-assisted degradation of bisphenol- A in aqueous TiO_2 dispersions revisited. Re-assessment of the microwave non-thermal effect. *J. Photochem. Photobiol. A: Chem.* **2007**, *188*, 1–4.

39. Horikoshi, S.; Tsuzuki, J.; Kajitani, M.; Abe, M.; Serpone, N. Microwave-enhanced radical reactions at ambient temperature Part 3: Highly selective radical synthesis of 3-cyclohexyl-1-phenyl-1-butanone in a microwave double cylindrical cooled reactor. *New J. Chem.* **2008**, *32*, 2257–2262.

40. Serpone, N.; Lawless, D.; Khairutdinov, R.; Pelizzetti, E. Subnanosecond relaxation dynamics in TiO_2 colloidal sols (Particle sizes Rp = 1.0–13.4 nm). Relevance to heterogeneous photocatalysis. *J. Phys. Chem.* **1995**, *99*, 16655–16661.

41. Ge, M.; Azouri, A.; Xun, K.; Sattler, K.; Lichwa, J.; Ray, C. Solar photocatalytic degradation of Atrazine in water by a TiO_2/Ag nanocomposite. In Proceedings of the ASME 2006 Multifunctional Nanocomposites International Conference, Honolulu, HI, USA, 20–22 September 2006; The American Society of Mechanical Engineers: New York, NY, USA; pp. 10016–5990.

42. Anpo, M. Utilization of TiO_2 photocatalysts in green chemistry. *Pure Appl. Chem.* **2000**, *72*, 1265–1270.

43. Horikoshi, S.; Hidaka, H.; Serpone, N. Hydroxyl radicals in microwave photocatalysis. Enhanced formation of •OH radicals probed by ESR techniques in microwave-assisted photocatalysis in aqueous TiO_2 dispersions. *Chem. Phys. Lett.* **2003**, *376*, 475–480.

44. Horikoshi, S.; Minatodani, Y.; Tsutsumi, H.; Uchida, H.; Abe, M.; Serpone, N. Influence of lattice distortion and oxygen vacancies on the UV-driven/microwave-assisted TiO_2 photocatalysis. *J. Photochem. Photobiol. A: Chem.* **2013**, *265*, 20–28.

45. Kuznetsov, V.N.; Serpone, N. On the origin of the spectral bands in the visible absorption spectra of visible-light-active TiO_2 specimens. Analysis and assignments. *J. Phys. Chem. C.* **2009**, *113*, 15110–15123.

46. Horikoshi, S.; Abe, M.; Serpone, N. Novel designs of microwave discharge electrodeless lamps (MDEL) in photochemical applications. Use in advanced oxidation processes. *Photochem. Photobiol. Sci.* **2009**, *8*, 1087–1104.

47. Giesy, J.P.; Kannan, K. Global distribution of perfluorooctane sulfonate in wildlife. *Environ. Sci. Technol.* **2001**, *35*, 1339–1342.

48. Houde, M.; Martin, J.W.; Letcher, R.; Solomon, K.; Muir, D.C.G. Biological monitoring of perfluoroalkyl substances: A review. *Environ. Sci. Technol.* **2006**, *40*, 3463–3473.

49. Key, B.D.; Howell, R.D.; Criddle, C.S. Fluorinated organics in the biosphere. *Environ. Sci. Technol.* **1997**, *31*, 2445–2454.

50. Schultz, M.M.; Barofsky, D.F.; Field, J.A. Fluorinated alkyl surfactants. *Environ. Eng. Sci.* **2003**, *20*, 487–501.

51. Hori, H.; Hayakawa, E.; Einaga, H.; Kutsuna, S.; Koike, K.; Ibusuki, T.; Kitagawa, H.; Arakawa, R. Decomposition of environmentally persistent perfluorooctanoic acid in water by photochemical approaches. *Environ. Sci. Technol.* **2004**, *38*, 6118–6124.

52. Hori, H.; Yamamoto, A.; Hayakawa, E.; Taniyasu, S.; Yamashita, N.; Kutsuna, S. Efficient decomposition of environmentally persistent perfluorocarboxylic acids by use of persulfate as a photochemical oxidant. *Environ. Sci. Technol.* **2005**, *39*, 2383–2388.

53. Chen, J.; Zhang, P.; Liu, J. Photodegradation of perfluorooctanoic acid by 185 nm vacuum ultraviolet light. *J. Environ. Sci.* **2007**, *19*, 387–390.

54. Horikoshi, S.; Tsuchida, A.; Sakai, H.; Abe, M.; Serpone, N. Microwave discharge electrodeless lamps (MDELs). VI. Performance evaluation of a novel microwave discharge granulated electrodeless lamp (MDGEL)—Photoassisted defluorination of perfluoroalkoxy acids in aqueous media. *J. Photochem. Photobiol. A: Chem.* **2011**, *222*, 97–104.

55. Horikoshi, S. Sophia University, Tokyo, Japan. Unpublished work, 2014.

Photocatalytic Solar Tower Reactor for the Elimination of a Low Concentration of VOCs

Nobuaki Negishi and Taizo Sano

Abstract: We developed a photocatalytic solar tower reactor for the elimination of low concentrations of volatile organic compounds (VOCs) typically emitted from small industrial establishments. The photocatalytic system can be installed in a narrow space, as the reactor is cylindrical-shaped. The photocatalytic reactor was placed vertically in the center of a cylindrical scattering mirror, and this vertical reactor was irradiated with scattered sunlight generated by the scattering mirror. About 5 ppm toluene vapor, used as representative VOC, was continuously photodegraded and converted to CO_2 almost stoichiometrically under sunny conditions. Toluene removal depended only on the intensity of sunlight. The performance of the solar tower reactor did not decrease with half a year of operation, and the average toluene removal was 36% within this period.

Reprinted from *Molecules*. Cite as: Negishi, N.; Sano, T. Photocatalytic Solar Tower Reactor for the Elimination of a Low Concentration of VOCs. *Molecules* **2014**, *19*, 16624-16639.

1. Introduction

Photochemical oxidants, mainly volatile organic compounds (VOCs), are currently a serious global environmental problem, particularly in large cities. In Japan, approximately 10% of VOCs are emitted from automobiles, while the rest are of fixed origin [1,2]. Another major source of VOC emissions (approximately 73%) is solvent from paint and printing industries [2]. Installation of a VOC elimination system is certainly possible at some fixed emission sources, and a few industry groups are undertaking its implementation. On the other hand, smaller industrial establishments with VOC emissions below the average level in Japan are under no such obligations, although self-imposed regulation is recognized. However, most establishments cannot install a VOC elimination system, mainly for economic reasons. Small industrial establishments are the main contributors to VOC emissions in Japan; for instance, these are the source of around 70% of VOC emissions in Tokyo [3]. Thus, minimization of VOC emissions from these facilities would be the most effective approach to eliminate this photochemical oxidant.

Practical application of photocatalysis by TiO_2 to self-cleaning materials has begun [4]; on the other hand, application to air purification is still not widely familiar, with the exception of a few examples, such as photocatalytic pavement blocks for the removal of nitrogen oxides (NOx) [5,6]. However, the application of TiO_2 photocatalysts for VOC elimination is also of interest. This is a simple process compared to NOx removal, since the final products are generally CO_2 and H_2O, which do not poison the surface of the photocatalyst. In addition, a few methods for the elimination of VOCs by photocatalytic materials have already been standardized (ISO 22197-2, ISO 22197-3) [7,8].

Photocatalysis can be activated by solar irradiation, which would enable continuous elimination of VOCs. Therefore, the combination of photocatalysis and solar energy would reduce the electric energy required for the elimination of VOCs and, consequently, the operational costs. While this method would be unsuitable for high concentrations of VOCs and/or high airflow, its application to small industrial establishments would be advantageous. That is, the amount of VOC emissions from these establishments is rather small; thus, the installation of a photocatalytic system would be suitable for such an emission source with a low VOC level and flow rate.

Various types of photocatalytic solar reactors have been recently proposed. Standard types of solar reactors are mainly parabolic trough or flat panel, and a few researchers have already investigated the optimum design to achieve maximum efficiency [9–13]. However, the design of solar reactors is limited to water treatment thus far, and that for air treatment has not received as much attention [14]. The discussion above explained the state of VOC emissions in Japan. In the case of a megalopolis, such as Tokyo or Osaka, small industrial establishments are cramped within a narrow space; therefore, a flat- or trough-type photoreactor would be difficult to set up for these establishments. In addition, expensive instruments, such as a solar tracking system for a solar photoreactor, are difficult to install [15]. Therefore, we developed a unique photocatalytic system consisting of a tower-like photoreactor with a lens barrel that would be suitable for installation in a narrow space, such as the gap between buildings. The performance of this photocatalytic solar tower reactor per unit area of photocatalyst is possibly lower than that of a flat- or trough-type photoreactor, because the surface exposed to solar radiation is limited to the cross-sectional area of the tower. However, the structure of this reactor is very simple, and this will reduce the costs of production, installation, operation and maintenance. We expect that it will be possible in the near future to eliminate photochemical oxidant generation under sunny conditions and to decrease VOC emission by extensively installing solar tower reactors in the areas populated by small industrial establishments. In this study, we describe the design and construction of the solar tower reactor and evaluate the VOC elimination performance through long-term experiments.

2. Results and Discussion

2.1. Optimization of the Height of the Modular Solar Tower Reactor

To determine the optimum height of the solar tower reactor, the change of the concentration of toluene (and formed CO_2) under a certain amount of UV radiation was measured as a function of the height of the reactor, which was adjusted by changing the number of photocatalyst modules. Toluene removal and CO_2 formation exhibited a staircase-like increase with increasing number of units. Photocatalytic degradation of toluene is a well-known reaction, which has been studied through various experiments. Toluene is simply converted to CO_2 when photocatalysis proceeds stoichiometrically [16]. Figure 1 shows the relationship between the amount of CO_2 formed and the number of photocatalyst modules. In this experiment, the number of modules was increased from two to five. As shown in Figure 1, the amount of CO_2 formed as a function of the number of modules was fitted to a second-order polynomial curve. Therefore, a maximum height of 2100 mm (equivalent to seven modules) for the tower was expected to be the best for this system. This estimation shows that UV light, which initiates photocatalytic degradation, reaches the interior of the solar tower reactor to a depth of 2100 mm. Therefore, a height of 2100 mm was chosen for the solar tower reactor.

Figure 1. Plot of the amount of CO_2 formed as a function of the number of photocatalyst units.

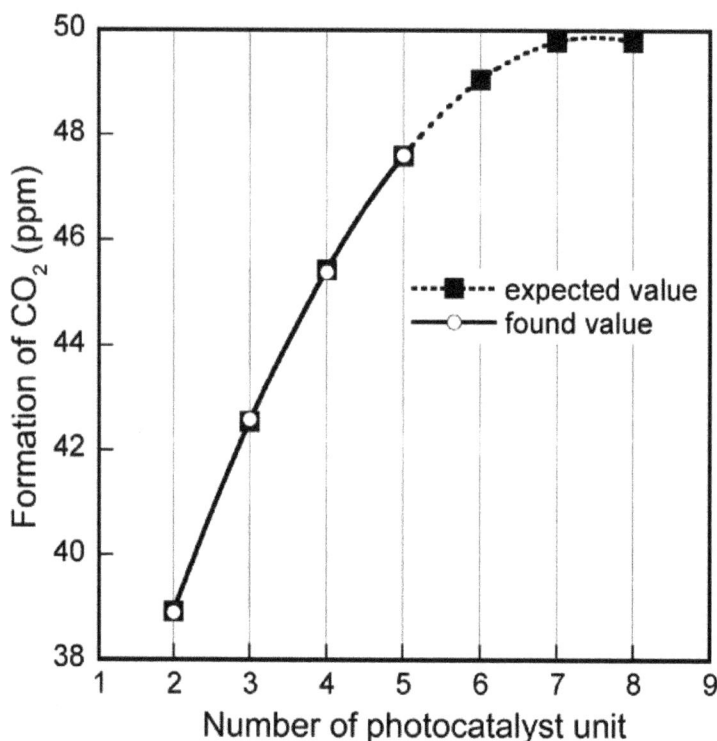

2.2. *Elimination of Toluene in the Solar Tower Reactor Installed with a TiO₂-Coated Ceramic Tube by Solar Irradiation*

The data for toluene vapor elimination under sunny and rainy conditions are shown in Figure 2. Under rainy conditions, the initial concentration of toluene (5 ppm) did not change after sunrise. However, UV light from the Sun arrives at the Earth's surface in spite of rainy conditions. Therefore, CO_2 formation resulting from the photodegradation of toluene was observed, and its oscillation profile was almost similar to the time profile of solar intensity.

Under sunny conditions, CO_2 formation began from sunrise, and a decrease in toluene concentration was observed. Within a day, approximately 100% of the toluene was removed, and an almost stoichiometric formation of CO_2 was observed. Based on these results, CO_2 formation was considered to depend principally on the intensity of solar radiation, since the oscillation profile of CO_2 formation was almost similar to the time profile of solar intensity. Under this condition, there was a time lag between toluene degradation and CO_2 formation, as shown in Figure 2. We assumed that this time lag could be attributed to toluene adsorption onto the TiO_2 surface before sunrise (dark condition). Toluene introduced to the solar tower reactor was adsorbed onto the TiO_2

surface at night. This adsorbed toluene was preferentially photodecomposed by solar irradiation instead of the toluene fed to the reactor after sunrise; therefore, the decrease in toluene concentration was not observed at the same time as sunrise. With the increase of solar radiation, the adsorbed toluene was removed by photocatalysis, and the continuously fed toluene started to photodecompose.

Figure 2. Elimination of toluene in the solar tower reactor installed with a TiO_2-coated ceramic tube under rainy and sunny conditions.

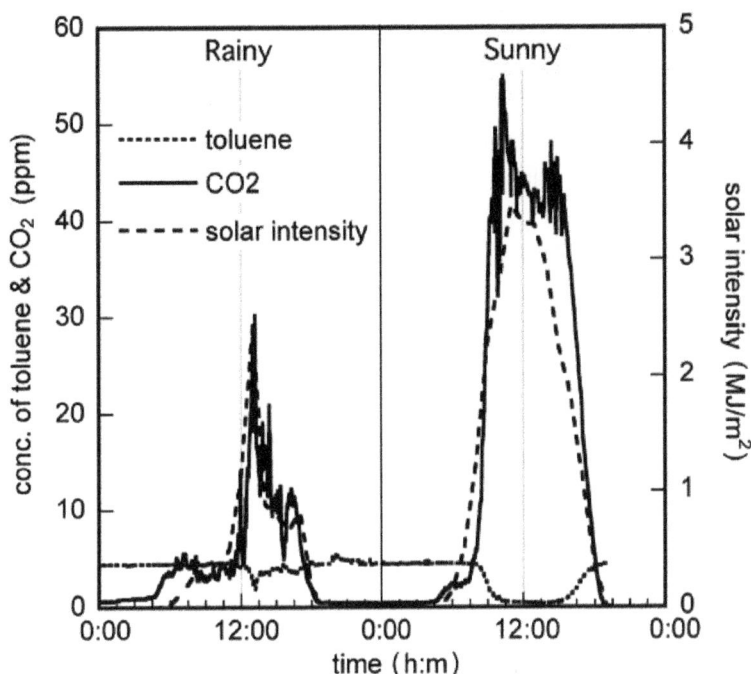

The long-term experiment was carried out from 1 August 2006 to 31 December 2006. Figure 3 shows the experimental data for September as representative of the long-term experiment. It is clear that the toluene removal ratio simply depended on the intensity of solar radiation, as also indicated in Figure 2. Figure 4 shows the plots of the integral values for the toluene removal ratio, the concentration of CO_2 product and the solar intensity per day for five months. As shown in this figure, the plots of the integrated values for CO_2 formation and the time profile of solar intensity are almost similar. In addition, the corresponding integrated values are proportional. This shows that there was no decrease in photocatalytic activity during the five months, and a total of 27.7% of the toluene introduced to the solar tower reactor was photocatalytically removed.

Figure 3. Experimental data for September 2006, obtained using the reactor installed with a TiO₂-coated ceramic tube and the changes of the temperature.

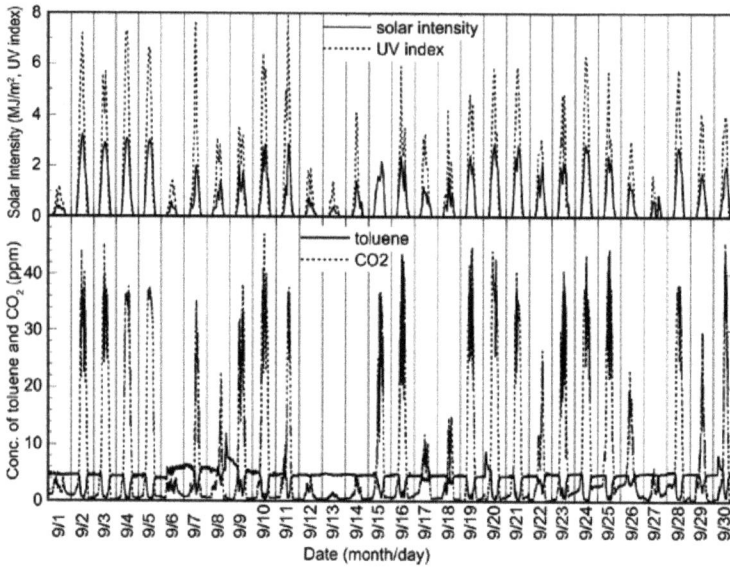

Figure 4. Plots of the integrated values for toluene removal ratio, the concentration of CO₂ product and the solar intensity per day for five months obtained using the reactor installed with a TiO₂-coated ceramic tube.

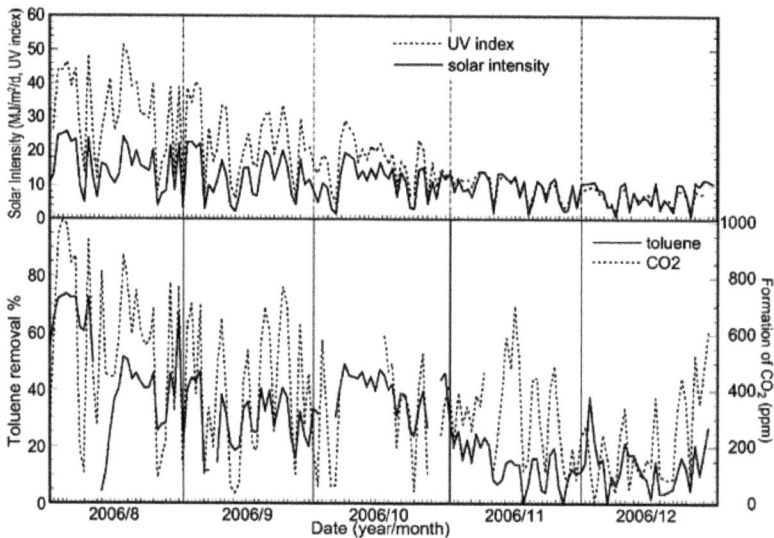

2.3. Elimination of Toluene in the Solar Tower Reactor Installed with a HQC21-Packed Glass Tube by Solar Irradiation

The data for toluene elimination by the solar tower reactor under sunny and rainy conditions are shown in Figure 5. HQC21 (TiO₂-coated silica gel produced by Shinto V Cerax) had strong adsorption efficiency, since it was unused, and the silica gel has a wide surface area. Therefore, 5 ppm toluene vapor was easily adsorbed onto HQC21, and the toluene concentration at the outlet of the reactor was constant at 0 ppm until sunrise. The toluene concentration after sunrise reached 30 ppm until 9 AM, after which the concentration began to decrease and was almost 0 ppm at 12 PM. This concentration was maintained until midnight. After 12 AM, the adsorption of toluene began again, and the concentration returned to its initial value at 6 AM, as shown in Figure 5.

Figure 5. Elimination of toluene by the solar tower reactor installed with a HQC21-packed glass tube under rainy and sunny conditions, and the changes of the temperature.

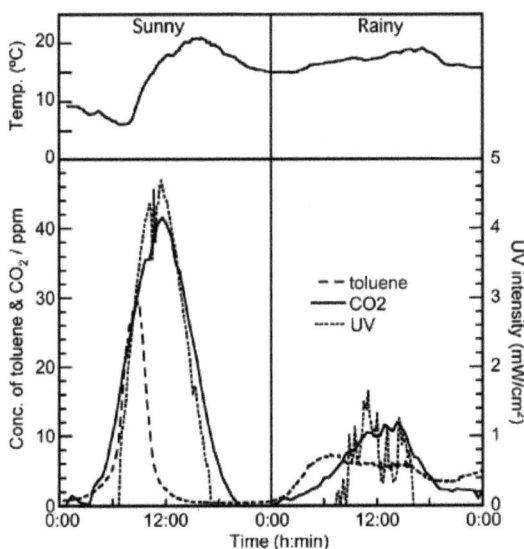

The sharp increase in toluene concentration under sunny conditions occurred around sunrise; on the other hand, a correlation between temperature and toluene concentration was not observed, as shown in Figure 5. It was assumed that this sharp increase in toluene concentration indicates desorption of a number of adsorbed toluene molecules from the TiO_2 surface by solar irradiation. That is, toluene molecules accumulated on the TiO_2 surface under dark conditions, and desorption of excess toluene began when the adsorbed toluene on the surface of TiO_2 photodecomposed just after sunrise. Toluene desorption was considerable, and the amount of toluene was higher than the photocatalytic capacity. However, after photocatalysis by solar irradiation, the TiO_2 surface became free, and consequently, the strong adsorption efficiency of HQC21 was steadily recovered. As shown in Figure 2, strong adsorption of toluene onto the TiO_2-coated ceramic tube was not observed. It was assumed that this resulted from the difference in the absolute surface area of the photocatalyst between the TiO_2-coated ceramic tube and the HQC21-packed glass tube. The absolute surface area of the TiO_2-coated ceramic tube with a total height of 2100 mm was around 11,000 m^2 (45 m^2/g × 37 mmD × 300 mm × 7 units). On the other hand, the absolute surface area of the HQC21-packed glass tube with a total height of 1500 mm was around 49,200 m^2 (200 m^2/g × 41 g/glass tube × 6 glass tubes) [17].

Under rainy conditions, the adsorption efficiency of toluene onto HQC21 was similar to that under sunny conditions until saturation. However, the concentration of toluene at the outlet of the system did not exceed the inlet concentration, which was constant the whole day. On the other hand, a slight amount of CO_2 was formed, which indicates that UV irradiation under rainy conditions influenced the photodegradation of adsorbed toluene on the photocatalyst surface.

This experiment was conducted for about half a year from 1 January 2008, and the data for February shown in Figure 6 were chosen as representative of this period, because this month had regular, relatively fine weather. As shown in this figure, the outlet concentration of toluene was higher than the inlet one, and it decreased to around 0 ppm in sunny conditions. This trend was observed throughout the half-year experimental period. A higher toluene concentration at the outlet indicates the increase of VOC emissions, and this excess emission to the environment is only a problem in the short term. However, the ratio of the total amount of toluene eliminated to that introduced by the solar tower reactor for the half year reached 36%, which shows that the solar tower reactor is well suited to the long-term reduction of total VOC emission.

Figure 6. Experimental data for February 2008, obtained using the reactor installed with a HQC21-packed glass tube.

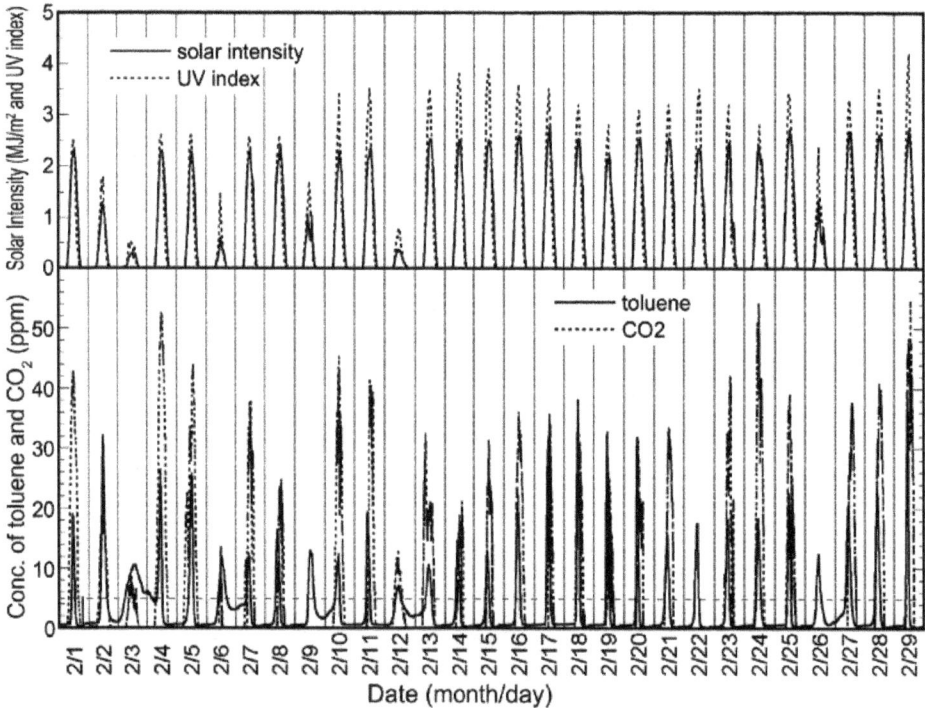

Figure 7 shows the changes in integral values for the toluene removal ratio, the amount of CO_2 and the solar intensity per day for a half year. In this experiment, the CO_2 data between March to the beginning of April were missing, owing to a malfunction of the CO_2 monitor. As shown in Figure 7, the profiles of solar intensity, UV index and the amount of CO_2 are similar. In addition, the emission of CO_2 also increased proportionally with the progress of the seasons from winter to summer. Therefore, the photocatalytic activity did not decrease during the long trial run of the solar tower reactor.

Figure 7. Plots of integrated values for the toluene removal ratio, the concentration of CO_2 product and the solar intensity per day for five months obtained using the reactor installed with a HQC21-packed glass tube.

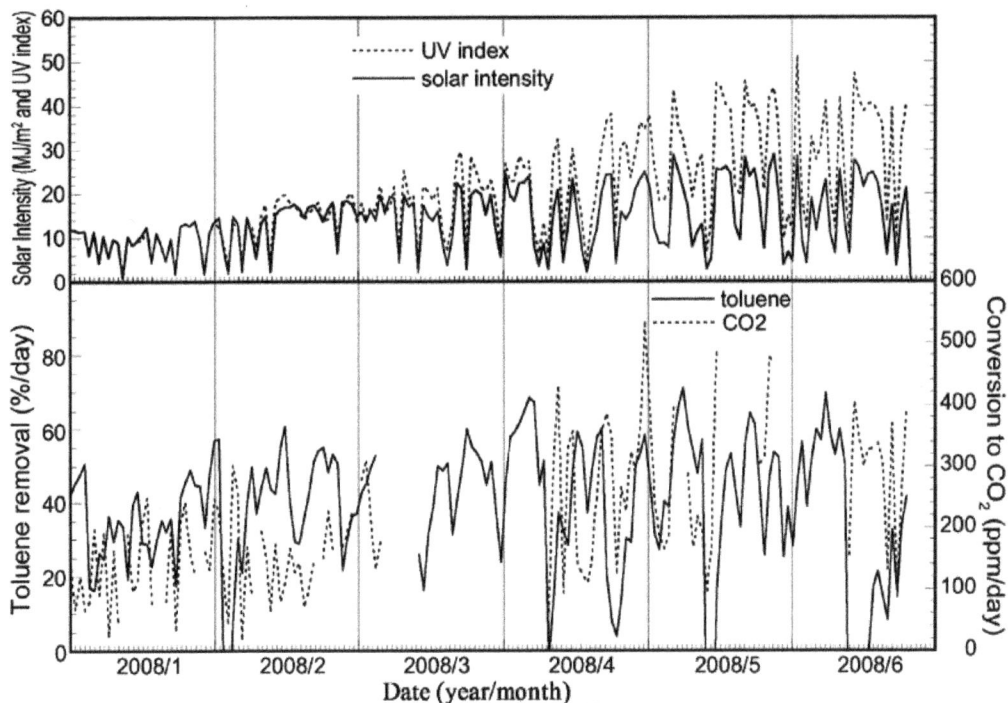

From the results of these experiments, it is clear that the toluene vapor introduced to the solar tower reactor was eliminated by solar irradiation. On the other hand, the flow rate of toluene vapor in these experiments was still lower than the actual level. Therefore, we will continuously improve the performance of the solar tower reactor by increasing the amount of VOCs eliminated, decreasing the pressure drop and reducing the production cost.

3. Experimental Section

3.1. Design of the Solar Tower Reactor

In this section, the structure and operating procedure of the solar tower reactor is described. The tower is shaped like a chimney, and the cylindrical mirror was assembled vertically, as shown in Figure 8. The tubular photocatalyst reactor was placed in its center. The cylindrical mirror was modularized, thus enabling adjustment of the height of the tower to the optimum value by changing the number of modules. Sunlight enters the interior of the cylindrical mirror through an opening at the top. A section of the tubular photocatalyst reactor protrudes above the top of the cylindrical mirror, as shown in Figure 8. The surface of the cylindrical mirror was embossed to scatter incoming solar radiation, and this scattered light reaches the bottom of the tower without concentrating in the center of the cylindrical mirror, the mechanism of which is different from a parabolic trough-type photocatalytic reactor [9]. Simulated VOC gas was supplied once to the top of the photocatalytic reactor, as shown in Figure 8. The top of the reactor was irradiated with direct sunlight, initiating photocatalysis and subsequent reduction of the initial concentration of the VOCs. The intensity of the light incident through the opening at the top decreases as repeated scattering and reflections go downward. However, the VOC concentrations are also continuously decreased by photocatalysis during the downward gas flow; therefore, the weakened light intensity is strong enough for the photocatalysis of residual VOCs. This mechanism allows continuous photocatalysis of the VOCs by the solar tower reactor. The top of the tubular mirror was diagonally cut for effective collection of solar radiation for the whole day. The angle of the cylindrical mirror of the sunlight entrance was determined from the following equation,

$$\text{Angle of the cylindrical mirror of the sunlight entrance } (°) = 45° + 0.5 \times \text{latitude} \quad (1)$$

There is a small deviance between the Earth's equatorial plane and the ecliptic plane; thus, this equation is not precisely applicable. However, this simple equation is sufficient for calculating the maximum sunlight intensity in a whole year.

Figure 8. Schematic diagram of the solar tower reactor.

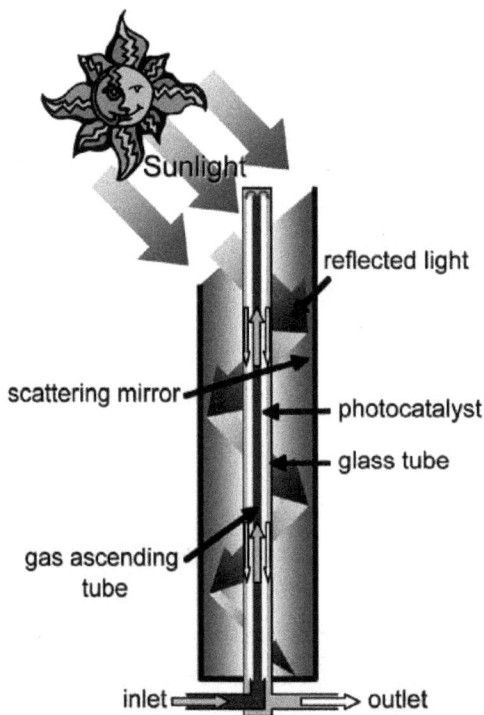

3.2. Photocatalysts

Two types of photocatalytic materials were used in the experiment. One was a TiO_2-coated ceramic tube with an external diameter of 37 mm and a length of 300 mm, and the other was a TiO_2-coated silica gel (HQC-21; purchased from Shinto V Cerax, Aichi, Japan) packed in a Pyrex® glass tube with an internal diameter of 10 mm and a length of 1500 mm. A single flow pass was employed for both types, as shown in Figure 8.

The preparation of the TiO_2-coated ceramic tube was carried out using the dip coating method [18]. A γ-alumina tube with an external diameter of 37 mm, an internal diameter of 30 mm and a length of 300 mm was used as the substrate for TiO_2 coating. The TiO_2 precursor was dip-coated onto this ceramic tube, and the TiO_2 thin film photocatalyst was obtained by calcination at 480 °C for 1 h. The coating procedure was repeated 18 times to obtain the final photocatalytic material. The film thickness of the TiO_2 layer on the ceramic tube was impossible to measure, except through destructive methods; therefore, the thickness was estimated from data obtained in our previous study [18]. It was shown that a single coating step yields a 200 nm-thick film. Therefore, the estimated film thickness of the TiO_2 film on the ceramic tube was approximately 3.6 μm for the 18 cycles of dip coating. Seven TiO_2-coated ceramic tubes were prepared, and each was placed inside a Pyrex® glass tube with an internal diameter of 49 mm, as shown in Figure 9a. This photocatalytic module was fitted with metal molding adapters to enable a connection with other modules. The photocatalytic material is placed in the center of the cylindrical mirror of the solar tower reactor. The external view is shown in Figure 9b. A diagram of the solar tower reactor is shown in Figure 10a. The cylindrical mirror that constitutes the reactor was modularized, as shown in Figure 10b, and the height of the tower can be adjusted to the optimum value by changing the number of modules.

Figure 9. (a) Module of the TiO_2-coated ceramic tube inside a Pyrex® glass tube; (b) external view of the solar tower reactor assembled from these modules.

(a)

Figure 9. *Cont.*

(b)

Figure 10. (a) Diagram of the solar tower reactor. 1: support rod; 2: cylindrical mirror; 3: embossed scattering mirror; 4: photocatalyst; 5: top view; 6: shaft carrier; 7: connector; 8: anchor; 9: gas return cap; 10: entry point of solar light; 11: inlet/outlet separator; 12: embossed scattering mirror; 13: air inlet; 14: air outlet. **(b)** Module of the solar tower reactor. A: cylindrical mirror; B: embossed scattering mirror; C: glass cover connector; D: TiO$_2$-coated ceramic tube; E: Pyrex glass cover.

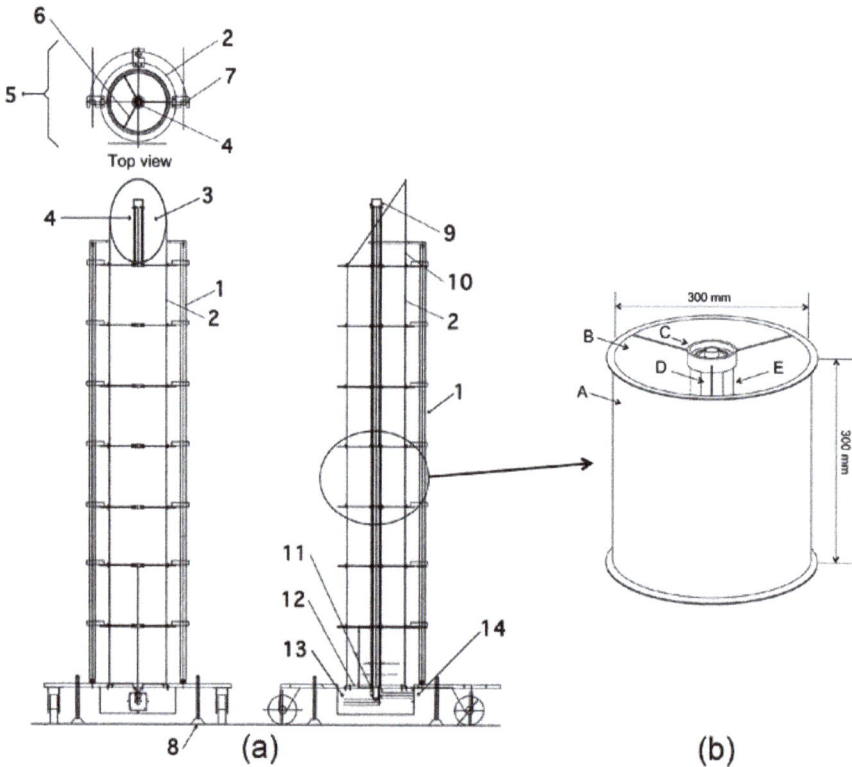

(a)

(b)

The glass tube packed with TiO$_2$-coated silica gel was prepared as follows. HQC21 (average diameter = 4 mm, specific surface area = 200 m^2/g, average TiO$_2$ deposition onto silica gel = around 20 wt %) [19] was used as TiO$_2$-coated silica gel. The dimensions of the Pyrex® glass tube were 12 mm external diameter, 10 mm internal diameter and 1500 mm length. Six glass tubes packed with HQC21 were prepared, each containing 41 g of HQC21. The HQC21-packed glass tubes were arranged at the center of the solar tower reactor, as shown in Figure 11.

Figure 11. Top view of the HQC21-packed glass tube reactor.

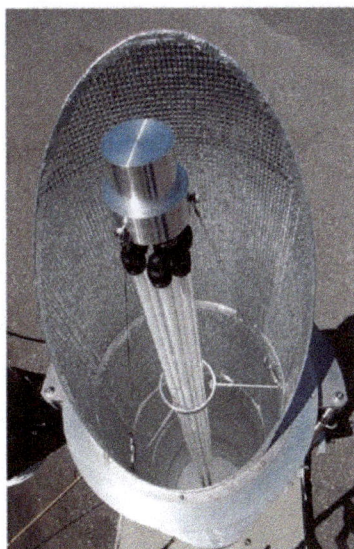

3.3. Construction of the Gas Flow System for Evaluating the Performance of the Solar Tower Reactor

Figure 12 shows the gas flow system constructed to evaluate the performance of the solar tower reactor. The air pump pulled in ambient air, which was then separated into two directions. One route led to the permeater for the generation of toluene vapor, while the other one led directly to the mass flow controller, where the normal air acted as the carrier gas. Toluene vapor and normal air were mixed through the mass flow controller, and 5 ppm toluene vapor was obtained. A different flow rate for the toluene vapor was used for each type of photocatalyst reactor. A flow rate of 1.5 L/min was used for the TiO$_2$-coated ceramic tube reactor; on the other hand, a flow rate of 10 L/min was used for the HQC21-packed glass tube reactor. To measure the toluene vapor concentration, the VM500 online VOC monitor (Yokogawa Inc., Tokyo, Japan) was used. It is well known that the humidity in air is a principal factor for photocatalysis, *i.e.*, photocatalysis of toluene requires humidity [20–23]. However, the humidity was not controlled in this experiment. One of the reasons is that we needed to simulate the actual operation condition. Obee *et al.* reported that the photocatalytic oxidation rate of toluene under low humidity was extremely low, and the high oxidation rate of toluene was indicated at around 20% relative humidity. The rate was gradually decreased with the increase in relative humidity until 90% [20]. The concentration of CO$_2$ produced from the photodecomposition of toluene was measured by the Thermo 41C CO_2 monitor (Thermo Fisher Scientific Inc., Waltham, MA, USA). When the flow rate used was 1.5 L/min, CO_2 in the ambient air was removed by 500 g soda lime (Wako Chemical, Osaka, Japan). On the other hand, an additional CO_2 monitor (LX-720, Iijima-Denshi Co., Aichi, Japan) was connected to the outlet of the solar tower reactor instead of the CO_2 adsorbent when the flow rate was 10 L/min. In this case, the concentration of CO_2 was calculated from the difference between the outlet and inlet concentrations at the solar tower reactor. Data were collected using a personal computer (PC) connected to both CO_2 monitors. During the experiment, some CO_2 data could not be collected, owing to malfunctions of either the CO_2 monitor or the PC.

Figure 12. Scheme of the gas flow system for the evaluation of the performance of the solar tower reactor.

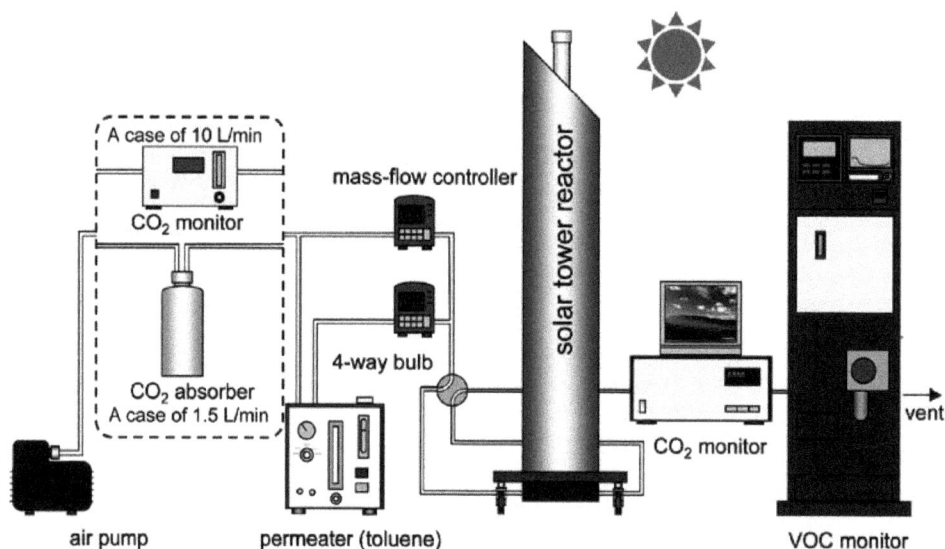

The pressure drop within the Pyrex® tube packed with TiO_2-coated silica gel was quite high compared with that of the TiO_2-coated ceramic tube; however, the actual values were not measured in the experiment. On the other hand, the residence time of toluene vapor in the reactor was calculated. The photocatalytic module made from the TiO_2-coated ceramic tube was annular, with an internal volume of 243 mL (Figure 9b). As the flow rate in this module was 1.5 L/min, the residence time of toluene vapor was around 9.7 s. The total residence time in the solar tower reactor was obtained by multiplying this value with the number of modules. For example, for 7 modules (*i.e.*, height of 2100 mm), the total residence time of the toluene vapor was around 68 s. In the case of the HQC21-packed glass tube, the free volume was around 64 mL, assuming the closest packing of particles with a 4 mm diameter into the glass tube having an internal diameter of 10 mm and a length of 1500 mm. There were 6 HQC21-packed glass tubes; therefore, the total free volume in the solar tower reactor was around 385 mL. The total residence time of the toluene vapor was around 2.3 s at a flow rate of 10 L/min.

3.4. Optimization of the Height of the Solar Tower Reactor

The optimum height of the solar tower reactor was determined from the variation of the toluene removal ratio with the number of TiO_2-coated ceramic tube modules. For this experiment, an artificial 400-W UV light (H400L-BL, Iwasaki Electric Co., Tokyo, Japan) was used (Figure 13). This procedure was not applied for the HQC21-packed glass tube reactor, because the corresponding experiment was only performed using the standard length of the Pyrex® glass (1500 mm). The intensity of the UV light at the top of the solar tower reactor was 3.5 mW/cm^2 (λ = 365 nm).

Figure 13. Artificial UV-A lamp.

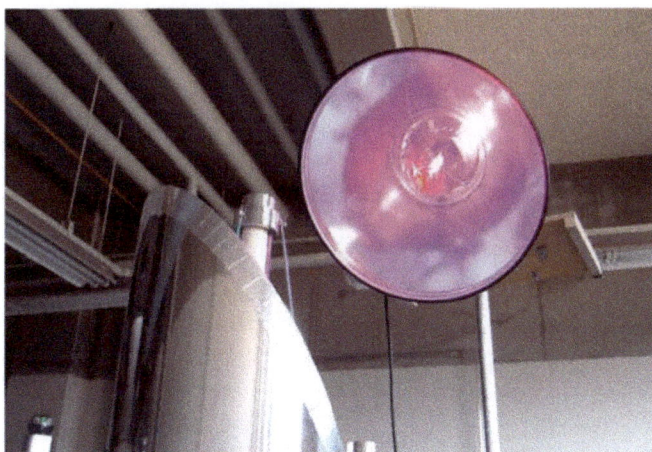

3.5. Evaluation of the Performance of the Solar Tower Reactor

Evaluation of the solar tower reactor using the TiO_2-coated ceramic tube was performed for 5 months from 1 August 2006 to 31 December 2006. Evaluation of the solar tower reactor using the HQC21-packed glass tube was performed for 6 months from 1 January 2008 to 24 June 2008. For the experiments done in 2006 and 2008, the intensity of sunlight was expressed in terms of the insolation and UV index. Insolation data from the Aerological Observatory/Japan Meteorological Agency (JMA) [24] were used, because JMA is adjacent to the Advanced Industrial Science and Technology (AIST), Tsukuba. For the 2008 experiment, the temperature was measured by Vantage Pro 2 (Davis Instruments Corp., Hayward, CA, USA).

4. Conclusions

We constructed a solar tower reactor for the elimination of low concentrations of VOCs, such as toluene. Two types of photocatalytic units were used: (1) a TiO_2-coated ceramic tube optimized for a low pressure drop and low flow rate (1.5 L/min); and (2) a glass tube packed with TiO_2-coated silica gel (HQC21) optimized for a high pressure drop and high flow rate (10 L/min). Both reactors effectively eliminated toluene vapor (5 ppm); however, the toluene removal profile was different between the TiO_2-coated ceramic tube and HQC21-packed glass tube reactor. The time profile of toluene removal by the TiO₂-coated ceramic tube reactor was almost similar to that of solar intensity. On the other hand, the toluene concentration at the outlet was temporarily higher than the inlet concentration at sunrise in the case of the HQC21-packed glass tube reactor. This difference between the two reactors was attributed to the difference in the actual surface area of the photocatalyst. Toluene vapor introduced to the reactor adsorbed and accumulated onto the TiO₂ surface at night, and this adsorbed toluene desorbed from the TiO₂ surface once it was by irradiated by sunlight. The difference between the specific surface areas of TiO₂ in the two reactors was not considerable; however, the total amount of photocatalyst in the HQC21-packed glass tube reactor was higher than that in the TiO₂-coated ceramic tube reactor. In either case, the long-term experiment conducted for a half year demonstrated that toluene vapor introduced into the solar tower reactor was eliminated by solar irradiation by an amount proportional to the integral solar power intensity.

Acknowledgments

This work was funded by the Ministry of the Environment of Japan (No. 13054-2125-14).

Author Contributions

N. N. and T. S. contributed to all of the reported research and writing of the paper.

Conflicts of Interest

The authors declare no conflict of interest.

References

1. Ministry of the Environment, Government of Japan. Available online: http://www.env.go.jp/air/osen/voc/materials/102.pdf (accessed on 10 June 2014).
2. Wakamatsu, S. VOC. *NIES Res. Bookl.* **2002**, *5*, 4–13. Available online: http://www.nies.go.jp/kanko/kankyogi/05/5.pdf (accessed on 14 October 2014).
3. Bureau of Environment, Tokyo Metropolitan Government. Available online: https://www.kankyo.metro.tokyo.jp/basic/attachement/tokyokankyo2011.pdf (English version: https://www.kankyo.metro.tokyo.jp/en/attachement/VOC%20Emissions%20Control.pdf) (accessed on 10 June 2014).
4. Fujishima, A.; Rao, T.N.; Tryk, D.A. Titanium dioxide photocatalysis. *J. Photochem. Photobiol. C* **2000**, *1*, 1–21.
5. Murata, Y.; Tawara, H.; Obata, H.; Takeuchi, K. Air purifying pavement: Development of photocatalytic concrete blocks. *J. Adv. Oxid. Technol.* **1999**, *4*, 227–230.
6. Negishi, N. Photocatalytic air purification—Roadside NOx removal by photocatalyst. *Bull. Ceram. Soc. Jpn.* **2004**, *39*, 504–506.

7. *ISO 22197-2:2011: Fine Ceramics (Advanced Ceramics, Advanced Technical Ceramics)—Test Method for Air-Purification Performance of Semiconducting Photocatalytic Materials—Part 2: Removal of Acetaldehyde*; International Organization for Standardization: Geneva, Switzerland, 2011.

8. *ISO 22197-3:2011: Fine Ceramics (Advanced Ceramics, Advanced Technical Ceramics)—Test Method for Air-Purification Performance of Semiconducting Photocatalytic Materials—Part 3: Removal of Toluene*; International Organization for Standardization: Geneva, Switzerland, 2011.

9. Alfano, O.M.; Bahnemann, D.; Cassano, A.E.; Dillert, R.; Goslich, R. Photocatalysis in water environments using artificial and solar light. *Catal. Today* **2000**, *58*, 199–230.

10. Malato, S.; Blanco, J.; Cáceres, J.; Fernández-Alba, A.R.; Agüera, A.; Rodríguez, A. Photocatalytic treatment of water-soluble pesticides by photo-Fenton and TiO_2 using solar energy. *Catal. Today* **2002**, *76*, 209–220.

11. Herrmann, J.-M.; Guillard, C.; Disdier, J.; Lehaut, C.; Malato, S.; Blanco, J. New industrial titania photocatalysts for the solar detoxification of water containing various pollutants. *Appl. Catal. B* **2002**, *35*, 281–294.

12. Malato Rodríguez, S.; Blanco Gálvez, J.; Maldonado Rubio, M.I.; Fernández Ibáñez, P.; Alarcón Padilla, D.; Collares Pereira, M.; Farinha Mendes, J.; Correia de Oliveira, J. Engineering of solar photocatalytic collectors. *Sol. Energy* **2004**, *77*, 513–524.

13. Braham, R.J.; Harris, A.T. Review of major design and scale-up considerations for solar photocatalytic reactors. *Ind. Eng. Chem. Res.* **2009**, *48*, 8890–8905.

14. Suárez, S.; Hewer, T.L.R.; Portela, R.; Hernández-Alonso, M.D.; Freire, R.S.; Sánchez, B. Behaviour of TiO_2–$SiMgO_x$ hybrid composites on the solar photocatalytic degradation of polluted air. *Appl. Catal. B* **2011**, *101*, 176–182.

15. Sano, T.; Negishi, N.; Takeuchi, K.; Matsuzawa, S. Degradation of toluene and acetaldehyde with Pt-loaded TiO_2 catalyst and parabolic trough concentrator. *Sol. Energy* **2004**, *77*, 543–552.

16. Muñoz-Batista, M.J.; Kubacka, A.; Gómez-Cerezo, M.N.; Tudela, D.; Fernández-García, M. Sunlight-driven toluene photo-elimination using CeO_2-TiO_2 composite systems: A kinetic study. *Appl. Catal. B* **2013**, *140–141*, 626–635.

17. Nishikawa, H.; Ihara, T. Active properties of thermally excited titanium dioxide/silica composite material for the decomposition of gaseous toluene. *Mater. Chem. Phys.* **2011**, *125*, 319–321.

18. Negishi, N.; Matsuzawa, S.; Takeuchi, K.; Pichat, P. Transparent micrometer-thick TiO_2 films on SiO_2-coated glass prepared by repeated dip-coating/calcination: Characteristics and photocatalytic activities for removing acetaldehyde or toluene in air. *Chem. Mater.* **2007**, *19*, 3808–3814.

19. Takeuchi, S. Shinto no Hikarishokubai silica gel series. *Toso Toryo* **2004**, *658*, 15–19.

20. Obee, N.T.; Brown, T.R. TiO_2 photocatalysis for indoor air applications—Effects of humidity and trace contaminant levels on the oxidation rates of formaldehyde, toluene, and 1,3-butadiene. *Environ. Sci. Technol.* **1995**, *29*, 1223–1231.

21. Marcí, G.; Addamo, M.; Augugliaro, V.; Coluccia, S.; García-Lopez, E.; Loddo, V.; Martra, G.; Palmisano, L.; Sciavello, M. Photocatalytic oxidation of toluene on irradiated TiO_2: Comparison of degradation performance in humidified air, in water and in water containing a zwitterionic surfactant. *J. Photochem. Photobiol. A Chem.* **2003**, *160*, 105–114.

22. Stavrakakis, C.; Raillard, C.; Hequet, V.; Le Cloirec, P. TiO_2-based materials for toluene photocalytic oxidation: Water vapor influence. *J. Adv. Oxid. Technol.* **2007**, *10*, 94–100.

23. Akly, C.; Chadik, P.A.; Mazyck, D.W. Photocatalysis of gas-phase toluene using silica-titania composites: Performance of a novel catalyst immobilization technique suitable for large-scale applications. *Appl. Catal. B Environ.* **2010**, *99*, 329–335.

24. Japan Meteorological Agency. Available online: http://www.data.jma.go.jp/obd/stats/etrn/index.php (accessed on 10 June 2014).

Sample Availability: Not available.

Field Performance Test of an Air-Cleaner with Photocatalysis-Plasma Synergistic Reactors for Practical and Long-Term Use

Tsuyoshi Ochiai, Erina Ichihashi, Naoki Nishida, Tadashi Machida, Yoshitsugu Uchida, Yuji Hayashi, Yuko Morito and Akira Fujishima

Abstract: A practical and long-term usable air-cleaner based on the synergy of photocatalysis and plasma treatments has been developed. A field test of the air-cleaner was carried out in an office smoking room. The results were compared to previously reported laboratory test results. Even after a treatment of 12,000 cigarettes-worth of tobacco smoke, the air-cleaner maintained high-level air-purification activity (98.9% ± 0.1% and 88% ± 1% removal of the total suspended particulate (TSP) and total volatile organic compound (TVOC) concentrations, respectively) at single-pass conditions. Although the removal ratio of TSP concentrations was 98.6% ± 0.2%, the ratio of TVOC concentrations was 43.8% after a treatment of 21,900 cigarettes-worth of tobacco smoke in the field test. These results indicate the importance of suitable maintenance of the reactors in the air-cleaner during field use.

Reprinted from *Molecules*. Cite as: Ochiai, T.; Ichihashi, E.; Nishida, N.; Machida, T.; Uchida, Y.; Hayashi, Y.; Morito, Y.; Fujishima, A. Field Performance Test of an Air-Cleaner with Photocatalysis-Plasma Synergistic Reactors for Practical and Long-Term Use. *Molecules* **2014**, *19*, 17424-17434.

1. Introduction

Photocatalysis-plasma synergistic reactors have been recently proposed for use in air-cleaners [1–4]. The synergistic effects of photocatalysis and plasma excitation achieve significant oxidative decomposition of gaseous compounds in laboratory tests. The coil-shaped reactor (Figure 1), using plasma-assisted catalytic technology (PACT) [5] and a TiO_2 impregnated Ti-mesh filter (TMiP^{TM}) [6], shows long-term capability of removing tobacco smoke compounds. High-level air-purification activity was maintained in the air-cleaner with the coil-shaped reactor (≥98, 98.9% ± 0.1%, and 88% ± 1% removal of the odour concentration, total suspended particulate (TSP), and total volatile organic compound (TVOC) concentrations, respectively) even after the treatment of 12,000 cigarettes-worth of tobacco smoke, which is the equivalent of using the air-cleaner in the smoking room for 6 months [4]. In this study, a field test of the air-cleaner using the coil-shaped reactor was carried out for 84 days in a functioning smoking room (Figure 2) of a typical office building. The air-purification ability and the long-term usability of the air-cleaner in the smoking room were discussed by comparison with previously reported laboratory test results.

Figure 1. Image and schematic illustration of the coil-shaped PACT-TMiP synergistic reactor (**left**) and the air-cleaner (**right**). Reproduced from Ochiai *et al.* [4], published by Scientific Research Publishing Inc., 2014.

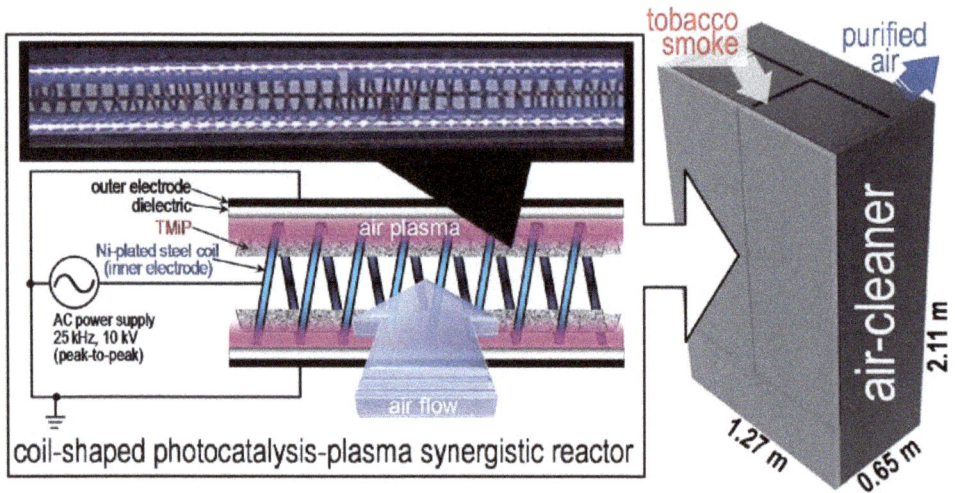

2. Results and Discussion

2.1. Evaluations of the Photocatalysis-Plasma Synergistic Air-Cleaner

Figure 3 shows the temperature distributions of the coil-shaped reactors in the air-cleaner 1, 5, and 10 min after the device is switched on without air flow. The temperature of the reactors reached an almost steady state at around 100 °C for 10 min (Figure 3c). This temperature is lower than the anatase-to-rutile transformation temperature of TiO_2 [7]. Therefore, the TiO_2 photocatalyst on the TMiP surface cannot be affected by the air-plasma. Conversely, ozone and NO_x concentrations near the reactors in the air-purifier were 8–9 and 0.7–0.8 ppm, respectively under a 1200 m^3/h flow rate. These species can accelerate the decomposition of TVOCs [1]. However, ozone and NO_x concentrations at the air-outlet of the air-cleaner were below the detection limit. Thus, the ozone-cut and activated carbon filters shown in Figure 2 can reduce excess ozone and NO_x.

2.2. Usage of the Smoking Room

The field test was carried out in a smoking room with a volume of approximately 31 m^3 (5.0 m × 2.5 m × 2.5 m), which was used by several smokers (Figure 2). The number of cigarettes burned every 30 min, from 11:00 to 17:00 during the working day, was counted and the results summarised in Figure 4. On average, 32 ± 11 and 350 cigarettes are burned on an hourly and daily basis, respectively. Generally, an estimated 12,000 cigarettes are burned in the smoking room every six months [4]. Therefore, the number of cigarettes burned in the smoking room in this study is around six times higher than in an average smoking room.

Figure 2. Schematic illustration of the test method in the smoking room for evaluating the air-purification ability of the air-cleaner (unit: cm). Sampling point (1) is near the air inlet of the air-cleaner, and (2) is inside the duct. Detailed experimental procedures are included in Section 3.

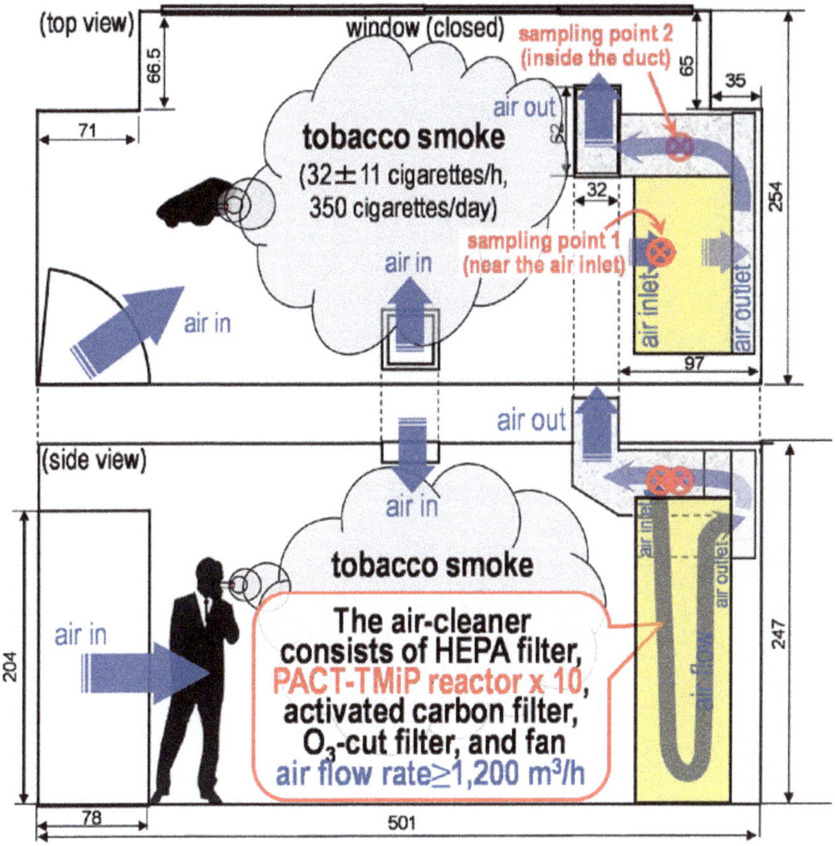

Figure 3. Temperature distribution of the coil-shaped reactors in the air-cleaner (**a**) 1; (**b**) 5; and (**c**) 10 min after the device is switched on without air flow.

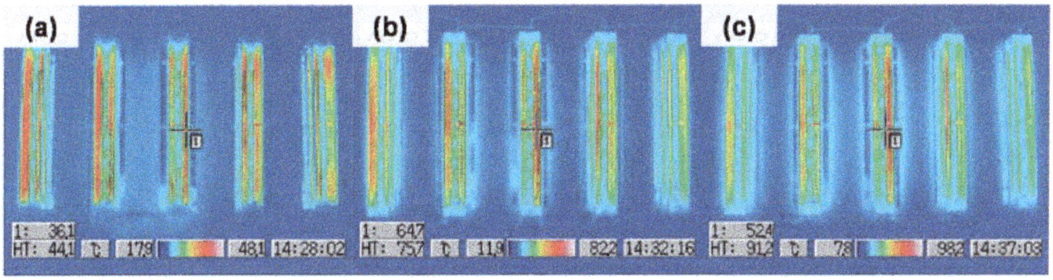

Figure 4. Summary of the number of cigarettes burned every 30 min during the working day (11:00–17:00, 1 May 2014).

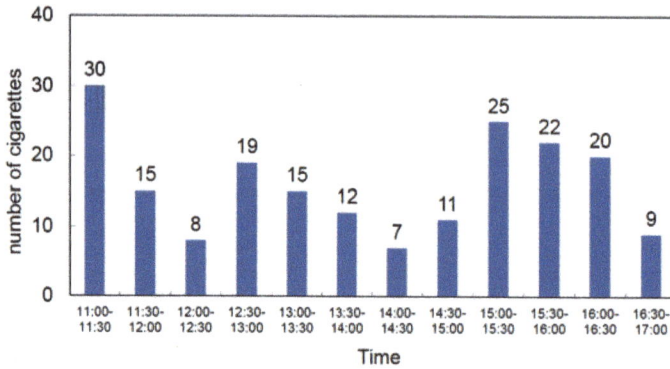

2.3. TSP Removal

Figure 5a shows TSP concentrations at sampling points 1 (near the air inlet of the air-cleaner, Figure 2) and 2 (inside the duct, Figure 2) after the treatment of 2300 cigarettes-worth of tobacco smoke. TSP concentrations at sampling point 1 fluctuated from 0.026 to 1.24 mg/m^3 with changes in the number of cigarettes burned. The TSP concentrations at sampling point 2 fluctuated from below the detection limit (0.0008 mg/m^3) to 0.014 mg/m^3. Thus, the average removal ratio of TSPs was 98.7% ± 0.4%. Moreover, the removal ratio maintained high-levels (98.6% ± 0.2%) after the treatment of 21,900 cigarettes (Figure 5b). These values indicate that the air-cleaner is able to remove TSPs efficiently, in agreement with previously reported air-cleaner trends observed in laboratory test [4].

Figure 5. TSP concentrations at sampling points 1 (black filled diamonds, Figure 2) and 2 (black open diamonds, Figure 2) and the removal ratios of TSPs (red filled diamonds) after the treatment of (**a**) 2300 and (**b**) 21,900 cigarettes-worth of tobacco smoke.

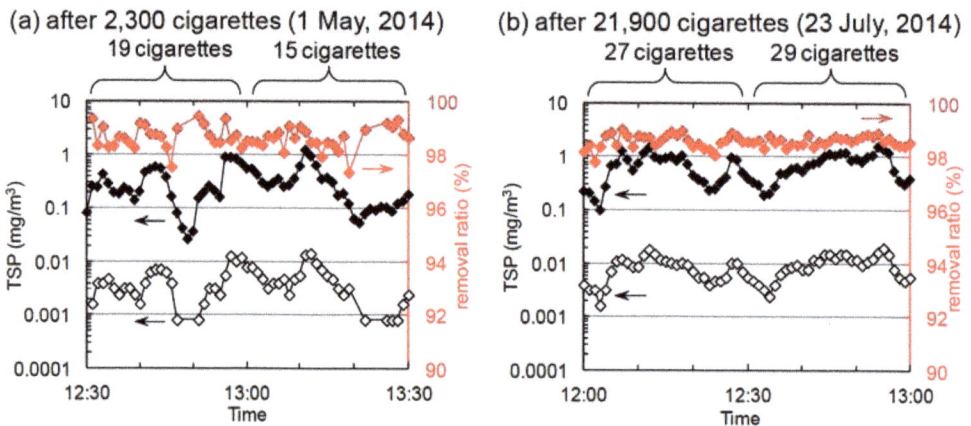

2.4. TVOC Removal

Figure 6a shows the normalised GC-MS chromatograms of the air samples at sampling points 1 and 2 after the treatment of 2300 cigarettes-worth of tobacco smoke. Many distinctive VOC peaks were observed in the chromatogram of the air sample at sampling point 1, which had almost disappeared at point 2. TVOC concentrations were calculated for all peaks between *n*-hexane (16.1 min) and *n*-hexadecane (54.7 min); they were then calibrated and converted to toluene peak (28.5 min) equivalents. In Figure 6a, the TVOC concentrations at sampling points 1 and 2 were 128.1 and 3.4 μg/m³, respectively (corresponding to a 97.3% removal ratio of TVOCs). However, several peaks remained or were amplified in the chromatogram of the sample at point 2 following the treatment of 21,900 cigarettes-worth of tobacco smoke (Figure 6b), especially between the *n*-hexane (16.1 min) and toluene (28.5 min) peaks. The TVOC removal ratio calculated from Figure 6b dramatically decreased to 43.8%, while the ratio for TSPs did not decrease (Figure 5). These data indicate that the TVOC removal efficiency of the air-cleaner under the present conditions is more easily affected than TSP removal efficiency by catalyst poisoning and the adsorption/desorption of VOCs on the filters during long-term use [8–11].

Figure 6. Normalised GC-MS chromatograms of the air samples from the Figure 2 sampling points 1 (black) and 2 (blue) after the treatment of (**a**) 2300 and (**b**) 21,900 cigarettes-worth of tobacco smoke.

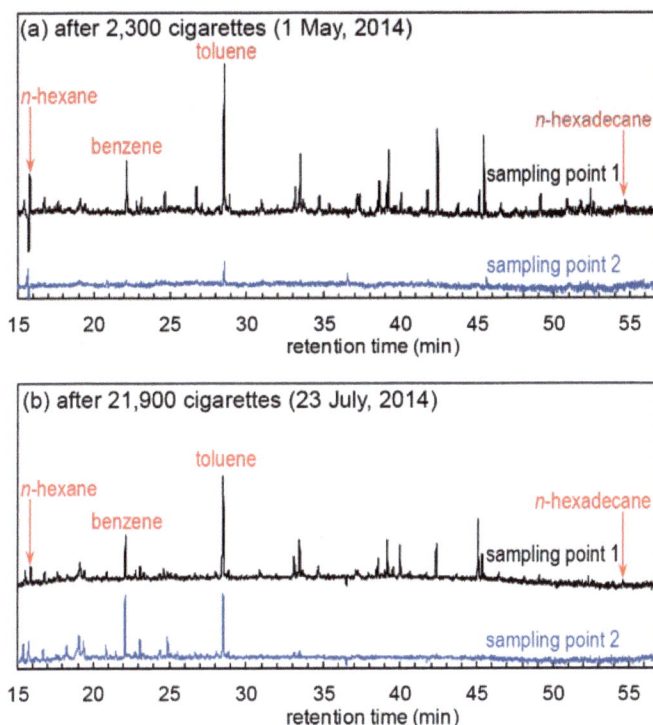

2.5. Comparison of the Field and Laboratory Tests; The Problems and Future Directions

The removal ratios of the TSPs and TVOCs from tobacco smoke by the air-cleaner in the field laboratory tests are summarised in Figure 7. In both the field and the laboratory tests, TSP removal ratios continued at high-levels (around 98.5%) throughout the experimental period. However, the TVOC removal ratios decreased with increases in the number of cigarettes. Intriguingly, the TVOC removal ratios in the field test decreased sooner than in the laboratory test. In this case, TSPs were removed by the HEPA filter and electrostatic precipitation in the plasma treatment [12,13]. However, as mentioned in Section 2.4, TVOC removal was easily affected by catalyst poisoning and adsorption/desorption of VOCs on the filters during long-term field use. To improve long-term usability, suitable maintenance methods must be developed such as plasma ashing of the reactor surfaces.

Figure 7. Removal ratios of TSPs (blue) and TVOCs (red) in tobacco smoke by the air-cleaner (**a**) in the field and (**b**) in laboratory tests. Panel (b) is reproduced from Ochiai *et al.*, [4], published by Scientific Research Publishing Inc., 2014.

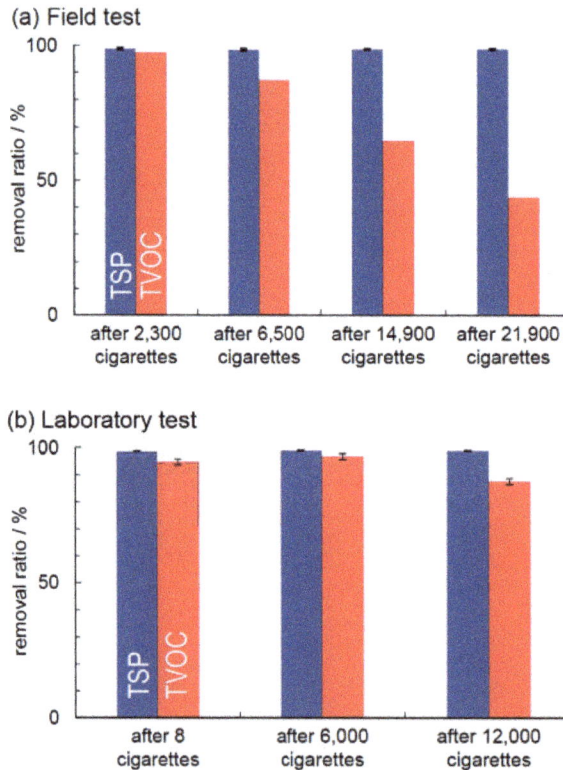

Another issue that must be considered is CO removal. Figure 8 shows CO and CO_2 concentrations at sampling points 1 and 2. The concentrations fluctuated with changes in the number of burning cigarettes, as was observed with the TSP concentrations (Figure 5). However, there are no clear differences between the concentrations at points 1 and 2 after the treatment of 2300 and 21,900 cigarettes-worth of tobacco smoke. These data indicate that the present experimental conditions of the photocatalysis-plasma synergistic reactor were not adequate for CO removal, despite the success with the TSP and TVOC removal. Several studies have investigated the synergistic effects of catalysis-plasma or photocatalysis-plasma systems. They found that synergism existed extensively but only under specific conditions [14–18]. Hence, a number of factors have been suggested that can affect efficiency such as catalyst loading level, input power, temperature, adsorption process, *etc.*

These factors can be easily influenced by the smoking room usage in this study. This may be the biggest drawback for developing a versatile and effective air-cleaner with a photocatalysis-plasma synergistic reactor. Currently, the oxidation of CO to CO_2 in the presence of noble metals is being studied for the development of effective catalytic converters and fuel cells [19–21]. In this study, the causes of the decrease in TVOC removal efficiency and the poor CO removal efficiency are still unclear. However, more suitable experimental conditions and combinations of catalysis, photocatalysis, and the plasma treatment for effective TVOC and CO removal in the field are being tested.

Figure 8. CO (black) and CO_2 (red) concentrations from the Figure 2 sampling points 1 (filled diamonds) and 2 (open diamonds) after the treatment of (**a**) 2300 and (**b**) 21,900 cigarettes-worth of tobacco smoke.

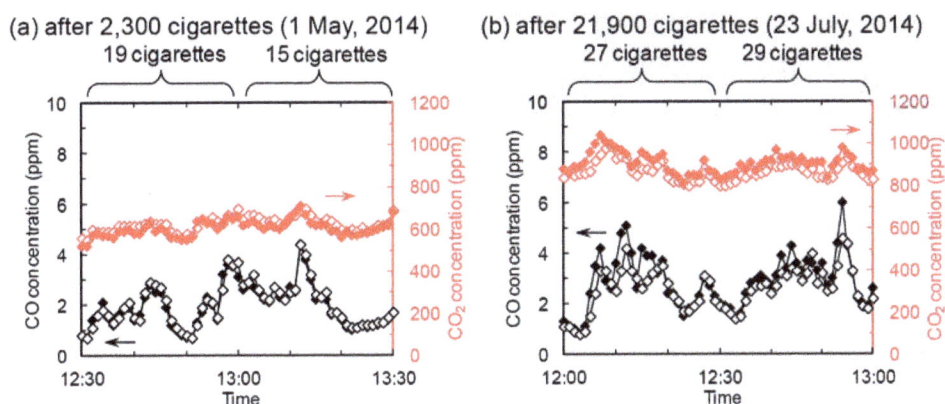

3. Experimental Section

3.1. Fabrication and Evaluation of the Photocatalysis-Plasma Synergistic Reactor and the Air-Cleaner

The image and schematic of the coil-shape PACT-TMiP synergistic reactor and air-cleaner are shown in Figure 1. The basic design and fabrication methods of the reactor and the air-cleaner have been previously described [4]. In this study, a voltage of 10 kV (peak-to-peak), a frequency of 25 kHz, and a power of 45 W were used. Air can be drawn through the gaps of the reactor while maintaining high surface contact with TMiP and air-plasma. A high efficiency particulate air (HEPA) filter, ten coil-shape reactors, two ozone-cut filters, an activated carbon filter, and a fan were arranged inside the air-cleaner. When the fan is turned on, air flow is generated inside the casing from the air inlet towards the air outlet, passing through the filters and the PACT-TMiP reactor. Temperature distributions of the coil-shaped reactors in the air-cleaner were measured by thermography using a Handy Thermo TVS-200EX (Nippon Avionics Co., Ltd., Tokyo, Japan). Ozone and NO_x concentrations were monitored using a Model 106-L ozone monitor (2B Technologies, Boulder, CO, USA) and MODEL42 I NO_x Analyser (Thermo Fisher Scientific, Waltham, MA, USA), respectively.

3.2. The Evaluation Method of the Air-Purification Activity by the Air-Cleaner in the Smoking Room

The schematic of the test method for evaluating the air-purification activity of the air-cleaner is shown in Figure 2. Air flow is generated inside the smoking room from the door and air inlet to the duct, passing through the air-cleaner. Under these conditions, the smoking room was filled with tobacco smoke from the sequential burning of cigarettes by the smokers (32 ± 11 cigarettes/h, 350 cigarettes/d). The concentrations of TSPs, TVOCs, carbon monoxide (CO), and carbon dioxide (CO_2) were measured after the treatments of 2300, 6500, 14,900, and 21,900 cigarettes-worth of tobacco smoke at sampling points 1 (near the air inlet) and 2 (inside the duct) defined in Figure 2. TSP concentrations were monitored using a digital real-time LD-3K2 dust monitor (Sibata Scientific Technology Ltd., Saitama, Japan) every minute for an hour. TVOC concentrations were calculated by qualitative and quantitative analysis using GC-MS analysis. A GC-17A-GCMS-QP5050A combination (Shimadzu, Kyoto, Japan) was used at an ionization voltage of 70 eV and a mass range of 35–200. The system was equipped with a 60 m × 0.25 mm internal diameter × 1.4 μm DB-624 fused silica capillary column (Agilent Technologies, Santa Clara, CA, USA) with split injection (split ratio 11:1). The oven was programmed to start at 35 °C (for 15 min) reaching 240 °C (for 8 min) at a rate of 6 °C/min. Samples were collected by drawing 60 L (0.6 L/min) of air through a charcoal tube, desorbed with 1 mL of carbon disulphide, and analysed by GC-MS. TVOC concentrations were calculated for all compounds eluted between *n*-hexane and *n*-hexadecane, they were then calibrated and converted to toluene equivalents. The removal ratios were calculated using the formula $(A_1 - A_2)/A_1$, where A_1 and A_2 are the amounts at sampling points 1 and 2, respectively. An important point to note is that the tobacco smoke was treated by the air-cleaner once, *i.e.*, this was a single-pass system. CO and CO_2 concentrations were also measured using a COX-3 CO/CO_2 analyser (Sibata Scientific Technology Ltd., Saitama, Japan) every minute for an hour.

4. Conclusions

The photocatalysis-plasma synergistic air-cleaner and its long-term usability in the field were investigated. Compared with previously reported laboratory test results for the air-cleaner, TSP removal ratios remained at high-levels (around 98.5%) throughout the experimental period in both the field and laboratory tests. However, the TVOC removal ratios in the field test decreased three times sooner than in the laboratory test. Additionally, the CO removal ability of the air-cleaner was almost negligible. In conclusion, these results indicate that the photocatalysis-plasma synergistic air-cleaner was effective in the long-term removal of TSPs, given the tobacco smoke conditions in the smoking room investigated. Suitable maintenance methods for the reactor surfaces would improve the long-term TVOC removal ability of the air-cleaner in the field.

Acknowledgments

We are grateful to N. Uchiyama, H. Inoue, H. Kasagi (I'm PACT World Co., Ltd.), Y. Abe, I. Kurihara, R. Yokoyama, H. Shigeno, and S. Arai (TANASHIN DENKI Co. Ltd.) for the experiments and discussions. We would also like to thank the staff at the SGS, Inc. offices for providing the opportunity for field testing and their support.

Author Contributions

Tsuyoshi Ochiai, Tadashi Machida, and Yoshitsugu Uchida participated in study design. Tadashi Machida and Yoshitsugu Uchida conducted the study. Data was collected and analysed by Tadashi Machida and Erina Ichihashi. The manuscript was written by Tsuyoshi Ochiai, Naoki Nishida, Tadashi Machida, and Yoshitsugu Uchida. Yuko Morito designed and made the TMiP. Yuji Hayashi, Yuko Morito, and Akira Fujishima provided valuable discussions and advice on the manuscript.

Conflicts of Interest

The authors declare no conflicts of interest.

References

1. Ochiai, T.; Nakata, K.; Murakami, T.; Morito, Y.; Hosokawa, S.; Fujishima, A. Development of an air-purification unit using a photocatalysis-plasma hybrid reactor. *Electrochemistry* **2011**, *79*, 838–841.

2. Ochiai, T.; Fujishima, A. Photoelectrochemical properties of tio_2 photocatalyst and its applications for environmental purification. *J. Photochem. Photobiol. C* **2012**, *13*, 247–262.

3. Ochiai, T.; Hayashi, Y.; Ito, M.; Nakata, K.; Murakami, T.; Morito, Y.; Fujishima, A. An effective method for a separation of smoking area by using novel photocatalysis-plasma synergistic air-cleaner. *Chem. Eng. J.* **2012**, *209*, 313–317.

4. Ochiai, T.; Hayashi, Y.; Ichihashi, E.; Machida, T.; Uchida, Y.; Tago, S.; Morito, Y.; Fujishima, A. Development of a coil-shape photocatalysis-plasma synergistic reactor for a practical and long-term usable air-cleaner. *Am. J. Anal. Chem.* **2014**, *5*, 467–472.

5. Chen, X.; Rozak, J.; Lin, J.-C.; Suib, S.L.; Hayashi, Y.; Matsumoto, H. Oxidative decomposition of chlorinated hydrocarbons by glow discharge in pact (plasma and catalyst integrated technologies) reactors. *Appl. Catal. A* **2001**, *219*, 25–31.

6. Ochiai, T.; Hoshi, T.; Slimen, H.; Nakata, K.; Murakami, T.; Tatejima, H.; Koide, Y.; Houas, A.; Horie, T.; Morito, Y.; *et al.* Fabrication of tio_2 nanoparticles impregnated titanium mesh filter and its application for environmental purification unit. *Catal. Sci. Technol.* **2011**, *1*, 1324–1327.

7. Hofer, M.; Penner, D. Thermally stable and photocatalytically active titania for ceramic surfaces. *J. Eur. Ceram. Soc.* **2011**, *31*, 2887–2896.

8. Knudsen, H.N.; Kjaer, U.D.; Nielsen, P.A.; Wolkoff, P. Sensory and chemical characterization of VOC emissions from building products: Impact of concentration and air velocity. *Atmos. Environ.* **1999**, *33*, 1217–1230.

9.   Wolkoff, P.; Wilkins, C.K.; Clausen, P.A.; Nielsen, G.D. Organic compounds in office environments—Sensory irritation, odor, measurements and the role of reactive chemistry. *Indoor Air* **2006**, *16*, 7–19.

10. Uhde, E.; Salthammer, T. Impact of reaction products from building materials and furnishings on indoor air quality—A review of recent advances in indoor chemistry. *Atmos. Environ.* **2007**, *41*, 3111–3128.

11. Batterman, S.; Godwin, C.; Jia, C. Long duration tests of room air filters in cigarette smokers' homes. *Environ. Sci. Technol.* **2005**, *39*, 7260–7268.

12. Cho, M.S.; Ko, H.J.; Kim, D.; Kim, K.Y. On-site application of air cleaner emitting plasma ion to reduce airborne contaminants in pig building. *Atmos. Environ.* **2012**, *63*, 276–281.

13. Molaei Najafabadi, M.; Basirat Tabrizi, H.; Aramesh, A.; Ehteram, M.A. Effects of geometric parameters and electric indexes on performance of a vertical wet electrostatic precipitator. *J. Electrost.* **2014**, *72*, 402–411.

14. Klett, C.; Duten, X.; Tieng, S.; Touchard, S.; Jestin, P.; Hassouni, K.; Vega-González, A. Acetaldehyde removal using an atmospheric non-thermal plasma combined with a packed bed: Role of the adsorption process. *J. Hazard. Mater.* **2014**, *279*, 356–364.

15. Lee, H.; Lee, D.-H.; Song, Y.-H.; Choi, W.C.; Park, Y.-K.; Kim, D.H. Synergistic effect of non-thermal plasma-catalysis hybrid system on methane complete oxidation over Pd-based catalysts. *Chem. Eng. J.* **2015**, *259*, 761–770.

16. Sano, T.; Negishi, N.; Sakai, E.; Matsuzawa, S. Contributions of photocatalytic/catalytic activities of tio2 and γ-al2o3 in nonthermal plasma on oxidation of acetaldehyde and co. *J. Mol. Catal. A* **2006**, *245*, 235–241.

17. Assadi, A.A.; Bouzaza, A.; Merabet, S.; Wolbert, D. Modeling and simulation of vocs removal by nonthermal plasma discharge with photocatalysis in a continuous reactor: Synergetic effect and mass transfer. *Chem. Eng. J.* **2014**, *258*, 119–127.

18. Assadi, A.A.; Bouzaza, A.; Vallet, C.; Wolbert, D. Use of dbd plasma, photocatalysis, and combined dbd plasma/photocatalysis in a continuous annular reactor for isovaleraldehyde elimination—Synergetic effect and byproducts identification. *Chem. Eng. J.* **2014**, *254*, 124–132.

19. Chan, D.; Tischer, S.; Heck, J.; Diehm, C.; Deutschmann, O. Correlation between catalytic activity and catalytic surface area of a pt/al2o3 doc: An experimental and microkinetic modeling study. *Appl. Catal. B* **2014**, *156–157*, 153–165.

20. Caballero-Manrique, G.; Velázquez-Palenzuela, A.; Brillas, E.; Centellas, F.; Garrido, J.A.; Rodríguez, R.M.; Cabot, P.-L. Electrochemical synthesis and characterization of carbon-supported Pt and Pt–Ru nanoparticles with Cu cores for CO and methanol oxidation in polymer electrolyte fuel cells. *Int. J. Hydrog. Energy* **2014**, *39*, 12859–12869.

21. Reshetenko, T.V.; Bethune, K.; Rubio, M.A.; Rocheleau, R. Study of low concentration CO poisoning of Pt anode in a proton exchange membrane fuel cell using spatial electrochemical impedance spectroscopy. *J. Power Sources* **2014**, *269*, 344–362.

Sample Availability: Not available.

Selective Reduction of Cr(VI) in Chromium, Copper and Arsenic (CCA) Mixed Waste Streams Using UV/TiO$_2$ Photocatalysis

Shan Zheng, Wenjun Jiang, Mamun Rashid, Yong Cai, Dionysios D. Dionysiou and Kevin E. O'Shea

Abstract: The highly toxic Cr(VI) is a critical component in the Chromated Copper Arsenate (CCA) formulations extensively employed as wood preservatives. Remediation of CCA mixed waste and discarded treated wood products is a significant challenge. We demonstrate that UV/TiO$_2$ photocatalysis effectively reduces Cr(VI) to less toxic Cr(III) in the presence of arsenate, As(V), and copper, Cu(II). The rapid conversion of Cr(VI) to Cr(III) during UV/TiO$_2$ photocatalysis occurs over a range of concentrations, solution pH and at different Cr:As:Cu ratios. The reduction follows pseudo-first order kinetics and increases with decreasing solution pH. Saturation of the reaction solution with argon during UV/TiO$_2$ photocatalysis had no significant effect on the Cr(VI) reduction demonstrating the reduction of Cr(VI) is independent of dissolved oxygen. Reduction of Cu(II) and As(V) does not occur under the photocatalytic conditions employed herein and the presence of these two in the tertiary mixtures had a minimal effect on Cr(VI) reduction. The Cr(VI) reduction was however, significantly enhanced by the addition of formic acid, which can act as a hole scavenger and enhance the reduction processes initiated by the conduction band electron. Our results demonstrate UV/TiO$_2$ photocatalysis effectively reduces Cr(VI) in mixed waste streams under a variety of conditions.

Reprinted from *Molecules*. Cite as: Zheng, S.; Jiang, W.; Rashid, M.; Cai, Y.; Dionysiou, D.D.; O'Shea, K.E. Selective Reduction of Cr(VI) in Chromium, Copper and Arsenic (CCA) Mixed Waste Streams Using UV/TiO$_2$ Photocatalysis. *Molecules* **2015**, *20*, 2622-2635.

1. Introduction

The practice of using chemicals to protect wood products from biological degradation and deterioration has been effectively used for nearly a century. One of the most widely used wood preservatives is Chromated Copper Arsenate (CCA), a mixture of oxyanions of chromium, copper and arsenic [1]. A number of different proportions of Cr:As:Cu have been employed as wood preservatives. The role of copper in CCA treated wood is to protect the wood from the attack of bacteria and fungi; arsenic plays the role of insecticide while chromium acts as a binder of arsenic and copper to the wood surface [2]. Human exposure to these toxic metals can occur from direct contact with the wood (dermal sorption), inhalation of wood dust [3] or from CCA leachate or waste streams leading to contaminated water and soil [4]. These metals/metalloids pose significant threats to human health and the environment. The highly toxic and carcinogenic nature of arsenic is well documented [5]. An estimated 50 to 100 million people suffer or are threatened from the negative health effects caused by the ingestion of arsenic contaminated water. In aqueous media, arsenic typically exists as arsenate, As(V) and arsenite, As(III). While both As(III) and As(V) are stable and toxic, As(III) is more poisonous and more mobile than As(V).Chromium, in the third oxidation state, Cr(III), is an essential micronutrient for human health however, it is highly toxic and carcinogen in the Cr(VI) state [6]. Although less pronounced, copper can also exert adverse effects on human health at high levels of exposure despite its important function as a nutrient at trace level concentrations [7]. The co-existence of these metals can be more toxic compared to the toxicity associated with each metal individually [8].

In response to the concern over the leaching of CCA from the treated wood products, such as playground equipment, patios, picnic tables, residential uses of CCA woods were banned in USA and Canada in 2004. However, the use of CCA treated wood is still allowed for a variety of commercial, industrial and agricultural purposes. Approximately, 300 million cubic meters of wood are treated with CCA, annually, consuming nearly 500,000 tons of CCA preservative formulations [9]. The extensive and sustained use of CCA preserved wood has led to the generation of significant quantities of CCA treated woodwaste which is predicted to reach approximately 9 million m^3/year by 2015 in USA and 2.5 million m^3/year by 2025 in Canada [10]. Various extraction methods including supercritical fluid extraction [11], hydrogen peroxide extraction [12], EDTA extraction [13], oxalic acid extraction [14], have been reported in the literature for the separation of Cr, Cu and As from CCA treated woodwaste. Unfortunately, these methods can suffer from limitations of low extraction efficiencies, long treatment times, interference from other metals, and high cost [11–15]. Proper handling and disposal of CCA woodwaste is a serious challenge as ultimately these toxic metals can leach from wood through aging and weathering processes leading to discharge of the toxic metals into the environment. Landfill is still the most common fate of CCA treated woodwaste in many countries including USA and Canada as CCA wood products are not currently categorized as hazardous waste [16].Once deposited into the landfills, CCA treated products leach toxic arsenic and chromium species into the environment through aqueous runoffs and pose a serious threat to human life and to the surrounding environment [17]. Thus there is an urgent need to identify an environmentally sound and cost effective method to treat CCA contaminated leachates and explore possible strategies to recover the metals for future use.

The toxicity of these metals can be dramatically influenced by their oxidation state. In CCA, Cu and As exist as oxides of Cu(II) and As(V) [18]. Cr is present as Cr(VI) in CCA before fixation, which involves a reductive process leading to partial conversion to Cr(III) during binding with wood lignin [19]. Cr(VI) is toxic and mobile in the environment while Cr(III) is less toxic and a number of reductive processes including photocatalysis can effectively transform Cr(VI) to Cr(III) [20]. Chromium species can be removed and recovered from aqueous media employing adsorption or precipitation processes [21,22]. While semiconductor mediated photocatalysis can be used to oxidize an extensive number of pollutant and toxins, a relatively limited number of examples have appeared on effective applications for reductive transformations of pollutants and toxins [23]. Among semiconductor photocatalysts, TiO_2 is most extensively studied for water purification. Upon photo-excitation of TiO_2 with wavelengths ≤385 nm, an electron/hole pair is produced, Equation (1), the hole can initiate oxidation through electron transfer with water, Equation (2) or a metal at the surface Equation (3), while the electron can act as a reducing agent to molecular oxygen, Equation (4) or metal ions, Equation (5):

$$TiO_2 \xrightleftharpoons{h\nu} h^+ + e^- \qquad (1)$$

$$h^+ + H_2O \rightarrow HO^\bullet \qquad (2)$$

$$HO^\bullet + m^{n+} \rightarrow m^{n+2} \qquad (3)$$

$$e^- + O_2 \rightarrow O_2^{\bullet-} \qquad (4)$$

$$e^- + m^{n+} \rightarrow m^{n-1} \qquad (5)$$

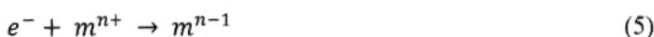

A number of reports have appeared in the literature on the TiO_2 adsorption and photocatalytic reduction of Cr(VI) [24–32]. TiO_2 materials can have strong metal ion adsorption properties and are thus attractive for adsorption and photo-initiated treatment as simultaneous processes for the removal of heavy metals from the contaminated waste streams and leachates. Since adsorption of Cr(VI) at the surface of the semiconductor is critical for effective reduction during photocatalysis, it is important to examine the effects of As and Cu in the treatment strategies for CCA waste streams. In this study, we have demonstrated the effective reduction of Cr(VI) to Cr(III) using TiO_2 photocatalyst in presence of As(V) and Cu(II). We have also investigated the influence of solution pH and CCA formulations on the reduction of Cr (VI) to Cr (III).

2. Results and Discussion

2.1. The Reduction of Cr(VI)

In a typical experiment, a suspension of TiO_2 was prepared with a CCA stock solution and transferred to a Pyrex reaction vessel. The reaction solution was purged with oxygen, air, or argon, and irradiated with UV light. Analyses were performed at specific time intervals. In the presence of TiO_2, oxygen and UV irradiation, Cr(VI) is readily converted to Cr(III) which is measured from the difference of initial and final concentration of Cr(VI). Control experiments established the negligible reduction (<3%) of Cr(VI) in the absence of TiO_2 and very little adsorption (<1%) of Cr(VI) on TiO_2 in the dark. The purging of the CCA reaction solution with different gases (Ar, air, O_2) prior and during irradiation had minimal effect on the photocatalytic reduction of Cr(VI) shown in Figure 1. These results are consistent with previous reports on the adsorption and reduction of Cr(VI) during TiO_2 photocatalysis [33]. The pKa and speciation of arsenic and chromium metals are shown in Table 1. Copper exists as non-protonated copper oxide form.

The solution pH has a pronounced effect on the Cr(VI) species and surface charge of TiO_2. The pK_{a1} and pK_{a2} of chromic acid (shown in Table 1) are 0.7 and 6.5, respectively. The zero point of charge (ZPC) of P25 TiO_2 is 6.8 [34]. Hence, the reduction of Cr(VI) was investigated as a function of solution pH. When the solution pH is less than 6.8, the TiO_2 is positively charged and the predominant Cr(VI) species $HCrO_4^-$ and CrO_4^{2-} are negatively charged. The photocatalytically initiated reactions mainly occur at or near the surface of TiO_2 and thus electrostatic attraction between charged substrate and oppositely changed surface can lead to strong adsorption and enhance the reactivity or reduction of Cr(VI) at solution pH below 6.8. As the solution pH increases above 6.8, the TiO_2 surface becomes progressively more negatively charged while the predominant Cr(VI) species is CrO_4^{2-}. The electrostatic repulsion between the negatively charged TiO_2 surface and CrO_4^{2-} inhibits Cr(VI) adsorption and retards reduction.

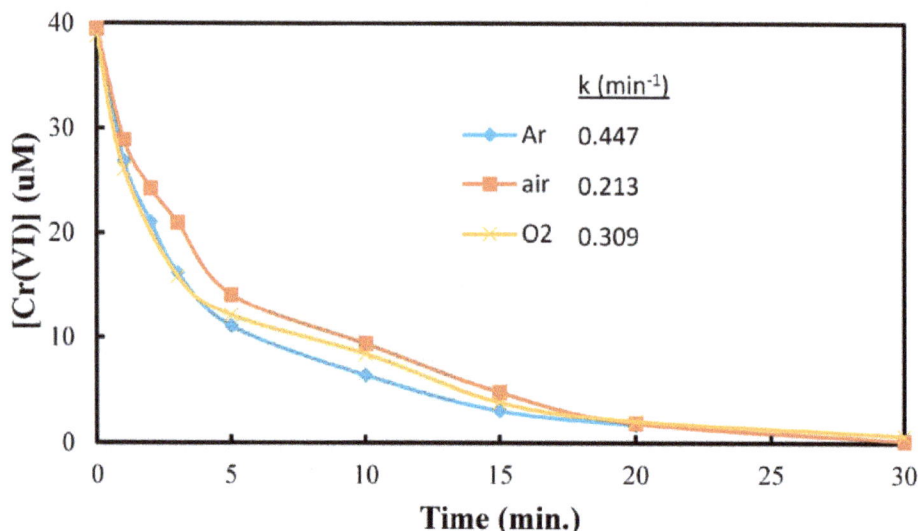

Figure 1. The effect of dissolved argon, oxygen and air on the reduction of Cr(VI). $[Cr(VI)] = 40\ \mu M$, $[TiO_2] = 0.10$ g/L, pH = 3, 350 nm.

Table 1. pKa and speciation of chromium and arsenic.

Arsenic (As)		Chromium (Cr)
OH $\mathsf{HO^-As^{(III)}_{\ \ OH}}$	O $\mathsf{HO^-As^{(V)}_{OH}OH}$	O (VI) $\mathsf{HO-Cr-OH}$ O
Arsenite	Arsenate	Chromate
pKa = 9.1, 12.1, 13.4	pKa = 2.1, 6.7, 11.2	pKa = 0.7, 6.5

UV/TiO_2 processes often following pseudo-first order kinetics according to Equation (6), below.

$$\ln(C_0/C) = kt \tag{6}$$

where k is the rate constant of pseudo-first order model (min^{-1}), t is time (min), C_0 is the initial concentration of Cr(VI) and C is the concentration of Cr(VI) at the specific time.

The reduction of Cr(VI) was monitored as a function of irradiation time at several different initial Cr(VI) concentrations. The UV/TiO_2 photocatalytic reduction of Cr(VI) was measured over a range of solution pH from ~3.0 to 10.0 with the fastest occurring under acidic conditions (Figure 2). The pH effect has been rationalized based on adsorption and the role of electrostatic interactions between TiO_2 surface and chromate species which are attractive at low pH and repulsive at high pH [33].

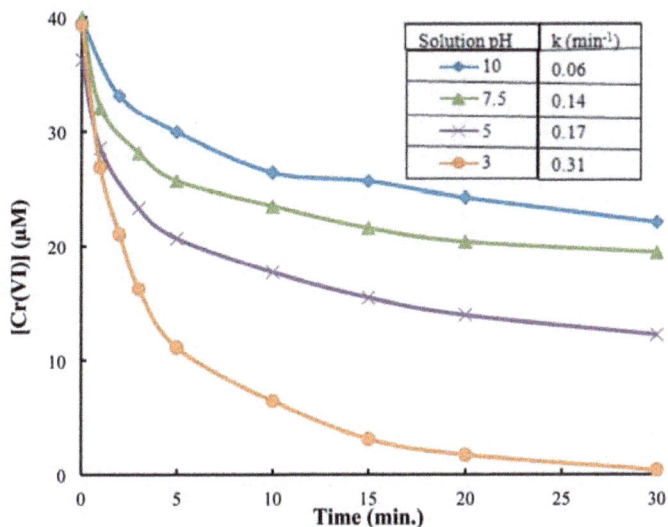

Figure 2. Investigation of reaction kinetics of Cr(VI) reduction as a function of solution pH. [Cr(VI)] = 40 µM; [TiO₂] = 0.10 g/L, 350 nm.

2.2. The Reduction of Cr(VI) in Presence of As(V) at pH 3.0

While transformation of Cr(VI) to Cr(III) can be achieved by employing UV and visible light-activated TiO₂ materials, the effects of As and Cu on the reduction process have not been reported which is especially important for the treatment of CCA waste streams. The effect of As(V) presence on the reduction rate of Cr(VI) at pH 3.0 is illustrated in Figure 3. While significant reduction of Cr(VI) is observed, the rate of reduction decreased with increasing As(V) concentration. A number of literature reports describe the oxidation of As(III) to less toxic As(V) under TiO₂ photocatalysis [34–36]. These reports also established TiO₂ photocatalysis does not lead to the reduction of As(V) and thus under our experimental conditions As(V) did not compete with Cr(VI) for the conduction band electrons. We did not detect the presence of As(III) in the reaction solution following the TiO₂ photocatalysis treatment. A modest inhibition of Cr(VI) reduction observed upon addition of As(V) and slight decrease of As(V) detected in solution relative to the initial concentration (Figure 4). These results are attributed to the competition between As(V) and Cr(VI) for adsorption at surface active sites where the conduction band electrons are available for reduction of the adsorbate.

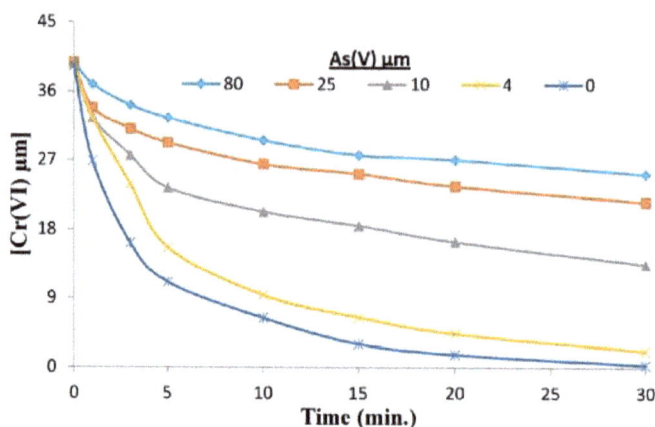

Figure 3. The reduction of Cr(VI) in the presence of As(V) at pH 3.0. [Cr(VI)] = 40 μM, [As(V)] = 0–80 μM, [TiO$_2$] = 0.10 g/L, 350 nm.

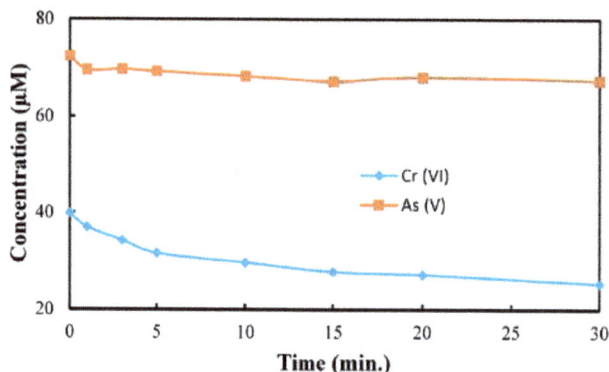

Figure 4. The reduction of As(V) in presence of Cr(VI). [Cr(VI)]$_0$ = 40 μM, [As(V)]$_0$ = 80 μM, pH = 3, [TiO$_2$] = 0.10 g/L, 350 nm.

2.3. The Reduction of Cr(VI) in Presence of Cu(II)

The effect of Cu(II), the third component in CCA on the reduction of Cr(VI) is shown in Figure 5. The addition of copper had no significant effect on the UV/TiO$_2$ photocatalytic reduction of Cr(VI) over a range of Cu(II) concentrations. While Cu(II) ions have been used as conduction band electron scavengers during TiO$_2$ photocatalysis, the presence of Cu(II) ions had no effect on the rate of reduction of Cr(VI) under our experimental conditions as illustrated in Figure 6. The concentration of Cu(II) was monitored using atomic absorption throughout the TiO$_2$ photocatalytic reduction of Cr(VI). The changes in the concentration of Cu(II) as a function of irradiation time were insignificant, as shown in Figure 6, indicating no Cu(II) reduction occurred during our UV/TiO$_2$ photocatalysis experiments [32]. These results demonstrate that Cu(II) as included in the CCA formulation does not compete with Cr(VI) for conduction band electrons or critical surface adsorption sites during TiO$_2$ photocatalytic reduction of Cr(VI).

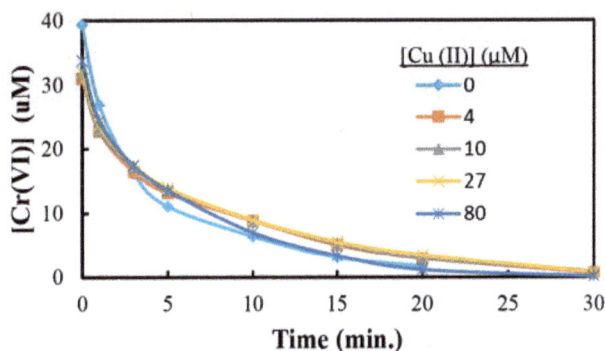

Figure 5. The reduction of Cr(VI) in the presence of Cu(II). [Cr(VI)] = 40 μM, [Cu (II)] = 0 μM–80 μM, pH = 3, [TiO₂] = 0.10 g/L, 350 nm.

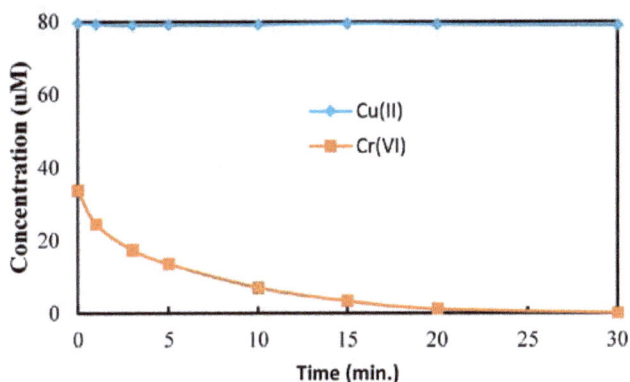

Figure 6. The reduction of Cu (II) in presence of Cr(VI). [Cr(VI)]₀ = 40 μM, [Cu(II)]₀ = 80 μM, pH = 3, [TiO₂] = 0.10 g/L, 350 nm.

2.4. Cr(VI) Reduction in Presence of As(V) and Cu(II)

Treatment of CCA as a mixture is much more practical compared to the processes requiring separation of the metal ions and multiple treatment trains. With this in mind the cooperative effect of Cu(II) and As(V) on the TiO₂ photocatalysis reduction of Cr(VI) to Cr(III) was investigated mimicking typical CCA formulations. TiO₂ photocatalysis experiments were conducted in presence of 25 μM As(V), 25 μM Cu(II) and 40 μM Cr(VI) under 350 nm irradiation. The reaction suspension was collected and filtered at specific time intervals. Analysis of the solution as a function of treatment time showed that Cr(VI) is reduced to Cr(III) at only a slightly slower rate than the reduction of Cr(VI) alone, in the absence of As and Cu (Figure 7). While the initial concentration of Cu(II) remained unchanged throughout the photocatalytic treatment, the small decrease in the initial As(V) concentration relative to the starting concentration is assigned to strong adsorption on TiO₂ surface. Although the addition of Cu(II) alone showed no effect on the rate of Cr(VI) reduction, in presence of 40 μm of Cr(VI) and 25 μm of As(V), addition of Cu(II) (≥25 μm) lead to a modest increase in the Cr(VI) reduction rate (Figure 8). The small enhancement may be due to interactions between

As(V) and Cu(II) and/or competitive adsorption processes which reduces the adsorption of As for active sites on the surface of the TiO$_2$ thus effectively improving the adsorption and subsequent reduction of Cr(VI).

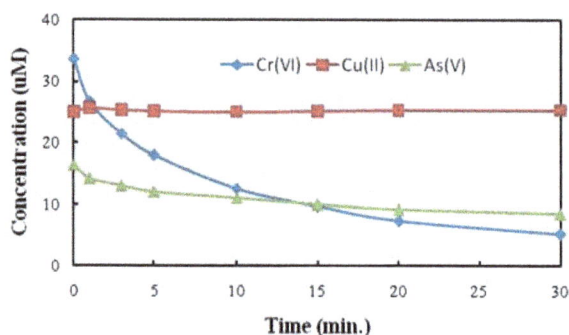

Figure 7. TiO$_2$ photocatalysis of Cr(VI) in presence of As(V) and Cu(II). [Cr(VI)] = 40 μM, [As(V)] = 25 μM, [Cu(II)] = 25 μM, pH = 2, [TiO$_2$] = 0.10 g/L, 350 nm.

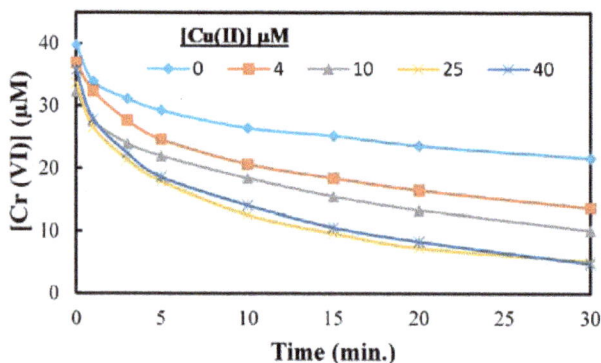

Figure 8. The effect of different Cu(II) concentration on Cr(VI) reduction in presence of As(V). [Cr(VI)] = 40 μM, [As(V)] = 25 μM, [Cu(II)] = 0–40 μM, pH = 3, [TiO$_2$] = 0.10 g/L, 350 nm.

The ratio of the initial concentrations of Cr(VI), As(V) and Cu(II) was varied during TiO$_2$ photocatalysis to model common CCA formulations. At a 0.5:1:1 ratio of Cr(VI):As(V):Cu(II) the fastest and most extensive reduction of Cr(VI) was observed (Figure 9). These changes as a function of CCA formulation are attributed to competitive adsorption among the metals and not for the reactions with the conduction band electrons. The effect of pH on the reduction of 40 μm of Cr(VI) in presence of 25 μm of As(V) and 25 μm of Cu(II) is illustrated in Figure 10. This result suggests that the UV/TiO$_2$ photocatalytic reduction of Cr(VI) is faster underacidic pH conditions both in the presence and absence of As(V) and Cu(II).

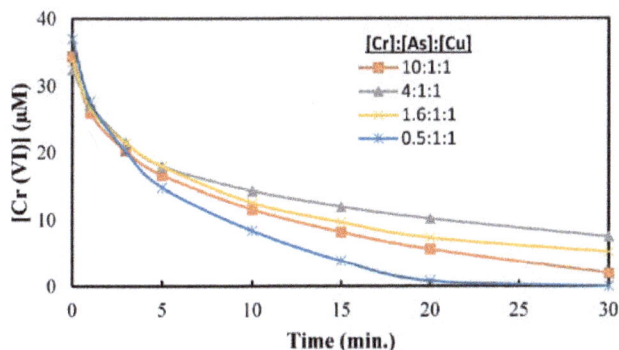

Figure 9. Cr(VI) reduction at different Cr(VI):As(V):Cu(II) ratios. pH = 3, [TiO₂] = 0.10 g/L, 350 nm.

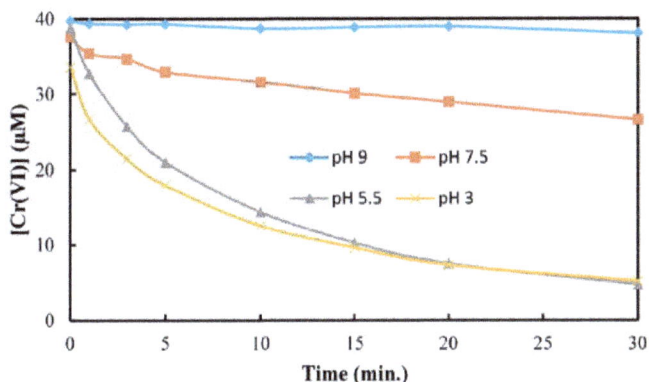

Figure 10. The pH effect on Cr(VI) reduction in presence of As (V) and Cu (II). Cr(VI):As(V):Cu(II) = 1.6:1:1. [TiO₂] = 0.10 g/L, 350 nm.

2.5. Effect of Hole Scavenger on Cr(VI) Reduction

Upon irradiation, TiO₂ can absorb photons of ≤385 nm promoting an electron from the valence band to the conduction band forming an electron-hole pair. Rapid recombination of the electron and hole pair prevents redox transformation of adsorbates, and typically an electron scavenger is required to trap the electron and inhibit recombination. UV/TiO₂ photocatalysis has been extensively used for oxidative transformation of pollutants and toxins with molecular oxygen purged through the system as an electron scavenger. Molecular oxygen acts as trap for the conduction band electron prolonging the lifetime of the hole and dramatically improving the probability to achieve oxidation of a surface adsorbed species. In the absence of a suitable electron trap little or no oxidative transformation is observed as photo-excitation energy is converted to thermal energy upon recombination.

For the case of Cr(VI), we chose to investigate the influence of adding an electron donor in an attempt to trap the hole, to inhibit recombination and extend the lifetime of the conduction band electron as the desired species to facilitate the reduction of Cr(VI). Formic acid can scavenge the valence band hole and is readily adsorbed onto the surface under neutral and acidic solution pH leading to an inhibition of recombination of electron and hole pairs on the TiO₂ surface Equations (7)–(9).

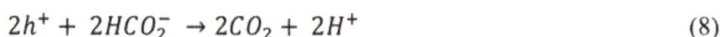

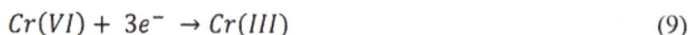

$$TiO_2 \underset{}{\overset{hv}{\rightleftharpoons}} h^+ + e^- \tag{7}$$

$$2h^+ + 2HCO_2^- \rightarrow 2CO_2 + 2H^+ \tag{8}$$

$$Cr(VI) + 3e^- \rightarrow Cr(III) \tag{9}$$

With this in mind formic acid was added to the reaction mixture prior to TiO₂ photocatalytic treatment of Cr(VI).We observed a dramatic enhancement in the reduction rate of Cr(VI), indicating that formic can trap the hole to extend the lifetime of the electron enhancing the rate of Cr(VI) reduction (Figure 11). Since molecular oxygen does not affect the rate it does not appear reduction to superoxide anion radical and subsequent reaction with Cr(VI) change the observed rates of reduction.

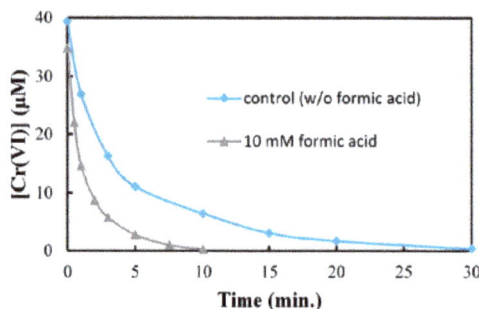

Figure 11. Effect of formic acid as hole scavenger on Cr(VI) reduction. [Cr (VI)] = 40 μM, [HCOOH] = 10 mM, [TiO₂] = 0.10 g/L, 350 nm.

3. Experimental

3.1. Materials

Degussa P25 TiO₂ (CAS NO. 13463-67-7), a mixture of 80% anatase and 20% rutile with an average surface area of 50 m²/g, was donated by Degussa (Ridgefield, NJ, USA). Copper(II) nitrate trihydrate (99%–104%), sodium arsenate dibasic heptahydrate (99%) and sodium *meta*-arsenite (98%) were obtained from Sigma (St. Louis, MO, USA), while potassium chromate (99.8%), sodium hydroxide (99.4%), ammonium hydroxide (29.15%), trace metal grade nitric acid (67%–70%) and formic acid (88%) were purchased from Fisher (Waltham, MA, USA). All the chemicals were used without further purification. Millipore filtered water (18 M*Ω cm) was obtained from a nanopure diamond lab water system (Barnstead Thermolyne Corporation, Dubuque, IA, USA) at room temperature. Volumetric glasswares were used to prepare all solutions.

3.2. Sample Preparation and Analysis

The concentration of TiO₂ was 0.10 g/L, and the initial concentration of Cr(VI) was 40 μmol/L unless otherwise stated. As(III), As(V) and/or Cu(II) were added into TiO₂ suspension based on experimental design. One hundred mL of TiO₂ suspension was subject to an ultrasonic bath for

15 minutes to achieve a homogeneous suspension of the catalyst in the reaction vessel (Pyrex tube 12×1 in, 160 mL capacity, with a vented Teflon screw top). The suspension was magnetically stirred for 1.0 hour in the dark to allow equilibrium adsorption between target compound and TiO_2. The solution pH was adjusted with 0.1 M HNO_3 and 0.1 M NaOH and measured using a calibrated Pinnacle 530 pH meter (Corning, Vernon Hills, IL, USA).The solution was gently purged with the desired gas (air, oxygen or argon) for 15 min prior to and during irradiation. Irradiation of TiO_2 suspension was conducted in a Rayonet photochemical reactor (model RPR-100, Southern New England Ultraviolet, Branford, Connecticut, USA), equipped with a cooling fan and 15 phosphor coated low-pressure mercury lamps (λ_{max} = 350 nm, light intensity = $5.2 \pm 0.1 \times 10^6$ photon/s/cm³). Five mL samples were taken at each specified time intervals, filtered through a 0.45 μm PTFE filter and subsequently analyzed.

The total copper and chromium were measured using an Analyst 600 atomic absorption spectrophotometer (Perkin–Elmer, Waltham, MA, USA) at 324.8 nm and 357.9 nm as recommended by the manufacturer. Determination of total arsenic was conducted using inductively coupled plasma mass spectroscopy (ICP-MS, HP 4500, Agilent, Santa Clara, CA, USA). The plasma and auxiliary gas flow rates of the ICP-MS were maintained at 15.4 and 1.0 L/min, respectively. Arsenic speciation of arsenate and arsenite and Cr(VI) were performed using a high-performance liquid chromatograph (HPLC) instrument equipped with a Lab Alliance series III digital pump (Fishersci, Waltham, MA, USA) and an ion exchange column coupled to ICP-MS detection. A PRP X-100 (250 mm × 4.6 mm, 10 μm particle size) anion-exchange column(Hamilton, North Quincy, MA, USA) was used to separate Cr(VI) and arsenic species with the mobile phase of 40 mM (pH 9.0) ammonium nitrate at the flow rate of 1.0 mL/min.

4. Conclusions

TiO_2 photocatalysis is effective for the complete reduction of Cr(VI) to Cr(III) in the presence and absence of Cu(II) and As(V). The reduction process is effective over a range of solution pH and under oxygen, air, or argon saturated conditions. The pH effects indicate that the reduction is favorable under acidic solution conditions due to enhanced adsorption through electrostatic attraction between positively charged TiO_2 surface and anionic chromate species. The addition of As(V) in the reaction mixture slightly reduces the reduction rate of Cr(VI) which is assigned to competitive adsorption between As(V) and Cr(VI) and competition for reduction processes since As(V) is not reduced under the experimental conditions. Although Cu(II) can be reduced during TiO_2 photocatalysis, we observe no reduction of Cu(II) and a slight enhancement in Cr(VI) reduction under very specific conditions. The effective reduction of Cr(VI) is achieved in the presence of As(V) and Cu(II) at different ratios. The addition of formic acid to the reaction solution shows a significant acceleration in the reduction of Cr(VI). Our study clearly demonstrates UV/TiO_2 photocatalysis is a promising single treatment process for the remediation of Cr(VI) mixed wastes under a variety of conditions.

Acknowledgments

KEO and DDD gratefully acknowledge the National Science Foundation for partial support of the project (Grant # CBET–1235803).

Author Contributions

Yong Cai, Dionysios D. Dionysiou, and Kevin E. O'Shea were critical to the conception, design, oversight of and data analysis for the project. Shan Zheng did the majority of the experimental work and provided a report of the results. Wenjun Jiang and Mamun Rashid ran follow-up experiments and revised the manuscript.

Conflicts of Interest

The authors declare no conflict of interest.

References

1. Cooper, P.A. Leaching of CCA: Is it a problem? In *Environmental Considerations in the Manufacturing, Use and Disposal of Preservative-Treated Wood*; Forest Products Society, Madison, WI, USA, 1994.
2. Joanna, S.W.; KongHwa, C. Extraction of chromated copper arsenate from wood wastes using green solvent supercritical carbon dioxide. *J. Hazard. Mater.* **2008**, *158*, 384–391.
3. Gordon, T.; Spanier, J.; Butala, J.H.; Li, P.; Rossman, T.G. *In vitro* bioavailability of heavy metals in pressure treated wood dust. *J. Toxicol. Sci.* **2002**, *67*, 32–37.
4. Stillwell, D.; Gorny, K. Contamination of soil with copper, chromium, and arsenic under decks built from pressure treated wood. *Bull. Environ. Contam. Toxicol.* **1997**, *58*, 22–29.
5. WHO. *Arsenic in Drinking-Water, Background Document for Development of WHO Guidelines for Drinking-Water Quality*; World Health Organization (WHO): Geneva, Switzerland, 2003.
6. WHO. *Chromium in Drinking-Water, Background Document for Development of WHO Guidelines for Drinking-Water Quality*; Geneva, Switzerland, 2003.
7. ATSDR. *Toxicological Profile for Copper, Background and Environmental Exposures to Copper in the United States*; Agency for Toxic Substances and Disease Registry (ATSDR): Atlanta, GA, USA, 2004.
8. Mason, R.W.; Edwards, I.R. Acute toxicity of combinations of sodium dichromate, sodium arsenate and copper sulphate in the rat. *Comp. Biochem. Physiol. Part C* **1989**, *93*, 121–125.
9. Humphrey, D.G. The chemistry of chromated copper arsenate wood preservatives. *Rev. Inorg. Chem.* **2002**, *22*, 1–40.
10. Cooper, P.A. A review of issues and technical options for managing spent CCA treated wood. In Proceedings of the AWPA Annual Meeting, Boston, MA, USA, 27–29 April 2003.
11. El-Fatah, S.M.; Goto, M.; Kodama, A.; Hirose, T. Supercritical fluid extraction of hazardous metals from CCA wood. *J. Supercrit. Fluids* **2004**, *28*, 21–27.

12. Gyu-Hyeok, K.; Jong-Bum, R.; li-Gon, K.; Yun-Sang, S. Optimization of hydrogen peroxide extraction conditions for CCA removal from treated wood by response surface methodology. *For. Prod. J.* **2004**, *54*, 141–144.

13. Kartal, S.N. Removal of copper, chromium, and arsenic from CCA-C treated wood by EDTA extraction. *Waste Manag.* **2003**, *23*, 537–546.

14. Clausen, C.A.; Smith, R.L. Removal of CCA from treated wood by oxalic acid extraction, steam explosion, and bacterial fermentation. *J. Ind. Microbiol. Biotechnol.* **1998**, *20*, 251–257.

15. Janin, A.; Blais, J.F.; Mercier, G.; Drogui, P. Optimization of a chemical leaching process for decontamination of CCA-treated wood. *J. Hazard. Mater.* **2009**, *169*, 136–145.

16. Khan, B.I.; Solo-Gabriele, H.M.; Townsend, T.G.; Cai, Y. Release of arsenic to the environment from CCA-treated wood.1.Leaching and speciation during service. *Environ. Sci. Technol.* **2006**, *40*, 988–993.

17. Townsens, T.; Solo-Gabriele, H.; Tolaymat, T.; Stook, K.; Hosein, N.; Chromium, Copper, and Arsenic Concentrations in Soil Underneath CCA-Treated Wood Structures. *Soil Sediment Contam.* **2003**, *12*, 779–798.

18. Janin, A.; Zaviska, F.; Drogui, P.; Blais, J.; Mercier, G. Selective recovery of metals in leachate from chromated copper arsenate treated wastes using electrochemical technology and chemical precipitation. *Hydrometallurgy* **2009**, *96*, 318–326.

19. Nygren, O.; Nilsson, C.A. Determination and speciation of chromium, copper and arsenic in wood and dust from CCA-impregnated timber. *Analusis* **1993**, *21*, 83–89.

20. Jiang, W.; Cai, Q.; Xu, W.; Yang, M.; Cai, Y.; Dionysiou, D.D.; O'Shea, K.E. Cr(VI) Adsorption and Reduction by Humic Acid Coatedon Magnetite. *Environ. Sci. Technol.* **2014**, *48*, 8078–8085.

21. Gupta, V.K.; Agarwal, S.; Saleh, T.A. Chromium removal by combining the magnetic properties of iron oxide with adsorption properties of carbon nanotubes. *Water Res.* **2011**, *45*, 2207–2212.

22. Kongsricharoern, N.; Polprasert, C. Chromium removal by a bipolar electro-chemical precipitation process. *Water Sci. Technol.* **1996**, *34*, 109–116.

23. Aarthi, T.; Madras, G. Photocatalytic reduction of metals in presence of combustion synthesized nano-TiO_2. *Catal. Commun.* **2008**, *9*, 630–634.

24. Chenthamarakshan, C.R.; Rajeshwar, K. Heterogeneous Photocatalytic Reduction of Cr(VI) in UV-Irradiated Titania Suspensions: Effect of Protons, Ammonium Ionsand, Other Interfacial Aspects. *Langmuir* **2000**, *16*, 2715–2721.

25. Lin, W.Y.; Wei, C.; Rajeshwar, K. Photocatalytic reduction and immobilization of hexavalent chromium at titanium dioxide in aqueous basic media. *J. Electrochem. Soc.* **1993**, *140*, 2477–2482.

26. Parida, K.; Mishra, K.G.; Dash, S.K. Adsorption of toxic metal ion Cr(VI) from aqueous state by TiO_2-MCM-41: Equilibrium and kinetic studies. *J. Hazard. Mater.* **2012**, *241–242*, 395–403.

27. Zhang, L.; Zhang, Y. Adsorption characteristics of hexavalent chromium on HCB/TiO2. *Appl. Surf. Sci.* **2006**, *316*, 649–656.

28. Zhang, Y.C.; Yang, M.; Zhang, G.; Dionysiou, D.D. HNO₃-involved one-step low temperature solvothermal synthesis of N-doped TiO_2 nanocrystals for efficient photocatalytic reduction of Cr(VI) in water. *Appl. Catal. B* **2013**, *142–143*, 249–258.

29. Yoon, J.; Shim, E.; Joo, H. Photocatalytic reduction of hexavalent chromium (Cr(VI)) using rotating TiO_2 mesh. *Korean J. Chem. Eng.* **2009**, *26*, 1296–1300.

30. Wang, Q.; Shang, Q.; Zhu, T.; Zhao, F. Efficient photoelectrocatalytic reduction of Cr(VI) usingTiO_2 nanotube arrays as the photoanode and a large-area titanium mesh as the photocathode. *J. Mol. Catal. A* **2011**, *335*, 242–247.

31. Xu, X.R.; Li, H.B.; Gu, J.D. Simultaneous decontamination of hexavalent chromium and methyl *tert*-butyl ether by UV/TiO_2 process. *Chemosphere* **2006**, *63*, 254–260.

32. Liu, W.; Ni, J.; Yin, X. Synergy of photocatalysis and adsorption for simultaneous removal of Cr(VI) and Cr(III) withTiO_2 and titanate nanotubes. *Water Res.* **2014**, *53*, 12–25.

33. Prairie, M.R.; Evans, L.R.; Stange, B. Martinez, S.L. An investigation of titanium dioxide photocatalysis for the treatment of water contaminated with metals and organic chemicals. *Environ. Sci. Technol.* **1993**, *27*, 1776–1782.

34. Xu, T.; Cai, Y.; O'Shea, K.E. Adsorption and Photocatalyzed Oxidation of Methylated Arsenic Species in TiO_2Suspensions. *Environ. Sci. Technol.* **2007**, *41*, 5471–5477.

35. Zheng, S.; Cai, Y.; O'Shea, K.E. TiO_2photocatalytic degradation of phenylarsonic acid. *J. Photochem. Photobiol. A* **2010**, *210*, 61–68.

36. Zheng, S.; Jiang, W.; Cai, Y.; Dionysiou, D.D.; O'Shea, K.E. Adsorption and photocatalytic degradation of aromatic organoarsenic compounds in TiO_2suspension. *Catal. Today* **2014**, *224*, 83–88.

Sample Availability: Samples of the compounds are not available from the authors.

Modeling the Photocatalytic Mineralization in Water of Commercial Formulation of Estrogens 17-β Estradiol (E2) and Nomegestrol Acetate in Contraceptive Pills in a Solar Powered Compound Parabolic Collector

José Colina-Márquez, Fiderman Machuca-Martínez and Gianluca Li Puma

Abstract: Endocrine disruptors in water are contaminants of emerging concern due to the potential risks they pose to the environment and to the aquatic ecosystems. In this study, a solar photocatalytic treatment process in a pilot-scale compound parabolic collector (CPC) was used to remove commercial estradiol formulations (17-β estradiol and nomegestrol acetate) from water. Photolysis alone degraded up to 50% of estradiol and removed 11% of the total organic carbon (TOC). In contrast, solar photocatalysis degraded up to 57% of estrogens and the TOC removal was 31%, with 0.6 g/L of catalyst load (TiO$_2$ Aeroxide P-25) and 213.6 ppm of TOC as initial concentration of the commercial estradiols formulation. The adsorption of estrogens over the catalyst was insignificant and was modeled by the Langmuir isotherm. The TOC removal via photocatalysis in the photoreactor was modeled considering the reactor fluid-dynamics, the radiation field, the estrogens mass balance, and a modified Langmuir–Hinshelwood rate law, that was expressed in terms of the rate of photon adsorption. The optimum removal of the estrogens and TOC was achieved at a catalyst concentration of 0.4 g/L in 29 mm diameter tubular CPC reactors which approached the optimum catalyst concentration and optical thickness determined from the modeling of the absorption of solar radiation in the CPC, by the six-flux absorption-scattering model (SFM).

Reprinted from *Molecules*. Cite as: Colina-Márquez, J.; Machuca-Martínez, F.; Li Puma, G. Modeling the Photocatalytic Mineralization in Water of Commercial Formulation of Estrogens 17-β Estradiol (E2) and Nomegestrol Acetate in Contraceptive Pills in a Solar Powered Compound Parabolic Collector. *Molecules* **2015**, *20*, 13354-13373.

1. Introduction

Pharmaceuticals and metabolites residues in the aquatic environment are cause of concern to many agencies, institutions and governments worldwide. Actions for monitoring their occurrence, preventive measures and novel technologies for their containment are currently being evaluated at national and international level [1–6]. Among many pharmaceuticals, endocrine disruptors in water are contaminants of emerging concern due to the risk they pose to the aquatic ecosystems and to the environment. Compounds with estrogenic, progestagenic and/or androgenic activities can have significant effect on humans and wildlife [7–10]. For example, the disruptive impact of 17-α ethynylestradiol (EE2) to the fish population in an experimental lake was demonstrated in a seven-year study, which showed near extinction of fish after four years of EE2 dosing, due to reproductive failure [11]. The female contraceptive pill active compounds 17-β estradiol (E2) and nomegestrol have recently been formulated as an alternative to pills containing the more common synthetic estrogen EE2, since these hormones are structurally identical to endogenous estrogen in women [12].

Current municipal wastewater treatment plants are unable to completely remove or destroy pharmaceuticals from domestic wastewater [13,14]. Advanced oxidation processes (AOPs), which are based on the generation of highly powerful reactive oxidative species (e.g., hydroxyl radicals), have been proposed as alternative processes to inactivate the biological and physiological effect of pharmaceuticals in water. Among AOPs, photocatalytic oxidation over an irradiated semiconductor photocatalyst (often titanium dioxide (TiO_2)) has proven to be effective in the removal of pharmaceuticals including estrogens [15–18]. Most studies that investigate the photocatalytic degradation of estrogens deal with idealized systems, using ultrapure water, synthetic chemicals and laboratory reactors [19–24]. However, there is a little information on the effectiveness of photocatalysis for the destruction of commercial estrogens formulations at pilot-scale and using real water. These waters may be the effluents from pharmaceutical manufacturing processes.

In this study, we investigate the treatment of commercial estradiols (17-β estradiol and nomegestrol acetate, Figure 1) aqueous solutions obtained from female contraceptive pills, in a pilot-scale compound parabolic collector (CPC) operated using a solar photocatalytic treatment process and titanium dioxide (TiO_2) suspensions. The radiation field in the CPC was modeled and the spatial distribution of the volumetric rate of photon absorption (VRPA) was evaluated by applying the six-flux photon absorption-scattering model (SFM) [25–27]. This model tracks scattered photons along the six directions of the Cartesian coordinates. The optical parameters of the catalyst suspension in water were averaged across the spectrum of the incident solar light to simplify the modeling methodology. The time-dependent degradation profiles of the commercial estrogens formulation in tap water were determined following the explicit consideration of the volumetric rate of photon absorption in the reaction kinetics and a mass balance across the CPC. The dependence of the treatment of the commercial formulation of estradiols on catalyst concentration and optical thickness was correlated to the rate of photon absorption in the reactor.

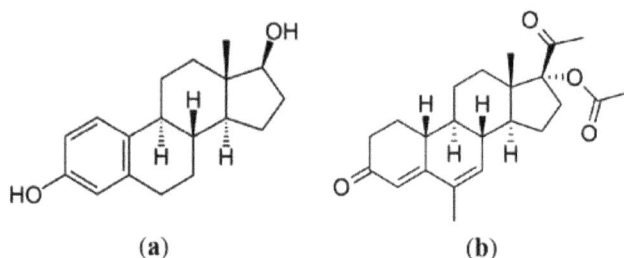

(a)   (b)

Figure 1. Chemical structures of the estrogens investigated: (a) 17-β estradiol and (b) Nomegestrol acetate.

2. Results and Discussion

2.1. Solar Photolysis

Figure 2 shows the rate of degradation and mineralization of the commercial estrogen mixture in the CPC by solar photolysis, in the absence of catalyst. The estrogens concentration and TOC removal after 42 min of irradiation and at a photon irradiance of 30 W/m², was 49% and 11%, respectively.

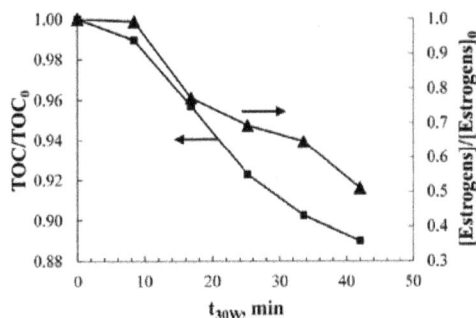

Figure 2. Solar photolytic degradation (**right axis**) and mineralization (**left axis**) of the commercial formulation of E2 and nomegestrol acetate mixture with an initial concentration of 5 ppm of estrogens and 388 ppm of TOC at pH 4.5 and with 97.2 kJ/m² of accumulated solar radiation (measured as UV-A).

The observed behavior is similar to the rate of photolysis reported in previous studies [23]. The estrogen mixture shows significant degradation under UV radiation exposure, especially with UV-A and UV-C radiation [23]. Nonetheless, the TOC removal was not as fast as the removal of estrogens. This may result from the presence of excipients compounds in the commercial estradiols formulation, consisting of lactose monohydrate, microcrystalline cellulose, magnesium stearate and polyvinyl alcohol, which are not easily mineralized by UV photolysis. In addition, the transformation products by photolysis are expected to degrade at slower rates, since the aliphatic derivatives of the photolytic degradation of estrogens can be more stable than the aromatic rings under UV irradiation conditions [28]. Although solar radiation only contains a small proportion of UV-C, the collective contribution of the other UV components can be sufficient for breaking chemical bonds in the estrogens molecules [29]. The effect of solar photolysis is usually significant for the degradation of the estrogens parent compound in natural waters. However, in heterogeneous photocatalysis, the absorption of UV photons by the photocatalyst, in well designed reactors, is several orders of magnitude higher than the absorption of photons by the molecules in solution [23], therefore the effect of photolysis can often be neglected when the contaminants and TOC removals are modeled.

2.2. Adsorption of Contaminants on the TiO₂ Catalyst Surface

The results of adsorption of the commercial estrogens mixture onto TiO_2 are presented in Figure 3 in terms of residual estrogens and TOC concentrations. The catalyst loadings selected in this study were limited to those concentrations that maximized the rate of photon absorption in the CPC and that yielded the fastest rate of contaminants degradation. After 12 h of stirring, under dark conditions at pH 4.5, approximately 13.5% of the initial estrogens concentration was adsorbed onto the surface of TiO_2. The role of adsorption can be significant in photocatalysis, since it can be rate-limiting to the contaminant degradation kinetics. The adsorption of hydroxyl anions onto the TiO_2

surface promotes the generation of free hydroxyl radicals via electron exchange, which in turn oxidize water contaminants [30,31]. However, hydroxyl anions and the contaminant molecules compete for adsorption on the catalyst surface sites, therefore, if the fractional coverage of OH is lower than that of estrogens and drug additives, this might have an adverse effect on the rate of generation of oxidative species. Contrasting to this effect, if the adsorption of the chemical species of interest is weak, the observed photocatalytic degradation kinetics of the pollutant may be affected negatively and controlled by the rate of mass transfer of contaminants from the bulk to the catalyst surface.

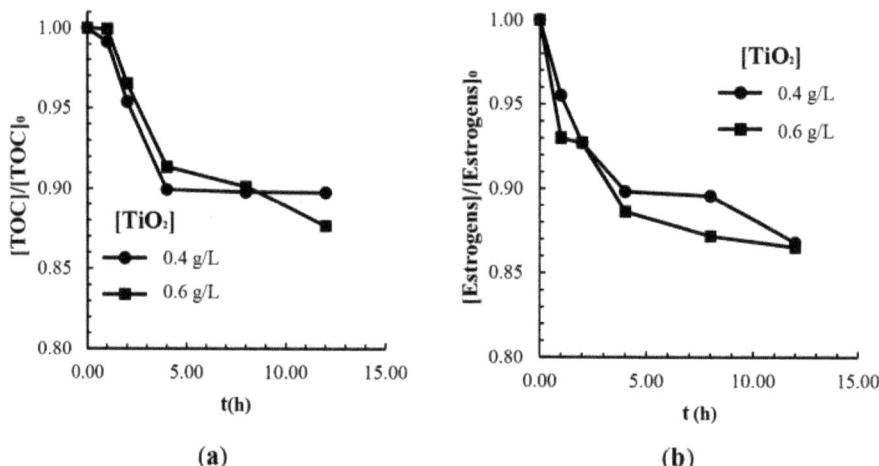

(a)

(b)

Figure 3. Adsorption of commercial estradiol at 0.4 and 0.6 g/L of catalyst loading with an initial concentration of 7 ppm of estrogens and 429 ppm of TOC: **(a)** TOC removal and **(b)** Estrogens removal.

The TOC of the estrogenic commercial drug was considered for modeling the adsorption phenomena onto the TiO$_2$ photocatalyst. The observed adsorption of estrogens in terms of TOC removal shows a typical monolayer behavior, which can be described by a Langmuir isotherm, according to the adsorption equation:

$$q = \frac{q_0 K_{ads}[TOC]}{1 + K_{ads}[TOC]} \tag{1}$$

where q is the amount of adsorbate per amount of adsorbent (mg$_{(TOC)}$/g TiO$_2$), q_0 is the maximum amount of TOC adsorbed, [TOC] is the concentration of estrogens in solution (mg$_{(TOC)}$/L) and K_{ads} is the adsorption equilibrium constant (L/mg$_{(TOC)}$). After rearrangement Equation (1), can be written as a linear expression, Equation (2), which allows the estimation of the adsorption parameters by linear fitting of the TOC adsorption results (Figure 4).

$$\frac{[TOC]}{q} = \frac{1}{q_0 K_{ads}} + \frac{1}{q_0}[TOC] \tag{2}$$

$$y = 0.6592x + 149.17$$
$$R^2 = 0.9217$$

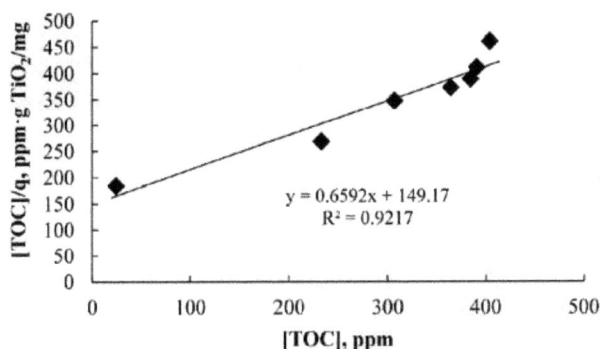

Figure 4. Linear regression of the Langmuir isotherm obtained from experiments at 30 °C, pH 6.9 and with 0.6 g/L of catalyst load.

The corresponding q_0 was 1.52 mg$_{(TOC)}$/g TiO$_2$, whereas K_{ads} was 4.42×10^{-3} ppm^{-1}. The relatively small value of the equilibrium constant, K_{ads}, suggest weak estrogens adsorption and pseudo first-order adsorption kinetics, as also shown in Figure 3. It is also possible that multilayer adsorption can be relevant at higher substrate concentrations above 400 ppm TOC since the Langmuir isotherm appears to deviate from the experimental data.

2.2.1. Photocatalytic Oxidation of a Commercial Formulation of Estrogens

The experiments performed in the pilot scale photoreactor, in the presence of suspended TiO$_2$, shows that the overall degradation rate of the commercial estrogens mixture increased as the initial concentration of the hormone disruptor decreased (Figure 5).

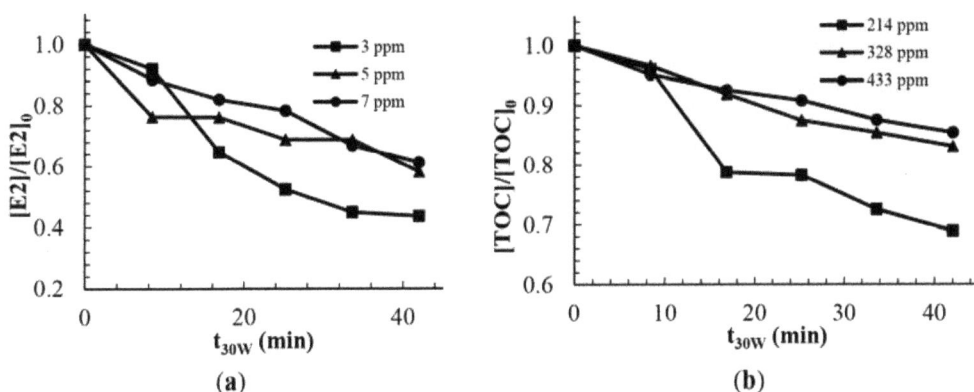

(a)

(b)

Figure 5. Effect of the initial concentration of the contaminant on the degradation rate with a catalyst load of 0.6 g/L; (**a**) Estrogens degradation; (**b**) TOC removal.

The estrogens removal was 56.7%, 41.7% and 38.7% for 3, 5 and 7 ppm of initial estrogens concentration, respectively, which means that the removal rates (ppm/min) increased as the initial estrogen concentration increased, whereas the TOC removal was 31.1%, 16.9% and 14.6% for 214,

328 and 433 ppm of initial TOC concentration, respectively. This behavior is consistent with the prevalent trend observed in other studies involving the photocatalytic oxidation of contaminants in water [23,32–34]. In heterogeneous photocatalysis, the apparent contaminants oxidation rates can be limited by adsorption, therefore, the initial concentration of contaminants can have a significant effect on the observed degradation rate. According to the Langmuir kinetic model, lower initial concentrations lead to first-order rate law, whereas higher concentrations lead to zero-order rate law. Figure 5 shows faster removal of both estrogens and TOC at the lowest initial concentrations, which is consistent with a pseudo first-order behavior.

It should be observed that the rates of degradation of estrogens in the presence of TiO_2 (Figure 5a) are not too dissimilar to the rate observed in the absence of catalyst (Figure 2), which contrast with other studies that have shown faster estrogens removal in the presence of TiO_2 photocatalyst [23]. This apparent coincidence may erroneously suggest that photolysis alone may be the most important factor responsible for the degradation of the contaminants in the aqueous solutions. However, it should be observed that at the same time, the absorption of UV photons by the photocatalyst is several orders of magnitude higher than the absorption of photons by the molecules in solution [23], therefore, the effective photon irradiance available for the photolysis of the molecules in solution is also several orders of magnitude smaller in comparison to the case in which the photocatalyst is absent. Since the rate of contaminants photolysis is first-order on the photon irradiance (Beer–Lambert law), it can be concluded that the contribution of photolysis alone should be insignificant in the observed degradation and mineralization of the contaminants in the presence of a photocatalyst. This conclusion is further supported by the analysis of the transformation products of estrogens (EE2) observed under UVA photocatalysis, which show that oxidation of estrogens occurs by hydroxyl radical attack [19,35]. The mineralization results (Figure 5b) show faster TOC removal in the presence of TiO_2 in comparison to photolysis alone (Figure 2).

2.2.2. Modeling and Experimental Validation of the Solar Photocatalytic Degradation of Contraceptive Pills Formulations

Due to the complex nature of the commercial contraceptive pill formulation, the photocatalytic treatment process was followed and modeled in terms of TOC removal.

The modified Langmuir–Hinshelwood (L–H) rate model proposed by Li Puma *et al.* [27], was adopted to describe the photocatalytic degradation of the commercial estradiol mixture. The radiation field in the CPC was analyzed using the six-flux absorption scattering model (SFM) [26]. The set of equations are summarized as follows:

(1) Reactor mass balance (expressed as differential equation in polar coordinates):

$$\frac{d[TOC]_{r,\theta}}{dt_{30W}} = r_{TOC} \tag{3}$$

(2) Contaminant rate law:

$$r_{TOC} = -k_T \frac{K_R[TOC]}{1 + K_R[TOC]} \int (LVRPA)^m dV_R \tag{4}$$

(3) Mass balance in the batch recirculation system:

$$[TOC]_{i+1}^{in} = \frac{[TOC]_i^{in}(V_T - V_R) + [TOC]_i^{out}V_R}{V_T} \tag{5}$$

(4) Hydrodynamic model for turbulent flow in the CPC, which was operated in the turbulent flow regime:

$$\frac{v_z}{v_{z,max}} = \left(1 - \frac{r}{R}\right)^{\frac{1}{n}} \tag{6}$$

$$n = 0.41\sqrt{\frac{8}{f}} \tag{7}$$

$$v_{z,average} = \frac{Q}{\pi r^2} \tag{8}$$

$$\frac{v_{z,max}}{v_{z,average}} = \frac{(n+1)(2n+1)}{2n^2} \tag{9}$$

(5) Optical properties of the catalyst and SFM parameters:

$$\omega = \frac{\sigma}{\sigma + \kappa} \tag{10}$$

$$a = 1 - \omega p_f - \frac{4\omega^2 p_s^2}{1 - \omega p_f - \omega p_b - 2\omega p_s} \tag{11}$$

$$b = \omega p_b - \frac{4\omega^2 p_s^2}{1 - \omega p_f - \omega p_b - 2\omega p_s} \tag{12}$$

$$\omega_{corr} = \frac{a}{b} \tag{13}$$

$$\tau = (\sigma + \kappa)\,\delta\,C_{cat} \tag{14}$$

$$\tau_{app} = a\,\tau\,\sqrt{1 - \omega_{corr}^2} \tag{15}$$

$$\gamma = \frac{1 - \sqrt{1 - \omega_{corr}^2}}{1 + \sqrt{1 - \omega_{corr}^2}}e^{-2\tau_{app}} \tag{16}$$

$$\lambda_{\omega corr} = \frac{1}{a(\sigma + \kappa)C_{cat}\sqrt{1 - \omega_{corr}^2}} \tag{17}$$

(6) The local volumetric rate of photon absorption (LVRPA) calculated from the SFM:

$$LVRPA = \frac{I_0}{\lambda_{\omega corr}\omega_{corr}(1 - \gamma)}\left[\left(\omega_{corr} - 1 + \sqrt{1 - \omega_{corr}^2}\right)e^{-\frac{r_p}{\lambda_{\omega corr}}} + \gamma\left(\omega_{corr} - 1 + \sqrt{1 - \omega_{corr}^2}\right)e^{\frac{r_p}{\lambda_{\omega corr}}}\right] \tag{18}$$

The parameters were fitted from the experimental results obtained with 0.6 g/L of TiO$_2$ using the optical properties of the Aeroxide P-25, shown in Table 1 [36].

Table 1. Optical parameters of TiO₂ Aeroxide P-25 in water averaged across the solar radiation spectrum up to the maximum wavelength that can photoactivate TiO₂ ($\lambda = 385$ nm) [36].

Parameter	Value
Absorption coefficient (κ)	174.7 m²/kg
Extinction coefficient (β)	1470.5 m²/kg
Scattering coefficient (σ)	1295.8 m²/kg
Forward scattering probability (p_f)	0.110
Backward scattering probability (p_b)	0.710
Side scattering probability (p_s)	0.045

The scattering albedo ω, calculated from Equation (10), was 0.88, and the corrected albedo, ω_{corr}, was 0.75, which was estimated from Equations (10)–(13).

The optical thickness (τ (Equation (14)) in the CPC reactor, estimated for a TiO₂ catalyst loading of 0.6 g/L and a reactor diameter of 33 mm, was equal to 29.0, whereas the apparent maximum optical thickness $\tau_{app.max}$ from Equation (15), at this catalyst concentration, was equal to 17.1. The volumetric rate of photon absorption per unit length of the solar CPC reactor ($VRPA/H$) could then be estimated using the modeling results of the CPC solar reactor previously reported [37], shown in Figure 6. These results were calculated for the solar irradiation conditions of Cali (Colombia), which, however, were very similar to the prevalent irradiation conditions of this study (latitude 3.5°). At $\tau_{app.max} = 17.1$, the $VRPA/H$ for $\omega = 0.88$ equals 0.405 W/m. Since the total length of the CPC reactor used in the experiments was 12 m, the corresponding $VRPA$ was 4.86 W. Combining Equations (3) and (4) and inverting yields Equation (19),

$$\frac{1}{\left(-V_T \frac{d[TOC]}{dt_{30W}}\right)} = \frac{1}{K_R k_T (VRPA)^{0.5}} \left(\frac{1}{[TOC]}\right) + \frac{1}{k_T (VRPA)^{0.5}} \tag{19}$$

and the kinetic parameters k_T and K_R were calculated by performing a linear fitting of the experimental data, with the TOC removal rates determined at time zero (Figure 7). The dimensionless parameter m in Equation (4) is related to the probability of electron–hole recombination and can take values within the 0.5–1.0 range, however, when the UV irradiance is significant and is not the limiting step of the photocatalytic reaction, m can be fixed as 0.5 [38].

From the slope and the intercept with the vertical axis (Equations (20) and (21)), the reaction kinetics and adsorption constants were estimated:

$$slope = \frac{1}{K_R k_T (VRPA)^{0.5}} = 348{,}947 \tag{20}$$

$$intercept = \frac{1}{k_T (VRPA)^{0.5}} = 25{,}773 \tag{21}$$

The estimated kinetic constants values were $k_T = 1.76 \times 10^{-5}$ ppm·m$^{1.5}$/s·W$^{0.5}$ and $K_R = 7.386 \times 10^{-2}$ ppm^{-1}. The adsorption constant K_{ads} estimated with the modified L–H kinetic model was one order of magnitude greater than the equilibrium adsorption constant K_{ads}, determined from the dark adsorption experiments. This apparent discrepancy results from the modification of the physical and chemical properties of the catalyst surface when this is irradiated with UV photons and when electron-hole pairs are generated [39]. When the TiO_2 surface is irradiated, there is an electron exchange with the hydroxyl anions with pH decrease, therefore, the net electric charge of the catalyst surface becomes positive and this effect might favor the physical adsorption of the anionic chemical species from the contaminants in solution. Nonetheless, the adsorption constant K_R estimated still suggests a weak adsorption of the contaminants.

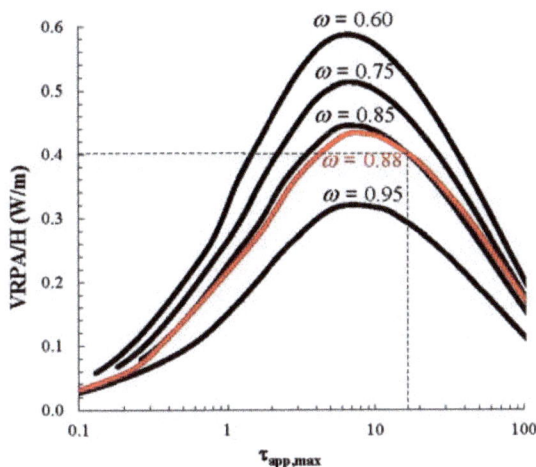

Figure 6. Effect of the maximum optical thickness on the VRPA per unit of length of a CPC solar reactor. Adapted from Colina-Marquez et al. [37].

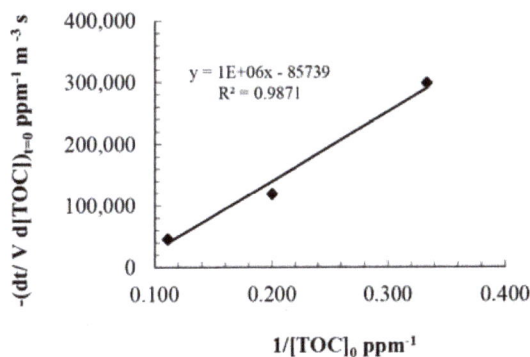

Figure 7. Linear regression for fitting parameters of the kinetic model.

The degradation of the commercial estradiol formulation was modeled in terms of TOC removal, using the proposed model and the kinetic parameters determined for the commercial estradiol formulation. Since the experimental pilot-scale CPC reactor was a flow-through reactor with external recirculation, the simulations and the latter validation were carried out considering the concentration changes after multiple passes through the solar photoreactor. The total number of passes in each simulation was estimated with Equation (22):

$$n_{pass} = \frac{Q t_{30W}}{V_R} \qquad (22)$$

where t_{30W} corresponds to the total time of the simulations. One hundred small sub-reactors ($j = 1$ to 100) of equal length ($L_j = L/100$) were considered to model the flow-through the CPC reactor during each pass, and the calculations were made considering that each sub-reactor behaved as a turbulent flow reactor. Considering the velocity profiles trough the cross-area of the reactor and the rate law (Equation (4)), the mass balance equation was expressed as follows:

$$Q \frac{d[TOC]_{r,\theta}}{dV_R} = -\frac{k_T K_R [TOC]_{r,\theta}}{(1 + K_R [TOC]_j^{in})} (LVRPA)_{r,\theta}^m \qquad (23)$$

where TOC_j^{in} is the feed TOC concentration to each sub-reactor. Since the LVRPA varies in both the radial and the angular directions, the local reaction rate also varies across the cross sectional area of the tube. The TOC concentration in the reactor further changes along the z-axis, therefore, the initial condition was set as follows:

$$z = 0, TOC_{j=1}^{in} = TOC_{i+1}^{in} \qquad (24)$$

where $[TOC]_{i+1}^{in}$ is the TOC concentration from the overall mass balance after each pass (Equation (5)). Obviously, for the first pass $[TOC]_{i=1}^{in}$ equals the initial TOC concentration in the reactor feed stream.

The local TOC concentration at the exit cross section of each sub-reactor was estimated for each point of the polar grid by solving the differential equation (Equation (23)):

$$[TOC]_{r,\theta}^{out} = exp\left[ln([TOC]_j^{in}) - \frac{k_R k_T}{v_z(1 + K_R[TOC]_j^{in})} \int_0^{L_j} (LVRPA)_{r,\theta}^m dz\right] \qquad (25)$$

where v_z is the average velocity of the fluid, which is a function of the radial coordinate (Equation (6)). The average TOC concentration at the exit of each sub-reactor was calculated by integrating the TOC concentration profile along the radial and angular directions taking into account the liquid flow rate:

$$[TOC]_j^{out} = \frac{\int_0^{2\pi} \int_0^R r v_z [TOC]_{r,\theta}^{out} dr d\theta}{Q} \qquad (26)$$

which is equivalent to mixing the fluid at the exit of each sub-reactor.

Finally, Equations (25) and (26) are solved iteratively for each sub-reactor after setting the initial condition:

$$[TOC]_{j+1}^{in} = [TOC]_j^{out} \qquad (27)$$

until the reactor exit at $z = L$ is reached ($j = 100$).

The reactor analysis presented above neglects the axial mixing of the fluid. Under turbulent flow conditions, it was necessary to assume complete radial and angular mixing, resembling the behavior of a continuous-stirred tank reactor (CSTR) across each cross section.

The calculation of the TOC concentration was made for each pass through the reactor until the number of passes established in Equation (22) was completed ($i = 1$ to n_{pass}).

The solid lines illustrated in Figure 8 correspond to the data generated by the model with the kinetic parameters k_T and K_R determined. The model described the experimental data satisfactorily, although it slightly overestimated the concentration profile at $t_{30W} > 15$ min, when the initial TOC concentration was 213.6 ppm. At lower TOC concentration, the catalyst surface may not be fully saturated with the substrates, and as a result a higher fraction of water adsorption yielded a higher rate of hydroxyl radical generation, ultimately contributing to a higher overall removal of the substrate. Whereas for the higher concentrations, the observed behavior could resemble to zero-order kinetics, for lower concentrations this behavior was closer to a first-order kinetics.

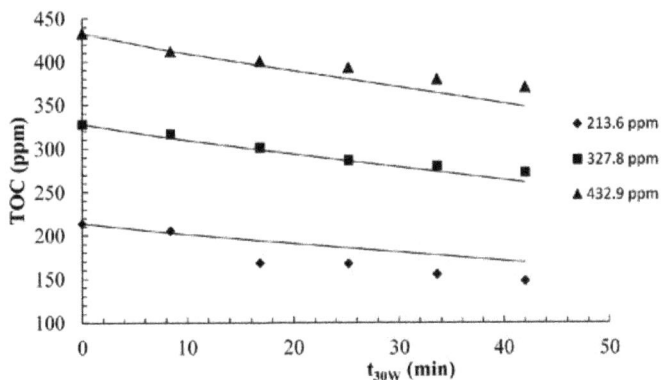

Figure 8. Modeling of the photocatalytic mineralization of commercial estradiol. The initial pH was 4.6 and the catalyst loading was 0.6 g/L.

The photoreactor model was further validated by comparing the experimental and model results of the the mineralization of the commercial estradiol formulation, using a different catalyst loading (0.4 g/L) and four initial TOC concentrations. The value of the $VRPA/H$ for the model simulations was determined from Figure 6, using the new value of the apparent maximum optical thickness ($\tau_{app,max}$), since this parameter is a function of the catalyst loading. Using Equation (15) $\tau_{app,max}$ was equal to 11.43, and the $VRPA$ was 5.18 W, which is slightly greater than the value corresponding to the catalyst loading of 0.6 g/L. In consequence, a greater TOC removal using 0.4 g/L of catalyst was expected, since the new $\tau_{app,max}$ approaches the optimum value (Figure 6). The results shown in Figure 9 demonstrate that the model described the photocatalytic degradation of estrogens satisfactorily, for both 387.7 and 409.5 ppm of TOC as initial concentrations.

The model appeared to underestimate the experimental results at higher TOC concentrations. One explanation for this behavior is that the underlying assumption of monolayer adsorption described by the L–H modeling approach may begin to fail, and that at these high TOC concentrations, the adsorption phenomena may be of multilayer nature, with complete saturation of the catalyst surface. The results shown in Figure 4 also suggest that a multilayer coverage of the catalyst at TOC concentrations higher than 400 ppm may be approached. Both cases lead to a decrease of the rate of •OH radical generation due to a higher fractional coverage of the surface with estradiols and, in consequence, to a reduction of the rate of TOC mineralization.

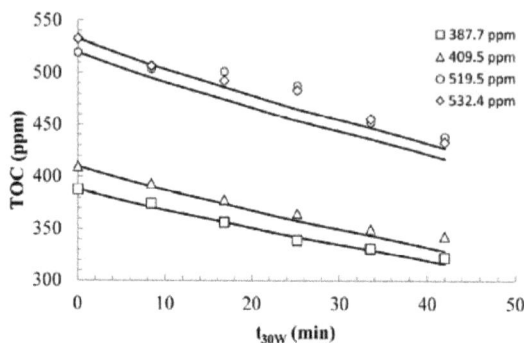

Figure 9. Validation of the TOC photocatalytic degradation with 0.4 g/L of catalyst load.

Figure 10 shows the fitting of the TOC values predicted by the model with the experimental data obtained in the validation tests. The R^2 value points to a satisfactory fitting in general, with low dispersion of the data. As shown in Figure 10, the deviation becomes more significant at higher concentrations of TOC, depicting the possible multilayer adsorption of substrates.

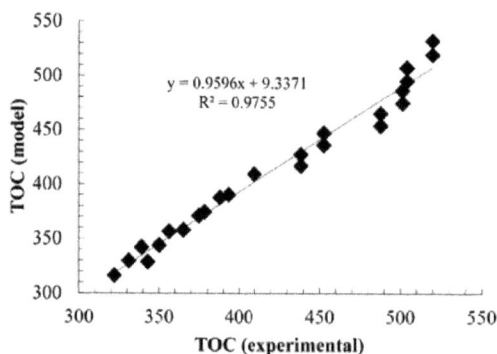

Figure 10. Comparison of the TOC values obtained by the model and the experimental TOC data.

The effect of the catalyst loading is evidenced by the increase of the TOC removal, which was slightly greater with 0.4 g/L in comparison to the results with 0.6 g/L (21.6% vs. 19.7%, respectively, for the case of 400 ppm of initial TOC concentration). Our previous study [37] established an optimal catalyst loading of 0.33 g/L for the CPC reactor under the same operating conditions and optical parameters used in this study. With a catalyst loading of 0.4 g/L, which is closer to this optimum, the model predicted a higher rate of contaminant mineralization than with 0.6 g/L, as predicted by the greater value of the *VRPA*. The existence of an optimum in the volumetric rate of photon absorption per unit reactor length (*VRPA/H*) is explained by a lower rate of photon adsorption at low catalyst concentrations when the catalyst surface area is insufficient for the absorption of the incident photons, and by a high scattering and "clouding" effects at catalyst concentrations much higher than the optimum, since under this situation the absorption of radiation is effective in a cross sectional area which is smaller than the actual physical cross section of the tube.

3. Experimental Section

3.1. Materials and Methods

The commercial contraceptive pill selected for this study contained 1.5 mg of 17-β estradiol (E2, as hemihydrate), 2.5 mg of nomegestrol acetate and excipients compounds (lactose monohydrate, microcrystalline cellulose, magnesium stearate and polyvinyl alcohol of unknown concentration) in each caplet. To prepare the estradiol aqueous solution, the caplets were pulverized and 30 mL of ethanol (Merck®) was added to the powdered caplets to extract the estrogens from the powder. The solution was filtered and the extract was lately dissolved in 40 L of tap water to perform the solar experimental tests. The concentration of the estradiols in each experiment was adjusted based on the content of multiple caplets. TiO$_2$ Aeroxide® P-25 (Evonik, primary particle size, 20–30 nm by TEM; specific surface area 52 m^2·g^{-1} by BET; composition 78% anatase and 22% rutile by X-ray diffraction) was used in the experiments. The catalyst was added in the reactor and circulated under dark conditions overnight before exposing the reactor to sunlight the next day. Sampled collected from the reactor containing TiO$_2$ were immediately filtered through a 0.45 μm Nylon filter (Millipore, Billerica, MA, USA) prior to immediate further quantitative analysis. The residual concentration of estrogens in the water was determined by UV-spectroscopy (Shimadzu UV1800 spectrophotometer) by measuring the absorbance in the UV region. The calibration curve and the absorbance spectra of estrogens are shown in Figure 11. The total organic carbon (TOC) of the samples was measured using a Shimadzu TOC-VCPH.

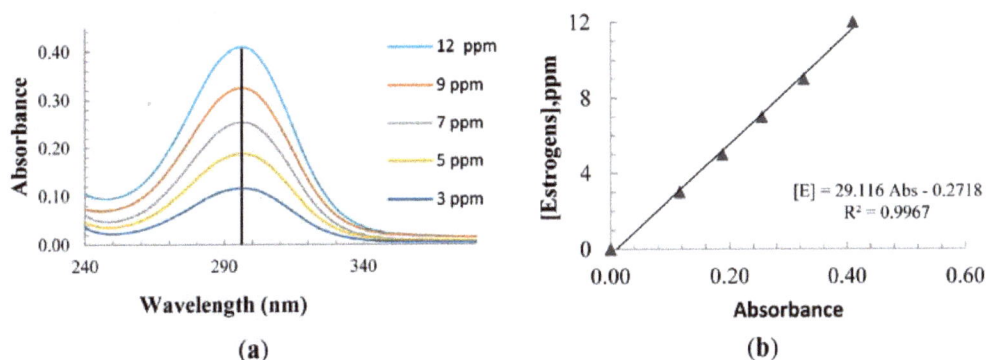

Figure 11. (**a**) UV absorbance of estrogens at different concentrations, showing a peak absorption wavelength at 297 nm. (**b**) Calibration curve for estrogens in tap water.

The adsorption tests were performed in sealed 500 mL-beakers containing 300 mL of an aqueous solutions of estrogens and suspended TiO$_2$ Aeroxide® P-25 at known concentrations. The samples were kept in a chamber under dark conditions and continuous magnetic stirring for 12 h. The concentration of estrogens in the water prior to adding the catalyst and after reaching equilibrium in the presence of catalyst was measured by UV-spectroscopy and TOC analysis.

The incident solar UV radiation accumulated in the photoreactor was measured by a UV radiometer (Delta OHM 210.2) with an UV-B probe, which covers the wavelength range between

280 and 315 nm. The effect of UV-A radiation from the solar radiation spectrum up to the absorption band edge of TiO_2 (384 nm) was evaluated with a common extrapolation by considering that the UV-B is approximately the 10% of the UV A+B radiation. The dissolved oxygen concentration in the water was monitored with a Spectroquant Pharo 3000. To account for variation in solar irradiance during the day, the photocatalytic treatment time was standardized based on the t_{30W} time, which considers that the average UV photon irradiance during a clear sunny day is 30 W/m^2 [36].

3.2. Photocatalytic Reactor

The photolytic and photocatalytic tests were performed in a solar CPC photoreactor shown in Figure 12. It consisted of ten Duran$^®$ glass tubes (1200 mm in length, 32 mm OD, 1.4 mm wall thickness), supported by a metal structure. The reactor was operated in a recirculation mode using a 40 L recycle feed tank, a recycling centrifugal pump (½ hp of nominal power) that delivered 30.2 L/min. The Reynolds number in the CPC for these operating conditions was 19,400, therefore the flow regime was fully developed and turbulent. The flow rate was measured by a calibrated flow meter. The pipeline and accessories used in the pilot plant were made of PVC, 1-inch diameter.

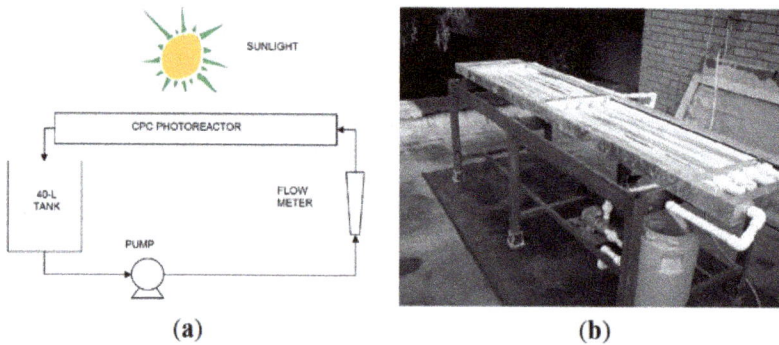

(a)   (b)

Figure 12. Pilot-scale solar CPC: (a) scheme of experimental setup and (b) photocatalytic reactor (Universidad del Valle, Cali-Colombia).

The geometry of the CPC is described in Figure 13. The curvature of the CPC involute (Figure 13) is described mathematically by the equations:

$$\rho = r\theta \text{ for } |\theta| \leq \theta_a + \pi/2 \quad \text{part AB of the curve} \tag{28}$$

$$\rho = r\frac{\theta + \theta_a + \pi/2 - \cos(\theta - \theta_a)}{1 - \sin(\theta - \theta_a)} \text{ for } \theta_a + \pi/2 \leq |\theta| \leq \frac{3\pi}{2} - \theta_a \text{ part BC of the curve} \tag{29}$$

where ρ is the radial coordinate of the involute, r is the reactor radius, θ is the angular coordinate of the involute and θ_a is the acceptance angle of the collector. For this case, the chosen acceptance angle was 90°.

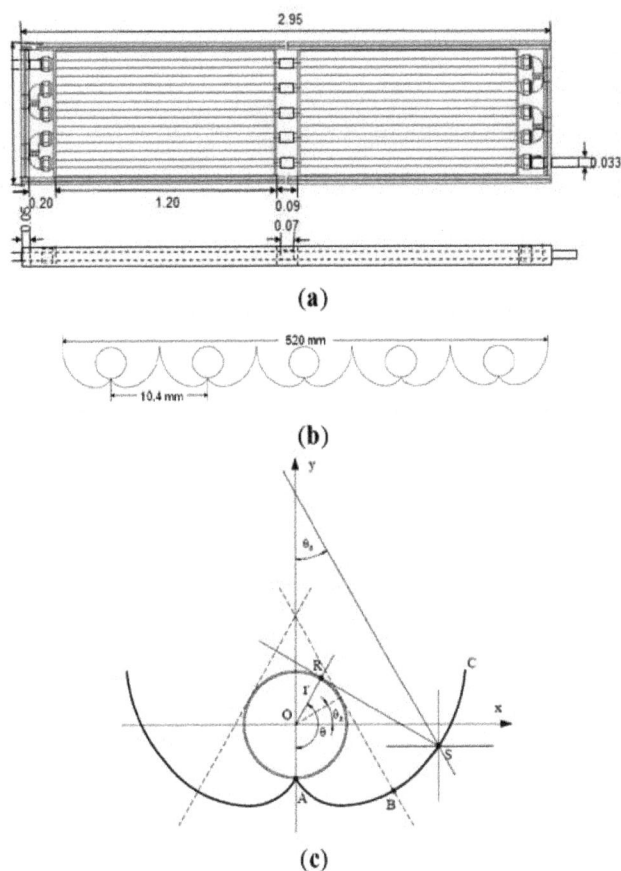

Figure 13. Solar CPC Photoreactor details: **(a)** superior and lateral views and **(b)** frontal view showing the curvature of the compound parabolic collectors; **(c)** CPC involute design.

4. Conclusions

The modified Langmuir–Hinshelwood (L–H) rate model, including the effect of photon adsorption and scattering represented by the SFM, modeled satisfactorily the removal of the total organic carbon of commercial estradiols (17-β estradiol and nomegestrol acetate) aqueous solutions, in a solar CPC photoreactor with relative errors of 5%. The application of solar photolysis without catalyst demonstrated a low rate of mineralization, although the rate of estrogens degradation was faster. The effect of solar photolysis could be neglected in the modeling of the TOC removal and the modified L–H rate model accounting for photocatalysis allowed the estimation of the kinetic parameters (reaction kinetics and adsorption constants) independent of the radiant field in the CPC photoreactor. It is important to note that the adsorption constant estimated with the Langmuir isotherm model under dark conditions, was one-order of magnitude greater than the adsorption constant obtained from the modified L–H kinetic model during the photocatalytic experiments. This demonstrates that the adsorption of estrogens over the TiO₂ catalyst surface is significantly affected under photon irradiation, as a result of the oxidation of the adsorbed estrogens. It was also demonstrated that at catalyst concentrations closer to the optimum predicted from radiation modeling consideration, the rate of TOC removal increased, and that the model could follow these trends. This study highlights the importance of kinetics and reactor modeling, which should always include the effect of the radiation field as a fundamental step, since without photon absorption there cannot be photocatalysis, as well as without reactants, there cannot be reactions. The current literature too often neglects this aspect in kinetic modeling.

Acknowledgments

Colina-Marquez and Machuca-Martinez thank Colciencias grant No. 117-521-28546 for the financial support and the chemical engineers Jaime Gomez and Leandro Quevedo for gathering the experimental data and running the simulations. Colina-Marquez also thanks the British Council for awarding the Travel Grant for Researcher Links, which made the collaboration between the research groups from Colombia and the United Kingdom possible.

Author Contributions

J.C.M., F.M.M. and G.L.P. wrote the paper and modeled the CPC photoreactor. In addition, J.C.M. developed the pilot-scale photoreactor, performed the experiments and analyzed the results.

Conflicts of Interest

The authors declare no conflict of interest.

Nomenclature

a	SFM parameter, dimensionless
b	SFM parameter, dimensionless
f	friction factor, dimensionless
I	UV radiation intensity, $W \cdot m^{-2}$
k_T	kinetic constant, $mol \cdot L^{-1} \cdot s^{-1} \cdot W^{-0.5} \cdot m^{1.5}$
K_{ads}	adsorption equilibrium constant, $L \cdot mol^{-1}$
K_R	reaction binding constant, $L \cdot mol^{-1}$
L	reactor length, m
$LVRPA$	local volumetric rate of photon absorption, $W \cdot m^{-3}$
m	reaction order respect to the LVRPA, dimensionless
n	parameter of velocity profile for turbulent regime, dimensionless
n_{pass}	number of pass of the fluid through the reactor space, dimensionless
p_b	probability of backward scattering, dimensionless
p_f	probability of forward scattering, dimensionless
p_s	probability of side scattering, dimensionless
Q	flow rate, $m^3 \cdot s^{-1}$

r	radial coordinate, m
r_p	auxiliary coordinate in the photon flux direction, m
R	reactor radius, m
t	time, s
t_{30W}	standardized 30 W time, s
TOC	TOC concentration, mol$\cdot$L^{-1}
v	fluid velocity, m$\cdot$s^{-1}
V	volume, m^3
$VRPA$	overall volumetric rate of photon absorption, W$\cdot$m^{-3}

Greek Letters

δ	reactor thickness, m
γ	SFM parameter, dimensionless
κ	specific mass absorption coefficient, m$^2\cdot$kg^{-1}
λ	radiation wavelength, nm
θ	polar coordinate, radians
σ	specific mass scattering coefficient, m$^2\cdot$kg^{-1}
τ	optical thickness, dimensionless
ϖ	declination angle, radians
ω	scattering albedo, dimensionless

Subscripts

app	apparent
$average$	average
$corr$	corrected
max	maximum
min	minimum
r	radial coordinate
R	reactor
$total$	total radiation
T	total
z	axial coordinate
0	relative to incident radiation, or initial condition
λ	radiation wavelength
θ	angular coordinate
$\omega corr$	corrected albedo

Superscripts

in	reactor inlet
out	reactor outlet

References

1. Thomaidi, V.S.; Stasinakis, A.S.; Borova, V.L.; Thomaidis, N.S. Is there a risk for the aquatic environment due to the existence of emerging organic contaminants in treated domestic wastewater? Greece as a case-study. *J. Hazard. Mater.* **2015**, *283*, 740–747.

2. Ramirez, A.J.; Brain, R.A.; Usenko, M.A.; Mottaleb, M.A.; O'Donnell, J.G.; Stahl, L.L.; Wathen, J.B.; Snyder, B.D.; Pitt, J.L.; Perez-Hurtado, P.; *et al*. Occurrence of pharmaceuticals and personal care products in fish: Results of a national pilot study in the United States. *Environ. Toxicol. Chem.* **2009**, *28*, 2587–2597.

3. Loos, R.; Gawlik, B.M.; Locoro, G.; Rimaviciute, E.; Contini, S.; Bidoglio, G. EU-wide survey of polar organic persistent pollutants in European river waters. *Environ. Pollut.* **2009**, *157*, 561–568.

4. Yang, Y.; Fu, J.; Peng, H.; Hou, L.; Liu, M.; Zhou, J.L. Occurrence and phase distribution of selected pharmaceuticals in the Yangtze Estuary and its coastal zone. *J. Hazard. Mater.* **2011**, *190*, 588–596.

5. Lindberg, R.H.; Östman, M.; Olofsson, U.; Grabic, R.; Fick, J. Occurrence and behaviour of 105 active pharmaceutical ingredients in sewage waters of a municipal sewer collection system. *Water Res.* **2014**, *58*, 221–229.

6. Bu, Q.; Wang, B.; Huang, J.; Deng, S.; Yu, G. Pharmaceuticals and personal care products in the aquatic environment in China: A review. *J. Hazard. Mater.* **2013**, *262*, 189–211.

7. Liu, Z.H.; Ogejo, J.A.; Pruden, A.; Knowlton, K.F. Occurrence, fate and removal of synthetic oral contraceptives (SOCs) in the natural environment: A review. *Sci. Total. Environ.* **2011**, *409*, 5149–5161.

8. Landrigan, P.; Garg, A.; Droller, D.B.J. Assessing the effects of endocrine disruptors in the National Children's Study. *Environ. Health Perspect.* **2003**, *111*, 1678–1682.

9. Colborn, T.; Vom Saal, F.S.; Soto, A.M. Developmental effects of endocrine-disrupting chemicals in wildlife and humans. *Environ. Health Perspect.* **1993**, *101*, 378–384.

10. Corcoran, J.; Winter, M.J.; Tyler, C.R. Pharmaceuticals in the aquatic environment: A critical review of the evidence for health effects in fish. *Environ. Health Perspect.* **2010**, *40*, 287–304.

11. Kidd, K.A.; Blanchfield, P.J.; Mills, K.H.; Palace, V.P.; Evans, R.E.; Lazorchak, J.M.; Flick, R.W. Collapse of a fish population after exposure to a synthetic estrogen. *PNAS* **2007**, *104*, 8897–8901.

12. Yang, L.P.H.; Plosker, G.L. Nomegestrol acetate/estradiol: In oral contraception. *Drugs* **2012**, *72*, 1917–1928.

13. Fernández, M.; Fernández, M.; Laca, A.; Laca, A.; Díaz, M. Seasonal occurrence and removal of pharmaceutical products in municipal wastewaters. *J. Environ. Chem. Eng.* **2014**, *2*, 495–502.

14. Collado, N.; Rodriguez-Mozaz, S.; Gros, M.; Rubirola, A.; Barceló, D.; Comas, J.; Rodriguez-Roda, I.; Buttiglieri, G. Pharmaceuticals occurrence in a WWTP with significant industrial contribution and its input into the river system. *Environ. Pollut.* **2014**, *185*, 202–212.

15. Prieto-Rodriguez, L.; Miralles-Cuevas, S.; Oller, I.; Agüera, A.; Puma, G.L.; Malato, S. Treatment of emerging contaminants in wastewater treatment plants (WWTP) effluents by solar photocatalysis using low TiO_2 concentrations. *J. Hazard. Mater.* **2012**, *211–212*, 131–137.

16. Kanakaraju, D.; Glass, B.D.; Oelgemöller, M. Titanium dioxide photocatalysis for pharmaceutical wastewater treatment. *Environ. Chem. Lett.* **2014**, *12*, 27–47.

17. Choi, J.; Lee, H.; Choi, Y.; Kim, S.; Lee, S.; Lee, S.; Choi, W.; Lee, J. Heterogeneous photocatalytic treatment of pharmaceutical micropollutants: Effects of wastewater effluent matrix and catalyst modifications. *Appl. Catal. B Environ.* **2014**, *147*, 8–16.

18. Yang, H.; An, T.; Li, G.; Song, W.; Cooper, W.J.; Luo, H.; Guo, X. Photocatalytic degradation kinetics and mechanism of environmental pharmaceuticals in aqueous suspension of TiO_2: A case of β-blockers. *J. Hazard. Mater.* **2010**, *179*, 834–839.

19. Mboula, V.M.; Héquet, V.; Andrès, Y.; Gru, Y.; Colin, R.; Doña-Rodríguez, J.M.; Silva, A.M.T.; Pastrana-Martínez, L.M.; Leleu, M.; Tindall, A.J.; *et al.* Photocatalytic degradation of estradiol under simulated solar light and assessment of estrogenic activity. *J. Hazard. Mater.* **2010**, *179*, 834–839.

20. Koutantou, V.; Kostadima, M.; Chatzisymeon, E.; Frontistis, Z.; Binas, V.; Venieri, D.; Mantzavinos, D. Solar photocatalytic decomposition of estrogens over immobilized zinc oxide. *Catal. Today* **2013**, *209*, 66–73.

21. Frontistis, Z.; Fatta-Kassinos, D.; Mantzavinos, D.; Xekoukoulotakis, N.P. Photocatalytic degradation of 17α-ethynylestradiol in environmental samples by ZnO under simulated solar radiation. *J. Chem. Technol. Biotechnol.* **2012**, *87*, 1051–1058.

22. Zhang, W.; Li, Y.; Su, Y.; Mao, K.; Wang, Q. Effect of water composition on TiO_2 photocatalytic removal of endocrine disrupting compounds (EDCs) and estrogenic activity from secondary effluent. *J. Hazard. Mater.* **2012**, *215–216*, 252–258.

23. Li Puma, G.; Puddu, V.; Tsang, H.K.; Gora, A.; Toepfer, B. Photocatalytic oxidation of multicomponent mixtures of estrogens (estrone (E1), 17β-estradiol (E2), 17α-ethynylestradiol (EE2) and estriol (E3)) under UVA and UVC radiation: Photon absorption, quantum yields and rate constants independent of photon absorption. *Appl. Catal. B Environ.* **2010**, *99*, 388–397.

24. Benotti, M.J.; Stanford, B.D.; Wert, E.C.; Snyder, S.A. Evaluation of a photocatalytic reactor membrane pilot system for the removal of pharmaceuticals and endocrine disrupting compounds from water. *Water Res.* **2009**, *43*, 1513–1522.

25. Li Puma, G.; Brucato, A. Dimensionless analysis of slurry photocatalytic reactors using two-flux and six-flux radiation absorption-scattering models. *Catal. Today* **2007**, *122*, 78–90.

26. Brucato, A.; Cassano, A.E.; Grisafi, F.; Montante, G.; Rizzuti, L.; Vella, G. Estimating radiant fields in flat heterogeneous photoreactors by the six-flux model. *AICHE J.* **2006**, *52*, 3882–3890.

27. Li Puma, G.; Gora, A.; Toepfer, B. Photocatalytic oxidation of multicomponent solutions of herbicides: Reaction kinetics analysis with explicit photon absorption effects. *Appl. Catal. B Environ.* **2006**, *68*, 171–180.

28. Chowdhury, R.R.; Charpentier, P.A.; Ray, M.B. Photodegradation of 17β-estradiol in aquatic solution under solar irradiation: Kinetics and influencing water parameters. *J. Photochem. Photobiol. A* **2011**, *219*, 67–75.

29. Leech, D.M.; Snyder, M.T.; Wetzel, R.G. Natural organic matter and sunlight accelerate the degradation of 17β-estradiol in water. *Sci. Total Environ.* **2009**, *407*, 2087–2092.

30. Dillert, R.; Cassano, A.; Goslich, R.; Bahnemann, D. Large scale studies in solar catalytic wastewater treatment. *Catal. Today* **1999**, *54*, 267–282.

31. Alfano, O.M.; Bahnemann, D.; Cassano, A.E.; Dillert, R.; Goslich, R. Photocatalysis in water environments using artificial and solar light. *Catal. Today* **2000**, *58*, 199–230.

32. Konstantinou, K.; Albanis, T. Degradation pathways and intermediates of photocatalytic transformation of major pesticide groups in aqeous TiO₂ suspensions using artificial and solar light: A review. *Appl. Catal. B Environ.* **2003**, *42*, 319–335.

33. Méndez-Arriaga, F.; Maldonado, M.; Giménez, J.; Esplugas, S.; Malato, S. Abatement of ibuprofen by solar photocatalysis process: Enhancement and scale up. *Catal. Today* **2009**, *144*, 112–116.

34. Minero, C.; Pelizzetti, E.; Malato, S.; Blanco, J. Large solar plant photocatalytic water decontamination: Effect of operational parameters. *Sol. Energy* **1996**, *56*, 421–428.

35. Ohko, Y.; Iuchi, K.I.; Niwa, C.; Tatsuma, T.; Nakashima, T.; Iguchi, T.; Kubota, Y.; Fujishima, A. 17β-estradiol degradation by TiO₂ photocatalysis as a means of reducing estrogenic activity *Environ. Sci. Technol.* **2002**, *36*, 4175–4181.

36. Colina-Márquez, J.; Machuca-Martínez, F.; Li Puma, G. Photocatalytic mineralization of commercial herbicides in a pilot-scale solar CPC reactor: Photoreactor modeling and reaction kinetics constants independent of radiation field. *Environ. Sci. Technol.* **2009**, *43*, 8953–8960.

37. Colina-Márquez, J.; Machuca, F.; Li Puma, G. Radiation Absorption and Optimization of Solar Photocatalytic Reactors for Environmental Applications. *Environ. Sci. Technol.* **2010**, *44*, 5112–5120.

38. Turchi, C.S.; Ollis, D.F. Photocatalytic degradation of organic water contaminants: Mechanisms involving hydroxyl radical attack. *J. Catal.* **1990**, *122*, 178–192.

39. Valente, J.P.S.; Padilha, P.M.; Florentino, A.O. Studies on the adsorption and kinetics of photodegradation of a model compound for heterogeneous photocatalysis onto TiO₂. *Chemosphere* **2006**, *64*, 1128–1133.

Sample Availability: Samples of the compounds are not available from the authors.

Chapter 4:
Photocatalysis and Photoelectrochemistry for
Production of Energy and Chemicals

Advances and Recent Trends in Heterogeneous Photo(Electro)-Catalysis for Solar Fuels and Chemicals

James Highfield

Abstract: In the context of a future renewable energy system based on hydrogen storage as energy-dense liquid alcohols co-synthesized from recycled CO_2, this article reviews advances in photocatalysis and photoelectrocatalysis that exploit solar (photonic) primary energy in relevant endergonic processes, viz., H_2 generation by water splitting, bio-oxygenate photoreforming, and artificial photosynthesis (CO_2 reduction). Attainment of the efficiency (>10%) mandated for viable techno-economics (USD 2.00–4.00 per kg H_2) and implementation on a global scale hinges on the development of photo(electro)catalysts and co-catalysts composed of earth-abundant elements offering *visible-light-driven* charge separation and surface redox chemistry in high quantum yield, while retaining the chemical and photo-stability typical of titanium dioxide, a ubiquitous oxide semiconductor and performance "benchmark". The dye-sensitized TiO_2 solar cell and multi-junction Si are key "voltage-biasing" components in hybrid photovoltaic/photoelectrochemical (PV/PEC) devices that currently lead the field in performance. Prospects and limitations of visible-absorbing particulates, e.g., nanotextured crystalline α-Fe_2O_3, g-C_3N_4, and TiO_2 sensitized by C/N-based dopants, multilayer composites, and plasmonic metals, are also considered. An interesting trend in water splitting is towards hydrogen peroxide as a solar fuel and value-added green reagent. Fundamental and technical hurdles impeding the advance towards pre-commercial solar fuels demonstration units are considered.

Reprinted from *Molecules*. Cite as: Highfield, J. Advances and Recent Trends in Heterogeneous Photo(Electro)-Catalysis for Solar Fuels and Chemicals. *Molecules* **2015**, *20*, 6739-6793.

1. Introduction

In 2004 Nobel Laureate Richard Smalley, discoverer of fullerenes and pioneer of modern nanoscience, gave a talk on the 10 great challenges facing humanity in the 3rd millenium. Placing *energy* at the top of the list, he proposed divestiture of fossil fuels in favour of sustainable and environmentally benign alternatives [1]. He also envisioned that nanotechnology, *i.e.*, the design and assembly of nanometer scale structures, would play a key role in our future prosperity. In fact, it can be argued that nanomaterials have already been exploited for many years in the form of heterogeneous catalysts in industrial chemical processing, responsible for the rapid growth in civilized life during the 20th century. Much of this was done with only marginal understanding of their workings prior to the advent of *in-situ* surface characterization, high-resolution imaging tools, and theoretical (computational) modeling [2,3]. Nowadays, the development of nanocatalysts with functionality optimized "by rational design" is a popular theme [4–8] but it remains an uphill challenge due to the myriad complexity of catalytic phenomena [9]. Nonetheless, recent articles focusing on the prospects for such materials in renewable energy applications give some ground for optimism [10–12]. They also introduce the main topic of this report, viz., the prodigious growth in research into heterogeneous photocatalysis, as attested by the drastic increase in literature citations over the last 20 years. While many of these concern environmental or "advanced oxidation" applications [13], multitudinous examples related to energy topics can be located under key (search) phrases like solar fuels, photo-catalytic hydrogen, photo-reforming, water photo-splitting, CO_2 photo-reduction, *etc.* A photocatalyst can be defined as "a solid material that accelerates a chemical reaction by light absorption while itself remaining unchanged" [14]. Just as in a thermal heterogeneous catalyst, a chemical process is made faster due to a significant lowering of the energy barrier of the associated transition state, e.g., by providing surface sites that activate unique modes of adsorption. On a typical semiconductor photocatalyst like titanium dioxide (TiO_2), the adsorbed (dark) state is further activated (or entirely new states are created) by surface interaction with highly energetic photo-induced charges, viz., electrons (e^-) and holes (h^+). Band-gap excitation of TiO_2 ($\lambda \leq 400$ nm) creates photons with an energy ≥ 300 kJ/mol. In principle, the use of photocatalysts enables substitution of expensive process heat by cheap solar (photon) energy, leading to reduced operating costs. However, an added impetus for research is that they can also drive *endergonic* (thermodynamically uphill) processes by converting light energy into stored chemical energy (bond enthalpy). In other words, the reaction selectivity can be steered towards more useful products. By analogy with the natural process, when the reactants are water and carbon dioxide, this is sometimes referred to as *artificial photosynthesis* [15], in which H_2 and/or its reaction product with CO_2 is isolated as a *solar fuel*.

To better appreciate the future impact of any viable solar fuel technology, it is helpful to consider it in the broader context of renewable energy schemes and their current limitations. In principle, the ideal energy carrier from an energetic and environmental viewpoint is hydrogen. On a weight basis, it has the highest energy density of any fuel (143 MJ/kg, or 3× the value of gasoline) and it burns cleanly and efficiently to water, producing heat and/or electrical power in a fuel cell [16]. Unfortunately, H_2 also has the lowest volumetric energy density under ambient conditions (0.011 MJ/L), making its storage in physical form (compressed, liquefied, adsorbed, *etc.*) impractical and expensive, especially for applications in the transportation sector [17]. Any renewable fuel should retain the positive attributes of gasoline but offer the environmental benefits of H_2. It must be a liquid (for ease of handling) and have a practical energy density (≥ 20 MJ/L). It should also contain a substantial level of "incipient hydrogen" (≥ 12 wt %) and be carbon-neutral in the long term. The recent (transitional) strategy of fuel "decarbonization" aims to exploit the clean energy associated with the H-component in fossil fuels as these become depleted [18].

A previous review by this author [19] dealt almost exclusively with *thermal* heterogeneous catalytic processes in a future renewable hydrogen energy system schematized in Figure 1.

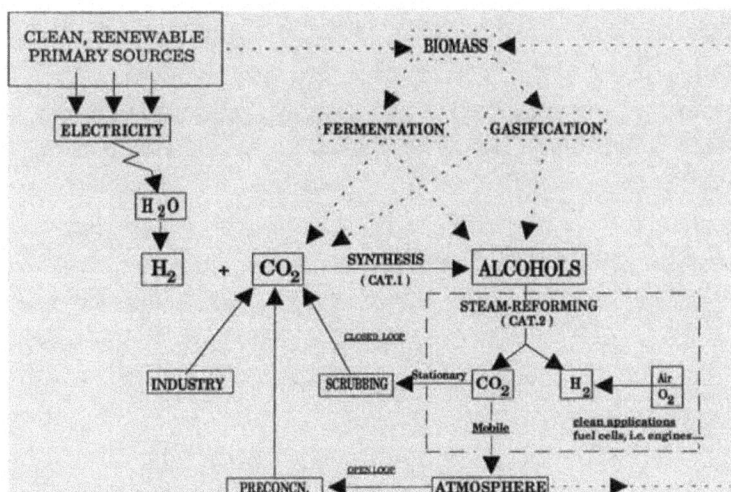

Figure 1. Idealized energy scheme based on renewable H_2 stored in simple alcohols (reproduced from [19] with permission).

In a gas phase catalytic process, H_2 derived from renewable primary sources (solar-electric, hydro-electric, *etc.*) is converted to alcohol(s) by a synthesis reaction with CO_2, itself recycled from industrial sources in concentrated form (closed loop). In the near future, the more challenging (open loop) process, *i.e.*, direct capture of CO_2 from the atmosphere, may become technical reality [20,21]. Simple alcohols are ideal H-carriers as they can be synthesized quite selectively while H_2 can be released at relatively low temperature. Insofar as biomass can supply renewable H_2, alcohols and oxygenates (even sugars) by gasification, fermentation, carbohydrate hydrolysis, *etc.* [22], this natural renewable (CO_2-neutral) energy source is integrated into the overall scheme. Essentially the same scientific precepts have since been advanced in a monograph by Olah *et al.*, which focuses on storage solely as methanol since this is already technically feasible [23]. The current status of *The Methanol Economy* has been the subject of a recent review, which re-emphasizes that the price of the alcohol will be strongly dependent on the cost of renewable electricity for H_2 generation [24]. Cost analyses show that "renewable" methanol will probably be 2–3× more expensive than the methane-based commodity [25] (see also Section 7). As regards the main catalytic cycle, the alcohol(s) synthesis can be written in general form as:

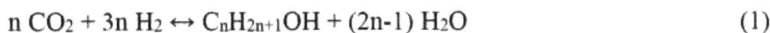

$$n\ CO_2 + 3n\ H_2 \leftrightarrow C_nH_{2n+1}OH + (2n-1)\ H_2O \tag{1}$$

Under high pressure, reaction (1) is favoured ($\Delta G < 0$) at ambient temperature but synthesis catalysts generally operate above 220 °C due to kinetic limitations. Product selectivity in ethanol synthesis is a major challenge [26,27]. Upon demand, H_2 is released catalytically from aqueous alcohol vapours to generate heat and power via steam-reforming (SR):

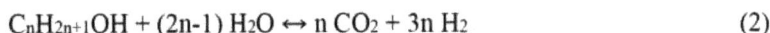

$$C_nH_{2n+1}OH + (2n-1)\ H_2O \leftrightarrow n\ CO_2 + 3n\ H_2 \tag{2}$$

SR is an endothermic process ($\Delta H \geq 0$) but reaction (2) is theoretically feasible ($\Delta G \leq 0$) near ambient for methanol, and above 230 °C for ethanol due to the molar volume increase (entropy factor). Since SR is simply the reverse of synthesis, catalysts for both reactions are similar.

Formulations based on Cu (modified by Co, Ni, Pd, Rh, *etc.*) on various oxide supports are quite effective. The main technical hurdle is the need for "fuel processing" because low levels of carbon monoxide in the reformate poison the fuel cell (Pt) anode and must be eliminated [28,29]. In addition, although it has been a growing area of research in the last 20 years, ethanol SR still suffers from catalyst deactivation due to carbon deposition, probably linked to the high temperatures necessary to establish good conversion rates [30,31]. Nevertheless, the scope for bio-hydrogen resources now encompasses a range of oxygenates and polyols, e.g., glycerol (a waste product from bio-diesel synthesis [32,33]), in which coking can be minimized via aqueous-phase-reforming (APR) under mild conditions [34].

For maximum technical impact, photocatalysis should logically be applied to the most energy-demanding steps in the scheme under consideration. Artificial photosynthesis to create a fuel such as methanol from aqueous CO_2:

$$CO_2 + 2\ H_2O \leftrightarrow CH_3OH + 1.5\ O_2 \tag{3}$$

is highly endothermic ($\Delta H = +727$ kJ/mol) but can be driven by solar photons in the visible/near IR region because it is 6 e^- process. Since reaction (3) involves water splitting (H abstraction) implicitly, it effectively couples two stages in the above scheme (H_2 generation and methanol synthesis), offering a potential process simplification. Supplying the energy needed for (endothermic) steam-reforming of alcohols ($\Delta H^\circ \approx +130$ kJ/mol CH_3OH or $+175$ kJ/mol C_2H_5OH) by means of photons is another obvious prospect, giving rise to intensive recent interest in *photo-reforming* [35]. In the review that follows, space limitations and a plethora of recent review articles [10,11,36–56] act as major con-straints on detail and impose a necessarily high degree of selectivity in given examples. Similarly, the emerging field of "renewable fuels from CO_2 and H_2O by solar-*thermal* processes" is beyond its scope [57]. As the title implies, this review deals only with materials that respond to (absorb) a significant fraction of the solar power spectrum, over 90% of which lies in the visible and near infrared region. Thus, recent advances in "sensitization" methods for TiO_2, the benchmark photocatalyst, are covered in some depth along with exploration of stable and non-toxic semiconductors of more suitable bandgap, either for independent use or in tandem (composite) arrangement for improved efficiency. The author has endeavoured to strike a balance between topicality and novelty, and apologizes in advance to the many authors who are not cited directly. Advances in selected materials are given regardless of their testing configuration, be it as dispersed nanoparticles in suspension or as "wired" electrodes in a photoelectrochemical cell (PEC). The question of the degree of complexity of any solar fuel "device", its future amenability to scale-up, and ultimate impact on process techno-economics (fuel cost) is still at an early stage of evaluation [58–62]. This is considered near the end of the review, along with the inevitable trend towards devices composed of "earth-abundant" elements [62–68], as necessitated by the ultimate (global) scale of fossil fuels replacement (see Section 7).

2. Water Splitting and Target Efficiency in Solar Hydrogen Generation

The fundamental groundwork in evaluating the maximum efficiency achievable in a solar photonic device was done by Bolton and co-workers [69–71], taking water splitting as the model reaction:

$$H_2O \rightarrow H_2 + 0.5\, O_2 \tag{4}$$

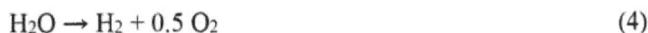

This two-electron, or effectively two-photon (one photon per H atom), reduction process is highly energetic and thermodynamically uphill by 237 kJ/mol H_2 or 119 kJ/mol photons, corresponding to a wavelength of 1008 nm (1.23 eV—all redox potentials are given *vs.* the normal hydrogen electrode at pH = 0). Inspection of the solar spectrum in Figure 2 shows that photons of greater energy ($\lambda \leq 1008$ nm) constitute only one half of the total incident solar flux.

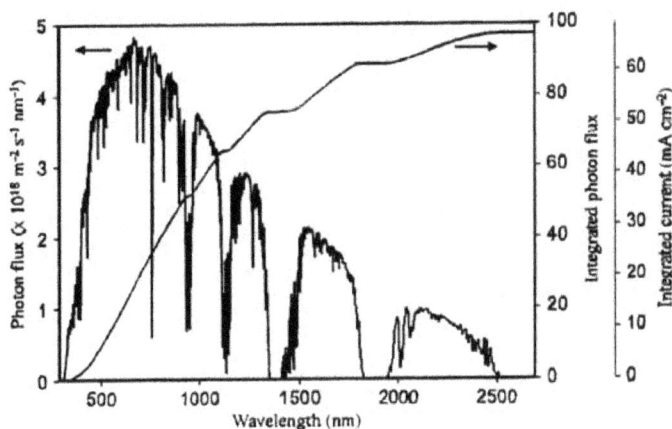

Figure 2. Solar photon flux at the earth's surface *vs.* wavelength and integrated current density obtainable from an ideal PV cell (adapted from [72] with permission of Elsevier).

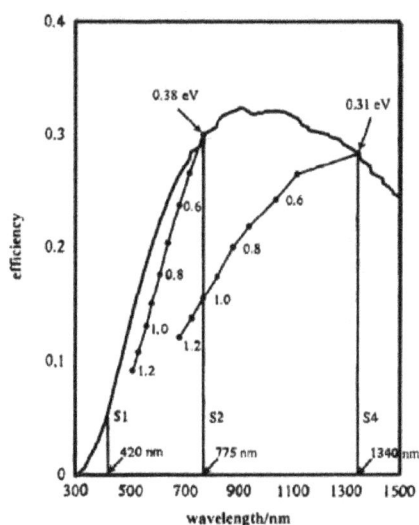

Figure 3. Solar photon-to-chemical energy conversion efficiency in a single bandgap current device (adapted from [71] with permission of Elsevier).

Furthermore, not all the energy in these photons is available for chemical energy storage after absorption due to intraband *thermalization, i.e.*, ultra-fast relaxation from excited vibrational states of the first electronic excited state (or conduction band in a semiconductor). To compensate for this energy loss the wave-length threshold must shift to 775 nm in this so-called S2 process, as shown in Figure 3. When entropic losses (non-equivalence of internal energy and Gibbs energy) are also factored in, the maximum efficiency (η_{STH}) of a device for solar-to-hydrogen (STH) energy conversion based on a single photo-system is around 30%. Also shown in Figure 3 are plots with more realistic entropic losses (>0.4 eV) and their progressive erosion of efficiency. In principle, a 4 e^- (S4) process extends the useful spectral range to 1340 nm based on successive absorption of two low energy photons to drive a single electron event like proton discharge ($H^+ + e^- \rightarrow H\cdot$). For a semiconductor, this is a futuristic concept because it would require hypothetical long-lived mid-gap states to be populated in the first photon absorption event, as shown in Figure 4. This will require a major advance in "bandgap engineering", but may yield so-called 3rd generation photocatalysts (*vide infra*) [73,74].

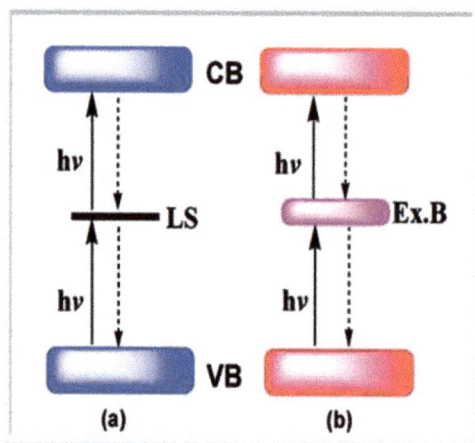

Figure 4. Successive two-photon excitation **(a)** via localized dopant states (LS) **(b)** via delocalized states in an extrinsic band (Ex.B) (reproduced from [74] with permission of ACS.

On the other hand, coupling of two absorbers of complementary bandgap and band-edge position (type II -see Figure 5) each absorbing two photons in a dual (D4) or tandem device is a viable water splitting configuration and offers better matching to the solar spectrum [75,76]. In practice, the mechanistic complexity (kinetic barrier) in the water oxidation half-cell reaction has kept conversion efficiencies below 2% until recently [60–62,77,78]. Efficiencies in the more challenging process of CO_2 photo-reduction are still below 0.5% [79]. The minimum workable efficiency for implementation of any solar-to-hydrogen (STH) process is taken as 10%, leading to a H_2 price approaching the DOE target of $4.00 per kg [60]. Immediately obvious from Figures 2 and 3 is that pristine TiO_2, the prototypical semiconductor photocatalyst [80,81] with a band-gap of ~3 eV ($\lambda \leq 400$ nm), absorbs just 4% of solar light. However, since this oxide is cheap and non-toxic and has otherwise excellent material properties, suitable energetics (band edge positions) to drive both proton reduction and water oxidation, stability in aqueous environment, *etc.*, ways of *sensitizing* TiO_2

to visible light are being studied intensively. At the same time, other non-TiO₂-based systems with intrinsic visible absorption are urgently sought.

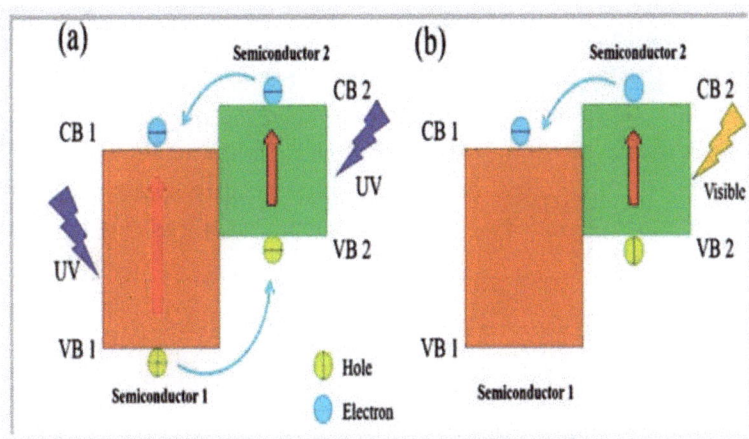

Figure 5. Dual absorbers with complementary bandgap/band edge positions (type II heterojunction) under (**a**) UV; and (**b**) Visible excitation (reproduced from [82] with permission of PCCP Owner Societies).

Incorporation of materials that perform efficient solar light harvesting is a fundamental (thought not the only) pre-condition for an effective energy conversion device. While it is generally assumed that absorption of one photon of energy exceeding the bandgap creates one electron/hole pair (exciton) with 100% efficiency, the fate of these excited states is less clear. One advantage of screening photocatalysts in "wired" mode, e.g., as photoanodes in an electrochemical cell, is that it promotes better charge separation via "band-bending", viz., the space-charge (depletion) layer formed spontaneously at the semiconductor/liquid junction. It also facilitates measurement of the combined efficiency of the two key initiatory processes (light absorption and charge collection) in terms of the resulting photocurrent. The incident photon-to-current conversion efficiency (IPCE) is a figure of merit for any photovoltaic (PV) cell. The photocurrent density limit shown in Figure 2 (~ 65 mA·cm^{-2}) is the ideal value achievable under so-called AM1.5G (1 Sun) illumination, corresponding to an incident optical power close to 1000 W·m^{-2} or 100 mW·cm^{-2} [72,83]. In solar fuel generation, these photo-generated charges are required instead to drive surface redox chemistry, in which the presence of suitable co-catalysts and rapid turnover of the substrate (diffusion) are crucial to performance. Due to these kinetic limitations, STH efficiencies lie below the PV (IPCE) value.

3. Advances in Absorber Materials with Improved Solar Spectral Matching

3.1. Modified TiO₂

The greatest success has been achieved in the so-called mesoscopic Dye-Sensitized Solar Cell (DSSC), or Grätzel cell as named after its inventor [84]. The basic principle is illustrated in

Figure 6. Upon photo-excitation of the chemically-anchored dye monolayer, electrons are injected to the conduction band of the mesoporous TiO₂ substrate indirectly via an excited state (S*). Back electron transfer to the oxidized dye is prevented by rapid dye regeneration (reduction) by a donor species (I⁻) present at high concentration in the electrolyte. To complete the circuit, the oxidized form of the mediator (I₃⁻) is discharged by electron flow through the external circuit to the Pt cathode. The energy difference between the TiO₂ conduction band and the redox level of the mediator (I⁻/I₃⁻) determines the maximum open circuit voltage ($V_{oc} \approx 0.8$ V), while the IPCE generally exceeds 70% up to 700 nm, leading to photocurrent densities $\approx$ 15 mA·cm⁻². With a typical "fill factor" (non-ideality in the power curve) of around 0.75, the resulting overall efficiency is ~10%.

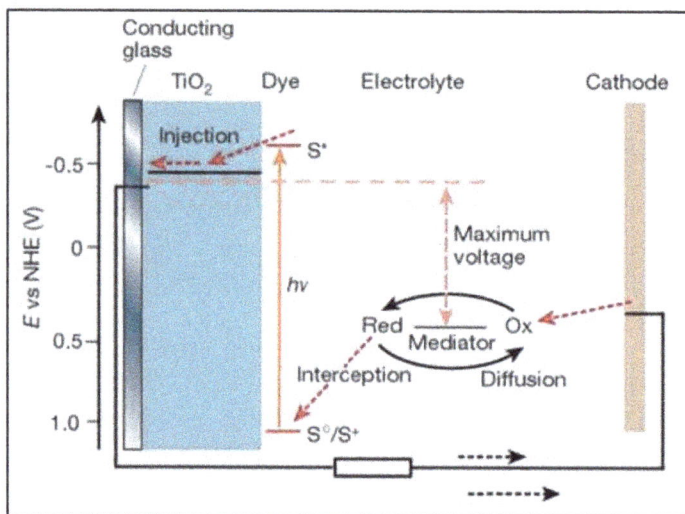

Figure 6. Working principle of dye-sensitized solar cell (DSSC) (reproduced from [76] with permission of Nature Publishing Group).

Most recently, this has been extended to 13% while at the same time reducing costs, substituting the previous Ru-based panchromatic (black) dyes with more absorptive (Zn-based) donor-π-acceptor porphyrins and the slightly corrosive I⁻/I₃⁻ redox mediator by Co²⁺/Co³⁺-based electrolytes [85]. Although the DSSC is strictly speaking a photovoltaic (PV) device rivaling conventional Si, their recent coupling into water splitting PEC cells has led to remarkably high efficiencies of H₂ generation (*vide infra*) [61,62]. Future advances can be expected from co-sensitization utilizing mixed dye systems, one of which absorbs in the near infrared region [86–91]. Alternatively, by varying its surface binding configuration, a single dye may achieve charge injection by both *direct* (type II—as exemplified by catechol, EDTA, *etc.* (see Figure 7b [92]) and *indirect* (type I—see Figure 7a) modes, thereby rendering it more panchromatic in response [93,94]. Although it has not yet come to fruition, pure type II sensitization should increase DSSC efficiency by eliminating the electron injection overpotential, *i.e.*, the energy loss due to thermalization from the excited state of the dye (S*) in the conventional (type I) process [93]. The general prospects for exploiting direct ligand-to-metal (Ti⁴⁺) charge transfer (LMCT) absorption in photocatalysis over TiO₂ have been reviewed [95]. It can be recognized by the appearance of a new absorption band absent in either of the free components. One serendipitous example is self-activation of hydrogen peroxide (which forms the yellow peroxotitanate complex on adsorption) for visible-driven environmental applications of TiO_2 [95,96]. The most exciting recent discovery in the DSSC field is the advantage of utilizing panchromatic semiconductor alkyl-ammonium lead (tin) trihalide perovskites as solid-state (layer) sensitizers [97–100], simultaneously replacing both the conventional dye and redox mediator. This technology-disruptive advance should ease fabrication costs and accelerate the development of a cheaper alternative solid-state PV cell of similar durability and efficiency to Si.

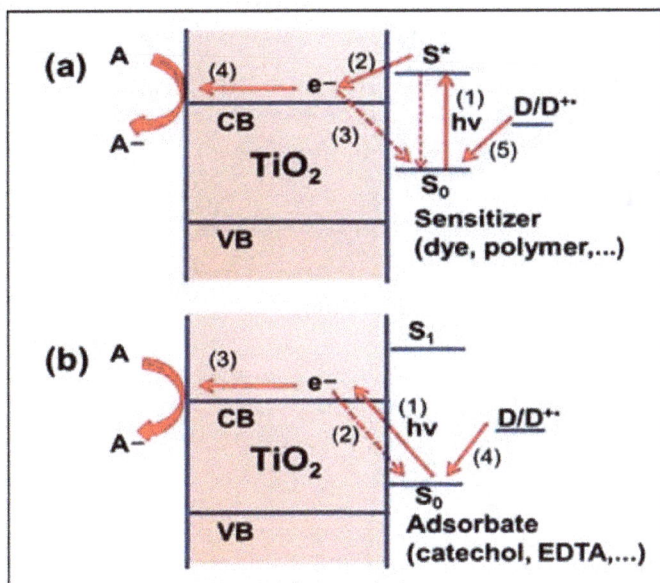

Figure 7. Scheme of charge transfer modes from sensitizer (S) to the TiO_2 conduction band (CB): **(a)** indirect (type I); **(b)** direct (type II or LMCT) (reproduced from [95] with permission of the Royal Society of Chemistry).

In particulate systems, bulk doping of TiO_2 with inorganic (metallic or non-metallic) elements has been a major strategy for visible sensitization in the last decade. Although this approach offers better long-term photostability (as compared to sensitization by organic dyes), the complexities of the solid-state chemistry have been decidedly more challenging. This is linked primarily to restricted solubility of dopants when introduced individually (1–2 at %), leading to insufficient visible light absorption. Furthermore, the concomitant introduction of defects that act as recombination centres has often led to efficiency losses. These are commonly O-vacancies but the dopant site itself can act deleteriously if present above a certain threshold concentration [101]. It is inadequate merely to impart colour to TiO_2, e.g., from transition metal ions (TMI) with a d^1–d^9 electronic configuration. Such optical transitions, being mostly localized ($d \leftrightarrow d$) type, do not involve charge transfer and merely act as parasitic absorbers competing with genuine (delocalized) charge injection. In a few cases, intervalence charge transfer may be effective provided the energy state of the photo-reduced acceptor lies above the bottom of the conduction band, e.g., Ti^{3+}, V^{2+}, or Fe^{2+} [102]. Early interest in TMI doping in TiO_2 was intended to improve the efficiency of charge separation and/or to extend the lifetime of surface-trapped carriers for photochemical action [103,104]. Any advantage to be gained by applying the TMI strategy to visible-light sensitization remains debatable as attested by more recent literature [105,106]. Indeed, it is increasingly recognized that incorporation of TMIs with empty (d^0) or filled (d^{10}) d-shells gets better results [107,108]. The recent flurry of excitement over the discovery of "black" TiO_2, obtained by high-pressure hydrogenation [109], is fading since it was confirmed that little or no visible photoactivity is generated [110], even though UV activity can be dramatically increased. Nevertheless, it has led to theoretical modeling [111] and renewed interest in defect engineering in pure TiO_2 [82,101,112].

The advent of what are now considered "2nd generation" photocatalysts [82,113] was triggered by independent reports in 2001 and 2002 that doping with electron acceptors, *i.e.*, elements forming anionic species such as N [114] and C [115], was the most effective way to impart visible-light sensitization. This was soon corroborated [116–119], and studies were extended to include F [120], S [121], and P [122]. Awareness of the benefits (synergies) of anionic *co-doping* gradually followed [123–125]. Thanks to good underpinning by DFT (calculational) modeling [117,126,127], most studies have focused on modified N-TiO_2 [128–130]. However, the visible sensitization effect is limited. N-TiO_2 generally appears pale yellow due to the low level of dopant achievable (<2 at % N), conferring only weak absorption in the blue-green region (see Figure 8).

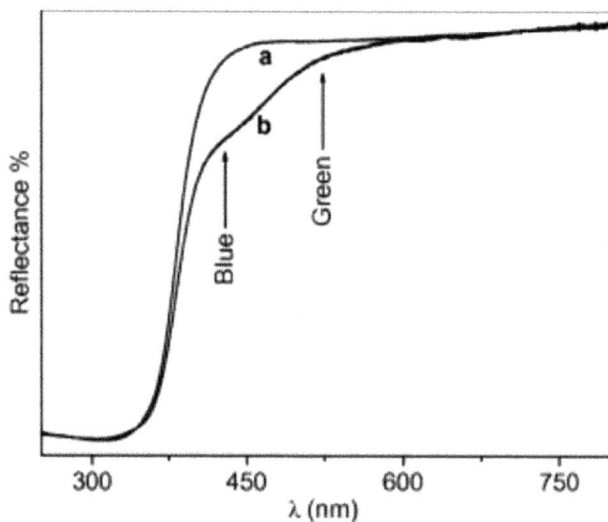

Figure 8. UV-visible reflectance spectra of (a) pristine TiO_2 (b) N-doped TiO_2 (reproduced from [126] with permission of Elsevier).

Furthermore, a number of studies have found a loss in oxidative power upon illumination within this visible band, and have attributed this mainly to faster charge carrier recombination [131–133]. Di Valentin *et al.* [126] have shown that the most stable doping configuration depends on the chemical potential of ambient oxygen during preparation. Since the most common reagent, NH_3, has reducing properties O-poor conditions prevail, favouring combination of two substitutional nitrogen sites compensated by one O-vacancy (2N$_s$ + V$_o$). The implied diamagnetic material of formula TiO$_{(2-3x)}$N$_{2x}$ has been affirmed as most likely by more recent modeling [127]. In an O-rich environment, a species consisting of an interstitial nitrogen atom associated with O (N$_i$-O) may also be stable, interacting with lattice Ti atoms through its π-bonding states, as shown in Figure 9. In either case, new (N$_{2p}$) energy states predicted to lie slightly above the valence band could be responsible for the observed visible absorption band around 450 nm. EPR spectroscopy has shown that irradiation within this band transfers an electron from the bulk diamagnetic N centre to the TiO$_2$ surface leading to formation of superoxide species (O$_2$·$^-$), a key activation process in environmental photocatalysis.

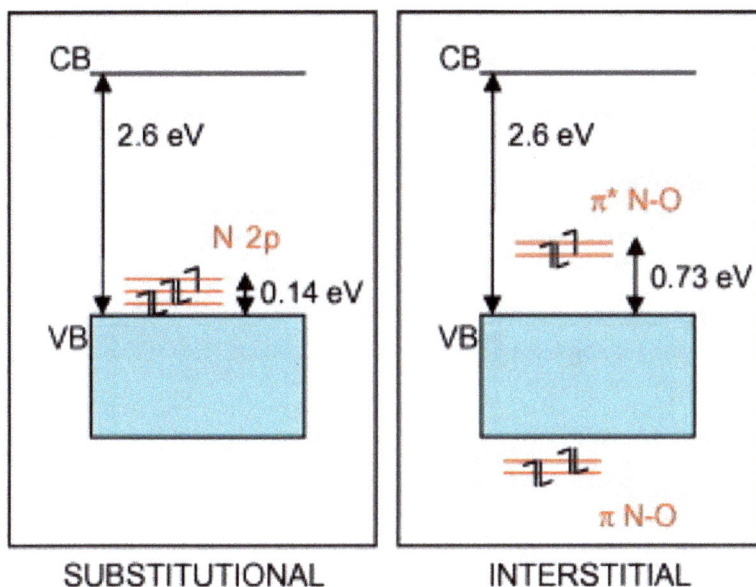

Figure 9. New energy states introduced into TiO$_2$ by substitutional and interstitial nitrogen. (reproduced from [126] with permission of Elsevier).

The current theoretical viewpoint on bulk doping is to take a donor/acceptor cooperative approach. This can be electro-neutral, e.g., "B + N" [134] or "Mo + 2N" [135], combinations that form intra-gap states extending from the bottom of the conduction band and top of the valence band, respectively. Charge compensation reduces the risk of introducing detrimental structural defects (interstitials, vacancies, *etc.*), while enabling the incorporation of higher dopant levels. Alternatively, even non-compensated p,n- type co-doping has been proposed. Using an excess of donor, e.g., Cr > N, the creation of states of intermediate energy results in a quasi-continuum visible absorption and an apparent bandgap energy of 1.5 eV [136]. The extrinsic states responsible for enhanced visible photoresponse often involve paramagnetic centres that can be explored by EPR [136,137].

One caveat on bulk doping should be mentioned. While re-affirming that N-TiO$_2$ has visible activity in formic acid mineralization, the same study claimed IR spectroscopic evidence for defective Ti≡N bonds and correlated this with weaker UV photoactivity due to related loss of crystallinity [138]. It is not clear if this trade-off in performance is inevitable [130]. Finally, it should be recognized that the "band narrowing" strategy vis-à-vis solar fuel generation may ultimately be constrained by high overpotentials associated with key redox processes, e.g., water oxidation (*vide infra*). The corollary is that *co-catalysts* will have a more vital role to play in lowering kinetic barriers in visible-active semiconductors, materials of intrinsically lower redox power than pristine TiO_2 (see Section 3.3).

A new class of more intensely coloured "multilayer-sensitized" titanias related to N-doped TiO_2 has emerged recently. These are obtained via mild preparative routes like sol-gel, hydrothermal, *etc.*, where the N-source is usually organic instead of ammonia, the preferred reagent for bulk doping. Starting from urea, the organic moiety transforms stepwise during calcination into the yellow-brown melon structure based on tri-*s*-triazine (heptazine) rings [139–141], a process catalyzed by acidic (H)-titanates [142]. Insofar as melon is structurally related to the more-condensed (fully dehydrogenated) graphitic carbon nitride, g-C_3N_4, a visible-absorbing semiconductor *per se* [143], these N-modified materials resemble nanocomposites. Figure 10 shows examples of these structural tectons (building blocks) and their inter-relationship. Starting from amines or alkyl-ammonium salts, the material appears more intensely coloured (brown) already below 200 °C due to a strong absorption tail extending across most of the visible region [96,144–147]. However, unlike the case of melon, the exact identity of the chromophore is uncertain and it is thermally labile. Calcination weakens both visible absorption and photoactivity [96,148]. Representative UV-Vis spectra of various C,N-based sensitizers loaded onto biphasic anatase/titanates (A/T) are shown in Figure 11.

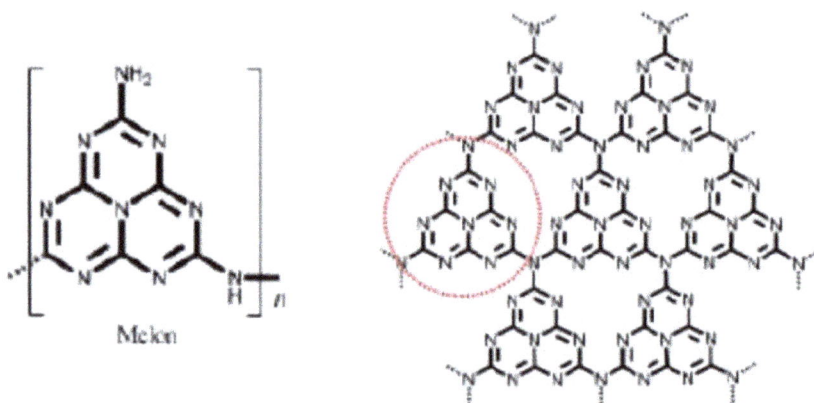

Figure 10. Melon and related tri-*s*-triazine unit (ringed) as building block for g-C_3N_4 (adapted from [143] with permission of Wiley).

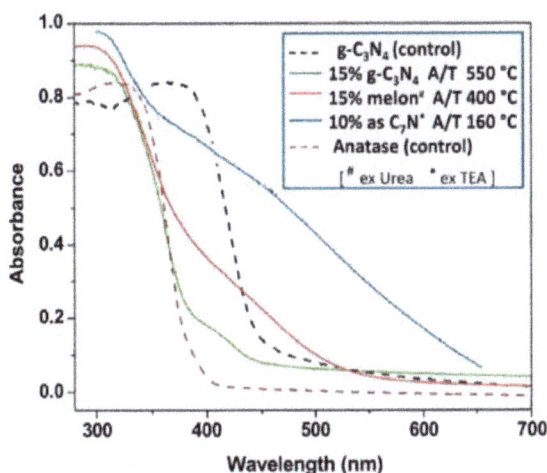

Figure 11. UV-Vis spectra of C/N-based sensitizers on biphasic anatase/titanates (adapted from [96,142] with permission).

Visible sensitization of TiO₂ via "plasmonics" is another rapidly intensifying field that may yield 3rd generation photoactive materials [73,74]. Localized Surface Plasmon Resonance (LSPR) is responsible for the now familiar intense coloration of mono-dispersed colloidal noble metals like Au and Ag, in which the absorption band may be "tuned" by varying the particle size and shape. The first convincing report that visible-light-induced metal-to-semiconductor electron transfer can be induced in Au/TiO₂ appeared in 2005 [149]. The action spectrum (IPCE) for photo-oxidation of ethanol was found to match the Au optical absorption peaking at ~550 nm (see Figure 12).

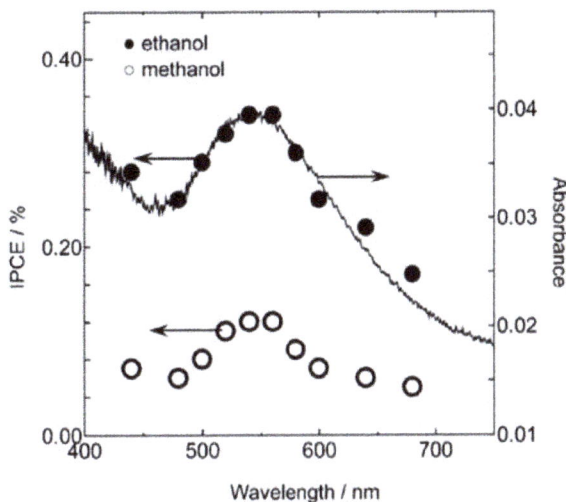

Figure 12. IPCE action spectrum in ethanol (methanol) photo-oxidation *vs.* LSPR (visible absorption) spectrum of gold in Au/TiO₂ (adapted from [149]; copyright (2005) American Chemical Society).

540

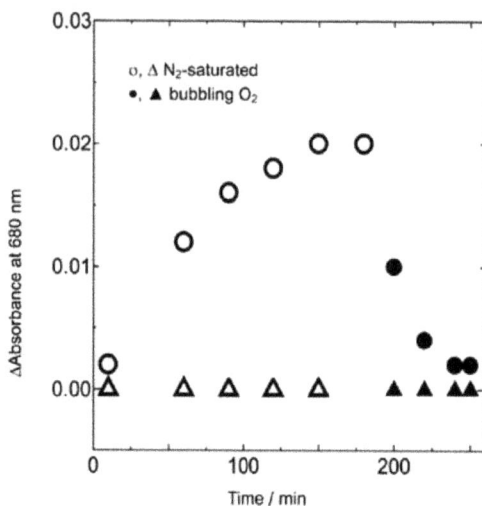

Figure 13. "Electron" absorption in TiO₂ @ 680 nm induced by visible illumination (λ > 500 nm) of Au deposits and its quenching by ambient O₂ (adapted from [149]; copyright (2005) American Chemical Society).

In addition, the characteristic spectrum of self-trapped electrons in TiO₂ ($\lambda_{max} \approx 680$ nm) developed under N₂. As shown in Figure 13, this was quenched in the presence of O₂ as electron acceptor. More extensive studies on Au/TiO₂ and Ag/TiO₂ have reached similar conclusions but efficiencies still need improvement [150–154]. The theoretical basis of plasmonics is developing rapidly, but two distinct types of interaction have already been identified [155–159]. The first is "hot electron" transfer from the metal to the semiconductor, which is the LSPR-induced analogue of indirect charge injection from a sensitizer dye in the DSSC, as described above. A second mechanism is also operative if their respective absorption bands overlap. This is termed plasmon resonant energy transfer (PRET), but its directionality may be reversed under UV irradiation (semiconductor → metal) via Forster resonant energy transfer (FRET). These "near-field" effects do not even require electrical contact at the interface. Notable examples are long-lived hot electron injection from Au to TiO₂ [160], PRET from insulated Ag nanocubes to N-doped TiO₂ [161]; or PRET from Au to α-Fe₂O₃ nanoplatelets [162]. Despite the promise of plasmonic sensitization, the long term economic outlook dictates a shift towards more earth abundant (but inevitably less stable) elements such as Cu [163–165], Al [166] and doped oxides [167].

3.2. Individual Alternatives to TiO₂

Recent reviews affirm that applied photocatalysis research is still largely (>80%) based on TiO₂, albeit in increasingly sophisticated (modified) forms [168–170]. Being amenable to nano-architecturing [170], and providing multi-phase heterojunctions for improved charge separation [171,172], this benchmark material has close to ideal properties as a photocatalyst [173,174] excepting its poor solar light response. One overdue task in TiO₂ research is a more quantitative evaluation of the importance of trapping states that do not lead to fast recombination but, on the contrary, extend charge carrier lifetime into the seconds or minutes time domain [175–181]. Recent modeling studies on photoexcited anatase show that the energetics (site stability—surface *vs.* bulk) favour surface-trapping of both the hole and the electron [182], with beneficial implications for surface redox chemistry.

The search for visible-light active semiconductors that also satisfy other key criteria for practical photocatalysis on a large scale is a difficult task. While various alternative semiconductors exist with suitable bandgap ($E_g = 1.5$–3.0 eV), most are inferior to TiO_2 in other respects, e.g., in having lower majority carrier conductivity, shorter minority carrier diffusion length (faster recombination), a less positive valence band edge (lower oxidizing power), instability under illumination (photo-corrosion), toxicity and/or high cost [183]. For these reasons, emphasis is now shifting to the development of type II (staggered bandgap) composites or tandem arrangements that perform complementary functions, coupled by directional electron transfer at the common heterojunction to "close the photochemical circuit" (*vide infra*) [111,183].

One notable exception that has emerged recently is graphitic carbon nitride (g-C_3N_4), which has a similar bandgap to N-TiO_2 ($E_g \approx 2.7$ eV—see also Figure 11), suitable energetics (band edge positions) for water splitting [142], and can be doped (bandgap tuned) and nano-textured to promote efficient charge migration [184–186]. However, a recent modeling study has identified a major kinetic constraint (large overpotential) linked to oxidative dissociation of water [187], helping to rationalize why co-catalysts are urgently needed for O_2 evolution [188]. In contrast, modeling studies have shown that N-doping of anatase TiO_2 may actually promote water dissociation [189].

Elsewhere, research into prospects for hematite (α-Fe_2O_3) has undergone a strong revival in the last decade, mainly due to efforts by the EPFL (Lausanne, Switzerland) group [162,183,190,191]. Pristine hematite is a cheap and stable indirect n-type semiconductor that absorbs visible light up to ~600 nm ($E_g = 1.9$–2.2 eV), offering a maximum photocurrent density of 12.6 mA cm^{-2}, or a solar-to-hydrogen (STH) conversion efficiency (η_{STH}) of ~16%. One limitation is its relatively low majority carrier (electron) conductivity, but this is readily overcome by incorporation of suitable dopants, e.g., Ti^{4+}, Sn^{4+}, Nb^{5+}, *etc.* [192]. However, it also suffers from several more challenging (deleterious) properties as a photo(electro)catalyst. Having a conduction band edge too low in energy for proton reduction and a large overpotential for O_2 evolution means water splitting over α-Fe_2O_3 will only work under external bias [183,191,193]. Furthermore, its high optical absorption depth (~400 nm [194]), coupled with a very short minority carrier (hole) diffusion length (~4 nm [195]) translates into a very low quantum efficiency for charge collection (IPCE). Nevertheless, structuring highly-crystalline deposits on the 20–30 nm scale has already raised the IPCE to over 30% with photocurrent densities exceeding 3 mA·cm^{-2} [190,191,196,197]. Reinforcing the contention that long-lived charge carriers are of key importance (*vide ultra*), application of transient optical and electrochemical techniques on α-Fe_2O_3 photo-electrodes held under positive (anodic) bias has shown a quantitative correlation between accumulated surface-trapped holes and photocurrent (electron) density [198]. The hole lifetime ($\tau = 0.1$–1 s) is sufficient for photo-oxidation of water, which has a rate constant in the range 0.1–10 s^{-1} [199,200]. The holes are reported to be of two distinct types, O_{2p} (O^-) and Fe_{3d} [Fe^{n+} ($n > 3$)], but both have similar activity [201].

3.3. Tandem (D4) Photoelectrochemical Cells, Composites, and the Role of Co-catalysts in Water Splitting

It is 30 years ago now that visible light-driven electron transfer from CdS to TiO_2 in an aqueous suspension of aggregated nanoparticles was first demonstrated [202]. The STH efficiency for the composite was very low but better than either of the pure components due to effective spatial separation and localization of electrons into TiO_2 (for proton reduction), the holes remaining on CdS (for H_2S oxidation) due to the relative energetics of the respective band edges (type II—see Figure 5 [82]). Here, the particles act as complementary self-biasing "photoelectrochemical diodes" in an efficient S2 mechanism (2 photons per H_2 molecule). Since that time, remarkable progress has been made in the visible-driven reduction half-reaction of water splitting:

$$2 H^+ + 2 e^- \rightarrow H_2 \ [E° \ (V) = 0.00 - (0.059 \times pH)] \tag{5}$$

A quantum yield of 93% was recorded over Pt/PdS/CdS at 420 nm, evolving 9 mmol/h H_2 with sulphide/sulfite as sacrificial donors [203]. Conventional wisdom has it that the low levels of Pt (0.3%) and PdS (0.13%) act as cocatalysts [51,53,203,204]. PdS is believed to promote oxidation of S^{2-} and SO_3^{2-} and transfer electrons to CdS. However, recent literature suggests that the combination of PdS and CdS may also be classified as an optical tandem system. PdS is an n-type semiconductor with a bandgap of ~1.6 eV and under investigation as a photovoltaic material *per se* [205,206]. It confers extra absorption in CdS composites that extends into the near IR region [207,208]. Unfortunately, the rarity of Pt and Pd, the tendency of sulphides to photocorrode, and the toxicity of cadmium ion, all militate against their use on a large scale. As shown in Figure 14, Pt as cocatalyst traps electrons from the semi- conductor to discharge protons, forming Pt-H bonds of ideal (intermediate) strength for H-H combination and desorption as molecular H_2 (Sabatier Principle). While often black in appearance, co-catalysts are strictly not photoactive and promote only dark elementary steps in the reaction. As shown in Figure 15, there is an optimum amount due to competing (beneficial and deleterious) effects. In practice loadings fall below 1%, fortuitously mitigating costs (many are precious metal-based, e.g., IrO_2, RuO_2, *etc.*) while minimizing parasitic light absorption. Prospects for alternative earth-abundant cocatalysts in photocatalytic water splitting have been reviewed [209]. Promising substitutes for Pt in H_2 evolution under neutral or alkaline conditions are Ni nanoclusters [210], Ni/Mo alloy [211,212], and $Cu(OH)_2$ [213].

The water oxidation half reaction [or oxygen evolution reaction (OER)]:

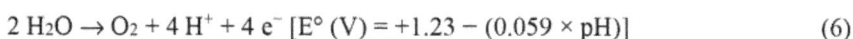

$$2 H_2O \rightarrow O_2 + 4 H^+ + 4 e^- \ [E° \ (V) = +1.23 - (0.059 \times pH)] \tag{6}$$

is the main obstacle to efficient water splitting as it suffers from a large activation energy barrier (overpotential > 0.4V) due to the necessary transfer of 4 charges per O_2 molecule in a complex proton-coupled electron transfer mechanism [46,214]. Nature's catalyst in photosynthetic water oxidation is the $CaMn_4O_x$ cubane-related molecular complex [215]. Traditionally, simple though expensive oxides like RuO_2 and IrO_2 have been used at the anode of PEM (acid) water electrolyzers [216], while mixed oxide Ni and Co spinels or perovskites (with inclusion of Cu) are favoured in alkali electrolyzers [217]. In heterogeneous particulate systems, the earth-abundant oxide CoO_x loaded onto $LaTiO_2N$ had an OER quantum efficiency (Φ_{O_2}) of 27% at 440 nm [218], while

CoOx or MnOx on TiO2 nanosheets achieved $\Phi_{O2} \approx 15\%$ at 365 nm [219]. In overall water splitting, only composite (tandem) absorbers, each optimized (with cocatalysts) for a single half-cell reaction have achieved quantum efficiencies greater than 5% (*vide infra*). Most progress has been made with oxysulphides or oxynitrides of d⁰ or d¹⁰ metal cations [107,220]. It should also be noted that quantum efficiency (Φ) values are not to be mistaken for η_{STH}. The last is theoretically ~10% for an absorber with a 500 nm absorption cut-off even at 100% quantum efficiency ($\Phi = 1$). In reality, despite intensive efforts over the last three decades η_{STH} values for particulate systems have yet to exceed 1% [220,221].

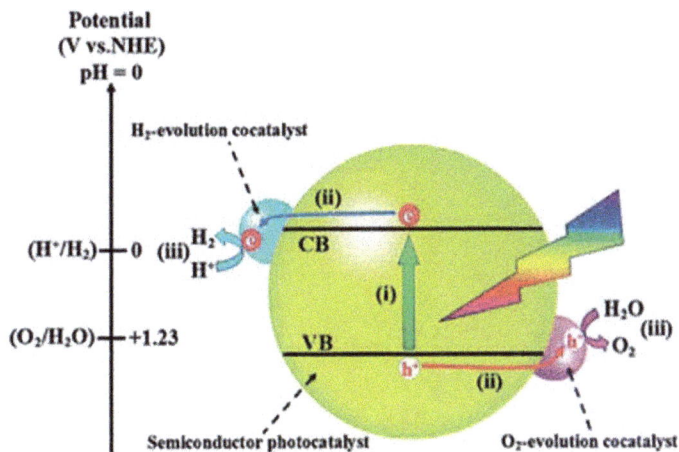

Figure 14. Photocatalytic water splitting over a visible-absorbing semiconductor loaded with H2- & O2-evolution co-catalysts (reproduced from [209] with permission of the Royal Society of Chemistry).

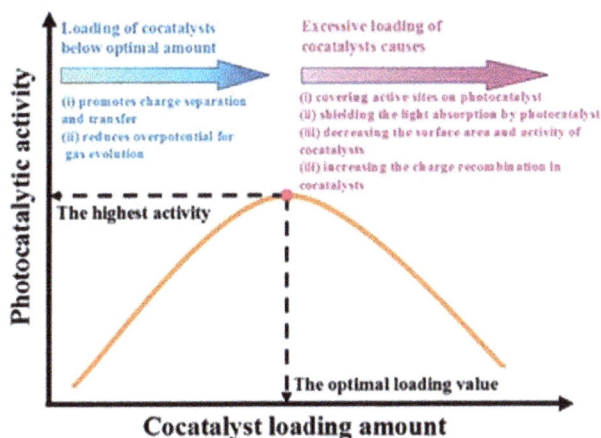

Figure 15. Principle of optimum loading of cocatalyst on a visible-absorbing semiconductor photocatalyst (reproduced from [209] with permission of the Royal Society of Chemistry).

544

Prospects for PEC cells look promising due to the efficient separation, collection and transport of photo-separated charges in a wired system. For most photoanodes, e.g., α-Fe$_2$O$_3$, WO$_3$, or other materials of more suitable bandgap (E$_g$ = 1.4–2.0 eV), the conduction band energy is so positioned that any injected electrons can only thermalize into the valence band of the photocathode, e.g., TiO$_2$, where they effectively neutralize holes created by direct photo-excitation (of TiO$_2$). By analogy with photosynthesis, such a configuration is generally referred to as a Z-scheme as originally proposed for spatially separated photoelectrodes [75]. Since most electrons reaching the photocathode conduction band are from the photoanode sensitizer and undergo two successive excitation steps, the mechanism is said to be of type D4 (4 photons per H$_2$), As shown in Figure 16 for a visible-absorbing WO$_3$ photoanode, if bare TiO$_2$ is replaced by a photoactive cathode (or a solar cell whose cathode is configured to evolve H$_2$) the theoretical combined efficiency can rise substantially due to wider light harvesting. This is most notably so (η_{STH} > 40%) in a series arrangement with an "in- front" photoanode (E$_g$ ~1.8 eV) that absorbs visible wavelengths to evolve O$_2$ from water. The transmitted near-IR light is incident on the solar cell (E$_g$ ~0.95 eV), which provides a voltage bias for H$_2$ evolution at the cathode (see Figure 17). The Z scheme principle has been extended to particulate systems but with limited success [50,107]. An added complexity here is the need to promote interparticle electron transfer using a redox mediator, e.g., IO$_3^-$/I$^-$, in solution. However, this suffers from "chemical short-circuiting", i.e., competitive reactions between water and the mediator, and especially reaction of its oxidized form, necessarily present in excess, with product H$_2$. Nevertheless, Maeda et al. [222] have reported a respectable overall quantum efficiency of 6.3% at 420 nm for a Pt-doped ZrO$_2$-protected TaON "cathode" (for H$_2$ evolution) suspended with PtO$_x$-loaded WO$_3$ as "anode" (for O$_2$ evolution). Better prospects may lie in elimination of mediators and the development of "all-solid-state" Z-scheme analogues, i.e., composite particles with heterojunctions [223–225].

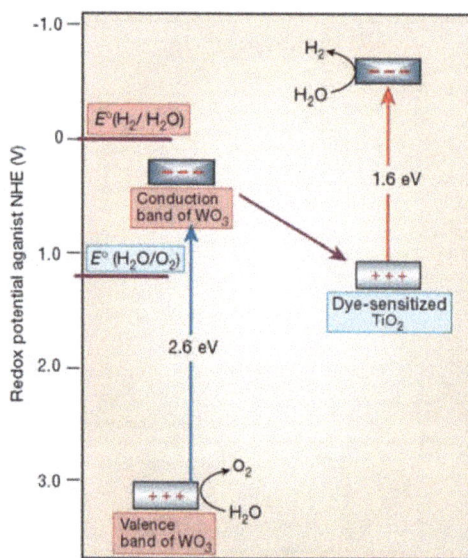

Figure 16. Example of Z-scheme (4D): WO$_3$ photoanode (E$_{cb}$ < E$^\circ_{H^+/H_2}$) coupled to dye-sensitized TiO$_2$ (reproduced from [76] with permission of the Nature Publishing Group).

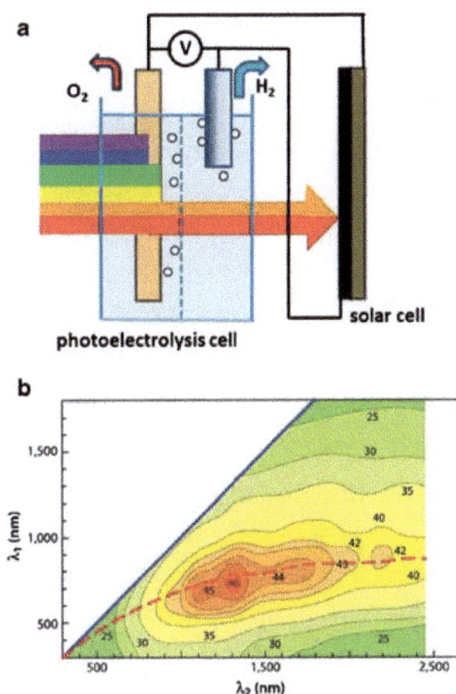

Figure 17. (a) Water splitting tandem cell: photoanode passes NIR light to solar cell, giving cathodic bias for H₂ evolution; **(b)** Dual absorber efficiency curve ($\eta_{STH} > 40\%$ at $\lambda_a \leq 750$nm (1.6 eV), $\lambda_{sc} = 750$–1300 nm (0.95 eV) (reproduced from [200] with permission from Springer).

The choice of materials for practical tandem PEC cells is restricted. Figure 18 shows the bandgap and bend edge position of representative semiconductors. They must be cheap (earth-abundant) and stable ideally in strongly acidic and/or alkaline conditions for good electrolyte conductance.

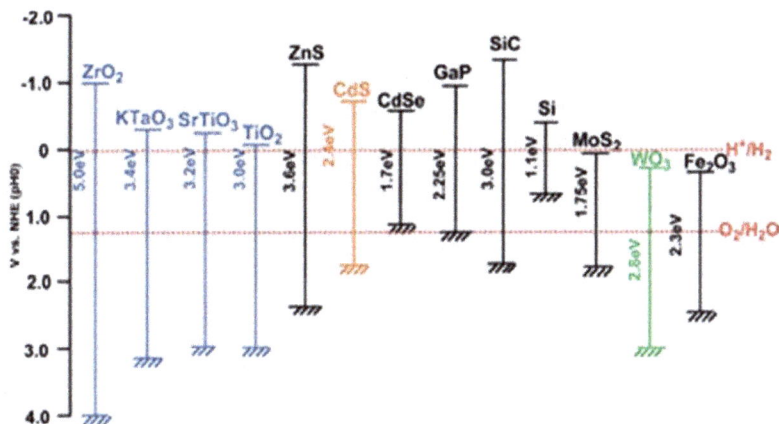

Figure 18. Bandgaps and band edge positions of representative semiconductors in relation to the redox potentials for water splitting at pH = 0. (Reproduced from [220] with permission of the Royal Society of Chemistry).

Bandgaps must also be higher to compensate for inevitable voltage losses, e.g., device series (ohmic) resistance, and electrode overpotentials. Nevertheless, the efficiencies of these tandem devices are expected to exceed 25% [226,227]. Major advances towards this goal were reported by Nocera et al. (η_{STH} = 4.7%, 2.5% wireless) [62] and Grätzel et al (η_{STH} = 3.1%) [61], but with a quite different system approach. Nocera's design consisted of a triple-junction (T6 or 6-photon) amorphous Si absorber loaded with cobalt phosphate (for the oxygen evolution reaction—OER) [228]) and Ni/Mo alloy (for H_2 evolution [211,212]) on a Ni mesh in phosphate or borate/nitrate buffer. The wireless layout suffered from an added resistance loss due to the longer migratory path imposed on proton transport from the front (anode) to the rear (cathode) of the cell. Grätzel's design was based on a single WO_3 (Fe_2O_3) photoanode coupled to a single DSSC, as shown in Figures 16 and 17. It is not clear if only Fe_2O_3 or both were loaded with IrO_2 co-catalyst. Continuing their exploration of a buried junction PV configuration with loaded electrocatalysts (EC), in which the Si absorber is protected from the electrolyte, the Nocera team most recently achieved η_{STH}= 10% with these so-called PV-EC tandem devices [78]. It consists of 4 single-junction crystalline Si solar cells connected in series to a NiMo cathode and a nickel borate anode, all immersed in a borate buffer (pH 9.2). The Ni-based anode is comparable in performance and cheaper than cobalt phosphate, but needs prior anodization (after deposition) to create the mixed-oxidation $Ni^{III/IV}$ state responsible for OE activity [229]. However, both Ni and Co salts can be electrodeposited, conveniently forming the anode in-situ from divalent ions in the appropriate buffer. The latest efficiency advance reported by the Grätzel team is η_{STH} = 12.3% in an analogous PV-EC device [77]. This was achieved with two DSSCs connected in series, each providing a short-circuit photocurrent density of 21.3 mA·cm^{-2}, open-circuit voltage (V_{oc}) = 1.06 V, and a fill factor of 0.76. A combined solar-to-electric power conversion efficiency of 15.7% was attained with superior light harvesting by lead iodide perovskite ($CH_3NH_3PbI_3$) sensitizers, prepared using a simple two-step spin-coating method at 100 °C. A cheap Ni-foam supported Ni/Fe layered double hydroxide, obtained by one-step hydrothermal growth, served as both cathode and anode in 1M NaOH electrolyte. It should be noted that a comparable performance may be obtained in PV-driven ("brute force") electrolysis, i.e., by combining state-of-the-art PV modules and electrolyzers optimized independently. For example, η_{STH} ≈ 12% has been reported in standalone systems where the voltage was maintained at ~1.7 V per cell (in a 20 cell PEM electrolyzer stack) with a DC-DC converter [230,231]. The PEC cell is a less expensive single integrated unit, provides a higher open circuit photovoltage, and reduces potential loss channels. However, it is susceptible to electrolyte resistance and polarization losses [227,232], especially under neutral conditions needed for operational stability of many earth-abundant co-catalysts. In view of such complications and exciting results with hybrid (PV-EC) devices, co-development of PV-electrolyzers and PEC water splitting cells may offer the best prospects [233,234].

4. Hydrogen Peroxide as Solar Fuel and Sustainable Chemical

Water splitting to H_2 and O_2 has been considered the "Holy Grail" of chemists working in the energy field. However, the co-production of H_2 and H_2O_2 is arguably a yet more valuable process:

$$2 H_2O \rightarrow H_2 + H_2O_2 \ [\Delta G° = +342 \ kJ/mol \ H_2O_2] \tag{7}$$

and may actually be easier, *i.e.*, the kinetic barrier may be lower, because it is just a 2 e⁻ process:

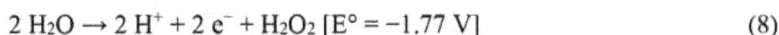

$$2 \, H_2O \rightarrow 2 \, H^+ + 2 \, e^- + H_2O_2 \; [E^\circ = -1.77 \; V] \tag{8}$$

This requires two photons at 171 kJ/mol (photons), corresponding to wavelengths below ~690 nm, which comprises ~50% of the solar power spectrum. Hydrogen peroxide is a valuable commodity chemical serving as a green oxidant in environmental clean-up, pulp bleaching, detergents, *etc.* [235]. It is now made largely by the *Anthraquinone Process* but research has intensified in recent years into direct catalytic synthesis from H_2 and O_2:

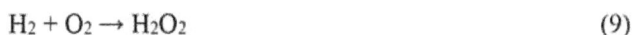

$$H_2 + O_2 \rightarrow H_2O_2 \tag{9}$$

which is exothermic ($\Delta H^\circ = -136$ kJ/mol) and a competitive option for small-scale on-site production ($< 10^4$ t/y) [236,237]. However, it deals with potentially explosive mixtures and only works efficiently over expensive rare metal (Pd or Pd/Au) catalysts. It is also an example of a highly selective partial oxidation reaction in which reaction with a second H_2 molecule:

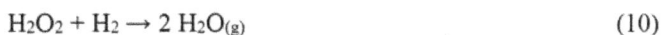

$$H_2O_2 + H_2 \rightarrow 2 \, H_2O_{(g)} \tag{10}$$

is even more exothermic (Equation (10) is the reverse of Equation (7)) and must be kinetically inhibited [238]. A PtHg₄/C electrocatalyst was shown to be active and highly selective for H_2O_2 synthesis, as predicted by DFT modeling [239]. Alloying leaves an isolated surface Pt atom for hydroperoxide (HOO*) stabilization in the on-top position while eliminating the hollow adsorption sites preferred by activated oxygen (O*) species, thereby inhibiting water as product. Unfortunately, it is still a rare-metal based formulation. The cathodic half-cell reacton for peroxide synthesis from water (Equation (8)) can be written in two ways depending on the electron acceptor.

For co-production of H_2 (Equation (7) overall) this is:

$$2 \, H^+ + 2 \, e^- \rightarrow H_2 \; [E^\circ = 0.00 \; V] \tag{11}$$

However, peroxide can also be synthesized by O_2 reduction:

$$O_2 + 2 \, H^+ + 2 \, e^- \rightarrow H_2O_2 \; [E^\circ = 0.68 \; V] \tag{12}$$

Summing the two half reactions (8) and (12), each yielding one peroxide molecule, gives Equation (13):

$$2 \, H_2O + O_2 \rightarrow 2 \, H_2O_2 \; [E^\circ = -1.09 \; V] \tag{13}$$

This is still a potentially visible-driven endergonic process so that H_2O_2 alone can be considered as an energy carrier derivable from cheap reactants. Supplied commercially as 30% aqueous solution (~9M) it is already in an energy dense form, unlike H_2 gas, and this underlines recent interest in peroxide as a solar fuel. Having acceptor and donor properties (reverse of Equations (8) and (12), respectively), the theoretical voltage of a "direct H_2O_2" fuel cell based on its own dismutation (reverse of Equation (13)) is 1.09 V. Although it enables a simplified fuel cell design (use of a single compartment is possible), the current densities are not yet of technical interest [240–246]. Alternatively, as a more powerful oxidant (than O_2) supplied to the cathode, it increases the operating voltage in cells based on H_2 [247] or liquid fuels such as aqueous NaBH₄ [248,249] and ethanol [250]. However, its powerful oxidizing properties and susceptibility to decomposition by traces of metal (ions) and redox-active surfaces can also lead to under-performance (mixed potentials) and introduces more stringent material compatibility issues, especially concerning long-term stability of polymer membranes and possibly carbon as an electrocatalyst support. Potential cathode materials explored to date include PbSO₄ [240], Au on Vulcan [249], and LaNiO₃ on N-doped graphene, which promotes the ORR [251].

In energetic terms, the photosynthetic route involving co-generation of hydrogen (Equation (7)) would be preferred as long as the H_2O_2/H_2 mixture remains stable. This has been reported for platinized calcium niobate [252] and Pt/TiO₂ [253]. Both showed relatively low quantum efficiencies (Φ_{uv} < 1%) but the 1:1 product stoichometry of Equation (7) was confirmed. A liquid water environ-ment would probably impede further reaction because any H_2O_2 dissolves selectively. However, little peroxide was found in solution, most remaining associated with the catalyst in a form seemingly immune to any back-reaction with H_2. This is consistent with previous literature showing that, in the absence of O_2, a stable form of peroxide (O_2^{2-}, HO_2^-, etc.) builds up and deactivates the photo-process in a few hours, typically affording only micromoles of products. Although the intermediate(s) can be decomposed easily to yield O_2 (the "missing" product in early studies of water splitting [254,255]), a way must be found to displace it as intact H_2O_2, e.g., by exploring more weakly-adsorbing (non-oxide) semiconductors and/or loading an oxidation co-catalyst. A viable system needs to sustain *milli*moles per hour productivity indicative of a broad spectral response and quantum yields exceeding 10%. This may be achievable but not easily recognized in practice because the peroxide, once formed, can decompose adventitiously, e.g., in the presence of trace metal ions, or excited by the UV component of a solar simulator, etc. Such H_2O_2 production rates (Φ_{uv} = 1%–30%) have been reported over quantum-sized ZnO and TiO₂ particles but only under oxygenated conditions, implying that photo-reduction of O_2 (Equation (12)) is the main source of peroxide [256]. Fluorination of TiO₂ improved dramatically the yield by weakening the surface adsorption of peroxide, a precursor step in self-decomposition [257].

As shown in Figure 19, the same effect was seen in a reduced graphene oxide (rGO)/TiO₂ composite in which the TiO₂ surface was phosphated to prevent simultaneous degradation, which is otherwise responsible for the attainment of photostationary product levels [258]. Alternatively, as shown in Figures 20 and 21, alloying Au (on TiO₂) with Ag suppresses selectively the intrinsic tendency of Au to simultaneously decompose its own product, thereby raising photostationary yields [259]. In contrast, it has been reported that TiO₂ *per se* is a poor catalyst in H_2O_2 decomposition due to its low affinity for the OH· radical [260,261]. This powerful but non-selective oxidant:

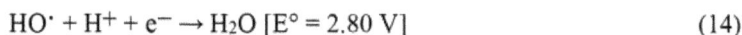

$$HO^{\boldsymbol{\cdot}} + H^+ + e^- \rightarrow H_2O \ [E° = 2.80 \text{ V}] \tag{14}$$

is produced from hydrogen peroxide by an electron donor:

$$H_2O_2 + H^+ + e^- \rightarrow HO^{\boldsymbol{\cdot}} + H_2O \ [E° = 0.72 \text{ V}] \tag{15}$$

uch as Fe^{2+} in the Photo-Fenton process, one of a variety of advanced oxidation processes (AOP) [13,96,262]. Unfortunately, it also promotes the autocatalytic decomposition of peroxide [263]:

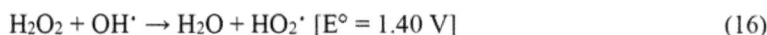

$$H_2O_2 + OH^{\boldsymbol{\cdot}} \rightarrow H_2O + HO_2^{\boldsymbol{\cdot}} \ [E° = 1.40 \text{ V}] \tag{16}$$

$$HO_2^{\boldsymbol{\cdot}} \rightarrow O_2 + H^+ + e^- \ [E° = -0.05 \text{ V}] \tag{17}$$

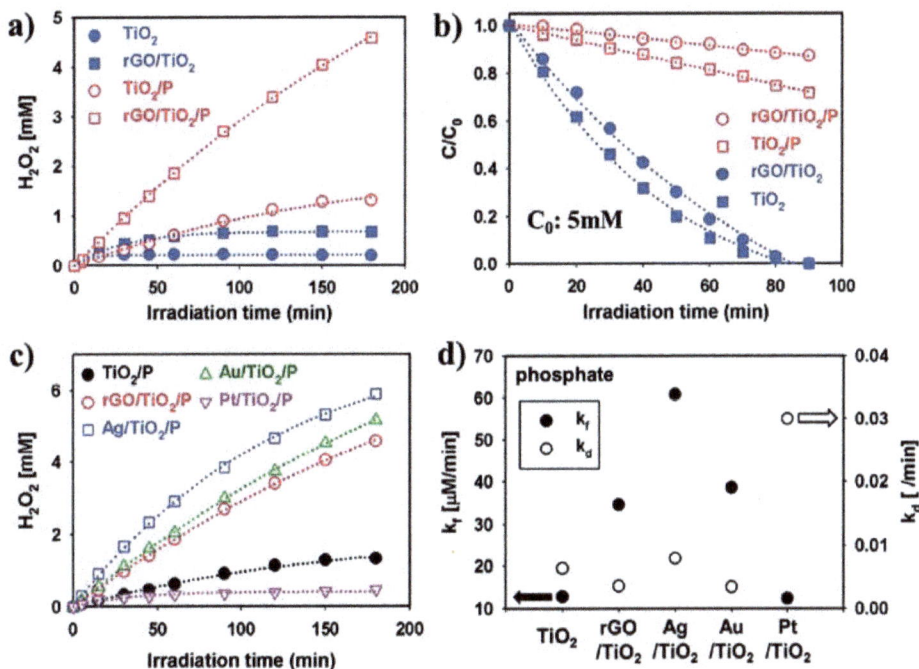

Figure 19. (a) Effect of pre-treating TiO₂ with phosphate on H₂O₂ photosynthesis ($\lambda > 320$ nm), and (b) 5 mM H₂O₂ decomposition over 6 wt % rGO/TiO₂ (0.5 g/L) in O₂-saturated aqueous buffer (pH 3) containing 5 vol % 2-propanol as hole scavenger; (c) Comparison of H₂O₂ synthesis rates over rGO/TiO₂(P) and 1 wt % metalized TiO₂(P) samples; (d) Rate constants of H₂O₂ synthesis (k_f) and decomposition (k_d) over TiO₂(P)-supported composites (reproduced from [258] with permission of the Royal Society of Chemistry).

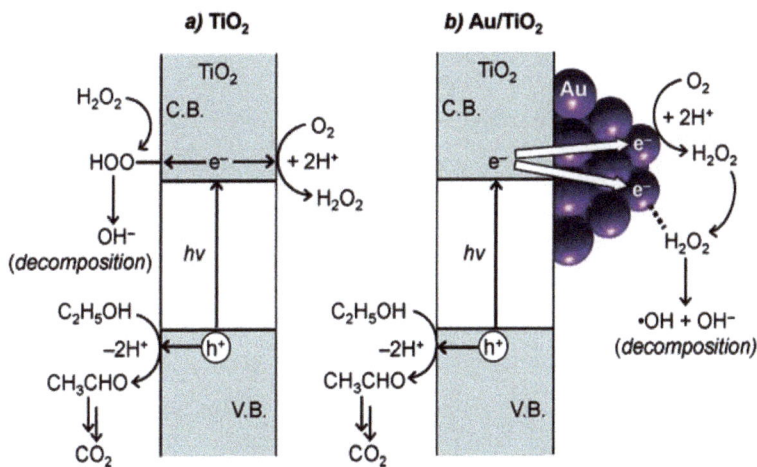

Figure 20. Scheme of synthesis and decomposition of H₂O₂ on (a) TiO₂; and (b) Au/TiO₂ photocatalysts. (Reproduced from [259]. Copyright (2012) American Chemical Society).

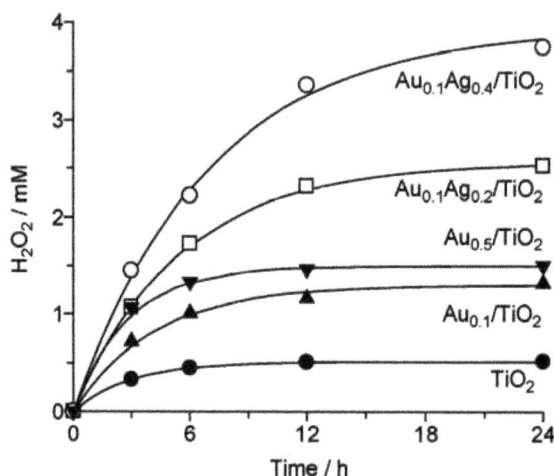

Figure 21. Build-up of photostationary levels of H_2O_2 on Au-Ag/TiO$_2$ catalysts (5 mg) in 5 mL aerated 4% aqueous ethanol ($\lambda > 280$ nm, $I_{uv} \approx 14$ mW)). (Reproduced from [259]. Copyright (2012) American Chemical Society).

Hole scavengers (H· donors) always increase decomposition rates, suggesting that the 2 e$^-$ water oxidation half-reaction (Equation (8)) is rate-determining in the synthesis. Under these conditions, *i.e.*, with little H_2 or H_2O_2 produced from water, the process is of no interest as an energy conversion scheme. Nevertheless, the peroxide can still be considered a value-added green reagent obtained efficiently and cheaply by photo-oxidation of organic wastes. Evidence has just been reported for visible-driven H_2O_2 photosynthesis from oxygenated ethanol over pristine g-C$_3$N$_4$ [264], whose conduction band minimum ($E° \approx -1.3$ V [265]) exceeds the reduction potential for the O_2/O_2^- couple ($E° \approx -0.3$ V [266]). A 1,4-endoperoxide intermediate stabilized by the g-C$_3$N$_4$ surface was identified by Raman spectroscopy. Addition of Pt had a deleterious effect on yield due to its tendency to break the O–O bond [239].

5. Photoreforming of Bio-Oxygenates

In the renewable energy scheme under consideration (*cf.* Figure 1), bio-oxygenates like sugars, alcohols, and polyols, all serve as CO_2-neutral energy carriers provided that their incipient H_2 can be extracted efficiently by catalytic reforming with steam [31,32,55,56]:

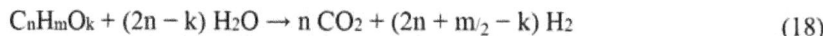

$$C_nH_mO_k + (2n - k)\, H_2O \rightarrow n\, CO_2 + (2n + m/_2 - k)\, H_2 \tag{18}$$

Despite being highly endothermic, Equation (18) is favoured thermodynamically above a threshold temperature due to the large volume expansion (entropy factor). Input thermal energy is converted into chemical energy (H_2) and represents a significant gain in exergy (20%–30%), as can be seen by comparing the heats of combustion of reactant and product. This is known as "chemical recuperation" [267,268], and pre-reforming of natural gas is likely to be incorporated into future gas turbine technologies [269]. Methanol is not currently made on a large scale from biomass or renewable H_2 and can be readily reformed by conventional (thermal) catalysis [23–25,28]. In contrast, ethanol comprises 90% of biofuel production and, due to its high growth forecast [270], is now being considered as a renewable platform chemical, e.g., for butadiene synthesis [271]. Ethanol is also a good model oxygenate as it is one of the simplest compounds containing C–C, C–H, and C–O bonds. However, its catalytic conversion in high activity and selectivity poses a major challenge [272,273].

Bio-ethanol obtained by fermentation of glucose:

$$C_6H_{12}O_6 \rightarrow 2\ CO_2 + 2\ C_2H_5OH \tag{19}$$

is an excellent energy carrier since almost the entire heating value of the original sugar (~2800 kJ/mol) is retained in the product (two moles liquid ethanol at 1365 kJ/mol). Ethanol steam reforming (ESR):

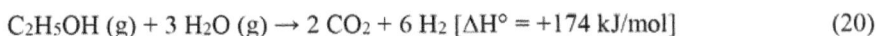

$$C_2H_5OH\ (g) + 3\ H_2O\ (g) \rightarrow 2\ CO_2 + 6\ H_2\ [\Delta H° = +174\ kJ/mol] \tag{20}$$

raises the fuel value substantially (six moles H_2 at 286 kJ/mol). In addition, Equation (20) has a crossover ($\Delta G° \leq 0$) temperature as low as 210 °C [19], suggesting it could be driven by "low-quality" heat provided a suitable catalyst can be found. This explains the major interest in ESR in recent years, e.g., as an on-board source of H_2 for PEM fuel cell (electric) vehicular propulsion [274,275]. However, the low rates encountered over many oxide-supported transition metals (Pt, Ni, Co, Rh, Ru) necessitate working above 400 °C, where rapid deactivation by coking ensues, possibly linked to acetic acid intermediate [30,276–278]. DFT modeling supports experimental data showing that the rate-determining step in ESR is initial dehydrogenation to acetaldehyde [279]:

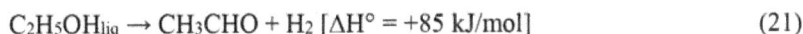

$$C_2H_5OH_{liq} \rightarrow CH_3CHO + H_2\ [\Delta H° = +85\ kJ/mol] \tag{21}$$

a modestly endothermic reaction but with an apparent activation energy (E_{app}) as high as +150 kJ/mol [280,281]. This is typically followed by decarbonylation of the aldehyde [281,282]:

$$CH_3CHO \rightarrow CO + CH_4\ [\Delta H° = +7\ kJ/mol] \tag{22}$$

an almost thermoneutral process giving an undesirable alkane product. Photocatalysis may circumvent these activity/selectivity issues because it works by an alternate mechanism (lowering E_{app}) and at low temperature where deleterious side reactions are inhibited. The pioneering work of Pichat et al. [283,284] established that photo-dehydrogenation (PDH) of alcohols (see Equation (21)) proceeds in high quantum efficiency ($\Phi_{uv} \geq 0.1$) over Pt/TiO$_2$. More recently, photo-assisted water-gas shift (WGS) [285], photoreforming of methanol [286,287], and combined photo-/thermal reforming of methanol (or glycerol) to H_2/CO_2 co-products over Pt/TiO$_2$ [288] (or Pd/TiO$_2$ [289]) have been studied. As shown in Figures 22 and 23, mild heating is a useful adjunct in photocatalysis when dark processes are rate-controlling. However, despite the sharp rise (×2–×5) in quantum efficiency ($\Phi_{uv} \approx 7\%$ for CH$_3$OH at 65 °C [288]) and better reaction stability (due to more complete product recovery [288,290]), this synergism has still not been widely exploited.

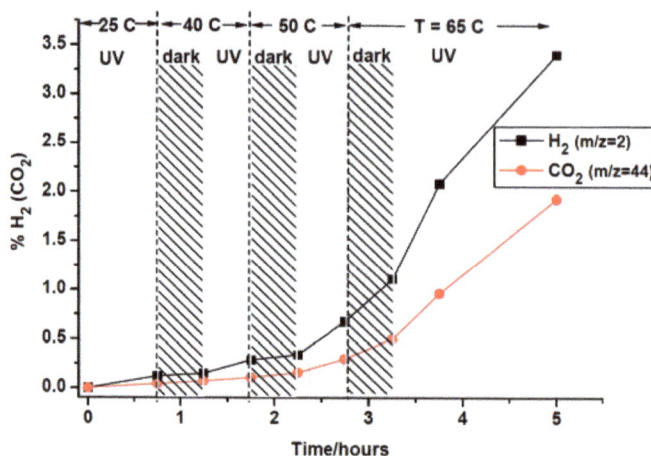

Figure 22. MS response (H_2/CO_2) showing effect of temperature on rate of vapour-phase CH_3OH photoreforming over Pt/TiO_2 (reproduced from [288] with permission of RSC).

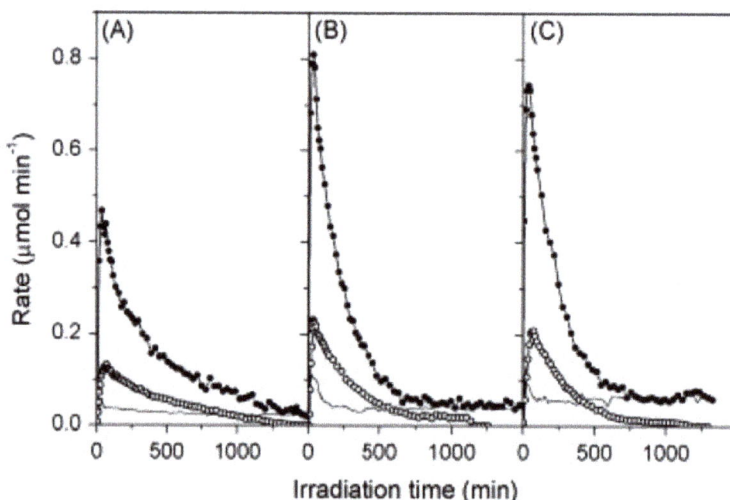

Figure 23. H_2 (full circles) & CO_2 (closed circles) evolution during liquid-phase (0.37 mM) aqueous glycerol reforming at (**A**) 40 °C; (**B**) 60 °C; (**C**) 80 °C (reproduced from [291] with permission of Elsevier).

Photoreforming *per se* has been extended to ethanol [290–294], various alcohol mixtures [35,294–298], glycerol [33,55,56,289,299], sugars [300,301], and acetic acid [302]. In the last case, no CH_4 product was seen, an encouraging result in view of previous claims of a novel Photo-Kolbé process over a similar catalyst [303]. All these studies used TiO_2-supported precious metals responding only to UV light at high efficiency. For 80% aqueous ethanol over a well-dispersed Pt/TiO_2 film at low light intensities ($I_{uv} \approx 0.8$ mW/cm², or 0.2 suns), a remarkable quantum efficiency ($\Phi_{uv} \approx 74\%$) was estimated [294]. Close to solar intensities, these were rather lower ($\Phi_{uv} = 10\%–30\%$) for ethanol [290,295,297], but several groups have reported mass-specific H_2 evolution rates of technical interest (>2 mmol/h/g_{cat}) [33,291,293]. As shown in Figures 24 and 25, bio-oxygenates are reformed at comparable rates [297,298], while quantum efficiencies can be raised significantly (to $\Phi \geq 30\%$) due to the superior illumination geometry available in an optical fibre honeycomb reactor [297].

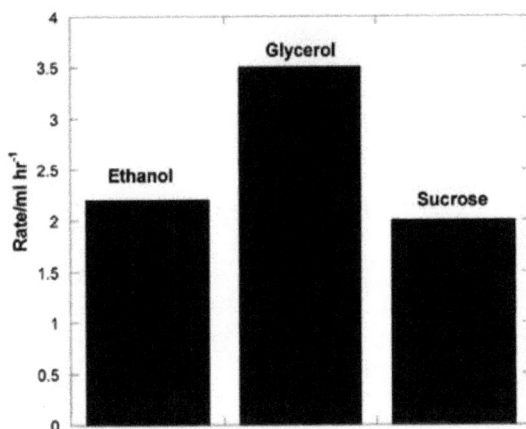

Figure 24. Rate of H_2 evolution by photo reforming of 3 important bio-oxygenates (1% in H_2O) over 0.3 wt % Pd/TiO$_2$ (P25) (reproduced from [298] with permission of RSC).

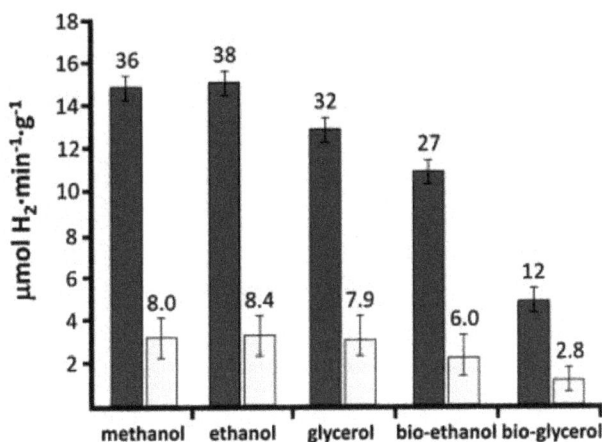

Figure 25. H_2 evolution rates and quantum efficiencies of bio-alcohols (1:1 H_2O) in a catalytic wall reactor (dark) or a slurry reactor (light) over 1 wt % Au/TiO$_2$. (reproduced from [297] with permission of Elsevier).

Early investigations of photo-reforming over visible-active catalysts include SPR metals on TiO$_2$ [304], Cd$_{1-x}$Zn$_x$S/ZnO [305], CdS/TiO$_2$ [306], ε-Fe$_2$O$_3$ [307,308], and a potentially self-degrading glucose-TiO$_2$ charge transfer complex [309]. However, only the chalcogenide-based system gave a respectable quantum efficiency, $\Phi_{>420nm} = 9.6\%$ [305]. Glycerol photoreforming over earth-abundant co-catalysts CuO$_x$/TiO$_2$ [310,311] and NiO$_x$/TiO$_2$ [312] has also been reported.

6. Photoreduction of Carbon Dioxide: Artificial Photosynthesis

As mentioned in the *Introduction*, a valuable (solar) energy-storing artificial photosynthetic process which couples two stages in the energy scheme is methanol synthesis from carbon dioxide and water:

$$CO_2 + 2\ H_2O \leftrightarrow CH_3OH + 1.5\ O_2 \tag{23}$$

although ethanol synthesis:

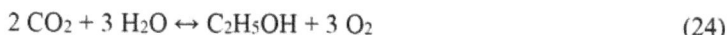

$$2\ CO_2 + 3\ H_2O \leftrightarrow C_2H_5OH + 3\ O_2 \tag{24}$$

would be even better due to its higher energy density, lower toxicity and volatility. However, Equation (24) is probably unrealistic due to its mechanistic complexity and associated selectivity issues, at least as reported in CO_2 hydrogenation [26,27]. Methanol synthesis *per se* (Equation (23)) is already complex as it subsumes the water splitting process (see Equation (6)) but is 50% more demanding energetically (ΔH = +727 kJ/mol) due to the evolution of proportionately more O_2, as shown by the relevant half-cell reactions:

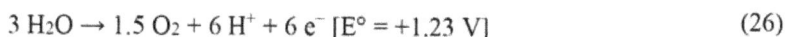

$$CO_2 + 6\ H^+ + 6\ e^- \rightarrow CH_3OH + H_2O\ [E° = -0.32\ V] \tag{25}$$

$$3\ H_2O \rightarrow 1.5\ O_2 + 6\ H^+ + 6\ e^-\ [E° = +1.23\ V] \tag{26}$$

and comparison of Equations (6) and (26). While this 6 e^- process can be driven by near-IR photons, it is mechanistically more complex than water splitting as two elementary steps (proton reduction and H atom coupling) at the photocathode are replaced by activation of a stable gas molecule of low aqueous solubility, its multi-step reductive de-oxygenation, and progressive hydrogenation (Equation (25)). It perhaps comes as no surprise that carbon-based solar-to-fuel conversion efficiencies lag far behind those of water splitting at < 1%, with turnover rates (10–100 μmol/h/g$_{cat}$) over pristine semiconductors still too low for technical exploitation [43,45,79]. Achieving high selectivity to methanol is also a challenge insofar as CH_4 is preferentially obtained over hydrated anatase TiO_2 [313,314], unless the surface Ti^{4+} (electron trap) centre is highly-dispersed or isolated [315].

Nonetheless, as shown in Figure 26, there is general agreement that the photocatalyst plays an indispensible role in activating CO_2 (the probable rate-determining step) by electron transfer and stabilizing the highly energetic $CO_2^{·-}$ radical ion in coordinated form(s) [43,316,317]. Possible sequences of proton-coupled electron transfer in CO_2 conversion to formic acid are shown in Figure 27. DFT-modeled energy barriers favor the green route via bidentate coordination mode B1 (0.87 eV) and the red route via the linear monodentate mode A1 (0.82 eV) although this is less likely as it requires a simultaneous two e^- transfer. Routes (black) via carbonato-type complexes A2 and B2 proceeding through carboxyl (COOH) intermediate have much higher energy barriers, 2.25 eV and 1.73 eV, respectively.

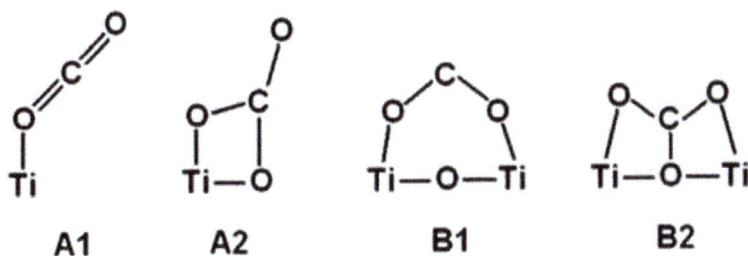

Figure 26. DFT-modeled states of neutral/anionic CO_2 adsorbed on TiO_2 anatase (101) (reproduced from [316] with permission of the Royal Society of Chemistry).

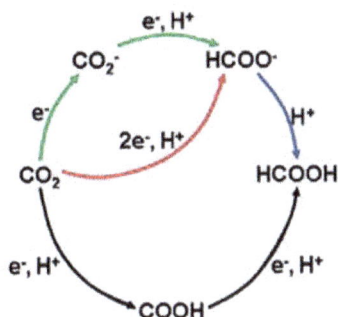

Figure 27. Pathways of $2e^-/2H^+$ photoreduction of CO_2 to formic acid. DFT energy barriers (<1 eV) favour the green route (via B1) and the red route (via A1) (reproduced from [316] with permission of the Royal Society of Chemistry).

It is evident from the literature [24,43,317] that earth-abundant Cu is a ubiquitous co-catalyst in CO_2 reduction, yielding mainly methanol in gas-solid photocatalysis or methane by electrocatalysis. This selectivity effect has been rationalized [318] as due to reaction of the surface-bound methoxy (CH_3O-Cu) intermediate either with a co-adsorbed H atom (favouring CH_3OH by a lateral surface mechanism) as in the industrial synthesis [319], or with a proton from aqueous solution (favouring CH_4 via attack on the protruding CH_3 moiety). The prevalence of Cu (or CuO_x) in composite photocatalysts [41,43–45,320–323] and electrodes [319,324–326] suggests that the dark mechanism "post-formate" is still operative but in which water photo-oxidation provides the electron/proton pairs in Equation (25), i.e., the half-cell equivalents of H atoms from H_2 dissociation. If this is the case, the thermal mechanism remains important [19] and an approach toward industrial synthesis conditions is worthy of study, e.g., mild heating/pressurization of CO_2/H_2O vapour under illumination. Publications have recently proliferated on the role of nanocarbons and graphenes as cocatalysts in composite photocatalysts [327–329], including photo- [40,330–332] and electro-reduction of CO_2 [333]. This is linked mainly to their effectiveness in promoting charge separation [327–329,334], although there are tentative claims for visible sensitization, possibly due to adventitious C doping [40,327–329]. There is also growing evidence that graphene oxide can act as a photocatalyst *per se* [335] although CH_3OH synthesis activity is improved by addition of Cu [336] and/or molecular sensitizers [337]. Otherwise, visible response has been conferred to TiO_2 nanocomposites by incorporating CdS/Bi_2S_3 [338], CdSe quantum dots [339], or plasmonic metal deposits [340]. While promising advances have been made in "self-biasing" particulates, the tendency of methanol to act as an efficient hole scavenger may ultimately militate against a gas-phase photocatalytic approach in favour of a PEC (membrane-separated photo-electrode) arrangement, as shown in Figure 28. This is despite the known limitations of the latter, viz., CO_2 solubility/mass transfer issues in the liquid phase [79], and a higher probability of obtaining CH_4 on a photoelectrode [318,326].

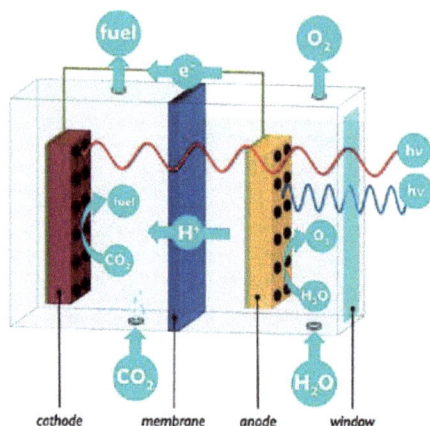

Figure 28. Liquid-phase 2-compartment PEC cell for CO_2 photo-reduction. (reproduced from [79] with permission of the Royal Society of Chemistry).

As a potential solution, and borrowing concepts from PEM fuel cell technology, gas-phase electrocatalytic studies using a gas diffusion membrane electrode (GDM) configuration have given improved CO_2 conversion rates (Faradaic efficiencies) and a selectivity shift towards oxygenates over carbon nanotube-supported Pt and Fe [333,341,342], as shown in Figure 29. The beneficial effect of mild heating is also evident.

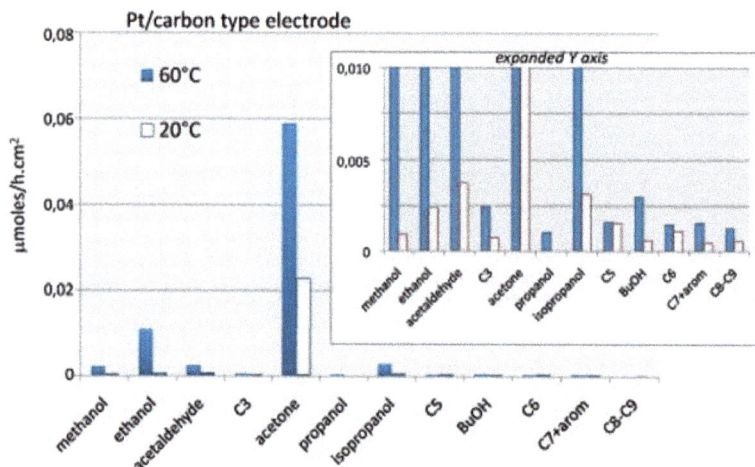

Figure 29. Temperature effect in gas phase CO_2 electro-reduction on 10% Pt/Carbon/Nafion117 GDM electrode (reproduced from [333] with permission of Elsevier).

Regarding the dark mechanism in industrial methanol synthesis (from CO_2/CO) over supported Cu, a recent microkinetic/DFT modeling paper [343] casts doubt on the formate (HCOO) → dioxomethylene (H_2COO) route in favour of carboxyl (COOH) → formic acid [HC(OH)O] → hydroxymethoxy [$H_2C(OH)O$] → formaldehyde (hydroxyl) [CH_2O (OH)] → methoxy (CH_3O). The hydroxymethoxy species is rightly considered the key intermediate but its identity as the conjugate base of methanediol or methylene glycol (formaldehyde hydrate) has seemingly not been recognized [343,344]. This species is implicated in the Cannizzaro disproportionation involving a hydride or proton-coupled 2 e⁻ transfer reaction [345–348]:

$$2 \ H_2C(OH)O \rightarrow CH_3O + HCOO + H_2O \qquad (27)$$

thereby providing a simpler (direct) path to methoxy.

7. Summary and Outlook

From this brief overview of a burgeoning field of research over the last two to three decades, it is somewhat puzzling and disappointing to admit that a "solar fuels" industry is still no more than a futuristic concept [349] with no clear indication as to when to expect its realization [58], or at least the emergence of (pre-commercial) demonstration systems. A recent review of the proliferating patent literature reveals that they are almost exclusively based on incremental advances at the fundamental level, many being filed by academic scientists [42]. However, looked at *sub specie aeternatitis,* one must recognize the enormity of the challenge facing Mankind, viz., a paradigm shift towards a sustainable global economy based on an entirely new foundation, or what could be termed "renewable petrochemistry" [350]. The expanding role of industrial catalysis in such a future will require a thorough evaluation of scalability issues, not least in global elemental resources [68] for a sector established largely on the exploitation of rare metals. Photocatalysis and the maturation of commercial photoreactor design [351] in environmental detoxification is certainly gaining an industrial foothold [13], but this reviewer has only come across one pertinent reference to H_2 photogeneration on a pilot scale, and even this did not involve water splitting (O_2 co-generation) but photoreforming of aqueous organic contaminants over Pt/N-TiO_2 and Pt/CdS/ZnS [352]. Otherwise, nanoparticulate suspensions have demonstrated very low efficiencies to date (≤1%) and offer merely simplicity and convenience in operation. Disadvantages include unwieldy reactor size/catalyst charge and an associated explosion hazard in the absence of a H_2/O_2 separation stage. In contrast, implementation of a PEC-membrane-integrated tandem system that yields pure H_2 at STH efficiencies already exceeding 10% in the laboratory (mandated to meet the US DOE cost target of USD 2.00–4.00 per kg H_2 [353]), is hampered by complexities associated with device design and scale-up, especially geometric factors and their role in loss (optical and overvoltage) minimization [354–356]. Indeed, the question as to whether an integrated (PV-PEC) design will ultimately outperform coupled PV-electrolyzers (with independently optimized components), or if their co-development has advantages [230,233,234,357], is still open to debate [58,59,226,358,359]. Certainly the former raises more materials compatibility issues since electrolyzers work best in acid or alkaline environments. According to McKone *et al.* [58], priority needs in fundamental work include: (a) Higher efficiency electrolysis in buffered pH-neutral electrolyte (for greater durability of earth-abundant absorbers and catalysts); (b) More conductive anion (alkaline) exchange membranes (thinner separators lead to higher efficiency electrolysis and faster pH equilibration); (c) More earth-abundant OER catalysts stable in acidic media (a major expansion in PEM-based devices will otherwise be limited by their present dependence on precious metal catalysts); (d) Optical transparency in lower mass-specific activity co-catalysts (minimal parasitic light absorption); and (e) A wider range of stable (acid/alkali -resistant) visible light absorbers and transparent (ultra-thin protective) layers for Si.

Two excellent articles have appeared recently that address the technoeconomic feasibility of centralized facilities for solar hydrogen [60] and the more complex case of solar methanol, which includes the problem of CO_2 sourcing [231]. Pinaud *et al.* [60] considered four types of reactor systems of increasing complexity designed to generate 1 ton per day of pure H_2 at 20 bar from 0.1 M KOH solution/electrolyte. Type 1 was an array of 18 shallow plastic "baggies" ($323 \times 12 \times 0.1$ m) housing a single-bed photocatalyst particle suspension working at an assumed STH efficiency (η) of 10% (theoretical $\eta_{max} \approx 23\%$). Type 2 was similar but had two (types of photocatalyst) beds for separate generation of H_2 and O_2 coupled by redox mediators (Z-scheme)—$\eta = 5\%$ ($\eta_{max} \approx 15\%$). Type 3 was a fixed panel array of ~27,000 monolithic tandem absorber PEC cells—$\eta = 10\%$ ($\eta_{max} \approx 30\%$), while Type 4 was similar but with a drastically reduced number of cells (~2,000, generating equivalent power) due to coupling with a tracking concentrator ($\times 10$) assembly—$\eta = 15\%$. The estimated cost of H_2 (per kg) for the four cases was $1.60, $3.20, $10.40, and $4.00, respectively. These are encouraging figures but it should be recognized that the efficiencies assumed for the particulate systems were "target" values several times higher than the current state-of-the-art. Clearly, the most promising and realistic system for early demonstration would be Type 4. However, the inclusion of low-power concentrators adds further complexity [360] and raises the question as to how PEC-based systems actually respond to light intensification and to what extent it is influenced by device configuration. While it is well-known that efficiencies in PV arrays are maintained (or even improved) by concentration of sunlight [361,362], the progressive drop in photochemical quantum yields with increasing incident power in suspended particulate systems is notorious [363], explaining the virtual absence of concentrating optics in advanced oxidation photoreactors, excepting perhaps the compound parabolic reflector for improved collection of diffuse sunlight [351]. The problem is generally attributed to kinetic limitations in surface redox processes and/or O_2 supply due to low aqueous solubility. Some of these "chemical" constraints are shared by PEC cells, such that a strong case can be made for a systematic evaluation of the sensitivity of water splitting efficiency to solar light concentration. The only positive evidence in the open literature is the NREL claim of a record-breaking 12% efficiency (in 1998) using a prototype hybrid PV/PEC system comprising a $GaInP_2$ photocathode voltage-biased by an underlying GaAs (PV) absorber under 11 Suns illumination [364].

Herron *et al.* [231] have taken a broader approach in designing a "transitional" solar refinery that produces H_2 as an intermediate in the generation of liquid fuels but with sub-systems still dependent to some degree on fossil fuel energy input, as represented schematically in Figure 30. First, a feasibility study was made by assessing the energy balance for the *indirect* route based on existing (sub-system) efficiencies. The CAMERE process [365] was selected and modeled, viz., methanol synthesis (at 1 kg/s ≡ 22.7 MW$_{HHV}$) from CO_2 (captured from a fossil fuel power station) with H_2 (produced increasingly from solar energy), and defining the solar or primary "energy incorporation efficiency" (EIE) as being positive (viable) only when the methanol energy content exceeded the sum of all fossil energy inputs. This rather severe criterion was not satisfied even when all the H_2 was derived from solar energy. Thus, there is a need for diversification and greater implementation of solar technologies in early demonstration systems. A good candidate (energy-intensive) process for fossil fuel substitution is solar heat-driven CO_2 recovery in amine scrubbers and, ultimately, atmospheric trapping devices [20,21].

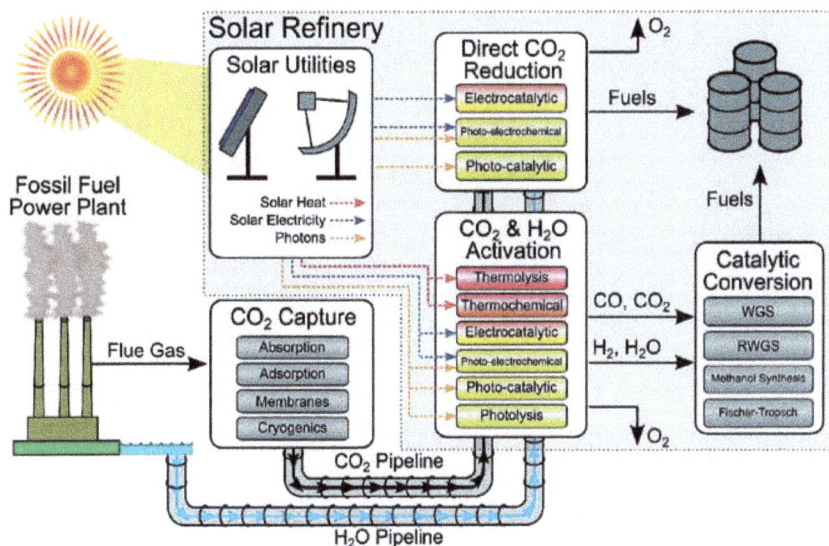

Figure 30. Scheme of a solar refinery based on CO_2 reduction to methanol *indirectly* by renewable H_2 (ex solar photon- or heat-driven electrolysis) or *directly* ("one pot") with H_2O using solar-electricity (PEC) or solar photons (photocatalysis) (reproduced from [231] with permission of the Royal Society of Chemistry).

A case study estimating the threshold CO_2 single-pass conversion for positive EIE based on a *direct* (photocatalytic) route was also made, assuming a reaction selectivity of 40% as below:

$$CO_2 + 2\ H_2O \rightarrow 0.6\ CH_4 + 0.4\ CH_3OH + 1.8\ O_2 \tag{28}$$

and a catalyst mass-specific rate of ~1 μmol/g$_{cat}$/h (Φ_{uv} = 0.28% over Ti-containing porous SiO_2 [366]). As shown in Figure 31, it is primarily the CO_2 capture stage (from dilute flue gas) that impedes the energy efficiency break-even point (EIE = 0), such that present costs of around 5.5 MJ/kg$_{CO2}$ must be further reduced. If the enthalpy from burning the renewable CH_4 co-product is valorized, positive energy incorporation is feasible below 50% conversion. Better still, if a more selective catalyst can be developed, it would simplify the process and have a dramatic effect on minimum one-pass conversion (EIE > 0), as shown in Figure 32. However, by the authors' own admission, this case is quite impractical due to the low activity of the photocatalyst, roughly 3 orders of magnitude below that required as discussed in Sections 4 and 5. Even at 1 mmol/g$_{cat}$/h, for an overall production rate of 1 kg/s, it would need 100 metric tons of catalyst and a reactor volume of 100 m³ severely constrained in one dimension by optical factors.

In closing, the author feels obliged to point out that the earlier review [19] was written on the premise that future market penetration by the renewable energy sector would be substantial and already impacting (slowing the rise of, if not stabilizing) global CO_2 levels in the atmosphere. Unfortunately, governments have not provided sufficient incentives or regulatory measures to curb our dependence on fossil fuels and the oil industry has conducted very much a "business-as-usual" policy. In the last 20 years, the rate of emissions has increased by 50% and a cumulative level of 400 ppm CO_2 has been reached, *i.e.*, 50 ppm in excess of the threshold considered necessary to avoid a mean global temperature rise of more than 2 °C with probable catastrophic effects. While the penetration of CO_2-neutral energy systems will help to ease the burden on natural sinks for CO_2, it now appears essential to augment these with artificial (man-made) disposal methods. The author's directions in research have broadened accordingly in recent years, favouring mineralization as the only technology that will guarantee CO_2 sequestration on the requisite (geological) time-scale [367–371].

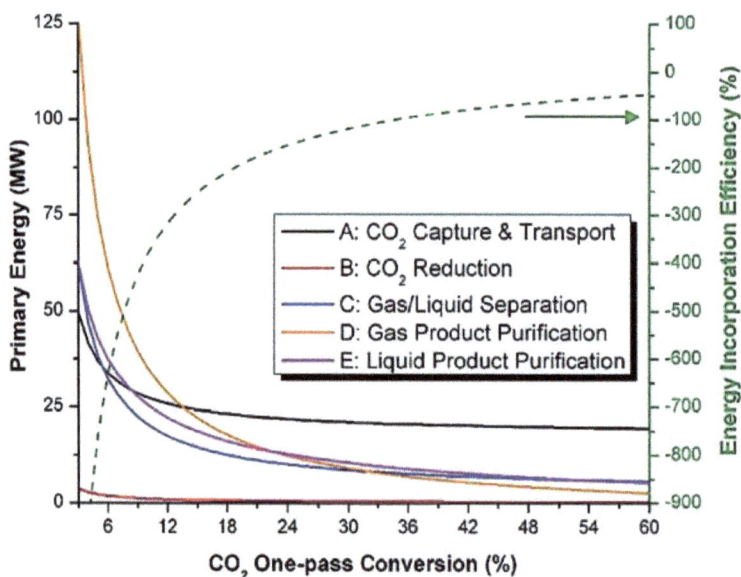

Figure 31. Dominance of CO_2 capture stage on primary energy costs in photocatalytic CH_3OH synthesis from CO_2/H_2O (reproduced from [232] with permission of the Royal Society of Chemistry).

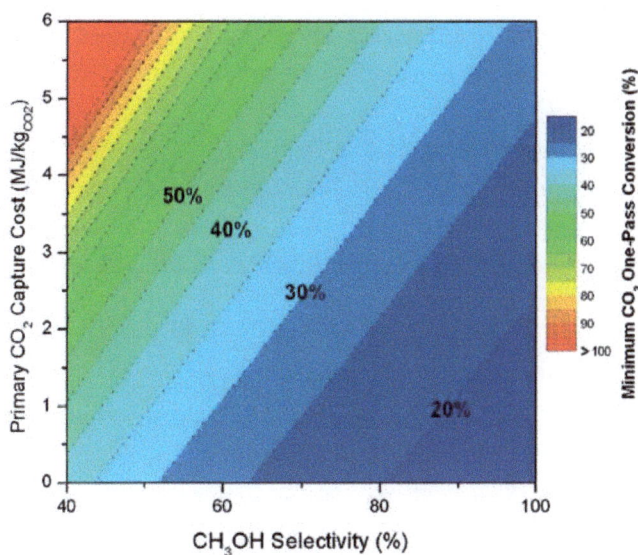

Figure 32. Effect of photocatalyst selectivity and CO₂ capture cost on minimum one-pass conversion for positive energy incorporation efficiency (EIE > 0) (reproduced from [231] with permission of the Royal Society of Chemistry).

Acknowledgments

The author is grateful to Eric Claude, Innovation Director at *Axane* (Air Liquide, Grenoble, France), for valuable advice on the hydrogen peroxide fuel cell.

Conflicts of Interest

The author declares no conflict of interest.

References

1. Smalley, R.E. Future Global Energy Prosperity: The Terawatt Challenge. *MRS Bull.* **2005**, *30*, 412.
2. Bell, A.T. The Impact of Nanoscience on Heterogeneous Catalysis. *Science* **2003**, *299*, 1688–1691.
3. Rolison, D.R. Catalytic Nanoarchitectures—The Importance of Nothing and the Unimportance of Periodicity. *Science* **2003**, *299*, 1698–1701.
4. Nørskov, J.K.; Bligaard, T.; Rossmeisl, J.; Christensen, C.H. Towards the Computational Design of Solid Catalysts. *Nat. Chem.* **2009**, *1*, 37–46.
5. Semagina, N.; Kiwi-Minsker, L. Recent Advances in the liquid-phase synthesis of metal nanostructures with controlled shape and size for catalysis. *Catal. Rev. Sci. Eng.* **2009**, *51*, 147–217.
6. Cuenya, B.R. Synthesis and Catalytic Properties of Metal Nanoparticles: Size, Shape, Support, Composition, and Oxidation State Effects. *Thin Solid Films* **2010**, *518*, 3127–3150.

7.   Yuan, Y.; Yan, N.; Dyson, P.J. Advances in the Rational Design of Rhodium Nanoparticle Catalysts: Control via Manipulation of the Nanoparticle Core and Stabilizer. *ACS Catal.* **2012**, *2*, 1057–1069.

8.   Guo, Z.; Liu, B.; Zhang, Q.H.; Deng, W.P.; Wang, Y.; Yang, Y.H. Recent Advances in Heterogeneous Selective Oxidation Catalysis for Sustainable Chemistry. *Chem. Soc. Rev.* **2014**, *43*, 3480–3524.

9.   Schlögl, R.; Abd Hamid, S.B. Nanocatalysis: Mature Science Revisited or Something Really New? *Angew. Chem. Int. Ed.* **2004**, *43*, 1628–1637.

10. Chen, X.; Li, C.; Grätzel, M.; Kostecki, R.; Mao, S.S. Nanomaterials for Renewable Energy Production and Storage. *Chem. Soc. Rev.* **2012**, *41*, 7909–7937.

11. Thomas, J.M. Heterogeneous Catalysis and the Challenges of Powering the Planet, Securing Chemicals for Civilised Life, and Clean Efficient Utilization of Renewable Feedstocks. *ChemSusChem* **2014**, *7*, 1801–1832.

12. Kondratenko, E.V.; Mul, G.; Baltrusaitis, J.; Larrazabal, G.O.; Perez-Ramirez, J. Status and Perspectives of CO_2 Conversion into Fuels and Chemicals by Catalytic, Photocatalytic and Electrocatalytic Processes. *Energy Environ. Sci.* **2013**, *6*, 3112–3135.

13. Lu, M. *Photocatalysis and Water Purification: From Fundamentals to Recent Applications*; Pichat, P., Ed.; Wiley-VCH Verlag: Weinheim, Germany, 2013.

14. Ohtani, B. Photocatalysis A to Z—What We Know and What We Do Not Know in a Scientific Sense. *J. Photochem. Photobiol. C Photochem. Rev.* **2010**, *11*, 157–178.

15. Kalyanasundaram, K.; Grätzel, M. Artificial Photosynthesis: Biomimetic Approaches to Solar Energy Conversion and Storage. *Curr. Opin. Biotechnol.* **2010**, *21*, 298–310.

16. Wang, Y.; Chen, K.S.; Mishler, J.; Cho, S.C.; Adroher, X.C. A review of polymer electrolyte membrane fuel cells: Technology, applications, and needs on fundamental research. *Appl. Energy* **2011**, *88*, 981–1007.

17. Durbin, D.J.; Malardier-Jugroot, C. Review of Hydrogen Storage Techniques for On-Board Vehicle Applications. *Int. J. Hydrogen Energy* **2013**, *38*, 14595–14617.

18. Muradov, N.Z.; Veziroglu, T.N. "Green" Path from Fossil-Based to Hydrogen Economy: An Overview of Carbon-Neutral Technologies. *Int. J. Hydrogen Energy* **2008**, *33*, 6804–6839.

19. Highfield, J.G. The Central Role of Catalysis in a Future Energy Cycle based on Renewable Hydrogen and Carbon Dioxide as Reactive Liquefier. *Trends Phys. Chem.* **1995**, *5*, 91–156.

20. Goeppert, A.; Czaun, M.; Prakash, G.K.S.; Olah, G.A. Air as the Renewable Carbon Source of the Future: An Overview of CO_2 Capture from the Atmosphere. *Energy Environ. Sci.* **2012**, *5*, 7833–7853.

21. Lackner, K.S. The Thermodynamics of Direct Air Capture of Carbon Dioxide. *Energy* **2013**, *50*, 38–46.

22. Alonso, D.M.; Bond, J.Q.; Dumesic, J.A. Catalytic Conversion of Biomass to Biofuels. *Green Chem.* **2010**, *12*, 1493–1513.

23. Olah, G.A.; Goeppert, A.; Prakash, G.K.S. *Beyond Oil and Gas: The Methanol Economy*, 1st ed.; Wiley-VCH: Weinheim, Germany, 2006.

24. Goeppert, A.; Czaun, M.; Jones, J.P.; Prakash, G.K.S.; Olah, G.A. Recycling of Carbon Dioxide to Methanol and Derived Products—Closing the Loop. *Chem. Soc. Rev.* **2014**, *43*, 7995–8048.

25. Galindo Cifre, P.; Badr, O. Renewable Hydrogen Utilisation for the Production of Methanol. *Energy Convers. Manag.* **2007**, *48*, 519–527.

26. Wang, W.; Wang, S.; Ma, X.; Gong, J. Recent Advances in Catalytic Hydrogenation of Carbon Dioxide. *Chem. Soc. Rev.* **2011**, *40*, 3703–3727.

27. Velu, S.; Gangwal, S.K. A Review of Recent Literature to Search for an Efficient Catalytic Process for the Conversion of Syngas to Ethanol. *Energy Fuels* **2008**, *22*, 814–839.

28. Palo, D.R.; Dagle, R.A.; Holladay, J.D. Methanol Steam Reforming for Hydrogen Production. *Chem. Rev.* **2007**, *107*, 3992–4021.

29. Huang, L.; Choong, C.; Chen, L.W.; Wang, Z.; Zhong, Z.Y.; Campos-Cuerva, C.; Lin, J.Y. Monometallic Carbonyl-Derived CeO_2-Supported Rh and Co Bicomponent Catalysts for CO-Free, High-Yield H_2 Generation from Low-Temperature Ethanol Steam-Reforming. *ChemCatChem* **2013**, *5*, 220–234.

30. Mattos, L.V.; Jacobs, G.; Davis, B.H.; Noronha, F.B. Production of Hydrogen from Ethanol: Review of Reaction Mechanism and Catalyst Deactivation. *Chem. Rev.* **2012**, *112*, 4094–4123.

31. De la Piscina, P.R.; Homs, N. The Use of Biofuels to Produce Hydrogen (Reformation Processes). *Chem. Soc. Rev.* **2008**, *37*, 2459–2467.

32. De Rogatis, L.; Fornasiero, P.; Catalyst Design for Reforming of Oxygenates. In *Catalysis for Sustainable Energy Production*; Barbaro, P., Bianchini, C., Eds.; Wiley-VCH Verlag GmbH: Weinheim, Germany, 2009; Chapter 6, pp. 173–232.

33. Tran, N.H.; Kannangara, G.S.K. Conversion of Glycerol to Hydrogen Rich Gas. *Chem. Soc. Rev.* **2013**, *42*, 9454–9479.

34. Chheda, J.N.; Huber, G.W.; Dumesic, J.A. Liquid-Phase Catalytic Processing of Biomass-Derived Oxygenated Hydrocarbons to Fuels and Chemicals. *Angew. Chem. Int. Ed.* **2007**, *46*, 7164–7183.

35. Bowker, M. Photocatalytic Hydrogen Production and Oxygenate Photoreforming. *Catal. Lett.* **2012**, *142*, 923–929.

36. Jiang, Z.; Xiao, T.; Kuznetsov, V.L.; Edwards, P.P. Turning Carbon Dioxide into Fuel. *Philos. Trans. R. Soc. A* **2010**, *368*, 3343–3364.

37. Li, C.; Domen, K.; Hutchings, G.J. Special Issue on Photocatalysis and Photoelectrolysis. *J. Catal.* **2014**, *310*, 1–108.

38. Dasgupta, S.; Brunschwig, B.S.; Winkler, J.R.; Gray, H.B. Solar Fuels Editorial. *Chem. Soc. Rev.* **2013**, *42*, 2205–2472.

39. Qu, Y.; Duan, X. Progress, Challenge and Perspective of Heterogeneous Photocatalysts. *Chem. Soc. Rev.* **2013**, *42*, 2568–2580.

40. Sun, H.; Wang, S. Research Advances in the Synthesis of Nanocarbon-based Photocatalysts and Their Applications for Photocatalytic Conversion of Carbon Dioxide to Hydrocarbon Fuels. *Energy Fuels* **2014**, *28*, 22–36.

41. Li, K.; An, X.; Park, K.H.; Khraisheh, M.; Tang, J. A Critical Review of CO_2 Photoconversion: Catalysts and Reactors. *Catal. Today* **2014**, *224*, 3–12.

42. Protti, S.; Albini, A.; Serpone, N. Photocatalytic Generation of Solar Fuels from the Reduction of H_2O and CO_2: A Look at the Patent Literature. *Phys. Chem. Chem. Phys.* **2014**, *16*, 19790–19827.

43. Habisreutinger, S.N.; Schmidt-Mende, L.; Stolarczyk, J.K. Photocatalytic Reduction of CO_2 on TiO_2 and Other Semiconductors. *Angew. Chem. Int. Ed.* **2013**, *52*, 7372–7408.

44. Izumi, Y. Recent Advances in the Photocatalytic Conversion of Carbon Dioxide to Fuels with Water and/or Hydrogen Using Solar Energy and Beyond. *Coord. Chem. Rev.* **2013**, *257*, 171–186.

45. Dhakshinamoorthy, A.; Navalon, S.; Corma, A.; Garcia, H. Photocatalytic CO_2 Reduction by TiO_2 and Related Titanium Containing Solids. *Energy Environ. Sci.* **2012**, *5*, 9217–9233.

46. Nocera, D.G. The Artificial Leaf. *Acc. Chem. Res.* **2012**, *45*, 767–776.

47. Hoffmann, M.R.; Moss, J.A.; Baum, M.M. Artificial Photosynthesis: Semiconductor Photocatalytic Fixation of CO_2 to Afford Higher Organic Compounds. *Dalton Trans.* **2011**, *40*, 5151–5158.

48. Ismail, A.A.; Bahnemann, D.W. Photochemical Splitting of Water for Hydrogen Production by Photocatalysis: A Review. *Sol. Energy Mater. Sol. Cells* **2014**, *128*, 85–101.

49. Hisatomi, T.; Kubota, J.; Domen, K. Recent Advances in Semiconductors for Photocatalytic and Photoelectrochemical Water Splitting. *Chem. Soc. Rev.* **2014**, *43*, 7520–7535.

50. Maeda, K. Z-Scheme Water Splitting Using Two Different Semiconductor Photocatalysts. *ACS Catal.* **2013**, *3*, 1486–1503.

51. Yang, J.; Wang, D.; Han, H.; Li, C. Roles of Cocatalysts in Photocatalysis and Photoelectrocatalysis. *Acc. Chem. Res.* **2013**, *46*, 1900–1909.

52. Horiuchi, Y.; Toyao, T.; Takeuchi, M.; Matsuoka, M.; Anpo, M. Recent Advances in Visible-Light-Responsive Photocatalysts for Hydrogen Production and Solar Energy Conversion—From Semiconducting TiO_2 to MOF/PCP Photocatalysts. *Phys. Chem. Chem. Phys.* **2013**, *15*, 13243–13253.

53. Yang, J.; Yan, H.; Zong, X.; Wen, F.; Liu, M.; Li, C. Roles of Cocatalysts in Semiconductor-Based Photocatalytic Hydrogen Production. *Philos. Trans. R. Soc. A* **2013**, *371*, 20110430.

54. Cargnello, M.; Diroll, B.T. Tailoring Photocatalytic Nanostructures for Sustainable Hydrogen Production. *Nanoscale* **2014**, *6*, 97–105.

55. Cargnello, M.; Gasparotto, A.; Gombac, V.; Montini, T.; Barreca, D.; Fornasiero, P. Photocatalytic H_2 and Added-Value By-Products—The Role of Metal Oxide Systems in Their Synthesis from Oxygenates. *Eur. J. Inorg. Chem.* **2011**, *28*, 4309–4323.

56. Navarro, R.M.; Sánchez-Sánchez, M.C.; Alvarez-Galvan, M.C.; del Valle, F.; Fierro, J.L.G. Hydrogen Production from Renewable Sources: Biomass and Photocatalytic Opportunities. *Energy Environ. Sci.* **2009**, *2*, 35–54.

57. Smestad, G.P.; Steinfeld, A. Review: Photochemical and Thermochemical Production of Solar Fuels from H_2O and CO_2 using Metal Oxide Catalysts. *Ind. Eng. Chem. Res.* **2012**, *51*, 11828–11840.

58. McKone, J.R.; Lewis, N.S.; Gray, H.B. Will Solar-Driven Water-Splitting Devices See the Light of Day? *Chem. Mater.* **2013**, *26*, 407–414.

59. Nielander, A.C.; Shaner, M.R.; Papadantonakis, K.M.; Francis, S.A.; Lewis, N.S. A Taxonomy for Solar Fuels Generators. *Energy Environ. Sci.* **2015**, *8*, 16–25.

60. Pinaud, B.A.; Benck, J.D.; Seitz, L.C.; Forman, A.J.; Chen, Z.; Deutsch, T.G.; James, B.D.; Baum, G.N.; Ardo, S.; Wang, H.; *et al.* Technical and Economic Feasibility of Centralized Facilities for Solar Hydrogen Production via Photocatalysis and Photoelectrochemistry. *Energy Environ. Sci.* **2013**, *6*, 1983–2002.

61. Brillet, J.; Yum, J.H.; Cornuz, M.; Hisatomi, T.; Solarska, R.; Augustynski, J.; Graetzel, M.; Sivula, K. Highly Efficient Water Splitting by a Dual-Absorber Tandem Cell. *Nat. Photon.* **2012**, *6*, 824–828.

62. Reece, S.Y.; Hamel, J.A.; Sung, K.; Jarvi, T.D.; Esswein, A.J.; Pijpers, J.J.H.; Nocera, D.G. Wireless Solar Water Splitting Using Silicon-Based Semiconductors and Earth-Abundant Catalysts. *Science* **2011**, *334*, 645–648.

63. McKone, J.R.; Marinescu, S.C.; Brunschwig, B.S.; Winkler, J.R.; Gray, H.B. Earth-Abundant Hydrogen Evolution Electrocatalysts. *Chem. Sci.* **2014**, *5*, 865–878.

64. Bozic-Weber, B.; Constable, E.C.; Housecroft, C.E. Light Harvesting with Earth-Abundant d-Block Metals: Development of Sensitizers in Dye-Sensitized Solar Cells (DSCs). *Coord. Chem. Rev.* **2013**, *257*, 3089–3106.

65. Thoi, V.S.; Sun, Y.J.; Long, J.R.; Chang, C.J. Complexes of Earth-Abundant Metals for Catalytic Electrochemical Hydrogen Generation under Aqueous Conditions. Chem. Soc. Rev. **2013**, *42*, 2388–2400.

66. Tran, P.D.; Pramana, S.S.; Kale, V.S.; Nguyen, M.; Chiam, S.Y.; Batabyal, S.K.; Wong, L.H.; Barber, J.; Loo, J. Novel Assembly of an MoS_2 Electrocatalyst onto a Silicon Nanowire Array Electrode to Construct a Photocathode Composed of Elements Abundant on the Earth for Hydrogen Generation. *Chem. Eur. J.* **2012**, *18*, 13994–13999.

67. Du, P.W.; Eisenberg, R. Catalysts Made of Earth-Abundant Elements (Co, Ni, Fe) for Water Splitting: Recent Progress and Future Challenges. *Energy Environ. Sci.* **2012**, *5*, 6012–6021.

68. Vesborg, P.C.K.; Jaramillo, T.F. Addressing the Terawatt Challenge: Scalability in the Supply of Chemical Elements for Renewable Energy. *RSC Adv.* **2012**, *2*, 7933–7947.

69. Bolton, J.R.: Strickler, S.J; Connolly, J.S. Limiting and Realizable Efficiencies of Solar Photolysis of Water. *Nature* **1985**, *316*, 495–500.

70. Archer, M.D; Bolton, J.R. Requirements for Ideal Performance of Photochemical and Photovoltaic Solar Energy Converters. *J. Phys. Chem.* **1990**, *94*, 8028–8036.

71. Bolton, J.R. Solar Photoproduction of Hydrogen: A Review. *Sol. Energy* **1996**, *57*, 37–50.

72. Smestad, G.P; Krebs, F.C; Lampert, C.M; Granqvist, C.G.; Chopra, K.L.; Mathew, X.; Takakura, H. Reporting Solar Cell Efficiencies in Solar Energy Materials and Solar Cells. *Sol. Energy Mater. Sol. Cells* **2008**, *92*, 371–373.

73. Emeline, A.V.; Kuznetsov, V.N.; Ryabchuk, V.K.; Serpone, N. On the Way to the Creation of Next Generation Photoactive Materials. *Environ. Sci. Pollut. Res.* **2012**, *19*, 3666–3675.

74. Serpone, N.; Emeline, A.V. Semiconductor Photocatalysis—Past, Present, and Future Outlook. *J. Phys. Chem. Lett.* **2012**, *3*, 673–677.

75. Bard, A.J. Photoelectrochemistry and Heterogeneous Photocatalysis at Semiconductors. *J. Photochem.* **1979**, *10*, 59–75.

76. Grätzel, M. Photoelectrochemical Cells. *Nature* **2001**, *414*, 338–344.

77. Luo, J.; Im, J.H.; Mayer, M.T.; Schrieier, M.; Nazeeruddin, K.; Park, N.G.; Tilley, S.D.; Fan, H.J.; Grätzel, M. Water Photolysis at 12.3% Efficiency via Perovskite Photovoltaics and Earth-Abundant Catalysts. *Science* **2014**, *345*, 1593–1596.

78. Cox, C.R.; Lee, J.Z.; Nocera, D.G.; Buonassisi, T. Ten-Percent Solar-to-Fuel Conversion with Nonprecious Materials. *Proc. Natl. Acad. Sci. USA* **2014**, *111*, 14057–14061.

79. Rongé, J.; Bosserez, T.; Martel, D.; Nervi, C.; Boarino, L.; Taulelle, F.; Decher, G.; Bordiga, S.; Martens, J.A. Monolithic Cells for Solar Fuels. *Chem. Soc. Rev.* **2014**, *43*, 7963–7981.

80. Fujishima, A.K.; Honda, K. Electrochemical Photolysis of Water at a Semiconductor Electrode. *Nature* **1972**, *238*, 37–38.

81. Borgarello, E.; Kiwi, J.; Pelizzetti, E.; Visca, M.; Graetzel, M. Photochemical Cleavage of Water by Photocatalysis. *Nature* **1981**, *289*, 158–160.

82. Wang, Z.; Liu, Y.; Huang, B.; Dai, Y.; Lou, Z.; Wang, G.; Zhang, X.; Qin, X. Progress on Extending the Light Absorption Spectra of Photocatalysts. *Phys. Chem. Chem. Phys.* **2014**, *16*, 2758–2774.

83. Varghese, O.K.; Grimes, C.A. Appropriate Strategies for Determining the Photoconversion Efficiency of Water Photoelectrolysis Cells: A Review with Examples using Titania Nanotube Array Photoanodes. *Sol. Energy Mater. Sol. Cells* **2008**, *92*, 374–384.

84. Grätzel, M. Conversion of Sunlight to Electric Power by Nanocrystalline Dye-Sensitized Solar Cells. *J. Photochem. Photobiol. A* **2004**, *164*, 3–14.

85. Mathew, S.; Yella, A.; Gao, P.; Humphry-Baker, R.; Churchod, B.F.E.; Ashari-Astani, N.; Tavernelli, I.; Rothlisberger, U.; Nazeeruddin, M.K.; Grätzel, M. Dye-Sensitized Solar Cells with 13% Efficiency Achieved through the Molecular Engineering of Porphyrin Sensitizers. *Nat. Chem.* **2014**, *6*, 242–247.

86. Hardin, B.E.; Snaith, H.J.; McGehee, M.D. The Renaissance of Dye-Sensitized Solar Cells. *Nat. Photon.* **2012**, *6*, 162–169.

87. Li, L.L.; Diau, E.W. Porphyrin-Sensitized Solar Cells. *Chem. Soc. Rev.* **2013**, *42*, 291–304.

88. Balasingam, S.K.; Lee, M.; Kang, M.G.; Jun, Y. Improvement of Dye-Sensitized Solar Cells Toward the Broader Light Harvesting of the Solar Spectrum. *Chem. Commun.* **2013**, *49*, 1471–1487.

89. Macor, L.; Fungo, F.; Tempesti, T.; Durantini, E.N.; Otero, L.; Barea, E.M.; Fabregat-Santiago, F.; Bisquert, J. Near-IR Sensitization of Wide Band Gap Oxide Semiconductor by Axially Anchored Si-Naphthalocyanines. *Energy Environ. Sci.* **2009**, *2*, 529–534.

90. Ono, T.; Yamaguchi, T.; Arakawa, H. Study on Dye-Sensitized Solar Cell using Novel Infrared Dye. *Sol. Energy Mater. Sol. Cells* **2009**, *93*, 831–835.

91. Altobello, S.; Argazzi, R.; Caramori, S.; Contado, C.; da Fré, S.; Rubino, P.; Chon, C.; Larramona, G.; Bignozzi, C.A. Sensitization of Nanocrystalline TiO_2 with Black Absorbers based on Os and Ru Polypyridine Complexes. *J. Am. Chem. Soc.* **2005**, *127*, 15342–15343.

92. Persson, P.; Bergstrom, R.; Lunell, S. Quantum Chemical Study of Photoinjection Processes in Dye-sensitized TiO_2 Nanoparticles. *J. Phys. Chem. B* **2000**, *104*, 10348–10351.

93. Ooyama, Y.; Yamada, T.; Fujita, T.; Harima, Y.; Ohshita, J. Development of D-π-Cat Fluorescent Dyes with a Catechol Group for Dye-Sensitized Solar Cells based on Dye-to-TiO_2 Charge Transfer. *J. Mater. Chem. A* **2014**, *2*, 8500–8511.

94. Pratik, S.M.; Datta, A. Computational Design of Concomitant Type-I and Type-II Porphyrin Sensitized Solar Cells. *Phys. Chem. Chem. Phys.* **2013**, *15*, 18471–18481.

95. Zhang, G.; Kim, G.; Choi, W. Visible-light driven photocatalysis mediated via ligand-to-metal charge transfer (LMCT): An alternative approach to solar activation of Titania. *Energy Environ. Sci.* **2014**, *7*, 954–966.

96. Cheng, Y.H.; Subramaniam, V.P.; Gong, D.G.; Tang, Y.; Highfield, J.; Pehkonen, S.O.; Pichat, P.; Schreyer, M.K.; Chen, Z. Nitrogen-Sensitized Dual Phase Titanate/Titania for Visible-Light Driven Phenol Degradation. *J. Solid State Chem.* **2012**, *196*, 518–527.

97. Grätzel, M. The Light and Shade of Perovskite Solar Cells. *Nat. Mater.* **2014**, *13*, 838–842.

98. Green, M.A.; Ho-Baillie, A.; Snaith, H.J. The Emergence of Perovskite Solar Cells. *Nat. Photon.* **2014**, *8*, 506–514.

99. Gao, P.; Grätzel, M.; Nazeeruddin, M.K. Organohalide lead perovskites for photovoltaic applications. *Energy Environ. Sci.* **2014**, *7*, 2448–2463.

100. Kazim, S.; Nazeeruddin, M.K.; Grätzel, M.; Ahmad, S. Perovskite as Light Harvester: A Game Changer in Photovoltaics. *Angew. Chem. Int. Ed.* **2014**, *53*, 2812–2824.

101. Nowotny, M.K.; Sheppard, L.R.; Bak, T.; Nowotny, J. Defect Chemistry of Titanium Dioxide. Application of Defect Engineering in Processing of TiO_2-Based Photocatalysts. *J. Phys. Chem. C* **2008**, *112*, 5275–5300.

102. Mizushima, K.; Tanaka, M.; Asai, A.; Iiga, S.; Goodenough, J.B. Impurity levels of Iron-Group Ions in TiO_2 (II). *J. Phys. Chem. Solids* **1979**, *40*, 1129–1140.

103. Choi, W.; Termin, A.; Hoffmann, M.R. The Role of Metal Ion Dopants in Quantum-Sized TiO_2: Correlation between Photoreactivity and Charge Carrier Recombination Dynamics. *J. Phys. Chem.* **1994**, *98*, 13669–13679.

104. Moser, J.; Grätzel, M.; Gallay, R. Inhibition of Electron-Hole Recombination in Substitution-ally Doped Colloidal Semiconductor Crystallites. *Helv. Chim. Acta* **1987**, *70*, 1596–1604.

105. Choi, J.; Park, H.; Hoffmann, M.R. Effects of Single Metal-Ion Doping on the Visible-Light Photoreactivity of TiO_2. *J. Phys. Chem. C* **2010**, *114*, 783–792.

106. Kernazhitsky, L.; Shymanovska, V.; Gavrilko, T.; Naumov, V.; Kshnyakin, V.; Khalyavka, T. A Comparative Study of Optical Absorption and Photocatalytic Properties of Nanocrystalline Single-Phase Anatase and Rutile TiO$_2$ Doped with Transition Metal Cations. *J. Solid State Chem.* **2013**, *198*, 511–519.

107. Maeda, K. (Oxy)nitrides with d°-Electronic Configuration as Photocatalysts and Photoanodes that Operate under a Wide Range of Visible Light for Overall Water Splitting. *Phys. Chem. Chem. Phys.* **2013**, *15*, 10537–10548.

108. Teoh, W.Y.; Scott, J.A.; Amal, R. Progress in Heterogeneous Photocatalysis: From Classical Radical Chemistry to Engineering Nanomaterials and Solar Reactors. *J. Phys. Chem. Lett.* **2012**, *3*, 629–639.

109. Chen, X.B.; Liu, L.; Yu, P.Y.; Mao, S.S. Increasing Solar Absorption for Photocatalysis with Black Hydrogenated Titanium Dioxide Nanocrystals. *Science* **2011**, *331*, 746–750.

110. Pesci, F.M.; Wang, G.; Klug, D.R.; Li, Y.; Cowan, A.J. Efficient Suppression of Electron-Hole Recombination in Oxygen-Deficient Hydrogen-Treated TiO$_2$ Nanowires for Photoelectro-chemical Water Splitting. *J. Phys. Chem. C* **2013**, *117*, 25837–25844.

111. Liu, L.; Yu, P.Y.; Chen, X.; Mao, S.S.; Shen, D.Z. Hydrogenation and Disorder in Engineered Black TiO$_2$. *Phys. Rev. Lett.* **2013**, *111*, 065505-(1–5).

112. Su, J.; Zou, X.; Chen, J.S. Self-Modification of Titanium Dioxide Materials by Ti^{3+} and/or Oxygen Vacancies: New Insights into Defect Chemistry of Metal Oxides. *RSC Adv.* **2014**, *4*, 13979–13988.

113. Fresno, F.; Portela, R.; Suárez, S.; Coronado, J.M. Photocatalytic Materials: Recent Achievements and Near Future Trends. *J. Mater. Chem. A* **2014**, *2*, 2863–2884.

114. Asahi, R.; Morikawa, T.; Ohwaki, T.; Aoki, K.; Taga, Y. Visible-Light Photocatalysis in Nitrogen-Doped Titanium Oxides. *Science* **2001**, *293*, 269–271.

115. Khan, S.U.M.; al-Shahry, M.; Ingler, W.B. Efficient Photochemical Water Splitting by a Chemically Modified n-TiO$_2$. *Science* **2002**, *297*, 2243–2245.

116. Diwald, O.; Thompson, T.L.; Goralski, E.G.; Walck, S.D.; Yates, J.T. The Effect of Nitrogen Ion Implantation on the Photoactivity of TiO$_2$ Rutile Single Crystals. *J. Phys. Chem. B* **2004**, *108*, 52–57.

117. Di Valentin, C.; Pacchioni, G.; Selloni, A. Origin of the Different Photoactivity of N-Doped Anatase and Rutile TiO$_2$. *Phys. Rev. B* **2004**, *70*, 085116.

118. Nakamura, R.; Tanaka, T.; Nakato, Y. Mechanism for Visible Light Responses in Anodic Photocurrents at N-Doped TiO$_2$ Film Electrodes. *J. Phys. Chem. B* **2004**, *108*, 10617–10620.

119. Choi, Y.; Umebayashi, T.; Yoshikawa, M. Fabrication and Characterization of C-Doped Anatase TiO$_2$ Photocatalysts. *J. Mater. Sci.* **2004**, *39*, 1837–1839.

120. Yu, J.C.; Yu, J.G.; Ho, W.K.; Jiang, Z.T.; Zhang, L.Z. Effects of F-Doping on the Photocatalytic Activity and Microstructures of Nanocrystalline TiO$_2$ Powders. *Chem. Mater.* **2002**, *14*, 3808–3816.

121. Ohno, T.; Akiyoshi, M.; Umebayashi, T.; Asai, K.; Mitsui, T.; Matsumura, M. Preparation of S-Doped TiO$_2$ Photocatalysts and Their Photocatalytic Activities under Visible Light. *Appl. Catal. A Gen.* **2004**, *265*, 115–121.

122. Lin, L.; Lin, W.; Zhu, Y.X.; Zhao, B.Y.; Xie, Y.C. Phosphor-Doped Titania—A Novel Photocatalyst Active in Visible Light. *Chem. Lett.* **2005**, *34*, 284–285.

123. Yin, S.; Ihara, K.; Aita, Y.; Komatsu, M.; Sato, T. Visible-Light Induced Photocatalytic Activity of $TiO_{(2-x)}A_{(y)}$ (A = N, S) Prepared by Precipitation Route. *J. Photochem. Photobiol. A Chem.* **2006**, *179*, 105–114.

124. Chen, D.M.; Jiang, Z.Y.; Geng, J.Q.; Wang, Q.; Yang, D. Carbon and Nitrogen Co-Doped TiO_2 with Enhanced Visible-Light Photocatalytic Activity. *Ind. Eng. Chem. Res.* **2007**, *46*, 2741–2746.

125. In, S.; Orlov, A.; Berg, R.; Garcia, F.; Pedrosa-Jimenez, S.; Tikhov, M.S.; Wright, D.S.; Lambert, R.M. Effective Visible Light-Activated B-Doped and B,N-Co-Doped TiO_2 Photocatalysts. *J. Am. Chem. Soc.* **2007**, *129*, 13790–13791.

126. Di Valentin, C.; Finazzi, E.; Pacchioni, G.; Selloni, A.; Livraghi, S.; Paganini, M.C.; Giamello, E.; N-Doped TiO_2: Theory and Experiment. *Chem. Phys.* **2007**, *339*, 44–56.

127. Harb, M.; Sautet, P.; Raybaud, P. Origin of the Enhanced Visible-Light Absorption in N-Doped Bulk Anatase TiO_2 from First-Principles Calculations. *J. Phys. Chem. C* **2011**, *115*, 19394–19404.

128. Marschall, R.; Wang, L.Z. Non-Metal Doping of Transition Metal Oxides for Visible-Light Photocatalysis. *Catal. Today* **2014**, *225*, 111–135.

129. Gomathi Devi, L.; Kavitha, R. Review on Modified N-TiO_2 for Green Energy Applications under UV/Visible Light: Selected Results and Reaction Mechanisms. *RSC Adv.* **2014**, *4*, 28265–28299.

130. Asahi, R.; Morikawa, T.; Irie, H.; Ohwaki, T. Nitrogen-Doped Titanium Dioxide as Visible-Light-Sensitive Photocatalyst: Designs, Developments, and Prospects. *Chem. Rev.* **2014**, *114*, 9824–9852.

131. Mrowetz, M.; Balcerski, W.; Colussi, A.J.; Hoffmann, M.R. Oxidative Power of Nitrogen-Doped TiO_2 Photocatalysts under Visible Illumination. *J. Phys. Chem. B* **2004**, *108*, 17269–17273.

132. Lindgren, T.; Lu, J.; Hoel, A.; Granqvist, C.G.; Torres, G.R.; Lindquist, S.E. Photoelectrochemical Study of Sputtered Nitrogen-Doped Titanium Dioxide Thin Films in Aqueous Electrolyte. *Sol. Energy Mater. Sol. Cells* **2004**, *84*, 145–157.

133. Tang, J.; Cowan, A.J.; Durrant, J.R.; Klug, D.R. Mechanism of O_2 Production from Water Splitting: Nature of Charge Carriers in Nitrogen Doped Nanocrystalline TiO_2 Films and Factors Limiting O_2 Production. *J. Phys. Chem. C* **2011**, *115*, 3143–3150.

134. Di Valentin, C.; Pacchioni, G. Trends in Non-Metal Doping of Anatase TiO_2: B, C, N and F. *Catal. Today* **2013**, *206*, 12–18.

135. Yin, W.J.; Tang, H.W.; Wei, S.H.; Al-Jassim, M.M.; Turner, J.; Yan, Y.F. Band Structure Engineering of Semiconductors for Enhanced Photoelectrochemical Water Splitting: The Case of TiO_2. *Phys. Rev. B* **2010**, *82*, 045106-(1–6).

136. Zhu, W.; Qiu, X.; Iancu, V.; Chen, X.Q.; Pan, H.; Wang, W.; Dimitrijevic, N.M.; Rajh, T.; Meyer, H.M., III; Paranthaman, M.P.; *et al.* Band Gap Narrowing of Titanium Oxide Semiconductors by Non-Compensated Anion-Cation Codoping for Enhanced Visible Light Photoactivity. *Phys. Rev. Lett.* **2009**, *103*, 226401:1–226401:4.

137. Czoska, A.M.; Livraghi, S.; Paganini, M.C.; Giamello, E.; di Valentin, C.; Pacchioni, G. The Nitrogen-Boron Paramagnetic Center in Visible Light-Sensitized N-B-Codoped TiO_2. Experimental and Theoretical Characterization. *Phys. Chem. Chem. Phys.* **2011**, *13*, 136–143.

138. Balcerski, W.; Ryu, S.Y.; Hoffmann, M.R. Visible-Light Photoactivity of Nitrogen-Doped TiO_2: Photo-Oxidation of HCO_2H to CO_2 and H_2O. *J. Phys. Chem. C* **2007**, *111*, 15357–15362.

139. Mitoraj, D.; Kisch, H. The Nature of Nitrogen-Modified Titanium Dioxide Photocatalysts Active in Visible Light. *Angew. Chem. Int. Ed.* **2008**, *47*, 9975–9978.

140. Mitoraj, D.; Kisch, H. On the Mechanism of Urea-Induced Titania Modification. *Chem. Eur. J.* **2010**, *16*, 261–269.

141. Ang, T.P.; Chan, Y.M. Comparison of the Melon Nanocomposites in Structural Properties and Photocatalytic Activities. *J. Phys. Chem. C* **2011**, *115*, 15965–15972.

142. Gong, D.; Highfield, J.G.; Zhong, E.N.; Tang, Y.; Ho, W.C.J.; Tay, Q.; Chen, Z. Poly Tri-s-triazines as Visible Light Sensitizers in Titania-Based Composite Photocatalysts: Promotion of Melon Development from Urea over Acid Titanates. *ACS Sustain. Chem. Eng.* **2014**, *2*, 149–157.

143. Wang, Y.; Wang, X.; Antonietti, M. Polymeric Graphitic Carbon Nitride as a Heterogeneous Organocatalyst: From Photochemistry to Multipurpose Catalysis to Sustainable Chemistry. *Angew. Chem. Int. Ed.* **2012**, *51*, 68–89.

144. Gole, J.L.; Stout, J.D.; Burda, C.; Lou, Y.; Chen, X. Highly Efficient Formation of Visible-Light Tunable $TiO_{2-x}N_x$ Photocatalysts and Their Transformation at the Nanoscale. *J. Phys. Chem. B* **2004**, *108*, 1230–1240.

145. Cantau, C.; Pigot, T.; Dupin, J.C.; Lacombe, S. N-Doped TiO_2 by Low Temperature Synthesis: Stability, Photo-Reactivity, and Singlet Oxygen Formation in the Visible Range. *J. Photochem. Photobiol. A Chem.* **2010**, *216*, 201–208.

146. Spadavecchia, F.; Cappelletti, G.; Ardizzone, S.; Bianchi, C.L.; Cappelli, S.; Oliva, C.; Scardi, P.; Leoni, M.; Fermo, P. Solar Photoactivity of Nano-N-TiO_2 from Tertiary Amine: Role of Defects and Paramagnetic Species. *Appl. Catal. B: Environ.* **2010**, *96*, 314–322.

147. Cheng, Y.H.; Huang, Y.; Kanhere, P.D.; Subramaniam, V.P.; Gong, D.; Zhang, S.; Highfield, J.; Schreyer, M.K.; Chen, Z. Dual-Phase Titanate/Anatase with Nitrogen Doping for Enhanced Degradation of Organic Dye under Visible Light. *Chem. Eur. J.* **2011**, *17*, 2575–2578.

148. Zhao, Y.; Qiu, X.; Burda, C. The Effects of Sintering on the Photocatalytic Activity of N-doped TiO_2 Nanoparticles. *Chem. Mater.* **2008**, *20*, 2629–2636.

149. Tian, Y.; Tatsuma, T. Mechanisms and Applications of Plasmon-Induced Charge Separation at TiO_2 Films Loaded with Gold Nanoparticles. *J. Am. Chem. Soc.* **2005**, *127*, 7632–7637.

150. Kowalska, E.; Mahaney, O.O.P.; Abe, R.; Ohtani, B. Visible-Light-Induced Photocatalysis through Surface Plasmon Excitation of Gold on Titania Surfaces. *Phys. Chem. Chem. Phys.* **2010**, *12*, 2344–2355.

151. Gong, D.; Ho, W.C.J.; Tang, Y.; Tay, Q.; Lai, Y.; Highfield, J.G.; Chen, Z. Silver-Decorated Titanate/Titania Nanostructures for Efficient Solar Driven Photocatalysis. *J. Solid State Chem.* **2012**, *189*, 117–122.

152. Sellapan, R.; Nielsen, M.G.; Gonzalez-Posada, F.; Vesborg, P.C.K.; Chorkendorff, I.; Chakarov, D. Effects of Plasmon Excitation on Photocatalytic Activity of Ag/TiO_2 and Au/TiO_2 Nanocomposites. *J. Catal.* **2013**, *307*, 214–221.

153. Mubeen, S.; Lee, J.; Singh, N.; Krämer, S.; Stucky, G.D.; Moskovits, M. An autonomous photosynthetic device in which all charge carriers derive from surface Plasmons. *Nat. Nanotechnol.* **2013**, *8*, 247–251.

154. Ma, X.; Dai, Y.; Yu, L.; Huang, B. New Basic Insights into the Low Hot Electron Injection Efficiency of Gold-Nanoparticle-Photosensitized Titanium Dioxide. *ACS Appl. Mater. Interfaces* **2014**, *6*, 12388–12394.

155. Clavero, C. Plasmon-Induced Hot-Electron Generation at Nanoparticle/Metal-Oxide Interfaces for Photovoltaic and Photocatalytic Devices. *Nat. Photon.* **2014**, *8*, 95–103.

156. Xu, H.; Ouyang, S.; Liu, L.; Reunchan, P.; Umezawa, N.; Ye, J. Recent Advances in TiO_2-based Photocatalysis. *J. Mater. Chem. A* **2014**, *2*, 12642–12661.

157. Warren, S.C.; Thimsen, E. Plasmonic Solar Water Splitting. *Energy Environ. Sci.* **2012**, *5*, 5133–5146.

158. Linic, S.; Christopher, P.; Ingram, D.B. Plasmonic-Metal Nanostructures for Efficient Conversion of Solar to Chemical Energy. *Nat. Mater.* **2011**, *10*, 911–921.

159. Zhou, X.M.; Liu, G.; Yu, J.G.; Fan, W.H. Surface Plasmon Resonance-Mediated Photocatalysis by Noble Metal-Based Composites under Visible Light. *J. Mater. Chem.* **2012**, *22*, 21337–21354.

160. DuChene, J.S.; Sweeny, B.C.; Johnston-Peck, A.C.; Su, D.; Stach, E.A.; Wei, W.D. Prolonged Hot Electron Dynamics in Plasmonic-Metal/Semiconductor Heterostructures with Implications for Solar Photocatalysis. *Angew. Chem. Int. Ed.* **2014**, *53*, 7887–7891.

161. Ingram, D.B.; Christopher, P.; Bauer, J.L.; Linic, S. Predictive Model for the Design of Plasmonic Metal/Semiconductor Composite Photocatalysts. *ACS Catal.* **2011**, *1*, 1441–1447.

162. Thimsen, E.; le Formal, F.; Grätzel, M.; Warren, S.C. Influence of Plasmonic Au Nanoparticles on the Photoactivity of Fe_2O_3 Electrodes for Water Splitting. *Nano Lett.* **2011**, *11*, 35–43.

163. Logar, M.; Bračko, I.; Potočnik, A.; Jančar, B. Cu and CuO/Titanate Nanobelt Based Network Assemblies for Enhanced Visible Light Photocatalysis. *Langmuir* **2014**, *30*, 4852–4862.

164. Nogawa, T.; Isobe, T.; Matsushita, S.; Nakajima, A. Preparation and Visible-Light Photo-catalytic Activity of Au- and Cu-modified TiO_2 Powders. *Mater. Lett.* **2012**, *82*, 174–177.

165. Kazuma, E.; Yamaguchi, T.; Sakai, N.; Tatsuma, T. Growth Behaviour and Plasmon Resonance Properties of Photocatalytically Deposited Cu Nanoparticles. *Nanoscale* **2011**, *3*, 3641–3645.

166. Knight, M.W.; King, N.S.; Liu, L.; Everitt, H.O.; Nordlander, P.; Halas, N.J. Aluminum for Plasmonics. *ACS Nano* **2014**, *8*, 834–840.

167. Lounis, S.D.; Runnerstrom, E.L.; Llordés, Milliron, D.J. Defect Chemistry and Plasmon Physics of Colloidal Metal Oxide Nanocrystals. *J. Phys. Chem. Lett.* **2014**, *5*, 1564–1574.

168. Henderson, M.A. A Surface Science Perspective on TiO_2 Photocatalysis. *Surf. Sci. Rep.* **2011**, *66*, 185–297.

169. Hernández-Alonso, M.D.; Fresno, F.; Suárez, S.; Coronado, J.M. Development of Alternative Photocatalysts to TiO_2: Challenges and Opportunities. *Energy Environ. Sci.* **2009**, *2*, 1231–1257.

170. Chen, X.; Mao, S.S. Titanium Dioxide Nanomaterials: Synthesis, Properties, Modifications, and Applications. *Chem. Rev.* **2007**, *107*, 2891–2959.

171. Zhang, J.; Xu, Q.; Li, M.; Li, C. Importance of the Relationship between Surface Phases and Photocatalytic Activity of TiO_2. *Angew. Chem. Int. Ed.* **2008**, *47*, 1766–1769.

172. Li, G.; Gray, K.A. The Solid-Solid Interface: Explaining the High and Unique Photocatalytic Reactivity of TiO_2-Based Nanocomposite Materials. *Chem. Phys.* **2007**, *339*, 173–187.

173. Chen, X.; Selloni, A. Introduction: Titanium Dioxide (TiO_2) Nanomaterials. *Chem. Rev.* **2014**, *114*, 9281–9282.

174. Kapilashrami, M.; Zhang, Y.; Liu, Y.S.; Hagfeldt, A.; Guo, J. Probing the Optical Property and Electronic Structure of TiO_2 Nanomaterials for Renewable Energy Applications. *Chem. Rev.* **2014**, *114*, 9662–9707.

175. Schneider, J.; Matsuoka, M.; Takeuchi, M.; Zhang, J.; Horiuchi, Y.; Anpo, M.; Bahnemann, D.W. Understanding TiO_2 Photocatalysis: Mechanisms and Materials. *Chem. Rev.* **2014**, *114*, 9919–9986.

176. Di Iorio, Y.; Aguirre, M.E.; Brusam M.A.; Grela, M.A. Surface Chemistry Determines Electron Storage Capabilities in Alcoholic Sols of Titanium Dioxide Nanoparticles. A Combined FTIR and Room Temperature EPR Investigation. *J. Phys. Chem. C* **2012**, *116*, 9646–9652.

177. Murakami, N.; Mahaney, O.O.P.; Abe, R.; Torimoto, T.; Ohtani, B. Double-Beam Photo-acoustic Spectroscopic Studies on Transient Absorption of Titanium(IV) Oxide Photocatalyst Powders. *J. Phys. Chem. C* **2007**, *111*, 11927–11935.

178. Chen, T.; Feng, Z.; Wu, G.; Shi, J.; Ma, G.; Ying, P.; Li, C. Mechanistic Studies of Photocatalytic Reaction of Methanol for Hydrogen Production on Pt/TiO_2 by In-Situ Fourier-Transform IR and Time-Resolved IR Spectroscopy. *J. Phys. Chem. C* **2007**, *111*, 8005–8014.

179. Yamakata, A.; Ishibashi, T.; Onishi, H. Effects of Water Addition on the Methanol Oxidation on Pt/TiO_2 Photocatalyst Studied by Time-Resolved Infrared Absorption Spectroscopy. *J. Phys. Chem. B* **2003**, *107*, 9820–9823.

180. Szczepankiewicz, S.H.; Moss, J.A.; Hoffmann, M.R. Slow Surface Charge Trapping Kinetics on Irradiated TiO_2. *J. Phys. Chem. B* **2002**, *106*, 2922–2927.

181. Highfield, J.G.; Grätzel, M. Discovery of Reversible Photochromism in Titanium Dioxide using Photoacoustic Spectroscopy: Implications for the Investigation of Light-Induced Charge Separation and Surface Redox Processes in Titanium Dioxide. *J. Phys. Chem.* **1988**, *92*, 464–467.

182. Di Valentin, C.; Selloni, A. Bulk and Surface Polarons in Photoexcited Anatase. *J. Phys. Chem. Lett.* **2011**, *2*, 2223–2228.

183. Peter, L.M.; Wijayantha, K.G.U. Photoelectrochemical Water Splitting at Semiconductor Electrodes: Fundamental Problems and New Perspectives. *ChemPhysChem* **2014**, *15*, 1983–1995.

184. Wang, X.C.; Biechert, S.; Antonietti, M. Polymeric Graphitic Carbon Nitride for Heterogeneous Photocatalysis. *ACS Catal.* **2012**, *2*, 1596–1606.

185. Martin, D.J.; Qiu, K.P.; Shevlin, S.A.; Handoko, A.D.; Chen, X.W.; Guo, Z.X.; Tang, J.W. Highly Efficient Photocatalytic H_2 Evolution from Water Using Visible Light and Structure-Controlled Graphitic Carbon Nitride. *Angew. Chem. Int. Ed.* **2014**, *53*, 9240–9245.

186. Hollmann, D.; Karnahl, M.; Tschierlei, S.; Kailasam, K.; Schneider, M.; Radnik, J.; Grabow, K.; Bentrup, U.; Junge, H.; Beller, M.; *et al.* Structure-Activity Relationships in Bulk Polymeric and Sol-Gel Derived Carbon Nitrides During Photocatalytic Hydrogen Production. *Chem. Mater.* **2014**, *26*, 1727–1733.

187. Wirth, J.; Neumann, R.; Antonietti, M.; Saalfrank, P. Adsorption and Photocatalytic Splitting of Water on Graphitic Carbon Nitride: A First Principles and Semi-Empirical Study. *Phys. Chem. Chem. Phys.* **2014**, *16*, 15917–15926.

188. Maeda, K.; Wang, X.; Nishihara, Y.; Lu, D.; Antonietti, M.; Domen. K. Photocatalytic Activities of Graphitic Carbon Nitride Powder for Water Oxidation and Reduction under Visible Light. *J. Phys. Chem. C* **2009**, *113*, 4940–4947.

189. Wang, X.; Pehkonen, S.O.; Rämö, J.; Väänänen, M.; Highfield, J.G.; Laasonen, K. Experimental and Computational Studies of Nitrogen-Doped Degussa P25 TiO_2: Application to Visible-Light Driven Photo-Oxidation of As (III). *Catal. Sci. Technol.* **2012**, *2*, 784–793.

190. Sivula, K.; le Formal, F.; Grätzel, M. Solar Water Splitting: Progress Using Hematite (α-Fe_2O_3) Photoelectrodes. *ChemSusChem* **2011**, *4*, 432–449.

191. Warren, S.C.; Voitchovsky, K.; Dotan, H.; Leroy, C.M.; Cornuz, M.; Stellacci, F.; Hébert, C.; Rothschild, A.; Grätzel, M. Identifying Champion Nanostructures for Solar Water-Splitting. *Nat. Mater.* **2013**, *12*, 842–849.

192. Shinar, R.; Kennedy, J.H. Photoactivity of Doped α-Fe_2O_3 Electrodes. *Sol. Energy Mater.* **1982**, *6*, 323–335.

193. Dare-Edwards, M.P.; Goodenough, J.B.; Hamnett, A.; Trevellick, P.R. Electrochemistry and Photoelectrochemistry of Iron(III) Oxide. *J. Chem. Soc. Faraday Trans. I* **1983**, *79*, 2027–2041.

194. Itoh, K.; Bockris, J.O. Thin Film Photoelectrochemistry: Iron Oxide. *J. Electrochem. Soc.* **1984**, *131*, 1266–1271.

195. Kennedy, J.H.; Frese, K.W. Photooxidation of Water at α-Fe_2O_3 Electrodes. *J. Electrochem. Soc.* **1978**, *125*, 709–714.

196. Wang, K.X.Z.; Wu, Z.F.; Liu, V.; Brongersma, M.L.; Jaramillo, T.F.; Fan, S.H. Nearly Total Solar Absorption in Ultrathin Nanostructured Iron Oxide for Efficient Photoelectrochemical Water Splitting. *ACS Photon.* **2014**, *1*, 235–240.

197. Wheeler, D.A.; Wang, G.M.; Ling, Y.C.; Li, Y.; Zhang, J.Z. Nanostructured hematite: Synthesis, characterization, charge carrier dynamics, and photoelectrochemical properties. *Energy Environ. Sci.* **2012**, *5*, 6682–6702.

198. Pendlebury, S.R.; Cowan, A.J.; Barroso, M.; Sivula, K.; Ye, J.; Grätzel, M.; Klug, D.R.; Tang, J.; Durrant, J.R. Correlating Long-Lived Photogenerated Hole Populations with Photocurrent Densities in Hematite Water Oxidation Photoanodes. *Energy Environ. Sci.* **2012**, *5*, 6304–6312.

199. Le Formal, F.; Pendlebury, S.R.; Cornuz, M.; Tilley, S.D.; Grätzel, M.; Durrant, J.R. Back Electron-Hole Recombination in Hematite Photoanodes for Water Splitting. *J. Am. Chem. Soc.* **2014**, *136*, 2564–2574.

200. Peter, L.M. Energetics and Kinetics of Light-Driven Oxygen Evolution at Semiconductor Electrodes: The Example of Hematite. *J. Solid State Electrochem.* **2013**, *17*, 315–326.

201. Braun, A.; Sivula, K.; Bora, D.K.; Zhu, J.; Zhang, L.; Grätzel, M.; Guo, J.; Constable, E.C. Direct Observation of Two Electron Holes in a Hematite Photoanode During Photoelectro-chemical Water Splitting. *J. Phys. Chem. C* **2012**, *116*, 16870–16875.

202. Serpone, N.; Borgarello, E.; Grätzel, M. Visible Light Induced Generation of H_2 from H_2S in Mixed Semiconductor Dispersions: Improved Efficiency through Interparticle Electron Transfer. *J. Chem. Soc. Chem. Commun.* **1984**, 342–344.

203. Yan, H.; Yang, J.; Ma, G.; Wu, G.; Lei, Z.; Shi, J.; Li, C. Visible-Light-Driven Hydrogen Production with Extremely High Quantum Efficiency on Pt-PdS/CdS Photocatalyst. *J. Catal.* **2009**, *266*, 165–168.

204. Yang, J.; Yan, H.; Wang, X.; Wen, F.; Wang, Z.; Fan, D.; Shi, J.; Li, C. Roles of Cocatalysts in Pt-PdS/CdS with Exceptionally High Quantum Efficiency for Photocatalytic Hydrogen Production. *J. Catal.* **2012**, *290*, 151–157.

205. Ehsan, M.A.; Ming, H.N.; McKee, V.; Peiris, T.A.N.; Gamage, U.W.K.; Arifin, Z.; Mazhar, M.; Vysotskite Structured Photoactive Palladium Sulphide Thin Films from Dithiocarbamate. Derivatives. *New J. Chem.* **2014**, *38*, 4083–4091.

206. Ferrer, I.J.; Díaz-Chao, P.; Sánchez, C. An Investigation on Palladium Sulphide (PdS) Thin Films as a Photovoltaic Material. *Thin Solid Films* **2007**, *515*, 5783–5786.

207. Meng, J.; Yu, Z.; Li, Y.; Li, Y. PdS-Modified CdS/NiS Composite as an Efficient Photocatalyst for H_2 evolution in Visible Light. *Catal. Today* **2014**, *225*, 136–141.

208. Zhang, S.; Chen, Q.; Jing, D.; Wang, Y.; Guo, L. Visible Photoactivity and Anticorrosion Performance of PdS-CdS Photocatalysts Modified by Polyaniline. *Int. J. Hydrog. Energy* **2012**, *37*, 791–796.

209. Ran, J.R.; Zhang, J.; Yu, J.G.; Jaroniec, M.; Qiao, S.Z. Earth-Abundant Cocatalysts for Semiconductor-Based Photocatalytic Water Splitting. *Chem. Soc. Rev.* **2014**, *43*, 7787–7812.

210. Dinh, C.; Pham, M.; Kleitz, F.; Do, T. Design of Water-Soluble CdS-Titanate-Nickel Nanocomposites for Photocatalytic Hydrogen production Under Sunlight. *J. Mater. Chem. A* **2013**, *1*, 13308–13313.

211. McKone, J.R.; Warren, E.L.; Bierman, M.J.; Boettcher, S.W.; Brunswig, B.S.; Lewis, N.S.; Gray, H.B. Evaluation of Pt, Ni, and Ni-Mo Electrocatalysts for Hydrogen Evolution on Crystalline Si Electrodes. *Energy Environ. Sci.* **2011**, *4*, 3573–3583.

212. Highfield, J.G.; Claude, E.; Oguro, K. Electrocatalytic Synergism in Ni/Mo Cathodes for Hydrogen Evolution in Acid Medium: A New Model. *Electrochim. Acta* **1999**, *44*, 2805–2814.

213. Yu, J.G.; Ran, J.R. Facile Preparation and Enhanced Photocatalytic H_2-Production Activity of $Cu(OH)_2$ Cluster Modified TiO_2. *Energy Environ. Sci.* **2011**, *4*, 1364–1371.

214. Gagliardi, C.J.; Vannucci, A.K.; Concepcion, J.J.; Chen, Z.; Mayer, T.J. The Role of Proton Coupled Electron Transfer in Water Oxidation. *Energy Environ. Sci.* **2012**, *5*, 7704–7717.

215. Armstrong, F.A. Why Did Nature Choose Manganese to Make Oxygen? *Philos. Trans. R. Soc. B* **2008**, *363*, 1263–1270.

216. Carmo, M.; Fritz, D.L.; Mergel, J.; Stolten, D. A Comprehensive Review on PEM Water Electrolysis. *Int. J. Hydrog. Energy* **2013**, *38*, 4901–4934.

217. Fabbri, E.; Habereder, A.; Waltar, K.; Kötz, R.; Schmidt, T.J. Developments and Perspectives of Oxide-Based Catalysts for the Oxygen Evolution Reaction. *Catal. Sci. Technol.* **2014**, *4*, 3800–3821.

218. Zhang, F.; Yamakata, A.; Maeda, K.; Moriya, Y.; Takata, T.; Kubota, J.; Teshima, K.; Oishi, S.; Domen, K. Cobalt-Modified Porous Single-Crystalline $LaTiO_2N$ for Highly Efficient Water Oxidation Under Visible Light. *J. Am. Chem. Soc.* **2012**, *134*, 8348–8351.

219. Liu, L.C.; Li, Z.Y.; Zou, W.X.; Gu, X.R.; Deng, Y.; Gao, F.; Tang, C.J.; Dong, L. *In Situ* Loading Transition Metal Oxide Clusters on TiO_2 Nanosheets as Co-Catalysts for Exceptional High Photoactivity. *ACS Catal.* **2013**, *3*, 2052–2061.

220. Kudo, A.; Miseki, Y. Heterogeneous Photocatalytic Materials for Water Splitting. *Chem. Soc. Rev.* **2009**, *38*, 253–278.

221. Maeda, K.; Domen, K. Photocatalytic Water Splitting: Recent Progress and Future Challenges. *J. Phys. Chem. Lett.* **2010**, *1*, 2655–2661.

222. Maeda, K.; Higashi, M.; Lu, D.; Abe, R.; Domen, K. Efficient Non-Sacrificial Water Splitting through Two-Step Photoexcitation by Visible Light Using a Modified Oxynitride as a Hydrogen Evolution Photocatalyst. *J. Am. Chem. Soc.* **2010**, *132*, 5858–5868.

223. Sasaki, Y.; Nemoto, H.; Saito, K.; Kudo, A. Solar Water Splitting Using Powdered Photocatalysts Driven by Z-Schematic Interparticle Electron Transfer without an Electron Mediator. *J. Phys. Chem. C* **2009**, *113*, 17536–17542.

224. Zhou, P.; Yu, J.; Jaroniec, M. All-Solid-State Z-Scheme Photocatalytic Systems. *Adv. Mater.* **2014**, *26*, 4920–4935.

225. Wang, Y.; Wang, Q.; Zhan, X.; Wang, F.; Safdar, M.; He, J. Visible Light Driven Type II Heterostructures and Their Enhanced Photocatalysis Properties: A Review. *Nanoscale* **2013**, *5*, 8326–8339.

226. Prévot, M.S.; Sivula, K. Photoelectrochemical Tandem Cells for Solar Water Splitting. *J. Phys. Chem. C* **2013**, *117*, 17879–17893.

227. Hu, S.; Xiang, C.; Haussener, S.; Berger, A.D.; Lewis, N.S. An Analysis of the Optimal Band Gaps of Light Absorbers in Integrated Tandem Photoelectrochemical Water-Splitting Systems. *Energy Environ. Sci.* **2013**, *6*, 2984–2993.

228. Kanan, M.W.; Nocera, D.G. *In Situ* Formation of an Oxygen-Evolving Catalyst in Neutral Water Containing Phosphate and Co^{2+}. *Science* **2008**, *321*, 1072–1075.

229. Bediako, D.K.; Lasalle-Kaiser, B.; Surendranath, Y.; Yano, J.; Yachandra, V.K.; Nocera, D.G. Structure-Activity Correlations in a Nickel-Borate Oxygen Evolution Catalyst. *J. Am. Chem. Soc.* **2012**, *134*, 6801–6809.

230. Gibson, T.L.; Kelly, N.A. Optimization of Solar Powered Hydrogen Production Using Photovoltaic Electrolysis Devices. *Int. J. Hydrog. Energy* **2008**, *33*, 5931–5940.

231. Herron, J.A.; Kim, J.; Upadhye, A.A.; Huber, G.W.; Maravelias, C.T. A General Framework for the Assessment of Solar Fuel Technologies. *Energy Environ. Sci.* **2015**, *8*, 126–157.

232. Hernández-Pagán, E.A.; Vargas-Barbosa, N.M.; Wang, T.H.; Zhao, Y.; Smotkin, E.S. Resistance and Polarization Losses in Aqueous Buffer-Membrane Electrolytes for Water-Splitting Photoelectrochemical Cells. *Energy Environ. Sci.* **2012**, *5*, 7582–7589.

233. Jacobsson, T.J.; Fjällstrom, V.; Edoff, M.; Edvinsson, T. Sustainable Solar Hydrogen Production: From Photoelectrochemical cells to PV-Electrolyzers and Back Again. *Energy Environ. Sci.* **2014**, *7*, 2056–2070.

234. Tributsch, H. Photovoltaic Hydrogen Generation. *Int. J. Hydrog. Energy* **2008**, *33*, 5911–5930.

235. Campos-Martin, J.M.; Blanco-Brieva, G.; Fierro, J.L.G. Hydrogen Peroxide Synthesis: An Outlook beyond the Anthraquinone Process. *Angew. Chem. Int. Ed.* **2006**, *45*, 6962–6984.

236. García-Serna, J.; Moreno, T.; Biasi, P.; Cocero, M.J.; Mikkola, J.P.; Salmi, T.O. Engineering in Direct Synthesis of Hydrogen Peroxide: Targets, Reactors and Guidelines for Operational Conditions. *Green Chem.* **2014**, *16*, 2320–2345.

237. Samanta, C. Direct Synthesis of Hydrogen Peroxide from Hydrogen and Oxygen: An Overview of Recent Developments in the Process. *Appl. Catal. A Gen.* **2008**, *350*, 133–149.

238. Edwards, J.K.; Solsona, B.; Ntainjua, E.; Carley, A.F.; Herzing, A.A.; Kiely, C.J.; Hutchings, G.J. Switching Off Hydrogen Peroxide Hydrogenation in the Direct Synthesis Process. *Science* **2009**, *323*, 1037–1041.

239. Siahrostami, S.; Verdaguer-Casadevall, A.; Karamad, M.; Deiana, D.; Malacrida, P.; Wickman, B.; Escudero-Escribano, M.; Paoli, E.A.; Frydendal, R.; Hansen, T.W.; *et al.* Enabling Direct H_2O_2 Production Through Rational Electrocatalyst Design. *Nat. Mater.* **2013**, *12*, 1137–1143.

240. Sanli, A.E. A Possible Future Fuel Cell: the Peroxide/Peroxide Fuel Cell. *Int. J. Energy Res.* **2013**, *37*, 1488–1497.

241. Kato, S.; Jung, J.; Suenobu, T.; Fukuzumi, S. Production of Hydrogen Peroxide as a Sustainable Solar Fuel from Water and Dioxygen. *Energy Environ. Sci.* **2013**, *6*, 3756–3764.

242. Fukuzumi, S.; Yamada, Y.; Karlin, K.D. Hydrogen Peroxide as a Sustainable Energy Carrier: Electrocatalytic Production of Hydrogen Peroxide and the Fuel Cell. *Electrochim. Acta* **2012**, *82*, 493–511.

243. Sanli, A.E.; Aytaç, A. Response to Disselkamp: Direct Peroxide/Peroxide Fuel Cell as a Novel Type Fuel Cell. *Int. J. Hydrog. Energy* **2011**, *36*, 869–875.

244. Disselkamp, R.S. Can Aqueous Hydrogen Peroxide be used as a Standalone Energy Source? *Int. J. Hydrogen Energy* **2010**, *35*, 1049–1053.

245. Yamazaki, S.; Siroma, Z.; Senoh, H.; Iroi, T.; Fujiwara, N.; Yasuda, K. A Fuel Cell with Selective Electrocatalysts Using Hydrogen Peroxide as Both an Electron Acceptor and a Fuel. *J. Power Sources* **2008**, *178*, 20–25.

246. Disselkamp, R.S. Energy Storage Using Aqueous Hydrogen Peroxide. *Energy Fuels* **2008**, *22*, 2771–2774.

247. Bussayajarn, N.; Harrington, D.A.; Therdthianwong, S.; Therdthianwong, A.; Djilali, N. The Cathodic Polarization Prediction of H_2/H_2O_2 Fuel Cells using EIS Spectra. In *Paper A-012*, Proceedings of the 2nd Joint International Conference on Sustainable Energy & Environment, Bankok, Thailand, 21–23 November 2006.

248. Ponce de León, Walsh, F.C.; Rose, A.; Lakeman, J.B.; Browning, D.J.; Reeve, R.W. A Direct Borohydride—Acid Peroxide Fuel Cell. *J. Power Sources* **2007**, *164*, 441–448.

249. Miley, G.H.; Luo, N.; Mather, J.; Burton, R.; Hawkins, G.; Gu, L.; Byrd, E.; Gimlin, R.; Shrestha, P.J.; Benavides, G.; *et al.* Direct $NaBH_4/H_2O_2$ Fuel Cells. *J. Power Sources* **2007**, *165*, 509–516.

250. An, L.; Zhao, T.S.; Zhou, X.L.; Wei, L.; Yan, X.H. A High Performance Ethanol—Hydrogen Peroxide Fuel Cell. *RSC Adv.* **2014**, *4*, 65031–65034.

251. Amirfakhri, S.J.; Meunier, J.L.; Berk, D. Electrocatalytic Activity of $LaNiO_3$ toward H_2O_2 Reduction Reaction: Minimization of Oxygen Evolution. *J. Power Sources* **2014**, *272*, 248–258.

252. Compton, O.C.; Osterloh, F.E. Niobate Nanosheets as Catalysts for Photochemical Water Splitting into Hydrogen and Hydrogen Peroxide. *J. Phys. Chem. C* **2009**, *113*, 479–485.

253. Daskalaki, V.M.; Panagiotopoulou, P.; Kondarides, D.I. Production of Peroxide Species in Pt/TiO_2 Suspensions under Conditions of Photocatalytic Water Splitting and Glycerol Photo-Reforming. *Chem. Eng. J.* **2011**, *170*, 433–439.

254. Arakawa, H. Water Photolysis by TiO_2 Particles—Significant Effect of Na_2CO_3 Addition on Water Splitting. In *Photocatalysis Science and Technology*; Kaneko, M., Okura, I., Eds.; Springer: New York, NY, USA, 2002; p. 235.

255. Yesodharan, E.; Yesodharan, S.; Grätzel, M. Photolysis of Water with Supported Noble Metal Cluster. The Fate of Oxygen in Titania Based Water Cleavage Systems. *Sol. Energy Mater.* **1984**, *10*, 287–302.

256. Hoffman, A.J.; Carraway, E.R.; Hoffmann, M.R. Photocatalytic Production of H_2O_2 and Organic Peroxides on Quantum-Sized Semiconductor Colloids. *Environ. Sci. Technol.* **1994**, *28*, 776–785.

257. Maurino, V.; Minero, C.; Mariella, G.; Pelizzetti, E. Sustained Production of H_2O_2 on Irradiated TiO_2-Fluoride Systems. *Chem. Commun.* **2005**, 2627–2629.

258. Moon, G.M.; Kim, W.; Bokare, A.; Sung, N.; Choi, W. Solar Production of H_2O_2 on Reduced Graphene Oxide-TiO_2 Hybrid Photocatalysts Consisting of Earth-Abundant Elements Only. *Energy Environ. Sci.* **2014**, *7*, 4023–4028.

259. Tsukamoto, D.; Shiro, A.; Shiraishi, Y.; Sugano, Y.; Ichikawa, S.; Tanaka, S.; Hirai, T. Photocatalytic H_2O_2 Production from Ethanol/O_2 System Using TiO_2 Loaded with Au-Ag Bimetallic Alloy Nanoparticles. *ACS Catal.* **2012**, *2*, 599–603.

260. Diesen, V.; Jonsson, M. Formation of H_2O_2 in TiO_2 Photocatalysis of Oxygenated and Deoxygenated Aqueous Systems: A Probe for Photocatalytically Produced Hydroxyl Radicals. *J. Phys. Chem. C* **2014**, *118*, 10083–10087.

261. Lousada, C.M.; Johansson, A.J.; Brinck, T.; Jonsson, M. Mechanism of H_2O_2 Decomposition on Transition Metal Oxide Surfaces. *J. Phys. Chem. C* **2012**, *116*, 9533–9543.

262. Bokare, A.D.; Choi, W. Review of Iron-Free Fenton-like Systems for Activating H_2O_2 in Advanced Oxidation Processes. *J. Hazard. Mater.* **2014**, *275*, 121–135.

263. Koppenol, W.H. The Haber-Weiss Cycle—70 Years Later. *Redox Rep.* **2001**, *6*, 229–234.

264. Shiraishi, Y.; Kanazawa, S.; Sugano, Y.; Tsukamoto, D.; Sakamoto, H.; Ichikawa, S.; Hirai, T. Highly Selective Production of Hydrogen Peroxide on Graphitic Carbon Nitride (g-C_3N_4) Photocatalyst Activated by Visible Light. *ACS Catal.* **2014**, *4*, 774–780.

265. Zhang, J.; Chen, X.; Takanabe, K.; Maeda, K.; Domen, K.; Epping, J.D.; Fu, X.; Antonietti, M.; Wang, X. Synthesis of a Carbon Nitride Structure for Visible-Light Catalysis by Copolymerization. *Angew. Chem. Int. Ed.* **2010**, *49*, 441–444.

266. Abe, R.; Takami, H.; Murakami, N.; Ohtani, B. Pristine Simple Oxides as Visible Light Driven Photocatalysts: Highly Efficient Decomposition of Organic Compounds over Platinum-Loaded Tungsten Oxide. *J. Am. Chem. Soc.* **2008**, *130*, 7780–7781.

267. Rostrup-Nielsen, J.R. Steam Reforming and Chemical Recuperation. *Catal. Today* **2009**, *145*, 72–75.

268. Hodoshima, S.; Shono, A.; Saito, Y. Chemical Recuperation of Low-Quality Waste Heats by Catalytic Dehydrogenation of Organic Chemical Hydrides and Its Exergy Analysis. *Energy Fuels* **2008**, *22*, 2559–2569.

269. Poullikkas, A. An Overview of Current and Future Sustainable Gas Turbine Technologies. *Renew. Sustain. Energy Rev.* **2005**, *9*, 409–443.

270. Sun, J.; Wang, Y. Recent Advances in Catalytic Conversion of Ethanol to Chemicals. *ACS Catal.* **2014**, *4*, 1078–1090.

271. Makshina, E.V.; Dusselier, M.; Janssens, W.; Degrève, J.; Jacobs, P.A.; Sels, B.F. Review of Old Chemistry and New Catalytic Advances in the On-Purpose Synthesis of Butadiene. *Chem. Soc. Rev.* **2014**, *43*, 7917–7953.

272. Alcalá, R.; Mavrikakis, M.; Dumesic, J.A. DFT Studies for Cleavage of C−C and C−O Bonds in Surface Species Derived from Ethanol on Pt(111). *J. Catal.* **2003**, *218*, 178–190.

273. Ferrin, P.; Simonetti, D.; Kandoi, S.; Kunkes, E.; Dumesic, J.A.; Nørskov, J.K.; Mavrikakis, M. Modeling Ethanol Decomposition on Transition Metals: A Combined Application of Scaling and Brønsted-Evans-Polanyi Relations. *J. Am. Chem. Soc.* **2009**, *131*, 5809–5815.

274. Salemme, L.; Menna, L.; Simeone, M. Thermodynamic Analysis of Ethanol Processors—PEM Fuel Cell Systems. *Int. J. Hydrog. Energy* **2010**, *35*, 3480–3489.

275. Ni, M.; Leung, D.Y.C.; Leung, M.K.H. A Review on Reforming Bio-Ethanol for Hydrogen Production. *Int. J. Hydrog. Energy* **2007**, *32*, 3238–3247.

276. Xu, W.Q.; Liu, Z.Y.; Johnston-Peck, A.C.; Senanayake, S.D.; Zhou, G.; Stacchiloa, D.; Stach, E.A.; Rodriguez, J.A. Steam Reforming of Ethanol on Ni/CeO$_2$: Reaction Pathway and Interaction between Ni and the CeO$_2$ Support. *ACS Catal.* **2013**, *3*, 975–984.

277. Sanchez-Sanchez, M.C.; Yerga, R.M.N.; Kondarides, D.I.; Verykios, X.E.; Fierro, J.L.G. Mechanistic Aspects of the Ethanol Steam Reforming Reaction for Hydrogen Production on Pt, Ni, and Pt/Ni Catalysts Supported on γ-Al$_2$O$_3$. *J. Phys. Chem. A* **2010**, *114*, 3873–3882.

278. Highfield, J.; Geiger, F.; Uenala, E.; Schucan, T. Hydrogen Release by Steam-Reforming of Ethanol for Efficient & Clean Fuel Applications. In Proceedings of the 10th World Hydrogen Energy Conference, Cocoa Beach, FL, USA, 20–24 June 1994; Hydrogen Energy Progress X; Block, D.L., Veziroglu, T.N., Eds.; Volume 2, p. 1039.

279. Sutton, J.E.; Panagiotopoulou, P.; Verykios, X.E.; Vlachos, D.G. Combined DFT, Microkinetic, and Experimental Study of Ethanol Steam Reforming on Pt. *J. Phys. Chem. C* **2013**, *117*, 4691–4706.

280. Maihom, T.; Probst, M.; Limtrakul, J. Density Functional Theory Study of the Dehydrogenation of Ethanol to Acetaldehyde over the Au-Exchanged ZSM-5 Zeolite: Effect of Surface Oxygen. *J. Phys. Chem. C* **2014**, *118*, 18564–18572.

281. Morgenstern, D.A.; Fornango, J.P. Low-Temperature Reforming of Ethanol over Copper-Plated Raney Nickel: A New Route to Sustainable Hydrogen for Transportation. *Energy Fuels* **2005**, *19*, 1708–1716.

282. Bowker, M.; Holroyd, R.; Perkins, N.; Bhantoo, J.; Counsell, J.; Carley, A.; Morgan, C. Acetaldehyde Adsorption and Catalytic Decomposition on Pd (110) and the Dissolution of Carbon. *Surf. Sci.* **2007**, *601*, 3651–3660.

283. Pichat, P.; Mozzanega, M.N.; Disdier, J.; Herrmann, J.M. Pt Content and Temperature Effects on the Photocatalytic H$_2$ production from Aliphatic Alcohols over Pt/TiO$_2$. *New J. Chem. (Nouv. J. Chim.)* **1982**, *6*, 559–564.

284. Pichat, P.; Hermann, J.M.; Disdier, J.; Courbon, H.; Mozzanega, M.N. Photocatalytic Hydrogen Production from Aliphatic Alcohols over a Bifunctional Platinum on Titanium Dioxide Catalyst. *New J. Chem. (Nouv. J. Chim.)* **1981**, *5*, 627–636.

285. Millard, L.; Bowker, M. Photocatalytic Water-Gas Shift Reaction at Ambient Temperature. *J. Photochem. Photobiol. A Chem.* **2002**, *148*, 91–95.

286. Bowker, M.; James, D.; Stone, P.; Bennett, R.; Perkins, N.; Millard, L.; Greaves, J.; Dickinson, A. Catalysis at the Metal-Support Interface: Exemplified by the Photocatalytic Reforming of Methanol on Pd/TiO$_2$. *J. Catal.* **2004**, *217*, 427–433.

287. Al-Mazroai, L.S.; Bowker, M.; Davies, P.; Dickinson, A.; Greaves, J.; James, D.; Millard, L. The Photocatalytic Reforming of Methanol. *Catal. Today* **2007**, *122*, 46–50.

288. Highfield, J.G.; Chen, M.H.; Nguyen, P.T.; Chen, Z. Mechanistic Investigations of Photo-Driven Processes over TiO$_2$ by in-situ DRIFTS-MS: Part 1: Platinization and Methanol Reforming. *Energy Environ. Sci.* **2009**, *2*, 991–1002.

289. Daskalaki, V.M.; Kondarides, D.I. Efficient Production of Hydrogen by Photo-Induced Reforming of Glycerol at Ambient Conditions. *Catal. Today* **2009**, *144*, 75–80.

290. Taboada, E.; Angurell, I.; Llorca, J. Dynamic Photocatalytic Hydrogen Production from Ethanol—Water Mixtures in an Optical Fiber Honeycomb Reactor Loaded with Au/TiO₂. *J. Catal.* **2014**, *309*, 460–467.

291. Puga, A.V.; Forneli, A.; García, H.; Corma, A. Production of H₂ by Ethanol Photoreforming on Au/TiO₂. *Adv. Funct. Mater.* **2014**, *24*, 241–248.

292. Murdoch, M.; Waterhouse, G.I.N.; Nadeem, M.A.; Metson, J.B.; Keane, M.A.; Howe, R.F.; Llorca, J.; Idriss, H. The Effect of Gold Loading and Particle Size on the Photocatalytic Hydrogen Production from Ethanol on Au/TiO₂ Nanoparticles. *Nat. Chem.* **2011**, *3*, 489–492.

293. Gallo, A.; Marelli, M.; Psaro, R.; Gombac, V.; Montini, T.; Fornasiero, P.; Pievo, R.; Dal Santo, V. Bimetallic Au-Pt/TiO₂ Photocatalysts Active under UV-A and Simulated Sunlight for H₂ Production from Ethanol. *Green Chem.* **2012**, *14*, 330–333.

294. Strataki, N.; Bekiari, V.; Kondarides, D.I.; Lianos, P. Hydrogen production by photocatalytic alcohol reforming employing highly efficient Nanocrystalline Titania films. *Appl. Catal. B Environ.* **2007**, *77*, 184–189.

295. Languer, M.P.; Scheffer, F.R.; Feil, A.F.; Baptista, D.L.; Migowski, P.; Macahdo, G.J.; de Moraes, D.P.; Dupont, J.; Teixeira, S.R.; Weibel, D.E. Photo-Induced Reforming of Alcohols with Improved Hydrogen Apparent Quantum Yield on TiO₂ Nanotubes Loaded with Ultra-Small Pt Nanoparticles. *Int. J. Hydrog. Energy* **2013**, *38*, 14440–14450.

296. Bowker, M.; Morton, C.; Kennedy, J.; Bahruji, H.; Greves, J.; Jones, W.; Davies, P.R.; Brookes, C.; Wells, P.P.; Dimitratos, N. Hydrogen Production by Photoreforming of Biofuels using Au, Pd and Au-Pd/TiO₂ Photocatalysts. *J. Catal.* **2014**, *310*, 10–15.

297. Taboada, E.; Angurell, I.; Llorca, J. Hydrogen Photoproduction from Bio-derived Alcohols in an Optical Fiber Honeycomb Reactor Loaded with Au/TiO₂. *J. Photochem. Photobiol. A Chem.* **2014**, *281*, 35–39.

298. Bowker, M. Sustainable Hydrogen Production by the Application of Ambient Temperature Photocatalysis. *Green Chem.* **2011**, *13*, 2235–2246.

299. Bowker, M.; Davies, P.R.; al-Mazroai, L.S. Photocatalytic Reforming of Glycerol over Gold and Palladium as an Alternative Fuel Source. *Catal. Lett.* **2009**, *128*, 253–255.

300. Luo, N.; Jiang, Z.; Shi, H.; Cao, F.; Xiao, T.; Edwards, P.P. Photo-Catalytic Conversion of Oxygenated Hydrocarbons to Hydrogen over Heteroatom-Doped TiO₂ Catalysts. *Int. J. Hydrog. Energy* **2009**, *34*, 125–129.

301. Fu, X.; Long, J.; Wang, X.; Leung, D.Y.C.; Ding, Z.; Wu, L.; Zhang, Z.; Li, Z.; Fu, X. Photocatalytic Reforming of Biomass: A Systematic Study of Hydrogen Evolution from Glucose Solution. *Int. J. Hydrog. Energy* **2008**, *33*, 6484–6491.

302. Zheng, X.J.; Wei, L.F.; Zhang, Z.H.; Jiang, Q.J.; Wei, Y.J.; Xie, B.; Wei, M.B. Research on Photocatalytic H₂ Production from Acetic Acid Solution by Pt/TiO₂ Nanoparticles under UV Irradiation. *Int. J. Hydrog. Energy* **2009**, *34*, 9033–9041.

303. Kraeutler, B.; Bard, A.J. Heterogeneous Photocatalytic Synthesis of Methane from Acetic Acid—New Kolbe Reaction Pathway. *J. Am. Chem. Soc.* **1978**, *100*, 2239–2240.

304. Tanaka, A.; Sakaguchi, S.; Hashimoto, K.; Kominami, H. Preparation of Au/TiO₂ with Metal Cocatalysts Exhibiting Strong Surface Plasmon Resonance Effective for Photoinduced Hydrogen Formation under Irradiation of Visible Light. *ACS Catal.* **2013**, *3*, 79–85.

305. Lyubina, T.P.; Markovskaya, D.V.; Kozlova, E.A.; Parmon, V.N. Photocatalytic Hydrogen Evolution from Aqueous Solutions of Glycerol under Visible Light Irradiation. *Int. J. Hydrog. Energy* **2013**, *38*, 14172–14179.

306. Melo, M.D.; Silva, L.A. Visible-Light-Induced Hydrogen Production from Glycerol Aqueous Solution on Hybrid Pt-CdS-TiO₂ Photocatalysts. *J. Photochem. Photobiol. A Chem.* **2011**, *226*, 36–41.

307. Carraro, G.; Gasparotto, A.; Maccato, C.; Gombac, V.; Rossi, F.; Montini, T.; Peeters, D.; Bontempi, E.; Sada, C.; Barreca, D.; *et al.* Solar H₂ Generation via Ethanol Photoreforming on ε-Fe₂O₃ Nanorod Arrays Activated by Ag and Au Nanoparticles. *RSC Adv.* **2014**, *4*, 32174–32179.

308. Carraro, G.; Maccato, C.; Gasparotto, A.; Montini, T.; Turner, S.; Lebedev, O.I.; Gombac, V.; Adami, G.; van Tendeloo, G.; Barreca, D.; *et al.* Enhanced Hydrogen Production by Photoreforming of Renewable Oxygenates through Nanostructured Fe₂O₃ Polymorphs. *Adv. Funct. Mater.* **2014**, *24*, 372–378.

309. Kim, G.; Lee, S.H.; Choi, W. Glucose-TiO₂ Charge Transfer Complex-Mediated Photocatalysis under Visible Light. *Appl. Catal. B Environ.* **2015**, *162*, 463–469.

310. Clarizia, L.; Spasiano, D.; di Somma, I.; Marotta, R.; Andreozzi, R.; Dionysiou, D.D. Copper Modified- TiO₂ Catalysts for Hydrogen Generation through Photoreforming of Organics. A Short Review. *Int. J. Hydrog. Energy* **2014**, *39*, 16812–16831.

311. Ampelli, C.; Passalacqua, R.; Genovese, C.; Perathoner, S.; Centi, G.; Montini, T.; Gombac, V.; Jaen, J.J.D.; Fornasiero, P. H₂ Production by Selective Photo-Dehydrogenation of Ethanol in Gas and Liquid Phase on CuOₓ/TiO₂ Nanocomposites. *RSC Adv.* **2013**, *3*, 21776–21788.

312. Liu, R.X.; Yoshida, H.; Fujita, S.; Arai, M. Photocatalytic Hydrogen Production from Glycerol and Water with NiOₓ/TiO₂ Catalysts. *Appl. Catal. B Environ.* **2014**, *144*, 41–45.

313. Shkrob, I.A.; Dimitrijevic, N.M.; Marin, T.W.; He, H.; Zapol, P. Heteroatom-Transfer Coupled Photoreduction and Carbon Dioxide Fixation on Metal Oxides. *J. Phys. Chem. C* **2012**, *116*, 9461–9471.

314. Shkrob, I.A.; Marin, T.W.; He, H.; Zapol, P. Photoredox Reactions and the Catalytic Cycle for Carbon Dioxide Fixation and Methanogenesis on Metal Oxides. *J. Phys. Chem. C* **2012**, *116*, 9450–9460.

315. Mori, K.; Yamashita, H.; Anpo, M. Photocatalytic Reduction of CO₂ with H₂O on Various Titanium Oxide Photocatalysts. *RSC Adv.* **2012**, *2*, 3165–3172.

316. He, H.; Zapol, P.; Curtiss, L.A. Computational Screening of Dopants for Photocatalytic Two-Electron Reduction of CO₂ on Anatase (101) Surfaces. *Energy Environ. Sci.* **2012**, *5*, 6196–6205.

317. Tu, W.; Zhou, Y.; Zou, Z. Photocatalytic Conversion of CO_2 into Renewable Hydrocarbon Fuels: State-of-the-Art Accomplishment, Challenges, and Prospects. *Adv. Mater.* **2014**, *26*, 4607–4626.

318. Peterson, A.A.; Abild-Pedersen, F.; Studt, F.; Rossmeisl, J.; Nørskov, J.K. How Copper Catalyzes the Electroreduction of Carbon Dioxide into Hydrocarbon Fuels. *Energy Environ. Sci.* **2010**, *3*, 1311–1315.

319. Behrens, M.; Studt, F.; Kasatkin, I.; Kühl, S.; Hävecker, M.; Abild-Pedersen, F.; Zander, S.; Girgsdies, F.; Kurr, P.; Kniep, B.L.; *et al.* The Active Site of Methanol Synthesis over $Cu/ZnO/Al_2O_3$ Industrial Catalysts. *Science* **2012**, *336*, 893–897.

320. Ola, O.; Maroto-Valer, M.M. Copper-based TiO_2 Honeycomb Monoliths for CO_2 Photoreduction. *Catal. Sci. Technol.* **2014**, *4*, 1631–1637.

321. Nunez, J.; O'Shea, V.A.D.; Jana, P.; Coronado, J.M.; Serrano, D.P. Effect of Copper on the Performance of ZnO and $ZnO_{1-x}N_x$ Oxides as CO_2 Photoreduction Catalysts. *Catal. Today* **2013**, *209*, 21–27.

322. Mao, J.; Li, K.; Peng, T.Y. Recent Advances in the Photocatalytic CO_2 Reduction over Semiconductors. *Catal. Sci. Technol.* **2013**, *3*, 2481–2498.

323. Richardson, P.L.; Perdigoto, M.L.N.; Wang, W.; Lopes, R.J.G. Manganese- and Copper-Doped Titania Nanocomposites for the Photocatalytic Reduction of Carbon Dioxide into Methanol. *Appl. Catal. B Environ.* **2012**, *126*, 200–207.

324. Kuhl, K.P.; Hatsukade, T.; Cave, E.R.; Abram, D.N.; Kibsgaard, J.; Jaramillo, T.F. Electrocatalytic Conversion of Carbon Dioxide to Methane and Methanol on Transition Metal Surfaces. *J. Am. Chem. Soc.* **2014**, *136*, 14107–14113.

325. Constentin, C.; Robert, M.; Saveant, J.M. Catalysis of the Electrochemical Reduction of Carbon Dioxide. *Chem. Soc. Rev.* **2013**, *42*, 2423–2436.

326. Kuhl, K.P.; Cave, E.R.; Abram, D.N.; Jaramillo, T.F. New Insights into the Electrochemical Reduction of Carbon Dioxide on Metallic Copper Surfaces. *Energy Environ. Sci.* **2012**, *5*, 7050–7059.

327. Tu, W.; Zhou, Y.; Zou, Z. Versatile Graphene-Promoting Photocatalytic Performance of Semiconductors: Basic Principles, Synthesis, Solar Energy Conversion, and Environmental Applications. *Adv. Funct. Mater.* **2013**, *23*, 4996–5008.

328. Fan, W.; Zhang, Q.; Wang, Y. Semiconductor-Based Nanocomposites for Photocatalytic H_2 Production and CO_2 Conversion. *Phys. Chem. Chem. Phys.* **2013**, *15*, 2632–2649.

329. Leary, R.; Westwood, A. Carbonaceous Nanomaterials for the Enhancement of TiO_2 Photocatalysis. *Carbon* **2011**, *49*, 741–772.

330. Liang, Y.T.; Vijayan, B.K.; Gray, K.A.; Hersam, M.C. Minimizing Graphene Defects Enhances Titania Nanocomposite-Based Photocatalytic Reduction of CO_2 for Improved Solar Fuel Production. *Nano Lett.* **2011**, *11*, 2865–2870.

331. Tu, W.G.; Zhou, Y.; Liu, Q.; Tian, Z.P.; Gao, J.; Chen, X.Y.; Zhang, H.T.; Liu, J.G.; Zou, Z.G. Robust Hollow Spheres Consisting of Alternating Titania Nanosheets and Graphene Nanosheets with High Photocatalytic Activity for CO_2 Conversion into Renewable Fuels. *Adv. Funct. Mater.* **2012**, *22*, 1215–1221.

332. Xia, X.H.; Jia, Z.H.; Yu, Y.; Liang, Y.; Wang, Z.; Ma, L.L. Preparation of Multi-Walled Carbon Nanotube Supported TiO_2 and Its Photocatalytic Activity in the Reduction of CO_2 with H_2O. *Carbon* **2007**, *45*, 717–721.

333. Centi, G.; Perathoner, S. Opportunities and Prospects in the Chemical Recycling of Carbon Dioxide to Fuels. *Catal. Today* **2009**, *148*, 191–205.

334. Yao, Y.; Li, G.; Ciston, S.; Lueptow, R.M.; Gray, K.A. Photoreactive TiO_2/Carbon Nanotube Composites: Synthesis and Reactivity. *Environ. Sci. Technol.* **2008**, *42*, 4952–4957.

335. Hsu, H.C.; Shown, I.; Wei, H.Y.; Chang, Y.C.; Du, H.Y.; Lin, Y.G.; Tseng, C.A.; Wang, C.H.; Chen, L.C.; Lin, Y.C.; *et al.* Graphene Oxide as a Promising Photocatalyst for CO_2 to Methanol Conversion. *Nanoscale* **2013**, *5*, 262–268.

336. Shown, I.; Hsu, H.C.; Chang, Y.C.; Lin, C.H.; Roy, P.K.; Ganguly, A.; Wang, C.H.; Chang, J.K.; Wu, C.I.; Chen, L.C.; *et al.* Highly Efficient Visible Light Photocatalytic Reduction of CO_2 to Hydrocarbon Fuels by Cu-Nanoparticle Decorated Graphene Oxide. *Nano Lett.* **2014**, *14*, 6097–6103.

337. Kumar, P.; Kumar, A.; Sreedhar, B.; Sain, B.; Ray, S.S.; Jain, S.L. Cobalt Phthalocyanine Immobilized on Graphene Oxide: An Efficient Visible-Active Catalyst for the Photoreduction of Carbon Dioxide. *Chem. Eur. J.* **2014**, *20*, 6154–6161.

338. Li, X.; Liu, H.; Luo, D.; Li, J.; Huang, Y.; Li, H.; Fang, Y.; Xu, Y.; Zhu, L. Adsorption of CO_2 on Heterostructure $CdS(Bi_2S_3)/TiO_2$ Nanotube Photocatalysts and Their Photocatalytic Activities in the Reduction of CO_2 to Methanol under Visible Light Irradiation. *Chem. Eng. J.* **2012**, *180*, 151–158.

339. Wang, C.; Thompson, R.L.; Baltrus, J. Matranga, Visible Light Photoreduction of CO_2 Using $CdSe/Pt/TiO_2$ Heterostructured Catalysts. *J. Phys. Chem. Lett.* **2010**, *1*, 48–53.

340. Hou, W.B.; Hung, W.H.; Pavaskar, P.; Goeppert, A.; Aykol, M.; Cronin, S.B. Photocatalytic Conversion of CO_2 to Hydrocarbon Fuels via Plasmon-Enhanced Absorption and Metallic Interband Transitions. *ACS Catal.* **2011**, *1*, 929–936.

341. Centi, G.; Perathoner, S. Catalysis: Role and Challenges for a Sustainable Energy. *Top. Catal.* **2009**, *52*, 948–961.

342. Gangeri, M.; Perathoner, S.; Caudo, S.; Centi, G.; Amadou, J.; Bégon, D.; Pham-Huu, C.; Ledoux, M.J.; Tessonier, J.P.; Su, D.S.; *et al.* Fe and Pt Carbon Nanotubes for the Electrocatalytic Conversion of Carbon Dioxide to Oxygenates. *Catal. Today* **2009**, *143*, 57–63.

343. Grabow, L.C.; Mavrikakis, M. Mechanism of Methanol Synthesis on Cu through CO_2 and CO Hydrogenation. *ACS Catal.* **2011**, *1*, 365–384.

344. Cheng, D.; Negreiros, F.R.; Aprà, E.; Fortunelli. A. Computational Approaches to the Chemical Conversion of Carbon Dioxide. *ChemSusChem* **2013**, *6*, 944–965.

345. Lazo, N.D.; Murray, D.K.; Kieke, M.L.; Haw, J.F. *In-Situ* C-13 Solid-State NMR Study of the $Cu/ZnO/Al_2O_3$ Methanol Synthesis Catalyst. *J. Am. Chem. Soc.* **1992**, *114*, 8552–8559.

346. Idriss, H.; Kim, K.S.; Barteau, M.A. Surface-Dependent Pathways for Formaldehyde Oxidation and Reduction on $TiO_2(001)$. *Surf. Sci.* **1992**, *262*, 113–127.

347. Busca, G.; Lamotte, J.; Lavalley, J.C.; Lorenzelli, V. FTIR Study of the Adsorption and Transformation of Formaldehyde on Oxide Surfaces. *J. Am. Chem. Soc.* **1987**, *109*, 5197–5202.

348. Rzepa, H.S.; Miller, J. An MNDO SCF-MO Study of the Mechanism of the Cannizzaro Reaction. *J. Chem. Soc. Perkin Trans. II* **1985**, 717–723, doi:10.1039/P29850000717.

349. Harriman, A. Prospects for Conversion of Solar Energy into Chemical Fuels: The Concept of a Solar Fuels Industry. *Philos. Trans. R. Soc. A* **2013**, *371*, 20110415.

350. Thomas, J.M. Reflections on the Topic of Solar Fuels. *Energy Environ. Sci.* **2014**, *7*, 19–20.

351. Malato, S.; Fernández-Ibáñez, P.; Maldonado, M.I.; Oller, I.; Polo-López, M.I. Solar Photocatalytic Pilot Plants: Commercially Available Reactors. In *Photocatalysis and Water Purification: From Fundamentals to Recent Applications*; Pichat, P., Ed.; Wiley-VCH Verlag: Weinheim, Germany, 2013; Chapter 15.

352. Villa, K.; Domènech, X.; Malato, S.; Maldonado, M.I.; Peral, J. Heterogeneous Photocatalytic Hydrogen Generation in a Solar Pilot Plant. *Int. J. Hydrog. Energy* **2013**, *38*, 12718–12724.

353. James, B.D.; Baum, G.N.; Perez, J.; Baum, K.N. *Technoeconomic Analysis of Photoelectrochemical (PEC) Hydrogen Production*; Directed Technologies Inc.: Arlington, VA, USA, 2009; US DOE Contract no. GS-10F-009.

354. Döscher, H.; Geisz, J.F.; Deutsch, T.G.; Turner, J.A. Sunlight Absorption in Water—Efficiency and Design Implications for Photoelectrochemical Devices. *Energy Environ. Sci.* **2014**, *7*, 2951–2956.

355. Seitz, L.C.; Chen, Z.; Forman, A.J.; Pinaud, B.A.; Benck, J.D.; Jaramillo, T.F. Modeling Practical Performance Limits of Photoelectrochemical Water Splitting Based on the Current State of Materials Research. *ChemSusChem* **2014**, *7*, 1372–1385.

356. Haussener, S.; Xiang, C.; Spurgeon, J.M.; Ardo, S.; Lewis, N.S.; Weber, A.Z. Modeling, Simulation, and Design Criteria for Photoelectrochemical Water-Splitting Systems. *Energy Environ. Sci.* **2012**, *5*, 9922–9935.

357. Rodriguez, C.A.; Modestino, M.A.; Psaltis, D.; Moser, C. Design and Cost Considerations for Practical Solar-Hydrogen Generators. *Energy Environ. Sci.* **2014**, *7*, 3828–3835.

358. Chandross, E.A. Shining a Light on Solar Water Splitting. *Science* **2014**, *344*, 469.

359. Turner, J.A. Response to Chandross. *Science* **2014**, *344*, 469–470.

360. Chen, Y.; Xiang, C.; Hu, C.; Lewis, N.S. Modeling the Performance of an Integrated Photoelectrolysis System with 10× Solar Concentrators. *J. Electrochem. Soc.* **2014**, *161*, F1101–F1110.

361. Khamooshi, M.; Salati, H.; Egelioglu, F.; Faghiri, A.H.; Tarabishi, J.; Babadi, S. A Review of Solar Photovoltaic Concentrators. *Int. J. Photoenergy* **2014**, doi:10.1155/2014/958521.

362. Swanson, R.M. The Promise of Concentrators. *Prog. Photovolt. Res. Appl.* **2000**, *8*, 93–111.

363. Hoffmann, M.R.; Martin, S.T.; Choi, W.; Bahnemann, D.W. Environmental Applications of Semiconductor Photocatalysis. *Chem. Rev.* **1995**, *95*, 69–96.

364. Khaselev, O.; Turner, J.A. A Monolithic Photovoltaic-Photoelectrochemical Device for Hydrogen Production via Water Splitting. *Science* **1998**, *280*, 425–427.

365. Joo, O.S.; Jung, K.D.; Jung, Y.S. CAMERE Process for Methanol Synthesis from CO_2 Hydrogenation. In *Carbon Dioxide Utilization for Global Sustainability*; Park, S.E., Chang, J.S., Lee, K.W., Eds.; Elsevier: Amsterdam, The Netherlands, 2004; Volume 153, pp. 67–72.

366. Ikeue, K.; Nozaki, S.; Ogawa, M.; Anpo, M. Photocatalytic Reduction of CO_2 with H_2O on Ti-Containing Porous Silica Thin-Film Photocatalysts. *Catal. Lett.* **2002**, *80*, 111–114.
367. Lackner, K.S. A Guide to CO_2 Sequestration. *Science* **2003**, *300*, 1677–1678.
368. Gerdemann, S.J.; O'Connor, W.K.; Dahlin, D.C.; Penner, L.R.; Rush, H. Ex Situ Aqueous Mineral Carbonation. *Environ. Sci. Technol.* **2007**, *41*, 2587–2593.
369. Khoo, H.H.; Sharratt, P.N.; Bu, J.; Yeo, T.Y.; Borgna, A.; Highfield, J.G.; Björklöf, T.G.; Zevenhoven, R. Carbon Capture and Mineralization in Singapore: Preliminary Environmental Impacts and Costs via LCA. *Ind. Eng. Chem. Res.* **2011**, *50*, 11350–11357.
370. Zevenhoven, R.; Fagerlund, J.; Nduagu, E.; Romão, I.; Bu, J.; Highfield, J. Carbon Storage by Mineralisation (CSM): Serpentinite Rock Carbonation via $Mg(OH)_2$ Reaction Intermediate without CO_2 Pre-Separation [GHGT-11]. *Energy Proced.* **2013**, *37*, 5945–5954.
371. Sanna, A.; Uibu, M.; Caramanna, G.; Kuusik, R.; Maroto-Valer, M.M. A Review of Mineral Carbonation Technologies to Sequester CO_2. *Chem. Soc. Rev.* **2014**, *43*, 8049–8080.

Energy and Molecules from Photochemical/Photocatalytic Reactions. An Overview

Davide Ravelli, Stefano Protti and Angelo Albini

Abstract: Photocatalytic reactions have been defined as those processes that require both a (not consumed) catalyst and light. A previous definition was whether such reactions brought a system towards or away from the (thermal) equilibrium. This consideration brings in the question whether a part of the photon energy is incorporated into the photochemical reaction products. Data are provided for representative organic reactions involving or not molecular catalysts and show that energy storage occurs only when a heavily strained structure is generated, and in that case only a minor part of photon energy is actually stored (ΔG up to 25 kcal·mol^{-1}). The green role of photochemistry/photocatalysis is rather that of forming highly reactive intermediates under mild conditions.

Reprinted from *Molecules*. Cite as: Ravelli, D.; Protti, S.; Albini, A. Energy and Molecules from Photochemical/Photocatalytic Reactions. An Overview. *Molecules* **2015**, *20*, 1527-1542.

1. Introduction: Photochemistry for Synthesis and Energy

More than a century ago, Giacomo Ciamician compared the way both Nature and chemistry practitioners produced complex molecules. Chemists had been extracting a variety of molecules from organisms and had demonstrated that, in most cases, the same structures could be built artificially. However, a key difference remained in the way such products were formed. In particular, plants appeared to be endowed by a "guarded secret" that made them able to synthesize complex structures under mild conditions, in strong contrast with the harsh conditions most often required for the generation of the same molecules in the laboratory. What could be the cause of such a difference? Ciamician thought this was the fact that they absorbed light. That is why he had embarked in a series of experiments and had found that solar light was actually able to cause a wealth of reactions. He was impressed by the smooth formation of new molecules under photochemical conditions. As an example, carbon-carbon bonds were formed upon irradiation and did not require a base, as it on the contrary occurred with the most typical C-C bond forming thermal reaction (known at that time), aldol condensation [1–4]. As a matter of fact, this issue was the subject of a dispute between Ciamician and the other "father" of organic photochemistry, Emanuele Paternò. Indeed, the latter scientist lamented that Ciamician had not grasped the potential of photochemistry in synthesis, and had limited himself to redox reactions of little synthetic significance [5]. Ciamician defended his ideas and evidenced the preparative significance of his work [6]. He also addressed the energetic aspect, with the consideration that fossil fuels are a sort of mineralized solar energy and are regenerated at a negligible rate with respect to the consumption by mankind, so that they are bound to finish in a few centuries. Therefore, according to Ciamician's idea, mankind should have learnt to use directly solar light, thus having at its disposal not only a boundless resource, but also a clean environment, abandoning the dirty world based on burning fuels [5].

From the energetic point of view, the key issue for classification was: in which way is the chemical equilibrium affected by light? A century ago, it was defined that the effect of light could be classed either as *photocatalytic* (photoaccelerating a process in the same direction in which it occurred spontaneously in the dark, $\Delta G < 0$) or *photochemical* in a full meaning (proceeding counter thermodynamically by incorporating part of the energy of the absorbed quantum in the product, $\Delta G > 0$; Scheme 1, where R: reagent, P: product) [7,8]:

$$R \; \underset{}{\overset{h\nu}{\rightleftharpoons}} \; P \qquad \begin{array}{ll} \Delta G < 0 & \text{Photocatalytic} \\ \Delta G > 0 & \text{Photochemical} \end{array}$$

Scheme 1. The effect of light on a chemical equilibrium and the historical definitions adopted.

The word photocatalytic reemerged much later with reference to a class of photochemical reactions, *viz.* to those processes where both light (that is consumed) and a catalyst (that is not) are required. Actually, the word photocatalysis designs a continuously developing area that since a couple of decades is the most extensively investigated subdiscipline of photochemistry.

The question is now, whether photocatalysis is a branch of the more general discipline of chemical catalysis. According to the IUPAC "Gold Book" [9], a catalyst is "*a substance that increases the rate of a reaction without modifying the overall standard Gibbs energy change*". Moreover, "*a catalyst is both a reagent and a product of the reaction*" (Equation (1)). It is further added that "*the words catalyst and catalysis should not be used when the added substance reduces the rate of reaction*" and that "*the term catalysis is also often used when the substance is consumed in the reaction (for example: base-catalysed hydrolysis of esters). Strictly, such a substance should be called an activator.*" This must be distinguished from a chain process:

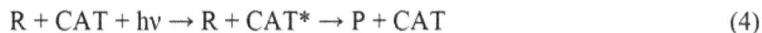

$$R + CAT \rightarrow P + CAT \tag{1}$$

$$R + h\nu \rightarrow R^* \rightarrow P \tag{2}$$

$$R + CAT + h\nu \rightarrow R + CAT^* \rightarrow R^* + CAT \rightarrow P + CAT \tag{3}$$

$$R + CAT + h\nu \rightarrow R + CAT^* \rightarrow P + CAT \tag{4}$$

In a light-induced reaction related definitions are used, but things are by necessity more complex. The chemical reaction involves either the reagent that has absorbed the photon (direct photochemistry, Equation (2)) or some not absorbing molecule (Equations (3) and (4)). In the latter case, if the light absorbing molecule is not consumed, it behaves like a catalyst and can be excited again. Two cases have to be distinguished, *sensitization* (Equation (3)) that involves a physical energy transfer between the absorbing molecule and the non-absorbing one (thus arriving again to an excited state of the reagent, though indirectly), and *photocatalysis* that involves any sort of chemical activation (such as the transfer of an electron or an atom; Equation (4)).

Sensitization is a simple and convenient way to carry out a photochemical reaction when the reactive excited state is not reached efficiently through the spectroscopic states (e.g., the triplet state of organic molecules or the singlet state of molecular oxygen). As for photocatalysis, according to the Gold Book [9] this is defined as a *"change in the rate of a chemical reaction or its initiation under the action of ultraviolet, visible or infrared radiation in the presence of a substance—the photocatalyst—that absorbs light and is involved in the chemical transformation of the reaction partners"*.

From the point of view of the mechanism of photocatalytic reactions, it is convenient to distinguish three cases. In the first case (Equation (5)), a thermal catalyst is produced photochemically and promotes a thermal reaction. This is simply a variation of thermal catalysis, in which an inactive precursor CAT_I is activated photochemically. The quantum yield of formation of the product may be $\gg 1$. An example is reported in Scheme 2, where the photoinduced generation of a thermally active Pt complex allows the hydrosilylation of olefins [10]:

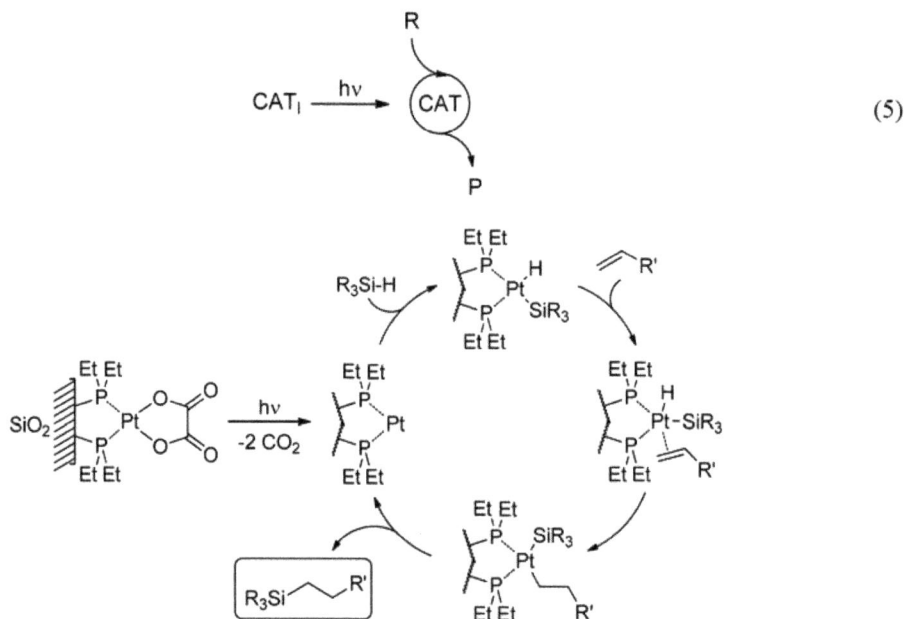

$$(5)$$

Scheme 2. Hydrosilylation of olefins catalyzed by a (supported) Pt complex photochemically generated *in situ* [10].

In the second class, an unstable complex is formed between catalyst and reagent (R'---CAT', Equation (6)). This is a different, ground-state species with its own absorption band. Irradiation in such a band causes the reaction and regenerates the catalyst. Examples are found in many fields with various mechanisms of complexation, including the interaction between Lewis/Brønsted acids or bases, charge-transfer complexes (see the example reported in Scheme 3), reversible condensations as in the case of the transient formation of enamines in organocatalysis by interaction between a carbonyl function and an amine [11], but also adsorption on the surface of a solid material, whether participating in the photochemical reaction (typically a semiconductor) [12] or not [13,14]. The unifying idea is that the irradiated species is in some way modified before turning the light on, and at least in principle, it is possible to irradiate selectively the complexed/adsorbed species. The quantum yield of the overall process is < 1. Semiconductor photocatalysis, by far the largest topic in this field, is in fact usually discussed in terms of electron/hole transfer to the *adsorbed* molecule [12].

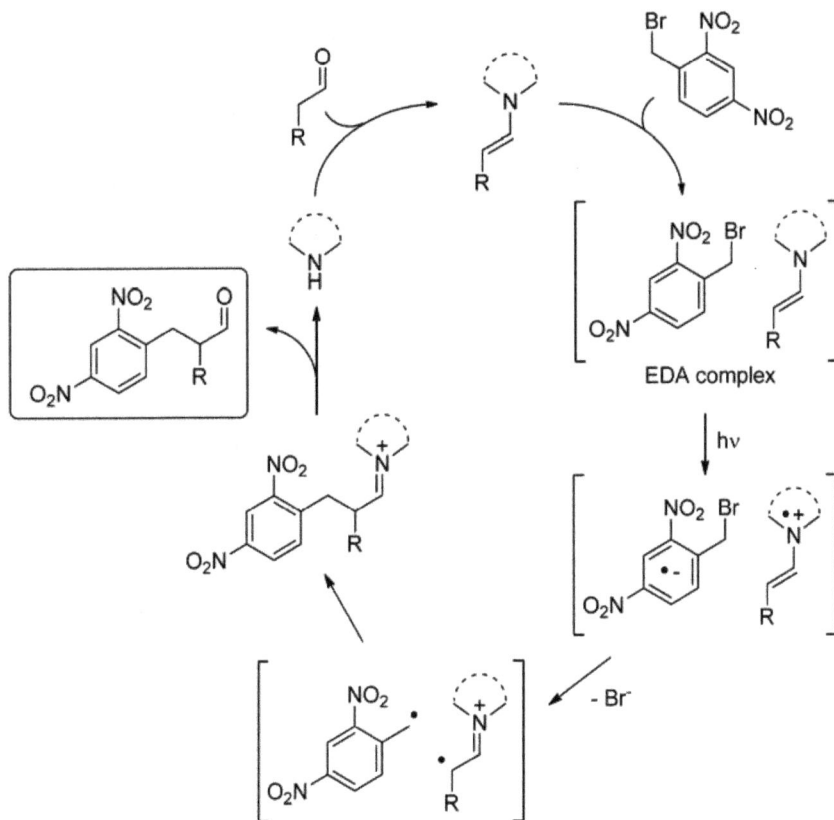

$$R + CAT \longrightarrow R'\text{-}CAT' \xrightarrow{h\nu} P'\text{-}CAT'' \longrightarrow P + CAT \tag{6}$$

Scheme 3. Organocatalytic (asymmetric) α-benzylation of aldehydes via an intermediate EDA (Electron Donor-Acceptor) complex under visible light irradiation [11].

In the third class, a catalyst is excited and interacts with a ground state reagent. Some chemical process causes the reaction of the reagent that is transformed into another species that gives the final product, thereby regenerating the catalyst in the original form (Equation (7)). A typical example (see Scheme 4) is the Hydrogen Atom Transfer (HAT) process from suitable H-donors (e.g., alkanes) promoted by excited ketones and polyoxometalates to give a radical species, then exploited in C-C bond forming reactions [15]. An alternative approach involves the interaction with some reagent Z (e.g., a sacrificial electron donor or acceptor) [16] that brings the photocatalyst into an "active" form CAT' then causing the reaction (Equation (8)). In a typical example, the excited state of a transition metal complex is reduced by a suitable electron donor (e.g., a tertiary amine) to give the "active" form that actually converts the reagent. In some way, the process then regenerates the photocatalyst (see the example reported in Scheme 5) [17]. The quantum yield is again < 1. A synthetically quite promising method involves the merging between organocatalysis and photocatalysis, particularly when activated by visible light. This path is enjoying an explosive development based on joining the experience on transition metal complexes as electron transfer sensitizers and organocatalysis, with the formation in equilibrium of active intermediates [18–23].

(7)

(8)

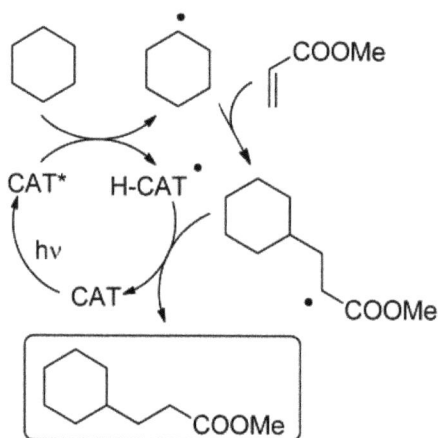

CAT = (Aromatic) Ketones
Polyoxometalates (e.g. $[W_{10}O_{32}]^{4-}$)

Scheme 4. Photocatalyzed conjugated radical addition to electron-poor olefins promoted by excited ketones and polyoxometalates [15].

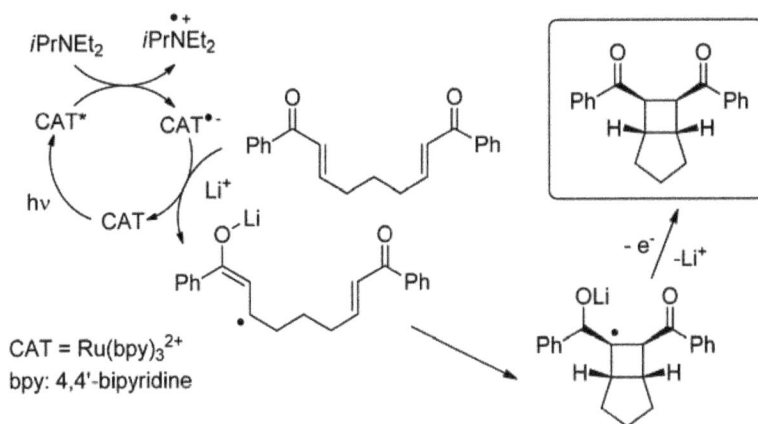

Scheme 5. Photocatalyzed cyclobutane ring formation exploiting a tertiary amine as a sacrificial electron donor [17].

After that attention in this field has been long concentrated on the degradation of pollutants and on the conversion of solar radiation into a convenient energy vector, as in semiconductor photocatalytic water splitting yielding hydrogen, synthetic approaches exploiting the methods above have now been conquering more attention. In this sense, the reaction of third class are simpler and easier to explore by physicochemical methods. However, those of the second class are more promising, because the photochemical reaction involves a prearranged, ground state complex, rather than having to count on the reaction of the short-lived excited states. The long contact time gives further possibilities of governing the reaction and its selectivity. These are the best conditions for "green" chemistry.

Apart from the first class, which is clearly something different, which is the fundamental difference between catalysis and photocatalysis? Some years ago, Ohtani observed that "*the most significant difference between photocatalysis and catalysis lies in their thermodynamics*" [24]. In a general definition, a catalyst reduces the activation energy of a given chemical reaction by changing the intermediate states and thereby accelerates the reaction which proceeds spontaneously with negative Gibbs energy change, that is, catalysis is limited to thermodynamically possible reactions. On the other hand, it is well known that photocatalysis can drive energy-storing reactions, for examples splitting water into hydrogen and oxygen. In this sense, "photocatalysis" must be recognized as a concept completely different from that of "catalysis".

If this criterion is important, then one should assess for every light-induced process, whether it occurs by direct irradiation (P), photosensitization (PS), or photocatalysis (PC), the position of the thermal equilibrium and the effect of a photon on it (see Scheme 1 above). With few exceptions, these data seem not to be available at present [25]. In order to explore in which way light induced processes occur from the energetic point of view, whether storing or consuming energy, we carried out a series of calculations through a uniform approach for typical photochemical, photosensitized and photocatalytic reactions (see Scheme 6 for the definition of the data reported) at the G3(MP2)B3 level (see Supplementary Materials for the computational details).

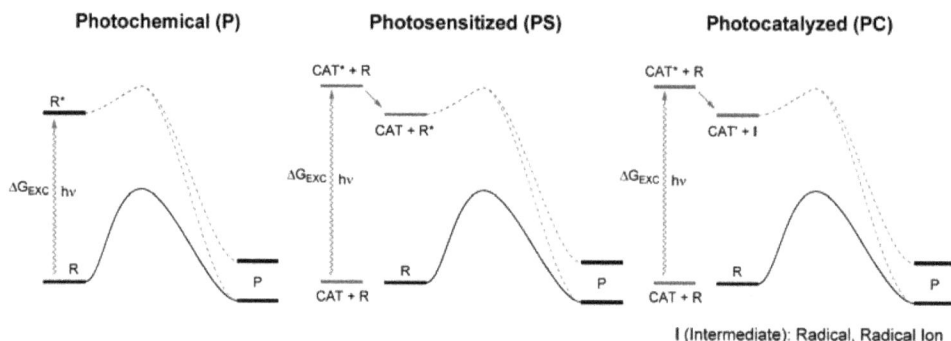

Scheme 6. Definition of the different light-induced processes examined in Table 1.

2. Results and Discussion

The obtained results are gathered in Table 1, where the Gibbs free energy changes (ΔG) between ground states of reagents and products, along with the energy of the activating radiation (ΔG_{EXC}), are listed. The energy of the activating light is meant as the minimum required, that is the less energetic radiation that is absorbed and makes the reaction run. As for photocatalytic reactions, processes involving a molecular photocatalyst soluble in the reaction medium have been chosen rather than those where the reaction occurred on the surface of a solid semiconductor material [26,27]. As a consequence, the reactions can be safely compared with each other since in all of them only molecular species in solution are involved, independently from the class they belong to, avoiding to have to treat adsorption phenomena. A perusal at the Table evidences different groups with reference to the type of process and the thermodynamic of the overall reaction.

Oxidations. These are strongly exergonic reactions, whichever is the mechanism, via electron transfer or singlet oxygen (Table 1, entries 1, 2) [28–30]. The irradiation wavelength used is that absorbed by the sensitizing dye.

Table 1. Calculated thermodynamic parameters at the G3(MP2)B3 level of theory for exemplificative photochemical reactions. These are classified as occurring either via direct irradiation (P), or under photosensitized (PS) or photocatalytic (PC) conditions (see Scheme 6).

Entry	Reaction	Class of Reaction	ΔG_{EXC} [a] (λ_{EXC}) [b]	ΔG [a] (ΔH) [a]
1 [28]		PS/PC	53 (540)	−80.24 (−80.74)
2 [30]		PS	53 (540)	−11.52 (−25.16)

Table 1. Cont.

Entry	Reaction	Class of Reaction	ΔG_{EXC} [a] (λ_{EXC}) [b]	ΔG [a] (ΔH) [a]
3 [31]	94%	P	94 (304)	−3.05 (−13.99)
4 [32]	65% TBADT: $(nBu_4N)_4[W_{10}O_{32}]$	PC	73 (392)	−11.22 (−22.05)
5 [15,33]	hv (310 nm) 57% Sunlight, 72% TBADT: $(nBu_4N)_4[W_{10}O_{32}]$	PC	73 (392)	−17.00 (−28.89)
6 [34]	87%	PS	92 (311)	−18.34 (−31.13)
7 [35]	63%	P	95 (301)	−19.31 (−32.89)
8 [36]	90%	P	77 (371)	−18.55 (−32.75)
9 [37]	100%	P	95 (301)	−17.91 (−6.15)
10 [38]	84%	PC [c]	58 (493)	−6.91 (−20.86)
11 [39]	9.4%	P	119 (240)	3.74 (0.27)
12 [40]	$\Phi_{-1} = 0.18$	PS	87 (330)	24.63 (24.28)

Table 1. *Cont.*

Entry	Reaction	Class of Reaction	ΔG_{EXC} [a] (λ_{EXC}) [b]	ΔG [a] (ΔH) [a]
13 [41]	MeO OMe → MeO OMe; hv (UV), Acetone; 90% (after hydrolysis)	PS	92 (311)	19.86 (18.43)
14 [42]	hv (UV), Cu(OTf), Et₂O; 70%	PC	89 (321)	−8.72 (−12.90)
15 [43]	HN(C=O)₂ Bu + Bu alkyne; hv (400 W Hg lamp), MeCN; 83%	P	–	−17.92 (−31.43)
16 [44]	furanone + PhCHO; hv (UV); 80%	P	94 (304)	20.75 (7.53)
17 [45]	CHO; 4-PBB, hv, 400 W Hg Lamp, CH₂Cl₂; 94%; 4-PBB = 4-phenylbenzophenone	PS	84 (340)	8.71 (8.16)
18 [46]	Me...Me; hv (300 nm), Acetone; 68%	PS	92 (311)	9.60 (9.57)
19 [47]	t-Bu aryl + N–Br succinimide; hv (Vis), Water; 86%; + N–H succinimide, Br	P	–	−24.25 (−24.71)
20	2 H₂ + O₂ → 2 H₂O	–	–	−107.2 (−113.5)
21	CH₄ + 2 O₂ → CO₂ + 2 H₂O	–	–	−193.6 (−193.9)
22 [48]	EtOOC...COOEt dihydropyridine; [Pt], hv (> 450 nm), Solvent, Φ₋₁ = 0.38; → EtOOC...COOEt pyridine + H₂ + [Pt] complex, Cl⁻	PC	67 (427)	−5.77 (1.40)

[a] Values expressed in kcal·mol⁻¹; [b] Values expressed in nm; [c] In the examined reaction, compounds iPr₂NEt and LiBF₄ act as sacrificial reductant and Lewis acid, respectively.

Carbon-Carbon (or C-H) bond forming reactions. When occurring via coupling of stabilized radicals, such reactions are weakly exergonic, as shown in the case of the pinacolic dimerization (entry 3) [31]. Reactions involving addition across a C=C double bond are the most exergonic, either when a carbon based radical (compare alkyl [32], acyl [33], and carbamoyl radicals [34], entries 4–6) or a photogenerated carbocation [35] (compare the phenyl cation-based arylation of nucleophilic alkenes, see entry 7) are involved. C-C bond formation is strongly exergonic also when acylation of a quinone ring is accompanied by conversion to hydroquinone (entry 8) [36]. Reduction of an aggressive, oxidant intermediate, such as a phenyl cation, is likewise strongly exergonic (entry 9) [37].

Small ring-forming reactions. Three and four-membered rings are notorious for the high energy connected to their strained structures. Thus, formation of cyclobutanes is exothermic when starting from two conjugated ketones and conserves both C=O groups (entry 10) [38], but slightly endothermic when only one of the C=C bonds is conjugated (entry 11) [39] and markedly so when both are isolated (entries 12, 13) [40,41], although the reaction turns exergonic when the molecule is not prearranged in a convenient conformation (entry 14) [42]. Notice, however, that cyclobutene formation from an alkyne is exergonic (entry 15) [43]. Oxetane formation is strongly endergonic (entry 16) [44]. Three-membered ring-formation likewise brings in endothermicity, as in the case of di-π-methane (entry 17) [45] and oxa-di-π-methane (entry 18) [46] rearrangements.

Other reactions. Formation of a compound containing a bond between carbon and an electron-withdrawing atom are strongly exergonic (entry 19) [47], as are of course combustions, included for the sake of comparison with high-energy thermal reactions (entries 20, 21). Notice that when the two last reactions are considered in the contrary direction, one has water-splitting or CO_2 reduction to methane (a 8-electron process), that are much more strongly endergonic than any of the above organic reactions and are typically studied by heterogeneous photocatalysis. In contrast, the almost thermoneutral oxidation of Hantzsch dihydropyridine (entry 22) [48] may be taken as a simplified example of processes occurring in living cells, where energy and "reducing power" are exchanged through small steps. It is apparent that at most a small fraction of the overall energy furnished to the system (ΔG_{EXC}) is stored as chemical energy in the products. Thus, in the most studied case, that of the rearrangement of norbornadiene (entry 12), most of the energy impinging is directly converted to heat during the conversion to quadricyclane.

Thermodynamics. As it is well known, comparing photochemical and thermal reactions is inappropriate, because excited states are different species, with a thermodynamic behavior of their own, not related to that of the ground states. Overall, photochemical reactions occur through different Potential Energy Surfaces (PESs). Molecules are promoted to the excited state with no geometric change and from there chemistry begins, until at some point they return to the ground state on a different PES and at a different atom configuration, that is to the product. The final product energy may lie either below or above the starting state of the reagent. It may be useful wondering at which point the system drops down. This may be illustrated by monitoring the bimolecular reduction of an aromatic carbonyl, an almost thermoneutral reaction. As it appears from Figure 1, formation of the intermediates, C-centered radicals, from triplet benzaldehyde and isopropanol involves only a moderate energy drop (steps *a*, *b*), while most of the decrease occurs at the bond formation stage (step *c*). In other words, most of the light energy absorbed is used in the bond breaking reaction, and then the high energy intermediates lose their energy by forming new bonds.

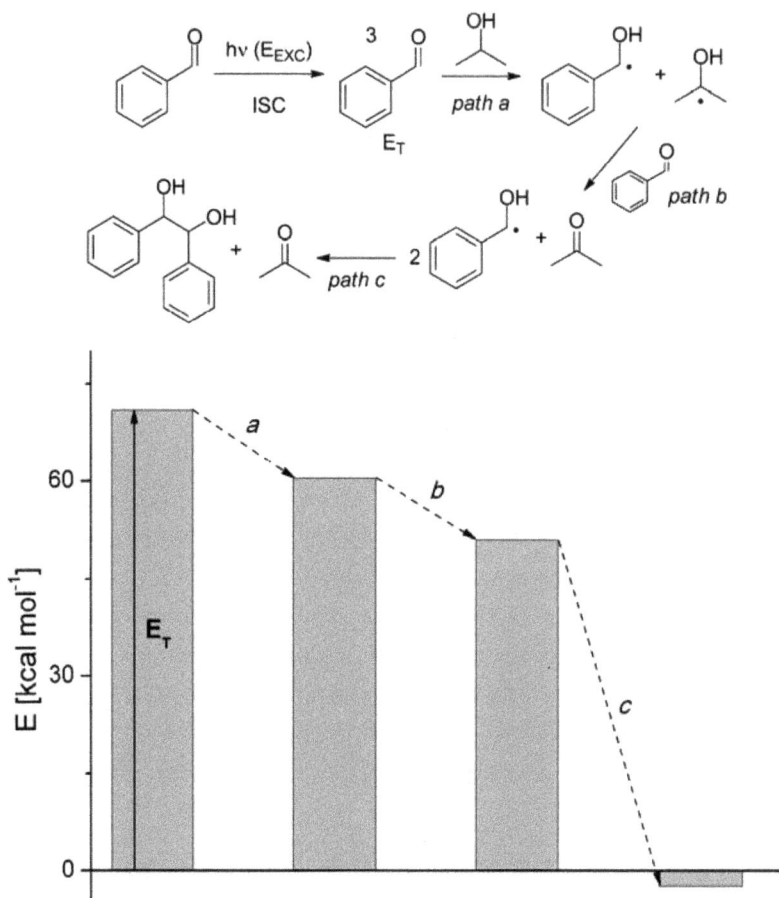

Figure 1. The energetic profile of the steps involved in the reaction reported in Table 1, entry 3. Adapted from Ref. [49].

Photocatalysis. The activation by the photocatalyst occurs by either of the two processes, atom transfer and electron transfer. A typical C-C bond forming reaction would be the addition of an organic compound R-H across a π-bond (see compound T, standing for "trap" in Figure 2; a well known example is an electron-poor olefin, such as acrylonitrile). The overall process is largely exothermic (see the related case reported in entry 4, Table 1), but requires the generation of an alkyl radical and thus the cleavage of a C-H bond, a process involving too high an energy for being amenable to laboratory conditions. The common way out is inserting a radicofugal group, as in the well known Giese method from alkyl halides via stannyl radicals as chain carriers [50]. The alkyl radicals are generated by transfer of a halogen atom X (Br, I) to a R₃Sn˙ radical (path *a* in Figure 2) and the radical adduct abstracts a hydrogen atom from a trialkylstannane R₃SnH (path *b*). The two processes are exergonic, *viz.* it is more convenient to make a Sn-X than a C-X bond, and a C-H than a Sn-H bond. In other words, the bond formed is always stronger than that broken (accordingly, black bars are smaller than grey bars for steps *a* and *b* in Figure 2). Thus, every cycle is exergonic and the process goes on. Photocatalysis has however more possibilities [51]. In fact, hydrogen transfer to an excited state that has a radical character is viable, in particular with triplet carbonyls, polyoxometalates and related species, where the excited state is similar to an alkoxy radical and thus hydrogen transfer forms a strong O-H bond (step *a'* in Figure 2). In the typical embodiment, alkylation by the radical formed by hydrogen abstraction follows. Thus, a catalytic, non-chain process is possible, with a quantum yield < 1, but occurring under very mild conditions.

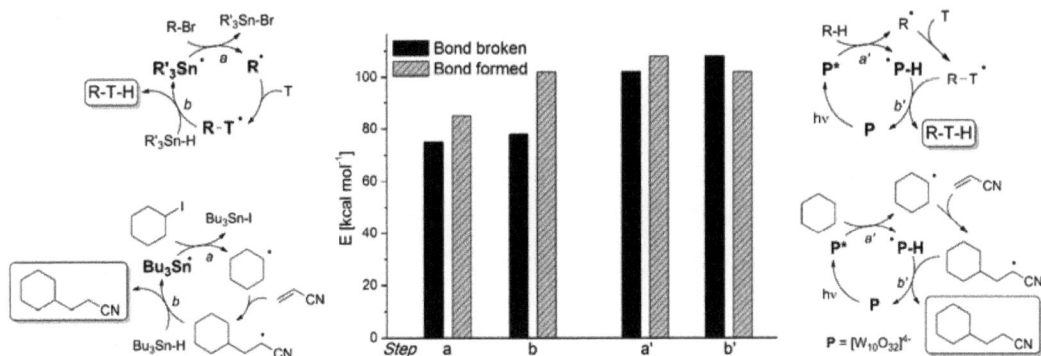

Figure 2. Chain *vs.* non-chain, photocatalytic process. The energy change associated to each step is considered. In the thermal chain process [50], both steps *a* and *b* involve the formation of a bond that is more energetic than the one that is cleaved, and the overall process involves relatively low energies. On the contrary, steps *a'* and *b'* in the photocatalytic process [15] involve the cleavage of a strong bond and the endergonic nature of the latter step precludes a chain process. Adapted from Ref. [49].

3. Experimental Section

The computational investigation has been carried out in order to recognize the energy change associated to the reaction under consideration. All of the simulations have been carried out with the Gaussian09 Software [52]. See Supplementary Materials for further details.

4. Conclusions

Both "catalytic" and "non-catalytic" photochemical reactions offer a wide-scope potential for a contribution to sustainability, that is green chemistry, by generating high-energy intermediates with the least possible exertion, coming down from above, from the high-lying excited state, not confronting a difficult ascending path from the ground state (as it occurs in thermal processes). These reactions appear to be a topic of great promise, although expensive from the energetic point of view. However, the mild conditions of the photochemical approach offer an unparalleled advantage and often allow to start from little-functionalized reagents via shorter paths [53,54]. As long as solar/visible light irradiation becomes used (a few examples are reported in Table 1), the potential of photocatalysis for organic synthesis should be fully developed [18–23,27].

As for energy storage, it is difficult to build molecules of very high energy content. The covalent bond is synonym of stability and even in the more favorable cases, in molecules specifically built for this aim, only a fraction of the energy of the photon (often around 10%–15%; as much as 25% in the most favorable instances) can be stored as chemical energy. If then one takes into account that exploiting such an energy requires overcoming a further barrier, and thus wasting part of the gain, it can be concluded that organic reactions (characterized by a small positive ΔG) are unlikely to become the main contribution of photochemistry to a sustainable world. The real energetic challenge remains limited to simple systems, such as water splitting and CO_2 reduction, where molecules not absorbing directly solar light are involved. Thus, these systems require a photocatalyst, allowing for the energy from an appropriate number of photons to be transferred into a molecule of a convenient energy vector, such as H_2 or CH_4 [55].

Supplementary Materials

Supplementary materials can be accessed at: http://www.mdpi.com/1420-3049/20/01/1527/s1.

Acknowledgments

Financial support from the Fondazione Cariplo (Grant No. 2009–2579) is gratefully acknowledged. S.P. acknowledges MIUR, Rome (FIRB-Futuro in Ricerca 2008 project RBFR08J78Q) for financial support. This work was funded by the CINECA Supercomputer Center, with computer time granted by ISCRA project HETFRA (code: HP10CSWWDQ).

Author Contributions

DR, SP and AA planned the research; DR and SP performed calculations and analyzed the data; DR, SP and AA wrote the paper. All authors read and approved the final manuscript.

Conflicts of Interest

The authors declare no conflict of interest.

References

1. Ciamician, G. Actions chimiques de la lumière. *Bull. Soc. Chim. Fr.* **1908**, *3–4*, i–xxvii.
2. Albini, A.; Fagnoni, M. Green chemistry and photochemistry were born at the same time. *Green Chem.* **2004**, *6*, 1–6.
3. Albini, A.; Fagnoni, M. 1908: Giacomo Ciamician and the concept of green chemistry. *ChemSusChem* **2008**, *1*, 63–66.
4. Albini, A.; Dichiarante, V. The belle époque of photochemistry. *Photochem. Photobiol. Sci.* **2009**, *8*, 248–254.
5. Paternò, E. Sintesi in chimica organica per mezzo della luce. Nota I. Introduzione. *Gazz. Chim. Ital.* **1914**, *44*, 31.
6. D'Auria, M. Una polemica fra Paternò e Ciamician. *Chim. Ind. (Milan)* **2009**, 106–109.

7. Warburg, E. The transformation of energy in photochemical processes in gases. II. *Sitzb. Preuss. Akad. Wiss.* **1912**, *746*, 216–225.

8. Benrath, A. On pure and combined photochemical reactions. *Z. Phys. Chem.* **1910**, *74*, 115–124.

9. IUPAC. Compendium of Chemical Terminology. The Gold Book. Available online: http://goldbook.iupac.org/ (accessed on 5 November 2014).

10. Prignano, A.L.; Trogler, W.C. Silica-Supported Bis(trialkylphosphine)platinum Oxalates. Photogenerated Catalysts for Hydrosilation of Olefins. *J. Am. Chem. Soc.* **1987**, *109*, 3586–3595.

11. Arceo, E.; Jurberg, I.D.; Alvarez-Fernàndez, A.; Melchiorre, P. Photochemical activity of a key donor-acceptor complex can drive stereoselective catalytic α-alkylation of aldehydes. *Nat. Chem.* **2013**, *3*, 750–756.

12. Minero, C.; Catozzo, F.; Pelizzetti, E. Role of Adsorption in Photocatalyzed Reactions of Organic Molecules in Aqueous TiO$_2$ Suspensions. *Langmuir* **1992**, *8*, 481–486.

13. Kamat, P.V. Photochemistry on Nonreactive and Reactive (Semiconductor) Surfaces. *Chem. Rev.* **1992**, *93*, 267–300.

14. Gerischer, H. Photochemistry of adsorbed species. *Faraday Discuss. Chem. Soc.* **1974**, *58*, 219–236.

15. Protti, S.; Ravelli, D.; Fagnoni, M.; Albini, A. Solar light-driven photocatalyzed alkylations. Chemistry on the window ledge. *Chem. Commun.* **2009**, 7351–7353.

16. Ravelli, D.; Protti, S.; Fagnoni, M.; Albini, A. Visible Light Photocatalysis. A Green Choice? *Curr. Org. Chem.* **2013**, *17*, 2366–2373.

17. Buckel, W. Radical and Electron Recycling in Catalysis. *Angew. Chem. Int. Ed.* **2009**, *48*, 6779–6787.

18. Zeitler, K. Photoredox Catalysis with Visible Light. *Angew. Chem. Int. Ed.* **2009**, *48*, 9785–9789.

19. Yoon, T.P.; Ischay, M.A.; Du, J. Visible light photocatalysis as a greener approach to photochemical synthesis. *Nat. Chem.* **2010**, *2*, 527–532.

20. Narayanam, J.M.R.; Stephenson, C.R.J. Visible light photoredox catalysis: applications in organic synthesis. *Chem. Soc. Rev.* **2011**, *40*, 102–113.

21. Teplý, F. Photoredox catalysis by [Ru(bpy)$_3$]$^{2+}$ to trigger transformations of organic molecules. Organic synthesis using visible-light photocatalysis and its 20th century roots. *Collect. Czech. Chem. Commun.* **2011**, *76*, 859–917.

22. Xuan, J.; Xiao, W.-J. Visible-Light Photoredox Catalysis. *Angew. Chem. Int. Ed.* **2012**, *51*, 6828–6838.

23. Prier, C.K.; Rankic, D.A.; MacMillan, D.W.C. Visible Light Photoredox Catalysis with Transition Metal Complexes: Applications in Organic Synthesis. *Chem. Rev.* **2013**, *113*, 5322–5363.

24. Ohtani, B. Photocatalysis by inorganic solid materials: Revisiting its definition, concepts, and experimental procedures. In *Advances in Inorganic Chemistry, Inorganic Photochemistry*; van Eldik, R., Stochel, G., Eds.; Academic Press, Elsevier Ltd.: London, UK, 2011; Volume 63, Chapter 10.

25. Bethke, S.; Drandm, S.; Treptow, B.; Geiter, R. Strained hydrocarbons from cyclic diynes, preparation and reactivity. *J. Phys. Org. Chem.* **2002**, *15*, 484–489.

26. Mills, A.; Le Hunte, S. An overview of semiconductor photocatalysis. *J. Photochem. Photobiol. A Chem.* **1997**, *108*, 1–35.

27. Kisch, H. Semiconductor Photocatalysis for Organic Synthesis. In *Advances in Photochemistry*; Neckers, D.C., von Bünau, G., Jenks, V.S., Eds.; John Wiley & Sons, Inc.: New York, NY, USA, 2007; Volume 26, p. 93.

28. Oelgemöller, M.; Jung, C.; Ortner, J.; Mattay, J.; Zimmermann, E. Green photochemistry: Solar photooxygenations with medium concentrated sunlight. *Green Chem.* **2005**, *7*, 35–38.

29. Oelgemöller, M.; Healy, N.; de Oliveira, L.; Jung, C.; Mattay, J. Green photochemistry: Solar-chemical synthesis of Juglone with medium concentrated sunlight. *Green Chem.* **2006**, *8*, 831–834.

30. Wootton, R.C.R.; Fortt, R.; de Mello, A.J. A microfabricated nanoreactor for safe, continuous generation and use of singlet oxygen. *Org. Proc. Res. Dev.* **2002**, *6*, 187–189.

31. Li, J.-T.; Yang, J.-H.; Han, J.-F.; Li, T.-S. Reductive coupling of aromatic aldehydes and ketones in sunlight. *Green Chem.* **2003**, *5*, 433–435.

32. Dondi, D.; Fagnoni, M.; Albini, A. Tetrabutylammonium Decatungstate-Photosensitized Alkylation of Electrophilic Alkenes: Convenient Functionalization of Aliphatic C-H Bonds. *Chem. Eur. J.* **2006**, *12*, 4153–4163.

33. Esposti, S.; Dondi, D.; Fagnoni, M.; Albini, A. Acylation of Electrophilic Olefins through Decatungstate-Photocatalyzed Activation of Aldehydes. *Angew. Chem. Int. Ed.* **2007**, *46*, 2531–2534.

34. Elad, D.; Rokach, J. The Light-Induced Amidation of Terminal Olefins. *J. Org. Chem.* **1964**, *29*, 1855–1859.

35. Lazzaroni, S.; Protti, S.; Fagnoni, M.; Albini, A. Photoinduced Three-Component Reaction: A Convenient Access to 3-Arylacetals or 3-Arylketals. *Org. Lett.* **2009**, *11*, 349–352.

36. Schiel, C.; Oelgemöller, M.; Ortner, J.; Mattay, J. Green photochemistry: The solar-chemical "Photo-Friedel-Crafts acylation" of quinones. *Green Chem.* **2001**, *3*, 224–228.

37. Dichiarante, V.; Fagnoni, M.; Albini, A. Eco-friendly hydrodehalogenation of electron-rich aryl chlorides and fluorides by photochemical reaction. *Green Chem.* **2009**, *11*, 942–945.

38. Du, J.; Yoon, T.P. Crossed Intermolecular [2+2] Cycloadditions of Acyclic Enones via Visible Light Photocatalysis. *J. Am. Chem. Soc.* **2009**, *131*, 14604–14605.

39. Büchi, G.; Goldman, I.M. Photochemical Reactions. VII. The Intramolecular Cyclization of Carvone to Carvonecamphor. *J. Am. Chem. Soc.* **1957**, *79*, 4741–4748.

40. Grutsch, P.A.; Kutal, C. Use of copper(I) phosphine compounds to photosensitize the valence isomerization of norbornadiene. *J. Am. Chem. Soc.* **1977**, *99*, 6460–6463.

41. Eaton, P.E.; Or, Y.S.; Branca, S.J. Pentaprismane. *J. Am. Chem. Soc.* **1981**, *103*, 2134–2136.

42. Langer, K.; Mattay, J. Stereoselective Intramolecular Copper(I)-Catalyzed [2+2]-Photocycloadditions. Enantioselective Synthesis of (+)- and (−)-Grandisol. *J. Org. Chem.* **1996**, *60*, 7256–7266.

43. Hook, B.D.A.; Dohle, W.; Hirst, P.R.; Pickworth, M.; Berry, M.B.; Booker-Milburn, K.I. A Practical Flow Reactor for Continuous Organic Photochemistry. *J. Org. Chem.* **2005**, *70*, 7558–7564.

44. D'Auria, M.; Racioppi, R.; Viggiani, L. Paternò-Büchi reaction between furan and heterocyclic aldehydes: Oxetane formation *vs.* metathesis. *Photochem. Photobiol. Sci.* **2010**, *9*, 1134–1138.

45. Armesto, D.; Ortiz, M.J.; Agarrabeitia, A.R.; El-Boulifi, N. Efficient photochemical synthesis of 2-vinylcyclopropanecarbaldehydes, precursors of cyclopropane components present in pyrethroids, by using the oxa-di-π-methane rearrangement. *Tetrahedron* **2010**, *66*, 8690–8697.

46. Hsu, D.-S.; Chou, Y.-Y.; Tung, Y.-S.; Liao, C.-C. Photochemistry of Tricyclo[5.2.2.0^{2,6}]undeca-4,10-dien-8-ones: An Efficient General Route to Substituted Linear Triquinanes from 2-Methoxyphenols. Total Synthesis of (±)-Δ$_9$(12)-Capnellene. *Chem. Eur. J.* **2010**, *16*, 3121–3131.

47. Podgoršek, A.; Stavber, S.; Zupan, M.; Iskra, J. Visible light induced "on water" benzylic bromination with *N*-bromosuccinimide. *Tetrahedron Lett.* **2006**, *47*, 1097–1099.

48. Zhang, D.; Wu, L.-Z.; Zhou, L.; Han, X.; Yang, Q.-Z.; Zhang, L.-P.; Tung, C.-H. Photocatalytic Hydrogen Production from Hantzsch 1,4-Dihydropyridines by Platinum(II) Terpyridyl Complexes in Homogeneous Solution. *J. Am. Chem. Soc.* **2004**, *126*, 3440–3441.

49. Fagnoni, M.; Albini, A. Photochemically-Generated Intermediates in Synthesis; Wiley: Hoboken, NJ, USA, 2013; Chapter 8.

50. Giese, B.; Gonzàlez-Gòmez, J.; Witzel, A. The Scope of Radical CC-Coupling by the "Tin Method". *Angew. Chem. Int. Ed.* **1984**, *23*, 69–70.

51. Fagnoni, M.; Dondi, D.; Ravelli, D.; Albini, A. Photocatalysis for the Formation of the C-C Bond. *Chem. Rev.* **2007**, *107*, 2725–2756.

52. Frisch, M.J.; Trucks, G.W.; Schlegel, H.B.; Scuseria, G.E.; Robb, M.A.; Cheeseman, J.R.; Scalmani, G.; Barone, V.; Mennucci, B.; Petersson, G.A.; *et al. Gaussian 09, Version D.01*; Gaussian, Inc.: Wallingford CT, USA, 2009.

53. Hoffmann, N. Photochemical Reactions as Key Steps in Organic Synthesis. *Chem. Rev.* **2008**, *108*, 1052–1103.

54. Dichiarante, V.; Protti, S. Photochemistry in ecosustanaible syntheses. Recent advances. In *CRC Handbook of Organic Photochemistry and Photobiology*, 3rd ed.; Griesbeck, A.G., Oelgemöller, M., Ghetti, A., Eds.; Taylor and Francis: Boca Raton, FL, USA, 2012; Chapter 9.

55. Protti, S.; Albini, A.; Serpone, N. Photocatalytic generation of solar fuels from the reduction of H$_2$O and CO$_2$: A look at the patent literature. *Phys. Chem. Chem. Phys.* **2014**, *16*, 19790–19827 and references therein.

Sample Availability: Not apply.

Formation of Combustible Hydrocarbons and H$_2$ during Photocatalytic Decomposition of Various Organic Compounds under Aerated and Deaerated Conditions

Sylwia Mozia, Aleksandra Kułagowska and Antoni W. Morawski

Abstract: A possibility of photocatalytic production of useful aliphatic hydrocarbons and H$_2$ from various organic compounds, including acetic acid, methanol, ethanol and glucose, over Fe-modified TiO$_2$ is discussed. In particular, the influence of the reaction atmosphere (N$_2$, air) was investigated. Different gases were identified in the headspace volume of the reactor depending on the substrate. In general, the evolution of the gases was more effective in air compared to a N$_2$ atmosphere. In the presence of air, the gaseous phase contained CO$_2$, CH$_4$ and H$_2$, regardless of the substrate used. Moreover, formation of C$_2$H$_6$ and C$_3$H$_8$ in the case of acetic acid and C$_2$H$_6$ in the case of ethanol was observed. In case of acetic acid and methanol an increase in H$_2$ evolution under aerated conditions was observed. It was concluded that the photocatalytic decomposition of organic compounds with simultaneous generation of combustible hydrocarbons and hydrogen could be a promising method of "green energy" production.

Reprinted from *Molecules*. Cite as: Mozia, S.; Kułagowska, A.; Morawski, A.W. Formation of Combustible Hydrocarbons and H$_2$ during Photocatalytic Decomposition of Various Organic Compounds under Aerated and Deaerated Conditions. *Molecules* **2014**, *19*, 19633-19647.

1. Introduction

Over the past thirty years increased concerns over emissions of greenhouse gases and the depletion of non-renewable resources of fossil fuels has caused the necessity to look for new methods of energy production. From both the ecological and economical point of view conversion of waste and wastewaters into energy is especially desirable. One of the most promising and popular approaches is biogas generation [1,2]. Biogas is a mixture of different gases, mainly methane and carbon dioxide. Its production during anaerobic digestion involves microorganisms, which results in some serious drawbacks of this technology, as the bacteria responsible for methane generation are very sensitive to the environmental conditions, such as oxygen content, pH or presence of certain organic and inorganic compounds [3]. Therefore, wastes or wastewaters containing substances which are toxic or recalcitrant to these microorganisms cannot be used in the traditional biogas production process.

Application of the photocatalytic process instead of the biological one could remove that restriction. Photocatalysis is not selective for any kind of substrates, therefore it might be used for treatment of all contaminants, even those which are toxic to the methanogenic bacteria [4].

Due to its significant activity, stability and low cost TiO$_2$ is widely used as a photocatalyst. Most investigations concerning the photocatalytic treatment of organic compounds in aqueous solutions are focused on their complete mineralization to CO$_2$ and H$_2$O. Usually, during these experiments the composition of the aqueous phase is only monitored. However, determination of the gas phase composition should be also of interest. There are some reports [5–11] showing that the process of a photocatalytic reduction of CO_2 may lead to methane formation.

The first papers concerning the photocatalytic generation of hydrocarbons from organics in liquid phase were published in the 1970s by Kraeutler and Bard [12–14]. These authors described a photocatalytic decarboxylation of acetic acid under UV light in the presence of Pt/TiO_2 photocatalyst. The reaction in which CH_4 and CO_2 were evolved as the products was named the "photo-Kolbe" reaction. A few years later Sakata *et al.* [15] reported methane and ethane formation during photodecomposition of acetic and propionic acids in the presence of bare and Pt modified TiO_2.

A possibility of hydrocarbon formation during photodegradation of C_1–C_3 alcohols in aqueous suspensions of TiO_2 was investigated by Dey and Pushpa [16]. They concluded that CH_4 and CO_2 were the main products of the reaction of methanol, ethanol and 2-propanol. Other hydrocarbons such as ethane, ethene and propene were also detected; however, at relatively low yields. Similar investigations were conducted by Bahruji *et al.* [17]. The authors used Pt–modified TiO_2 in order to increase H_2 formation. CH_4, CO_2, C_2H_6 and C_3H_8 were also identified in the gas phase.

Xu *et al.* [18] reported biomass reforming on Pt/TiO_2 (anatase-rutile structure) leading to H_2 generation. Methanol, propanetriol, formic acid and glucose were used as the model compounds and sacrificial agents. The possibility of hydrogen production from glucose, sucrose and starch over noble metal-loaded TiO_2 photocatalysts was also described by Fu *et al.* [19]. The results revealed an enhancement of H_2 production in case of Pd and Pt modified TiO_2 and an inhibition of the efficiency in aerated systems.

Recently, Klauson *et al.* [20] described the application of TiO_2 modified with Pt, Co, W, Cu or Fe for the production of hydrogen, oxygen and low molecular weight hydrocarbons from aqueous solutions of humic substances under anoxic conditions. In the presence of all the above materials the formation of CH_4 was observed, although the highest yield was found in case of Pt-TiO_2. That photocatalyst was also the most efficient when formation of C_2H_4, C_2H_6 and H_2 was taken into account.

In the present work an Fe-modified TiO_2 photocatalyst was applied for the photocatalytic generation of useful hydrocarbons and hydrogen which could be regarded as the potential source of "green energy". Different organics representing biomass-derived compounds, including an aliphatic acid (acetic acid), aliphatic alcohols (methanol and ethanol) and glucose were used in the experiments. In particular the influence of the reaction atmosphere on the products evolution was investigated. The Fe/TiO_2 photocatalyst was chosen on a basis of our previous investigations [21] during which we found that it exhibits high activity in the "photo-Kolbe" reaction using acetic acid as a substrate.

2. Results and Discussion

2.1. Photocatalytic Decomposition of Various Organic Compounds: The Influence of a Substrate on the Formation of the Gaseous and Liquid Products

Depending on the substrate, different gases were identified in the headspace volume of the reactor (Table 1). In case of acetic acid, the main products of its decomposition were CH_4 and CO_2. Low amounts of C_2H_6, C_3H_8 and H_2 were also identified. During the photocatalytic degradation of alcohols the following gaseous products were identified: CO_2, CH_4 and H_2 in case of CH_3OH and CO_2, CH_4, C_2H_6 and H_2 in case of C_2H_5OH. The gaseous products formed during photodegradation of $C_6H_{12}O_6$ were CH_4, CO_2 and H_2 (Table 1). The diversity of the products generated from the applied substrates resulted from their different photocatalytic decomposition pathways.

Table 1. Products identified in the gas and liquid phases after 27 h of irradiation over Fe/TiO_2.

Substrate	Composition of a Gas Phase	Composition of a Liquid Phase
CH_3COOH	CH_4, CO_2, C_2H_6, C_3H_8, H_2	CH_3COOH, CH_3OH, C_2H_5OH, $CO(CH_3)_2$, CH_3CHO, CH_3COOCH_3
CH_3OH	CH_4 [a], CO_2, H_2	CH_3OH, CH_3CHO
C_2H_5OH	CH_4, CO_2, C_2H_6, H_2	C_2H_5OH, CH_3CHO, CH_3OH
$C_6H_{12}O_6$	CH_4, CO_2, H_2	$C_6H_{12}O_6$, CH_3CHO, C_2H_5OH, CH_3COOCH_3

[a] in air atmosphere only.

Taking into consideration that some by-products of the organics' degradation must have been generated in the liquid phase, the composition of the reaction solution was also examined. The investigations revealed (Table 1) the presence of trace amounts of acetaldehyde (CH_3CHO) in all cases. Furthermore, methanol (CH_3OH) in the case of acetic acid and ethanol decomposition, and ethanol (C_2H_5OH) and methyl acetate (CH_3COOCH_3) in the case of acetic acid and glucose degradation were identified. In addition, small quantities of acetone ($CO(CH_3)_2$) were detected during the photodecomposition of acetic acid. The amounts of all the products in the liquid phase were very low and no clear dependence of the liquid phase composition on the reaction atmosphere used was found.

2.2. Effect of the Reaction Atmosphere on Gas Phase Composition during the Photodegradation of Various Organic Substrates

The concentrations of gaseous reaction products evolved with time of irradiation were continuously monitored during the experiments. Figures 1–4 present changes of the amounts of CO_2 and CH_4 in the gaseous phase during the processes conducted under either N_2 or air atmospheres. In Figures 5 and 6 a comparison of the amounts of C_2H_6 and H_2 evolved after 27 h of the decomposition of the model compounds is shown.

Figure 1. Evolution of CH₄ and CO₂ in time of irradiation during the photocatalytic degradation of CH₃COOH. Photocatalyst loading: 1g/dm³; CH₃COOH concentration: 1 mol/dm³; solution pH: 2.6; t = 25 °C.

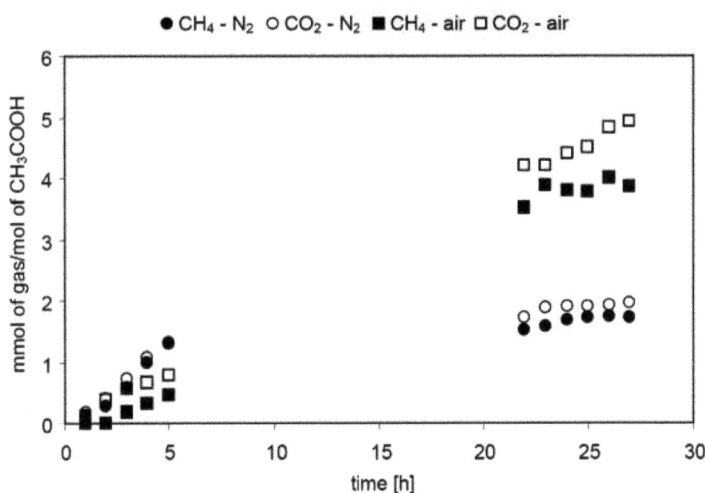

Figure 2. Evolution of CH₄ and CO₂ in time of irradiation during the photocatalytic degradation of CH₃OH. Photocatalyst loading: 1g/dm³; CH₃OH concentration: 1 mol/dm³; solution pH: 6.3; t = 25 °C.

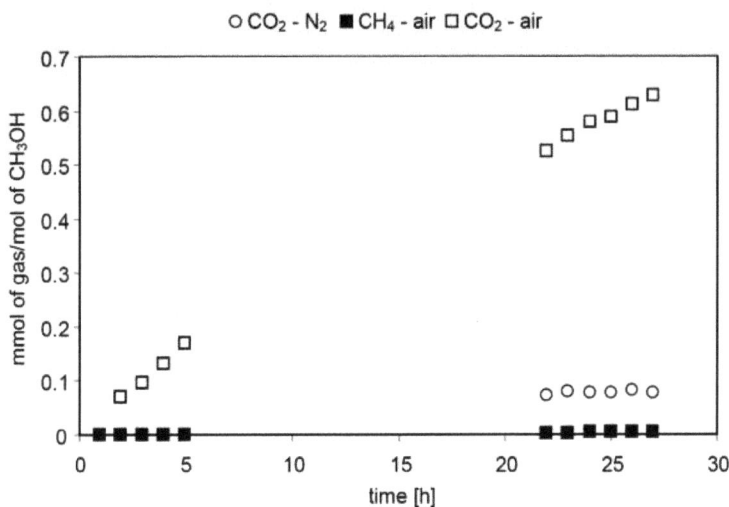

Figure 3. Evolution of CH₄ and CO₂ in time of irradiation during the photocatalytic degradation of C₂H₅OH. Photocatalyst loading: 1g/dm³; C₂H₅OH concentration: 1 mol/dm³; solution pH: 4.8; t = 25 °C.

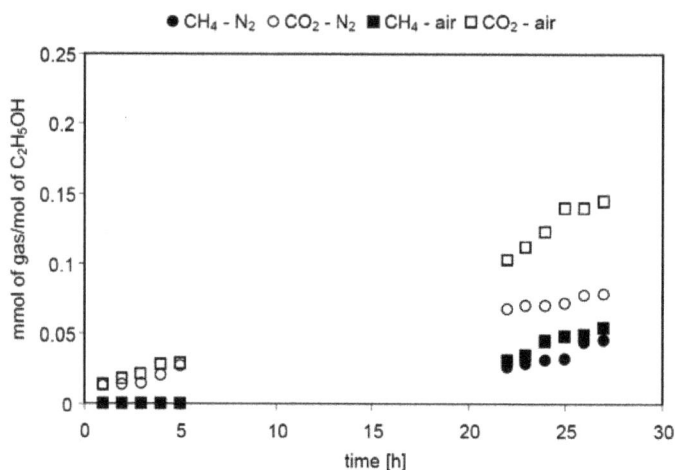

Figure 4. Evolution of CH_4 and CO_2 in time of irradiation during the photocatalytic degradation of $C_6H_{12}O_6$. Photocatalyst loading: 1 g/dm^3; $C_6H_{12}O_6$ concentration: 1 mol/dm^3; solution pH: 5.4; t = 25 °C.

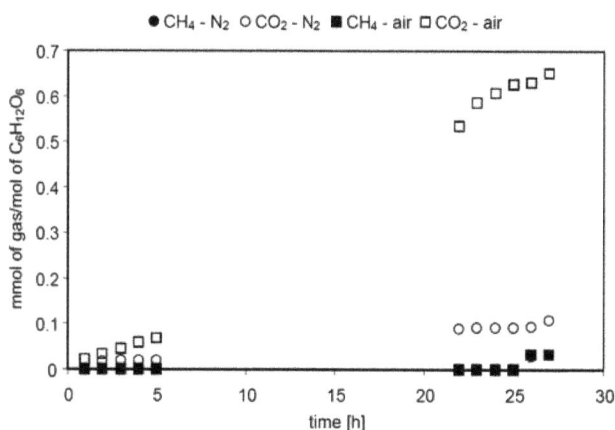

2.2.1. Acetic Acid

In general, the main mechanism responsible for a photocatalytic decomposition of CH_3COOH is its decarboxylation initiated by the photogenerated holes (h^+). This reaction, known as the "photo–Kolbe" reaction, leads to the production of one mole of CO_2 and one mole of CH_4 from one mole of CH_3COOH:

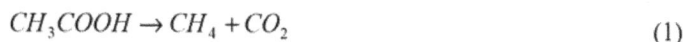

$$CH_3COOH \rightarrow CH_4 + CO_2 \tag{1}$$

Moreover, recombination of methyl radicals might take place, which results in a formation of C_2H_6, except from CH_4 [12–15,21–24]. Formation of C_2H_6 and H_2 can be written as follows:

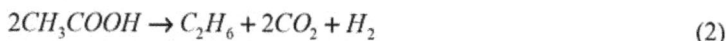

$$2CH_3COOH \rightarrow C_2H_6 + 2CO_2 + H_2 \tag{2}$$

Further, as can be seen in Table 1, formation of C_3H_8 can also occur. A possible mechanism of propane generation can be as follows [15]:

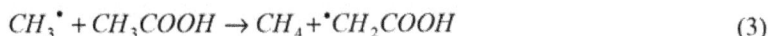

$$CH_3^\bullet + CH_3COOH \rightarrow CH_4 + {}^\bullet CH_2COOH \tag{3}$$

or:

$$OH^\bullet + CH_3COOH \rightarrow H_2O + {}^\bullet CH_2COOH \tag{4}$$

$${}^\bullet CH_2COOH + CH_3^\bullet \rightarrow C_2H_5COOH \tag{5}$$

$$C_2H_5COOH + h^+ \rightarrow {}^\bullet C_2H_5 + CO_2 + H^+ \tag{6}$$

$${}^\bullet C_2H_5 + CH_3^\bullet \rightarrow C_3H_8 \tag{7}$$

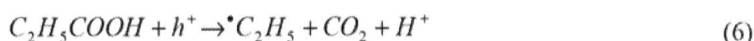

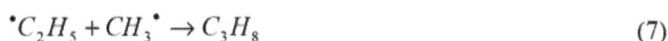

Nevertheless, the present results clearly show that the CH_4/CO_2 ratio after 27 h of irradiation was 0.88 when a N_2 atmosphere was applied and 0.78 when the process was conducted in the presence of air. This suggests that reaction (1) was not the only one proceeding in the system. From Table 1 it can be found that aside from methane, ethane was also formed. In this process methyl radicals are consumed. Therefore, the amount of ethane should be also taken into consideration. Assuming that two methyl radicals form one molecule of C_2H_6 the $CH_3^\bullet/CO_2$ ratio can be calculated. After 27 h of irradiation of acetic acid solution the amount of C_2H_6 evolved in a N_2 atmosphere was 0.05 mmol C_2H_6/mol CH_3COOH, whereas under aerated conditions it was 0.09 mmol C_2H_6/mol CH_3COOH. Thus, the $CH_3^\bullet/CO_2$ ratio was 0.93 and 0.82 for N_2 and air atmosphere, respectively. However, the values are still below 1. Incorporation of C_3H_8 in the calculations also does not allow one to get a ratio of 1, since the amount of propane was an order of magnitude lower than that of ethane. These results suggest that formation of carbon dioxide might also be due to the mineralization of CH_3COOH to H_2O and CO_2:

$$CH_3COOH + 2O_2 \rightarrow 2CO_2 + 2H_2O \tag{8}$$

Reaction (8) is understandable when the aerated conditions are considered; however, the obtained results revealed that it also proceeded in the N_2-purged system. In our previous paper [21] we have discussed higher evolution rate of CO_2 compared to CH_4 by the reaction of CH_3COOH with the photogenerated oxygen. This O_2 as well as the hydroxyl radicals might be responsible for the mineralization of CH_3COOH [21], which leads to higher CO_2 evolution.

The results shown in Figure 1 revealed that the amounts of CH_4 and CO_2 evolved under aerated conditions were more than two times higher compared to a N_2 atmosphere (1.72 vs. 3.85 mmolCH_4/molCH_3COOH and 1.95 vs. 4.93 mmolCO_2/molCH_3COOH, respectively, after 27 h). Higher efficiency of CH_4 evolution under the aerated conditions can be explained by more effective separation of e^-/h^+ pairs in the presence of O_2, being an efficient electron scavenger, and acetic acid, which is known as an effective hole scavenger. Therefore, in the presence of both oxygen and CH_3COOH the "photo–Kolbe reaction" should occur more easily, what was confirmed by the results presented in Figure 1. Moreover, it was found that the concentration of O_2 in the headspace volume of the reactor decreased from 21 to 12 vol.% after 27 h of irradiation, which confirms that oxygen was consumed in the process.

Figure 5. Comparison of the amounts of C_2H_6 evolved during the photocatalytic degradation of various organic substrates after 27 h of irradiation in the presence of Fe/TiO_2. Photocatalyst loading: $1 g/dm^3$; substrate concentration: $1 mol/dm^3$; $t = 25$ °C.

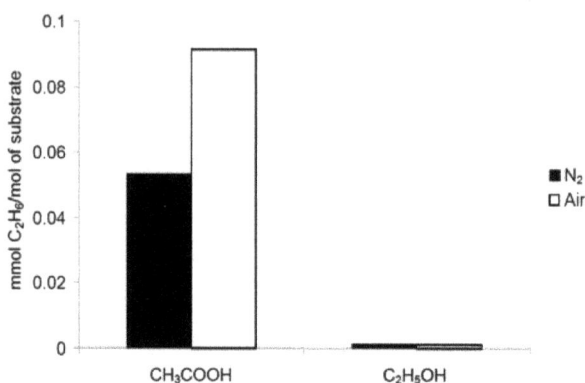

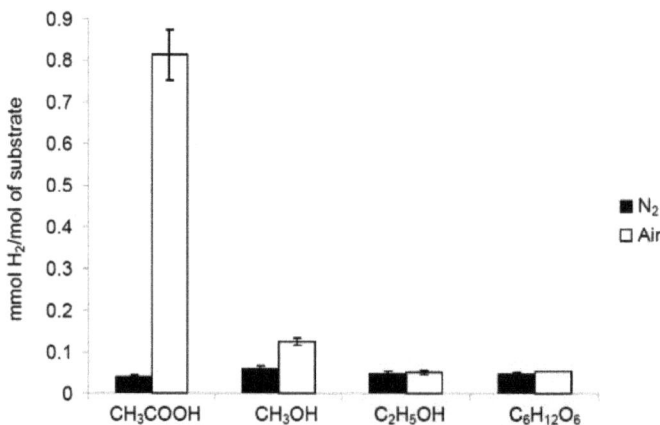

The obtained results (Figure 5) also revealed higher efficiency of C_2H_6 evolution in the aerated compared to the N_2 purged system. Ethane formation (Reaction (2)) is initiated by the photogenerated holes, therefore, can easily proceed under both deaerated and aerated conditions. However, like in case of methane, more efficient separation of e^-/h^+ pairs contributes to the enhancement of ethane formation. Moreover, the presence of O_2 can result in the increase of the amount of C_2H_6 by enabling of its formation according to the following equation [23,24]:

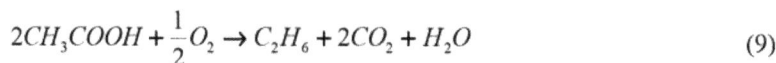

$$2CH_3COOH + \frac{1}{2}O_2 \rightarrow C_2H_6 + 2CO_2 + H_2O \tag{9}$$

Figure 6. Comparison of the amounts of H_2 evolved during the photocatalytic degradation of various organic substrates after 27 h of irradiation in the presence of Fe/TiO_2. Photocatalyst loading: $1 g/dm^3$; substrate concentration: $1 mol/dm^3$; $t = 25$ °C.

As shown in Table 1, amongst the products of CH_3COOH decomposition hydrogen was also present. As in case of other gases, evolution of H_2 was significantly higher in an air atmosphere compared to a N_2 one (Figure 6). After 27 h of irradiation the amounts of H_2 were 0.04 and 0.81 mmolH_2/mol CH_3COOH in N_2 and air purged system, respectively. The data discussed above show that the photocatalytic conversion of CH_3COOH into hydrocarbons and hydrogen was significantly more effective in the presence of air than in the N_2 purged system.

2.2.2. Methanol

The photocatalytic degradation of methanol under deaerated conditions can be written as [25]:

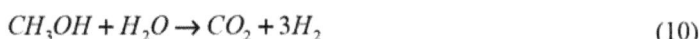

$$CH_3OH + H_2O \rightarrow CO_2 + 3H_2 \tag{10}$$

This reaction can also be represented as two half-reactions of oxidation and reduction, respectively:

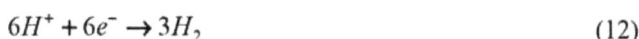

$$CH_3OH + H_2O + 6h^* \rightarrow CO_2 + 6H^+ \tag{11}$$

$$6H^+ + 6e^- \rightarrow 3H_2 \tag{12}$$

As reported by Chen et al. [25], Reaction (12) cannot occur easily in an aerated system because only few hydrogen atoms are formed in the presence of oxygen. Under such conditions, oxygen is more competitive in capturing the photogenerated electrons, which eventually leads to the formation of H_2O_2 and $OH^\cdot$.

The obtained results (Table 1) revealed formation of CO_2 and H_2 as the only gaseous products of CH_3OH decomposition in N_2 atmosphere, which confirms the mechanism presented by Equations (10)–(12). Nonetheless, if the only reaction occurring in the investigated system were Reaction (10), the H_2/CO_2 ratio should be equal to 3, but the experimental data show that the ratio is significantly lower (ca. 0.7–0.8). This suggests that some other reactions proceeded in the system. As in case of CH_3COOH, such a reaction can be mineralization of CH_3OH yielding CO_2 and H_2O as products [24]:

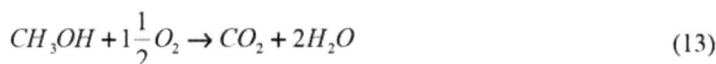

$$CH_3OH + 1\frac{1}{2}O_2 \rightarrow CO_2 + 2H_2O \tag{13}$$

During the experiments conducted in the air-purged system, the evolution of methane, except from CO_2 and H_2, was observed (Figure 2). Its concentration in the gaseous mixture was, however, very low and after 27 h of irradiation it only amounted to 4.26 μmol/molCH_3OH. Nonetheless, the observed formation of CH_4 might lead to a conclusion that the mechanism of methanol decomposition in the presence of air is not as simple as the one described by Equation (13). For example, a possibility of CO_2 photoreduction cannot be excluded here [16]. Dey and Pushpa reported that carbon dioxide, generated during mineralization of methanol, could undergo a methanation reaction by e^- and yield CH_4. In the case of greater amounts of CO_2 (as is the case in this work, when the system was aerated) there is a better chance of it being reduced, which can explain the results shown in Figure 2.

The amount of CO_2 evolved in the presence of air was at the end of the experiment about eight times higher compared to the N_2 atmosphere (0.627 vs. 0.078 mmolCO_2/mol CH_3OH, respectively). High CO_2 evolution was an effect of methanol mineralization (Equation (13)) and was accompanied by a decrease of O_2 concentration in the gaseous phase (from 21 to 16 vol.%). It was also observed that in the presence of N_2 no gaseous product evolved from the reaction mixture within the initial 5 h of the experiment. On the contrary, when the reaction was conducted under aerated conditions the evolution of CO_2 started after 2 h of irradiation.

Evolution of hydrogen was significantly lower compared to that of CO_2 (Figure 6). After 27 h of irradiation the amount of H_2 was 0.06 and 0.13 mmolH_2/mol CH_3OH in the N_2 and air purged systems, respectively.

2.2.3. Ethanol

In the case of ethanol, the main products identified in the gaseous mixture were CH_4 and CO_2 (Figure 3). Moreover, some amounts of C_2H_6 and H_2 were also identified (Figures 5 and 6). The CH_4/CO_2 ratio was higher in N_2 than in an air atmosphere and amounted to 0.83 and 0.05, respectively. This resulted from significantly higher CO_2 evolution in the presence of air compared to the N_2-purged system (1.45 $vs.$ 0.078 mmolCO_2/mol C_2H_5OH after 27 h). As in case of other substrates a decrease of O_2 concentration in the gaseous phase in case of the experiments conducted under aerated conditions was found (from 21 to 15 vol.%). It was also observed that the amount of methane obtained under both conditions was comparable (Figure 3).

Decomposition of ethanol is more complex compared to methanol due to the presence of the ethyl group in the C_2H_5OH structure. As a result, the range of intermediate degradation products is very wide [25]. In case of the deaerated conditions the overall reaction of ethanol decomposition can be written as follows [17,24]:

$$C_2H_5OH + H_2O \rightarrow CO_2 + 2H_2 + CH_4 \tag{14}$$

The reduction reaction can be represented by Equation (12), like in case of methanol [25]. However, the oxidation reactions are different. Generally, methane can be produced either by the reaction of free methyl radicals with H˙ or ethanol, or the reaction of acetic radicals with ethanol [24]. In case of the present research, since acetic acid was not identified in the liquid phase (Table 1), the most probable pathway of CH_4 formation was the one involving CH_3˙ and C_2H_5OH. Moreover, as in case of methanol [16], the reduction of CO_2 leading to the methane production cannot be ignored here. Furthermore, methyl radicals can also recombine yielding C_2H_6 as the product (Figure 5).

In the presence of oxygen the decomposition of ethanol can be described by the following equation [24]:

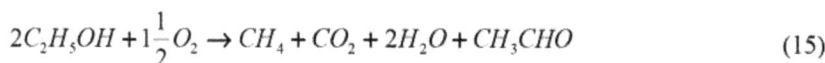

$$2C_2H_5OH + 1\frac{1}{2}O_2 \rightarrow CH_4 + CO_2 + 2H_2O + CH_3CHO \tag{15}$$

Equation (14) indicates that H_2 should be present amongst the ethanol decomposition products. Indeed, the analysis of the gaseous phase composition revealed evolution of hydrogen under both the aerated and deaerated conditions (Figure 6). Furthermore, H_2 could be produced by a degradation of the intermediate products present in the liquid phase (CH_3CHO, CH_3OH, Table 1). However, taking into account that their concentrations were very low, this pathway was of minor importance. The amount of H_2 formed in the N_2 purged system was comparable to that in the aerated system (0.049 vs. 0.051 mmolH$_2$/molC$_2$H$_5$OH, respectively). If we recall the methane evolution under aerated and deaerated conditions (Figure 3) we may find that the reaction atmosphere did not clearly influence the effectiveness of H$_2$ and CH$_4$ formation during ethanol decomposition.

2.2.4. Glucose

The photocatalytic reforming of C$_6$H$_{12}$O$_6$ is a very complex process which proceeds through numerous steps, in which intermediates such as carboxylic acids, aldehydes and hydrocarbons are formed [19,26]. A detailed probable mechanism of glucose degradation under anaerobic conditions leading to the formation of H$_2$ and CO$_2$ was recently discussed by Fu et al. [19]. The overall reaction can be written as:

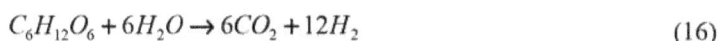

$$C_6H_{12}O_6 + 6H_2O \rightarrow 6CO_2 + 12H_2$$ (16)

The present research confirmed the formation of CO$_2$ and H$_2$ (Figures 4 and 6). In addition, small amounts of CH$_4$ were identified as well. From Figure 4 it can be found that the amount of CO$_2$ was higher in the presence of air compared to a N$_2$ atmosphere, which is consistent with the results observed for other substrates. After 27 h of irradiation the amount of CO$_2$ in the gaseous mixture was 0.65 mmolCO$_2$/molC$_6$H$_{12}$O$_6$ and 0.11 mmolCO$_2$/molC$_6$H$_{12}$O$_6$ for air and N$_2$, respectively. In the experiment conducted under aerated conditions the concentration of O$_2$ in the gaseous phase decreased from 21 to 15 vol.% which confirms its consumption during glucose decomposition.

No significant difference between the efficiency of hydrogen evolution in the two systems was observed. In case of the N$_2$-purged system the amount of H$_2$ was 0.048 mmolH$_2$/molC$_6$H$_{12}$O$_6$, whereas in case of the aerated system, it was 0.054 mmolH$_2$/molC$_6$H$_{12}$O$_6$ (Figure 6). Similarly, no difference in the amount of methane evolved in the presence and in the absence of oxygen was found. After 27 h of the reaction in both N$_2$ and air atmospheres, the amount of CH$_4$ reached 0.033 mmolCH$_4$/molC$_6$H$_{12}$O$_6$ (Figure 4). The observed evolution of methane can be explained by decomposition of by-products formed in the liquid phase (Table 1) as well as CO$_2$ photoreduction, as discussed earlier.

2.3. Hydrogen Evolution in the Presence of Oxygen: a Point of Discussion

The results discussed above revealed that the presence of oxygen at a concentration of 21 vol.% or less (i.e., oxygen in air) did not suppress hydrogen evolution during the photodegradation of organic compounds in the performed experiments. What is more, in the cases of acetic acid and methanol a significant enhancement of H$_2$ formation was even observed (Figure 6). This is somewhat unusual in view of the electron acceptability of O$_2$ and the competitiveness with H$^+$ for electron scavenging [16,19,24,25,27].

In order to investigate if the observed phenomenon resulted from the presence of Fe in the photocatalyst structure, an additional experiment was performed. A TiO$_2$ photocatalyst prepared in a similar way to the Fe/TiO$_2$, but without impregnation with Fe(NO$_3$)$_3$, was applied in a process of photocatalytic CH$_3$COOH degradation under N$_2$ and air atmosphere. After 27 h of irradiation it was found that the effectiveness of evolution of CH_4 and H_2 in the air purged system was higher by 65 and 45%, respectively, compared to the N_2 atmosphere. Therefore, it was concluded that the addition of iron was not responsible for the phenomenon described above.

There are very few papers reporting that O_2 does not affect negatively or could have a positive influence on hydrogen photogeneration [28–30]. Korzhak et al. [28] found that when a small amount of air was introduced to a photocatalytic system containing ethanol, the yield of H_2 formation increased. However, in case of mixtures saturated with oxygen or air, hydrogen formation was almost completely suppressed. The authors contributed the observed increase in hydrogen production to the fact that under such conditions the reactions of O_2 with active free organic radicals take place with high rate constants. Therefore oxygen is consumed mainly in the process leading to the evolution of additional amounts of hydrogen. Moreover, dissolved oxygen might be involved in stabilization of the radical intermediates thus could enhance the reaction efficiency [31]. Furthermore, organic substrates such as acids, alcohols or glucose, contribute to the improvement of charge separation by scavenging of photogenerated holes and consuming O_2 in diverse direct oxidation reactions, which leads to a decrease of the oxygen concentration [32–34].

Anyhow, the majority of the work on hydrogen generation with semiconductors dispersed in a solution is carried out in an oxygen–free atmosphere to avoid the back recombination processes, oxygen interferences with the photocatalyst which occurs while forming of superoxides and/or peroxides and the competition of O_2 and H^+ for the reduction sites [15,17–19,26,32,35–37]. Most of the papers which describe the photocatalytic degradation of organics in the presence of O_2 are focused on its total mineralization, thus the evolution of H_2 is not discussed. We have proved that the negative O_2 influence on the H_2 generation from different organic substrates is not so evident. In some cases (e.g., decomposition of acetic acid) an increase in H_2 evolution yield can even be obtained. Therefore, a broad and detailed discussion is needed in order to explain the discussed phenomenon.

3. Experimental Section

3.1. Photocatalyst

The photocatalyst used in this study was described in details in our previous paper [21]. In brief, the Fe/TiO_2 was prepared by an impregnation method using crude TiO_2 obtained from the Chemical Factory "Police" (Police, Poland) and $(Fe(NO_3)_3)$ as the Fe precursor. The sample was calcined at 500 °C. The amount of Fe introduced to the sample was 20 wt.%. The Fe/TiO_2 contained anatase, rutile and Fe_2O_3 phases. The crystallite size of anatase and the anatase over rutile ratio were equal to 9 nm and 87:13, respectively. The specific surface area S_{BET} was 82 m^2/g.

3.2. Photocatalytic Reaction

The photocatalytic reaction was conducted in a cylindrical quartz reactor (type UV-RS-2, Heraeus, Hanau, Germany) equipped with a medium pressure mercury vapour lamp (TQ-150, λ_{max} = 365 nm). The total volume of the reactor was 765 cm^3 (350 cm^3 of a liquid phase and 415 cm^3 of headspace). In the upper part of the reactor a gas sampling port was mounted. At the beginning of the experiment 0.35 dm³ of CH₃COOH, CH₃OH, C₂H₅OH or C₆H₁₂O₆ solution and 1 g/dm³ of the photocatalyst were introduced into the reactor. The concentration of the organic substrates was 1 mol/dm³ in all the experiments.

Before the photocatalytic reaction N₂ (in order to eliminate the dissolved oxygen) or air were bubbled through the reactor for 1 h. Then, the gas flow was stopped and UV lamp, positioned in the centre of the reactor, was turned on to start the photoreaction. The process was conducted for 27 h. The reaction mixture containing the photocatalyst in suspension was continuously stirred during the experiment by means of a magnetic stirrer. All the experiments were repeated at least twice in order to confirm the reproducibility of the results. Gaseous products of the reaction were analyzed using a SRI 8610C GC (SRI Instruments, Torrance, CA, USA) equipped with TCD and HID detectors, and Shincarbon (carbon molecular sieve; 2 m, 1 mm, 100–120 mesh), molecular sieve 5 Å (3 m, 2 mm, 80–100 mesh) and 13× (1.8 m, 2 mm, 80–100 mesh) columns. Helium was used as the carrier gas. The composition of the liquid phase was determined using a SRI 8610C GC equipped with a FID detector and a MXT®-1301 (60 m) column. Hydrogen was used as the carrier gas.

4. Conclusions

The possibility of photocatalytic generation of combustible hydrocarbons and hydrogen from various organic substrates, including an aliphatic acid (CH₃COOH), alcohols (CH₃OH, C₂H₅OH) and sugar (C₆H₁₂O₆) was demonstrated. The composition of the gaseous phase was influenced by both the applied substrate and the reaction atmosphere. In general, higher efficiency of hydrocarbon and hydrogen generation was obtained under aerated conditions, which is very advantageous from the point of view of possible future applications. In the presence of air, the gaseous phase contained CO₂, CH₄ and H₂, regardless of the substrate used. Moreover, formation of C₂H₆ and C₃H₈ in the case of acetic acid and C₂H₆ in the case of ethanol was observed.

The obtained results revealed that the presence of oxygen did not suppress hydrogen evolution during the photodegradation of organic compounds. In the cases of acetic acid and methanol a significant enhancement of H₂ formation was even observed. Further investigations concerning this issue as well as the improvement of the efficiency of the presented system are in progress.

Acknowledgments

This work has been supported by the Polish Ministry of Science and Higher Education as a scientific project N N523 413435 (2008–2011).

Author Contributions

Sylwia Mozia designed the study, managed the literature search and was involved in writing the first draft and data collection. Aleksandra Kułagowska performed measurements and was involved in manuscript writing. Antoni W. Morawski participated in analysis and data interpretation. All authors read and approved the final manuscript.

614

Conflicts of Interest

The authors declare no conflict of interest.

References

1. Demirel, B.; Scherer, P. Bio-methanization of energy crops through mono-digestion for continuous production of renewable biogas. *Renew. Energy* **2009**, *34*, 2940–2945.
2. Pöschl, M.; Ward, S.; Owende P. Evaluation of energy efficiency of various biogas production and utilization pathways. *Appl. Energy* **2010**, *87*, 3305–3321.
3. Hanaki, K.; Hirunmasuwan, S.; Matsuo, T. Protection of methanogenic bacteria from low pH and toxic materials by immobilization using polyvinyl alcohol. *Water Res.* **1994**, *28*, 877–885.
4. Augugliaro, V.; Litter, M.; Palmisano, L.; Soria, J. The combination of heterogeneous photocatalysis with chemical and physical operations: A tool for improving the photoprocess performance. *J. Photochem. Photobiol. C* **2006**, *7*, 127–144.
5. Tan, S.S.; Zou L.; Hu E. Photosynthesis of hydrogen and methane as key components for clean energy system. *Sci. Technol. Adv. Mater.* **2007**, *8*, 89–92.
6. Subrahmanyam, M.; Kaneco, S.; Alonso-Vante, N. A screening for the photo reduction of carbon dioxide supported on metal oxide catalysts for C_1–C_3 selectivity. *Appl. Catal. B* **1990**, *23*, 169–174.
7. Asi, M.A.; He, C.; Su, M.; Xia, D.; Lin, L.; Deng, H. Photocatalytic reduction of CO_2 to hydrocarbons using AgBr/TiO_2 nanocomposites under visible light. *Catal. Today* **2011**, *175*, 256–263.
8. Collado, L.; Jana, P.; Sierra, B.; Coronado, J.M.; Pizarro, P.; Serrano, D.P.; de la Peña O'Shea, V.A. Enhancement of hydrocarbon production via artificial photosynthesis due to synergetic effect of Ag supported on TiO_2 and ZnO semiconductors. *Chem. Eng. J.* **2013**, *224*, 128–135.
9. Tahir, M.; Saidina Amin, N. Recycling of carbon dioxide to renewable fuels by photocatalysis: Prospects and challenges. *Renew. Sustain. Energy Rev.* **2013**, *25*, 560–579.
10. Mei, B.; Pougin, A.; Strunk, J. Influence of photodeposited gold nanoparticles on the photocatalytic activity of titanate species in the reduction of CO_2 to hydrocarbons. *J. Catal.* **2013**, *306*, 184–189.
11. Tu, W.; Zhou, Y.; Zou, Z. Photocatalytic conversion of CO_2 into renewable hydrocarbon fuels: State-of-the-art accomplishment, challenges, and prospects. *Adv. Mater.* **2014**, *26*, 4607–4626.
12. Kraeutler, B.; Bard, A.J. Photoelectrosynthesis of ethane from acetate ion at an n-type TiO_2 electrode. The photo-Kolbe reaction. *J. Am. Chem. Soc.* **1977**, *99*, 7729–7731.
13. Kraeutler, B.; Bard, A.J. Heterogeneous photocatalytic synthesis of methane from acetic acid–new Kolbe reaction pathway. *J. Am. Chem. Soc.* **1978**, *100*, 2239–2240.
14. Kraeutler, B.; Jaeger, C.D.; Bard, A.J. Direct observation of radical intermediates in the photo–Kolbe reaction–Heterogeneous photocatalytic radical formation by electron spin resonance. *J. Am. Chem. Soc.* **1978**, *100*, 4903–4905.

15. Sakata, T.; Kawai, T.; Hashimoto, K. Heterogeneous photocatalytic reactions of organic acids in water. New reaction paths besides the photo-Kolbe reaction. *J. Phys. Chem.* **1984**, *88*, 2344–2350.

16. Dey, G.R.; Pushpa, K.K. Formation of different products during photo-catalytic reaction on TiO$_2$ suspension in water with and without 2-propanol under diverse ambient conditions. *Res. Chem. Intermed.* **2006**, *32*, 725–736.

17. Bahruji, H.; Bowker, M.; Davies, P.R.; Saeed Al-Mazroai, L.; Dickinson, A.; Greaves, J.; James, D.; Millard, L.; Pedrono, F. Sustainable H$_2$ gas production by photocatalysis. *J. Photochem. Photobiol. A* **2010**, *216*, 115–118.

18. Xu, Q.; Ma, Y.; Zhang, J.; Wang, X.; Feng, Z.; Li, C. Enhancing hydrogen production activity and suppressing CO formation from photocatalytic biomass reforming on Pt/TiO$_2$ by optimizing anatase–rutile phase structure. *J. Catal.* **2011**, *278*, 329–335.

19. Fu, X.; Long, J.; Wang, X.; Leung, D.Y.C.; Ding, Z.; Wu, L.; Zhang, Z.; Li, Z.; Fu, X. Photocatalytic reforming of biomass: A systematic study of hydrogen evolution from glucose solution. *Int. J. Hydrog. Energy* **2008**, *33*, 6484–6491.

20. Klauson, D.; Budarnaja, O.; Beltran, I.C.; Krichevskaya, M.; Preis, S. Photocatalytic decomposition of humic acids in anoxic aqueous solutions producing hydrogen, oxygen and light hydrocarbons. *Environ. Technol.* **2014**, *35*, 2237–2243.

21. Mozia, S.; Heciak, A.; Morawski, A.W. Photocatalytic acetic acid decomposition leading to the production of hydrocarbons and hydrogen on Fe-modified TiO$_2$. *Catal. Today* **2011**, *161*, 189–195.

22. Asal, S.; Saif, M.; Hafez, H.; Mozia, S.; Heciak, A.; Moszyński, D.; Abdel-Mottaleb, M.S.A. Photocatalytic generation of useful hydrocarbons and hydrogen from acetic acid in the presence of lanthanide modified TiO$_2$. *Int. J. Hydrog. Energy* **2011**, *36*, 6529–6537.

23. Muggli, D.; Falconer, J.L. Parallel pathways for hotocatalytic decomposition of acetic acid on TiO$_2$. *J. Catal.* **1999**, *197*, 230–237.

24. Blount, M.C.; Buchholz, J.A.; Falconer, J.L. Photocatalytic decomposition of aliphatic alcohols, acids, and esters. *J. Catal.* **2001**, *197*, 303–314.

25. Chen, J.; Ollis, D.F.; Rulkens, W.H.; Bruning, H. Photocatalyzed oxidation of alcohols and organochlorides in the presence of native TiO$_2$ and metallized TiO$_2$ suspensions. Part (II): Photocatalytic mechanisms. *Water Res.* **1999**, *33*, 669–676.

26. Li, Y.; Wang, J.; Peng, S.; Lu, G.; Li, S. Photocatalytic hydrogen generation in the presence of glucose over ZnS-coated ZnIn$_2$S$_4$ under visible light irradiation. *Int. J. Hydrog. Energy* **2010**, *35*, 7116–7126.

27. Dey, G.R.; Nair, K.N.R.; Pushpa, K.K. Photolysis studies on HCOOH and HCOO$^-$ in presence of TiO$_2$ photocatalyst as suspension in aqueous medium. *J. Nat. Gas Chem.* **2009**, *18*, 50–54.

28. Korzhak, A.V.; Kuchmii, S.Y.; Kryukow, A.I. Effects of activation and inhibition by oxygen of the photocatalytic evolution of hydrogen from alcohol-water media. *Theor. Exp. Chem.* **1994**, *30*, 26–29.

29. Borgarello, E.; Serpone, N.; Pelizzetti, E.; Barbeni, M. Efficient photochemical conversion of aqueous sulphides and sulphites to hydrogen using a rhodium-loaded CdS photocatalyst. *J. Photochem.* **1986**, *33*, 35–48.

30. Medrano, J.A.; Oliva, A.; Ruiz, J.; Garcia, L.; Arauzo, J. Catalytic steam reforming of acetic acid in a fluidized bed reactor with oxygen addition. *Int. J. Hydrog. Energy* **2008**, *33*, 4387–4396.

31. Rauf, M.A.; Meetani, M.A.; Hisaindee, S. An overview on the photocatalytic degradation of azo dyes in the presence of TiO_2 doped with selective transition metals. *Desalination* **2011**, *276*, 13–27.

32. Fu, X.; Wang, X.; Leung, D.Y.C.; Xue, W.; Ding, Z. Photocatalytic reforming of glucose over La doped alkali tantalate photocatalysts for H_2 production. *Catal. Commun.* **2010**, *12*, 184–187.

33. Rosseler, O.; Shankar, M.V.; Karkmaz-Le Du, M.; Schmidlin, L.; Keller, N.; Keller, V. Solar light photocatalytic hydrogen production from water over Pt and Au/TiO_2(anatase/rutile) photocatalysts: Influence of noble metal and porogen promotion. *J. Catal.* **2010**, *269*, 179–190.

34. Patsoura, A.; Kondarides, D.I.; Verykios, X.E. Photocatalytic degradation of organic pollutants with simultaneous production of hydrogen. *Catal. Today* **2007**, *124*, 94–102.

35. Strataki, N.; Bekiari, V.; Kondarides, D.I.; Lianos, P. Hydrogen production by photocatalytic alcohol reforming employing highly efficient nanocrystalline titania films. *Appl. Catal. B* **2007**, *77*, 184–189.

36. Wu, G.; Chen, T.; Su, W.; Zhou, G.; Zong, X.; Lei, Z.; Li, C. H_2 production with ultra-low CO selectivity via photocatalytic reforming of methanol on Au/TiO_2 catalyst. *Int. J. Hydrog. Energy* **2008**, *33*,1243–1251.

37. Fu, X.; Leung, D.Y.C.; Wang, X.; Xue, W.; Fu, X. Photocatalytic reforming of ethanol to H_2 and CH_4 over $ZnSn(OH)_6$ nanocubes. *Int. J. Hydrog. Energy* **2011**, *36*, 1524–1530.

Sample Availability: Samples are not available.

Photocatalysis for Renewable Energy Production Using PhotoFuelCells

Robert Michal, Stavroula Sfaelou and Panagiotis Lianos

Abstract: The present work is a short review of our recent studies on PhotoFuelCells, that is, photoelectrochemical cells which consume a fuel to produce electricity or hydrogen, and presents some unpublished data concerning both electricity and hydrogen production. PhotoFuelCells have been constructed using nanoparticulate titania photoanodes and various cathode electrodes bearing a few different types of electrocatalyst. In the case where the cell functioned with an aerated cathode, the cathode electrode was made of carbon cloth carrying a carbon paste made of carbon black and dispersed Pt nanoparticles. When the cell was operated in the absence of oxygen, the electrocatalyst was deposited on an FTO slide using a special commercial carbon paste, which was again enriched with Pt nanoparticles. Mixing of Pt with carbon paste decreased the quantity of Pt necessary to act as electrocatalyst. PhotoFuelCells can produce electricity without bias and with relatively high open-circuit voltage when they function in the presence of fuel and with an aerated cathode. In that case, titania can be sensitized in the visible region by CdS quantum dots. In the present work, CdS was deposited by the SILAR method. Other metal chalcogenides are not functional as sensitizers because the combined photoanode in their presence does not have enough oxidative power to oxidize the fuel. Concerning hydrogen production, it was found that it is difficult to produce hydrogen in an alkaline environment even under bias, however, this is still possible if losses are minimized. One way to limit losses is to short-circuit anode and cathode electrode and put them close together. This is achieved in the "photoelectrocatalytic leaf", which was presently demonstrated capable of producing hydrogen even in a strongly alkaline environment.

Reprinted from *Molecules*. Cite as: Michal, R.; Sfaelou, S.; Lianos, P. Photocatalysis for Renewable Energy Production Using PhotoFuelCells. *Molecules* **2014**, *19*, 19732-19750.

1. Introduction

Simultaneous conversion of solar energy into chemical energy and electricity is achieved by means of photoelectrochemical cells. A standard configuration for a photoelectrochemical cell involves a photoanode electrode carrying an n-type semiconductor photocatalyst, a counter electrode carrying an electrocatalyst and an electrolyte. Light is absorbed by the photocatalyst generating electron-hole pairs. Electrons are guided through an external circuit to the cathode (counter) electrode, where they take part in reduction reactions while holes are consumed through oxidation reactions. In regenerative solar cells, the electrolyte involves a redox couple, which is reduced at the counter electrode and oxidized at the photoanode electrode thus refilling holes and converting photon energy into electricity. Those cells act as photovoltaic devices. The present work will not deal with such cells but it will rather focus on photoelectrochemical procedures where both oxidation and reduction reactions involve consumption or production of chemical substances. In that case, holes are consumed by oxidizing an organic or inorganic substance, which acts as a sacrificial agent. Electrons are consumed by some standard reduction reactions, for example, water, oxygen or CO_2 reduction. A less common configuration for a photoelectrochemical cell involves a photocathode electrode carrying a p-type photocatalyst while the anode electrode carries an oxidation electrocatalyst. The present work will deal with a standard cell employing a photoanode and n-type semiconductor photocatalyst. In both types of cells, the energy of photons and the chemical energy of the sacrificial agent, *i.e.*, of the "fuel", are converted into electricity and/or stored in a useful form of chemical energy, for example, production of hydrogen. For this reason, photoelectrochemical cells operating by consumption of a fuel are called PhotoFuelCells (PFCs) [1,2].

The origins of photoelectrochemical cells can be traced back to the findings of Alexander-Edmond Becquerel in 1839, who reported for the first time the photovoltaic (in reality, the photoelectrochemical) conversion of light into electricity [3]. However, it was a work by Fujishima and Honda in 1972 [4] describing water splitting by using a TiO_2 photoanode that set the foundations of modern photoelectrochemistry. After more than 40 years, the conception of a photoelectrochemical cell remains the same but the new materials and procedures employed for their construction has kept the field alive with ever increasing popularity. Photoanode electrodes are usually made by depositing nanoparticulate titania (np-TiO_2) on transparent conductive glasses like Fluorine-doped Tin Oxide (FTO). The choice of a photocatalyst should be dictated by the position of its conduction and valence band, which define its oxidation and reduction capacities. A large variety of different semiconductor photocatalysts have been studied through the years and reported in several publications [5–8]. However, other factors finally become more important, like the ease of synthesis and deposition, the specific surface area of the obtained film, the abundance of materials and the effectiveness of their photocatalytic functionality. Thus nanoparticulate titania is the champion of all photocatalysts, despite of some known disadvantages. The most serious of these disadvantages is the fact that it only absorbs UV radiation. In order to utilize this photocatalyst and at the same time exploit visible light, np-TiO_2 must be combined with appropriate sensitizers. The present work, among others, reviews our approach for the use of visible-light-absorbing sensitizers. A second main issue to face in the study of PFCs and of all photoelectrochemical cells is the nature of the electrocatalyst. Nanoparticulate Pt is acknowledged as the best electrocatalyst for both oxidation and reduction reactions. There are two reasons for this quality. Pt can be easily obtained in fine nanoparticles and it has the highest work function, thus becoming an electron sink, which can exchange charges with the active liquid phase. However, Pt is rare and costly and this discourages large scale applications. Alternative materials employed as electrocatalysts should be also deposited in nanoparticulate form providing high specific surface area and should be good conductors facilitating transfer of charges. The present work will shortly describe our approach for the employment of alternative reduction electrocatalysts. Finally, the third principal issue involved in the use of PFCs is the choice of the fuel. Any organic substance can play the role of fuel and be photocatalytically oxidized during PFC operation. In fact, one of the main assets of a PFC is its capacity to oxidize any organic substance and produce usable energy. Thus water soluble wastes or products of biomass can be used as fuel, offering the multiple environmental benefit of water cleaning, of using biomass that may be too costly to refine and of producing renewable energy.

Small chain length alcohols have been used as model fuels to employ with PFCs. In the present work, we use ethanol as model organic fuel. On the contrary, the choice of inorganic fuels is very limited. Most published works use sulfur containing compounds, for example, a combination of Na_2S with Na_2SO_3 [9]. The present work will not deal with inorganic sacrificial agents.

2. Model of PFC Operation

The PFC operation reactions are summarized in Table 1 and are detailed as follows: reactions in Table 1 are designed for ethanol, which is presently employed, as already said, as a model fuel. When UV light is shined on the photocatalyst (np-TiO_2), it is excited generating e^-—h^+ pairs (reaction (I)). These photogenerated charges may partly recombine and dissipate their energy. Electrons escaping recombination may be partly retained in the presence of O_2 leading to formation of superoxide radical and then to $^{\bullet}OH$ radicals, which add to the degradation capacity towards the organic fuel (reaction (II)). The majority of free electrons flow through the external circuit. Of course, in the absence of oxygen all electrons escaping recombination flow through the external circuit.

The type of reactions induced by photogenerated holes depend on solution pH and on the presence or absence of O_2. In the absence of O_2 and at low pH, holes mediate the generation of hydrogen ions H^+ according to reaction III. Reaction III represents an overall scheme. In reality, the reaction proceeds by steps, where the following route usually prevails [2]:

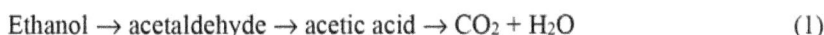

$$\text{Ethanol} \rightarrow \text{acetaldehyde} \rightarrow \text{acetic acid} \rightarrow CO_2 + H_2O \tag{1}$$

Table 1. Typical reactions taking place in a PFC employing ethanol as model fuel.

Photoanode
(1) *Reactions induced by absorption of UV photons (radiation mainly absorbed by the majority species, i.e., np-TiO₂)*
$$TiO_2 \xrightarrow{hv} e^- + h^+ \tag{I}$$
•Fate of the photogenerated electrons: Most electrons flow in the external circuit. Some may interact with O_2:
$$O_2 \xrightarrow{+e^-} O_2^{\bullet-} \xrightarrow{+H^+} HO_2^{\bullet} \xrightarrow{+e^-} HO_2^- \xrightarrow{+H^+} H_2O_2 \xrightarrow{+e^-} {}^{\bullet}OH + OH^- \tag{II}$$
•Fate of the photogenerated holes
$$C_2H_5OH + 3\ H_2O + 12\ h^+ \rightarrow 2\ CO_2 + 12\ H^+ \text{ (low pH)} \tag{III}$$
$$OH^- + h^+ \rightarrow OH^{\bullet} \text{ and } C_2H_5OH + 12\ OH^{\bullet} \rightarrow 2\ CO_2 + 9\ H_2O \text{ (high pH)} \tag{IV}$$
•Intermediate steps may involve interaction of the fuel with photogenerated holes in the valence band of titania or holes injected into the valence band of the sensitizer
$$C_2H_5OH + 2\ h^+ \rightarrow CH_3CHO + 2\ H^+ \text{ or } C_2H_5OH + h^+ \rightarrow C_2H_5O^{\bullet} + H^+ \tag{V}$$
(2) *Reactions induced by absorption of Visible light (exclusive excitation of the photosensitizer)*
•Excited electrons are rapidly injected into the conduction band of TiO_2. Some may interact with O_2 forming superoxide radical and following the scheme of reaction II. Most electrons flow through the

external circuit. Holes at the valence band of the photosensitizer may go through reactions III–V depending of the nature of the photosensitizer.
•Other schemes may also be possible.

Cathode	
Inert environment (no O_2 present)	Aerated electrolyte or cathode exposed to ambient air
•Low pH (potential: 0.00 V at pH = 0) $2 H^+ + 2 e^- \rightarrow H_2$ (VI) •High pH (potential: −0.83 V at pH = 14) $2 H_2O + 2 e^- \rightarrow H_2 + 2 OH^-$ (VII)	•Low pH (potential: 1.23 V at pH = 0) $2 H^+ + \frac{1}{2} O_2 + 2 e^- \rightarrow H_2O$ (VIII) •High pH (potential: 0.40 V at pH = 14) $H_2O + \frac{1}{2} O_2 + 2 e^- \rightarrow 2 OH^-$ (IX)

Overall Cell Reactions (combination of anode and cathode reactions)
•absence of oxygen (ethanol reforming): $\qquad C_2H_5OH + 3 H_2O \rightarrow 2 CO_2 + 6 H_2$ (X) •presence of oxygen (ethanol mineralization): $\qquad C_2H_5OH + 3 O_2 \rightarrow 2 CO_2 + 3 H_2O$ (XI)

Most of these intermediate reactions are endothermic. The Gibbs free energy change for reaction (III) is positive, therefore the overall balance is endergonic and is mediated by photogenerated holes. The same holds true for all reactions associated with chemical substances of the general composition $C_xH_yO_z$, which are generalized by the following scheme [2]:

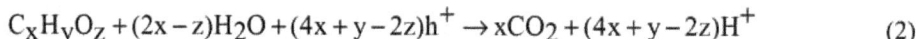

$$C_xH_yO_z + (2x - z)H_2O + (4x + y - 2z)h^+ \rightarrow xCO_2 + (4x + y - 2z)H^+ \qquad (2)$$

At high pH, a lot of $^\bullet$OH radicals are formed by interaction of OH^- with holes

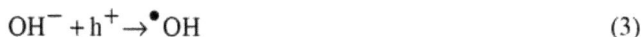

$$OH^- + h^+ \rightarrow {}^\bullet OH \qquad (3)$$

therefore, photocatalytic degradation should be visualized with their participation. In reality, ethanol photodegradation still follows the route of Equation (1) through intermediate acetaldehyde formation, as it has been previously detected and analyzed [10]. However, the number of hydrogen ions should be limited at high pH or completely eliminated by interacting with OH^- and producing water. Thus at high pH the photocatalytic degradation of ethanol should be represented by reaction IV if Table 1 while a more general scheme equivalent to Equation (2) should be defined by the following equation:

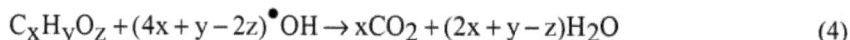

$$C_xH_yO_z + (4x + y - 2z){}^\bullet OH \rightarrow xCO_2 + (2x + y - z)H_2O \qquad (4)$$

where we assume that $H^+ + OH^- \rightarrow H_2O$ and $H_2O + h^+ \rightarrow H^+ + {}^\bullet OH$. In the presence of oxygen, alternative photocatalytic degradation routes may be activated, however, what is mainly affected by the presence of oxygen is the production of molecular hydrogen as will be seen in the next paragraph. We have previously found that the presence of O_2 accelerates mineralization of ethanol [10,11] at the expense of the current flowing in the external circuit. Special attention must be paid to the second reaction V of Table 1, showing ethanol radical formation by direct interaction with one hole. This radical is unstable and by interaction with the photocatalyst it injects an electron into its conduction band. This is responsible for the so-called "Current Doubling" effect [2], which will be further discussed in Section 3.1.

Reactions VI–IX taking place at the cathode electrode, as seen in Table 1, are less complicated and are independent of what is the fuel or the photocatalyst. It is seen that in the absence of oxygen, hydrogen is produced by reduction of hydrogen ions at low pH or by reduction of water at high pH. In the presence of O_2, reactions proceed with the participation of the latter. Of course, no hydrogen is produced in that case. The electrochemical potentials for the reduction reactions are given in Table 1. The variation of the potential with pH follows the general rule [8]:

$$\Delta V \,(Volts) = -0.059 \times \Delta pH \tag{5}$$

The values of these potentials define the conditions of PFC operation. When reactions VI–IX are combined with reactions III and IV they yield the characteristic schemes of cell operation, which are independent of the pH but distinguish themselves by the presence or not of oxygen. Thus electricity flows through the external circuit and hydrogen is produced in the absence of oxygen while in its presence only electricity is produced. These two overall operation schemes are represented by reactions X and XI of Table 1. The corresponding schemes for substances of the general composition $C_xH_yO_z$ are given by:

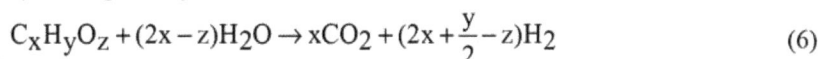

$$C_xH_yO_z + (2x-z)H_2O \rightarrow xCO_2 + (2x + \frac{y}{2} - z)H_2 \tag{6}$$

in the absence and by:

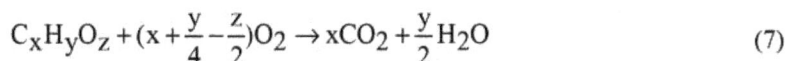

$$C_xH_yO_z + (x + \frac{y}{4} - \frac{z}{2})O_2 \rightarrow xCO_2 + \frac{y}{2}H_2O \tag{7}$$

in the presence of oxygen. The Gibbs free energy change ΔG^0 for Equation (7) has been calculated and listed in Table 2 for a few substances of the type $C_xH_yO_z$. The overall balance is exergonic ($\Delta G^0 < 0$). The corresponding standard potential, was calculated by the following equation:

$$E^0 = \frac{\Delta G^0}{-nF} \tag{8}$$

where n is the number of electron moles involved in the reaction and F is the Faraday constant 96.486 kC·mol^{-1}, while ΔG^0 is given in kJ·mol^{-1}. n and E^0 are also listed in Table 2. The number of electrons (and holes) is equal to $n = (4x + y - 2z)$, i.e., equal to the number of holes involved in reactions Equations (2) and (4) or to 4 electrons per oxygen molecule involved in reaction Equation (7). Thus all potentials are positive and their range is limited between 1.12–1.21 V with respect to the Standard Hydrogen Electrode (SHE). The corresponding value for ethanol is +1.14 V. Figure 1 shows energy levels for some semiconductors and redox reactions at pH 13. Alkaline pH is chosen because, as it will be discussed later, PFCs usually operate at high pH. In the diagram of Figure 1, the potential for ethanol mineralization in the presence of oxygen is depicted as EtOH/CO$_2$ level. Values in Figure 1 are referred to Reversible Hydrogen Electrode (RHE) to take account of the pH. Since pH = 13, it corresponds to a negative potential shift by 0.059 × 13 = 0.77 V, then the level of EtOH/CO$_2$ in Figure 1 is placed at 1.14 − 0.77 = +0.37 V. Furthermore, reaction (V), which is the first step of ethanol oxidation, with ΔG^0 equal to 34.8 kJ mol^{-1} and corresponding $E^0 = -0.18$ V is placed in the diagram of Figure 1 at $0.18 - 0.77 = -0.59$ V. The diagram of Figure 1 reveals that for holes found at the valence band level of titania, the oxidation power is high enough to carry out all listed reactions. However, in the presence of a sensitizer, holes are injected into the higher lying level of the sensitizer and then their oxidative power dramatically decreases. Thus all depicted sensitizers can oxidize ethanol to acetaldehyde, CdS and CdSe may catalyze ethanol mineralization and water oxidation but only CdS is oxidative enough to generate $^{\bullet}$OH radicals. Direct scavenging of holes by ethanol means that ethanol molecules are adsorbed on the semiconductor, but it is easier to react with $^{\bullet}$OH, which can take place in the liquid phase. Therefore, titania and CdS have higher chance to oxidize ethanol than CdSe and PbS. It is obvious that these interactions are fairly complicated. Obviously, mineralization of the fuel is encouraged by photocatalysts with high enough oxidation capacity.

Table 2. Calculated Gibbs free energy change and standard potential for reaction Equation (7) and for a few organic substances of the general composition $C_xH_yO_z$.

Name	Chemical Composition	Number of Electrons	$\Delta G^0(kJ \cdot mol^{-1})$	E^0 (V)
Methanol	CH_3OH	6	−702	1.21
Ethanol	C_2H_5OH	12	−1325	1.14
n-Propanol	C_3H_7OH	18	−1965	1.13
n-Butanol	C_4H_9OH	24	−2595	1.12
n-Pentanol	$C_5H_{11}OH$	30	−3249	1.12
Glycerol	$C_3H_8O_3$	14	−1656	1.22
Sorbitol	$C_6H_{14}O_6$	26	−3084	1.23
Acetic acid	CH_3COOH	8	−873	1.13

If titania is combined with a sensitizer then visible light absorption is possible. When visible light is shined on the photoanode, only the photosensitizer is excited, since titania does not absorb in the visible. Electrons are rapidly injected into the conduction band of titania (*cf.* Figure 1) and follow the routes related to reaction Equation (II) of Table 1. The fate of the holes in the valence band of the sensitizer depends on the valence band potential, as already discussed. Photocatalytic degradation of alcohols using oxide photocatalysts has been extensively studied, however, rare data or no data at all exist for combined titania-QD photocatalysts. The reason is that such systems are thought to be unstable. This is not the case for certain semiconductor sensitizers like CdS, as has been shown in previous publications [12,13].

The conduction band level of nanocrystalline titania lies at about −0.2 to −0.3 V *vs.* SHE [5–8,14] or about −1 V *vs.* RHE (pH 13). Then the potential difference with the cathode is ideally no more than 0.3 V in the absence of oxygen (*cf.* Figure 1 and Table 1). This difference, taking into account the inevitable losses, is not strong enough to provide drive for cell operation. Consequently, a bias is necessary in that case, which can be applied either externally or chemically. If the anode is in an alkaline environment and the cathode in acidic environment a chemical forward bias is provided [15] according to Equation (5). When the cathode electrode is in an aerated environment, a potential difference of around 1.4 V is established between the anode and the cathode electrode as seen in Figure 1 and Table 1. Indeed, experimental data have repeatedly fulfilled this expectation [12,16]. PFCs then operating in the presence of oxygen produce high voltage and operate without bias, like solar cells. Obviously, in that case a PFC can be used as a device to produce electricity consuming organic wastes and being activated by absorption of light. Despite the fact that in the absence of oxygen a bias is necessary to run the cell, as already said, it is still possible to produce hydrogen by minimizing the losses. This has been achieved by minimizing the resistance of electron flow between anode and cathode electrodes. An example of a functional cell of this type will be presented in the present work. The following sections present a short review on experimental data obtained by the above described PFC types including some unpublished data on electricity and hydrogen production using PFCs.

Figure 1. Approximate energy levels of a few semiconductors together with some characteristic redox potentials. The levels correspond to pH 13. They were negatively shifted with respect to the values at pH = 0, according to Equation (5). Thus at pH = 0 the potential of H_2/H_2O is zero. CdS, CdSe and PbS are shown as sensitizers by Fermi level alignment with TiO_2. The band gap values correspond to quantum dot nanoparticles and are larger than in bulk semiconductors.

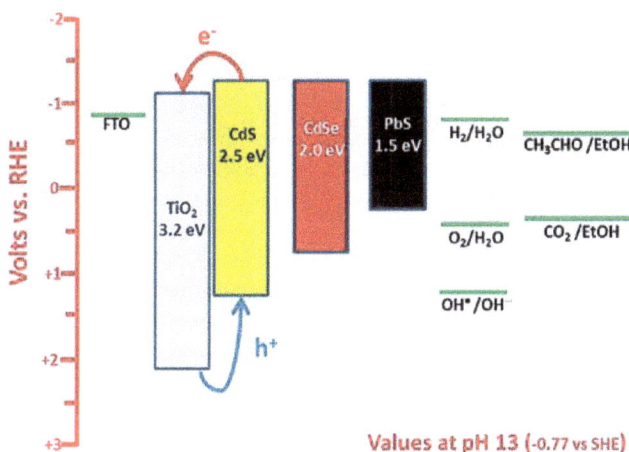

3. Results and Discussion

3.1. Study of PFCs Operating in the Presence of Oxygen and Producing Electricity

PFCs operating in the presence of oxygen with the purpose of producing electricity by photocatalytically consuming an organic fuel are schematically represented by Figure 2A. The photoanode carries np-TiO_2 photocatalyst with or without sensitizer. The cathode electrode is made of "air-breathing" carbon cloth on which a hydrophobic material (usually, carbon black and Teflon paste) is deposited. This hydrophobic paste prevents water leak through the carbon cloth. On this modified electrode a catalytic layer is deposited. The best catalyst is nanoparticulate Pt, as in the present case (see Experimental Section). However, satisfactory results were obtained in the past by employing carbon nanotubes or organic conductive polymers [17,18] as electrocatalysts. The fuel is added in the aqueous alkaline electrolyte. The cell can be also operated in the absence of fuel by oxidizing water, acting as purely water-splitting apparatus. Figure 3 shows JV characteristics obtained in a 2-electrode configuration using an air saturated PFC comprising a photoanode carrying bare titania and filled with NaOH electrolyte without fuel. The cell produced zero current in the dark but gave substantial current under illumination by a Xenon lamp, providing simulated solar radiation. Of course, in that case only the UV portion of the lamp is absorbed by the photocatalyst. In the presence of fuel (ethanol) both open-circuit voltage V_{oc} and short-circuit current density J_{sc} increased (curve 2 L of Figure 3). It is obvious that the presence of the fuel results in higher consumption of holes liberating more electrons. More electrons means higher current and, since the increase of electron population makes the conduction band of titania more negative, according to the discussion is Section 2, the potential difference between the anode and the cathode electrode increases, thus making V_{oc} higher. Indeed, in the absence of fuel J_{sc} and V_{oc} were 0.14 mA and 0.84 V, respectively but in its presence they became 0.61 mA and 1.03 V, respectively. However, part of the current recorded in the presence of fuel, may also come from "Current Doubling", as it will be discussed below. The above results show that it is possible to produce electricity by consumption of a fuel using a PFC. Water itself can play the role of fuel but the current produced is lower and so is the voltage. The current can be much higher if a sensitizer is used and a greater portion of the solar radiation is absorbed. This is shown in Figure 4, where it is seen that J_{sc} increased more than tenfold when CdS quantum dots were deposited on the np-TiO$_2$ film. Indeed, J_{sc} in that case became 6.6 mA. Interestingly, V_{oc} of the sensitized photoanode was the same as the non-sensitized one. This is expected. V_{oc} is approximately the difference between the conduction band level of nanocrystalline titania and the reduction level at the cathode electrode. In an alkaline environment, as in the present case, according to the discussion in Section 2, it is expected that V_{oc} may be as high as 1.4 V and it may be even higher in the presence of the fuel, which consumes holes and increases conduction band electronegativity, as already said. This value was never obtained due to inevitable losses. In addition, it is possible that reaction IX of Table 1 may be substituted by other reduction reactions with less positive potential, for example hydrogen peroxide formation (cf. [2]). Whatever is the case, the potential of the conduction band of titania defines the upper level. In the presence of CdS, if the latter acted as an independent semiconductor, this upper level would have been modified (cf. Figure 1). Because CdS acts as a sensitizer, its excited electrons are injected into the conduction band of titania, therefore, the latter defines the upper level. In fact the non-variation of V_{oc} in the presence of CdS is an index of sensitization.

It has been thus shown that CdS is an efficient sensitizer of titania in PFCs, applicable in aqueous alkaline environments. How stable is this sensitizer? A sensitizer remains stable as long as the photogenerated holes are consumed thus preventing sensitizer photo-corrosion by self-oxidation. This matter has been treated in previous publications [12,13]. It has indeed been found that if the oxidative power of the combined semiconductor-sensitizer is high enough, in order to be capable of producing hydroxyl radicals (cf. Figure 1), which are the main scavengers of photogenerated holes, then hole scavenging is ensured and the sensitizer is preserved. If the oxidative power is not sufficiently high, i.e., the valence band is not positive enough, as in the case of CdSe and PbS, then hole scavenging will rely on less efficient routes and the sensitizer becomes more vulnerable to oxidation. The best case certified by the previous findings is sensitization by moderately sized CdS quantum dots [13], as in the present case.

Figure 2. Design of reactors and electrodes: (**A**) Design of a PFC exclusively producing electricity; (**B**) Geometrical distribution of photocatalyst (small circles) and electrocatalyst (black area) on a "photoelectrocatalytic leaf"; and (**C**) Production of hydrogen using the "photoelectrocatalytic leaf".

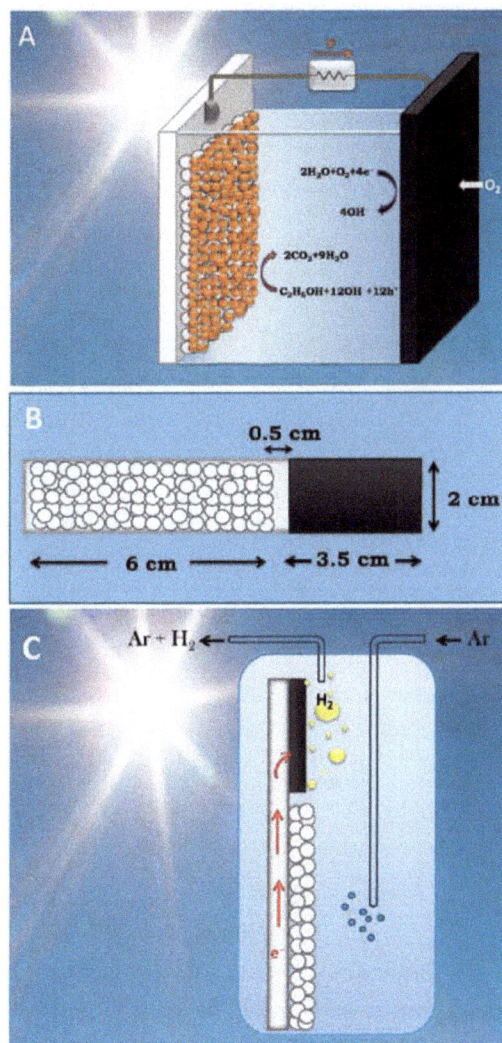

Figure 3. Current density-Voltage curves obtained with a PFC functioning in the presence of oxygen. The aqueous electrolyte always contained 0.5 mol·L^{-1} NaOH without (curve 1) or with (curve 2) 5 v % ethanol. L signifies irradiation (Light) and D dark. In both cases photoanode carried bare np-TiO$_2$ film. The curves were traced in a 2-eletrode configuration.

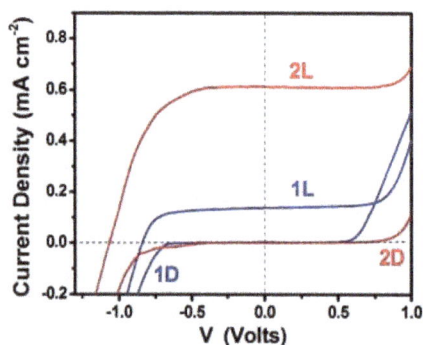

Figure 4. Current density-voltage curves obtained with a PFC functioning in the presence of oxygen. The aqueous electrolyte always contained 0.5 mol·L^{-1} NaOH and 5 v % ethanol. Both curves were obtained by illumination with simulated solar radiation. The lower curve corresponds to a photoanode carrying bare titania while the upper curve corresponds to a CdS-sensitized titania photoanode. The curves were traced in a 2-eletrode configuration.

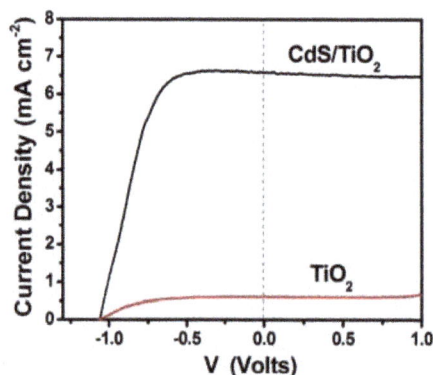

When the cell functions in the absence of organic fuel, photogenerated holes oxidize water. The data of Figure 3 show that it is easier to oxidize an organic fuel than to oxidize water. This is because organic substances can be oxidized by interaction with hydroxyl radicals, which, as already said, are efficient hole scavengers. Water cannot be easily oxidized because it takes two unit charges to split water, while an organic molecule can be oxidized by single steps, similar to that shown by the second reaction V of Table 1. This difficulty in oxidizing water precludes the employment of a sensitizer in the absence of a sacrificial agent. Thus CdS-sensitized photoanode was rapidly decomposed when used in the absence of ethanol [12]. Therefore, in the absence of fuel, titania must function alone without sensitizer. In this respect, it is interesting to also inspect the data presented in Figure 5, where current-voltage curves are plot in a 3-electrode configuration. It is seen that the saturation current in the presence of ethanol is to a rough approximation the double of the current in its absence. This finding calls for the "current doubling" effect frequently observed with such systems [2,19]. Current doubling is attributed to the photocatalytic formation of unstable organic radicals, as in the second reaction V of Table 1. The radical interacts with titania injecting an electron into its conduction band, according to the following reactions:

$$CH_3CH_2OH + h^+ \rightarrow CH_3CH_2O^\bullet + H^+ \tag{9}$$

$$CH_3CH_2O^\bullet \rightarrow CH_3CHO + H^+ + e^- \tag{10}$$

Figure 5. Current density-Voltage curves obtained with a PFC functioning in the presence of oxygen. The aqueous electrolyte always contained 0.5 mol·L^{-1} NaOH without (curve 1) or with (curve 2) 5 v % ethanol. L signifies irradiation (Light) and D dark. In both cases photoanode carried bare np-TiO$_2$ film. The curves were traced in a 3-electrode configuration using Ag/AgCl as reference electrode. The curves are presented *vs.* RHE by adding 0.97 V to the recorded potential.

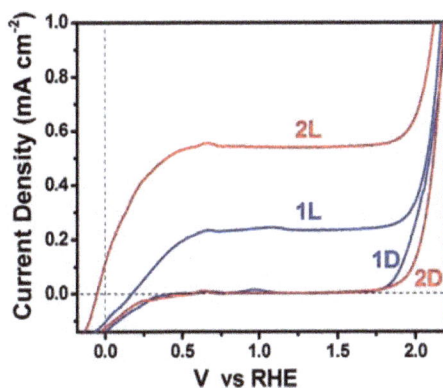

Thus for each photon absorbed, there are two electrons created in the conduction band, one photogenerated and one injected by the radical. This doubles the current, hence the term current doubling. We have not observed current doubling systematically [2,16] and it was not detected by the 2-electrode plots seen in Figure 3. We believe that current doubling is frequently shadowed by the normal photocatalytic mineralization route. That is, if the current recorded in the absence of fuel is low, the current observed in its presence may appear several times higher and not just the double [2,16]. The current observed in the absence of fuel may be small because of electron-hole recombination. In the presence of fuel, consumption of holes decreases recombination thus liberating more electrons and increasing current. Therefore, increase of current in the presence of an organic sacrificial agent may both come from the current doubling effect and from the decrease of electron-hole recombination. current doubling is surely responsible for unusually high currents. We have, for example, sometimes observed photon to electron conversion efficiencies, which are higher than 100% [12,20]. Even in the present case, CdS-sensitized photoanode gave high current, as seen in Figure 4, just above the theoretical limit. Indeed, for sensitizer band gap equal to 2.5 eV

(496 nm) the theoretical current density expected for 100% efficiency is 6.3 mA·cm^{-2}. The presently recorded short-circuit current density in Figure 4 was 6.6 mA·cm^{-2}. Such unexpected high current can only be justified if we accept that current doubling does take place during PFC operation.

3.2. Use of PFCs for Hydrogen Production

PFCs operating in an inert environment, *i.e.*, in the absence of oxygen, can produce hydrogen at the cathode electrode according to reaction VI or VII of Table 1. This has been verified in the past by using various cell configurations. As already analyzed in Section 2, a bias is necessary in order to produce a measurable quantity of hydrogen. Indeed, by using a two compartment cell separated by an ion transfer membrane, it is possible to operate the photoanode in an alkaline electrolyte and the cathode electrode in an acidic electrolyte. This introduces a chemical forward bias in the cell according to Equation (5). In addition, the acidic environment of the cathode electrode provides a reservoir of H$^+$ ions, which can be reduced to generate hydrogen molecules. Such cells have been previously studied and reported [15,21,22]. Replacement of the consumed hydrogen ions is supposed to be made by hydrogen ions produced during oxidation of the fuel. However, since the photoanode is immersed in an alkaline electrolyte, these hydrogen ions would have to struggle to survive against interaction with hydroxyl ions. For this reason, such cells do not function for a long period of time unless a continuous supply of acid is assured.

Another possibility is to use a one-compartment cell filled with a single (alkaline) electrolyte and apply an electric forward bias. We have reported this case in a previous publication [23]. Hydrogen was indeed produced but the quantity was disappointingly low. Production of hydrogen in an alkaline environment is foreseen by reaction VII of Table 1. However, this reaction does not take into account losses and the fact that it is hard for hydrogen ions to survive in a highly alkaline environment. Surprisingly, even though, there is a lot of work on water splitting this issue of the employed electrolyte is not sufficiently studied and it calls for further investigations. In any case, it is necessary to limit losses in order to produce a sufficient quantity of hydrogen by using PFCs functioning in an alkaline environment.

3.3. Short-Circuiting Anode and Cathode Electrode for Hydrogen Production in an Alkaline PFC. The "Photoelectrocatalytic Leaf"

In the previous subsection it was stated that it is necessary to keep losses to a minimum in order to detect hydrogen production by a PFC in an alkaline environment. Alkaline environment is preferred because it offers a higher supply of hydroxyl radicals and enhances the photocatalytic capacity of titania, as analyzed in Section 2. One way to limit losses is to short-circuit anode and cathode electrode facilitating electron transfer from anode to cathode. This is usually done by depositing photocatalyst and electrocatalyst on the two sides of a metal support [21,22]. A more practical configuration is presented in the present work by the "photoelectrocatalytic leaf" of Figures 2B,C, which supports oxidation and reduction reactions side by side, like in natural photosynthesis. By lying next to each other, the distance between anode and cathode is decreased.

Thus the route that ionic species have to run within the electrolyte is also decreased. In the present work, we present an easy to prepare and functional electrode using commercial materials deposited on a single FTO slide. As described in the Experimental section and depicted in Figure 2B, the photocatalyst covers an area of 6 cm × 2 cm and the electrocatalyst an area of 3.5 cm × 2 cm. Three possibilities of electrocatalyst have been tested (see Section 4.3). Pt casted directly on FTO, a commercial carbon paste called Elcocarb and Pt mixed with Elcocarb. All three were functional in producing molecular hydrogen, as can be seen in Figure 6. The curves of Figure 6 are structured in three parts: a rising part, which corresponds to hydrogen building up in the reactor and the tubing; the peak rate, which shows the relative efficiency of each electrocatalyst; and the falling part, showing exhaustion of the fuel. It is seen that Elcocarb paste alone gave very low hydrogen production rate. Pt alone gave much higher rate but the highest was obtained with the Pt-enriched Elcocarb paste. It must be underlined at this point that the quantity of Pt alone (0.6 mg) was much larger than the quantity of Pt mixed with Elcocarb paste (0.08 mg). Therefore, combination of these materials offers a very important advantage. As seen by its FESEM image shown in Figure 7, Elcocarb film is made of a nanostructured material that apparently provides a means for fine distribution of Pt nanoparticles. Thus much smaller quantity of platinum has a better electrocatalytic effect than pure Pt, which apparently loses its properties by aggregation.

Figure 6. Hydrogen production rate using the "photoelectrocatalytic leaf" (short-circuited electrodes) of Figure 2B,C. The photoanode was made of np-TiO_2 while the electrocatalyst was made of: (1) 4 mg of Elcocarb alone; (2) 0.6 mg Pt alone and (3) 0.08 mg of Pt dispersed in 4 mg of Elcocarb. The alkaline electrolyte contained 5 v % ethanol.

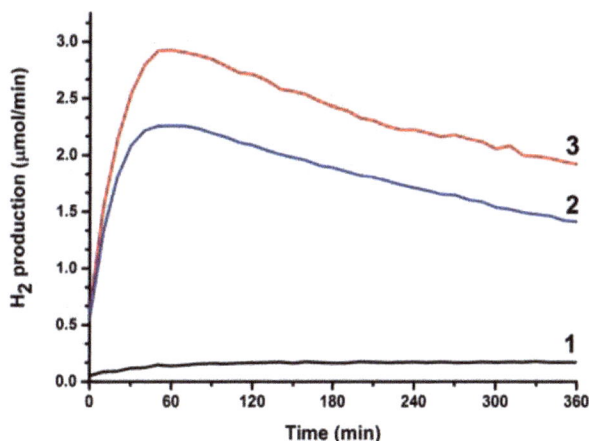

Figure 7. FESEM image of the Elcocarb film. The scale bar is 200 nm.

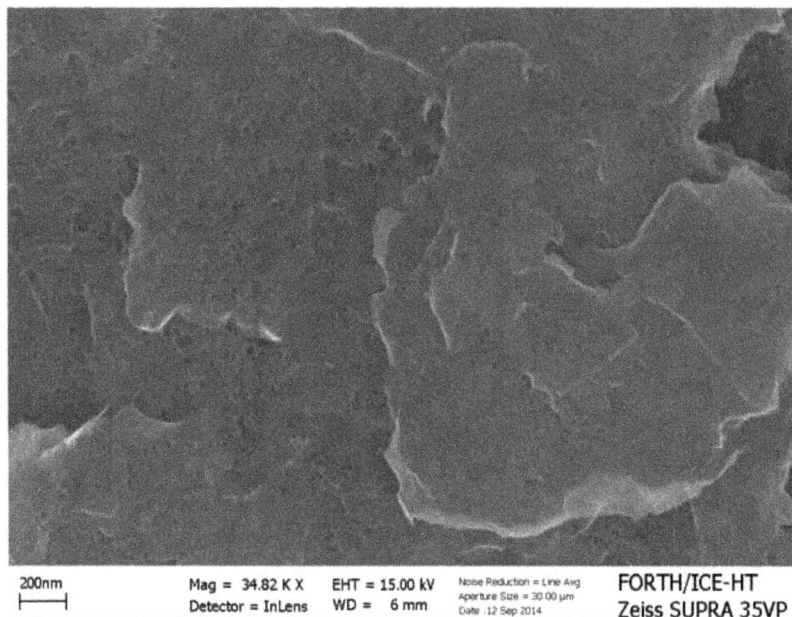

200nm	Mag = 34.82 K X	EHT = 15.00 kV	Noise Reduction = Line Avg	FORTH/ICE-HT
	Detector = InLens	WD = 6 mm	Aperture Size = 30.00 µm Date :12 Sep 2014	Zeiss SUPRA 35VP

The present "photoelectrocatalytic leaf" is one step apart from the combined np-TiO$_2$-Pt photocatalyst and the photocatalytic hydrogen production, which is the most popular and the most efficient route for hydrogen production by using photocatalysts. In that case, Pt (or other noble metal) nanoparticles are deposited on titania nanoparticles and thus each combined particle is a nano-size device capable of oxidizing the fuel and at the same time of reducing hydrogen ions to molecular hydrogen. In that case, no electrolyte is necessary to make the system function and thus the above pH-associated issues become less important. Photocatalytic hydrogen production is very popular and it has been treated in hundreds of publications [22,24–27]. What interest could then be ascribed to the above "photoelectrocatalytic leaf"? The present data show that by separating the photocatalyst from the electrocatalyst, it is possible to employ an alternative material for the latter and avoid the use of a rare metal like Pt, or at least reduce its quantity, as in the present case.

4. Experimental Section

4.1. Materials

Unless otherwise indicated, reagents were obtained from Aldrich (Taufkirchen, Germany) and were used as received. Commercial nanocrystalline titania Degussa P25 (specific surface area 50 m^2/g) was used in all cell constructions and Millipore (Merck-Millipore, Darmstadt, Germany) water was used in all experiments. SnO$_2$:F transparent conductive electrodes (FTO, Resistance 8 Ω/square) were purchased from Pilkington (Toledo, OH, USA).

4.2. Preparation of np-TiO₂ Films and Deposition of CdS by the SILAR Method

Nanoparticulate titania (np-TiO₂) films were deposited on FTO transparent electrodes by the following procedure: a FTO glass was cut in the appropriate dimensions and was carefully cleaned first with soap and then by sonication in isopropanol, water and acetone. A thin layer of compact titania was first sprayed over a patterned area by using 0.2 mol·L^{-1} diisopropoxytitanium bis(acetylacetonate) solution in ethanol and was calcined at 500 °C. Deposition of this bottom compact layer is a common practice with nanocrystalline titania photoanodes, since it enhances attachment of the top thick film, prevents short circuits and facilitates electron flow towards the electrode. On the top of this compact film, we applied a titania paste made of P25 nanoparticles by doctor blading. The film was calcined up to 550 °C at a rate of 20 °C/min. The final thickness of the film, as measured by SEM, was approximately 10 μm. The geometrical area of the film varied according to the particular application. CdS was deposited by Successive Ionic Layer Adsorption and Reaction (SILAR) [12,13,28]. 10 SILAR cycles were applied by using $Cd(NO_3)_2$ as Cd^{2+} and Na_2S as S^{2-} precursor. In all cases, we used 0.1 mol·L^{-1} aqueous solutions for both cations and anions. After SILAR deposition and final washing with tripled distilled water, the films were dried in an oven at 100 °C and were ready for use.

4.3. Construction of the Counter Electrode

Experimental data obtained for the purpose of the present work involve the following counter electrode constructions. Electrodes allowing O₂ diffusion for the case of oxygen saturated cell operation were based on commercial carbon cloth, which was functionalized with carbon black and Pt similarly to previous publications [12,13,16]. A hydrophobic layer was first applied as follows: carbon black (0.246 g) was mixed with distilled water (8 mL) by vigorous mixing in a mixer (about 2400 r.p.m.) until it became a viscous paste. This paste was further mixed with polytetrafluoroethylene (Teflon 60% wt. dispersion in water, 0.088 mL) and then applied on a carbon cloth cut to the necessary dimensions. This has been achieved by first spreading the paste with a spatula, preheating for a few minutes at 80 °C and finally heating also for a few minutes in an oven at 340 °C. Subsequently, the catalytic layer was prepared as follows: 1 g of Pt-carbon black electrocatalyst (30% on Vulcan XC72, Cabot, Leuven, Belgium) was mixed with Nafion perfluorinated resin (5 wt. % solution in lower aliphatic alcohols and water, 8 g) and a solution made of 7.5 g H₂O and 7.5 g isopropanol (15 g). The mixture was ultrasonically homogenized and then applied on the previously prepared carbon cloth bearing carbon black. The electrode was then heated at 80 °C for 30 min and the procedure was repeated as many times as necessary to load about 0.5 mg of Pt/cm². The thus prepared Pt-carbon black/carbon-cloth (Pt/CC) electrode was ready for use. Its dimensions were 2.25 cm² (1.5 cm × 1.5 cm). The above porous carbon cloth electrode allows contact with air while the deposited hydrophobic materials prevent water leak.

For the purpose of hydrogen production the counter electrode was based on an FTO glass. In fact, in that case photoanode and cathode were patterned on the same FTO slide, spatially separated as shown in Figure 2B,C. One part was covered with photocatalyst while the other part was covered with the electrocatalyst. In this way the electrons could be directly transferred from the photocatalyst to electrocatalyst through the FTO layer without wiring, thus minimizing losses. This is a "photoelectrocatalytic leaf" similar to the "quasi-artificial leaf" described in [29] and also previously presented by us [30,31]. The photocatalyst was deposited by the procedure described in Section 3.2. The electrocatalyst was deposited by using the following alternatives: (1) A solution of Diamminedinitritoplatinum (II) in ethanol was cast on warm FTO so that the solvent evaporated and the remaining material formed a dark film. It was annealed at 450 °C forming a film of clustered Pt nanoparticles. The total quantity of Pt in the film was calculated to be 0.6 mg; (2) A commercial carbon paste named Elcocarb C/SP (Solaronix, Aubonne, Switzerland), was applied on FTO by doctor blading and was annealed at 450 °C. This paste forms a uniform and very stable film. Its quantity was 4 mg; (3) Elcocarb was mixed with Pt. The mixed material was again applied by doctor blading and was annealed at 450 °C. The quantity of Elcocarb was again 4 mg and it contained 2 wt % (0.08 mg) of Pt. Therefore, in the mixed Pt-Elcocarb film, the quantity of Pt was 7.5 times lower than when Pt was cast alone.

4.4. Device (Reactor) Construction

Two reactors were used for the purposes of the present work. When PFC was employed as a device functioning in the presence of oxygen to produce electricity, it was made of a Plexiglas body with the two electrodes facing each other at a distance of 5 mm. As depicted in Figure 2A, the photoanode electrode played the role of cell window while the opposite side was closed with the Pt/CC electrode. The quantity of the electrolyte was 10 mL and it contained 0.5 mol·L^{-1} NaOH with or without 5 v % ethanol. The active area of the photoanode was 1 cm × 1 cm = 1 cm^2 and that of the counter electrode 1.5 cm × 1.5 cm = 2.25 cm^2. In the case of hydrogen production using the electrode of Figure 2B,C, the reactor was a Pyrex cylinder containing 100 mL of an aqueous solution of 0.5 mol·L^{-1} NaOH and 5 v % ethanol. The double electrode of Figure 2B,C was completely immersed in the electrolyte and was placed in an upright position. The cylinder was equipped with fittings allowing Ar gas to flow through the electrolyte. The latter was deoxygenated for 20 min by a vigorous flow of Ar and then the flow continued at a controlled rate of 20 mL·min^{-1}. Hydrogen production was monitored on-line by a gas chromatographer. Both reactors were illuminated by simulated solar light set at around 100 mW·cm^{-2}. When titania was alone without sensitizer it was excited by the UV portion of the incident radiation.

4.5. Measurements

Current-voltage curves were obtained with the help of an Autolab PGSTAT128N potentiostat (Metrohm-Autolab, Utrecht, Holland). They were all traced at a rate of 5 mV·s^{-1}. Ag/AgCl electrode was used as reference, when necessary, but corresponding plots were presented vs. Reversible Hydrogen Electrode (RHE) by adding 0.97 V to the recorded voltage values (i.e., +0.2 V for the potential of Ag/AgCl vs. SHE + 0.059 × 13, where 13 is the corresponding value of the pH). Hydrogen was detected on line by using a SRI 8610C gas chromatograph (Torrance, CA, USA). Calibration of the chromatograph signal was accomplished by comparison with a standard of 0.25% H$_2$ in Ar. FESEM images were obtained with a Zeiss SUPRA 35VP (Oberkochen, Germany).

5. Conclusions

The present work has examined PhotoFuelCells (PFCs) as devices which can convert solar light to electricity or chemical energy, for example hydrogen, at the same time consuming a fuel, which can be a water waste or a water pollutant. PFCs thus offer the double benefit of renewable energy production and environmental remediation. PFCs can function in the presence of oxygen producing electricity with a high open-circuit voltage. Photoanodes carrying nanoparticulate titania can be sensitized in the Visible by quantum dot metal sulfide semiconductor sensitizers. The choice of sensitizer is limited in the case of PFCs since the combined titania-sensitizer photocatalyst must retain its oxidative power to oxidize the fuel. Only moderately sized CdS nanoparticles have been so far demonstrated to be efficient and stable sensitizers of nanoparticulate titania. Titania alone can oxidize water producing a rather low current, which increases in the presence of a fuel. A cell can run alone without any bias just by shining light, if the cathode is aerated. If it is necessary to produce hydrogen, that is, in the absence of oxygen, a bias must be applied. A chemical bias does lead to hydrogen production but it necessitates a continuous supply of acid. In an alkaline environment it is difficult to produce hydrogen even under bias because of the scarcity of hydrogen ions. It is still possible to produce hydrogen in an alkaline environment by minimizing losses and this was achieved by the "photoelectrocatalytic leaf" where anode and cathode electrode were short-circuited by being deposited on the same FTO slide.

Acknowledgments

This project is implemented under the "ARISTEIA" Action of the "OPERATIONAL PROGRAMME EDUCATION AND LIFELONG LEARNING" and is co-funded by the European Social Fund (ESF) and National Resources (Project No. 2275). Robert Michal wishes to acknowledge a grant provided by the Scientific Grant Agency of the Slovak Republic (Project VEGA 1/0276/15) and National Scholarship Program of the Slovak Republic funded by Ministry of Education, Sport, Science and Research of the Slovak Republic managed by SAIA N.o., which allowed his stay in the University of Patras.

Author Contributions

RM and SS designed research and analyzed the data and PL wrote the paper.

Conflicts of Interest

The authors declare no conflict of interest.

References

1. Kaneko, M.; Nemoto, J.; Ueno, H.; Gokan, N.; Ohnuki, K.; Horikawa, M.; Saito, R.; Shibata, T. Photoelectrochemical Reaction of Biomass and Bio-Related Compounds With Nanoporous TiO_2 Film Photoanode and O_2-Reducing Cathode. *Electrochem. Commun.* **2006**, *8*, 336–340.

2. Lianos, P. Production of Electricity and Hydrogen by Photocatalytic Degradation of Organic Wastes in a Photoelectrochemical Cell: The Concept of the Photofuelcell: A Review of a Re-Emerging Research Field. *J. Hazard. Mater.* **2011**, *185*, 575–590.

3. Becquerel, E. Mémoire sur les effets électriques produits sous l'influence des rayons solaires. *Comptes Rendus* **1839**, *9*, 561–567.

4. Fujishima, A.; Honda, K. Electrochemical Photolysis of Water at a Semiconductor Electrode. *Nature* **1972**, *238*, 37–38.

5. Van de Krol, R.; Liang, Y.; Schoonman, J. Solar Hydrogen Production with Nanostructured Metal Oxides. *J. Mater. Chem.* **2008**, *18*, 2311–2320.

6. Kudo, A.; Miseki, Y. Heterogeneous Photocatalyst Materials for Water Splitting. *Chem. Soc. Rev.* **2009**, *38*, 253–278.

7. Gratzel, M. Photoelectrochemical Cells. *Nature* **2001**, *414*, 338–344.

8. Bak, T.; Nowotny, J.; Rekas, M.; Sorell, C.C. Photo-Electrochemical Hydrogen Generation from Water Using Solar Energy. Materials-Related Aspects. *Int. J. Hydrog. Energy* **2002**, *27*, 991–1022.

9. Antoniadou, M.; Sfaelou, S.; Dracopoulos, V.; Lianos, P. Platinum-Free Photoelectrochemical Water Splitting. *Catal. Commun.* **2014**, *43*, 72–74.

10. Panagiotopoulou, P.; Antoniadou, M.; Kondarides, D.I.; Lianos, P. Aldol Condensation Products During Photocatalytic Oxidation of Ethanol in a Photoelectrochemical Cell. *Appl. Catal. B* **2010**, *100*, 124–132.

11. Antoniadou, M.; Panagiotopoulou, P.; Kondarides, D.I.; Lianos, P. Photocatalysis and Photoelectrocatalysis Using Nanocrystalline Titania Alone or Combined with Pt, RuO2 or NiO Co-Catalysts. *J. Appl. Electrochem.* **2012**, *42*, 737–743.

12. Antoniadou, M.; Kondarides, D.I.; Dionysiou, D.D.; Lianos, P. Quantum Dot Sensitized Titania Applicable as Photoanode in Photoactivated Fuel Cells. *J. Phys. Chem. C* **2012**, *116*, 16901–16909.

13. Sfaelou, S.; Sygellou, L.; Dracopoulos, V.; Travlos, A.; Lianos, P. Effect of the Nature of Cadmium Salts on the Effectiveness of CdS SILAR Deposition and Its Consequences on the Performance of Sensitized Solar Cells. *J. Phys. Chem. C* **2014**, *118*, 22873–22880.

14. Yu, H.; Irie, H.; Hashimoto, K. Conduction Band Energy Level Control of Titanium Dioxide: Toward an Efficient Visible-Light-Sensitive Photocatalyst. *J. Am. Chem. Soc.* **2010**, *132*, 6898–6899.

15. Antoniadou, M.; Lianos, P. Near Ultraviolet and Visible Light Photoelectrochemical Degradation of Organic Substances Producing Electricity and Hydrogen. *J. Photochem. Photobiol. A* **2009**, *204*, 69–74.

16. Antoniadou, M.; Lianos, P. Production of Electricity by Photoelectrochemical Oxidation of Ethanol in a PhotoFuelCell. *Appl. Catal. B* **2010**, *99*, 307–313.

17. Sfaelou, S.; Antoniadou, M.; Trakakis, G.; Dracopoulos, V.; Tasis, D.; Parthenios, J.; Galiotis, C.; Papagelis, K.; Lianos, P. Buckypaper as Pt-Free Cathode Electrode in Photoactivated Fuel Cells. *Electrochim. Acta* **2012**, *80*, 399–404.

18. Balis, N.; Dracopoulos, V.; Antoniadou, M.; Lianos, P. One-Step Electrodeposition of Polypyrrole Applied as Oxygen Reduction Electrocatalyst in Photoactivated Fuel Cells. *Electrochim. Acta* **2012**, *70*, 338–343.

19. Fujishima, A.; Kato, T.; Maekawa, E.; Honda, K. Mechanism of the Current Doubling Effect. I. The ZnO Photoanode in Aqueous Solution of Sodium Formate. *Bull. Chem. Soc. Jpn.* **1981**, *54*, 1671–1674.

20. Antoniadou, M.; Daskalaki, V.M.; Balis, N.; Kondarides, D.I.; Kordulis, Ch.; Lianos, P. Photocatalysis and Photoelectrocatalysis Using (CdS-ZnS)/TiO₂ Combined Photocatalysts. *Appl. Catal. B* **2011**, *107*, 188–196.

21. Selli, E.; Chiarello, G.L.; Quartarone, E.; Mustarelli, P.; Rossetti, I.; Forni, L. A Photocatalytic Water Splitting Device for Separate Hydrogen and Oxygen Evolution. *Chem. Commun.* **2007**, 5022–5024.

22. Horiuchi, Y.; Toyao, T.; Takeuchi, M.; Matsuoka, M.; Anpo, M. Recent Advances in Visible-Light-Responsive Photocatalysts for Hydrogen Production and Solar Energy Conversion—From Semiconducting TiO₂ to MOF/PCP Photocatalysts. *Phys. Chem. Chem. Phys.* **2013**, *15*, 13243–13253.

23. Antoniadou, M.; Sfaelou, S.; Lianos, P. Quantum Dot Sensitized Titania for Photo-Fuel-Cell and for Water Splitting Operation in the Presence of Sacrificial Agents. *Chem. Eng. J.* **2014**, *254*, 245–251.

24. Nomikos, G.N.; Panagiotopoulou, P.; Kondarides, D.I.; Verykios, X.E. Kinetic and Mechanistic Study of the Photocatalytic Reforming of Methanol Over Pt/TiO₂ Catalyst. *Appl. Catal. B* **2014**, *146*, 249–257.

25. Chiarello, G.L.; Dozzi, M.V.; Scavini, M.; Grunwaldt, J.-D.; Selli, E. One Step Flame-Made Fluorinated Pt/TiO₂ Photocatalysts for Hydrogen Production. *Appl. Catal. B* **2014**, *160–161*, 144–151.

26. Ismail, A.A.; Bahnemann, D.W. Photochemical Splitting of Water for Hydrogen Production by Photocatalysis: A Review. *Sol. Energy Mater. Sol. Cells* **2014**, *128*, 85–101.

27. Puskelova, J.; Michal, R.; Caplovicova, M.; Antoniadou, M.; Caplovic, L.; Plesch, G.; Lianos, P. Hydrogen Production by Photocatalytic Ethanol Reforming Using Eu- and S-Doped Anatase. *Appl. Surf. Sci.* **2014**, *305*, 665–669.

28. Nicolau, Y.F. Solution Deposition of Thin Solid Compound Films by a Successive Ionic-Layer Adsorption and Reaction Process. *Appl. Surf. Sci.* **1985**, *22–23*, 1061–1074.

29. Trevisan, R.; Rodenas, P.; Gonzalez-Pedro, V.; Sima, C.; Sanchez, R.S.; Barea, E.M.; Mora-Sero, I.; Fabregat-Santiago, F.; Gimenez, S. Harnessing Infrared Photons for Photoelectrochemical Hydrogen Generation. A PbS Quantum Dot Based "Quasi-Artificial Leaf". *J. Phys. Chem. Lett.* **2013**, *4*, 141–146.

30. Strataki, N.; Antoniadou, M.; Dracopoulos, V.; Lianos, P. Visible-Light Photocatalytic Hydrogen Production from Ethanol-Water Mixtures Using a Pt-CdS-TiO₂ Photocatalyst. *Catal. Today* **2010**, *151*, 53–57.

31. Daskalaki, V.M.; Antoniadou, M.; Li Puma, G.; Kondarides, D.I.; Lianos, P. Solar Light-Responsive Pt/CdS/TiO$_2$ Photocatalysts for Hydrogen Production and Simultaneous Degradation of Inorganic or Organic Sacrificial Agents in Wastewater. *Environ. Sci. Technol.* **2010**, *44*, 7200–7205.

Sample Availability: Samples are available from the authors.

TiO₂ and Fe₂O₃ Films for Photoelectrochemical Water Splitting

Josef Krysa, Martin Zlamal, Stepan Kment, Michaela Brunclikova and Zdenek Hubicka

Abstract: Titanium oxide (TiO₂) and iron oxide (α-Fe₂O₃) hematite films have potential applications as photoanodes in electrochemical water splitting. In the present work TiO₂ and α-Fe₂O₃ thin films were prepared by two methods, e.g., sol-gel and High Power Impulse Magnetron Sputtering (HiPIMS) and judged on the basis of physical properties such as crystalline structure and surface topography and functional properties such as simulated photoelectrochemical (PEC) water splitting conditions. It was revealed that the HiPIMS method already provides crystalline structures of anatase TiO₂ and hematite Fe₂O₃ during the deposition, whereas to finalize the sol-gel route the as-deposited films must always be annealed to obtain the crystalline phase. Regarding the PEC activity, both TiO₂ films show similar photocurrent density, but only when illuminated by UV light. A different situation was observed for hematite films where plasmatic films showed a tenfold enhancement of the stable photocurrent density over the sol-gel hematite films for both UV and visible irradiation. The superior properties of plasmatic films could be explained by ability to address some of the hematite drawbacks by the deposition of very thin films (25 nm) consisting of small densely packed particles and by doping with Sn.

Reprinted from *Molecules*. Cite as: Krysa, J.; Zlamal, M.; Kment, S.; Brunclikova, M.; Hubicka, Z. TiO₂ and Fe₂O₃ Films for Photoelectrochemical Water Splitting. *Molecules* **2015**, *20*, 1046-1058.

1. Introduction

Thin films of titanium dioxide deposited on various supports are very useful photocatalysts in a number of applications, primarily in environmental protection. Another application is in alternative energy generation, e.g., the photoelectrochemical water splitting [1] where photogenerated holes act as an oxidant, in this particular case to evolve molecular oxygen, photogenerated electrons are transferred via an external circuit to the auxiliary electrode where they are used to evolve hydrogen.

Great advantages of TiO₂ are its low price, high stability and nontoxicity [2]. However, for practical applications, there is a huge disadvantage consisting in a large band gap energy resulting in utilization of very small part of sunlight (4%–5%). Iron oxide (α-Fe₂O₃) with hematite crystalline structure has recently attracted much attention as a potentially convenient material to be used for hydrogen production via photoelectrochemical water splitting. This is due to its favourable properties such as a band gap between 2.0–2.2 eV, which allows absorbing a substantial fraction of solar spectrum, chemical stability in aqueous environment, nontoxicity, abundance and low cost. For such band gaps and assuming standard solar illumination conditions (AM 1.5 G, 100 mW·cm⁻²) the theoretical maximal solar-to-hydrogen (STH) conversion efficiency can be calculated as 15% [3]. However not everything is ideal and hematite also has certain disadvantages. Among the most cited limitations are the nonideal position of hematite's conduction band, which is too low for spontaneous water reduction (this can be addressed by applying e.g., a PV cell to provide the additional energy needed); the low absorptivity (especially for longer wavelengths); and very short diffusion length of photogenerated holes. This creates a disaccord between the depth where charge carriers are photogenerated (in the bulk) and the distance they diffuse before recombining. Doping with elements such as Sn, Ti, Ge, Si, Nb, *etc.* can significantly increase the electronic conductivity by increasing the number of carriers. The negative effect of the short diffusion length of holes can be suppressed by using very thin films of hematite or their nanostructuring. Furthermore, remarkable overpotential of about 0.4–0.6 V for the onset of the water splitting photocurrent, which has been assigned to poor oxygen evolution kinetics on hematite surfaces and/or to the presence of surface defects acting as traps, is another crucial issue.

The aim of this work was the preparation of TiO_2 and α-Fe_2O_3 thin films of well-defined thicknesses by two methods, e.g., sol–gel and the advanced pulsed plasma deposition method of High Power Impulse Magnetron Sputtering (HiPIMS). Due to the potential applications for photoelectrochemical water splitting such prepared films were characterised by the measurement of photocurrent and open circuit potential in aqueous media containing an inert electrolyte.

The sol-gel technique enables photocatalyst films, e.g., TiO_2, to be prepared in a way that controls surface properties such as composition, thickness and morphology [4,5]. This technique consists of several steps: (i) precursor synthesis; (ii) precursor deposition (usually by dip-coating); (iii) drying; and (iv) calcination at elevated temperatures. Layer thickness can be controlled by viscosity of the sol-gel precursor and by the withdrawal rate during dip-coating and could be between 60–300 nm. Thicker (or multilayer) films can be obtained by repeating steps (ii), (iii) and (iv) [6]. The HiPIMS discharges are operated in pulse modulated regime with a low repetition frequency (typically about 100 Hz) and a short duty cycle (~1%) with applying high peak powers (~kW/cm^2) during the active part of the modulation cycle [7]. A distinguishing feature of HiPIMS is its high degree of ionization of the sputtered metal and a high rate of molecular gas dissociation due to very high plasma density near the target (order of 10^{13} ions cm^{-3}).

2. Results and Discussion

2.1. TiO_2 Films

XRD diffraction patterns of TiO_2 films of comparable thickness (65 nm) prepared by two different techniques are shown in Figure 1. Both types of films have an anatase crystalline structure. This is well documented in Figure 1a, where the substrate is quartz. Sol gel TiO_2 film deposited on quartz has similar patterns so it is not shown here and instead XRD patterns of such a film deposited on FTO glass is shown as Figure 1b. It can be seen that only two Bragg diffraction peaks corresponding to anatase can be identified (due to overlapping of other bands with those corresponding to fluorine doped SnO_2 layer), particularly for position $2\Theta = 25.2°$ (101) and $2\Theta = 48.0°$ (200).

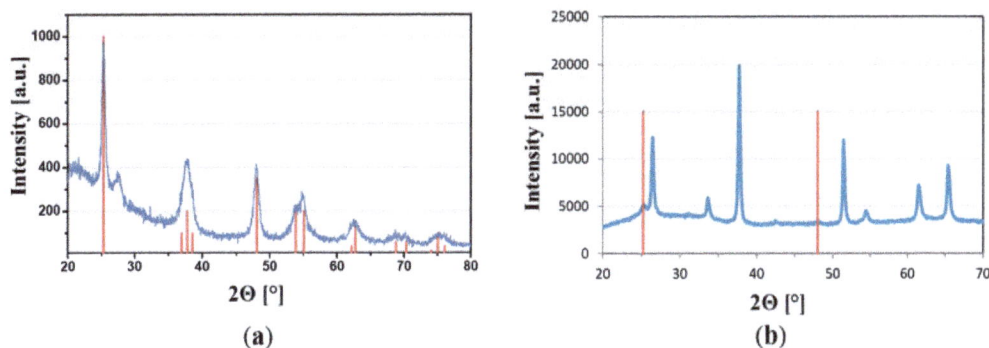

Figure 1. XRD patterns of (**a**) plasmatic TiO₂ thin films deposited on quartz and (**b**) sol gel TiO₂ thin films deposited on FTO substrate. Thickness of both films 65 nm ± 10 nm.

Figure 2. SEM surface morphology (**a**) plasmatic TiO₂ thin films deposited on quartz and (**b**) sol gel TiO₂ thin films deposited on FTO substrate.

The electron microscopy images (Figure 2) show that the morphologies of plasmatic (left image) and sol gel (right image) films are different. Morphology of sol gel film is smooth and due to homogeneous well transparent structure we can see only the grooves corresponding to the morphology of FTO glass substrate. On the other side the roughness of plasmatic film is significantly higher obviously due to the bigger grains.

Figure 3 shows the comparison of chopped light polarization curves for both types of films. The response of current to light on and light off is similar for both films, only the time interval for light on and off was two times higher for plasmatic films.

From the shape of polarization curve, for a photocurrent equal to zero we can estimate values of OCP under light. Thus obtained values of OCP differ for both films, for plasmatic it is around −0.15 V (Ag/AgCl) while for sol gel film it is around −0.40 V (Ag/AgCl). It was reported previously (in [6]) that the value of OCP for TiO₂ film decreased when the electrode is illuminated, which is due to the electron access generated in the nanocrystalline film. After some time of irradiation a steady state value is reached and upon turning the illumination off the photostationary potential decays and eventually reaches its initial dark value. The less negative value of OCP for plasmatic layers suggests a significantly lower amount of electrons trapped in the film upon irradiation. This can be supported by the shape of polarization curve in the vicinity of OCP where we can see fast response to light and subsequently fast decay of photocurrent resulting in very small difference between dark and light current. The possible explanation is in faster electron-hole recombination at OCP (zero current conditions) in comparison to the sol gel film.

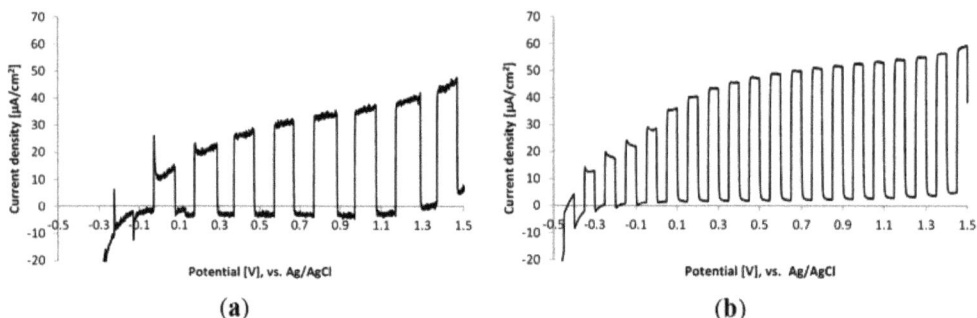

(a)   (b)

Figure 3. Chopped light polarization curve of (a) plasmatic TiO_2 thin films and (b) sol gel TiO_2 thin films. Irradiation wavelength 365 nm (1.46 mW/cm^2), electrolyte 0.1 M Na_2SO_4, scan rate 5 mV·s^{-1}.

Another difference in electrochemical properties of both films is in the potential value where the current plateau starts. For plasmatic film (even the plateau is not well developed) it starts around 0.56 V (Ag/AgCl). For sol gel films plateau starts around 0.3 V (Ag/AgCl). This indicates that for plasmatic film higher applied bias is necessary for the formation of space charge layer and sufficient electron hole separation. But when we discuss a different parameter, e.g., the difference between the open-circuit potential under light and the applied potential we can see that approximately 0.7 V is necessary for the achievement of the current plateau for both films.

2.2. (α-Fe₂O₃) Hematite Films

Due to the small thickness of Fe_2O_3 films the crystalline structure of deposited films was assessed by Raman spectroscopy (Figure 4). We have reported previously that the hematite crystalline phase can be achieved already during the HiPIMS deposition [8]. However in order to improve the charge transfer between the hematite and FTO substrate as well as among the hematite grains themselves, the films had to be calcined at 650 and 750 °C always for 40 min in air. The database hematite spectrum is presented for reference and it unambiguously matches the measured HiPIMS Raman spectra and no significant difference (see Figure 4) between the films was evident. On the contrary to HiPIMS deposition in the case of sol gel films, the hematite structure was only achieved after calcination of the deposited precursors at 400 °C. A further increase to 600 °C has no influence on the crystalline structure.

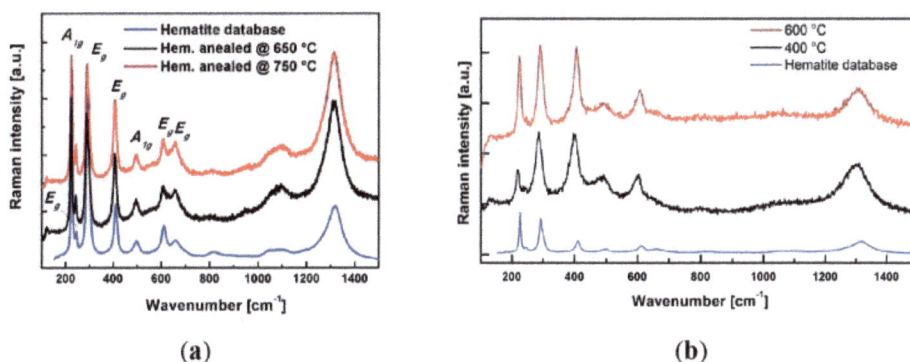

Figure 4. Raman spectra of (a) HiPIMS Fe₂O₃ films (thickness 40 ± 5 nm) and (b) sol gel Fe₂O₃ films (thickness 60 ± 5 nm) on FTO glass substrate.

With the help of SEM the surface morphology of the films was visualized (Figure 5). The morphology of sol gel films has a "worm like structure" suggesting a porous character of the film.

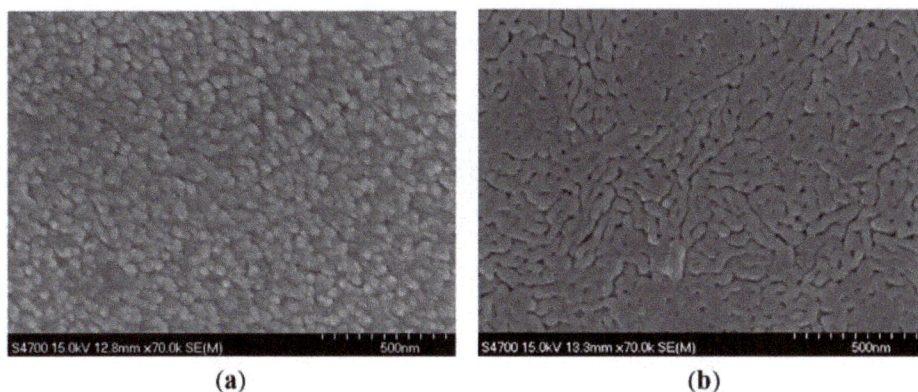

Figure 5. SEM surface morphology of (a) HiPIMS Fe₂O₃ films (calcined at 650 °C) and (b) sol gel Fc₂O₃ films (calcined at 600 °C) on FTO glass substrate.

On the other hand from the captured images of HiPIMS films it is clearly seen that the films consist of very small densely packed particles. Nevertheless, such organizations are a common feature of HiPIMS produced thin films in general. Furthermore the measured surface RMS roughness (AFM) was only ~2 nm for the HiPIMS films.

Sol gel hematite films calcined at 400 °C exhibit negligible electrochemical response so the data are not shown here, but the sol gel films calcined at 600 °C exhibit better performance as shown in Figure 6. The response to light is rapid and after the initial spike the photocurrent decreases due to the strong hole-electron recombination. Fast change of open circuit potential after light is switched on correlates with fast response of photocurrent to light. But the difference between the dark and light value of OCP is rather small (around −50 mV) which suggests small concentration of photogenerated electrons. Very important with respect to the PEC performance are the very sharp photocurrent spikes present in the chopped-light voltammogram when the light is turned on (see

Figure 6b) denoting a high density of undesirable surface states acting as recombination centres for photogenerated carriers [9]. The surface crystalline defects and oxygen vacancies are mostly attributed to these states [10].

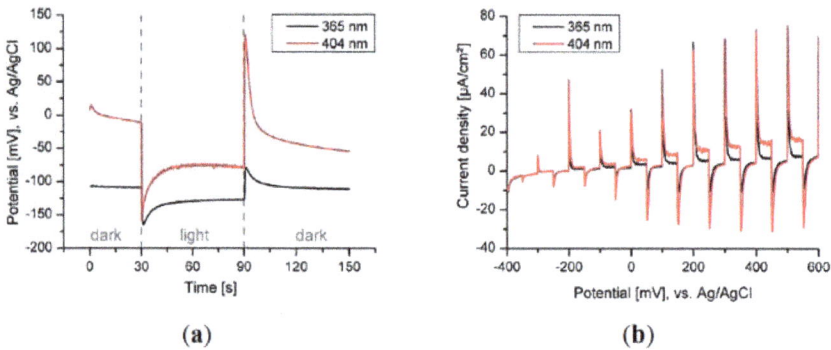

(a)   (b)

Figure 6. Open circuit potential (OCP) (**a**) and chopped light polarization curve (**b**) for sol gel hematite thin films calcined at 600 °C (FTO electrode). Irradiation wavelength 365 nm and 404 nm.

Although the as-deposited plasmatic hematite films were crystalline, they exhibit almost negligible photocurrent [11]. The high density of defects and imperfections in crystalline structure, and thus a high extent of backward electron-hole pair recombination of as-deposited hematite films, is probably the main reason for the poor photoactivity. Thus, in the next step the hematite films were annealed in air at 650 and 750 °C for 40 min. and the photoelectrochemical performance of hematite films under conditions simulating UV and visible light are shown in Figures 7 and 8. It was reported previously that such high temperature is needed for the diffusion of tin from the FTO substrate into hematite to occur. The diffused tin ions results in the extrinsic doping of hematite improving its electronic properties [11].

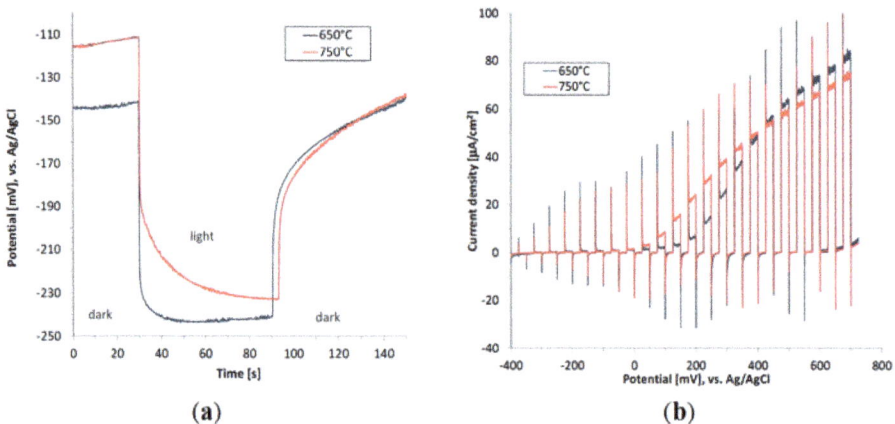

(a)   (b)

Figure 7. Open circuit potential (OCP) (**a**) and chopped light polarization curve (**b**) for hematite thin films calcined at 650 °C and 750 °C (FTO electrode). Irradiation wavelength 365 nm (1.46 mW/cm^2).

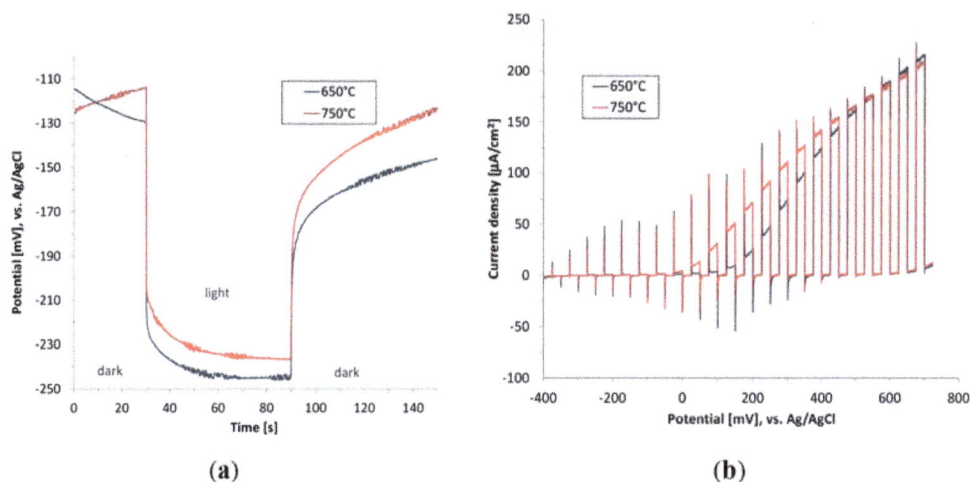

Figure 8. Open circuit potential (OCP) (**a**) and chopped light polarization curve (**b**) for hematite thin films calcined at 650 °C and 750 °C (FTO electrode). Irradiation wavelength 404 nm (6.1 mW/cm²).

Increase of the annealing temperature to 750 °C does not bring about any improvement in terms of the obtained photocurrent value. On the other hand the onset of photocurrent starts already at about 0 V (*vs.* Ag/AgCl), which is shifted 100 mV negatively in relation to the film annealed at 650 °C. This phenomenon is apparently due to enhanced oxygen evolution kinetics as a consequence of self-passivating surface states serving as recombination centres.

The observed photocurrent spikes when light is turned on and off (Figures 7b and 8b) are due to the surface recombination and are typical for n-type photoelectrode [12]. The anodic current spikes have their origin in the accumulated holes at the electrode/electrolyte interface. These holes are not injected to the electrolyte as a consequence of the slow water oxidation kinetics. Instead they have ability to oxidize trap states in the bulk and on the surface. Conversely, the cathodic current spikes are generated when the light illumination is off, denoting the recombination of the accumulated holes at the semiconductor/liquid junction by the electrons diffusing from the external circuit [13,14].

One of the positive effect of HiPIMS method enhancing the photoefficiency is apparently related to a very high energy of ions and sputtered particles (which can be higher than 20 eV [15]) bombarding the substrate during the deposition by means of HiPIMS. As a consequence the interface and electrical conductivity between the FTO and hematite film is significantly boosted. A second reason for the increased photoactivity of HiPIMS is apparently attributed to the much smaller grains, which are beneficial for the process in terms of reduced electron-hole recombination due to optimal matching of nanoparticle size with the hole diffusion length [3]. As stated in the introduction the hematite films struggle with short diffusion length of the holes. This can be demonstrated on the influence of photocurrent on layer thickness where very thin film of 25 nm shows much better performance than film of thickness 40 nm [16]. The short diffusion length of the holes can be moderated not only by fabrication of very thin films but also by a careful nanostructuring in various architectures such as wormlike structures [17], nanorods [18], nanowires [19], *etc.*

2.3. Comparison of TiO₂ and α-Fe₂O₃ Films

IPCE values for hematite and TiO₂ films are shown in Table 1. Due to different electrolytes and thus different pH values used for photocurrent measurement, photocurrent density and IPCE were compared for the same value of potential related to RHE (1.6 V).

Table 1. Photocurrents of hematite and TiO₂ films at 1.6 V (RHE).

	Wavelength [nm]	j_{ph} (at 0.98 V vs. Ag/AgCl)	P [10^{-9} Einstein·s·cm^{-2}]	IPCE [%]
TiO₂				
sol gel	365	50.8	6.40	8.2
plasmatic	365	37.1	5.79	6.6
		j_{ph} (at 0.57 V vs. Ag/AgCl)		
Fe₂O₃				
sol gel	365	2.1	4.45	0.5
	404	11	20.6	0.6
plasmatic	365	74.5	4.45	17.3
	404	186.7	20.6	9.4

The measured potentials *vs.* Ag/AgCl are related to those *vs.* reversible hydrogen electrode (RHE) according to the Nernst Equation:

$$E_{RHE} = E_{Ag/AgCl} + 0.059pH + E^0_{Ag/AgCl} \qquad (1)$$

where E$_{RHE}$ is the converted potential *vs.* RHE, E$^0_{Ag/AgCl}$ is 0.205 V at 25 °C, and E$_{Ag/AgCl}$ is the experimentally measured potential against Ag/AgCl reference electrode.

Unlike TiO₂ films for hematite films we observe significant photoresponse also for the 404 nm interference filter. Owing to the band gap energy, which is tabulated to be in the region of 2.0–2.2 eV, hematite absorbs photons also in the visible range of light spectra. The overall photocurrent value at 404 nm is almost doubled comparing to the value at 365 nm but it is due to the higher incident light intensity at 404 nm (four times higher than at 365 nm). IPCE of plasmatic hematite film is in fact for 365 nm almost two times higher than for 404 nm which is due to the increased light absorption at 365 nm. Comparing the best TiO₂ and Fe₂O₃ films, at 365 nm Fe₂O₃ has two times higher value of IPCE (17.3% *vs.* 8.2%), at 404 nm Fe₂O₃ reach 9.4% while IPCE for TiO₂ was negligible.

3. Experimental Section

3.1. Glass Substrates

Two types of glass substrates were used namely quartz (SiO₂) and fluorine doped tin oxide (FTO) transparent conducting glass (TCO) slides (TCO-7) supplied by Solaronix (Aubonne, Switzerland).

3.2. Preparation of TiO₂ Films

Sol gel films were prepared using titanium(IV) isopropoxide (97%, Sigma-Aldrich, Prague, Czech Republic) as TiO₂ precursor, absolute ethanol (p.a., Penta, Prague, Czech Republic) and ethyl acetylacetate (p.a. 99%, Fluka, Prague, Czech Republic) as solvent and nitric acid (p.a. 65%, Penta, Prague, Czech Republic) as catalyst based on the method reported previously [6,20]. Absolute ethanol was added drop wise under stirring to titanium isopropoxide. In the next step absolute ethanol was mixed with ethyl acetoacetate and nitric acid and then added to the isopropoxide mixture. Thus prepared sol was stirred under vigorous stirring for 24 h. FTO substrates were dip-coated with prepared TiO₂ sol (withdrawal speed 60 mm·min⁻¹). Deposited films were calcined at 500 °C for 1 h.

The preparation of plasmatic films was based on reactive magnetron sputtering using a commercial Z 550 M magnetron (Leybold-Heraeus, Cologne, Germany). The diameter of the Ti target was 151 mm with purity of Ti 99.5%. The Ar-O₂ working gases mixture with the flow rates of 20 and 10 standard cubic centimetres per minute (sccm), respectively, was used. The depositions were carried out under ambient conditions. The distance between the magnetron target and substrates was 10 cm. The films were deposited onto carefully cleaned quartz and FTO substrates. After the deposition the films were thermally treated at 450 °C for 3 h.

3.3. Preparation of Hematite Films

Sol gel hematite layers were manufactured as follows: the precursor was prepared by mixing Fe(NO₃)₃·9H₂O, absolute ethyl alcohol and propylene oxide. The layers were deposited on the glass substrates (SiO₂, FTO) by dip coating (withdrawal speed 60 mm·min⁻¹) and after drying, calcined at 400, 500 and 600 °C for 2 h [21,22].

The HiPIMS deposition employed a metallic target of pure iron (99.995%, Plasmaterials, Livermore, CA. USA) with outer diameter 50 mm and an Ar-O₂ atmosphere as working gases mixture in an ultra-high vacuum (UHV) reactor continuously pumped down by a turbo-molecular pump providing the base pressure of 10⁻⁵ Pa. Glass substrates were carefully cleaned before deposition. The working gases were fed to the reactor with the flow rates of 30 sccm and 12 sccm corresponding to argon and oxygen, respectively. The depositions were carried out at room temperature, operating pressure was 1 Pa. The pulsing frequency of DC HiPIMS discharge was in the range 70–1000 Hz with the "ON" time of 100 μs and the maximal current density achieved in a pulse was ≈5 A·cm⁻² at 70 Hz [23].

3.4. Film Characterisation

The structural, morphological, electronic and optical properties of the deposited films were determined using X-ray diffraction (Seifert-XRD 3000; Panalytical HighScore Plus, Ahrensburg, Germany), space resolved Raman spectroscopy (RM 1000 Raman Microscope, Renishaw, Wotton-under-Edge, UK), AFM (Explorer, ThermoMicroscopes, Sunnyvale, CA USA) and UV-Vis absorption spectroscopy (Cary 100, Santa Clara, CA, USA).

3.5. Photoelectrochemical Activity Measurement

Polarization curves and the open circuit potential (OCP) measurements seem to be suitable for the assessment of the photo-electrochemical efficiency of the TiO_2 and α-Fe_2O_3 thin films. Photoelectrochemical properties of films illuminated from the front side (from film/electrolyte interface) were tested in a three electrode arrangement [24,25]. The TiO_2 or α-Fe_2O_3 thin film on TCO served as the working electrode, the silver-silver chloride electrode (Ag/AgCl) as the reference and a platinum sheet as the counter electrode. 0.1 M solution of Na_2SO_4 for TiO_2 and 1 M NaOH for α-Fe_2O_3 thin films was used as an electrolyte. The exposed film area (1 cm^2) was defined by teflon tape. The electric contact was made by pressing stainless steel to upper part of FTO layer, not covered by TiO_2 (α-Fe_2O_3). The cell was connected to the optical bench (Melles Griot, Rochester, NY, USA) which included optical filters and a shutter. As the source of radiation the DC Arc polychromatic high pressure mercury lamp (LOT LSH201 Hg, Darmstadt, Germany) was employed with a characteristic broadband line source. Optical interference filters were utilized in order to ensure the layers exposure to the radiation of precise wavelength (365 $\pm$ 10 nm and 404 $\pm$ 10 nm). The photocurrents were measured using a Voltalab10 PGZ-100 potentiostat with software VoltaMaster 4 (Radiometer Analytical SAS, Villeurbanne Cedex, France). The starting potential was around the value of OCP (-0.1 V to -0.5 V (Ag/AgCl), sweeping was positive towards the end potential 1.5 V or 0.7 V (Ag/AgCl).

The incident intensity of the light when passing the filter was measured using a S1337-1010BQ photodiode (Hamamatsu Photonics K.K., Hamamatsu, Japan). The light intensity was influenced by the lamp age and thus decreased slightly during a few months period but were always in the range 1–2 mW·cm^{-2}. The light intensities were always recorded before each experiment and recalculated to the incident photon flux intensity P [Einstein s^{-1}·cm^{-2}]. Incident photon to current conversion efficiency (IPCE) was then calculated using Equation (2):

$$IPCE = \frac{j_{ph}}{F\,P} \tag{2}$$

where j_{ph} denotes for the photocurrent density [A·cm^{-2}], F is the Faraday constant (96.485 C·mol^{-1}) and P is the incident light intensity [Einstein cm^{-2}·s^{-1}].

4. Conclusions

TiO_2 and Fe_2O_3 represent two very promising materials for use as photoanodes for photoelectrochemical water splitting. In this study these materials were fabricated in the form of thin films by chemical sol-gel method and the plasma-assisted technique known as high power impulse magnetron sputtering (HiPIMS). The significant difference between the methods revealed here is that the plasmatic method provides the crystalline structure of anatase TiO_2 and hematite Fe_2O_3 already during the deposition, whereas to finalize the sol-gel route the as-deposited films must always be annealed to obtain the crystalline phase. It has been shown that in the case of TiO_2 films the sol-gel method produces very smooth surface of extremely small grains, while a more porous structure was observed in the case of hematite films. Slightly higher surface roughness was observed in the case of both materials fabricated by the plasmatic method.

Regarding the PEC activity of the TiO$_2$ films, a higher photoresponse (about 20%) was evident for sol-gel films when illuminated by UV light using the 365 nm interference filter. A different situation was observed for the hematite films. In this case the plasmatic films showed a tenfold enhancement of the stable photocurrent density over the sol-gel hematite films. A very high concentration of the surface states acting as the recombination centres might be the reason for the poor activity of the hematite sol-gel films. The large extent of the surface states is evident as the transient spike currents of the chopped-light polarization curves. As far as the plasmatic hematite films are concerned, some of the abovementioned drawbacks for its application for the PEC water splitting were addressed in this study by deposition of very thin films of hematite (25 nm) and by doping the hematite with Sn.

Acknowledgments

This work was supported by the Grant Agency of the Czech Republic (P108/12/2104).

Author Contributions

Josef Krysa managed the literature search and was involved in data collection and writing the manuscript. Martin Zlamal performed photoelectrochemical measurements, Stepan Kment participated in the literature search, data analysis and interpretation and was involved in manuscript writing. Michaela Brunclikova prepared iron oxide sol gel films. Zdenek Hubicka worked on plasmatic film deposition method. All authors read and approved the final manuscript.

Conflicts of Interest

The authors declare no conflict of interest.

References

1. Fujishima, A.; Honda, K. Electrochemical photolysis of water at a semiconductor electrode. *Nature* **1972**, *238*, 37–38.
2. Fujishima, A.; Rao, T.N.; Tryk, D.A. Titanium dioxide photocatalysis. *J. Photochem. Photobiol. C Photochem. Rev.* **2000**, *1*, 1–21.
3. Sivula, K.; Le Formal, F.; Gratzel, M. Solar Water Splitting: Progress Using Hematite (alpha-Fe$_2$O$_3$) Photoelectrodes. *Chemsuschem* **2011**, *4*, 432–449.
4. Liu, J.X.; Yang, D.Z.; Shi, F.; Cai, Y.J. Sol-gel deposited TiO$_2$ film on NiTi surgical alloy for biocompatibility improvement. *Thin Solid Films* **2003**, *429*, 225–230.
5. Choi, S.Y.; Mamak, M.; Coombs, N.; Chopra, N.; Ozin, G.A. Thermally stable two-dimensional hexagonal mesoporous nanocrystalline anatase, meso-nc-TiO$_2$: Bulk and crack-free thin film morphologies. *Adv. Funct. Mater.* **2004**, *14*, 335–344.
6. Krysa, J.; Baudys, M.; Zlamal, M.; Krysova, H.; Morozova, M.; Kluson, P. Photocatalytic and photoelectrochemical properties of sol-gel TiO$_2$ films of controlled thickness and porosity. *Catal. Today* **2014**, *230*, 2–7.

7. Lundin, D.; Sarakinos, K. An introduction to thin film processing using high-power impulse magnetron sputtering. *J. Mater. Res.* **2012**, *27*, 780–792.

8. Kment, S.; Hubicka, Z.; Krysa, J.; Olejnicek, J.; Cada, M.; Gregora, I.; Zlamal, M.; Brunclikova, M.; Remes, Z.; Liu, N. High-power pulsed plasma deposition of hematite photoanode for PEC water splitting. *Catal. Today* **2014**, *230*, 8–14.

9. Kment, S.; Kluson, P.; Hubicka, Z.; Krysa, J.; Cada, M.; Gregora, I.; Deyneka, A.; Remes, Z.; Zabova, H.; Jastrabik, L. Double hollow cathode plasma jet-low temperature method for the TiO2-xNx photoresponding films. *Electrochim. Acta* **2010**, *55*, 1548–1556.

10. Sivula, K. Metal Oxide Photoelectrodes for Solar Fuel Production, Surface Traps, and Catalysis. *J. Phys. Chem. Lett.* **2013**, *4*, 1624–1633.

11. Sivula, K.; Zboril, R.; Le Formal, F.; Robert, R.; Weidenkaff, A.; Tucek, J.; Frydrych, J.; Gratzel, M. Photoelectrochemical Water Splitting with Mesoporous Hematite Prepared by a Solution-Based Colloidal Approach. *J. Am. Chem. Soc.* **2010**, *132*, 7436–7444.

12. Berger, T.; Monllor-Satoca, D.; Jankulovska, M.; Lana-Villarreal, T.; Gomez, R. The Electrochemistry of Nanostructured Titanium Dioxide Electrodes. *ChemPhysChem* **2012**, *13*, 2824–2875.

13. Dotan, H.; Sivula, K.; Gratzel, M.; Rothschild, A.; Warren, S.C. Probing the photoelectrochemical properties of hematite (alpha-Fe2O3) electrodes using hydrogen peroxide as a hole scavenger. *Energy Environ. Sci.* **2011**, *4*, 958–964.

14. Le Formal, F.; Tetreault, N.; Cornuz, M.; Moehl, T.; Gratzel, M.; Sivula, K. Passivating surface states on water splitting hematite photoanodes with alumina overlayers. *Chem. Sci.* **2011**, *2*, 737–743.

15. Bohlmark, J.; Lattemann, M.; Gudmundsson, J.T.; Ehiasarian, A.P.; Gonzalvo, Y.A.; Brenning, N.; Helmersson, U. The ion energy distributions and ion flux composition from a high power impulse magnetron sputtering discharge. *Thin Solid Films* **2006**, *515*, 1522–1526.

16. Rioult, M.; Magnan, H.; Stanescu, D.; Barbier, A. Single Crystalline Hematite Films for Solar Water Splitting: Ti-Doping and Thickness Effects. *J. Phys. Chem. C* **2014**, *118*, 3007–3014.

17. Kim, J.Y.; Magesh, G.; Youn, D.H.; Jang, J.W.; Kubota, J.; Domen, K.; Lee, J.S. Single-crystalline, wormlike hematite photoanodes for efficient solar water splitting. *Sci. Rep. UK* **2013**, *3*, doi:10.1038/srep02681.

18. Mao, A.; Shin, K.; Kim, J.K.; Wang, D.H.; Han, G.Y.; Park, J.H. Controlled Synthesis of Vertically Aligned Hematite on Conducting Substrate for Photoelectrochemical Cells: Nanorods versus Nanotubes. *ACS Appl. Mater. Int.* **2011**, *3*, 1852–1858.

19. Li, L.S.; Yu, Y.H.; Meng, F.; Tan, Y.Z.; Hamers, R.J.; Jin, S. Facile Solution Synthesis of α-FeF3·3H2O Nanowires and Their Conversion to α-Fe2O3 Nanowires for Photoelectrochemical Application. *Nano Lett.* **2012**, *12*, 724–731.

20. Zita, J.; Maixner, J.; Krysa, J. Multilayer TiO2/SiO2 thin sol–gel films: Effect of calcination temperature and Na+ diffusion. *J. Photochem. Photobiol. Chem.* **2010**, *216*, 194–200.

21. Park, C.D.; Magana, D.; Stiegman, A.E. High-quality Fe and gamma-Fe2O3 magnetic thin films from an epoxide-catalyzed sol–gel process. *Chem. Mater.* **2007**, *19*, 677–683.

22. Park, C.D.; Walker, J.; Tannenbaum, R.; Stiegman, A.E.; Frydrych, J.; Machala, L. Sol-Gel-Derived Iron Oxide Thin Films on Silicon: Surface Properties and Interfacial Chemistry. *ACS Appl. Mater. Int.* **2009**, *1*, 1843–1846.

23. Hubicka, Z.; Kment, S.; Olejnicek, J.; Cada, M.; Kubart, T.; Brunclikova, M.; Ksirova, P.; Adamek, P.; Remes, Z. Deposition of hematite Fe_2O_3 thin film by DC pulsed magnetron and DC pulsed hollow cathode sputtering system. *Thin Solid Films* **2013**, *549*, 184–191.

24. Morozova, M.; Kluson, P.; Krysa, J.; Zlamal, M.; Solcova, O.; Kment, S.; Steck, T. Role of the template molecular structure on the photo-electrochemical functionality of the sol-gel titania thin films. *J. Sol Gel Sci. Technol.* **2009**, *52*, 398–407.

25. Kment, S.; Hubicka, Z.; Kmentova, H.; Kluson, P.; Krysa, J.; Gregora, I.; Morozova, M.; Cada, M.; Petras, D.; Dytrych, P. Photoelectrochemical properties of hierarchical nanocomposite structure: Carbon nanofibers/TiO_2/ZnO thin films. *Catal. Today* **2011**, *161*, 8–14.

Sample Availability: Samples of the compounds are not available from the authors.

Theoretical Verification of Photoelectrochemical Water Oxidation Using Nanocrystalline TiO₂ Electrodes

Shozo Yanagida, Susumu Yanagisawa, Koichi Yamashita, Ryota Jono and Hiroshi Segawa

Abstract: Mesoscopic anatase nanocrystalline TiO₂ (nc-TiO₂) electrodes play effective and efficient catalytic roles in photoelectrochemical (PEC) H₂O oxidation under short circuit energy gap excitation conditions. Interfacial molecular orbital structures of $(H_2O)_3$ &OH(TiO₂)₉H as a stationary model under neutral conditions and the radical-cation model of $[(H_2O)_3\&OH(TiO_2)_9H]^+$ as a working nc-TiO₂ model are simulated employing a cluster model OH(TiO₂)₉H (Yamashita/Jono's model) and a H₂O cluster model of $(H_2O)_3$ to examine excellent H₂O oxidation on nc-TiO₂ electrodes in PEC cells. The stationary model, $(H_2O)_3\&OH(TiO_2)_9H$ reveals that the model surface provides catalytic H₂O binding sites through hydrogen bonding, van der Waals and Coulombic interactions. The working model, $[(H_2O)_3\&OH(TiO_2)_9H]^+$ discloses to have a very narrow energy gap (0.3 eV) between HOMO and LUMO potentials, proving that PEC nc-TiO₂ electrodes become conductive at photo-irradiated working conditions. DFT-simulation of stepwise oxidation of a hydroxide ion cluster model of $OH^-(H_2O)_3$, proves that successive two-electron oxidation leads to hydroxyl radical clusters, which should give hydrogen peroxide as a precursor of oxygen molecules. Under working bias conditions of PEC cells, nc-TiO₂ electrodes are now verified to become conductive by energy gap photo-excitation and the electrode surface provides powerful oxidizing sites for successive H₂O oxidation to oxygen via hydrogen peroxide.

Reprinted from *Molecules*. Cite as: Yanagida, S.; Yanagisawa, S.; Yamashita, K.; Jono, R.; Segawa, H. Theoretical Verification of Photoelectrochemical Water Oxidation Using Nanocrystalline TiO₂ Electrodes. *Molecules* **2015**, *20*, 9732-9744.

1. Introduction

Photoelectrochemical (PEC) water (H₂O) oxidation on TiO₂ electrodes was qualitatively explained as due to downward band bending induced by depletion layer of TiO₂ rutile crystal electrodes by assuming the energy structure of TiO₂, e.g., conduction band potential (E_{cb}) of −0.65 V(SCE) and valence band potential (E_{vb}) of 2.35 V (SCE), energy gap 3.0 eV, and 1.23 eV of the equilibrium cell potential for H₂O electrolysis at 25 °C and 1 atmospheric pressure [1,2]. We noticed recently that efficient photoelectrochemical H₂O oxidation using anatase nc-TiO₂ electrodes was at first reported in 1987 by Sakka *et al.* [3]. Interestingly, they reported vigorous oxygen (O₂) and hydrogen (H₂) evolution using acidic aqueous solution (0.1 N H₂SO₄) at the PEC nc-TiO₂ cell. It is worth noting that the notable electron flow due to H₂O oxidation to O₂ becomes detectable when bias potential reaches at about 2.0 V (*vs.* SCE), and that under energy gap UV irradiation, photocurrent starts to flow at bias potential around −0.5~−0.3 V (*vs.* SCE), showing vigorous O₂ and H₂ evolution at bias potential around 0.5–1.0 V (*vs.* SCE). Such effective acidic H₂O oxidation on mesoporous nc-TiO₂ electrodes prompts us to understand the H₂O photooxidation on the basis of molecular orbital (MO) theory, because the band-bending concept is based on crystal-level physics, and nc-TiO$_2$ in PEC electrodes is too small to form depletion layer in nc-TiO$_2$ with average size of 25 nm.

Computational chemistry using density functional theory (DFT) well explains and predicts molecular energy structures and properties functioned by self-association of molecules where hydrogen bonding or van der Waals and Coulombic interactions as non-covalent bonding play an essential role [4–9]. DFT calculations using the nc-TiO$_2$ model of Ti$_9$O$_{18}$H-OH (Yamashita/Jono model) successfully verified that the surface complex between nc-TiO$_2$ and 7,7,8,8-tetracyanoquinodimenthane shows charge transfer transition [10]. We now report verification of PEC oxidation of H$_2$O on nc-TiO$_2$ electrodes on the basis of DFT simulation using Yamashita/Jono nc-TiO$_2$ model. Here, DFT-simulations verify that H$_2$O forms H$_2$O clusters via hydrogen bonding and that the H$_2$O clusters-associated nc-TiO$_2$ electrodes provide excellent H$_2$O oxidation sites. In addition, the working models of H$_2$O clusters-associated PEC-nc-TiO$_2$ electrodes are simulated as radical cations, and effective PEC H$_2$O oxidation is verified as well theoretically.

2. Results and Discussion

2.1. Yamashita/Jono Model for Simulation of PEC-nc-TiO$_2$ Electrodes

Yamashita and Jono's anatase nc-TiO$_2$ model (Ti$_9$O$_{18}$H-OH) consists of nine TiO$_2$ units (TiO$_2$)$_9$ derived from the packing unit of the crystalline anatase TiO$_2$, hydroxide on surface side of the TiO$_2$ and hydrogen on one side of the TiO$_2$. To optimize the non-covalent distance between hydroxyl group and nc-TiO$_2$ unit, all heavy atoms of the nc-TiO$_2$ unit are frozen and the model was simulated to the energetically optimized geometry (see Supplementary Figure S1). The distance (1.862 Å) become shorter, and the refined Yamashita/Jono model is abbreviated as OH(TiO$_2$)$_9$H hereafter and size of about 1 nm length, Mulliken charge, the energy structures and configurations of the highest occupied molecular orbital (HOMO) and the lowest unoccupied molecular orbital (LUMO) are shown in Figure 1.

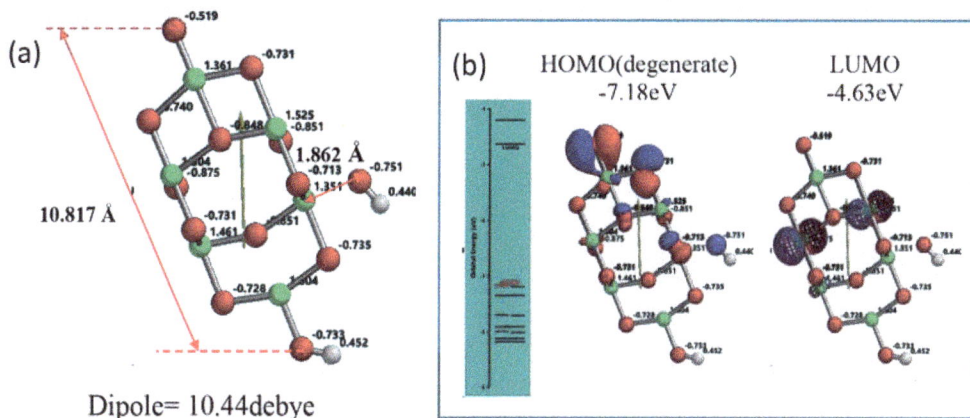

Figure 1. Yamashita/Jono model of OH(TiO$_2$)$_9$H as a model of PEC-nc-TiO$_2$ electrodes, **(a)** Size, distance and Mulliken charge; **(b)** the energy structure and configuration of HOMO and LUMO.

Mulliken charge in OH(TiO₂)₉H indicates that hydroxyl group is charged negative and the protonated nc-TiO₂ unit positive, and then we regard that Yamashita/Jono model is a kind of ion-dipole complex or weak charge transfer complex. The HOMO distributes exclusively on hydroxide ion and the LUMO inside of the nc-TiO₂ unit, giving energy gap 2.55 eV. To know self-association of Yamashita/Jono model, the OH and H-detached (TiO₂)₉ and (TiO₂)₉H units are both DFT-simulated as charge is neutral and cation, respectively (Supplementary Tables S1 and S2 and Figures S1 and S2).

The DFT simulation data reveals that Mulliken charge on the hydrogen-bearing side is positive and another side negative as well as the OH(TiO₂)₉H model. Further, the HOMO and LUMO are almost degenerate. The association through van der Waals and Coulombic interaction and the comparable dipole to that of the Yamashita/Jono model support that Yamashita/Jono model may self-associate with (TiO₂)₉ and (TiO₂)₉H to yield surface thin-film anatase TiO₂ electrodes as depicted in Figure 2.

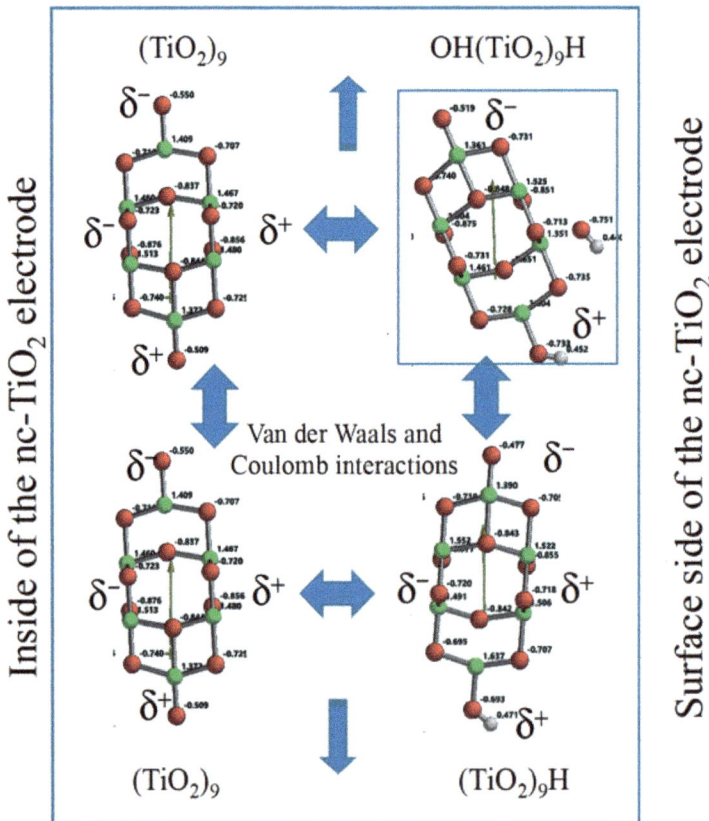

Figure 2. Association of Yamashita/Jono model via van der Waals and Coulomb interactions, explaining growth of the model to larger size PEC-nc-TiO₂ electrodes.

2.2. Water Cluster Models for Modeling of Interface Structures of PEC-nc-TiO$_2$ Electrodes

Water (H$_2$O) molecules aggregate each other via hydrogen bonding. The structures of H$_2$O clusters (H$_2$O)$_n$ (n = 1~3, 6) are simulated to understand their molecular orbital energy structures (Supplementary Table S3). One of the (H$_2$O)$_3$ trimer is simulated to have dipole and the highest HOMO potential −6.67 eV. The HOMO distributes on the H$_2$O, of which hydrogen atoms have hydrogen bond with oxygen atoms of other two H$_2$O, rationalizing the most positive HOMO potential among the examined H$_2$O clusters. In other words, the trimer (H$_2$O)$_3$ is the most oxidizable H$_2$O model.

Similarly, H$_2$O hydroxide ion clusters and H$_2$O hydronium ion clusters are simulated (Supplementary Tables S4 and S5). In general, H$_2$O hydroxide ion clusters have positive HOMO potential. With increase of H$_2$O molecules, the HOMO shifts to negative potential, which means that HOMO potential is controllable by number of associating H$_2$O as pH is controllable by dilution. In addition, all of H$_2$O hydronium ion clusters have very negative HOMO potential −20.2~−12.36 eV with large size of LUMO configurations (Supplementary Table S5). With increase of H$_2$O molecules, HOMO potential shifts to positive potential, verifying that the HOMO level is changeable as is the case of H$_2$O hydroxide ion clusters, and the hydrated hydronium ion clusters become oxidizable energetically.

With these simulations and considerations, the polar (H$_2$O)$_3$ and OH$^-$(H$_2$O)$_3$ are employed for DFT-modeling of neutral interface of PEC-nc-TiO$_2$ electrodes. As for hydronium ion cluster, H$_3$O$^+$(H$_2$O)$_2$, which is derived from (H$_2$O)$_3$, are introduced for the modeling of acidic interface of PEC-nc-TiO$_2$ electrodes (Figure 3). The HOMO configurations indicate that the model clusters have electron rich parts with a wide variety of potentials ranging from −1.07 to −13.5 eV.

2.3. Photoelectrochemical Oxidation of H$_2$O on PEC-nc-TiO$_2$

As an interface model of nc-TiO$_2$ electrodes, the Yamashita/Jono model, OH(TiO$_2$)$_9$H, is structurally frozen and three H$_2$O molecules are made manually to interact via hydrogen bonding with the frozen hydroxyl group in Yamashita/Jono model. The (H$_2$O)$_3$-hydrogen bonded model structure is optimized by molecular mechanics (MMFF operation in Spartan), and the molecular orbital is verified by DFT-single-point simulation of an interface structure of (H$_2$O)$_3$&OH(TiO$_2$)$_9$H as a stationary model (Figure 4). Mulliken charge and electrostatic potential map indicates that negative charge locates on surface oxygen atoms of TiO$_2$ and of the (H$_2$O)$_3$ cluster, and the more negative charge (stronger red color) on the H$_2$O molecule in the cluster is worth noting in the stationary state model structure.

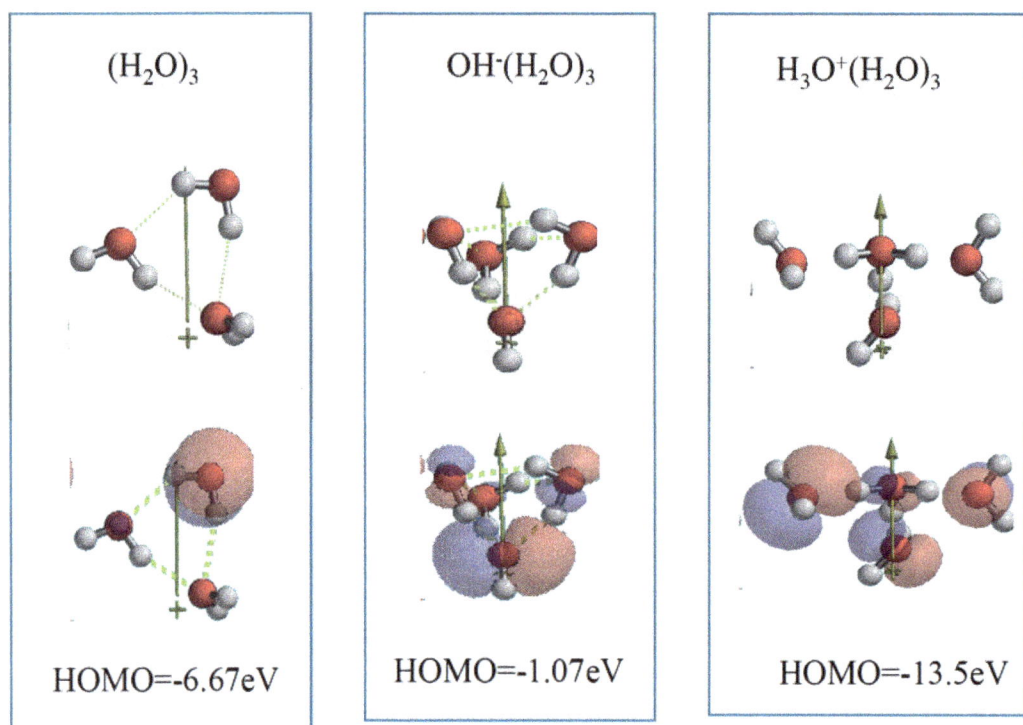

Figure 3. Water cluster models for DFT simulation of PEC oxidation of water molecules on PEC-nc-TiO₂ electrodes.

In the stationary model, configurations of HOMO distributes on one H_2O of the cluster $(H_2O)_3$ unit, and LUMO inside the nc-TiO₂ unit. Interestingly, the energy gap (0.73 eV) between HOMO and LUMO is smaller than the energy gap (2.55 eV) in the stationary state of Yamashita/Jono model. This fact suggests that nc-TiO₂ surface binds H_2O molecules via hydrogen bond, forming kinds of charge transfer complexes. In addition, HOMO(−1) (−7.4 eV) distributes on the whole $(H_2O)_3$ unit, implying more effective H_2O oxidation under negative potential of −7.4 eV (Figure 4).

For comparison, only one H_2O-associated Yamashita/Jono model, H_2O&$OH(TiO_2)_9H$, is simulated as the simplest structure of PEC-nc-TiO₂ electrodes (Supplementary Figure S3). The energy gap (1.75 eV) is given and the hydrogen-bonded H_2O molecule locates HOMO(−1) at −7.8 eV. The simplest orbital energy structure verifies that nc-TiO₂ electrodes intrinsically work as catalytic sites of H_2O oxidation through hydrogen bonding with H_2O molecules.

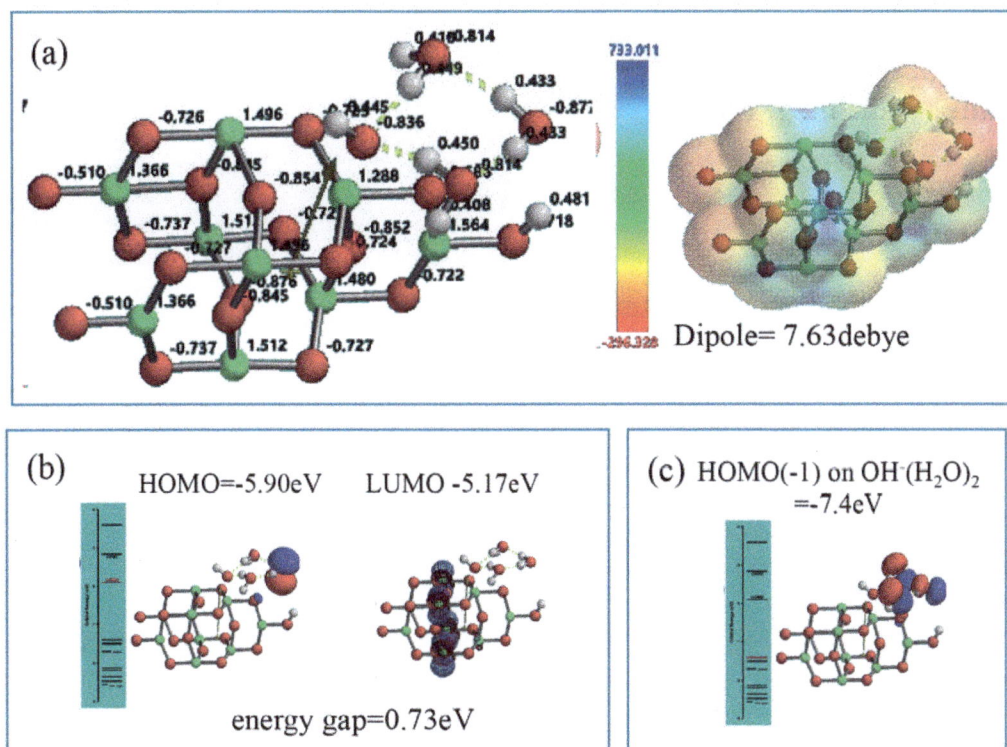

Figure 4. DFT-simulation of $(H_2O)_3\&OH(TiO_2)_9H$ as a stationary state of PEC-nc-TiO_2 electrodes, (**a**) Mulliken charge and electrostatic density map; (**b**) energy structures of HOMO and LUMO; (**c**) the configuration of HOMO(−1) on $(H_2O)_3$.

When nc-TiO_2 electrodes are energy-filled by UV-irradiation at short circuit PEC conditions, photoelectron on nc-TiO_2 is ejected to conducting grids instantly, and photocurrent become observable. The $(H_2O)_3$-associated PEC-nc-TiO_2 model is simulated as radical cation model of $[(H_2O)_3 \&OH(TiO_2)_9H]^{\cdot+}$ as the energy-filled structure under working conditions (Figure 5). Interestingly, the radical cation model as the working model is endothermically simulated (ΔE = 192.85 kcal/mol). The orbital energy analysis of the energy-filled model of $[(H_2O)_3\&OH(TiO_2)_9H]^{\cdot+}$ reveals that the energy gap between HOMO and LUMO potential becomes narrow as small as 0.3 eV with largely negative HOMO (−10.2 eV) and LUMO (−9.9 eV) potential. The HOMO and LUMO distribute in the model with almost the same configurations, and that the spin (unpaired electron) density distributes with the same configuration as the HOMO and LUMO.

The orbital energy analysis of the energy-filled Yamashita/Jono model of $[OH(TiO_2)_9H]^{\cdot+}$ reveals that the energy gap is pretty narrow (0.7 eV) (Supplementary Table S2). These facts suggest that self-organization of Yamashita/Jono model will give photoconductive nc-TiO_2 electrodes when the PEC cell is kept at negative oxidation potential at short circuit conditions and energized by band gap excitation. In fact, the sharp rise in photoconductivity of nc-TiO_2 electrodes was reported and discussed as an insulator-metal (Mott) transition in a donor band of anatase

TiO₂ [11]. It is also worth noting that in studies on dye-sensitized nc-TiO₂ solar cells (DSC), adsorption of cationic species like tetrabutylammonium cation and sensitizing dye molecules enhanced electron transport in nc-TiO₂ electrodes [12,13]. Accordingly, the photo-enhanced electron transport is now verified as a key function of nc-TiO₂ electrodes not only in DSC but also in PEC H₂O oxidation.

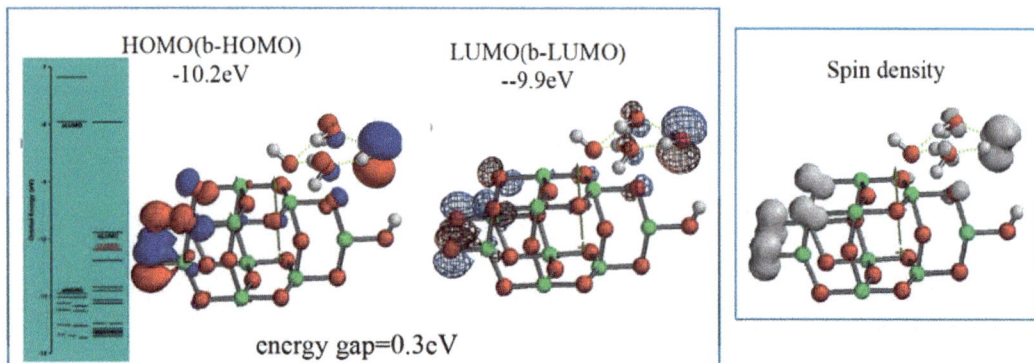

HOMO(b-HOMO)
-10.2eV

LUMO(b-LUMO)
--9.9eV

Spin density

cncrgy gap=0.3cV

Figure 5. Energy structures of $(H_2O)_3\&[OH(TiO_2)_9H]^+$ as a radical cation model of $(H_2O)_3$-interacted PEC-nc-TiO₂ electrode under UV-irradiated bias conditions of PEC-nc-TiO₂ electrodes, *i.e.*, a photo-energy-driven operational state model.

The HOMO configuration on the H₂O unit at the stationary state of the model, and the spin density distribution on the $(H_2O)_3$ unit at the energy-filled working state strongly suggest that PEC-nc-TiO₂ electrodes provides catalytic binding sites of H₂O. The same functions of PEC nc-TiO₂ electrodes are confirmed by the molecular orbital simulation of the energy-filled $[H_2O\&OH(TiO_2)_9H]^+$ model (Tables S1 and S2, Figure S4). The narrowed energy gap (0.3 eV), the comparable configurations of HOMO, LUMO and spin density are quite comparable with those of $[(H_2O)_3\&OH(TiO_2)_9H]^+$.

2.4. DFT Simulation of H₂O Oxidation to Hydrogen Peroxide

In PEC H₂O oxidation on nc-TiO₂ electrodes, bias potential is essential to start H₂O oxidation. The HOMO potential of Yamashita/Jono model, −7.18 eV and the average HOMO potential of H₂O clusters, −7.48 eV (Supplementary Table S3) verify that the bias potential >0.3V is at lease required for PEC H₂O oxidation under neutral working conditions. On the other hand, the HOMO potential (−10.2 eV) of the working model of $[(H_2O)_3\&OH(TiO_2)_9H]^+$ predicts that successive oxidation should occur under more negative bias potential (>2.72 eV) for H₂O oxidation to oxygen through formation of hydrogen peroxide.

In order to verify whether hydroxyl radical may form successively on PEC-nc-TiO₂ electrodes, the two-electron oxidation structure of $[H_2O\&OH(TiO_2)_9H]^{\cdot++}$ is simulated as dication-diradical model as another working interface model of PEC-nc-TiO₂ electrodes. The more energy-filled model is endothermically (ΔE = 458.43 kcal/mol) simulated as powerful working model (Figure 6). The energy gap (0.3 eV) and the configuration of HOMO and LUMO are confirmed to verify that such largely energized PEC-nc-TiO₂ electrodes keep photoconductivity with keeping high oxidation potential. The spin density distributes on the $(H_2O)_3$ unit, suggesting that two-electron H_2O oxidation occurs successively on the catalytic site on nc-TiO₂ electrodes.

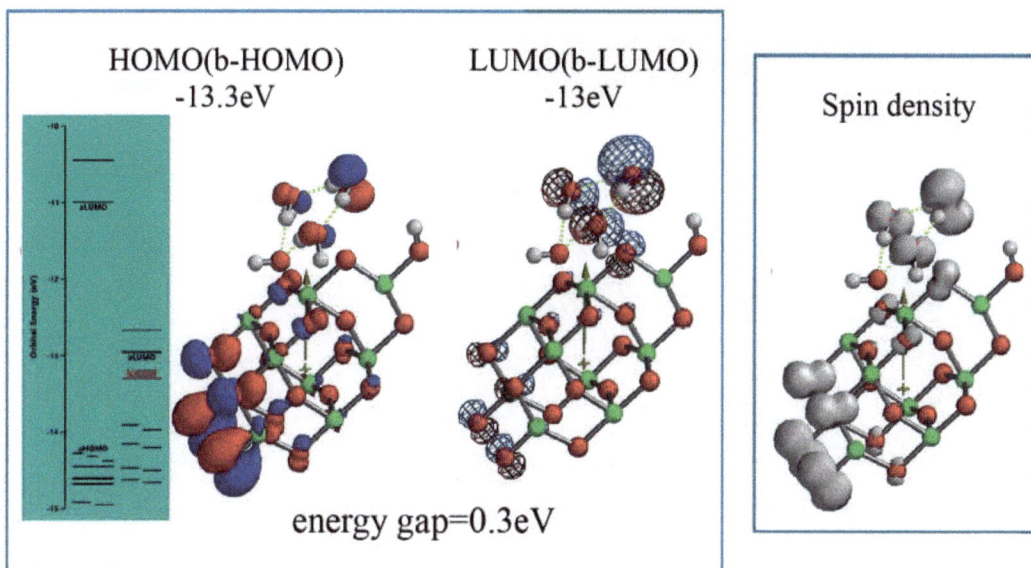

Figure 6. Energy structures of $[(H_2O)_3]\&OH(TiO_2)_9H]^{..++}$ as the model of two electron oxidation state of PEC-nc-TiO₂ electrodes, *i.e.*, the diradical-dication state of the electrodes.

With these simulation analyses, step-wise PEC-H_2O oxidation is shown in Scheme 1. One-electron oxidation of H_2O molecule yields radical cation of H_2O (H_2O^+) and the removable of proton from the radical cation (deprotonation) leads to hydroxyl radical (HO) (Equations (1) and (2) in Scheme 1). When H_2O hydroxide ion cluster of $OH^-(H_2O)_3$ undergoes further oxidation, another hydroxyl radical favorably forms in neighbor on PEC-nc-TiO₂ electrodes, and efficient and effective formation of hydrogen peroxide occurs (Equation (3) in Scheme 1).

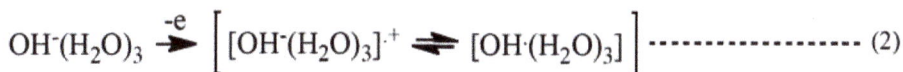

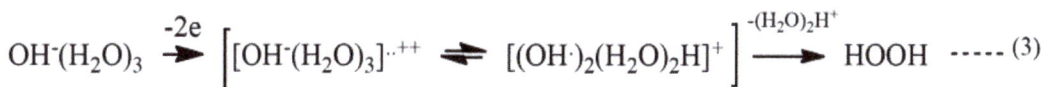

$$H_2O \xrightarrow{-e} [H_2O]^{.+} \xrightarrow{-H^+} HO^{.} \qquad \text{----------------------------- (1)}$$

$$OH^-(H_2O)_3 \xrightarrow{-e} \left[[OH^-(H_2O)_3]^{.+} \rightleftharpoons [OH^.(H_2O)_3] \right] \text{------------------ (2)}$$

$$OH^-(H_2O)_3 \xrightarrow{-2e} \left[[OH^-(H_2O)_3]^{..++} \rightleftharpoons [(OH^.)_2(H_2O)_2H]^+ \right] \xrightarrow{-(H_2O)_2H^+} HOOH \text{ ----- (3)}$$

Scheme 1. PEC-H_2O oxidation to hydrogen peroxide as a precursor of oxygen molecule.

Figure 7 shows DFT-simulation results of oxidation of $OH^-(H_2O)_3$ to $[OH^-(H_2O)_3]^{\cdot+}$ or $[OH^{\cdot}(H_2O)_3]$ as one-electron oxidation products (Equation (2)), and to $[OH^-(H_2O)_3]^{\cdot++}$ or $[(OH^{\cdot})_2(H_2O)_2H]^+$ as two-electron oxidation products. They are simulated endothermically, suggesting that they are in energy filled states (Supplementary Table S6). In the equilibrium geometry of $[OH^{\cdot}(H_2O)_3]^{\cdot+}$, the spin density distributes only on the hydroxyl group, and the Mulliken charge on hydroxyl group decreases largely from −0.900 to −0.404. Thus the one-electron oxidation product has the structure of $[OH^{\cdot}(H_2O)_3]$ rather than $[OH^-(H_2O)_3]^{\cdot+}$.

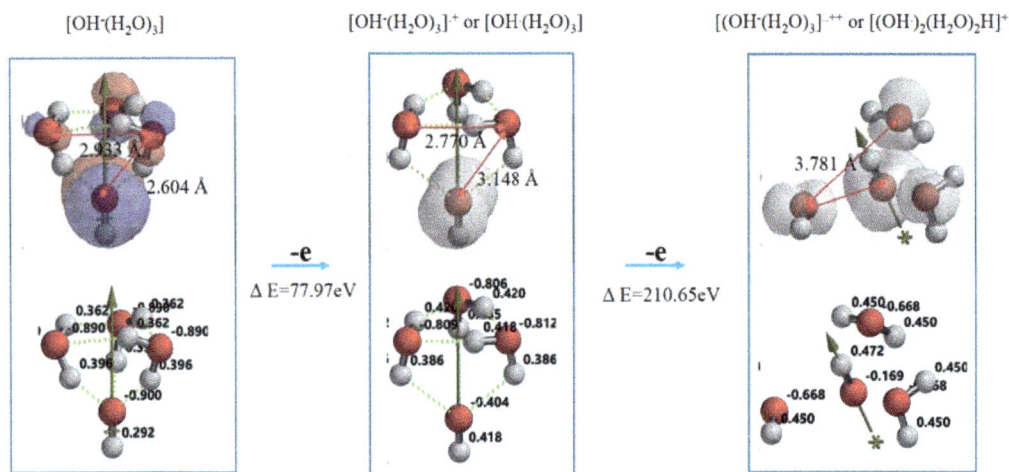

Figure 7. DFT-simulation of step-wise oxidation of the hydroxide ion cluster model of $[OH^-(H_2O)_3]$.

On the other hand, the two-electron oxidation product shows that the spin density distributes on hydroxyl group and the $(H_2O)_3$ units. The Mulliken charge on them decreases from −0.806 to −0.668. The distance between hydroxyl group and H_2O is shortened from 3.148 Å to 2.239 Å and the hydrogen bonds observed in $[OH^-(H_2O)_3]^{\cdot+}$ disappear. In addition, the hydrogen-oxygen bond distance of H_2O (0.976 Å) in $[OH^-(H_2O)_3]^{\cdot++}$ is quite comparable with that (0.988 Å) in $[OH^-(H_2O)_3]^{\cdot+}$. The H_2O components in the most energy-filled cluster $[OH^-(H_2O)_3]^{\cdot++}$ should be tightly aggregated one another. Detachment of $(H_2O)_3H^+$ from $[(OH^{\cdot})_2(H_2O)_2H^+]$ leaves two hydroxyl radical in neighbor, yielding hydrogen peroxide as a precursor of O_2 molecule (Equation (3) in Scheme 1). The oxidation of neutral H_2O is verified to occur initially in photocatalytic processes to give effectively hydrogen peroxide on PEC nc-TiO$_2$ electrodes.

2.5. Verification of the Sakka's PEC H2O Oxidation under Acidic Conditions

The stationary model of $H_3O^+(H_2O)$-associated structure of $H_3O^+(H_2O)\&OH(TiO_2)_9H$ and two kinds of energy filled models, $[H_3O^+(H_2O)\&OH(TiO_2)_9H]^{\cdot+}$ and $[H_3O^+(H_2O)\&OH(TiO_2)_9H]^{\cdot++}$ are simulated as an interface model of PEC-nc-TiO$_2$ electrodes under Sakka's acidic conditions (Supplementary Figures S5 and S6). However, the energy gap are rather wider and the spin density does not localize on $H_3O^+(H_2O)$ in the most energized state of $[H_3O^+(H_2O)\&OH(TiO_2)_9H]^{\cdot++}$.

The hydronium ion cluster, $H_3O^+(H_2O)_2$ represent less acidic than $H_3O^+(H_2O)$ (Supplementary Table S5). The stationary states of $H_3O^+(H_2O)_2$-associated structure of $H_3O^+(H_2O)_2\&OH(TiO_2)_9H$ is simulated as an interface model under Sakka's less acidic conditions, and analyzed as well in view of molecular orbital energy structure (Figure 8).

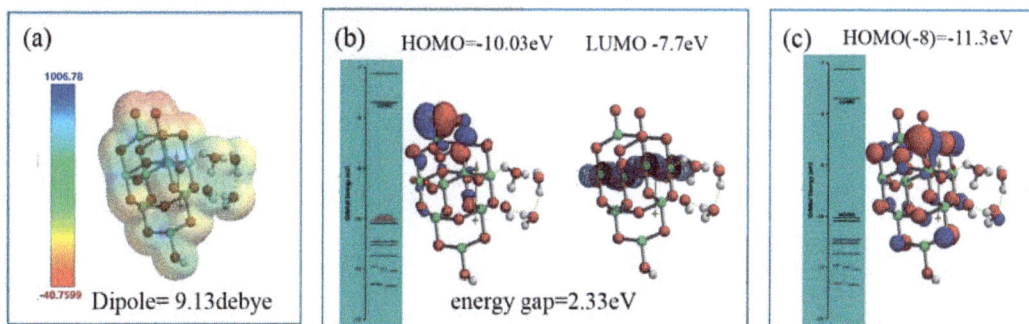

Figure 8. DFT-simulation of $H_3O^+(H_2O)_2\&OH(TiO_2)_9H$ as a model of hydronium ion clusters on PEC-nc-TiO$_2$ electrodes, **(a)** electrostatic potential map; **(b)** structures of HOMO and LUMO; **(c)** Configuration of HOMO(−8).

Differently from the modeling for the neutral PEC H_2O oxidation, electrostatic potential map indicates that negative charge locates much more on the nc-TiO$_2$ unit rather than the H_2O unit, and HOMO distributes only on oxygen atoms in the nc-TiO$_2$ unit. The energy gap 2.33 eV implies weak association of acidic H_2O on nc-TiO$_2$ electrodes. The orbital energy indicates that HOMO(−8) distributes slightly on the $H_3O^+(H_2O)_2$ unit with very negative potential of −11.3 eV.

The acidic interface model of $[H_3O^+(H_2O)_2\&OH(TiO2)_9H]$ is simulated to $[H_3O^+(H_2O)_2\&OH(TiO2)_9H]^{\cdot+}$ as the radical cation of the one-electron oxidation state, and to $[H_3O^+(H_2O)_2\&OH(TiO2)_9H]^{\cdot\cdot++}$ as the diradical-dication model of the two electron oxidation sate (Figure 9). The former radical cation model reveals that configurations of HOMO, LUMO and spin density distribute on the nc-TiO$_2$ unit and not on the $H_3O^+(H_2O)_2$, and the energy gap 0.7eV is not favorable in view of photoconductivity compared to that under neutral conditions. However, HOMO(−1) distributes on the $(H_2O)_2$ unit with orbital potential of −13.9 eV.

As for the latter dication-diradical model, the energy gap 0.5 eV and HOMO potential, −16.6 eV are given, and the spin density distributes on $H_3O^+(H_2O)_2$ unit. Accordingly, the DFT-based orbital energy structure verifies that PEC H_2O oxidation occurs even under acidic conditions, when nc-TiO$_2$ electrodes are energized by bias potential under energy-gap UV irradiation.

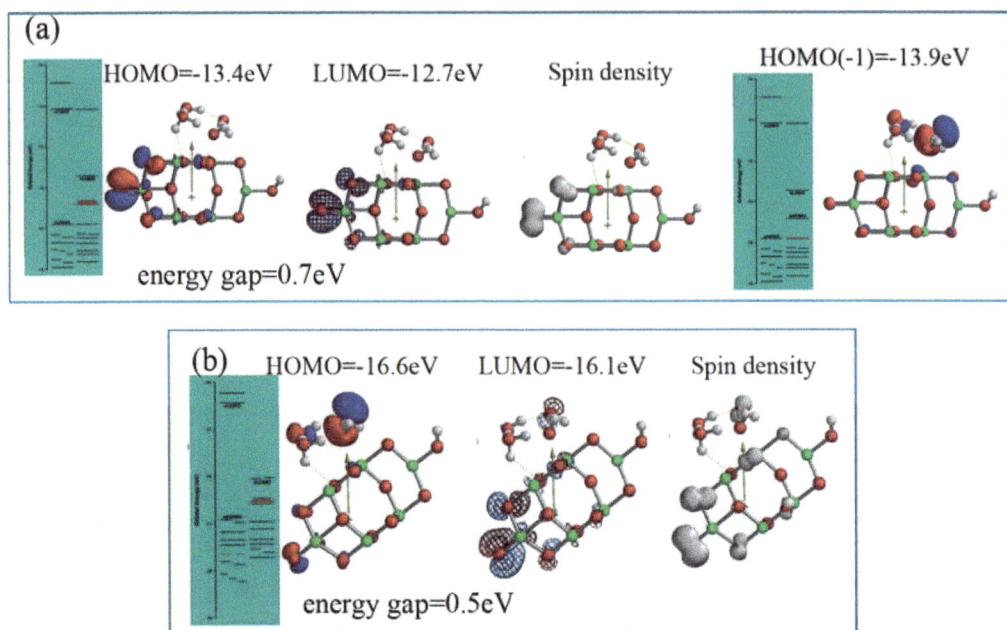

Figure 9. DFT-simulation of oxidation states of $H_3O^+(H_2O)_2\&OH(TiO_2)_9H$, (**a**) energy structures of the one-electron oxidation state, $[H_3O^+(H_2O)_2\&OH(TiO_2)_9H]^{·+}$; (**b**) energy structures of the two-electron oxidation state, $[H_3O^+(H_2O)_2\&OH(TiO_2)_9H]^{··++}$.

3. Experimental Section

DFT calculations were performed using the B3LYP exchange-correlation functional and the 6-31G(d) basis set with *Spartan'14* (Wavefunction, Inc. Irvine, CA, USA) installed on VAIO Model SVP132A1CN, Intel(R) core(TM)i7-4500U CPU and on VAIO PC-Z (Intel core 2 Duo processor T9900, system memory (RAM) 8G and hard disk drive, SSD 128, 2GB).

Molecular mechanic optimization (e.g., Merck Molecular Force Factor (MMFF) operation in Spartan program) and DFT (B3LYP 6-31G*) modeling determine molecular orbital structure of equilibrium geometry as an inter-atomic potential model [7,8]. In the case of Spartan program, molecular orbital energy structures (HOMO(0~9), LUMO(0~1)), their configurations, electrostatic potential map, spin (unpaired electron of radical) density are visualized by graphic conveniently.

As for interface energy-filled model structures with unpaired electron, *i.e.*, radical cations, orbital energy diagrams are shown in two ways, affixed 'a-' and 'b-' because the radical cations have two available wave functions. The 'b'-HOMO and 'b'-LUMO are employed, since 'β'-HOMO configurations are almost the same as spin density configurations. Electron energy gap of the radical cation components is obtained from energy difference between 'b'-HOMO and 'b'-LUMO. Mulliken charge, spin densities and their maps are informative for theoretical understandings of energy structures of energy-filled molecular orbitals of nc-TiO₂ interfaces.

The anatase nc-TiO₂ model structure has a pretty large size of $OH(Ti_9O_{18})H$ and is named as Yamashita/Jono model. The nc-TiO₂ model structure is refined as described in Figure S7. As for orbital configurations, HOMO is shown by solid or solid transparent, LUMO by mesh, and spin density by white solid or white solid transparent. As for color in electrostatic potential map, red is negative, green neutral and blue positive qualitatively. Formation energy (ΔE) of key model molecules is determined from total energy (E) of their related components in Supplementary Tables S1–S3.

4. Conclusions

DFT-based modeling enables to verify molecular orbital level interfacial structures of photo-electrochemically energized nc-TiO$_2$ electrodes. Although nc-TiO$_2$ electrodes are composed of nc-TiO$_2$ particles with average size 25 nm, Yamashita/Jono nc-TiO$_2$ model (length size = ~1 nm) is large enough to model nc-TiO$_2$ electrodes because the model may self-aggregate to larger sizes through hydrogen bond and van der Waals and Coulombic interactions. Water (H_2O) cluster models ($H_2O)_3$ and $H_3O^+(H_2O)_2$ are appropriate to bind Yamashita/Jono model cluster, providing interfacial PEC-nc-TiO$_2$ electrode structures at neutral and acidic H_2O conditions.

Molecular orbital analyses of the stationary and the working PEC-nc-TiO$_2$ cluster models reveal that H_2O clusters are adsorbed effectively (catalytically) via hydrogen bonding to PEC-nc-TiO$_2$ electrodes at stationary state, and that the conductivity of PEC-nc-TiO$_2$ electrodes is enhanced without loosing oxidation potential, leading to successive water oxidation to oxygen molecules through hydrogen peroxide in PEC cells. The molecular modeling of nc-TiO$_2$ electrodes in PEC cells verifies that the photo-induced conductivity is the most important driving force of PEC H_2O oxidation on nc-TiO$_2$ electrodes. The DFT-verified photoconductivity is true for understanding of photocatalysis of Pt-deposited nc-TiO$_2$ particles and dye-sensitized nc-TiO$_2$ solar cells, rationalizing their remarkable efficiencies and effectiveness.

Supplementary Materials

Supplementary materials can be accessed at: http://www.mdpi.com/1420-3049/20/06/9732/s1.

Acknowledgments

The authors thank W. J. Hehre (Wavefunction, Inc. Irvine, CA, USA) and N. Uchida and M. Takahashi (Wavefunction, Inc., Japan Branch Office, Kouji-machi, Chiyoda-ku, Tokyo, Japan) for discussion on DFT simulation using 'Spartan 14'.

Author Contributions

Shozo Yanagida, Susumu Yanagisawa, KY, RJ and HS designed research; Shozo Yanagida, Susumu Yanagisawa, KY, RJ and HS performed research, analyzed the data and discussed; Shozo Yanagida, Susumu Yanagisawa wrote the paper. All authors read and approved the ¯nal manuscript.

Conflicts of Interest

The authors declare no conflict of interest.

References

1. Grätzel, M. Photoelectrochemical cells. *Nature* **2001**, *414*, 338–344.
2. Hashimoto, K.; Irie, H.; Fujishima, A. Photoelectrochemical cells. *JSAP Int.* **2006**, *24*, 4–20.
3. Yoko, T.; Kamiya, K.; Sakka, S. Photoelectrochemical Properties of TiO_2 Films Prepared by the Sol-Gel Method. *Yogyo Kyoukai-Shi* **1987**, *95*, 150–155.
4. Kanemoto, M.; Hosokawa, H.; Wada, Y.; Murakoshi, K.; Yanagida, S.; Sakata, T.; Mori, H.; Ishikawa, M.; Kobayashi, H. Semiconductor photocatalysis. Part 20. Role of surface in the photoreduction of carbon dioxide catalysed by colloidal ZnS nanocrystallites in organic solvent. *J. Chem. Soc. Faraday Trans.* **1996**, *92*, 2401–2411.
5. Manseki, K.; Yu, Y.; Yanagida, S. A phenyl-capped aniline tetramer forZ907/*tert*-butylpyridine-based dye-sensitized solar cells and molecular modelling of the device. *Chem. Commun.* **2013**, *49*, 1416–1418.
6. Yanagisawa, S.; Yasuda, T.; Inagaki, K.; Morikawa, Y.; Manseki, K.; Yanagida, S. Intermolecular Interaction as the Origin of Red Shifts in Absorption Spectra of Zinc-Phthalocyanine from First-Principles. *J. Phys. Chem. A* **2013**, *117*, 11246–11253.
7. Hehre, W.J. Chapter 19. Application of Graphical models in book. In *A Guide to Molecular Mechanics and Quantum Chemical Calculations*; Wavefunction, Inc.: Irvine, CA, USA, 2003; p. 473.
8. Hoffmann, R. The frontier orbital perspective in book SOLID and SURFACES. In *A Chemist's View of Bonding in Extended Structures*; Wiley-VCH Inc.: Toronto, ON, Canada, 1988.
9. Agrawal, S.; English, N.J.; Thampi, K.R.; MacElroy, J.M.D. Perspectives on *ab initio* molecular simulation of excited-state properties of organic dye molecules in dye-sensitised solar cells. *Phys. Chem. Chem. Phys.* **2012**, *14*, 12044–12056.
10. Jono, R.; Fujisawa, J.; Segawa, H.; Yamashita, K. Theoretical Study of the Surface Complex between TiO_2 and TCNQ Showing Interfacial Charge-Transfer Transitions. *J. Phys. Chem. Lett.* **2011**, *2*, 1167–1170.
11. Wahl, A.; Augustynski, J. Charge Carrier Transport in Nanostructured Anatase TiO_2 Films Assisted by the Self-Doping of Nanoparticles. *J. Phys. Chem. B* **1998**, *102*, 7820–7828.
12. Kambe, S.; Nakade, S.; Kitamura, T.; Wada, Y.; Yanagida, S. Influence of the Electrolytes on Electron Transport in Mesoporous TiO_2-Electrolyte Systems. *J. Phys. Chem. B* **2002**, *106*, 2967–2972.
13. Nakade, S.; Saito, Y.; Kubo, W.; Kanzaki, T.; Kitamura, T.; Wada, Y.; Yanagida, S. Enhancement of electron transport in nano-porous TiO_2 electrodes by dye adsorption. *Electrochem. Commun.* **2003**, *5*, 804–808.

Sample Availability: Not available.

MDPI AG
Klybeckstrasse 64
4057 Basel, Switzerland
Tel. +41 61 683 77 34
Fax +41 61 302 89 18
http://www.mdpi.com/

Molecules Editorial Office
E-mail: molecules@mdpi.com
http://www.mdpi.com/journal/molecules

www.ingramcontent.com/pod-product-compliance
Lightning Source LLC
Chambersburg PA
CBHW051927190326
41458CB00026B/6430